P9-EFH-601

DURHAM COUNTY LIBRARY
OCT - - 2000
DURHAM, NC

Also by Richard Ellis

The Book of Whales

Dolphins and Porpoises

The Book of Sharks

Men and Whales

Monsters of the Sea

Great White Shark (with John McCosker)

Deep Atlantic

Imagining Atlantis

The Search for the Giant Squid

Encyclopedia of the Sea

Encyclopedia

of the Sea

Written and Illustrated by

Richard Ellis

THIS IS A BORZOI BOOK PUBLISHED BY ALFRED A. KNOPF

www.aaknopf.com

Library of Congress Cataloging-in-Publication Data

Ellis, Richard, [date]
Encyclopedia of the sea / by Richard Ellis. — 1st ed.
p. cm.
ISBN 0-375-40374-4 (alk. paper)
1. Oceanography Encyclopedias. 2. Ocean Encyclopedias.
3. Marine ecology Encyclopedias. I. Title.
GC9.E38 2000
551.46'003—dc21 99-42401
CIP

Manufactured in the United States of America

FIRST EDITION

Introduction

There are perhaps twenty thousand known species of fishes in the world, many of which live in freshwater, and about eight thousand species of birds, many of which have nothing whatever to do with the ocean. There are only seventy-odd species of whales, dolphins, and porpoises, and twenty-five seals and sea lions, however, all of which have a lot to do with the ocean; some of them never leave it from the moment they are born until the day they die. There are five sirenians (one extinct), one sea otter, and one polar bear. There are 350 species of sharks, about 700 species of squid, about 100 octopuses, and heaven knows how many worms, starfishes, clams, crabs, oysters, and sponges. (There are some items, like diatoms or plankton, that by their very numbers can be included as plural entries; no one would attempt to identify every microsopic member of these diverse fauna.) I have included all the cetaceans and all the pinnipeds, but for everything else I had to be extremely selective. Not all the fishes are all that different from one another; many of them are minor variations on a particular theme, like the 100-odd blennies, or the 250 species of lantern fishes, which look very much alike except for the different arrangement of photophores on their bodies. Since I could only describe a couple of them, I had to use some criteria for this choice: Were they unusual in some aspect? Were they particularly well known, even to nonspecialists? If not, did the specialists think they were special?

Scientists do not usually play favorites: an ornithologist writing a monograph about pelicans will probably not tell you which one he likes best; an ichthyologist will not choose one grouper over another. But an encyclopedia is a different kettle of fish, and he who composes it has to decide what to include and what to exclude. To include every gull or every shark is far beyond the scope of this book. The number of monographs on individual groups, genera, or even single species is enormous; I consulted many of them in the preparation of this book. I have been ruthlessly selective about sharks, for example, including most of the better-known species and some of the oddities and omitting many of the poorly known, rarely seen, or very-similar-to-others. I have done the same with seabirds, also throwing into the mix some particularly rare or even extinct species, and I did the same with shorebirds, cephalopods, and fishes. There are only eight kinds of sea turtles, however, so they are all here.

Even though all the arms of the "seven seas" are connected—one could, in principle, sail (or row, or paddle) from every one into every other one without touching land—they have been awarded names that allow us to differentiate the Kara Sea from the Tasman Sea, or the Gulf of Tonkin from Prince William Sound. This encyclopedia also includes as many islands as possible. You will find islands that range in size from Greenland, at 840,000 square miles, to Wake Island, a tiny coral atoll in the middle of nowhere. The "nowhere," of course, is the Pacific Ocean, the largest single feature on the planet by several orders of magnitude, and many of the islands in it, trenches under it, fishes that swim through it, and birds that fly over it can be found as individual entries.

The number of references required to write an encyclopedia essentially negates the possibility of a bibliography. To have listed the sources of the information on the myriad of subjects in this book would have produced a work as long or longer than the book itself. The very magnitude of the works consulted is itself the reason for this book: I consulted thousands of reference works so you won't have to.

I have had to make decisions about how the entries will be listed, and I decided to use the vernacular "Pacific giant squid" instead of "giant squid, Pacific" or the even more awkward "squid, giant Pacific." Many of the animals included in this book, mostly invertebrates but many fishes as well, have no common names. In many cases, therefore, I have listed the subject by its scientific name instead of an unfamiliar or nonexistent common name. No one would think of looking up "flowervase jewel squid" (the name assigned to *Histioteuthis dofleini* in the FAO *Species Catalog of Cephalopods of the World*), so it would be found under the scientific designation, *Histioteuthis.* Scientific names are awkward and troublesome to those who are not familiar with their usage, but they are necessary in many cases to identify the animal that is being discussed.

A note on the conventions of binomial nomenclature would not be out of place here, if only to make more palatable the repeated use of Latinized, often difficult-to-pronounce names. According to a system developed in the eighteenth century by the Swedish naturalist Carolus Linnaeus, every living thing (plants included) is to have a name identifying its genus (the generic name), followed by a specific name, to identify the species. The generic name is always capitalized; the specific name always begins with a lowercase letter. Scientific names are always supposed to be in italics. For the sake of brevity, when listing members of the same genus, the generic name can be abbreviated by its initial letter, as with the gulls *Larus argentatus, L. hyperboreus,* and *L. californicus*—respectively, the herring gull, the glaucous gull, and the California gull.

During the period I was working on this project, people often asked me what letter I was up to, as if I had a long list of entries, beginning with "abalone" and ending with "zooxanthellae," and I was going to plow through them in order. Not only did I not go from "abalone" to "abyssal plains," but I didn't even start with "abalone." I don't remember which entry was first, but I know that one entry led naturally to another, and often suggested the next one. When I was doing the penguins, for example, I researched and wrote up the Humboldt penguin, and then segued right into the Humboldt squid, followed by the Humboldt Current, but then I realized that I'd better include Alexander von Humboldt, for whom these various things were named. Thinking about currents aimed me in the direction of the Kuroshio, the Benguela, the Agulhas, and the Gulf Stream, and these in turn suggested other bodies of water, explorers, islands, and wildlife. I would be disappointed if people simply looked up "Christian, Fletcher" and were not inspired to go on to "Bligh, William," "*Bounty,*" and perhaps even "breadfruit." Or if they turned to "blue-footed booby" and did not find themselves heading for "red-footed booby" or "albatross."

This encyclopedia is supposed to be a reference work. It is designed to be consulted if one wants to know what a turbot looks like, how big

Sri Lanka is, or what happened to the *Andrea Doria* and when. But as I was assembling this collection of entries, I found that either the alphabetical entries or the links (the "see also"s) made me want to write about (or read about, if the entry had been completed) other things, either immediately after the entry I had come for, or wherever it was suggested that I look for some related topic.

After I had completed the initial group of entries (and been informed by my editor that the book would not be published until the fall of 2000, a year later than originally planned), I sent a copy of the index (which was for my own reference, and does not, for obvious reasons, appear in this book) to my friend Rick Martin in Vancouver. Because he was one of the few people I could think of who had as broad and diversified a range of marine interests as I did, I asked him to read this working index and see if he could think of anything I'd missed. Ha. He sent me a list of 372 suggestions, many of which I incorporated into the final product before publication. I am profoundly grateful for his help, especially for suggesting things that would be obvious only by their omission.

During a last-minute effort to proofread the manuscript, I found myself aboard the MV *Hanseatic* cruising Antarctic and sub-Antarctic waters. It was a pleasant surprise—and a godsend—to find that Don Walsh was aboard, someone who was mentioned in several entries. He took the manuscript to his cabin, intending to read only those entries about which he had firsthand knowledg, but like me, he became involved in the interconnections and read the whole thing. By now I feel pretty secure about *Trieste*, bathyscaph, Mariana Trench, etc., but I accept full responsibility for any mistakes that he didn't catch. Also aboard *Hanseatic* on this same cruise was Sylvia Stevens, out of San Diego by way of Glasgow, and she carefully read those entries that pertained to her specialty, the birds and mammals of the Southern Ocean. I can only hope that these two close friends and severe critics have saved me from a few of the pitfalls that I know await me.

Their willingness to read an entire encyclopedia from A to Z was an unexpected boon. That privilege is usually reserved for the author, his editor, and the copy editors and proofreaders who were given this assignment from hell. I would never ask that of anyone, not even someone who lived with me during the process. Throughout the research, writing, and drawing, Stephanie left me to flounder (see also "flatfish") through the doldrums, high winds, and depths of the world's oceans, more or less singlehandedly (see also "Slocum, Joshua"). Even so, this book is for her, as I am.

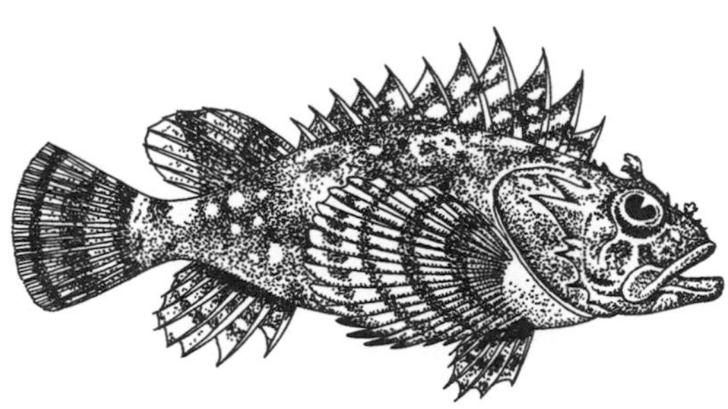

Encyclopedia of the Sea

abalone (*Haliotis* spp.) A large marine gastropod of the genus *Haliotis,* with a dishlike shell punctuated by a row of holes on the outer edge. Abalones cling to rocks with their powerful foot and breathe through gills, passing out water through the holes in the shell. They feed by browsing on sea lettuce and kelp and snaring passing plankton by means of tentacles projecting from the edge of the mantle. Abalones are found on the coasts of the northern Pacific, Australia, New Zealand, and South Africa, where they are pried from the rocks by divers and the large foot is prepared as a steak. The biggest of the abalones is the red, 1-foot-long *Haliotis rufescens* of the California coast. California abalones are consumed locally, but Australians ship most of their harvest to Southeast Asia, Japan, and China.

See also gastropoda, limpet

Abbott, R. Tucker (1919–1995) Born in Watertown, Massachusetts, and educated at Harvard, with a Ph.D. from George Washington University, Tucker Abbott was probably the best-known malacologist of his time, and a renowned popularizer of the art and science of shell collecting. During World War II, after a stint as a navy fighter pilot, he was attached to a medical unit in the Pacific and discovered that schistosomiasis was transmitted by a freshwater snail, *Oncolmelania.* He served at the Smithsonian from 1944 to 1954 and published the first edition of *American Seashells,* still the most important book on this subject. Then he went to the Academy of Natural Sciences in Philadelphia, where he was chairman of the Department of Mollusks and published several important books, including *Introducing Seashells* and *Van Nostrand's Standard Catalog of Shells.* In 1969, Abbott was named to the Du Pont Chair of Malacology at the Delaware Museum, and produced *Kingdom of the Seashell* in 1972. During his lifetime, he traveled to every conceivable remote location in search of shells. In 1976, he "retired" to Melbourne, Florida, but almost immediately became involved in the building of the Bailey-Matthews Shell Museum on Sanibel Island. From his earliest publications to his relationship with the Conchologists of America, Tucker Abbott was dedicated to the dissemination of information about snails and shells, and remains one of the most important conchologists in history.

abyss The depths of the ocean. No light reaches here, and the temperature is constant between 0°C and 2°C. There is sufficient oxygen to support life, but most abyssal forms are suspension feeders or scavengers and depend on particles of edible material that rain down from above. At the oceans' greatest depth—35,800 feet in the Mariana Trench—the most common animals are holothurians (sea cucumbers) and brittle stars. The world's oceans average 13,000 feet in depth, which means that deep ocean is the earth's predominant environment. Most of the creatures that live in the depths are still undiscovered. **See also benthos**

abyssal plains Although the floor of the ocean is far from the monotonous expanse that was originally imagined, it does have vast stretches that are exceedingly flat. The abyssal plains are characterized by a slope of no more than one part in one thousand, or 5 feet per mile—completely beyond the ability of the human eye to recognize as anything but perfectly flat. They are usually composed of silts that have accumulated over millions of years, filling in and leveling the previously irregular topography. Except for the Gulf of Alaska, the Pacific lacks abyssal plains, but they are present on both sides of the Mid-Atlantic Ridge in the North and South Atlantic. **See also Mid-Atlantic Ridge**

acorn worm (phylum Hemichordata) Named for the acornlike swelling at the anterior end—it has no eyes, no brain, and the most primitive mouth, so it would be a stretch to call it a head—acorn worms burrow into the sand or mud of seashores around the world. Some species have been found at depths of up to 10,000 feet. They have a three-part body, consisting of a muscular proboscis, attached by a stalk to a short collar, and a long, wormlike trunk, which contains the digestive tract. The mouth is located beneath the collar, and they feed by ingesting sediment. Along the anterior portion of the trunk, acorn worms have a series of gill slits, which are comparable only to those found in the higher vertebrates. Acorn worms are placed in their own phylum (Hemichordata), between the invertebrates and the primitive chordates (lancelets and tunicates), although they have only a short, rodlike growth (the buccal diverticulum) at the anterior end of the digestive tract. The commonest genus is *Balanoglossus,* which can range in size from 6 inches to 6 feet.

See also lancelet, tunicate

Adélie penguin (*Pygoscelis adeliae*) Reaching a length of about 27 inches, the Adélie penguin is easily recog-

Adélie penguin

nized by its black back, white front, and white eye-rings. Adélies nest in large colonies on the coast of the Antarctic continent, only rarely appearing as far north as South Georgia. Their nests are made of stones piled up in hollowed-out mounds. In the water, they swim with only their heads out, and they are capable of fantastic leaps out of the water, especially if pursued by their archenemy, the leopard seal. Adélie was the wife of the French admiral Dumont d'Urville, who explored the Antarctic in 1837 and also named Terre Adélie for his wife.

See also chinstrap penguin, Dumont d'Urville, gentoo penguin

Admiralty Islands The northwestern part of the Bismarck Archipelago in the southwest Pacific, belonging to Papua New Guinea. The islands may have been sighted by the Spanish navigator Álvaro de Saavedra in 1583, but they were recorded by the Dutchman Willem Schouten in 1616 and named by the British captain Philip Carteret in 1767. In 1884, the group became part of German New Guinea, but they were captured by Australia in 1914. The Admiralties were occupied by the Japanese from 1942 to 1944, but after the war they were made part of the UN Trust Territory of New Guinea. When Papua New Guinea attained its independence in 1975, the islands were included. The largest of the forty volcanic islands is Manus (sometimes called Great Admiralty Island), 50 by 20 miles in size and covering some 633 square miles. The islands produce copra and some coffee, but most of the 29,000 residents are engaged in subsistence farming and fishing. The U.S. Navy built a large base in Seeadler Harbor on Manus in 1944, and after the war a "cargo cult" was founded by the prophet Paliau, who encouraged believers to build replicas of airstrips and hangars to encourage the return of the wartime bounty of goods.

See also Bismarck Archipelago, New Guinea

Adriatic Sea That part of the Mediterranean between Italy and the Balkan Peninsula extending from the Gulf of Venice to the Strait of Otranto, between the heel of the Italian boot and the coast of Albania. (South of the Strait of Otranto is the Ionian Sea.) The sea is approximately 500 miles long and covers an area of 50,000 square miles. Venice, Ancona, and Bari are important ports on the Italian side, and the port city of Trieste, which has changed affiliations many times since Julius Caesar made it a colony of Rome, is now part of the Italian region of Friuli-Venezia Giulia. The Italian coast is relatively straight and continuous, but the Dalmatian coast, a favorite destination of yachtsmen, has many islands and inlets. Like most of the Mediterranean, the Adriatic is polluted and deficient in life, although there is fishing for lobsters, sardines, and tuna.

See also Ionian Sea, Mediterranean Sea

Aegean Sea An arm of the eastern Mediterranean between mainland Greece to the west and Turkey to the east, the Aegean is bordered on the south by the island of Crete. To the northeast, the Dardanelles (Hellespont) and the Sea of Marmara connect it with the Black Sea. It covers about 83,000 square miles and contains many islands, including the Thracian Sea group (Thasos, Samothrace, Lemnos); the Aegean group (Lesbos, Chios, Ikaria, Samos); the northern Sporades (Skyros); the Cyclades (Melos, Paros, Naxos, Santorini); the Saronic Islands (Salamis, Aegina, Poros, Hydra, Spetsai); the Dodecanese (whose principal island is Rhodes); and Crete, the largest of the Greek islands and the southernmost land in Europe. Bronze Age cultures flourished in the Aegean region between 3000 and 1000 B.C., and included the Minoan civilization of Crete and the Mycenaeans on mainland Greece. The so-called Aegean civilizations collapsed when they were attacked and conquered by invaders from the north, known colloquially as the Sea Peoples.

See also Dardanelles, Mediterranean Sea

Aeolian Islands Named for Aeolus, the Greek god of winds, this group of seven major islands and numerous smaller ones is sometimes called the Lipari Islands for the largest of the group. Lying in the Tyrrhenian Sea, off the northeast coast of Sicily, the group also includes Stromboli, Vulcano, Alicudi, Filicudi, Panarea, and Salina. Stromboli and Vulcano are active volcanoes and have erupted in recent times. (Stromboli is known as the lighthouse of the Mediterranean because it has been erupting constantly for five thousand years.) Lipari, with 8,500 inhabitants, is the administrative center of six of the islands; Salina is self-governing.

See also Sicily, Stromboli, Vulcano

African penguin (*Spheniscus demersus*) Like their close relatives in South America, African penguins are known as jackass penguins because of their loud, braying call. They are the only penguins native to southern Africa, but there have been occasional reports of stray macaronis, kings, and rockhoppers. Standing about 2 feet high, this species, like the other members of the genus *Spheniscus* from southern South America and the Galápagos, is strongly marked with a black band

African penguin

that runs from the chest to the feet, and white coloring that swoops up and over the eye, giving it the appearance of wearing a black mask. Oil spills and predation by sea lions, sharks, and men have greatly reduced the population, which was estimated at more than 1 million birds at the beginning of the twentieth century.

See also Galápagos penguin, Humboldt penguin, Magellanic penguin

Agassiz, Alexander (1835–1910) The son of Louis Agassiz, Alexander Agassiz was born in Neuchâtel, Switzerland. He came to the United States in 1849, but after studying at Harvard, he went west to become a mine superintendent in Calumet, Michigan. He turned an unprofitable company into the world's foremost copper mine, and when he returned to Massachusetts in 1869, he made large donations to the Museum of Comparative Zoology, where he served as curator from 1874 to 1885. He greatly advanced the knowledge of marine zoology and oceanography and led expeditions to South America, the Gulf Stream, the West Indies, Hawaii, the Bahamas, Bermuda, and the Great Barrier Reef.

Agassiz, Louis (1807–1873) Swiss naturalist, originally trained in philosophy and medicine. An associate of Alexander von Humboldt and Baron Georges Cuvier, he achieved fame in Europe for his publications on fossil mollusks and fishes and the glaciers of Switzerland, and came to the United States in 1846 to lecture at the Lowell Institute in Boston. The following year, he accepted a professorship of zoology at Harvard, and developed the collections that were to become the Museum of Comparative Zoology. He founded the Anderson School of Natural History on Penikese Island in Buzzards Bay, Massachusetts, for which he was recognized as the most influential science teacher in America. When he died, the school closed. He led expeditions to Brazil in 1865, to Cuba in 1869, and around Cape Horn to California in 1871. Agassiz was a lifelong opponent of Darwin's *Origin of Species*, maintaining to his death that only a Supreme Being could have created the variety of animal and plant life on earth. Each species, he wrote, was "a thought of God."

See also Humboldt

Agulhas Current Flowing off the east coast of southern Africa, the Agulhas Current is a part of the large-scale circulation of the southern Indian Ocean. (*Agulha* is Portuguese for "needle" and refers to the dangerous, saw-edged rocks that mark Cape Agulhas, Africa's southernmost point.) As with most ocean currents, it is wind-driven, but not by the winds in its immediate vicinity. The southeast trade winds set up an east-to-west current (the South Equatorial Current) that is deflected by the southeast coast of Africa, part of it becoming the Mozambique Current between Madagascar and the African continent and part of it becoming the Agulhas. It is largely a warm-water current, responsible for the warm, often humid weather of the Natal coast.

See also Benguela Current, current

air bladder: See gas bladder

aircraft carrier A naval vessel from which airplanes can take off and land. In 1910, Eugene Ely flew a plane off a specially built deck on the U.S. cruiser *Saratoga*, and the following year (in another plane) he landed on a platform built on the quarterdeck of the battleship *Pennsylvania*. During World War I, the British built the first aircraft carrier with an unobstructed flight deck, the HMS *Argus*, but the war was over before it could be put into action. The U.S. Navy's first carrier was the *Langley*, a converted collier, in 1922. In 1927, the U.S. Navy converted the *Lexington* and the *Saratoga* to carriers from battle cruisers. Carriers were used extensively in combat in World War II; the attack on Pearl Harbor in December 1941 was made by planes launched from Japanese carriers. The carrier played a major role in the battles of Midway and Leyte Gulf, and the Battle of the Coral Sea (May 1942), fought exclusively with naval aircraft, was the first sea battle where the opposing fleets never saw each other. In 1955, the *Forrestal* became the first carrier with a large extension to the flight deck, and in 1961 the *Enterprise*, the first nuclear-powered carrier, was built. Jet aircraft are stored belowdecks and brought up and down by elevators. Most are launched into the air by catapults, and caught when they land by arresting wires and tailhooks. The 102,000-ton Nimitz class carrier *John S. Stennis*, commissioned in 1995, is the largest warship ever built, with a flight deck that is 1,092 feet long and a complement of six thousand including aircrews.

See also battleship

albacore (*Thunnus alalunga*) Identifiable by its very long pectoral fins that reach beyond the anal fin, the albacore is a medium-sized tuna (maximum length: 4½ feet) that inhabits the temperate and subtropical waters of the world. The average albacore weighs between 10 and 20 pounds, but the present rod-and-reel record is 88 pounds. It is one of the world's most important food

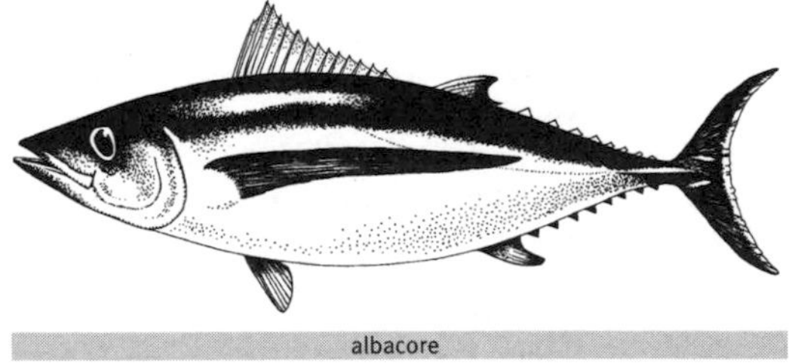
albacore

fishes, the only tuna that can be labeled "white meat" on the can. They are commercially caught by pole-and-line fishermen from bait boats, although there is some purse seining and long-lining.

See also skipjack, yellowfin tuna

albatross (*Diomedea* spp.) The common name for a group of large, long-winged seabirds found mostly in the Southern Hemisphere. (The waved albatross, *D. irrorata,* nests in the Galápagos, on the equator: and the Laysan albatross, *D. immutabilis,* nests on Laysan Island but circumnavigates the North Pacific.) They are famous for their sustained soaring abilities, and they have been known to follow ships for weeks at a time. The largest is the wandering albatross (*D. exulans*) with a wingspan that may exceed 12 feet, making it the biggest flying bird in the world. Because of their awkwardness on land, albatrosses are sometimes called mollymawks, from the Dutch *mallemuk,* which means "foolish gull." At the start of the breeding season, albatrosses perform elaborate displays, which include head raising, bill clacking, and intricate posturing that has been described as "dancing." A single egg is laid, usually in a simple nest in the grass, and the parents travel enormous distances to capture food for the chicks, which they pass along by regurgitating the semi-digested food into the chick's mouth. Their diet consists mostly of squid and small fishes that they catch at or near the surface of the water, but they will also eat garbage from ships. Like many other oceanic birds, albatrosses drink seawater. Seamen considered it bad luck to kill an albatross, a theme immortalized in Coleridge's "Rime of the Ancient Mariner" (1798). Along with the shearwaters, storm petrels, and diving petrels, albatrosses are tubenoses. Their nostrils extend on top of the bill through two horny tubes, the function of which is not fully understood but probably has something to do with oil secretion.

See also black-browed albatross, black-footed albatross, Laysan albatross, shy albatross, sooty albatross, wandering albatross, waved albatross

Alcatraz Island in San Francisco Bay that was first visited by the Spanish in 1769, who named it for the pelicans (*alcatraz* in Spanish) they found there. It was fortified by the Spanish, and came under U.S. control in 1852. Alcatraz was used as a military prison from 1859 to 1934, when it became a federal penitentiary. Known as the Rock, it was the most notorious maximum-security facility in the United States, symbolic of strict discipline, impregnability, and inescapability. (Two prisoners escaped, but they are not believed to have survived the swim across the bay.) All told, some 1,500 prisoners were incarcerated there, averaging 250 at any one time. Among the more notable inmates were Al Capone; "Machine-Gun" Kelly; Alvin Karpis; Robert Stroud, "the Birdman of Alcatraz"; and Morton Sobell, who had been tried for treason with the Rosenbergs in 1952. The prison was closed in 1963, and in 1970 the island was invaded and occupied by a force of Native Americans, asserting their rights to abandoned federal property. In 1972, Alcatraz was made a tourist destination, part of the Golden Gate National Recreation Area.

Alcidae The family of seabirds that is commonly known as auks. There are thirteen genera and twenty-two species. (There were twenty-three, but the great auk, the largest of the alcids, became extinct in 1844.) Alcids, which are confined to the Northern Hemisphere, are more or less the counterparts of the penguins in the Southern Hemisphere. (Living alcids can all fly, but the great auk, like the penguins, could not.) They have dense, waterproof plumage (usually black and white) and are awkward on land. They are skillful divers and swimmers that feed by diving from a sitting position and "flying" through the water. Depending on size, they feed on fishes or smaller planktonic organisms. The largest of the alcids is Brünich's guillemot (*Uria louvia*), with a 30-inch wingspan; the smallest is the least auklet (*Aethis pusillia*), with a wingspan of just 6 inches. Most alcids breed in colonies on sea cliffs and rocky slopes, but some nest in trees.

See also great auk, murre, puffin, razorbill

Aldabra An oval-shaped coral atoll in the Indian Ocean, north of the Comoros and Madagascar. Aldabra is 21 miles long and about 8 miles wide, enclosing a shallow lagoon. The island is now uninhabited, but in the 1950s it was mined for the guano produced by the numerous seabirds that nest there. With Farquhar, Desroches, and the Chagos Archipelago, Aldabra was bought from private owners in the Seychelles in 1965, and now forms the British Indian Ocean Territory (BIOT). A joint British-American plan to build an airstrip was thwarted by conservationists and Britain's Royal Society to protect the island's rare and endangered wildlife. Rare green turtles nest on Aldabra, but the island's most celebrated inhabitant is the giant tortoise (*Geochelone gigantea*), which was eliminated from many other Indian Ocean islands, including Réunion, Rodriguez, and Mauritius. The only other giant tortoises are found on the Galápagos Islands in the eastern Pacific, and they, too, are endangered.

See also Galápagos tortoise, Madagascar

Alderney Known as *Aurigny* in French, Alderney is one of the Channel Islands, just off the coast of Normandy, but officially a part of England. The nearest English coast is 55 miles away; it is separated from the French mainland by the 10-mile-wide Race of Alderney, a swift tidal current that makes the crossing difficult. The northernmost of the Channel Islands, Alderney is only

3 square miles in area and sustains a population of some 2,100 people. The island has been inhabited since the Bronze Age (about 3,500 years ago), and the Romans built a fort there. Along with the other Channel Islands, Alderney became part of England after the Norman invasion in 1066 and has been the subject of ownership disputes between Englishmen and Frenchmen ever since. It is now in the bailiwick of Guernsey, but has its own elected assembly. The populace was evacuated to England before the Germans arrived in 1940 and repatriated after the war.

See also Channel Islands, Guernsey

Aleutian Islands Chain of about seventy volcanic islands stretching westward for 1,200 miles from the Alaska Peninsula to the Commander Islands off Kamchatka. The four main groups of islands (Fox, Andreanof, Rat, and Near) separate the North Pacific Ocean from the Bering Sea. There are forty-six active volcanoes in the Aleutians, twenty-six of which have erupted since 1760. Deep under the Pacific where the Aleutians rise from the seafloor is the great Aleutian Trench, one of the most seismically active regions in the world. (The great tsunami of 1946, which devastated Hilo in Hawaii, was generated by an earthquake in the Aleutian Trench.) When Vitus Bering visited the islands in 1741, he saw groups of indigenous people living there; relatives of Siberian Eskimos called Aleuts, they hunted whales and seals with poisoned arrows. Bering did not survive the "Great Northern Expedition," but others reported the profusion of fur seals, sea otters, and foxes, and soon Russian hunters were setting up camps throughout the islands to kill these fur-bearing animals. (Steller's sea cows also lived in the Aleutians. They were totally eliminated by Russian sealers by 1768, only twenty-seven years after their discovery in the Commander Islands.) When the United States bought Alaska from Russia in 1867, the Aleutians were included in the purchase; they became part of the United States, but they brought their Russian Orthodox religion with them. During World War II, the Japanese occupied Attu, Kiska, and Amchitka islands at the western end of the chain, but the United States regained them in 1943. The conservation organization Greenpeace was formed in 1969 to protest nuclear testing on Amchitka Island. (The testing took place anyway.) Nowadays, the islands that have been developed—such as Unalaska, which has a population of about 4,000—serve as home to Aleuts who base their economy on support of the commercial fishing industry in the Bering Sea. Dutch Harbor in Unalaska is the number one port in the nation for seafood volume and value: in 1993, 52 percent of Alaska's entire commercial fish catch was landed there.

alewife (*Alosa pseudoharengus*) Sometimes known as grayback herrings, alewives can be differentiated from

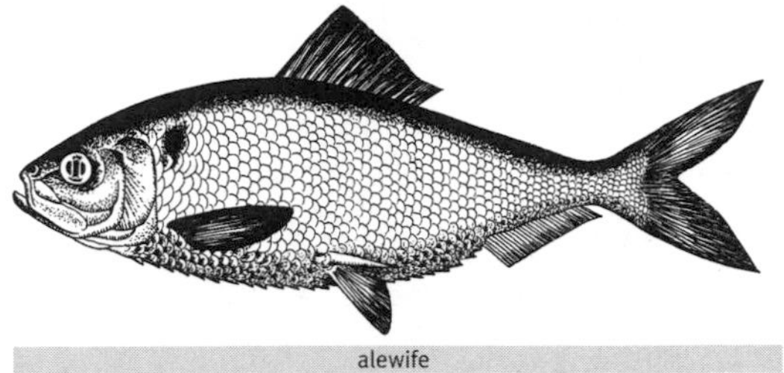

alewife

Atlantic herrings (to which they are closely related) by their deeper body shape and the presence of a small patch of teeth on the tongue. They are found along the Atlantic coast of North America, from Labrador to Florida, and throughout the St. Lawrence River drainage into Lakes Huron and Michigan. They reach a length of 15 inches.

See also Atlantic herring, menhaden, shad

Alexander Archipelago Group of islands off southeast Alaska, including Prince of Wales Island, Admiralty, Kupreanof, Baranof, Chichagof, Kuiu, Mitkof, Wrangell, Revillagigedo, and many smaller islands. (The city of Sitka is on Baranof, and Ketchikan is on Revillagigedo.) The entire archipelago is the Tongass National Forest, but the U.S. government has allowed lumber interests to clear-cut vast tracts of the 17 million acres of hemlocks, spruce, and cedar to the point where Prince of Wales Island, once the most heavily forested of all the islands, is virtually stripped bare of its original growth. Glacier Bay National Park is at the northern end of the archipelago. **See also Prince of Wales Island**

algae The most primitive plants, lacking true roots, stems, leaves, and flowers but containing chlorophyll, algae are worldwide in distribution and are the major aquatic plant life in both seawater and freshwater. The simplest forms are tiny and unicellular, but some of the more complex forms, like kelp, can reach a length of 100 feet. Blue-green algae (cyanobacteria) contain chlorophyll, but under certain conditions the cells can ingest food in an animal-like manner, and many scientists now classify the blue-green algae as neither plant nor animal but a separate phylum altogether. Algae is eaten by many of the smallest members of the zooplankton, and therefore forms the very base of the entire oceanic food pyramid. Most of the oceans are plantless, so the algae provide the oxygen required by animals that live in the sea. Ninety percent of the earth's photosynthesis is carried out by algae. Virtually all seaweeds are marine algae. In addition to providing food and oxygen for fishes, clams, crabs, shrimp, squid, octopuses, worms, and almost everything else that lives in the ocean, algae is an important food source for people. The red alga *Porphyra* is farmed in great quantities in Japan; other forms of "seaweed" are eaten in Canada and numerous countries in Europe and Asia. The green alga *Ulva,* which looks like lettuce, was eaten by early seafarers to ward off scurvy. A variety of seaweeds are used in the manufacture of prepared foods, such as cake mixes, ice creams, and puddings, as well as shampoos, cosmetics, and soaps. Agar, used as a laboratory culture medium, comes from algae. Algae can also be toxic. Ciguatera, a disease that can be transmitted by eating certain fishes, is algae-based. Algal saxitoxins that accumulate in the flesh of various shellfish are responsible for paralytic shellfish poisoning, and the "red tides" are caused by the alga *Gymnodinium breve.*

See also ciguatera, kelp, red tide, zooplankton

Almeida, Francisco de (1450–1510) Portuguese admiral; the first viceroy of Portuguese India. Sent to India in 1503 as a captain major, he was involved in the Portuguese conquest of Calicut. After returning to Lisbon, he set out again for India in 1505, rounded the Cape of Good Hope, and sailed up the east coast of Africa, where he destroyed Mombasa, then sailed on to Cochin to take up his position as viceroy. He built numerous forts to strengthen Portugal's position in India and sent his son Lourenço on various exploring missions, one of which resulted in the European discovery of the island of Ceylon. He repulsed attempts by Arabs and Egyptians to unseat him, and when Afonso de Albuquerque arrived from Lisbon to replace him, he had him thrown in prison. On his return voyage to Portugal, Almeida was killed in a skirmish with the Hottentots at Table Bay, South Africa.

Alvin Research submersible named for Allyn Vine and launched in 1964 at Woods Hole Oceanographic Institution in Massachusetts. It consists of a pressurized spherical compartment housed in a little submarine, which gives it the mobility lacking in the bathysphere. After many modifications, *Alvin* is now capable of dives to a depth of 15,000 feet. In addition to her immeasurable contributions to deep-sea biology and geology, *Alvin* has been involved in many of the more dramatic deep-sea discoveries in recent years, including the discovery of hydrothermal vents in the Galápagos, the mapping of the Mid-Atlantic Ridge (Project FAMOUS), and the locating of the *Titanic.* She also assisted in the retrieval of a lost hydrogen bomb off Palomares, Spain, in 1966 and was rammed by a swordfish in 1967. *Alvin* is now owned by the U.S. Navy.

See also Ballard, bathysphere, hydrothermal vents, Project FAMOUS, *Titanic, Trieste*

Amazon River dolphin (*Inia geoffrensis*) Also known as boutu, these 8-foot-long freshwater dolphins vary in color from dark gray in the juveniles to almost white in the adults, which often have a pinkish tinge. Their eyes are small but functional, and their acoustic senses are

Amazon River dolphin

particularly well developed, enabling them to navigate and hunt in the murky waters in which they live. Unique among all dolphins, boutus have two kinds of teeth in the jaws, conical ones in the front and molar-like ones toward the rear. They are found throughout the Amazon and Orinoco basins in northern South America. **See also echolocation, freshwater dolphins**

ambergris From the French for "amber" and "gray," this is a waxy, gray, or blackish peatlike substance that forms as an impaction in the lower intestinal tract of some sperm whales. It is not known how or why it forms, and it is encountered only rarely. Ambergris (pronounced "amber-*griss*" or "amber-*grease*") was sometimes vomited up by the whales in their death throes, sometimes found when the whale was cut open, and sometimes spotted floating on the surface of the sea. It was used as a fixative for perfumes, and because it served this purpose better than any known substance, it was extremely valuable. At times, it was worth more than three times its weight in gold. The largest ambergris "boulder" ever collected weighed 983 pounds; in 1912 it was sold for sixty thousand dollars, saving its finder, a small whaling company, from bankruptcy. **See also sperm whale, sperm whaling**

amberjack (*Seriola dumerili*) The largest of the jacks and the most common in the tropical western Atlantic, it is also found worldwide. Although most amberjacks are bluish above and silvery below, they are occasionally amber-colored. Sometimes mistaken for the bluefish, it has a more incavated tail and a dark stripe that runs from the corner of the mouth through the eye and

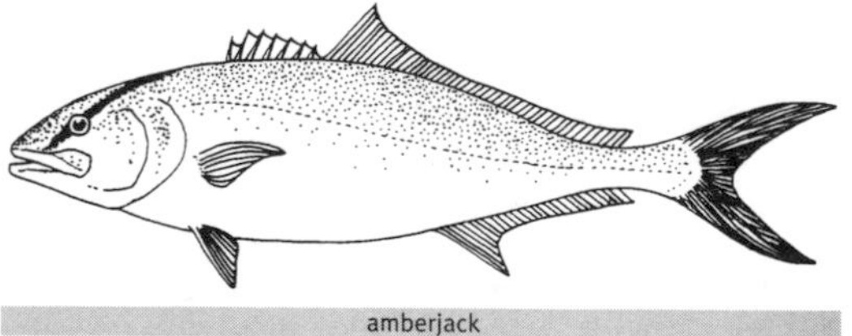

amberjack

to the base of the first dorsal fin. It is a popular game fish, but care has to be taken in eating it because it is second only to the barracuda in causing cases of ciguatera in the West Indies. The amberjack can reach a length of 6 feet and a weight of 155 pounds.

See also ciguatera, jack crevalle, jacks, permit, pompano

Ambon Also spelled "Amboine" or "Amboina," this is a fertile Indonesian island in the Moluccas, the group known as the Spice Islands for the nutmeg and cloves grown there. Discovered by the Portuguese in 1512, Ambon was captured by the Dutch in 1605. In 1623, when the Dutch governor of the island believed that certain English merchants were planning to kill him and overwhelm the Dutch garrison, he had the alleged plotters (ten English and ten Japanese) captured and executed. Known thereafter as the Amboina Massacre, this event contributed to the friction between the Dutch and the British not only in the East Indies but also on the whaling grounds of the Arctic, where they were also in competition. After the massacre, the British East India Company concentrated its efforts in India. Ambon is now the capital of the Indonesian province of Maluku.

See also British East India Company, Moluccas

American Virgin Islands Although geographically a part of the Lesser Antilles, the Virgin Islands are perched on the same continental shelf as Puerto Rico and the Greater Antilles to the west. When Columbus landed on St. Croix in 1493, he found nothing to keep him there and sailed on. (He named the islands after Saint Ursula and the eleven thousand martyred virgins.) St. Thomas and St. John, the other large islands that constitute the American Virgins, were discovered by Columbus on the same voyage, but Spain evinced little interest, and by 1625 English and French settlers were farming on St. Croix and it had become a base for pirates like Blackbeard (Edward Teach). In 1717, the Danish West India Company colonized the islands for tobacco and cotton plantations. The Danes and the French put down a slave rebellion on St. John in 1732, and the slaves, rather than go back to servitude, committed mass suicide. During the nineteenth century, Denmark and England vied for dominance of these islands, but Denmark won—at least until 1917, when the United States purchased the islands for $25 million to keep Germany away from the Panama Canal. The islands are popular tourist destinations; each of them has a notably different character. Two-thirds of St. John is a national park; there is one luxury resort, Caneel Bay, built by the Rockefellers. St. Croix, with its bustling capital of Christiansted, is more developed. (The world's largest petroleum refinery is also on St. Croix.) St. Thomas is the administrative capital of the USVI, and the duty-free port of Charlotte Amalie (population 12,500) has one of the best harbors in the Caribbean and is the popular destination of many cruise ships. Residents have all the rights and privileges of American citizens and elect a nonvoting representative to the House of Representatives; they do vote in presidential elections.

See also St. Croix, St. John

America's Cup Sailing trophy originally known as the Hundred Guineas Cup and won by the New York Yacht Club's *America* in an 1851 yacht race around the Isle of Wight. The American winners donated the cup to the New York Yacht Club (NYYC) for a perpetual international competition, but with the rules to be determined by the holders. In 1880, for example, the NYYC sent no fewer than seventeen boats to defend the cup, claiming that the *America* had won it against seventeen British yachts. When Sir Thomas Lipton was making his regular challenges (1899–1930), the rules stipulated that the challenger had to be sailed across the Atlantic to the race, which meant that it had to be much heavier than the defender, which could be a pure racing machine. Originally the race was held between the largest racing yachts, but after World War II the decision was made to race only 12-meter boats. (No 12-meter yacht actually measures 12 meters; there is a complicated mathematical formula that determines the actual allowable length.) The cup remained the property of the NYYC for 132 years—the longest unbroken winning streak in sports—until 1983, when *Australia II*, outfitted with a revolutionary "winged keel," won the seventh and deciding race from Dennis Conner, sailing *Liberty*. In 1987, the Americans took back the cup with *Stars & Stripes*. Then, in 1988, the New Zealanders entered a huge 130-foot monohull, and the Americans, again led by Dennis Conner, raced a catamaran, which ran away with the contest. The 1995 race, held in Conner's home waters off San Diego, was won by the New Zealand yacht *Black Magic*, only the second victory by a non-American challenger in the history of the competition.

Amistad Schooner that was carrying fifty-three slaves in 1839 from Sierra Leone to Cuba when the slaves, led by a man named Cinqué, mutinied and took the ship. Since they did not know how to navigate, they turned over the helm to one of the Spaniards who was planning to sell them in Cuba, and told him to take them back to Africa. By sailing west at night and eastward during the day, he tricked them into believing he was taking them home. Instead he got them to the eastern tip of Long Island, where the *Amistad* was apprehended by the U.S. Coast Guard and brought to New London, Connecticut. In the court cases that followed, abolitionists argued that Cinqué and his codefendants had been kidnapped and ought to be returned to Africa, and the Spanish authorities claimed that they were

Spanish property. Attorney Roger Baldwin, of New Haven, Connecticut, argued successfully that they had been illegally kidnapped, and the judge set them free, but President Martin Van Buren sided with the Spanish claims, and the case was appealed all the way to the Supreme Court. Former president John Quincy Adams argued in the slaves' defense, and his arguments persuaded the court to free the slaves and return them to Africa. In 1997, Steven Spielberg made a movie about the *Amistad* story.

ammonite Named for the Egyptian god Ammon, whose head was that of a ram, ammonites are extinct cephalopods that flourished from the Devonian (400 million years ago) to the Cretaceous period (144 to 66 million years ago). They lived in the forechamber of chambered shells, most of which were coiled like the horn of a ram, but others were straight. Some were tiny, but *Pachydiscus* had a great spiral shell that was 6 feet across. Most of the ammonites were extinct by the Triassic period, some 200 million years ago, but some lasted for another 130 million years, into the Cretaceous. They left no descendents. **See also cephalopod**

Amoco Cadiz Supertanker carrying 68.7 million gallons of crude oil that lost her steering and ran aground in a gale on the rocks off Portsall, Brittany (France), on March 17, 1978. Heavy seas made it impossible for a smaller tanker to come alongside the *Amoco Cadiz* to offload the remaining oil, and when a tug attempted to tow the tanker out of trouble, she broke in two. The entire cargo of the 220,000-ton ship was discharged into the ocean, covering 400 miles of the French coastline with oil up to a foot thick. A massive cleanup followed, but more than thirty thousand seabirds died, along with 230,000 tons of crabs, lobsters, and fish. The prized oyster beds of the area were completely destroyed. The U.S.-owned Amoco Oil Corporation was found guilty of negligence and failure to properly train the ship's crew, and in 1998 Amoco had to pay $85.2 million to ninety Breton communities.

See also *Castillo de Bellver*, *Exxon Valdez*, oil spill, supertanker, *Torrey Canyon*

amphipod The vast majority of the three thousand species of amphipods are marine, living in algae or swarming on the bottom near the coastal shelves. Where isopods—which they resemble and are related to—are *de*pressed (flattened from above and below), amphipods are *com*pressed (flattened from side to side). In most amphipods, the body is elongated and the thorax is fused to the head. The largest amphipod (*Cytosoma magna*) is 5 inches long, but most species are less than ¼ inch in length. With their arched backs, they resemble fleas, and many of the species known as beach fleas or sand hoppers can be seen hopping about in seaweed that has been washed ashore. Amphipods feed on decaying plant and animal matter, and in turn are fed upon by invertebrates, fishes, whales, and birds. The group includes the skeleton shrimp (family Caprellidae), which are amphipods with an elongated body composed of tubular sections; these are given to posing vertically on grasses while waiting for food. Whale lice are dorsiventrally flattened amphipods that are adapted to live on the skin of whales. **See also copepod, isopod, skeleton shrimp, whale lice**

amphipod

ampullae of Lorenzini Electrical receptor organs found on the underside of the snout of all sharks and rays. These small, jelly-filled pores (*ampulla* is Latin for "flask") were first described in detail in 1678 by Stefano Lorenzini, an Italian anatomist. In the 1970s, Adrianus Kalmijn and others showed that sharks have the greatest electrical sensitivity in the animal kingdom; they can detect electrical discharges as small as 0.005 microvolts, and can also determine the directionality of the charge. While the shark's smell and vision are useful at a distance, electrical sensitivity is probably its most effective short-range stimulus. Sharks and rays that swim close to the bottom searching for prey, such as the hammerhead and the goblin shark, have particularly numerous and well-developed ampullae.

Amsterdam and Saint-Paul Islands Part of the French Southern and Antarctic Territories (along with the Kerguelen and Crozet Islands, and Adélie Land on the Antarctic continent), Île Amsterdam and Île Saint-Paul are in the southern Indian Ocean, 1,000 miles northeast of Kerguelen and about midway between the southern tip of Africa and the southwest corner of Australia. The islands were discovered by Portuguese explorers in the sixteenth century, but attempts to colonize the islands from Réunion, far to the north, failed repeatedly. Sealers visited the islands during the nineteenth century, but the inhospitability of the islands (Amsterdam has no harbor) encouraged their quick departure after they had slaughtered and skinned the seals. When Abel Tasman visited Tonga in 1643, he named one of the islands Amsterdam and another Rotterdam. When Cook arrived in 1773, he named the archipelago the Friendly Islands, and although he noted that he had landed on "Amsterdam Island," it was not the island that currently bears the name. Saint-Paul is a volcanic crater that has been inundated by the sea. Île Amsterdam and Île Saint-Paul are uninhabited, but many birds breed there, including rockhopper penguins.

See also Crozet Islands, Kerguelen Islands, Réunion

Amundsen, Roald (1872–1928) Born in Oslo, Roald Amundsen was the most celebrated and successful of polar explorers. He studied medicine, but the call of the sea was too great and he chose to become an adventurer. He was a member of the first party to winter in the Antarctic (1898), and the first to successfully navigate the Northwest Passage, in the 47-ton sloop *Gjöa,* from 1903 to 1906. He had planned to drift across the North Pole in Fridtjof Nansen's *Fram,* but when he learned that Robert Peary had already conquered the Pole, he continued his preparations but sailed south instead. On October 20, 1911, he set out with four companions and fifty-two dogs, and reached the South Pole on December 14. In 1925, with the American explorer Lincoln Ellsworth, he flew to within 170 miles of the North Pole, and the following year he passed over it in a dirigible with Ellsworth and Italian engineer Umberto Nobile. In 1928, Amundsen died in an attempt to rescue Nobile, whose dirigible had crashed on the ice northeast of Spitsbergen. Nobile was rescued by others.

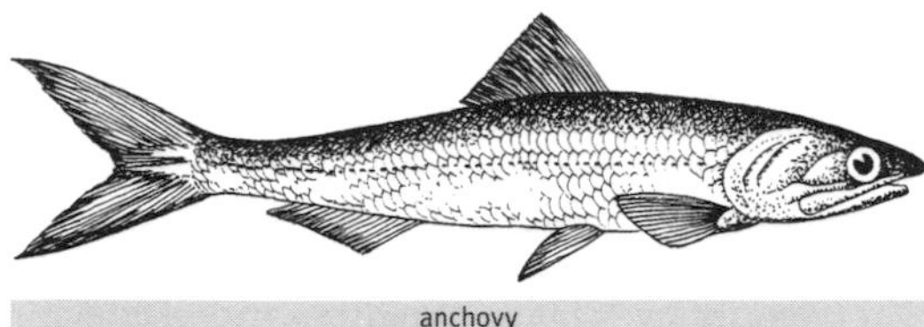

anchovy

anchovy (*Engraulis mordax*) There are many species of anchovies, all of which are small, long-snouted fishes with an underslung mouth that average about 4 inches in length. (Shown here is the northern anchovy of California, which gets somewhat larger.) Another species, *Engraulis encrasicholus,* is harvested in Spain, Portugal, and Italy, where it is canned or made into relishes and pastes. (Anchovies are not eaten fresh.) The Peruvian fishery for the anchoveta *E. ringens* was the world's largest in the 1970s, but overfishing, combined with adverse environmental conditions associated with the 1972–1973 El Niño, caused the fishery to collapse. From a 1970 high of 12.3 million tons, the number crashed to 4.5 million in 1972, and continued to decline until 1980, when the fishery shut down completely.

See also sardine

Andaman Islands Usually grouped with the Nicobar Islands to the south, the Andamans are part of a union territory of India in the Bay of Bengal. The Andamans consist of more than two hundred islands, but only twenty-six are inhabited. The main islands are North, Middle, and South Andaman. The islands were first visited by Lieutenant Archibald Blair of the British East India Company in 1789, and the capital city, on South Andaman, is named Port Blair. In the late eighteenth century, the British established a penal colony in the Andamans, and the Japanese occupied the islands from 1942 to 1945. They passed to India with that country's independence in 1947. The Nicobars consist of twelve inhabited and seven uninhabited islands, where coconuts, betel nuts, pandanus, and mangoes are grown.

Andrea Doria Named for the Genoese naval hero who lived from 1466 to 1560, the *Andrea Doria,* built in 1951, was Italy's finest postwar cruise ship. Fitted out in the most contemporary style, she had a Gothic chapel, a swimming pool on every deck for every class of passenger, and a promenade for dogs. Her owners believed she was virtually unsinkable because of her double hull and watertight compartments, but she had sixteen lifeboats anyway, with a maximum capacity of two thousand people. From 1952 to 1956, she functioned as a successful cruise ship, appealing especially to Europeans. On July 23, 1956, as she was passing through Nantucket Sound in a dense fog on her way to New York, she was struck broadside by the Swedish liner *Stockholm.* The bow of the *Stockholm* drove 35 feet into the starboard side of the *Andrea Doria,* causing her to take on such an enormous quantity of water that she began to sink. Half her lifeboats were useless because they were wedged against the superstructure of the boat deck. (Captain Piero Calamai would later testify that he didn't make any announcements because he didn't want the passengers to know about the shortage of lifeboats.) An SOS went out immediately, and a flotilla of Coast Guard cutters, oil tankers, and freighters came to the rescue. The French liner *Île de France,* more than two hours away, rushed to the scene, providing many of the necessary lifeboats. Taking in water, the Italian liner sank to the bottom in eleven hours. The *Stockholm,* with her bow caved in and many of the *Andrea Doria*'s passengers aboard, limped into New York Harbor. Miraculously, only fifty-one people died as a result of the collision, forty-four of them passengers on the *Andrea Doria.* In 1982, underwater explorer Peter Gimbel (who had been the first diver to photograph the *Andrea Doria* after she sank) led a salvage expedition to the sunken hull and retrieved the purser's safe, which was opened with great fanfare on national television. It contained a lot of waterlogged bills.

anemone fish (*Amphiprion* spp.) Damselfishes with a strange lifestyle: they live within the poisonous tentacles of sea anemones, made immune to the stinging nematocysts by a coating of protective mucous. These little fishes—also known as clown fishes—maintain a commensal relationship with the anemones, in which both benefit from the association. The fish receives pro-

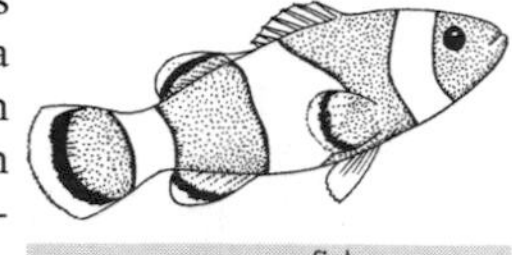

anemone fish

tection while cleaning debris and waste from the tentacles of the anemone. Found throughout the coral reefs of the Indian and Pacific Oceans, these 4-inch fishes are orange or yellow, with bold white vertical stripes.

See also damselfishes, sea anemone

angelfishes (family Pomacanthidae) Among the most strikingly colored of all tropical fishes, angelfishes are found throughout the world's coral reefs. Where angelfishes have a sharp, tailward-pointing spine on the lower margin of the gill cover (the operculum), their relatives the butterfly fishes do not. They are flattened, pancake-thin fishes with prominent eyes and a little, beaklike mouth. They nibble on small invertebrates, such as polychaete worms, corals, sponges, and algae. Quite often the juvenile forms are patterned completely differently than the adults, and when they sleep at night, the adults often change color too.

angelfish

See also butterfly fish, queen angelfish

angel shark (*Squatina dumerili*) This shark probably received its common name from its greatly expanded pectoral fins, but it is hard to imagine a more unangelic creature. Angel sharks, which reach a length of about 5 feet, have a flattened body, with the mouth at the end of the head and gill slits at the juncture of the head and the pectoral fins. They are classified as sharks, not rays, because the pectoral fins are not joined to the head. There are two very similar species in North American waters, the Atlantic angel shark and *S. californica* in the Pacific.

See also guitarfish

anglerfishes A group of fishes equipped with a lure at the end of a movable stalk that is used to attract prey items to the waiting angler. The lure is the modified first spiny ray of the dorsal fin. Anglerfishes are usually slow moving, and the shallow-water varieties spend a lot of time waiting on the bottom. This behavior, of course, is not possible for the deep-sea anglers, which are mostly mid- or deep-water forms and have to keep swimming. The deep-sea anglers have developed the additional enticement of a lure that lights up. In order to breed, the males, tiny compared to the females, attach themselves permanently to the females and deteriorate until they become nothing more than sexual appendages.

anglerfish

See also deep-sea anglerfish, frogfish

Anguilla The Carib Indians called this island Maliouhana, which means "eel," and in 1493 Columbus gave it the same name in Italian. The Spanish never colonized the island, but by 1650 English settlers had arrived from neighboring St. Kitts, and it has been affiliated with Britain ever since. In 1882, the St. Kitts, Nevis, and Anguilla federation was formed, and in 1958 the islands became a unit in the Federation of the West Indies. This was dissolved in 1962, and in 1969, chafing under St. Kitts control, Anguilla seceded from the St. Kitts–Nevis–Anguilla union. The British sent in troops to quell the rioting, and Anguilla was granted internal autonomy under British control. The most northerly of the Leeward Islands, Anguilla is 16 miles long by 3 miles wide; 35 square miles of sandy shoreline and low scrub, with a reputation for having some of the most beautiful beaches in the world. In order of economic importance, the major industries on Anguilla are tourism, boatbuilding, salt production (natural evaporation from salt ponds), and the export of spiny lobsters.

See also Columbus, St. Kitts and Nevis

Año Nuevo Island The Spanish navigator Sebastián Vizcaíno passed this location off the coast of northern California on January 3, 1603, and named it Punta Año Nuevo—"New Year's Point"—and the name was used for both the point and the island 850 yards offshore. The island, 12 acres in area at low tide, was occupied by Ohlone Indians, then ranchers, and, in 1870, a lighthouse to announce the hazardous nature of the coastline. In 1948, the manned light was replaced by an automatic light, sound, and radar reflector, and the island was abandoned by humans. The lighthouse crashed down in 1976, and the keeper's house has been completely colonized by California sea lions, who can be seen poking their heads out of the second-story windows. The marine mammals that bred along the shore, undeterred by the little channel, also occupied the island. Año Nuevo Island is the most important pinniped rookery in central and northern California, home to thousands of sea lions, harbor seals, Steller's sea lions, and elephant seals. The elephant seals (*Mirounga angustirostris*) were eliminated from the area by the beginning of the twentieth century, but the few remaining animals on Isla Guadaloupe (off Baja California) recolonized the mainland, and then the island, where more than three thousand of these ponderous giants now come ashore annually.

See also elephant seal, Steller's sea lion

Anson, George (1697–1762) British admiral and circumnavigator. After service in American, West Indian, and African waters, he was given command of a squadron of six ships and ordered to the Pacific to harass Spanish shipping. Five of his ships were lost and his crews depleted by scurvy, but in 1743, sailing in the

Centurion near the Philippines, he captured the *Nuestra Señora de Covadonga,* one of the fabled Manila galleons, loaded with more than a million silver pesos and 36,000 ounces of bullion. Brought to England, the treasure was evaluated at more than £500,000 and was paraded through the streets of London in thirty-two wagons. In 1745, his squadron defeated that of the French admiral Jacques de la Jonquière at the Battle of Cape Finisterre off northern Spain, and added to his personal fortune. Now a rich man, Anson was appointed admiral and made a peer, and married the daughter of the Lord Chancellor. He was made Lord of the Admiralty twice and reorganized British naval administration, including the introduction of standardized uniforms. Under his regime, England won the Seven Years' War (1756–1763), but he did not live to see England's triumph. **See also Manila galleon**

Antarctica Officially, the continental landmass that encompasses the South Pole, but known as the Antarctic when used to designate the entire region, including the islands of South Georgia, Heard, and Bouvetøya, as well as the South Shetlands, the South Orkneys, and the South Sandwich Islands. Early geographers used to postulate the existence of a southern continent, Terra Australis Incognita, to balance the landmasses of the north, but although James Cook sailed close and others approached the islands, the actual continent was not sighted until Thaddeus von Bellingshausen saw rocks (instead of ice) in January 1821. An otherwise obscure sealer named John Davis may have been the first person to *land* there, as shown in his log—discovered only in 1952—which indicates that he came ashore at Hughes Bay on the Antarctic Peninsula in February 1821. The Antarctic continent is roughly circular, with a peninsula pointing in the direction of the tip of South America. At an average elevation of 9,000 feet, it is the highest continent in the world, with a maximum height of 16,000 feet at the Vinson Massif in the Ellsworth Mountains. It is also the driest and coldest continent. The ice sheet covering the continent represents 90 percent of the world's ice, and more than half of its freshwater. The Ross Ice Shelf and the Ronne-Filchner Ice Shelf are each about the size of France. The tabular (flat-topped) icebergs of the Southern Ocean, some of which have been more than 100 miles long, break off from the ice sheets; the smaller bergs are calved from moving glaciers. There are no trees or terrestrial mammals in Antarctica, but penguins abound (and are found only in the Southern Hemisphere, except for the anomalous Galápagos penguins), seabirds are extremely numerous, and whales and seals used to be so abundant that their presence motivated the whalers and sealers to explore this region before anyone else. Explorers such as Robert Scott and Ernest Shackleton established their reputations by trying to reach the South Pole, but the Norwegian Roald Amundsen got there first—on December 14, 1911, a month before Scott's ill-fated expedition arrived—and perished on the return journey. Many countries have laid claims to portions of the Antarctic continent, and several nations have scientific stations there, but the Antarctic Treaty, an international agreement enacted in 1961 and finally ratified in 1997, designates the entire continent as a demilitarized and nonnuclear zone, and one where no commercial activity may be undertaken—except for controlled tourism.

See also Amundsen, Bellingshausen, Bouvet Island, Heard Island, Scott, Shackleton, South Georgia, South Orkney Islands, South Sandwich Islands, South Shetland Islands

Antarctic Convergence A boundary in the southern oceans in which cold, poorly saline Antarctic surface water sinks beneath warmer, southward-flowing subantarctic water. It is found at about 50° south latitude and marks a distinct change in the ocean's surface temperature and chemical composition, which in turn affects the creatures—fish, whales, seabirds, etc.—that live on either side of the convergence.

Antarctic Peninsula An S-shaped finger of land that points from the Antarctic continent toward the southern tip of South America, between 59° west and 67° west longitude and terminating in a cluster of rugged islands at about 62° south latitude. The northern portion of the peninsula is known as Graham Land and the southern is Palmer Land. It is separated from South America by the 600-mile-long Drake Passage, considered the worst body of water in the world because of the prevailing winds and high seas. The Antarctic Peninsula is bounded on the east by the Weddell Sea, and on the west by the Bellingshausen Sea. The Larsen Ice Shelf and the Ronne Ice Shelf form the eastern shore. Off the northwestern tip are the South Shetland Islands, which include Deception, an active volcano. After Ernest Shackleton's *Endurance* was crushed in the ice in 1915, he and his men made it to Elephant Island just beyond the northern tip of the peninsula, and Shackleton and five others began their incredible 800-mile voyage to South Georgia. The peninsula is the most accessible part of the Antarctic because it is the least beleaguered by pack ice and has the warmest climate. It is the location of many Antarctic scientific bases; it is also the most popular destination for tourist cruises. With its hundreds of offshore islands, the peninsula is a rich breeding ground for seabirds, seals, and penguins.

See also Drake Passage, *Endurance*, Shackleton, South Georgia

Antarctic tern (*Sterna vittata*) Since both the Antarctic and the Arctic terns are found around the Antarctic ice pack in summer, and since they look almost exactly alike, they ought to be difficult to tell apart, but they are not. When the Arctic tern arrives during the Southern Hemisphere summer, the Antarctic tern is in its breeding plumage, with an all-black cap. The Arctic tern, which breeds thousands of miles away in the high Arctic, is in its nonbreeding plumage, with a white forehead and underparts. Unlike its Arctic counterpart, which migrates thousands of miles every year, the Antarctic tern spends its entire life in the Southern Hemisphere. It breeds on subantarctic islands such as Kerguelen, the Crozets, Marion, the South Shetlands, the South Orkneys, and Tristan da Cunha, and disperses to the coasts of South America and South Africa, where from July to September flocks of five thousand or more are seen around Cape Town.

See also Arctic tern, tern

Anticosti Island Discovered by Jacques Cartier in 1534, this island in the Gulf of St. Lawrence was originally named Assumption. It is 135 miles long and covers some 3,000 square miles. The name "Anticosti," adopted in the seventeenth century, is believed to be derived from the Indian word *naticousti,* meaning "where bears are hunted." Louis XIV granted the island to Louis Joliet in 1680 as a reward for his discovery of the Mississippi, and it was annexed to Newfoundland (then a separate colony) in 1763. All attempts to settle this low, flat island failed, so in 1895 it was leased to a French chocolate factory owner named Henri Menier, who introduced various species of animals, including forty white-tailed deer. In 1926, the island was sold to Consolidated Bathurst for pulpwood lumbering, but forest fires and high transportation costs put an end to this endeavor. Anticosti is now inhabited by 250 people and 125,000 deer, but hunting permits have to be granted by Consolidated Bathurst.

Antigua and Barbuda Republic in the Leeward Islands of the Caribbean. At 108 square miles, Antigua is the largest of the three islands; Barbuda and Redonda are much smaller. The islands were discovered by Columbus in 1493, and Antigua (which means "ancient" in Spanish) was named for Santa Maria de la Antigua, a church in Seville. British colonists introduced sugarcane in 1632, and like many Caribbean islands, Antigua supplied sugar, rum, and molasses to Britain and her colonies. The sugar plantations were worked by slaves from Africa, but the abolition of slavery in England in 1833 applied to her overseas colonies (after a five- to seven-year period of apprenticeship to their owners), and without slaves, production declined. Sugar has not been commercially grown on the island since 1985; tourism is now the dominant industry. Antigua joined the Federation of the West Indies in 1958, and when the federation broke up, it became an associated state of the British Commonwealth in 1967. The little island nation achieved full independence in 1981.

See also Leeward Islands

aquaculture: See mariculture

Aran Islands A popular tourist attraction consisting of three islands 6 miles off the coast of western Ireland. The total population of the three islands—Inishmore (the largest), Inishmaan, and Inisheer (*inish* means "island" in Irish Gaelic)—is about two hundred people, most of whom speak Gaelic as their primary language. The prehistoric fort known as Dun Aengus is situated high on the cliffs of Inishmore, its makers and purpose unknown. In 1904, the Irish playwright John Millington Synge wrote the one-act *Riders to the Sea* about Aran fishermen, and in 1934 Robert Flaherty made the famous documentary *Man of Aran,* which showed the harsh life of the Aran Islanders, who are even today fishing in leather- or treated-canvas-covered boats known as curraghs. **See also curragh**

Arawak Indians American Indians who inhabited various islands in the Caribbean and the interior of South America. (Arawaks still survive in the interior of the Amazon rain forests.) They were agriculturalists who lived in villages of up to three thousand people, where they grew manioc and maize. They were peaceful, gentle people who were persecuted and driven out of the Lesser Antilles by the aggressive, cannibalistic Caribs even before the arrival of the Spanish. On other islands, such as the Bahamas, Cuba, Puerto Rico, Jamaica, and Trinidad, the Arawaks were enslaved by the Spaniards and often worked to death. Because they had no immunities to the diseases brought from Europe, those that were not killed outright died within a few years of the arrival of the Spanish conquerors. By the early decades of the fifteenth century, millions of Arawaks had been reduced to a few thousand. The word "buccaneer" comes from the Arawak word *buccan,* which was a grill made of green wood upon which meat was smoked.

Archelon A precursor of modern sea turtles, *Archelon* was a 12-foot-long giant from the Cretaceous period (150 million years ago). It lacked the distinct plates (technically known as scutes) that characterize most other sea turtles today; like the leatherback turtle, it had a thick, rubbery skin covering on its back. It had a narrow head, a hooked beak, and enormous, paddle-like flippers, and it probably fed on jellyfish. Most of the fossils of this gigantic turtle have been found in Kansas and South Dakota.

See also leatherback turtle, sea turtle

***Architeuthis:* See giant squid**

Arctic Circle A line of latitude around the earth, at approximately 66°30′ north, which marks the southern limit of the area where, for one day or more per year, the sun does not set (June 21) or rise (December 21). The length of continuous day or night increases northward from one day at the Arctic Circle to six months at the North Pole. Most of Greenland is above the Arctic Circle, as are a large portion of the Canadian Arctic Archipelago, northern Scandinavia, Alaska and Siberia, Wrangel Island, the New Siberian Islands, Severnaya Zemlya, Novaya Zemlya, Franz Josef Land, and Spitsbergen.

Arctic Ocean The smallest of the world's oceans; it is only one-sixth the volume of the next largest, the Indian Ocean. Its area is 5.44 million square miles, and its depth averages 3,240 feet. (The deepest point, on the Pole Abyssal Plain, is 14,800 feet below sea level.) The total includes the marginal Chukchi, East Siberian, Laptev, Kara, Barents, White, Greenland, and Beaufort Seas. The Arctic Ocean is almost completely surrounded by the landmasses of North America, Eurasia, and Greenland. While there are occasional small gaps in this encirclement—the Bering Strait, for example—the major part of the water inflow and outflow takes place in the Greenland Sea; roughly 80 percent of the water leaving and entering the Arctic passes through the narrow channel between Greenland and Spitsbergen. Within 80° north latitude, there are several islands and parts of islands (the northernmost part of Greenland is Cape Morris Jesup, at 83°40′ north), including several islands in the Canadian Arctic Archipelago, Severnaya Zemlya, Franz Josef Land, and northern Spitsbergen, but the North Pole is in the middle of the Arctic Ocean. Between 60° north and 75° north sea ice is seasonal (it appears in the fall, and remains through the winter and early spring), but most of the region above 75° north is permanently ice-covered. The ice of the northern Arctic Ocean is not solid like that of the Antarctic or even the Greenland ice cap; it floats in the ocean, put in motion by the prevailing currents and winds. (Sea ice is frozen salt water; it is different from icebergs, which are freshwater and have broken off from glaciers.) Because of the ice, and the continuous night during the winter, the Arctic Ocean is poorly explored. (Those who sought the Pole did very little exploration.) The exception is Fridtjof Nansen, whose *Fram* expedition in 1893–1895 was the first scientific study of the waters, contours, and conditions of the Arctic Ocean. In 1928, Sir Hubert Wilkins made the first airplane flight across the Arctic Ocean; in 1937 the U.S.S.R. set up the first floating scientific station; and now airplane and satellite observations have provided accurate images of this previously uncharted region. The first overland crossing of the Arctic Ocean was made by explorer Wally Herbert, who used dogsleds to traverse the ocean from Alaska to Spitsbergen (arriving at the Pole en route) in 1968–1969. **See also Barents Sea, Beaufort Sea, Chukchi Sea, East Siberian Sea, Franz Josef Land, Greenland Sea, iceberg, Kara Sea, Laptev Sea, Nansen, North Pole, sea ice, Spitsbergen**

Arctic skua (*Stercorarius parasiticus*) Also known as the parasitic skua or parasitic jaeger, this predatory bird resembles a large, brownish gull; it nests throughout the Arctic tundra regions and migrates south to a more pelagic habitat in winter. As with the other *Stercorarius* skuas (pomarine and long-tailed), the two central tail feathers are elongated, but less so than in the other species. The upper parts are brown, with characteristic white patches at the base of the primaries, and the undersides are lighter. There are dark and pale versions (morphs) but most of them have a black cap and yellowish cheeks. They feed by chasing other birds and forcing them to disgorge their food, which the skuas often catch before it hits the water. In the air, Arctic skuas are swift and agile flyers, which they have to be to chase the terns that are their principal victims. They are most common in the Arctic, from Greenland to Iceland, the Shetlands, the Orkneys, Norway, Sweden, Russia, and Spitsbergen, but they appear regularly along the coasts of South America and southwestern Africa. **See also long-tailed skua, McCormick's skua, pomarine skua, skua**

Arctic tern (*Sterna paradisaea*) The Arctic tern makes the longest annual migration of any animal in the world. Its breeding grounds are in Greenland (to about 83° north, farther north than any other bird), Iceland, the Faeroes, the Baltic, Russia, Alaska, and Canada; it winters in the Antarctic ice pack. The annual round-trip journey covers about 24,000 miles, a little less than the circumference of the earth. Summer in the north means twenty-four-hour daylight; when the terns arrive in the Antarctic, it is summer there, so this bird spends almost all its life in daylight. It is grayish white underneath, with white underwings, a gray back, a black cap, and a blood-red bill and feet. The wingspan is about 30 inches, and in breeding birds, the sharply forked tail streamers are particularly long. On their breeding grounds, where the females lay two eggs in a nest on the ground, Arctic terns are incredibly ferocious, attacking anything that comes within range, including people.

Arctic tern

See also Antarctic tern, common tern, tern

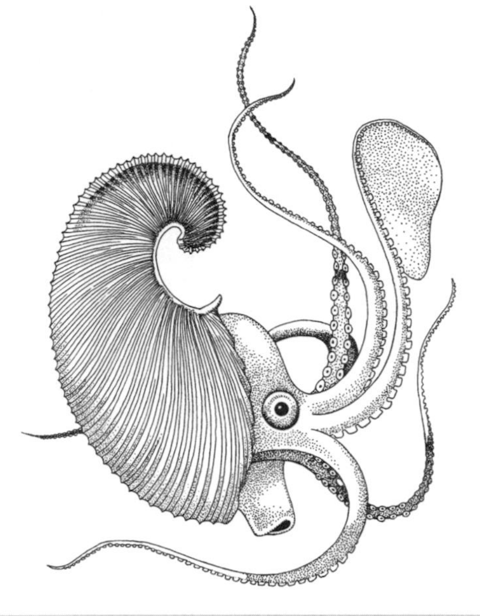

argonaut

argonaut (*Argonauta argo*) In Greek mythology, the Argonauts were the warriors who accompanied Jason on his quest for the Golden Fleece. In the ocean, the argonaut is an octopus. The female argonaut has wide webs on two of her arms; early naturalists believed she used these to "sail," but in fact, the webs produce the delicately ribbed, paper-thin shell and then hold it in place. A female argonaut in her shell looks something like a chambered nautilus, but the two are not related except that they are both cephalopods. The chambered nautilus is a primitive cephalopod with a chambered shell and as many as ninety tentacles; the argonaut is an octopus (eight arms) that builds a thin shell. It is difficult to tell which of the sexes is stranger—the 1-foot-long female that lives in a papery, ribbed shell that she constructs herself (hence the other name, "paper nautilus"); or the male, all of ½ inch long, with one arm that becomes detached and swims over to the female, who places it in her mantle cavity to fertilize her eggs. In 1829, Baron Georges Cuvier examined several argonauts and found what he believed to be a parasitic worm. Because the worm resembled the arm of a cephalopod, he named the genus *Hectocotylus*, "arm of a hundred suckers." It was not until 1853 that the German zoologist Heinrich Müller recognized that this "worm" was actually the sex organ of the tiny male argonaut. After the eggs are fertilized, the female broods them in the shell. There are several species of argonaut, which are found throughout the temperate and tropical waters of the world and differentiated by the construction of the shell. These delicate shells float, and they occasionally wash up on the beach. In 1936, for reasons unexplained, hundreds of thousands of argonauts appeared off Trieste in the northern Adriatic.

See also chambered nautilus, octopus

Armada, Spanish In 1588, the battle lines were drawn between Catholic Spain—the most powerful empire in Europe—and Protestant England. In 1587, as the Marqués de Santa Cruz, one of the heroes of the Battle of Lepanto, in 1571, was assembling his fleet for an invasion of England, Sir Francis Drake attacked the Spanish ships in the harbor at Cádiz, postponing the battle for almost a year. After Santa Cruz died, command of the Armada fell to the Duke of Medina-Sidonia. It consisted of 130 galleons and galleasses, carrying about thirty thousand men. They intended to go to Flanders, pick up the army of the Duke of Parma (who was then at Bruges), invade England, and place Philip of Spain on the English throne. The Armada left Lisbon in May 1588, but storms forced them back and they did not depart again until July. When they entered the English Channel, they were met by the British fleet led by Lord Howard of Effingham, consisting of some ninety-seven ships commanded by the likes of Sir Francis Drake, Sir Martin Frobisher, and Sir John Hawkins. (Frobisher and Hawkins were knighted during the battle; Drake had been knighted seven years earlier by the queen.) The British, with the weather gauge (the wind behind them), were in the superior position, and they followed the Spanish fleet eastward through the Channel. One distinct British advantage was their use of the culverin, a cannon with a longer range than the weapons used by the Spanish. No significant action occurred until the Spaniards anchored off Calais, where they hoped to make contact with Parma. The English sent fire ships into the Spanish fleet, causing the Armada to scatter, and then attacked. Unable to re-form, the Spanish suffered tremendous losses, but a fortuitous change of wind enabled them to escape northward through the North Sea and around the east coast of England. Rounding Scotland, the Spanish fleet was hit by storms, and many of the ships that were not already damaged in battle were wrecked on the rocks. Those Spaniards who landed in Scotland or Ireland were often killed or captured by local inhabitants or English troops. Only sixty-seven of the Spanish ships reached Spain; the British did not lose a single vessel.

Arnoux's beaked whale (*Berardius arnuxii*) Similar in every aspect to Baird's beaked whale of the North Pacific, this species is found only in the Southern Ocean, having been reported from Australia, New Zealand, Argentina, Tierra del Fuego, and the Antarctic Peninsula. Because almost all the information on this species has come from strandings, little is known of its habits. It has been measured at 32 feet in length. Like the northern variety, both sexes have teeth, and indeed they may be a single species, differentiated primarily by geography. **See also Baird's beaked whale, beaked whales**

arrowworm (phylum Chaetognatha) From *chaetae,* meaning "bristles," and *gnathos,* meaning "jaws," the scientific name of the phylum describes the animals accurately. Arrowworms are thin, transparent worms that range from 1 to 4 inches in length and have spines on the prehensile hooks that are the most conspicuous part of the head. On the sides of the tubelike body are narrow fins, like the vanes of an arrow, that give the worm its common name. Arrowworms resemble chordates but are not, because no skeletal elements ever develop, though they do have an internal digestive tract: a tube within a tube. They live in shallow water, often in vast numbers, but are rarely seen because they are almost totally transparent.

arthrodire: See placoderm

Aruba In 1499, the Spanish explorer Alonso de Ojeda, traveling with Amerigo Vespucci, landed on this little island off the coast of Venezuela and claimed it for Queen Isabella. The Spanish found the island too arid for cultivation, so they left it to the Caiquetio Indians (a tribe of Arawaks) who had been living there for more than two thousand years. Pirates and buccaneers used it as a base for attacking treasure ships bound for Europe. In 1634, when the Dutch had been expelled from St. Martin, they captured Aruba, Bonaire, and Curaçao (now known as the ABC Islands), and Curaçao became the administrative capital for the Dutch West India Company. Except for the period 1805–1815, when the island was taken by the British, it has been in Dutch hands since 1634. Gold was discovered in Aruba in 1824, and was mined until 1916, when the precious metal ran out. Oil was discovered in Lake Maracaibo (in Venezuela—only 35 miles away) in 1918, and massive refineries, among the largest in the world, were built on Aruba and Curaçao. The refineries closed in 1985, but have reopened since. Tourism is now Aruba's most important industry.

See also Netherlands Antilles

Ascension Island An isolated volcanic island in the South Atlantic Ocean, about midway between the easternmost bulge of Brazil and the coast of West Africa. It is a dependency of the British colony of St. Helena, 700 miles to the southwest. Located some 90 miles west of the midline of the Mid-Atlantic Ridge, Ascension is one of the few surface manifestations of the undersea mountain range that runs down the middle of the Atlantic Ocean. Discovered in 1501 by the Portuguese navigator João da Nova Castella, it remained uninhabited until the French stationed some troops there when Napoleon was exiled to St. Helena in 1815. The island was taken by the British in 1922 and used as a naval station, and the American space program has installed a tracking station there. Biologists do not know where they come from or how they get there, but green turtles (*Chelonia mydas*) have chosen Ascension Island as one of their primary nesting sites.

See also green turtle, St. Helena

Assateague Island Barrier island off the coasts of Maryland and Virginia, now a national park. The 37-mile-long island covers some 18,000 acres and is the home of more than three hundred species of birds, including snow geese, ospreys, herons, egrets, plovers, and sandpipers. The best-known "wildlife," however, are the wild horses (often called Chincoteague ponies) that inhabit the island; these are rounded up every July and driven across the sound to Chincoteague Island, where some of them are auctioned off. When he and his party landed on the island in 1524, Giovanni da Verrazano found Indians there and kidnapped one of them. Around 1680, settlers from Jamestown used the island for grazing their cattle, and in 1711 a Marylander named William Whittington obtained 1,000 acres. The U.S. Coast Guard used the island as a base for their lifesaving operations until 1962, when a violent storm destroyed most of their facilities. The island became part of the U.S. national park system in 1965, and hosts more than a million visitors a year. As a federally designated national seashore, the island offers opportunities for swimming, crabbing, fishing, and off-road vehicle use.

See also Chincoteague Island, Verrazano

astrolabe The *planispheric astrolabe,* on which a celestial sphere is projected onto a plane of the equator, was originally used for measuring the altitude of the sun and the stars and for the solution of other problems of astronomy and navigation. The *mariner's astrolabe* determines the latitude of a ship at sea by measuring the noon altitude of the sun or the meridian altitude of a star of known declination. The first devices that measured the sun's noon altitude were the quadrant, cross-staff, backstaff, and, later, the mariner's astrolabe. Popular in the fifteenth and sixteenth centuries, it consisted of a brass ring with the circumference marked off in degrees and a movable pointer called the alidade. The instrument was usually cast of brass, and was cut away to keep it from blowing in the wind. By sighting with the alidade and taking readings of the graduated circle, angular distances could be determined, but only latitude could be reckoned in this way; the calculation of longitude depended upon a clock that could accurately record the difference between local and Greenwich mean time, and this would not be invented until 1773. One of the French explorer Jean-François La Pérouse's ships was named *L'Astrolabe,* and there is an Astrolabe Bay in New Guinea.

See also compass, La Pérouse, quadrant, sextant

Atlantic, Battle of the The term given to the World War II campaign fought by the Allies against German and Italian submarine predation on shipping. Conflicts occurred from Murmansk to Trinidad, from Boston to Bremerhaven. The "battle," which began in September 1939 with the torpedoing of the British liner *Athenia* and the aircraft carrier *Courageous*, lasted for the entire duration of the war. In October, at Scapa Flow in the Orkneys, where the British fleet was anchored, the German submarine *U-14* slipped in undetected and sank the battleship *Royal Oak* with a loss of 833 lives. The U-boats, under the command of Admiral Karl Dönitz, sank 2,603 merchant ships and 175 naval vessels, with a loss of more than forty thousand lives. In response, the Allies destroyed 784 German and 85 Italian submarines. Twice, the crippling loss of merchant shipping brought Britain to the brink of starvation and Hitler to the threshold of victory. The turning point of the battle was around May 1943, when enough aircraft could be marshaled to supplement the surface escorts of convoys throughout their passage across the Atlantic. In America, shipbuilders created 2,770 Liberty ships at an astonishing rate (some were built in four days) to replace the ships that were being sunk by the enemy.

Atlantic Empress* and *Aegean Captain On July 19, 1979, northeast of Trinidad and Tobago, two supertankers, the *Atlantic Empress* and the *Aegean Captain*, collided. Both ships burst into flames, and the *Aegean Captain* spilled 4.31 million gallons of oil. Two weeks later, as the *Atlantic Empress* was being towed from the original spill site, another 41.5 million gallons of oil were spilled. In this tragedy, twenty-six men were killed and many more injured. **See also oil spill, supertanker**

Atlantic gray whale (*Eschrichtius robustus*) Although it is the same species as the Pacific—also known as the California—gray whale, the Atlantic gray whale is usually treated separately because it is extinct. The Atlantic whales probably followed a migration pattern similar to that of their Pacific counterparts, feeding in the north, probably around Greenland and Iceland, and then swimming south to breed in the warmer, protected waters of the Bay of Biscay, Sweden, and the Netherlands in the eastern Atlantic. Fossil remains found in the bays and inlets of the United States from New Jersey to South Carolina indicate that the whales also inhabited the inshore waters of the western North Atlantic. Atlantic gray whales were known to have been hunted by Native Americans and early Basque and Icelandic whalers, but the reason for the whales' extinction is not known. It is likely that the whalers contributed to the demise of an already depleted population. **See also gray whale, whaling**

Atlantic herring (*Clupea harengus*) Gathering in schools that numbered in the billions, the Atlantic herring was

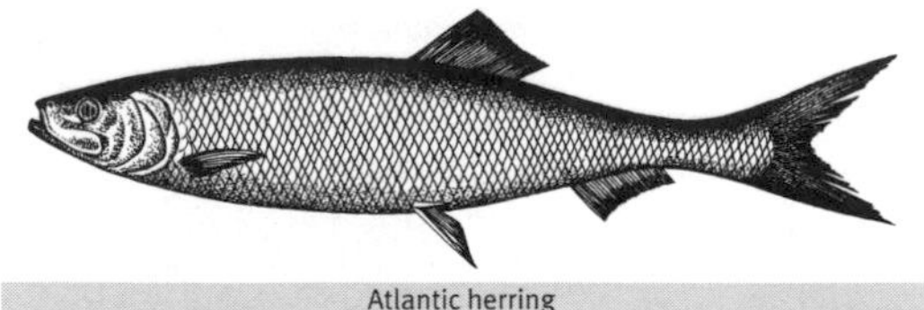
Atlantic herring

once believed to be the most numerous fish in the world. It is not—the title actually belongs to the bristlemouths—but herring is still the basis for one of the world's largest commercial fisheries. Herring is consumed in large quantities by the residents of those countries that ring the Baltic and North Seas: Norway, Sweden, Denmark, Germany, the Netherlands, and Britain (kippers are smoked herring). They are sold canned, fresh, pickled, or salted. Young herring are sometimes called sardines, but there are also other fishes with that name. Herring form the base of the North Atlantic food pyramid, as they are eaten by bluefish, mackerel, tuna, bonito, porpoises, dolphins, and some large whales. Herring become sexually mature when they are about 8 inches in length, and may grow to 20 inches. They feed on plankton. The vast number of herring made it almost unthinkable that they could be overfished, but no animal, marine or terrestrial, can withstand massive population destruction, and herring stocks around the world are in decline. In some areas, on Georges Bank, for instance, the numbers fell so low that the fishery was abandoned, and there has been no directed herring fishery there since 1976.

See also alewife, Georges Bank, menhaden, wolf herring

Atlantic humpback dolphin (*Sousa teuszii*) Named for the characteristic ridge on its back, on which is perched a small dorsal fin, this 8-foot-long, dark gray dolphin is found only in the waters off West Africa. (It is sometimes known as the West African humpbacked dolphin.) It is closely related to the Indo-Pacific humpback dolphins, of which there may be as many as three distinct species, or as few as one, with color variations. Because of confusion in the earliest descriptions, this species was believed to be the only vegetarian dolphin. It now appears that the seeds and grass thought to have been found in the stomach of the specimen described in 1892 were actually from the stomach of a manatee caught at the same time. *Sousa teuszii* feeds on fish, like all other dolphins.

See also Indo-Pacific humpback dolphin

Atlantic humpback dolphin

Atlantic mackerel (*Scomber scombrus*) Averaging about 1 pound in weight, the Atlantic mackerel is one of the most important food fishes in the world. Large schools of these streamlined, spindle-shaped scombroids roam the western Atlantic from Georges Bank to the Carolinas, and the eastern from Norway to Spain. It is characterized by the wavy bands on its back and the little finlets between the second dorsal and the tail fin. The mackerel feeds mostly on shrimps and other crustaceans, but also on smaller fishes such as herring and pilchards. Millions of tons of mackerel are caught every year, usually in purse seines, but also in pound nets. Their oily flesh is considered particularly healthy because it is rich in the omega-3 fatty acids.

See also Scombridae

Atlantic mackerel

Atlantic Ocean The body of water that separates the western coasts of Europe and Africa and the eastern coasts of North and South America, bounded on the north by the Arctic Basin and on the south by the Southern Ocean. It covers some 31,815,000 square miles, and its average depth is 10,390 feet. The volume of water in the Atlantic is 77,235,000 cubic miles. The Puerto Rico Trench, the Atlantic's lowest point, is 28,224 feet deep, more than a mile shallower than the deepest part of the Pacific. The Atlantic is higher than the adjacent oceans because most of the world's major rivers empty into it. Unlike the Pacific, the Atlantic has few islands: Iceland, the British Isles (including the Shetlands and Orkneys), the West Indies, the Azores, the Canaries, Madeira, Bermuda, and the Cape Verdes in the North Atlantic; and Fernando de Noronha, St. Helena, Tristan da Cunha, and the Falklands in the South. The floor of the Atlantic is bisected by the Mid-Atlantic Ridge, an S-shaped underwater volcanic mountain range that runs from Iceland to Bouvet Island in the Antarctic and produces the seafloor spreading believed to be responsible for continental drift. The Gulf Stream runs northward along the east coast of North America, bringing a large body of warmer water into the cooler north as far east as the British Isles. The Atlantic Ocean was the cradle of some of the world's oldest and most profitable fisheries, including herring, cod, haddock, and flounder.

See also Caribbean Sea, continental drift, Gulf Stream, Mid-Atlantic Ridge, Pacific Ocean, seafloor spreading

Atlantic salmon (*Salmo salar*) Found in the eastern North Atlantic from Greenland to the Bay of Biscay and in the western from Hudson Bay to New England, the Atlantic salmon is one of the most popular game fishes in the world. A spindle-shaped, silvery fish, it is blue-gray on the back, with little spots or cross-shaped markings. The world record salmon—caught in Norway in 1928—weighed 79 pounds, but most are smaller, around 10 to 20 pounds. With the exception of some landlocked populations, Atlantic salmon are anadromous: born in freshwater, they migrate to the sea, and then return to freshwater to spawn. Unlike Pacific salmon, which die when they reach their freshwater spawning grounds, Atlantic salmon may repeat the cycle as many as three or four times. At every stage of their growth, salmon acquire new names: the eggs hatch into *alevins,* which develop into *fry.* Two- or three-month-old fry are called *parrs;* after several years in freshwater, they head for the sea as *smolts* and return to the streams of their birth as *grilse.* Only the full-grown fish are called *salmon,* and these prizes are eagerly sought in freshwater streams and rivers by fly fishermen. Some of the most popular salmon rivers are in Scotland, Iceland, Norway, Nova Scotia, Quebec, and Maine. Regarded as one of the world's premier food fishes, salmon are eaten fresh, smoked, or canned, and in Japan, raw as sushi. Atlantic salmon are also caught in great numbers by commercial fishermen—particularly gillnetters—which has led not only to conflicts with sport fishermen but also to a catastrophic reduction in the number of fish. *S. salar* is classified with the trouts in the genus *Salmo* (the brown trout is *S. trutta*), while the Pacific salmons are placed in the genus *Oncorhyncus.*

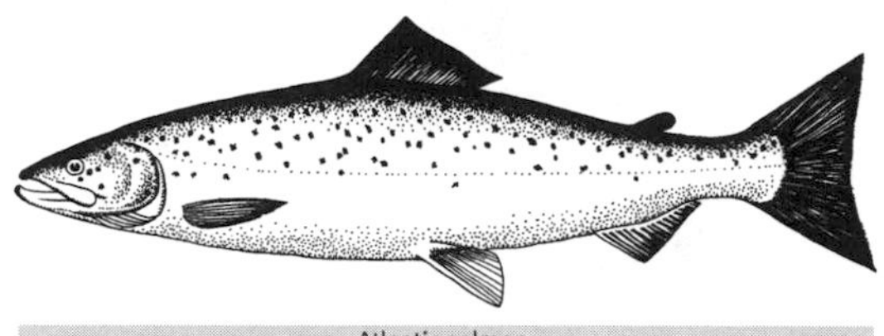

Atlantic salmon

See also chinook salmon, chum salmon, coho salmon, Pacific salmon, pink salmon, sockeye salmon

Atlantic white-sided dolphin (*Lagenorhynchus acutus*) One of the few dolphins that exhibits a color other than black, white, or gray, this species has a yellowish or orange patch on the tailstock that is often visible when the animal jumps, and it jumps so frequently that it is known as jumper by Newfoundland fishermen. This dolphin, which reaches a length of 8½ feet, is also characterized by greatly exaggerated vertical "keels" on the dorsal and ventral aspects of the caudal peduncle. It is found in the northern North Atlantic, but not as far north as its relative, the white-beaked dolphin (*L. albirostris*). Al-

Atlantic white-sided dolphin

though the ranges of the two species often overlap, they can be easily differentiated by the yellow flash of the white-beaked dolphin.

See also Pacific white-sided dolphin, white-beaked dolphin

Atlantis A mythical land that is said to have vanished into the sea as a result of a massive landslide or earthquake. The story was first told by the Greek philosopher Plato (428–347 B.C.) in the dialogues *Timaeus* and *Critias*, written around 360 B.C. Since Plato's time, there have been hundreds of books written about Atlantis, and their authors have tried to locate it in such places as the Aegean (on the Greek islands of Crete and Santoríni), the North Atlantic, the Sahara, Germany, Spain, the Bahamas, and the Antarctic. The city of Helice, which sank into the sea as a result of an earthquake in 373 B.C. (during Plato's lifetime), is believed to be one of the elements that inspired him to write the story. Although people have been searching for Atlantis for almost three thousand years, no legitimate evidence of its existence has ever turned up, and it is now believed that the "Lost City" existed only as a creation of Plato's fertile imagination. **See also Crete, Helice, Santoríni**

atoll A horseshoe-shaped or circular coral reef enclosing a lagoon. Atolls are believed to be formed by coral polyps on the tops of submerged volcanoes. The outer edges are kept alive by the continual surge of the sea, which brings nutrients, but they die on the inside edge because the still waters of the lagoon provide no nourishment. The largest atolls in the world are in the South Pacific: Kwajalein in the Marshall Islands and Rangiroa in French Polynesia.

See also coral, coral reef

Auckland Islands Six volcanic islands 290 miles south of South Island, New Zealand, discovered in 1806 by Abraham Bristow. The islands were named after the Earl of Auckland, Viceroy of India (as was the second capital of New Zealand). Elephant and fur seals were slaughtered in such numbers on the islands that sealing ended ten years after it began. On his epic voyage in the *Erebus* and *Terror* to the Antarctic, James Clark Ross set out from Hobart in 1840 and stopped off at the Aucklands. In the same year, British whaling entrepreneur Samuel Enderby sent his son Charles to establish a permanent whaling station there, but it failed after two years. The islands—Auckland is the largest, Adams is next in size, and the remaining islets are merely exposed rocks—are now uninhabited by people, but many species of petrels breed there, and it and Campbell Island are the only places where the royal albatross nests. **See also New Zealand**

Audouin's gull (*Larus audouini*) Once widely distributed throughout the Mediterranean, this species can now be found only on the coasts of Tunisia and Morocco, and on a few small islets near Sardinia, Corsica, and Cyprus. It looks like a smaller and slenderer herring gull, with its white head and tail and light gray back, but where the herring gull has a yellow bill with a red spot, the bill of Audouin's gull is red. It is one of the rarest gulls in the world, with fewer than 1,000 in existence. The reason for its decline is not clear, but it may have to do with egg thieving by fishermen and/or herring gulls.

See also gull, herring gull

auklet Any of a number of small alcids found in the Northern Hemisphere. These birds usually nest on cliff faces, but some species dig burrows. In pursuit of their prey, which consists of crustaceans and small fishes, they "fly" underwater, but unlike penguins, they can also fly in air. They are characterized by dense, waterproof plumage and relatively small wings. Most species are gregarious, but the quail-sized parakeet auklet (*Cyclorhynchus psittacula*) is usually seen only in pairs. The rhinoceros auklet (*Cerrorhinca monocerata*) is named for the little "horn" at the base of the beak that appears during the breeding season. Islands in the Bering Sea are home to millions of least auklets (*Aethis pusilla*), which are the smallest of the alcids at a body length of 6 inches. The crested auklet (*A. cristatella*), a native of northern Japan and coastal Siberia, is an all-black bird with an orange bill, a crest, and a thin white plume behind the eye. **See also Alcidae, Cassin's auklet, least auklet, murre, puffin**

Australian sea lion (*Neophoca cinerea*) Known as a hair seal in southern Australia, this sea lion is one of the few that is nonmigratory. They spend most of their lives on or near the beach where they were born. As with all of the eared seals, males are considerably larger and heavier than females. They are renowned for their ability to move about on land, and there is a record of an individual that was found 6 miles from the sea. In the waters of South Australia and Kangaroo Island, Australian sea lions are the preferred prey of great white sharks. Seal hunters massacred this species during the

Australian sea lion

eighteenth and nineteenth centuries, and it is believed that there are no more than 5,000 left.

See also great white shark, sealing, seals and sea lions

Avery, John (1653?–1697?) Pirate, known as Long Ben, who plied his trade in the Caribbean and the Indian Ocean. Details of his early life are not known (indeed, his name might actually have been Henry Every), but it is known that he was a slaver employed by the Royal Governor of Bermuda along Africa's Guinea Coast until around 1694, when he turned pirate. Aboard as sailing master, he took the Bristol ship *Charles II,* renamed her *Fancy,* and set out for Madagascar, where, alongside Thomas Tew, he plundered and ravaged every ship he encountered. In 1695, Avery captured the *Gang-i-Sawai,* the treasure ship of the Great Mogul Aurangzeb, which contained gold and silver pieces that the Indian owners valued at £600,000; the figure may have been exaggerated to improve their compensation from the East India Company. Long Ben Avery and his crew headed for the Caribbean, selling their booty and dispersing as pirates were wont to do if they lived long enough. Avery himself, known as the Arch Pirate, showed up in Ireland, but there the trail turns cold. He may have sold his gold and diamonds and retired to a life of bourgeois respectability, or he may have died a pauper. **See also Madagascar, Tew**

avocet

avocet (*Recurvirostra americana*) A large shorebird with contrasting plumage, long bluish legs, and a long black bill that turns up at the tip. Avocets inhabit marshes with open water and mudflats, where they feed on minnows and crustaceans. Avocets are found in Africa, central Asia, Europe, and the Americas.

avocet eel (*Avocettina infans*) A fish with a long, filamentous body with continuous dorsal and anal fins and dark coloration, like most of the snipe eels. They can reach a length of 24 inches. Avocet eels are named because of their upturned upper jaw, but the downturned lower jaw means that they cannot close their mouths. It is thought that they trap passing food items on the tiny teeth in the jaws. **See also snipe eel**

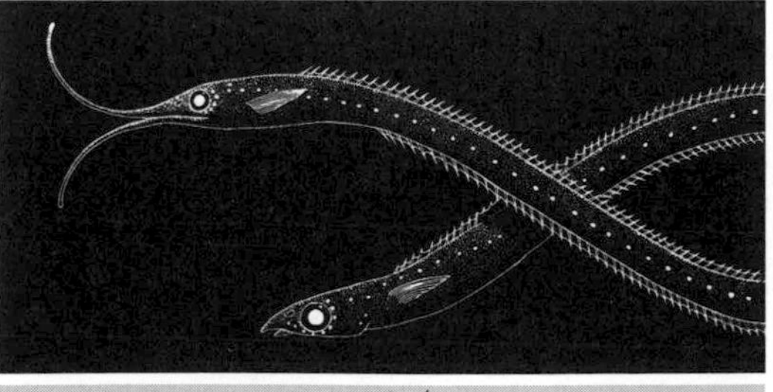
avocet eel

Axel Heiberg Island Uninhabited island in Nunavut (formerly part of the Northwest Territories) of Arctic Canada, west of Ellesmere Island. Axel Heiberg is 15,779 square miles of snow and ice, deeply indented by fjords. After Fridtjof Nansen drifted in the *Fram* from 1893 to 1896, Otto Sverdrup took the ship north again in 1898 and explored the region around Ellesmere Island on foot and by sledge, surveying and naming the islands now known as Axel Heiberg, Amund Ringnes, and Ellef Ringnes. (Heiberg and the Ringnes brothers were sponsors of the expedition, and the founders of the Ringnes brewery, which is still in existence.) Sverdrup claimed the islands for Norway, and although they later became part of Canada, the Norwegian names were retained. When Frederick Cook claimed to have reached the North Pole in 1908, he said he departed from the northern tip of Axel Heiberg; his claim has now been discounted. In 1985, a geological survey team flying over the island observed a field of large stumps, subsequently dated at 45 million years old, which indicated that an extraordinary change in climate had taken place. The wood was so well preserved that it could be cut like recent lumber and readily burned.

See also Cook (Frederick), Ellesmere Island, Nansen, Nunavut

Azores Originally Islas de Açores, or "Islands of Hawks," this North Atlantic group of nine islands lies about 900 miles due west of Lisbon. Discovered over a period of time by Portuguese explorers of the fifteenth century, they are now a Portuguese colony. The three groups of islands stretch over 340 miles and consist of the eastern group of São Miguel (on which the capital, Ponta Delgada, is located) and Santa Maria; the central group of Faial, Pico, São Jorge, Terceira, and Graciosa; and the western pair of Flores and Corvo. The islands are a surface manifestation of the Mid-Atlantic Ridge, and their volcanic nature is evidenced by frequent earthquakes, basaltic eruptions, and numerous hot springs. The islands sit on a microplate at the junction of the North American, Eurasian, and African plates. The volcanic activity is the result of the plate moving over a "hot spot," as in the Hawaiian chain. Yankee sperm whalers used to visit the Azores frequently; they imparted whaling technology to the Azoreans, who hunted whales from the late nineteenth century to around 1985, when the practice was suspended. Now whale watchers can venture out into the North Atlantic to see the very whales that were hunted from the shore stations at Faial and Pico.

See also Hawaii, Mid-Atlantic Ridge, whale watching, whaling

B

backstaff An instrument used to measure latitude by measuring the altitude of the sun. At noon, the sun is directly overhead at the equator, so however far the sun deviates from that position gives a pilot a measurement of how far north or south he has sailed. Used during the sixteenth century, it gained its name because, unlike the cross-staff, which it replaced, the measurer had the sun behind him when taking an observation.

See also astrolabe, compass, quadrant

bacteria, marine The smallest living things in the sea, bacteria are one-celled animals with no cell nucleus, therefore *prokaryotic.* (*Eukaryotic* cells have a nucleus containing DNA, organized into chromosomes; all your cells are eukaryotic.) Bacteria live in the sediments at all levels in the ocean and have even been discovered deep below the ocean floor. Some bacteria function as mineralizers, decomposing complex organic matter; others live in the intestinal tracts of various creatures; some are even capable of photosynthesis. Luminescent bacteria are responsible for the sometime "phosphorescence" of the ocean and for the glowing organs of certain fishes and shrimps. While most bacteria are microscopic, the largest bacteria known—large enough to be seen easily with the naked eye—were recently discovered living in the sediments off the coast of Namibia. Sulfide-oxidizing bacteria, requiring no oxygen, are an integral part of the chemosynthetic digestive system of hydrothermal vent animals such as tube worms, clams, crabs, and mussels.

See also bioluminescence, hydrothermal vents

Baffin, William (1584–1622) English Arctic explorer who first sailed to the north as pilot on Robert Bylot's 1615 and 1616 voyages aboard the *Discovery* in search of the Northwest Passage. (Bylot had accompanied Henry Hudson in 1607 and 1608 on earlier failed attempts.) Nothing is known about Baffin's early years; only the last decade of his life appears in the history books. During this period he made no fewer than five voyages to the Arctic, and although there are references to his map of the bay that now bears his name, the map has been lost. He penetrated 300 miles deeper into Baffin Bay than John Davis, and if he had only entered Lancaster Sound (which he named for a patron of his voyage), he would have discovered the elusive Northwest Passage. Baffin named Lancaster, Smith, and Jones Sounds; the bay and the island were named for him. After the Arctic, he sailed to the Red Sea for the British East India Company, and he was killed in an attack on Qeshm, the largest island in the Persian Gulf.

Baffin Bay Between Greenland and Baffin Island, Baffin Bay is a 900-mile-long, ice-choked body of water. It connects in the south to the Davis Strait, which in turn opens to the North Atlantic, and at its northern terminus is Lancaster Sound, which, when the ice breaks up, opens to a passage through to the Beaufort Sea—the Northwest Passage. Baffin Bay covers an area of some 266,000 square miles, and its deepest point is 7,000 feet down. Most of the icebergs that come down as far as Newfoundland (including the one that sank the *Titanic*) are calved from the glaciers that push into northern Baffin Bay. In 1587, John Davis sailed the *Sunshine* and the *Moonshine* well into Baffin Bay, but at what is now Godthaab Fjord, he came about and headed southwest, entering Cumberland Sound, which he believed might be the passage to the west. Then, in 1615, with William Baffin as his pilot, Robert Bylot sailed the *Discovery* to the northern terminus of the bay, but went no farther. During the eighteenth century, the bowhead whales (*Balaena mysticetus*) of Baffin Bay were the object of concentrated "ice fisheries" by the British and the Dutch. These fisheries—along with those in the region of Spitsbergen and the east coast of Greenland—virtually eliminated this species of whale from the eastern Arctic.

Baffin Island The fifth largest in the world, Baffin Island covers some 195,000 square miles, an area approximately the size of Spain. It is separated from the eastern Canadian mainland by Hudson Strait, and from Greenland by Davis Strait and Baffin Bay. To the north is Lancaster Sound, the route to the Northwest Passage. The eastern shore of the island (facing Greenland) is punctuated by numerous fjords and inlets, the most prominent of which are Frobisher Bay and Cumberland Sound. The island is believed to be Helluland ("Flat Rock Land"), the area Leif Ericsson discovered around A.D. 1000. Martin Frobisher landed there in 1578. Uninhabited except for some small Inuit villages on the southeastern coast, Baffin Island in its entirety was incorporated in 1999 into the new Canadian territory of Nunavut, to be administered by Inuit. The capital of the new territory is Iqaluit, on Frobisher Bay.

Bahamas A group of some seven hundred islands and more than two thousand islets 50 miles off the east coast of Florida, stretching for about 760 miles. The better-known islands are New Providence (on which the capital, Nassau, is located), Grand Bahama, Abaco, Andros, Eleuthera, Cat Island, and San Salvador (also known as Watlings Island), which was Columbus's first landfall in October 1492. The population on the twenty-two inhabited islands is about 260,000 people. The Bahamas were a British crown colony from 1717 to 1964, when they were granted self-government. They still belong to the British Commonwealth.

See also Columbus

Bahamonde's beaked whale (*Mesoplodon bahamondi*) In June 1986, a single beaked-whale skull was found on Robinson Crusoe Island in the Juan Fernández Archipelago off the Chilean coast. It was examined by Chilean cetologists, who compared it with the skulls of other species and concluded that it was different enough to warrant designation as a new species. That it differs from all the others is clear, but everything else about it is a complete mystery. It was named for Professor Nibaldo Bahamonde, a Chilean marine biologist.

See also beaked whales, *Mesoplodon*, Peruvian beaked whale

Baird's beaked whale

Baird's beaked whale (*Berardius bairdii*) The largest of the beaked whales; a specimen stranded in California in 1910 was 42 feet long. Both males and females have two pairs of teeth, one near the front and the second somewhat farther back. The lower jaw is longer than the upper, making the first pair of teeth clearly visible even when the mouth is closed. It has been recorded off the Aleutians, British Columbia, Washington, and California, and there is a fishery for these animals off Japan. Baird's beaked whale (named for Spencer F. Baird, the second secretary of the Smithsonian Institution) has what may be the longest gestation period of any cetacean: seventeen months.

See also Arnoux's beaked whale, beaked whales

Balboa, Vasco Núñez de (1475–1519) The first European to see the Pacific Ocean, Balboa arrived in the New World in 1501. He tried farming on the island of Hispaniola (Haiti), but when that didn't work out, he turned to more traditional conquistador practices, such as killing the natives and stealing their gold. In 1513, he led an expedition of 190 Spaniards (one of whom was Francisco Pizarro, the future conqueror of Peru) and 1,000 Indians across the isthmus of Panama to Darien. Upon sighting the great ocean, he took possession of it for Spain. (Although Keats wrote that it was Cortés who stood "silent, upon a peak in Darien," it was actually Balboa.) Balboa's rival in the New World was Pedro Arias Dávila (usually called Pedrarias), who had him imprisoned, tried for treason and rebellion, and beheaded in 1519.

bald eagle

bald eagle (*Haliaetus leucocephalus*) Best known as America's national bird, the bald eagle is one of the group known as fish eagles, large birds of prey most frequently found near water. (Benjamin Franklin favored the wild turkey as the national bird.) Adults are easily recognized by their brown plumage, white head and tail, and yellow beak and legs. Juveniles can be confused with golden eagles because both are large and brown, but the legs of bald eagles are bare, whereas those of the golden eagle are feathered to the feet. Despite their powerful and imposing appearance, bald eagles are primarily carrion eaters and often steal the prey of smaller birds, particularly ospreys. In the "lower forty-eight," bald eagles are rare—and earnestly protected—but in southeast Alaska, where they congregate to feed on salmon, they are quite common. Alaskan eagles are larger than their more southerly counterparts, and their wingspan can reach 98 inches.

See also osprey, Steller's sea eagle, white-tailed sea eagle

baleen The fibrous plates that hang from the roof of the mouth of certain filter-feeding whales. These plates, which are actually made of keratin (the same material that forms the basis of human hair and fingernails), are fringed on the inner surface, and it is these fringes that trap the food items upon which the whales feed. Those whales with baleen plates are known as baleen whales, and they are the bowhead, right whale, gray whale, humpback, blue whale, fin whale, sei whale, Bryde's whale, minke whale, and pygmy right whale.

Bali Indonesian island just east of Java, across the 1-mile-wide Bali Strait. It is a mountainous island with several active volcanoes, including the sacred Mount Agung. Rice is grown on intricately terraced hillsides; other crops include vegetables, fruits, and coffee. Klungklung was the capital until 1908, when the Dutch moved it to Denpasar. During Suharto's purge of Communists in 1965, more than forty thousand Balinese people were killed and entire villages were destroyed. The Balinese are renowned for their music, dance, crafts, and architecture, and the island is considered one of Asia's primary tourist attractions.

See also Indonesia

Ballard, Robert (b. 1942) American geologist, oceanographer, and explorer, born in Wichita, Kansas. He received his undergraduate degree from the University of California and his Ph.D. in chemistry and geophysics from the University of Rhode Island. While doing postgraduate work at the University of Hawaii, Ballard also worked as a porpoise trainer at Sea Life Park in Hawaii. He joined the U.S. Navy in 1967, and was assigned to the Deep Submergence Laboratory at Woods Hole, Massachusetts. He participated in Project FAMOUS, investigating the Mid-Atlantic Ridge from submersibles in 1973 and 1974, and in 1977, aboard the *Alvin,* he was among the first scientists to see the strange and completely unexpected animal life of the deep hydrothermal vents in the Galápagos Rift Zone. He has spent more time in submersibles investigating the deep oceans than any man alive. In 1985, working with French scientists, he discovered the resting place of the world's most famous shipwreck, the *Titanic.* In 1989, he located the wreck of the German battleship *Bismarck,* and in that same year developed the JASON Project, designed to allow children all over the world to investigate the underwater world by television—a process he calls telepresence. He is the author of numerous scientific papers, popular articles, the books *The Discovery of the Titanic, The Discovery of the Bismarck, The Lost Ships of Guadalcanal, Exploring Our Living Planet,* and the novel *Bright Shark.* After thirty years at Woods Hole, he was named director of the Institute for Exploration at the Mystic Marinelife Aquarium in Connecticut, dedicated to undersea research, using the U.S. Navy's nuclear research submarine *NR-1.* In May 1998, he located the American aircraft carrier *Yorktown,* sunk during the Battle of Midway in the central Pacific in May 1942. **See also *Alvin,* Project FAMOUS**

ballyhoo (*Hemiramphus balao*) One of the larger halfbeaks, the ballyhoo (also called balao) can reach a length of 2 feet. Found worldwide in warmer waters, this fish aggregates in large schools and is a preferred food item for many larger fishes, especially sailfish. Occasionally several sailfishes will surround a school of ballyhoo and force them into a ball so they can slash at them more efficiently and then feed on the dead or wounded ones. Like other halfbeaks, ballyhoos are often seen skittering along the surface.

See also halfbeak, sailfish

Baltic Sea Semienclosed, brackish extension of the North Atlantic that covers some 160,000 square miles and touches the shores of Denmark, Sweden, Finland, Russia, Estonia, Latvia, Lithuania, Poland, and Germany. The Baltic is connected to the North Sea by the Skagerrak, which separates southern Sweden and Norway from the Jutland Peninsula of Denmark; and, to the east of the Skagerrak, the Kattegat, between northeastern Denmark and Sweden. The Gulf of Bothnia, an arm of the Baltic Sea, extends past the Åland Islands (belonging to Finland) to the north, and separates eastern Sweden from Finland. The Gulf of Finland pushes eastward past the coasts of Latvia, Estonia, and Russia until it reaches St. Petersburg. So many rivers, including the Oder, Vistula, Neva, and Neman, drain into the Baltic that the water is nearly fresh. The most important ports on the Baltic are Copenhagen, Gdansk, Riga, St. Petersburg, Stockholm, and Helsinki. The Baltic was known in ancient times as a major source of amber, and in the Middle Ages commerce was dominated by the Hanseatic League. **See also Hanseatic League; Jutland, Battle of; North Sea**

bamboo shark (*Chiloscyllium punctatum*) A small, slender shark with eyes near the top of its head, short nasal barbels, and a rounded mouth. Three-foot-long bamboo sharks are found in shallow waters throughout the Australasian region, sometimes in tide pools. They eat small fishes and bottom-dwelling invertebrates and are harmless to people. Bamboo sharks are popular with saltwater aquarists. **See also epaulette shark**

Banks, Sir Joseph (1743–1820) British naturalist, explorer, and patron of the sciences. Born into a wealthy London family, he was educated at Eton, Harrow, and Christ Church, Oxford. As a naturalist, he participated in a voyage to Iceland in 1772, and another to Newfoundland and Labrador in 1776, where he collected numerous previously undescribed plants. He sailed as naturalist/botanist aboard the *Endeavour* on James Cook's first voyage from 1768 to 1771, accompanied by the Swedish naturalist Daniel Solander. *Endeavour* sailed into Botany Bay in April 1770, and Banks chose the location for Britain's first penal colony. In 1778, he was elected president of the Royal Society, a position he held until his death. He did not sail on Cook's second and third voyages, but he did propose that William

Bligh command two voyages to Tahiti to collect breadfruit seedlings for transplanting in the West Indies, an enterprise that ended with the mutiny of the *Bounty's* crew in 1789. Banks had a hand in choosing Arthur Phillip to command the "First Fleet" that sailed to Australia, but when they got to Botany Bay, Phillip rejected Banks's suggestion of Botany Bay and moved the settlement to Port Jackson (Sydney). It was Banks who recommended Bligh for the governorship of New South Wales in 1805. He was made a baronet in 1781 and invested with the Order of the Bath in 1795. Banks Island in the Canadian Arctic was named for him, as was the plant genus *Banksia,* a type of Australian honeysuckle.

See also Bligh, *Bounty*, Cook (James)

Barbados Because it is not in the immediate vicinity of the other Windward Islands (Dominica, Martinique, St. Lucia, St. Vincent, etc.), but about 100 miles to the east, this island was not colonized by the early Portuguese and Spanish explorers and was not "discovered" until 1627, when the British found it occupied by the peaceful Arawak Indians. When the Arawaks were gone, the British imported African slaves to work the sugar plantations, and most of today's Barbadians are descended from these slaves. Like many of its neighbors, Barbados was a member of the West Indian federation that was established in 1958 and dissolved in 1962. Barbados remained a British colony until it achieved independence in 1966, and it is now a member of the British Commonwealth. Some 255,000 people live on the 21-mile-long island. Removed from the "Caribbean" culture, Barbados has developed its own version of England, with an emphasis on pubs, cricket, and Anglican Christianity. The primary industry is tourism, but internationally acclaimed rums are also produced here. **See also Arawak Indians, Dominica, Martinique, St. Lucia, St. Vincent and the Grenadines**

Barbarossa "Red Beard," the name given to the sons of Yakub of Mitylene who took to piracy on the Barbary Coast of North Africa. From about 1510 to 1546, two of the redbeards, Khayr-al-Din and his brother Arouj, mercilessly plundered Mediterranean shipping and invaded the Spanish, Italian, and Greek coasts, burning and murdering as they went. Khayr-al-Din became admiral of the fleet under Sulayman the Magnificent (1494–1566), the Ottoman sultan who greatly expanded the Turkish empire and twice defeated Andrea Doria, the admiral of the Italian fleet. He retired a great Islamic hero to a vast palace in Constantinople, and died there in 1547. **See also Barbary pirates**

Barbary pirates Generations of corsairs who operated from the Barbary coast of North Africa, the Barbary pirates were notorious for their ferocity and skill. In the mid-seventeenth century, the coastal towns—Tripoli, Tunis, Algiers, Salé, and Bône—broke free of Turkish rule and became military republics, existing mostly on piracy. Originally working in oared galleys rowed by slaves, the pirates converted to sailing ships in the seventeenth century and greatly expanded their range, traveling as far afield as Iceland and Ireland. A major element in their plundering was the capture of slaves, and it is estimated that in the seventeenth century, more than twenty thousand Christians were sold in the market of Algiers alone. In 1800, the United States went to war with the Barbary pirates because they had been demanding tribute and taking American sailors as slaves. Many other nations mounted expeditions against the Barbary pirates, but their predations did not end until 1830, when a French expedition captured Algiers and annexed it. **See also Barbarossa, piracy, Tripolitan War**

Barents, Willem (1550–1597) A Dutch navigator and explorer, Willem Barents (Dutch: Barendsz) served as navigator on the expedition in search of a northerly route to India over the northern coast of Russia in 1594. In a ship commanded by Jacob van Heemskerck, they reached Novaya Zemlya on the first voyage, but they were turned back by the ice. In 1595, they tried and failed again, but in the following year they landed on Bear Island and proceeded on to Spitsbergen before arriving at Novaya Zemlya, where their ship was wrecked by the ice. They built a house from the ship's timbers and spent the winter in almost impossibly inhospitable conditions. When spring came, Barents and his crew abandoned their house and set out for the Kola Peninsula, 1,600 miles away. Heemskerck and most of the men survived, but Barents died after five days in the open boats.

See also Bear Island, Heemskerck, Northeast Passage, Novaya Zemlya, Spitsbergen

Barents Sea Bounded by the archipelagoes of Spitsbergen and Franz Josef Land in the north, the Norwegian and Russian mainland to the south, and Novaya Zemlya to the east, the Barents Sea covers some 542,000 square miles. Medieval Russians and Vikings knew it as the Murmean Sea, but it was named in 1853 for Willem Barents, the navigator who in 1594 accompanied Dutch captains Jacob van Heemskerck and Jan Cornelisz Rijp in search of the northeast passage to Asia. In 1596, with Barents again as navigator, they discovered Bear Island en route to the (accidental) discovery of Spitsbergen. The waters of the Barents Sea are warmed by the Norway Current, and although some ice forms, the Russian port of Murmansk and the Norwegian port of Vardö are navigable all year 'round.

See also Bear Island, Heemskerck, Franz Josef Land, Spitsbergen

bark A sailing vessel with three masts, square-rigged on the fore and main, and fore-and-aft rigged on the mizzen. Originally small sailing ships, they were built up to 3,000 tons for the grain and guano routes around Cape Horn in the late nineteenth century. Eventually, four- and even five-masted barks were built. A barkentine is square-rigged on the foremast only and is fore-and-aft rigged on the main and mizzen.

See also brigantine, ship

barnacle (class Cirripedia) "A barnacle," wrote Thomas Huxley, "is a little shrimplike animal standing on its head within a limestone house and kicking its food into its mouth with its feet." Technically, it is a sedentary marine crustacean that is permanently attached to a moving or nonmoving object. Barnacles have a shell resembling that of a mollusk, but they have six pairs of thoracic legs (the *cirri* of "Cirripedia") that protrude through the shell plates to filter food suspended in seawater. Barnacles are hermaphrodites, carrying out cross-fertilization between neighbors. After hatching, they are one-eyed animals with three pairs of legs, floating around in the plankton until they find some surface to adhere to. When they are permanently attached, they metamorphose into eyeless animals with cirripeds and a new, calcareous shell covering. There are stalked (or gooseneck) barnacles, rock (or acorn) barnacles, and others that are specific to humpback whales. Barnacles adhere to living objects, like crabs, turtles, sea snakes, and whales, but they are also a major problem for ships, slowing them down and therefore costing them more fuel.

barndoor skate (*Raja laevis*) One of the largest of the skates, this species is easily identified by its size, pointed snout, and smooth skin. These skates, which can be 6 feet across, are usually reddish brown on the back, and unlike most other skates, which are white below, the barndoor skate's ventral surface is pigmented. They are bottom feeders but spend time hovering above the gravel bottom. They used to be plentiful in northern Atlantic waters such as the Gulf of Maine and Georges Bank, and although they had no commercial value, sport fishermen liked the idea of hauling in such a large, heavy creature. Recently, the population has been diminished to the point where they are in danger of extinction. There was never a directed fishery for barndoors, but the bycatch for groundfishes and scallops has included so many—particularly juveniles—that their numbers have been drastically reduced. If protection is not afforded this species in the near future, it will have the dubious distinction of becoming the first marine fish whose extinction is scientifically documented in modern times. **See also rays, skates**

barracuda, great (*Sphyraena barracuda*) The largest of the twenty-odd species of barracudas, the great barracuda can reach a length of 6 feet. It is found worldwide in temperate and tropical waters, where it feeds primarily on fish. Because of its fearsome teeth, the barracuda is generally believed to be a threat to swimmers and divers, but stories of actual attacks are rare. Perhaps more dangerous to man is the barracuda's potential to carry ciguatera, a poison that can be transmitted when the fish is eaten. **See also ciguatera**

barramundi (*Lates calcifer*) Probably Australia's most popular sport fish, the barramundi is found in coastal rivers and mangrove estuaries, but not in the clear offshore waters of the Great Barrier Reef. They have been recorded at up to 100 pounds, but heavy fishing has greatly reduced their numbers, and the average fish nowadays does not exceed 20 pounds. *Barramundi* is an Aboriginal word meaning "fish with large scales." An unrelated species of Australian grouper (*Cromileptes altivelis*) is commonly known as the barramundi cod.

barrier island A narrow, usually sandy island that runs parallel to the mainland and separates a lagoon or bay from the open ocean. Barrier islands are usually separated from one another by narrow inlets, and are subject to rapid change from storms and currents. In the United States, the best-known barrier islands are found off the Atlantic coasts of Virginia, North Carolina, and Georgia. Whereas barrier islands are usually composed of sand, barrier *reef* islands are mostly coral.

barrier reef A partially submerged coral outcrop that follows the contour of the land. Inside the reef the water is usually shallow, permitting the growth of coral, but the reefs usually fall off sharply on the seaward side. The largest coral reef in the world is Aus-

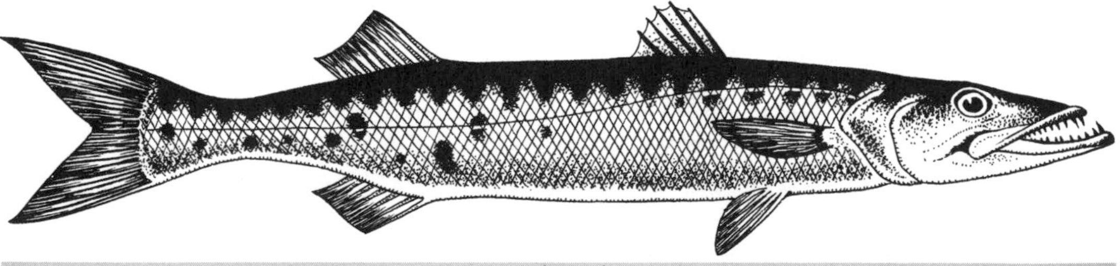

great barracuda

tralia's Great Barrier Reef; at 1,250 miles in length, it is the largest structure on earth built by living organisms. The second largest, 135 miles long, is off the Caribbean coast of the Central American country of Belize.

See also coral, coral reef

Barron, James (1768–1851) American naval officer; served in the Revolutionary War and in the Mediterranean. In 1807, in the *Chesapeake,* the flagship of the U.S. Mediterranean squadron, he sailed for Europe but was stopped off Virginia by the British frigate *Leopard,* which demanded the return of three British deserters. When Barron refused, the *Leopard* opened fire, and Barron, unable to return the fire, offered to surrender, but the British simply took the men and sailed away. Barron was court-martialed for cowardice but convicted only of neglecting to clear his ship for action. The "*Chesapeake* Affair" was considered one of the incidents that led to the War of 1812. Barron was suspended for five years, during which he served in the French navy. Upon his return, he applied for reinstatement but was refused. He believed that Stephen Decatur, a junior officer who had served as one of his judges, was responsible for his misfortune, and challenged him to a duel. On March 22, 1820, Barron shot and killed Decatur, one of America's most beloved naval heroes. Although he was reinstated to duty, Barron never regained his earlier status, and he retired from the navy in 1848. **See also Decatur**

Barrow, John (1764–1848) British statesman who served as second secretary of the Admiralty for forty years, and although his only visit to the Arctic was a whaling voyage to Spitsbergen when he was young, he was the moving force behind Britain's quest for the Northwest Passage, which he believed should be the highest goal of the Royal Navy. Whaling captain William Scoresby offered his services, but was turned down in favor of Commander John Ross and Lieutenant Edward Parry, who sailed in separate expeditions in 1818. Under Barrow's administration, the Royal Navy offered cash rewards for sailing the farthest north and for finding the Northwest Passage. The Ross and Parry voyages were followed by Franklin and Richardson (1819–1822), Parry and Liddon (1819–1820), Parry and Hoppner (1824–1825), Franklin and Richardson (1825–1827), Beechey (1826–1828), Parry (1827), and others, mostly by James Ross, all of which failed to achieve Barrow's goals but were critical to the mapping of the Canadian Arctic. Barrow's Arctic enthusiasm culminated in the Franklin expedition of 1845, which was lost and required a whole new series of voyages in search of the missing *Erebus* and *Terror.* Barrow died in 1848, before the fate of the Franklin expedition became known. Barrow Strait, between Cornwallis and Somerset Islands—which does lead to the Northwest Passage—was named for him.

Barton, Otis (1899–1992) Explorer, inventor, and visionary who designed the bathysphere. Otis Barton was raised in New York and Massachusetts, and graduated from Harvard in 1922. Inspired by Sir Arthur Conan Doyle's *Lost World,* Barton traveled on his own to Central America to look for prehistoric reptiles, and then went to Russia and China to search for fossils. He spent a lot of time and money drawing up plans for a self-contained underwater diving capsule. William Beebe (who needed Barton's bathysphere as much as Barton needed Beebe's leadership) put together an expedition to Bermuda (under the combined auspices of the National Geographic Society and the New York Zoological Society), where the two of them made the series of dives off Nonsuch Island (Bermuda) that culminated in the then-record descent of 3,028 feet in 1934. The record remained unbroken until 1949, when Barton, who by then had designed the benthoscope, another underwater vehicle, made a descent to 4,500 feet off the coast of Southern California and set the record again. During World War II, Barton served as a naval photographer in the Pacific. After the 1954 publication of his *World beneath the Sea,* Barton turned his attention to Africa, where he wanted to join Richard Leakey in the search for early man. (His plan to search for fossils by attaching a plow to a jeep was not received enthusiastically.) Barton spent the rest of his life fantasizing about exploration and the discovery of exotic animals, designed an elevator to take him to the tallest treetops, and died in obscurity.

basket star (*Gorgonocephalus arcticus*) Although basket stars have five arms, each arm branches, branches again, and then branches again and again until the arms form a dense tangle. *Gorgonocephalus* therefore resembles the Gorgons, three women in Greek mythology with snakes for hair. Basket stars use their tangled arms to snare passing organisms, which they transfer to the mouth on the underside of the disk. The North Atlantic *G. arcticus,* which can achieve an arm span of 20 inches, is found from the Arctic to Cape Cod at depths of 4,000 feet or more.

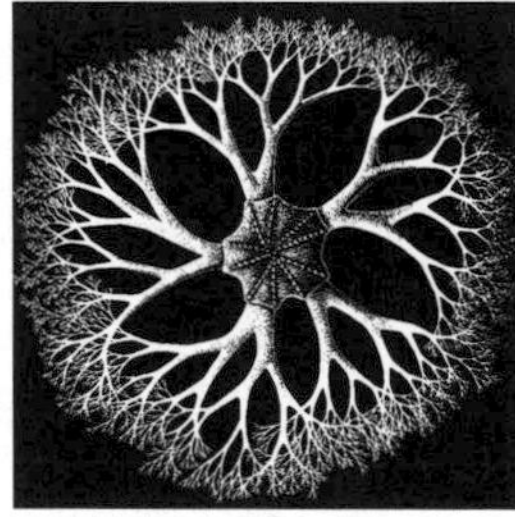
basket star

See also brittle star

basking shark (*Cetorhinus maximus*) At a maximum known length of 32 feet and a weight of 8,500 pounds, the basking shark is one of the largest sharks in the world. (Only the whale shark grows larger.) It is easily recognized by its size, its mottled brownish-gray color, its pointed snout, and its enormous gill slits, which extend almost all the way around the head. It is found in

basking shark

subpolar and temperate waters around the world, occasionally straying into warmer seas. They are often seen swimming in groups near the surface. Basking sharks are plankton eaters, and strain microorganisms from the water with their gill rakers. Despite their size, they are harmless to people. For the oil contained in their enormous livers, basking sharks have been fished commercially for centuries in Scotland and Ireland. Large specimens that wash ashore and deteriorate have been responsible for many tales of sea serpents. In 1952, Gavin Maxwell wrote *Harpoon Venture,* the story of his attempts to open a basking shark fishery on the island of Soay, just south of the Isle of Skye in the Hebrides.

See also Skye, Isle of; whale shark

Bass, George (1771–1802) English naturalist and explorer; sailed in 1795 for New South Wales as surgeon aboard the *Reliance,* on which Matthew Flinders was serving as midshipman. From Port Jackson (Sydney), Bass, Flinders, and a servant boy took an 8-foot boat (christened *Tom Thumb*) on a coastal survey to the south, and later Bass took a whaleboat with six rowers and discovered the strait that separates Van Diemen's Land (Tasmania) from the mainland; the strait now bears his name. In 1798, in the sloop *Norfolk,* Flinders and Bass made the first circumnavigation of the island, proving that it was not attached to the mainland. Bass took ill in New South Wales and returned to England in 1800, but when he tried to return to Australia, he got lost en route and was never heard of again.

Bass, George (b. 1932) Marine archaeologist, trained at the American School for Classical Studies in Athens and the University of Pennsylvania, who directed the excavation of a Bronze Age shipwreck off Cape Gelidonya, Turkey, the first complete excavation of an ancient wreck on the seafloor. After receiving a Ph.D. in 1964, Bass excavated two Byzantine wrecks near Yassaida, Turkey, developing a submersible, an underwater telephone booth, and the method of underwater mapping known as stereophotogrammetry. In 1973, he founded the Institute for Nautical Archaeology (INA), a private organization affiliated with Texas A&M University. Under Bass's leadership, the INA has been involved in research on four continents, including the excavation of the city of Port Royal on the island of Jamaica, which slid into the Caribbean during a 1692 earthquake; and a shipwreck from the fourteenth century B.C., the oldest known shipwreck, which contained a scarab from Queen Nefertiti of Egypt and the oldest ingots of tin and glass. Dr. Bass, considered the founder of scientific marine archaeology, has published seven books and more than one hundred relevant articles, and is the recipient of the National Geographic Society Gold Medal, among other awards.

bass, giant sea: See giant sea bass

Bataan Peninsula north of Manila Bay on Luzon Island in the Philippines. In 1941, after the Japanese invaded the islands and Manila was taken, American and Filipino forces withdrew to Bataan. On April 9, 1942, Bataan fell to the Japanese, and 55,000 Filipinos and 8,000 Americans were taken prisoner. On the torturous six-day "death march" from Bataan to the prison camp at Cabanatuan, given hardly any food or water by their captors, 2,300 Americans and 10,000 Filipinos died of beatings, wounds, starvation, dysentery, beriberi, malaria, and thirst. Under Lieutenant General Jonathan Wainwright, those who were not captured at Bataan escaped to heavily fortified Corregidor Island in the mouth of the bay, where they held off the Japanese for another month. In the Philippines, April 9 is known as Bataan Day, and in homage, Filipino men re-create the march every year. A 70,000-acre refuge, Bataan National Park, has been set aside as a memorial.

See also Corregidor; Leyte Gulf, Battle of; Philippines

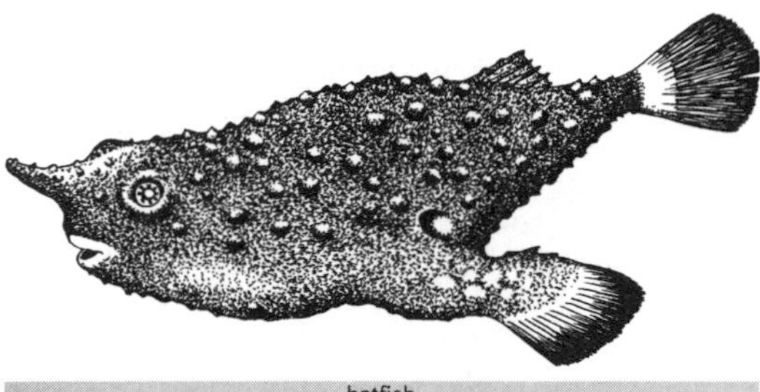
batfish

batfish (*Ogcocephalus* spp.) Flattened fishes with bat-like "wings," these peculiar creatures are also equipped with widely separated pelvic fins that they use for walking over the sandy bottom. A modified dorsal spine hangs down as a tentacle from beneath the snout and is used as a lure as the fish waits in the sand, and then bursts forth to capture its unwary prey. There are some thirty species of batfishes, found throughout the warm, shallow waters of the world. In South Africa, the name "batfish" is used for the spadefish (*Platax pinnatus*), a deep-bodied, compressed fish with large, erect dorsal and anal fins that give it a delta-winged shape.

See also spadefish

Bathurst Island (1) A 7,600-square-mile island in the Canadian Arctic Archipelago, between Melville and Devon Islands, currently the location of the magnetic North Pole. It was first described by William Edward Parry in 1819 and was named for Lord Henry Bathurst, Secretary of State for Colonial Affairs when British explorers were seeking the Northwest Passage. (Banjul, the capital city of the West African country of Gambia, was also formerly known as Bathurst.)

See also Melville Island (2)

Bathurst Island (2) Island in the Northern Territory of Australia, between the Timor and Arafura Seas, separated from Melville Island by the Apsley Strait. Like Melville Island, this 1,000-square-mile island was deeded to the Tiwi tribe under the 1976 Aboriginal Land Rights Act. **See also Melville Island (1)**

bathybius In 1868, Thomas Henry Huxley "discovered" a supposed primitive life-form on the floor of the ocean, consisting of muddy sediment containing foraminiferan shells and small, chalky disks that he called coccoliths. Huxley believed that this gelatinous material was the source of all life, a network of living protoplasm that was spontaneously generated from inorganic materials. He named it *Bathybius haeckeli* (after the German zoologist Ernst Haeckel) and believed that this "living ooze" would be found over all the ocean floors. Instead, bathybius was shown to be a precipitate of calcium sulfate in seawater that had been turned to jelly by the addition of the preserving alcohol. Huxley realized he had made a mistake, but Haeckel, who referred to the material as *Urschleim* ("original slime"), kept the idea alive until well into the 1880s. **See also foraminifera, sediments**

bathyscaph An untethered submersible that served from 1948 to 1982 as the primary device for manned exploration of the deep oceans. The first bathyscaph (*bathyscaphe* in French) was *FNRS-2* (1948), followed by the French navy's *FNRS-3* (1953) and then *Trieste*, designed by Auguste Piccard (1884–1962), and put into service in 1953. For *Trieste*, an underwater free balloon, buoyancy was provided by a float that was filled with gasoline, which is lighter than water. The crew was housed in a thick steel sphere suspended beneath the float. Vertical motion was achieved through gaining and losing weight. The French Navy's last bathyscaph was *Archemide*, built in 1964, and retired in 1978. The bathyscaph *Trieste*, piloted by Piccard's son Jacques and U.S. Navy lieutenant Don Walsh, set the unbreakable world's depth record of 35,800 feet in 1960 when they descended into the deepest spot on the ocean floor, the Challenger Deep in the Mariana Trench in the southwestern Pacific. The U.S. Navy continued to use *Trieste* until 1982. **See also Challenger Deep, Project FAMOUS, Pacific Ocean, *Trieste*, Walsh**

bathysphere A spherical submersible a little less than 5 feet in diameter, made of steel that was 1½ inches thick, with three quartz portholes. Commissioned by William Beebe and designed by Otis Barton for the New York Zoological Society, the bathysphere was lowered to a maximum depth of 3,000 feet off Nonsuch Island in Bermuda on several dives from 1930 to 1934. At the time, Beebe's "half mile down" was the record for submersible descents, and Beebe wrote several books and hundreds of descriptions of the fishes and other deep-sea life that he observed from the portholes and collected in nets. **See also Barton, Beebe**

battleship Late in the nineteenth century, the French began building iron and steel battleships, and the British followed with a class of ships that they called dreadnoughts, characterized by their size and heavy naval guns. In fact, HMS *Dreadnought*, built in 1906, was the first modern battleship equipped with all big guns. She was the first ship to have the primary gun batteries in turrets on the centerline instead of along each side, as they had been in all previous warships. Capable of speeds up to 21 knots, *Dreadnought* was 526 feet long and had a beam of 82 feet. The U.S. battleship

Maine, which blew up in Havana harbor in January 1898, precipitated the Spanish-American War. British and German fleets consisting primarily of battleships fought the Battle of Jutland (May 31, 1916), the largest naval battle of World War I. The Russian battleship *Potemkin* was the scene of a famous mutiny in 1905 and the subject of a famous film in 1925. The *Bismarck,* chased and sunk by the British in 1941, was a battleship. Part of the Japanese strategy in the surprise attack on Pearl Harbor was to cripple as much of the American Pacific fleet as possible; the *Arizona, Nevada, West Virginia, Tennessee, Oklahoma, Maryland,* and *California* were sunk or damaged on "battleship row" on December 7, 1941. The Japanese surrender was signed aboard the battleship *Missouri* on September 2, 1945. Although a small number of battleships remained usable (the *New Jersey* was used for shore bombardment during the Vietnam War), they are no longer considered a necessary part of naval warfare, and the last were decommissioned in 1992.

See also *Bismarck;* dreadnought; Jutland, Battle of; Pearl Harbor; *Potemkin*

Bay of Islands Formed when the sea inundated an old river valley at the northern end of South Island, the Bay of Islands is among New Zealand's most important historical locations. James Cook was the first European to enter the inlet, arriving in 1769, and he named the bay for the 150 small islands in it. He careened and overhauled the *Endeavour* at a little inlet that he named Queen Charlotte's Sound, passed through the strait, now known as Cook Strait, that separates the two main islands, then circumnavigated and mapped both islands. (On this voyage, Cook demonstrated that New Zealand was composed of two major islands, and that it was not part of the mysterious "southern continent.") By the beginning of the nineteenth century, whaling ships from England and America were using the Bay of Islands for provisioning, and even for recruiting Maoris as whalemen. The town of Kororareka was the scene of the first white settlement in New Zealand in 1809, and a hotbed of whalemen's debauchery. Renamed Russell, it became the first capital of New Zealand. The 1840 treaty between the Maoris and the British that annexed New Zealand to England was signed at Waitangi, also in the Bay of Islands. It is now a popular resort area, especially for big-game fishing, as popularized by western author Zane Grey, who wrote *Angler's Eldorado* in 1926.

See also Cook (James), Maori, New Zealand

Beagle Four-gun sloop built at Woolwich in 1820 and rigged as a brig, the HMS *Beagle* was allocated to the surveying service in 1825, and sailed to the Straits of Magellan under Robert FitzRoy, accompanied by Phillip Parker King in the *Adventure.* Under FitzRoy's command, she sailed around the world from 1830 to 1836 with Charles Darwin aboard as naturalist. In *The Voyage of the Beagle* (1836), Darwin described the voyage, which began on November 23, 1831, at Devonport, and headed for the Canary Islands, the Cape Verdes, and then the east coast of South America, stopping to explore at Bahia and Rio de Janeiro in Brazil, Montevideo in Uruguay, and various anchorages in Argentina. After landing on Tierra del Fuego, they sailed eastward to the Falklands, and then back to Tierra del Fuego. All told, the *Beagle* spent almost two years on the east coast of South America. They passed through the Straits of Magellan and entered the Pacific in June 1834. The *Beagle* then sailed along the Chilean coast and departed from Callao in Peru for the Galápagos. It was here that Darwin is supposed to have realized that the differentiation of various finches proved the mutability of species—a realization that led him to formulate the seminal *Origin of Species,* which would be published in 1869. After spending a month in the Galápagos, the *Beagle* headed for home, sailing westward through the Society Islands and the Marquesas, then heading for New Zealand and Australia. From there she went to the Keeling Islands, Mauritius, around the Cape of Good Hope to Ascension Island, and then back to Bahia in Brazil before returning to England with stops at the Cape Verdes and the Azores. Darwin and the *Beagle* anchored at Falmouth on October 2, 1836.

See also FitzRoy, Galápagos Islands, Tierra del Fuego

Beagle Channel Named for the ship *Beagle,* commanded by Robert FitzRoy, who, along with Phillip Parker King in the *Adventure,* had been commissioned by the British Admiralty to survey the Strait of Magellan. The Beagle Channel passes between the island of Tierra del Fuego, to the north, and Navarin, Hoste, and several smaller islands to the south. In decent weather, the channel can be a passage from east to west, avoiding the often difficult Strait of Magellan and the much more unwelcoming Cape Horn.

See also *Beagle;* Cape Horn; Magellan, Strait of; Tierra del Fuego

beaked coral fish (*Chelmon rostratus*) A butterfly fish with an extremely elongated snout, this species is found around the coral reefs of northern and western Australia, New Guinea, and the Indian and western Pacific Oceans. In certain areas of the Great Barrier Reef, it is replaced by *C. muelleri,* which has bronze-colored instead of yellow stripes. Beaked coral fishes are popular with saltwater aquarists. **See also angelfish, butterfly fish**

beaked whales A group of some twenty species, generally characterized by a spindle-shaped body, protruding upper and lower jaws, no central notch in the tail fin, and teeth only in the lower jaw of the males. Called Ziphiids or Mesoplodonts, very little is known about their habits, and some species are represented only by animals that have stranded. In those cases where the stomach contents have been examined, it was seen that beaked whales feed mostly on squid and some cetologists believe they are suction feeders. The Indo-Pacific beaked whale (*Indopacetus pacificus*) has never been seen in the flesh and is known only from two skeletons. Beaked whales may be the least-understood group of large animals in the world, and in recent years, two previously unsuspected new species have been described, the Peruvian beaked whale (*Mesoplodon peruvianus*) in 1991 and Bahamonde's beaked whale (*M. bahamondi*) in 1995. **See also Baird's beaked whale, bottlenose whale, goosebeak whale, *Mesoplodon***

bearded seal (*Erignathus barbatus*) The long and plentiful whiskers—technically known as vibrissae—of this Northern Hemisphere seal are obviously responsible for its name. Males, which grow larger than females, can reach a length of 12 feet, but 10 feet is more common. Throughout their ice-defined range, bearded seals are hunted for their meat, blubber, and skin, but they were never the object of a major sealing industry because they are usually solitary animals and do not assemble in large herds the way many other pinnipeds do. They have a circumpolar Arctic distribution, from Labrador and Greenland to Siberia and Alaska, and as far south as northern Norway. Bearded seals are very vocal, and they are among the few mammals that actually "sing."

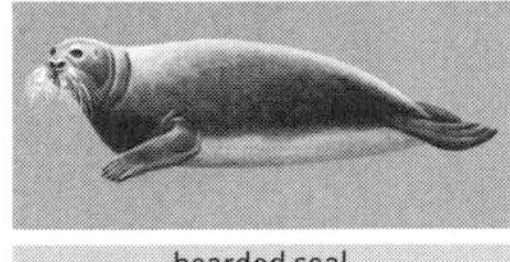
bearded seal

Bear Island Isolated little island in the Barents Sea in the high Arctic, between northern Norway and Spitsbergen. (In Norwegian, Bear Island is Bjornøya.) The 10-mile-long island was discovered in 1596 by the Dutch explorers Jacob van Heemskerck and Jan Cornelisz Rijp, on a voyage in search of a route to the east over the Eurasian continent. Willem Barents was the navigating officer on board. On their first landing, they killed a polar bear and named the island Beyren Eylandt. (Subsequently, the British claimed the island and named it Cherry Island, but the name was later changed back.) Although their quest was unsuccessful, the Dutch explorers continued northward, discovered Spitsbergen, and reported that walruses and whales were abundant in the waters of Spitsbergen and Bear Island, which led to the seventeenth- and eighteenth-century Dutch and British Arctic whale fisheries. The island is now part of Spitsbergen and belongs to Norway.
See also Barents, Barents Sea, Spitsbergen

Beaufort, Sir Francis (1774–1857) British admiral who served as hydrographer of the Royal Navy for a longer period of time than anyone before or since. After being wounded in battle twice, he worked on his surveys and scientific studies, and developed his scale for wind force at sea in 1805. (In 1874, with the addition of sea-state phenomena, it was adopted by the International Meteorological Committee for worldwide use.) From 1829 to 1835, in his office as hydrographer, he introduced many scientific innovations, including the development of an office to correlate compasses, and another to record and publish tide tables for all British coasts. In Beaufort's last years, the Crimean War broke out, and he saw that surveying officers were attached to all British ships to facilitate navigation in uncharted waters. The Beaufort Sea, north of Alaska and western Canada, is named for him. **See also Beaufort Scale, Beaufort Sea**

Beaufort Scale Properly known as the Beaufort Wind Force Scale, this scale that classified wind force at sea was devised in 1805 by Admiral (later Knight Commander of the Bath) Sir Francis Beaufort. As originally drawn up, the scale made no reference to the speed of the wind, but these figures were added during the twentieth century. The scale is rarely used by professional meteorologists or sailors today, but it can be used where there are no wind instruments.

BEAUFORT NUMBER	WIND	WIND SPEED (KNOTS)	DESCRIPTION OF SEA SURFACE
0	Calm	less than 1	Sea like a mirror
1	Light air	1–3	Ripples like scales
2	Light breeze	4–6	Small wavelets
3	Gentle breeze	7–10	Large wavelets
4	Moderate breeze	11–16	Small waves
5	Fresh breeze	17–21	Moderate waves
6	Strong breeze	22–27	Large waves
7	Moderate gale	28–33	Sea heaps up
8	Fresh gale	34–40	Moderately high waves
9	Strong gale	41–47	High waves
10	Whole gale	48–55	Very high waves
11	Storm	56–63	Exceptionally high waves
12–17	Hurricane	64+	Air completely filled with foam and spray

Beaufort Sea Section of the Arctic Ocean north of Canada and Alaska between Point Barrow, Alaska, and the Canadian Arctic Archipelago. It covers about 184,000 square miles and its average depth is 3,239 feet. It is under ice most of the year except for August and

September, when the ice breaks up, and then only near the shore. The Mackenzie River debouches into the Beaufort Sea, depositing some 15 million tons of alluvial sediment annually. The sea is rich in plankton, and it is the breeding ground of the bowhead whale (*Balaena mysticetus*). In the area of Point Barrow, at the western end of the Beaufort Sea, American Inuit hunt the bowhead under a strict, government-imposed quota system. Prudhoe Bay is a small inlet in the Beaufort Sea, 200 miles southeast of Point Barrow. Since the 1968 discovery of vast petroleum deposits on Alaska's North Slope, Prudhoe Bay has been the site of the origin of the Trans-Alaska Pipeline. Valdez, an ice-free port some 800 miles to the south on Prince William Sound, is where tankers are loaded—and where the 987-foot-long supertanker *Exxon Valdez* filled up before she ran aground, dumping 11 million gallons of crude oil into Prince William Sound, the largest oil spill ever to occur in the United States. **See also Arctic Ocean, bowhead whale, *Exxon Valdez***

Bedloe's Island: See Liberty Island

Beebe, C. William (1877–1962) Although he achieved fame as an explorer, oceanographer, and ornithologist, William Beebe was probably best known as an author and popularizer of science. He wrote more than eight hundred journal and magazine articles and twenty-four books. Among his works was a definitive monograph on pheasants, *Unseen Life in New York,* and *Half Mile Down,* the story of his 1930–1934 descents in the bathysphere. During these descents he claimed to have seen many species of fishes that no one has seen since. Beebe was curator of ornithology at the New York Zoological Gardens from 1899 to 1919, when he was appointed director of the Bronx Zoo's Department of Tropical Research Station in Trinidad. **See also Barton, bathysphere**

belemnite Extinct group of marine cephalopods, similar to modern squids and cuttlefishes in that they were dartlike, rapid swimmers, and could move tailward or tentacleward with equal facility. Belemnites (technically called belemnoids) had ten tentacles that were set with rows of little hooks rather than suckers. Like the modern cephalopods, they were dibranchiates, with two gills in the mantle cavity, and they had an ink sac. Belemnites had a squidlike body enclosing a bullet-shaped internal shell (the phragmocone), which consisted of gas-filled chambers and a bladelike forward extension (known as the pro-ostracum). Behind the phragmocone was the rostrum (also known as the guard), a strong, massive, calcareous shell, which is the part most usually found as a fossil. The belemnites were especially abundant in Jurassic and Cretaceous seas, 200 to 140 million years ago. The largest known belemnite was *Megateuthis,* which measured more than 11 feet from tentacle tips to tail tip. The state fossil of Delaware is a belemnite, *Belemnitella americana,* from the Cretaceous silts of the town of St. Georges. **See also ammonite, cephalopod, chambered nautilus, squid**

Bellamy, Sam (d. 1717) The story is told that Sam Bellamy (later known as Black Sam or Black Bellamy) became a pirate because he could not locate enough treasure legally to impress his lady love, Maria Hallett of Cape Cod. He took to piracy and was said to have plundered more than fifty ships. Among them was the *Whydah,* a slaver and treasure ship, which he was sailing in Massachusetts waters in April 1717 when a storm came up and sank the ship, killing everybody on board. The wreck of the *Whydah* was found in 1984 off Wellfleet, Massachusetts, by treasure hunter Barry Clifford.

Bellingshausen, Thaddeus Fabian von (1779–1852) Russian naval officer who was ordered by Tsar Alexander I to circumnavigate the world. In 1819 he sailed in the *Mirny,* with Mikhail Lazarev in command of the *Vostok.* They sighted the South Sandwich Islands and then proceeded farther south than any expedition at that time, and although they made no claim to have landed on the Antarctic landmass, they were probably the first to sight it, on January 27, 1820. (Sealing captain Nathaniel Palmer claimed to have been the first, but his sighting was on November 16, 1820, almost nine months later.) The Bellingshausen Sea in the Antarctic is named for him.

beluga (*Delphinapterus leucas*) Found only in cold Arctic and subarctic waters, the beluga is a chubby, 16-foot white whale with a rounded, prominent forehead (called the melon), a low ridge instead of a dorsal fin, and a permanent grin. Called sea canaries by early whalers, belugas (the name comes from the Russian word for "white") are among the noisiest of all whales, with a repertoire of whistles, squeaks, chirps, barks, whinnies, and snores. They do well in captivity and are often seen as the stars of oceanarium shows. In Canadian waters—particularly the St. Lawrence River—their populations have been heavily affected by chemical pollutants. **See also narwhal**

beluga

Benbecula The smallest main island in the Outer Hebrides of Scotland, Benbecula separates the islands of North and South Uist. It is connected to them by a causeway and a bridge, respectively. Like many of the islands in the Outer Hebrides (also known as the Western Isles), Benbecula is losing population because the crofters—farmers who lease narrow strips of land—cannot earn a living farming its moorlands. During the cold war, the British army trained on Benbecula, which added substantially to the island's economy. In 1830, so the story goes, a Benbecula woman was cutting seaweed when she came upon a mermaid the size of a small child. A cruel boy threw a stone at the mermaid and killed her, but the townspeople made a coffin for her and buried her close to where she had washed up. **See also Hebrides**

Benchley, Peter (b. 1940) American author, educated at Harvard; the author of several novels that featured sea creatures as their protagonists, including *Jaws,* the 1974 story of a vengeful great white shark that terrorizes the beaches of the fictional Amity Island. After the worldwide success of *Jaws* (book and movie), Benchley wrote *The Deep* (1976), *The Island* (1979), *The Girl of the Sea of Cortez* (1982), and in 1991, *Beast,* a story of a man-eating giant squid. Most of his nautical novels have been turned into Hollywood or television movies, and he is one of the best-selling novelists at work today. **See also *Jaws***

bends, the Also known as caisson disease or decompression sickness. When a diver descends to significant depths, nitrogen—which constitutes four-fifths of every breath we take—is forced into solution under the ambient pressure. As the pressure falls—that is, as the diver rises to the surface—the nitrogen turns back into bubbles of gas and, if not allowed to dissolve, can obstruct blood vessels. In arm and leg joints this can cause serious cramps and pain, but if the bubbles obstruct the blood flow to the heart or the brain, they can cause death. Divers who have been at depths for any amount of time have to decompress, either by waiting at certain stations for a prescribed amount of time or, in an emergency, by entering a decompression chamber. Deep-diving animals, such as certain whales, seals, and penguins, do not get the bends because, where human divers take more than one breath, the other animals submerge after taking only one breath, and the air is compressed into "safe" regions of the respiratory pathway, such as the air sinuses at the base of the skull. Rather than particularly large lung capacity, the deepest-diving whales, such as sperm whales and bottlenose whales, have relatively small lungs, and at depths of 300 feet or more, their lungs collapse completely, forcing all the air into the rigid respiratory dead spaces. Another problem with breathing nitrogen under pressure is nitrogen narcosis, or "rapture of the deep," where human divers begin to lose the power of rational thought and, if they do not ascend, are capable of removing their masks, diving deeper, doing an impromptu dance, or performing other irrational and potentially dangerous acts. **See also bottlenose whale, caisson, scuba, sperm whale**

Bengal, Bay of Arm of the Indian Ocean bordered by Sri Lanka and India on the west and Myanmar (formerly Burma) and Thailand on the east. The Bay of Bengal covers some 839,000 square miles, and its only major island groups are the Andamans and Nicobars. In the spring and summer, the monsoons cause the currents to move clockwise; they reverse direction in the autumn. The Irawaddy, Ganges-Brahmaputra, Mahanadi, and Cauvery Rivers empty into the bay, and their fertile deltas are the site of such cities as Madras, Calcutta, and Rangoon (now Yangon). Sediment from the rivers has reduced the depth of the waters near the shore and greatly reduced the salinity. The deepest part of the bay, 14,764 feet, is in the south, off the west coast of the island of Sumatra. **See also Andaman Islands, Indian Ocean**

Benguela Current Named for the Angolan city on the South Atlantic coast of Africa, the Benguela Current is a cool-water upwelling that flows northward off the coasts of Namibia and Angola. It is formed by westerly trade winds (the West Wind Drift) that are deflected off South America and set up the Brazil Current that flows southward, passing across the South Atlantic and picking up cold Antarctic water before turning north along the coast of Africa. At the Cape of Good Hope, where the cold Benguela Current meets the warm Agulhas Current, there are often crosscurrents and eddies that are dangerous to shipping. Upwelling of nutrients has made this area particularly attractive to fishermen and whalers, especially in the area of Walvis Bay in Namibia. Walvis Bay is also the scene of massive fish kills attributable to red tides. **See also Agulhas Current, red tide**

benthos From the Greek word meaning "depth of the sea," the benthos is the deep-sea environment; the adjectival form, "benthic," is usually used in reference to those animals or plants attached to the ocean floor or crawling on it. The smallest benthic organisms are protozoans and bacteria, while the largest are polychaete worms, corals, shellfish, and starfish. **See also abyss**

Bering, Vitus (1681–1741) Danish explorer in the service of Tsar Peter the Great of Russia. He was charged with exploring the Arctic, and in 1725 he traveled overland from St. Petersburg to Kamchatka, where he built the ship *Gabriel.* The sea and the strait that were named for him were actually discovered by Semyon Dezhnev in 1684, but Bering discovered Little and Big Diomede and St. Lawrence Island. In 1741, as part of Russia's "Great Northern Expedition," he made the first European landing in Alaska, but on the return voyage, his ship *St. Peter* was wrecked on the Commander Islands (which were later named for him), and he died there. A survivor of this voyage was Georg Steller, the naturalist who first described the sea otter, the sea lion, the eagle, and the sea cow that bear his name. **See also Dezhnev, Steller**

Bering Sea and Strait The Bering Sea is the northernmost part of the Pacific Ocean, separated from the Pacific by the Aleutian chain of islands. It is bordered on the west by the east coast of Asia and on the east by Alaska. The sea, roughly triangular in shape, is 890,000 square miles in area. Because of its severe weather, ice, fog, winds, and rain, the Bering Sea is considered one of the most hazardous of seas. The Bering Strait, some 53 miles across, separates the Chukchi Peninsula of Siberia from the Seward Peninsula of Alaska, and the Bering Sea from the Arctic Ocean. Big Diomede Island (the westernmost of the two islands in the Bering Strait) belongs to Russia; Little Diomede belongs to the United States.

Bermuda Located about 600 miles east of Cape Hatteras in the western North Atlantic, Bermuda is a self-governing British colony. It consists of four large islands—Somerset, St. George's, St. David's, and "Main" (Bermuda)—and many smaller ones, bunched together in a fishhook shape. Facts about the discovery of the islands are obscure, but the name is derived from that of Juan de Bermúdez, a Spanish navigator who is reputed to have sighted the islands in 1503. En route to the Virginia colony in 1609, the *Sea Venture,* commanded by Sir George Somers, ran aground on Bermuda's treacherous reefs, but the settlers found the islands uninviting and they left as quickly as they could. (This event was the inspiration for Shakespeare's *Tempest,* written around 1611.) In 1612, King James I of England authorized the Virginia Company to send sixty settlers to colonize the island. Bermuda has remained a British colony ever since, and the parliament is the oldest in the British Commonwealth, having been established in 1620. Tourism is the major industry, and the population is around 58,000. Bermuda's climate, its proximity to the United States and England, and its splendid hotels, beaches, and diving sites (it has the most northerly coral reef in the world) have established its preeminence as a resort and tourist destination. The white-roofed, pastel-colored houses and pink sand beaches are particularly photogenic. The depths of Nonsuch Island, on the perimeter of Castle Harbour, were the site of the record-breaking bathysphere dives of William Beebe and Otis Barton in 1930–1934.

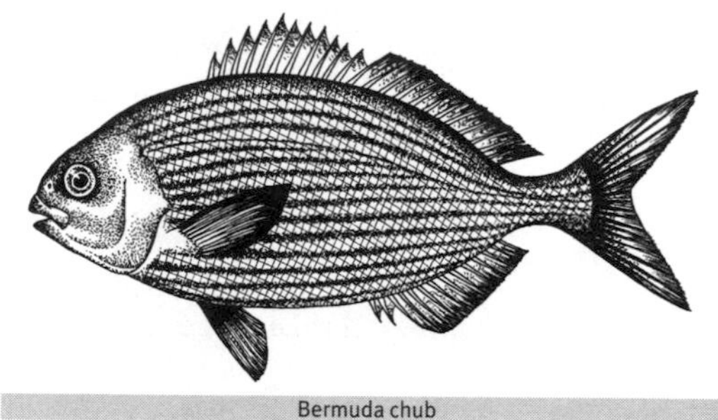

Bermuda chub

Bermuda chub (*Kyphosus sectatrix*) This horizontally striped fish, which reaches a length of about 30 inches, is common from Massachusetts to the Caribbean, and in the eastern Atlantic from England to the Canary Islands. Sea chubs have small, delicate mouths, and although they are prized as sport fish, they are difficult to hook. Throughout their range, they are known by many common names, including rudderfish for their habit of following ships. An Australian version (*K. sydneyanus*) is known as the silver drummer, and another Indo-Pacific species (*K. biggibus*) is called the gray chub. There is also a freshwater minnow (*Nocomis biguttatus*) that is known as a chub.

Bermuda Triangle An area of the western Atlantic, roughly drawn from Bermuda south to southern Florida, then east through the Bahamas to Puerto Rico, then north again to Bermuda. In 1974, Charles Berlitz published a book called *The Bermuda Triangle,* in which he listed "more than 100 planes and ships [that] literally vanished into thin air . . . and more than 1,000 lives [that] have been lost in the past twenty-six years, without a single body or even a piece of wreckage from the vanishing planes or ships having been found." The absolute lack of physical and documentary evidence should have shown that this book was utter hokum, but it was enormously popular, and many people believed Berlitz's unsubstantiated stories of time-space warps, UFOs, alien abductors, and ancient Egyptian space travel. **See also Atlantis**

bigeye (*Priacanthus* spp.) Small, usually red or orange fishes that live in shallow water on both sides of the North Atlantic but are best known from the Caribbean. They have rough, prominent scales, and, as befits their common name, their eyes are huge: 1 inch in diameter for the 10-inch-long *P. arenatus.* Also known as catalufa, the glasseye snapper (*Heteropriacanthus cruentatus*) is slightly larger and is found in the same Caribbean waters. Bigeyes are adept at color change and can go white or blotchy in seconds, usually to match their environment.

bigeye thresher shark (*Alopias superciliosus*) Recognizable by its enormous, upward-looking eyes and a

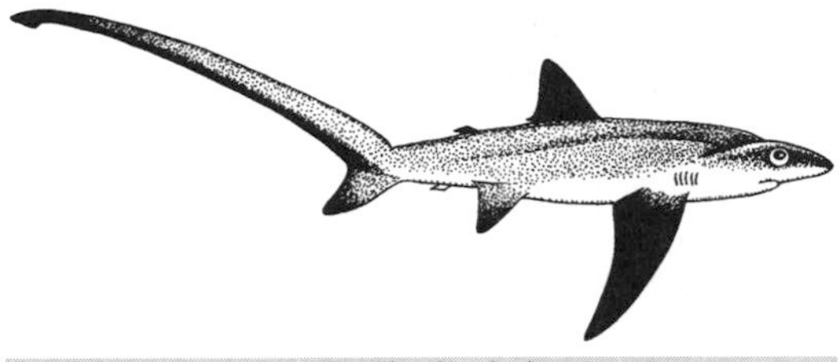
bigeye thresher shark

groove on the top of the head, the bigeye thresher is the deep-water version of the common thresher. Bigeyes can reach a length of 18 feet, approximately half of which is the greatly elongated upper lobe of the tail fin. Feeding at depth on squid, small tunas, lancet fishes, and hake, the bigeye thresher probably uses its long tail to herd or round up its prey. Bigeye threshers are often caught on swordfish longlines. **See also thresher shark**

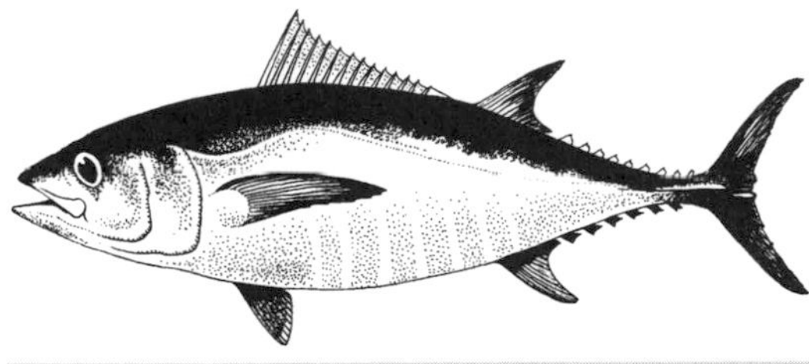
bigeye tuna

bigeye tuna (*Thunnus obesus*) The large eyes of this species suggest that it lives and hunts at greater depths than other tunas. Bigeyes reach a maximum length of 7½ feet and can weigh 400 pounds. They are among the most commercially important tunas, especially in Pacific Rim countries like Japan and Australia. They are usually caught on longlines, not in nets, and are sold fresh, not canned. The bright red meat, higher in fat than that of other large tunas, is prized for sashimi and grilling. Bigeyes are found all over the world in tropical and subtropical waters, and they are popular big-game fish on both coasts of North America and in Hawaii and Australia. **See also albacore, bluefin tuna, skipjack**

big-game fishing Fishing (usually from specially equipped boats) for any large fish, often those that put up a fight. Although there are many species of fishes that are recognized as game fishes by the International Game Fish Association (IGFA), those that are considered the premier big-game fishes are tunas, broadbill swordfish, various marlins, sailfish, and some sharks, such as the mako, tiger, and great white.

Bikini Atoll in the western (Ralik) chain of the Marshall Islands in the region of the central Pacific known as Micronesia. Bikini was first described by Otto von Kotzebue, who sailed past it without landing in 1825 and named it Escholtz Atoll. Unlike its neighbors, Eniwetok and Kwajalein, Bikini was not touched by the fighting as the United States fought to regain the islands that had been occupied by the Japanese during World War II. After the war, the Marshalls came under the administration of the U.S. Navy, and in 1946 Bikini became the centerpiece for Operation Crossroads, a project to determine the effect of atomic bombs on naval vessels. The third and fourth atomic explosions in history took place on Bikini in July 1946. The second of these, an underwater bomb, produced a 1-million-ton hollow column of water, 2,000 feet in diameter, which rose to a mile above Bikini Lagoon before crashing down in a storm of waves, debris, and radioactivity. Among the target ships were the American carrier *Saratoga* and the superbattleship *Nagato,* the only Japanese battleship to survive the war. Further tests—including the 1954 explosion of "Bravo," the world's first deliverable hydrogen bomb, one thousand times more powerful than the atomic bomb that had been dropped on Hiroshima—were conducted until 1958, spreading lethal radioactive fallout over an unexpectedly wide area. Bikini's 167 inhabitants, who had been moved to the island of Rongerik, were unable to sustain themselves there and were moved to Kwajalein and then to Kili Island. The Bikini Islanders, assured that the dangers of radioactivity were over, moved back to their island in 1974, but when further testing showed that the level of radioactivity was still too high, they were evacuated in 1978. The island is still dangerously radioactive, and nobody can live there.

See also Eniwetok, Kotzebue, Kwajalein, Marshall Islands

Bimini Islands in the northern Bahamas, also called Biminis, consisting of North Bimini, South Bimini, and several smaller cays to the south, including Cat Cay, a world-famous center for big-game fishing. The islands are flat, with an elevation of no more than 20 feet. A popular resort destination, North Bimini is also the home of the Lerner Marine Laboratory, administered by the American Museum of Natural History in New York. In 1513, Juan Ponce de León (1460–1521) led an expedition from Puerto Rico to Bimini in search of the fountain of youth, but he found Florida instead. When he returned to Spain in 1514, he was named military governor of Bimini and Florida.

See also Bahamas, Ponce de León

binomial nomenclature Developed by the Swedish naturalist Carolus Linnaeus (1707–1778), this is the system of giving living things two names, a generic name (the genus) and a specific name (the species). The system is used for every animal and plant, and enables scientists, no matter what language they are writing in, to identify the subject. Thus the bigeye tuna (see above) is known as *Thunnus* (all tunas) *obesus* (which means "fat" in Latin, and has nothing to do with big eyes). This differentiates this species from, say, the bluefin tuna, which belongs to the same genus (*Thunnus*) and is known as *Thunnus thynnus*. The generic name always begins with a capital letter, the specific name with a lowercase letter. When species in the same genus are listed, it is often convenient to use just the first letter of the generic name, e.g., *Stenella longirostris*, the spinner dolphin, and *S. coeruleoalba*, the striped dolphin.

bioluminescence The generation of light by living organisms. Many benthic creatures have bioluminescent capabilities, including the great majority of deep-sea fishes, some sharks, and many squids, shrimps, euphausiids, and even some sea cucumbers and starfishes. There are two categories of bioluminescence: that generated in special cells known as photophores (intracellular), and that generated by symbiont bacteria in the body of the host (extracellular). Bioluminescence serves many functions, the primary one of which seems to be concealment. If a fish can eliminate its silhouette by effectively matching the background, it is said to be counterilluminated, and since most light organs are found on the ventral surface of fishes, this seems a likely explanation. There are fishes equipped with bioluminescent lures that are used to attract prey, and others with lights that might be used to illuminate the immediate vicinity so that the fish can see in what would otherwise be total darkness. (The squid *Histioteuthius* has photophores around one of its eyes that act as a searchlight.) Some bioluminescent fishes use lights to attract potential mates, but these lights might also serve a negative function in attracting predators. For mobile animals such as shrimps and squids, bioluminescence probably serves the same purposes as it does for fishes, but why a sea cucumber or a starfish would need to light up is unclear.

See also anglerfishes, lantern fishes, photophore, viperfish

birdbeak dogfish (*Deania calcea*) A small deepwater dogfish with an elongated snout that reaches a known maximum length of 6 feet. Like many other dogfishes, it has spines in front of each dorsal fin. Little is known about this species, but it has been caught in pelagic trawls in the eastern Atlantic and on handlines in the Pacific, around South Africa, and western South America. **See also dogfish, spiny dogfish**

birdbeak dogfish

bireme A galley having two banks of oars, used for warlike purposes in the Mediterranean from ancient days until the seventeenth century. They were almost invariably fitted with a ram or beak fixed to the bow at or below sea level, since their primary means of attack was to ram an enemy vessel. **See also trireme**

Biscay, Bay of Part of the North Atlantic that forms a large indentation in the coast of western Europe, bound by the eastern coast of France and the northern coast of Spain and covering around 86,000 square miles. The French call it Golfe de Gascogne, and the Spanish Golfo de Vizcaya. There are several small islands off the French coasts of Brittany and Aquitaine, including Ushant, Belle-Île, Nourmoutier, Île de Ré, and Oléron, the last two of which are known for their saltworks. The Loire and the Garonne Rivers flow into the bay, and the major ports are Brest, Nantes, Bordeaux, Bayonne, and La Rochelle in France; and San Sebastián, Bilbao, and Santander in Spain. The Basque people, whose origins are unknown, were early inhabitants of the coastal areas in northeastern Spain, particularly the province of Guipúzcoa, and embarked on the earliest commercial whaling voyages (around A.D. 1000) from this region, eliminating the right whales of the Bay of Biscay before crossing the Atlantic to Newfoundland. Biarritz and Saint-Jean-de-Luz are popular French resorts on the bay, but Basque terrorists, fighting for independence from Spain, have made parts of the Spanish coast dangerous for tourists.

See also Atlantic gray whale, right whale, Ushant

Bismarck German battleship completed in 1940 and designated for action against British convoys in the North Atlantic. In May 1941, having sailed from the shipyard at Gdynia, in Poland, the *Bismarck* and the heavy cruiser *Prinz Eugen* were encountered by the British battle cruiser *Hood* and the battleship *Prince of Wales* in the Denmark Strait, between Iceland and Greenland. In the ensuing battle, the British ships fared badly: the *Hood* was sunk with the loss of all but three of her fourteen hundred crew members, and the *Prince of Wales* was badly damaged. The aircraft carrier *Victorious* was dispatched to carry out an air attack on the German battleship, but it was inconsequential. The British Admiralty then ordered the battle cruiser *Renown*, the aircraft carrier *Ark Royal*, and the cruiser *Sheffield* to close with the enemy ships. The *Bismarck* headed for port in occupied France, since she had lost 1,000 gallons of fuel in the battle with the *Hood*. Over a period of three days, she was followed and attacked by numerous ships and planes, and on May 26 a torpedo fired from a British biplane damaged her rudder so she could not maneuver. The British ships *Rodney* and *King George V* fired shell after shell at the wounded *Bismarck*; in all, 2,786 shells were fired, of which 400 struck the ship, but still she would not go down. Captain Ernst Lindemann gave the order to scuttle her so she would avoid the ignominy of falling into British hands, and on May 27 the great battleship went to the bottom, taking more than two thousand German sailors with her and leaving the survivors to be picked up by British ships. In June 1989, Robert Ballard, the man who had located the *Titanic* in 1985, found the *Bismarck* 600 miles west of Brest, France, sitting upright on the bottom in 15,500 feet of water.

Bismarck Archipelago Island group in the Bismarck Sea, northeast of New Guinea, including New Ireland, New Britain, and the Admiralty and Duke of York Islands. New Ireland was first sighted in 1616 by the Dutch navigator Jakob Le Maire, who believed it was contiguous with the island of New Britain; it is actually separated from New Britain by the 20-mile-wide St. George's Channel, which was discovered and named in 1767 by the Englishman Philip Carteret. (Carteret also named the island "New Hibernia," which became "New Ireland.") William Dampier, an Englishman, discovered and named New Britain during his voyage in 1700, and it became part of German New Guinea in 1884. During this period, the archipelago was named for Otto von Bismarck, the "Iron Chancellor" of the German republic from 1871 to 1890. After World War I, the islands were mandated to Australia by the League of Nations, but in 1941 they were taken by the Japanese. The battle for the Japanese air base on Rabaul (New Britain) was one of the turning points of the war in the Pacific. **See also Dampier, Le Maire, New Guinea, Rabaul**

Blackbeard (d. 1718) Probably the most famous pirate in history, Edward Teach (or perhaps Thatch) was nicknamed for his luxurious beard. He is believed to have been a privateer for the British during the War of the Spanish Succession (1701–1713); he turned pirate in 1717, when he captured a French warship and converted it to the 40-gun pirate ship *Queen Anne's Revenge.* When Stede Bonnet, the "gentleman pirate," ran into Blackbeard in Honduras, Blackbeard captured him and his ship. Blackbeard became notorious for raids on ships along the Virginia and Carolina coasts, and even ventured into the Caribbean, setting up shop on the island of St. Croix. Alexander Spotswood, the lieutenant governor of Virginia, dispatched a naval force under Robert Maynard, who succeeded in capturing Bonnet and killing Blackbeard. The latter was decapitated, and his head displayed on the bowsprit of Maynard's ship. If the pirate ever left any "buried treasure," its whereabouts are still a mystery.

See also Bonnet, buccaneer, Kidd, piracy, privateer, Roberts

black-browed albatross (*Diomedea melanophris*) With its white head and tail and dark upper wings, this albatross has the coloring of a gigantic black-backed gull, but its wings are narrower and much longer—about 7½ feet. The common name is derived from the dark eyebrow, which gives it a frowning or angry appearance. This species breeds on subantarctic islands and may be the commonest and most widespread of all the albatrosses. (There are fourteen species in the genus *Diomedea,* and two in *Phoebetria.*) It is an inveterate ship-follower, and the most likely species of albatross to be seen in the North Atlantic. A single individual associated with nesting gannets in the Faeroes from 1860 to 1894, thousands of miles from its regular range, and beginning in 1970, another began appearing on a ledge on the Shetland island of Unst. Nicknamed "Albert Ross," the bird has been seen intermittently every year during the summer, also in the middle of a gannet colony.

black-browed albatross

See also albatross, gannet, Laysan albatross, sooty albatross, wandering albatross, waved albatross

black-capped petrel (*Pterodroma hasitata*) Seabird with a 40-inch wingspan, dark back, white undersides, and a black cap. Once extremely abundant in the West Indies, the black-capped petrel (also known as diablotin, or "little devil") was believed to have been eliminated from its nesting islands of Guadeloupe, Jamaica, Hispaniola, Dominica, and Martinique by hunters who plucked the birds and eggs from the burrows. That they were not completely extinct was evidenced by occasional sightings at sea, but their only nesting site on the Morne la Selle in Haiti was not found until 1963. The introduction of the Burmese mongoose to Haiti now threatens the last known breeding colony. The Bermuda petrel or cahow (*P. cahow*) looks very much like the diablotin, but it lacks the white band behind the black cap. **See also cahow, petrel**

black coral (*Antipathes* spp.) Thorny corals that grow at depths of 300-plus feet in the Indo-Pacific and Caribbean. Black coral polyps grow on a treelike skeleton of hard, horny material. Black coral is one of the most desirable corals for jewelry because it polishes to a glossy black finish. For this reason it has been heavily harvested in those areas where it can be reached by divers, especially off the Hawaiian island of Maui.

See also coral

black dolphin (*Cephalorhynchus eutropia*) Sometimes called the Chilean dolphin, this small, stocky dolphin is known only from the waters of southern Chile. It is one

black dolphin

of several species of dolphins illegally hunted, killed, and used for bait in the Chilean crab fishery. Another is Commerson's dolphin (whose white flanks make confusion between the two species unlikely). The dolphins killed annually for this purpose are believed to number in the thousands. As befits its name, this species is largely black, but there is a lighter patch on the top of the head that resembles that of Hector's dolphin, a close relative. This animal can reach a length of 5 feet and a weight of 120 pounds, but hardly anything else is known of its natural history. **See also Commerson's dolphin, Hector's dolphin**

black drum (*Pogonias cromis*) A fish of the croaker family (Sciaenidae) that gets its name from the "drumming" noise produced by its air bladder. During the breeding season, the males emit particularly loud noises; those produced by the females are softer. Black drums are the largest members of the family, reaching a length of 4 feet and a weight of 125 pounds. They are found in the western North Atlantic, from Massachusetts to Argentina, in shallow inshore waters, near breakwaters, jetties, bridges, pilings, channels, and estuaries.

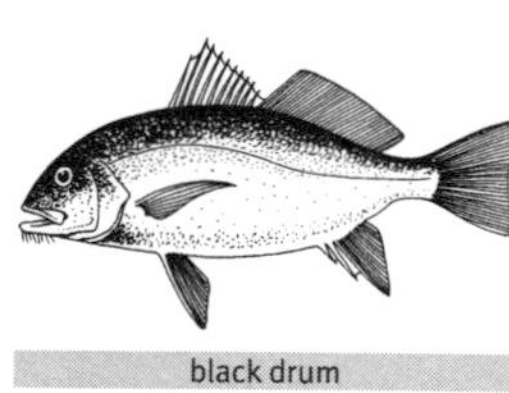
black drum

See also red drum

black-footed albatross (*Diomedea nigripes*) Juvenile black-footed albatrosses are dark brown all over except for a white area at the base of the bill; the adults are likewise, except for aberrant color phases (maybe aged birds), which have an almost white head and paler plumage. It breeds on central Pacific islands such as Laysan, Midway, Pearl and Hermes Reef, and Lisianski. It is the most abundant North Pacific albatross, following ships and fishing boats, but it is occasionally seen off the California and Oregon coasts. **See also albatross, black-browed albatross, Laysan albatross**

black-headed gull (*Larus ridibundus*) The most familiar and abundant small gull of Europe, the black-headed gull can be seen along the shore and in every river valley, moving inland in the morning and seaward in the evening. Despite its name, the head of the adult is not black but dark brown, and the coloring does not extend down the neck. The bill and feet are red. The very similar Mediterranean gull (*L. melanocephalus*) is larger and heavier, with a black head and no black on the wings. The smaller and lighter Bonaparte's gull (*L. philadelphia*), which is primarily North American (although an occasional visitor to Europe), has a black head and bill. The great black-headed gull (*L. ichthyaetus*) is an occasional visitor to western Europe, but this much larger bird with prominent black wing bands and a heavy bill is found primarily in temperate Asia, from the Black Sea to China.

black-headed gull

See also gull, herring gull

black marlin (*Makaira indica*) Probably the world's most sought-after game fish, the black marlin occurs in the tropical Indian and Pacific Oceans. (Some strays have rounded the Cape of Good Hope and visited the waters of West Africa, but these are rare occurrences.) It can be easily differentiated from the blue marlin by pectoral fins that cannot be folded flat against the body.

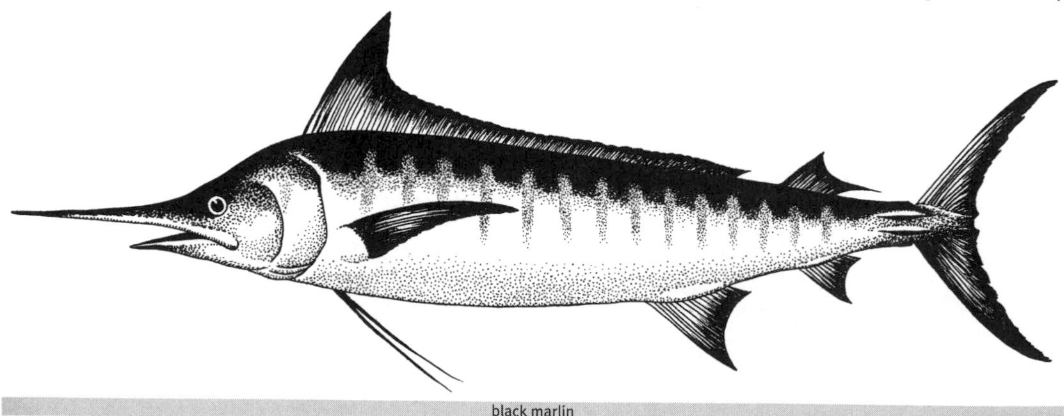
black marlin

It has proportionally the shortest dorsal fin of any billfish. The body is slate blue above, and silvery white below, with occasional pale vertical stripes. The black marlin is the largest of the billfishes, and commercial longliners have reported 2,500-pounders. The rod-and-reel record weighed 1,560 pounds and was caught off Cabo Blanco, Peru, in 1953. Most of the largest specimens have been females; a 500-pound male is considered large. **See also big-game fishing, blue marlin**

black-necked stilt (*Himantopus mexicanus*) Although it is most often found in the vicinity of brackish ponds and salt flats, this close relative of the avocet is also an inhabitant of shores and marshes. This delicate bird is surprisingly noisy and aggressive, and its alarm call sounds like the yapping of a terrier. The bill is black, but the long legs are bright red. Similar species are found around the world, differing only in the pattern of black on the back and head. **See also avocet, shorebirds**

black-necked stilt

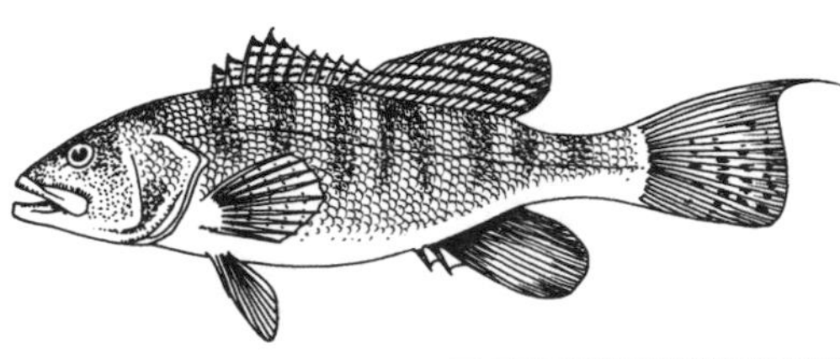
black sea bass

black sea bass (*Centropristis striatas*) Reaching a length of about 2 feet and a weight of 9 pounds, the black sea bass is found in the western North Atlantic from Massachusetts to the Gulf of Mexico. It is an inshore species, found around wrecks, reefs, and piers and in the vicinity of coral and rocks. It is fished commercially off Cape Hatteras, where approximately 1,000 tons are taken annually. Most begin life as females, and after about five years, they become males. Older males are sometimes known as humpbacks because they develop a prominent bulge behind the head. **See also giant sea bass**

black swallower (*Chiasmodon niger*) A deep-sea fish with fearsome fangs and jaws hinged at the front rather than the rear. This arrangement allows the 6-inch-long fish to expand its gape to the extent that it can swallow prey items considerably larger than its head. Swallowers are found throughout the world's temperate and tropical waters, at depths of up to 10,000 feet. **See also needletooth swallower**

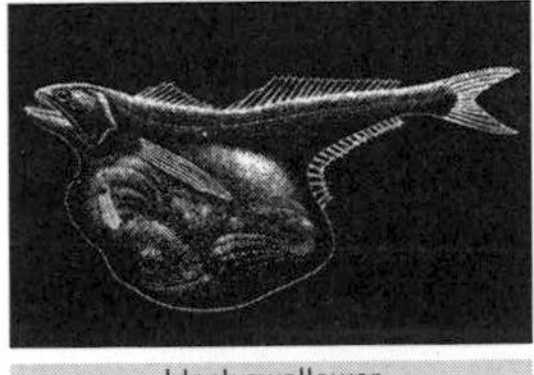
black swallower

blacktip reef shark (*Carcharhinus melanopterus*) A medium-sized brownish-gray shark with a rounded snout and sharply defined black tips on all fins. Rarely reaching 5 feet in length, the blacktip reef shark is mostly a creature of the inshore tropics of the Indo-Pacific region, but it is also found in the eastern Mediterranean, the Red Sea, and all long the east coast of Africa. With the white-tipped reef shark (*Trianodon obesus*) and the gray reef shark (*C. amblyrhynchos*), it is the commonest shark of the coral reefs of Oceania. Although it is usually not aggressive, so many people encounter this species—often while just walking in shallow water—that it has to be considered dangerous. **See also carcharhinid sharks, gray reef shark, white-tipped reef shark**

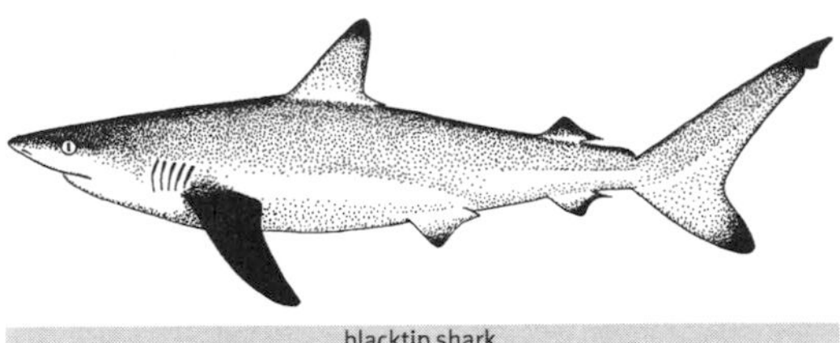
blacktip shark

blacktip shark (*Carcharhinus limbatus*) Blacktips are found in tropical and subtropical continental waters around the world, but are rarely encountered in waters more than 100 feet deep. Dark or bluish gray above and yellowish white below, this species gets its common name from the black tips on all of its fins except the anal fin, which is white. It reaches a maximum length of 6 feet and a weight of 270 pounds. A fast-moving shark, it has been seen leaping and spinning out of the water. Blacktips are considered game fishes, and are also the objects of commercial shark fisheries. **See also carcharhinid sharks, shark cartilage products**

blenny (family Blenniidae) A large family (more than three hundred species) consisting mostly of small, agile, bottom-dwelling fishes. Most are blunt-headed, with the mouth low on the head and not protractile. For fishes so small, many species have formidable canine teeth. They are found throughout the world's shallow waters, although there are a few deepwater species. One group is scaleless; another (known as klipfishes) has scales; and the pricklebacks have scales on the body, but the rays of the dorsal fin are completely spiny. The "eel-like blennies" (Pholidae) are elongated, 1-foot-long fishes found in the North Pacific, and the eelpouts (Zoarcidae) occur from the shallows to great depths. The giants of the blenny family are the wolffishes and wolfeels (Anarhichadidae), which can get to be 6 feet long; with their ferocious-looking teeth they are the delight of underwater photographers in the Pacific Northwest. **See also wolffish**

Bligh, William (1754–1817) British admiral, best known for his command of HMS *Bounty,* which culminated in the most famous mutiny in history. He began his naval career at the age of nine, when he was entered as a personal servant to the commander of a man-of-war. By 1770, he was master's mate aboard the HMS *Hunter,* and in 1776 he was appointed sailing master of the *Resolution* for Captain Cook's third voyage, a tribute to his skill as a navigator. As captain of the *Bounty,* he set sail from Tahiti for the West Indies with a cargo of breadfruit trees, but on April 28, 1789, the ship was seized by mutineers led by Fletcher Christian, and Bligh and eighteen loyal crew members were set adrift in a longboat. Demonstrating remarkable seamanship and fortitude, Bligh reached the island of Timor in the East Indies six weeks later, traveling a distance of 3,618 miles. He returned to England, where he was given several commands, one of which involved transporting breadfruit trees from Tahiti to the West Indies again. He served twice with distinction in military engagements alongside Admiral Horatio Nelson, and after several more voyages, he was appointed governor of New South Wales in 1805. Here his sailing skills availed him not, and he was such an unpopular administrator that in 1808 his officers "mutinied" and had him arrested and recalled to England.

See also *Bounty,* Christian, Pitcairn Island

Block Island Island in the Atlantic, first sighted in 1524 by Giovanni da Verrazano (who also discovered Aquidneck, or "Rode Island"), and visited in 1614 by the Dutch navigator Adriaen Block, for whom it was named. Its first settlers arrived in 1661. Block Island is about 7 miles long and 3.5 miles wide, and it is about 9 miles south of the Rhode Island mainland. Once dedicated to fishing and farming, the island is now a popular summer resort destination, reachable by ferry from New London, Point Judith, or Newport. It is part of the state of Rhode Island.

blowhole The common term for the nostril (or nostrils) of whales and dolphins. Most of the smaller cetaceans have a single blowhole, but the larger whales, e.g., the rorquals, the right whales, the gray whale, and the humpback, have paired blowholes, that look not unlike the nose of a human being, turned upside-down—that is, with the openings facing the rear. This arrangement enables the animal to inhale while moving forward, without taking water into its nose. The blowholes of almost every cetacean are located approximately above the eyes, but the sperm whale differs from the other large whales in that it has a single nostril, an S-shaped affair located at the end of the nose and off-center to the left. All cetaceans exhale through the blowhole, and when the exhalation is visible—as in the larger species—it is known as a spout.

See also dolphin, sperm whale, whale

blubber The fat layer of whales and dolphins; also used in discussions of seals, particularly elephant seals. Blubber insulates the animals from the heat-draining cold water in which they live, and also provides the whales with buoyancy. Right whales and bowheads have the thickest blubber, which can be 2 feet thick. Along with baleen, it was the reason whales were hunted. In the early whaling days, the blubber was peeled off the whale and boiled down in trypots on the decks of whaling vessels; the resulting oil was transferred to casks and stowed. With the introduction of more efficient methods, the blubber was boiled down, but the entire whale was chopped up and put into huge pressure-cookers, and the oil was extracted from the flesh and the bones as well. **See also baleen, whaling**

blue chromis (*Chromis cyanea*) One of about a dozen species of damselfishes (the common name is the same as the generic name) found throughout the world's coral reefs and adjacent waters. Of all the damselfishes, the blue chromis is one of the most brilliantly colored, with its electric blue body, black back, and black on the outer margins of its forked tail. It is the sharply forked tail that distinguished these fishes from the other damselfishes. **See also damselfishes**

blue crab (*Callinectes sapidus*) The most important commercial species of crab caught on the east coast of North and South America. In the United States, some 80 million pounds are taken every year. The blue crab's

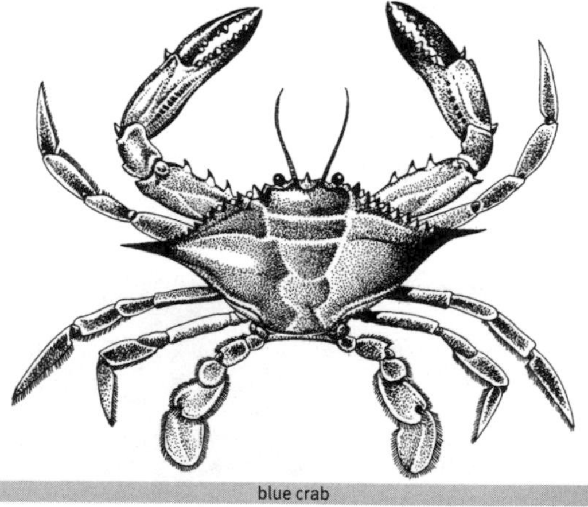

blue crab

preferred habitat is muddy shores, bays, and estuaries. Its claws and legs are blue, and the carapace is greenish gray on top and whitish on the bottom. The carapace of the largest ones can be 9 inches wide, enhanced by the presence of a distinctive sharp spine on each side. The so-called soft-shell crabs are blue crabs that are harvested shortly after molting, while the new shell is still soft. **See also crab, Dungeness crab, spider crab**

blue-eyed shag (*Phalacrocorax atriceps*) A strongly marked black-and-white cormorant found throughout southern South America, subantarctic islands, and the Antarctic Peninsula. Also known as the imperial or king cormorant, this species aggregates in large numbers, and there are times when thousands can be seen in dense concentrations. During the breeding season, the birds develop bright orange caruncles at the base of the bill, as well as the cobalt-blue eye-ring that gives the bird its common name. (Its eyes are actually brown.) Because of the climate, blue-eyed shags do not spread their wings to dry, in the manner so characteristic of temperate and tropical cormorants. **See also cormorant, shag**

bluefin tuna

bluefin tuna (*Thunnus thynnus*) The largest of the tunas—and one of the largest fishes in the world—the bluefin can reach a length of 12 feet and a weight of ¾ ton. (The International Game Fish Association record for a bluefin caught on rod and reel is 1,496 pounds.) They are immensely powerful and fast, capable of speeds up to 50 mph. The bluefin is a blue-water, schooling fish that makes some of the longest migrations of any fish. Specimens tagged in the Bahamas have been recaptured in Newfoundland and Norway, and even in Uruguay. (The southern bluefin, *T. maccoyii*, a very similar species, lives in the Southern Hemisphere.) Bluefins feed on mackerel, herring, mullet, whiting, squid, eels, and crustaceans. They are prized as big-game fishes, but it is the Japanese sashimi market that sets the astronomical prices on these fish. They are the source of *toro*, the fatty belly meat that sells in Tokyo restaurants for the equivalent of $50 an ounce. When a big, top-quality tuna is caught in New England, Australia, or New Zealand waters, it may sell for $80,000 on the dock, and by the time it is served in a restaurant, its value may have increased tenfold. As a result, bluefins have been overfished and their populations are threatened, but few protection measures have been taken, because this would require unprecedented international cooperation.

See also albacore, big-game fishing, bonito, yellowfin tuna

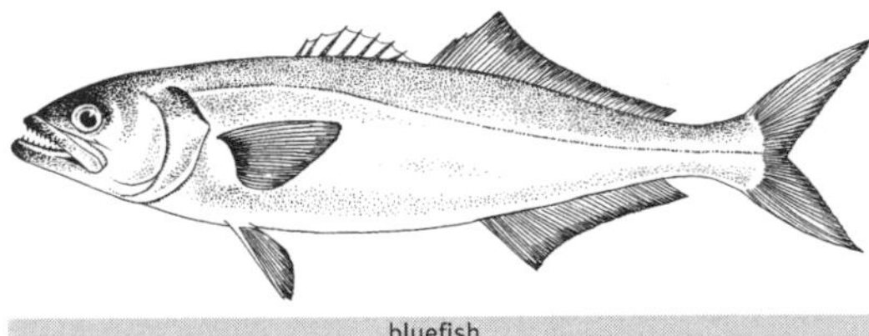

bluefish

bluefish (*Pomatomus saltatrix*) Found in scattered locations in the eastern United States, the Mediterranean and Black Seas, the South Pacific, and Africa, the bluefish is a favorite food and sport fish. (It is known as tailor in Australia and elf in South Africa.) Bluefish average about 2 feet in length and 10 pounds in weight, but larger ones have been caught. They are extremely aggressive and will swim through schools of bait fish, slashing and destroying everything in their path, including other bluefish. In this "feeding frenzy," they have also been known to attack swimmers, biting viciously with their sharp teeth.

blue-footed booby (*Sula nebouxii*) A mottled, cinnamon-brown bird from the western coasts of South and Central America and the Galápagos. Its range overlaps that of the brown booby, but the two species can be differentiated because the adult brown is a dark, chocolate-brown above, and the blue-footed—not surprisingly—has bright blue feet. The wingspan averages about 5 feet and the eye of the adult is straw-colored. Boobies are so named because they appear to be stupid; they do not fly away at the approach of men, and can be easily caught by hand.

See also brown booby, masked booby, red-footed booby

blue-green algae Primitive prokaryotic (cells without a nucleus) organisms that resemble the stromatolites, the earliest known form of life on earth. Blue-green algae contain chlorophyll-*a*, which enables the organisms to trap light and synthesize organic molecules. Lacking a nucleus, the blue-green algae most closely resemble bacteria, and the earlier name *cyanophyta* has been replaced by *cyanobacteria*, classifying them as animals rather than plants. They thrive in anaerobic environments, like the black mud of a lagoon or the polluted sediments of a harbor. Blue-green algae form threads about the thickness of a human hair; one group is *Spirulina*, touted as a high-protein wonder food.

See also bacteria, marine

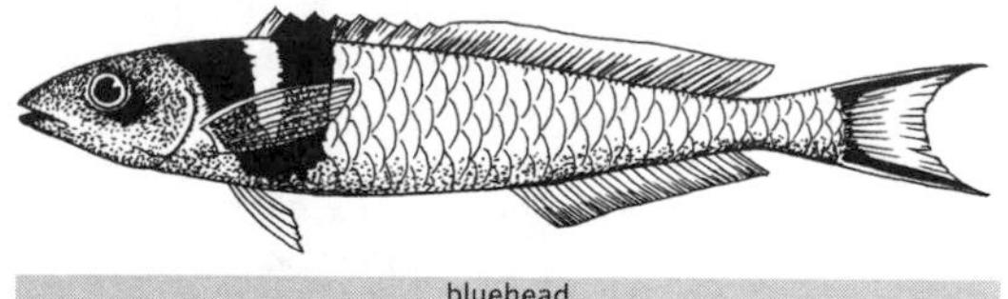

bluehead

bluehead (*Thalassoma bifasciatum*) One of the commonest fishes in the reefs of the Caribbean, Florida, and the Bahamas, the bluehead (also known as the bluehead wrasse) is 4 to 5 inches long and comes in two completely different color schemes, depending on sex. Juveniles and young adults of both sexes are yellow with a broad black stripe along the side, dark backs on the upper and lower edges of the tail, and a prominent spot in front of the dorsal fin. Breeding males—known as supermales—have a blue head separated from the green body by two thick black bars with a pale blue band between them. Blueheads sometimes act as cleaner fish. **See also cleaner fishes, wrasse**

blue holes Underwater caves caused by the deterioration of limestone on various Caribbean islands, particularly in the Bahamas. Many of the caverns, such as those on Andros and Grand Bahama Islands, are 200 to 300 feet deep and have been explored by divers, but others remain unexplored. Perhaps the most famous of these is "Tongue of the Ocean," immediately east of Andros Island, which drops precipitously to a depth of 6,500 feet. **See also Bahamas, giant octopus**

blue marlin (*Makaira nigricans*) Because they are found in both the Atlantic and the Pacific Oceans, some taxonomists have separated them into two distinct species. Most authorities, however, believe there is only a single species. Blue marlins are among the world's premier game fishes, being fast, powerful, and spectacular jumpers, and among the largest of the bony fishes. The rod-and-reel-record blue marlin (caught in 1992 off Vitoria, Brazil) weighed 1,402 pounds, but commercial fishermen have hauled in 2,000-pounders. True to its name, the blue marlin is dark blue above, with white undersides and pale blue vertical stripes along the back that quickly fade as the fish dies. Blue marlins are found in the waters of the Virgin Islands, the Gulf of Mexico, the Caribbean, West Africa, and Georges Bank in the Atlantic; and Hawaii, Baja California, and Ecuador in the Pacific. **See also big-game fishing, black marlin, white marlin**

blue parrot fish (*Scarus coeruleus*) Unlike many species of parrot fishes, the blue parrot does not change very much between the juvenile and adult phases. (However, like all parrot fishes, they are born female and change into males.) The "initial phase" of the blue parrot is a grayish blue; in the "terminal phase," the fish becomes deep blue and develops a vertical profile. Blue parrots, which reach a length of 3 feet, are found from Maryland to Rio de Janeiro. **See also parrot fish**

blue penguin (*Eudyptula minor*) In New Zealand, it is known as the blue penguin, but on Phillip Island (off Melbourne, Australia), where a study has been ongoing since 1967, it is still known by an earlier name, "fairy penguin." The smallest of all the penguins, the blue penguin stands only 17 inches tall. Blue penguins are bluish gray on the back and have light-colored eyes. They are the only nocturnal penguins, entering and leaving the water under cover of darkness. They form rafts offshore until dusk, then come out of the surf to walk to their burrows, where they remain for the day before returning to the water to feed. Blue penguins are nonmigratory, but they do make long excursions in search of food. **See also penguin, Phillip**

blue penguin

blue-ringed octopus

blue-ringed octopus (*Hapalochlaena maculosa*) A 5-inch-long Indo-Pacific and Australasian shallow-water octopus with venom powerful enough to kill a grown man in under an hour. It is a pretty little thing, with a span of only about 8 inches, with blue rings and spots that become more iridescent as the animal gets excited or irritated. **See also octopus**

blue shark (*Prionace glauca*) A pelagic, deepwater species, found throughout the world's temperate oceans. It is a slender, graceful shark, characterized by

blue shark

long, sickle-shaped pectoral fins; large, white-rimmed eyes; and dark blue dorsal coloration. The maximum known length is 12½ feet, but most are smaller. Blue sharks are viviparous (giving birth to live young), and can have as many as 135 pups per litter. They feed on small schooling fishes and squid, and although they have been implicated in attacks on victims of air or sea disasters, they are not otherwise considered dangerous to man. **See also mako shark**

blue tang (*Acanthurus coeruleus*) A common surgeonfish in the West Indies, Florida, the Bahamas, and Brazil, the blue tang is bright yellow as a juvenile, changing to its familiar deep blue when it reaches a length of about 3 inches. At maturity they can reach a length of 14 inches. The spine at the base of the tail is clearly marked in white. **See also surgeonfishes**

blue whale (*Balaenoptera musculus*) Even though there are some recently discovered dinosaurs that may have been longer, for sheer bulk the blue whale is the largest animal ever to have lived on earth. The biggest specimens are more than 100 feet long and weighed 150 tons (300,000 pounds). Blue whales feed on shrimplike creatures known as euphausiids, which they ingest by the millions. They take in and expel huge mouthfuls of water, trapping the food items on the fringes of their baleen plates and then swallowing them. Blue whales were hunted nearly to extinction in the Antarctic by twentieth-century commercial whalers, and although hunting them has been illegal since 1966 and they are fully protected around the world, the total population of around 5,000 animals does not appear to be increasing. **See also baleen, euphausiids, whaling**

boat A generic name for any small open craft without decking that is usually propelled by oars or an outboard engine. Another definition: "You can put a boat on a ship, but you cannot put a ship on a boat." Some exceptions to these general definitions are fishing boats, which may be almost any size and powered by any kind of engine, and submarines, which are almost always referred to as boats, probably because of their early designation as "submarine boats." Other exceptions are patrol boats and ferryboats. **See also ship, submarine, U-boat**

Bombay duck (*Harpodon nehereus*) Found in the estuaries of the Ganges and Brahmaputra Rivers, this 1-foot-long fish looks like a short-snouted lizardfish with long pelvic and pectoral fins and a spike that extends beyond the fork of the tail fin. Dried and salted, it is a common ingredient in curries, which accounts for its unusual common name. Closely related species are found in Indonesian and Japanese waters. **See also lizardfish**

Bonaire The "B" of the ABC Islands (the others are Aruba and Curaçao) of the Netherlands Antilles, located off the coast of Venezuela. Bonaire is a dry, dusty island of some 111 square miles. Amerigo Vespucci and Alonso de Ojeda landed on the island in 1499, but found nothing of interest except the Caiquetio Indians (a tribe of Arawaks) who were rounded up and exported to Hispaniola as slaves. In 1634, the Dutch, having lost the island of St. Martin to the Spanish, retaliated by capturing Aruba, Bonaire, and Curaçao.

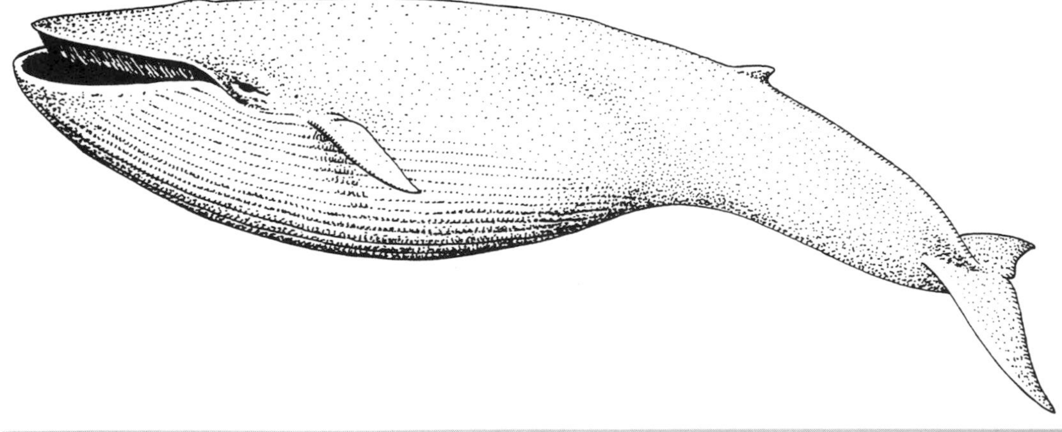
blue whale

Settlers representing the Dutch West India Company arrived on the island and tried to make a living raising corn and cattle, and harvesting salt. The British occupied the island twice (1800–1803 and 1807–1815), but Bonaire was returned to the Dutch in 1816, and has been part of the Kingdom of the Netherlands ever since. As with its neighbors, the lingua franca of Bonaire is Papiamentu, a creole amalgam of Dutch, Portuguese, Spanish, English, French, African, and Arawak. Tourism sustains the island's economy, especially for naturalists who come to view the wildlife above and below the water. There are fifteen thousand flamingos that congregate at Goto Meer, a saltwater lake in the north, but Bonaire's principal attractions can be found in the offshore reefs, where the diving is said to be the best in the Caribbean. In 1979, Bonaire established one of the world's first marine parks, encompassing the shelving reefs all around the island and providing a haven for a myriad of fish species and corals; it is a preferred destination for underwater photographers.

See also Aruba, Curaçao, Netherlands Antilles

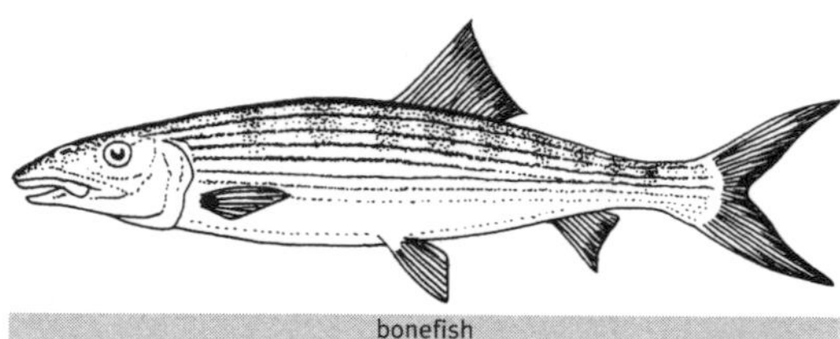

bonefish

bonefish (*Albula vulpes*) The family Albulidae contains a single species, found in warm shallows throughout the world, where its members forage for small crabs, shrimps, worms, and mollusks in the muddy or sandy bottom. Bonefishes are popular sport fishes, especially in the Florida Keys, where fishermen stalk them by poling flat-bottomed boats along the mudflats. Like eels, tarpon, and ladyfishes, bonefishes pass through a transparent, small-headed, ribbonlike leptocephalus stage before metamorphosing into miniature bonefishes. The world-record bonefish weighed 19 pounds and was caught off Zululand, South Africa, in 1962.

See also ladyfish, leptocephalus, tarpon

Bonin Islands Known also as the Ogasawara Islands, a group of small, volcanic islands about 500 miles south of Tokyo. The largest island is Chichi-jima, with the capital city of Omura. (In some accounts, Iwo Jima is considered one of the Bonin Islands.) The first European report of these islands was made by the Spanish navigator Ruy Lopez de Villalobos in 1543. They were rediscovered in 1824 by Josiah Coffin, captain of the Nantucket whaler *Transit,* who vaguely claimed them for the United States, but they were "officially" claimed the following year by Britain. Prior to the Sino-Japanese War of 1894–1895, the Japanese took over the islands and made them a part of Tokyo Prefecture. During World War II, the islands were a major Japanese stronghold until they were occupied by the U.S. Navy in 1945. The waters around the islands are particularly popular with whales, and therefore with whalers, but in recent years the moratorium on commercial whaling has encouraged whale watchers to come to the Bonin Islands for views of humpbacks and other species.

bonito (*Sarda* spp.) There are several species of bonito found throughout the world's tropical and temperate oceans. They reach a length of about 2½ feet and are characterized by stripes on the back and, like the other scombrids, the presence of little finlets aft of the second dorsal fin. The Atlantic bonito (*S. sarda*) is a fast, schooling fish that preys on smaller fishes and squid, usually near the surface. Other species are the California bonito (*S. orientalis*), the Australian bonito (*S. australis*), and the Chilean bonito (*S. chilensis*).

See also albacore, bluefin tuna, Scombridae, skipjack, yellowfin tuna

Bonnet, Stede (d. 1718) A retired British officer and wealthy plantation owner in Barbados when he decided to turn pirate, Major Stede Bonnet had a small sloop built that he claimed was to be used in interisland trading. The *Revenge* had ten guns and flew the Jolly Roger, which might have given a clue to its captain's intentions. Wearing a satin waistcoat, snow-white breeches, and a powdered wig, the "gentleman pirate" slipped out of Barbados; headed north, where he captured several prizes; and then made for Honduras. There he encountered Edward Teach (Blackbeard) who promptly captured him and his ship, but later restored the *Revenge* to Bonnet and allowed him participate in more voyages of piracy. They were both captured in 1718, but Bonnet was released by the governor of North Carolina. Bonnet changed the name of his ship to *Royal James* and continued his pirating until he was taken in South Carolina by Colonel William Rhett. Despite his pleas for clemency, he was hanged.

See also Blackbeard

bony fishes The "true" fishes (Osteichthyes), including all but the sharks, rays, and chimaeras (Chondrichthyes), which have skeletons of cartilage. (The "jawless fishes"—hagfishes and lampreys—are classified separately.) They are subdivided into two groups, the fleshy-finned fishes (Sarcopterygii), which includes the freshwater lungfishes and the coelacanth; and the ray-finned fishes (Actinopterygii), which includes all other freshwater fishes and the great majority of marine species. They have an ossified (bony) skeleton, true scales on the skin, a flap (the operculum) covering the gills, and movable rays in the fins and tail. They come in all sizes, from the ¼-inch-long pygmy goby to the 30-

foot-long oarfish; and all shapes, from the snakelike eels and the upright sea horses to the typically "fishlike" salmon and sardines. The heaviest of the bony fishes is probably the black marlin, which can weigh 2,500 pounds. They can be found at every depth, from the shallowest surface waters to the deepest trenches. Some, like the marlins and tunas, are swift ocean rangers, while others, like flounders, are adapted to a life spent lying on the bottom. Approximately half of the known vertebrates in the world are bony fishes; there are perhaps 20,000 species, compared to some 550 species of cartilaginous fishes.

See also *individual fish species;* chimaera, rays, shark

Bora-Bora Regarded as the quintessential South Pacific island because it consists of a dramatic mountain peak rising from an aquamarine lagoon, Bora-Bora is one of the leeward group of the Society Islands in French Polynesia. It is 161 miles northwest of Papeete, the capital of Tahiti, the most populous of the islands. The main island is a little more than 6 miles long, and the towering volcanic plug that dominates it is 2,400 feet high. Although it was long occupied by Polynesians, it was "discovered" by the Dutch navigator Jacob van Roggeveen in 1722. James Cook visited the island in 1777, and by 1895 it was annexed to France. During World War II, the Americans fortified the island against a Japanese attack that never came. One of the Americans stationed on Bora-Bora was James Michener, who fictionalized the island as "Bali Hai" in his *Tales of the South Pacific.* The population is around 4,500, but tourism has overtaken this island paradise, and its roads, beaches, streets, and hotels are disagreeably overcrowded. **See also Cook (James), French Polynesia, Tahiti**

Borneo The third largest island in the world, after Greenland and New Guinea, Borneo is occupied mostly by Indonesian Borneo (called Kalimantan by the Indonesians) and, on the north coast, by the Malaysian states of Sabah and Sarawak and the sultanate of Brunei. The island consists of dense, tropical jungle and mountains that tower more than 13,000 feet high. Because of the inhospitable climate and impenetrable forests, Borneo is one of the least densely populated countries in the world, with approximately 7 million people spread out over 287,000 square miles. The two major ethnic groups are the Dyaks and the coastal Malays. The fauna consists of elephants, deer, orangutans, gibbons, crocodiles, and many varieties of snakes. Rhinoceroses have been hunted practically to extinction. In 1983, some 13,000 square miles of rain forest were destroyed by fire. The island was discovered by Portuguese navigators in 1521 and shortly thereafter by the Spanish. The Dutch arrived in the early 1600s, and the English around 1665. British adventurer James Brooke became the "white rajah" of Sarawak in 1841, and present-day Sabah, Sarawak, and Brunei were declared British protectorates in 1880. During World War II, the island was occupied by the Japanese from 1942 to 1945. Dutch Borneo became part of the republic of Indonesia in 1950, and Sabah and Sarawak joined the Malay Federation in 1963.

Borough, Stephen (1525–1584) British navigator who, as sailing master of the *Edward Bonaventure,* accompanied Hugh Willoughby and Richard Chancellor in 1553 on a voyage to discover the Northeast Passage to China. Borough (sometimes spelled "Burrow" or "Borrows") was the first Englishman to sight and name the North Cape (Norway), and to reach the White Sea. Of the other senior officers on the 1553 expedition (Chancellor, Willoughby, and Cornelius Durforth), only Borough survived to lead the next voyage. In 1556, in the *Searchthrift,* a pinnace so small that its crew numbered only eight, he sailed to Russia, and traversed 600 miles of the Siberian coast before he reached the southern end of Novaya Zemlya, where he was stopped by the impassable ice. He visited Russia again in 1561 in the *Swallow,* and in 1663 was named chief pilot for Queen Elizabeth's ships. His younger brother William (1536–1599), who was aboard the *Edward Bonaventure* as an ordinary seaman, went on to a distinguished naval career himself, culminating in his command of the ship *Lion* during Sir Francis Drake's raid on Cádiz in 1587 and the *Bonvolia* during the battle against the Spanish Armada in 1588.

See also Chancellor, Novaya Zemlya, White Sea, Willoughby

Botany Bay Inlet on the east coast of Australia, 6 miles south of Sydney. It was first discovered by James Cook in 1770 during his first circumnavigation, and was named by Sir Joseph Banks, the official botanist of Cook's expedition. In 1787, the "First Fleet" left England for Australia with a cargo of convicts to found a penal colony there. They landed at Botany Bay, but Captain Arthur Phillip found the site unsuitable, and they moved northward until they reached the site of Sydney, one of the best natural harbors in the world. Botany Bay is now a major oil-tanker port. **See also Banks**

bottlenose dolphin (*Tursiops truncatus*) The most familiar of all the dolphins, *T. truncatus* is the species most often seen performing in oceanarium shows. "Flipper" was a bottlenose, and *The Day of the Dolphin* was written about this species. Echolocation (the ability to find things underwater by broadcasting sounds and then receiving the echoes) was first observed in captive bottlenoses, and most work with captive cetaceans is done with this species. Reaching a maximum length of

bottlenose dolphin

10 feet and a weight of more than 600 pounds, the bottlenose is a powerful animal, capable of prodigious leaps out of the water and measured dives to 1,500 feet. In the wild it is usually found within 100 miles of shore, and there are recognizable populations that remain in a particular bay or estuary all their lives. Of all the dolphins, bottlenoses have proven the most adaptable to training. They breed readily; most oceanarium animals are not caught in the wild, but born and raised in captivity. The life span of a captive bottlenose is about thirty years.

See also dolphin, echolocation

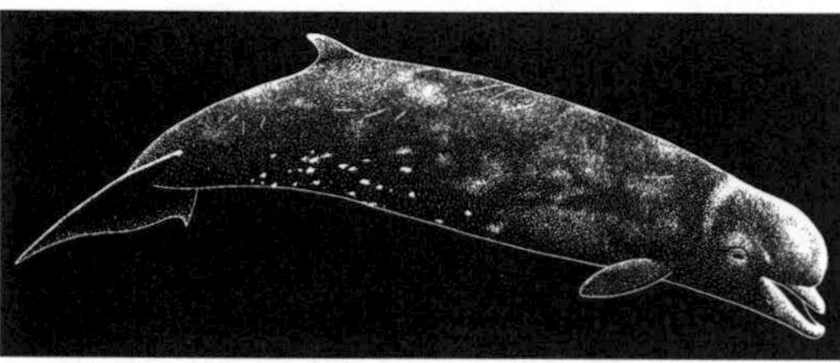
bottlenose whale

bottlenose whale (*Hyperoodon* spp.) Not to be confused with the bottlenose dolphin, which is much smaller and different in many particulars. There are two types of bottlenose whales, the Northern Hemisphere version (*H. ampullatus*) and the Southern (*H. planifrons*). Both have bulging foreheads, and although they are born dark gray or brown, they become lighter with age. Males differ so markedly from females (they are considerably larger and have a much more pronounced forehead, known as the melon) that early cetologists classified them as two different species. Maximum length for a male is 30 feet, for a female 23 feet. The northern bottlenose whale was hunted intensely by Norwegian and Scots whalers in the late nineteenth century, and although its current population is not known, it is considered endangered.

See also beaked whales, bottlenose dolphin

Bougainville, Louis-Antoine de (1729–1811) First recognized as a French army officer, Bougainville served as aide-de-camp to General Montcalm in Canada during the Seven Years' War (1856–1863). In 1766, accompanied by astronomers and naturalists, he organized a colonizing settlement to the Falkland Islands, but the Spanish, having their own claims to these islands, drove him off. Passing through the Strait of Magellan, Bougainville then sailed westward into the South Pacific and annexed Tahiti, even though it had been discovered some eight months earlier by the Englishman Samuel Wallis. From there he proceeded to Espiritu Santo, and found himself stopped from further exploration by the Great Barrier Reef. He then passed through the Solomon Islands, one of which now bears his name, and after stops at New Britain, the Moluccas, and Batavia (now Djakarta), he returned to France, arriving in March 1769. For his services during the French Revolution, Napoleon made him a count of the empire, a senator, and a member of the Legion of Honor. The colorful climbing plant bougainvillea is named for him.

Bougainville Island Although it is the largest of the Solomon Islands, Bougainville, with neighboring Buka, is part of Papua New Guinea. It was discovered (along with Choiseul, which he named for the French foreign minister who had supported his expedition), by Louis-Antoine de Bougainville during his 1766–1769 'round-the-world voyage. The island became a part of German New Guinea in 1884 and was mandated to Australia by the League of Nations in 1920. This rugged, densely forested island was the scene of some of the bitterest jungle fighting of World War II. After they lost Guadalcanal, Japanese troops dug in at Bougainville, their last stronghold in the Solomons, and a massive effort by air, sea, and land forces was required to rout them. The Solomon Islands campaign began with the taking of Guadalcanal in December 1942, was successful in cutting off Japan's forward air and naval base at Rabaul on the island of New Britain, and ended in the resounding American victory at Bougainville in late 1943. In 1989, the Bougainville Revolutionary Army was formed to protest the injustices of the Panguna copper mine, which was draining the resources of the island without returning anything to the islanders. After much bloodshed, which included the taking of the island by South African mercenaries and the forced resignation of the prime minister of Papua New Guinea, a stalemate was reached. In 1998, a cease-fire was signed, ending almost ten years of bloodshed, during which more than twenty thousand islanders died. **See also Guadalcanal, New Guinea, Rabaul, Solomon Islands**

Bounty British armed transport ship, famous for the mutiny of her crew on April 28, 1789, off the Friendly Islands (Tonga). Under the command of William Bligh, the 94-foot-long ship set out from Spithead to pick up a cargo of breadfruit trees in Tahiti and carry them to the West Indies, where they would be used as food for slaves. The mutiny, led by master's mate Fletcher Christian, took place near Tonga on the return voyage, and is believed to have been inspired by Bligh's harsh treatment of the crew and his unwillingness to allow his subordinates to fraternize with Tahitian women. After putting Bligh and eighteen loyalists over in a longboat, Christian and the mutineers returned to Tahiti to pick up the women and several men, and then sailed to uninhabited Pitcairn Island, where they removed everything that might be useful and burned the *Bounty*. The colony they established was not discovered until 1808, when a passing sealer stopped for water. The story of the mutiny on the *Bounty* was told in a famous novel by Charles Nordhoff and James Norman Hall, and has been made into several motion pictures. The roles of Christian and Bligh have been played by Mayne Lynton and Errol Flynn (1933); Clark Gable and Charles Laughton (1935); Marlon Brando and Trevor Howard (1962); and Mel Gibson and Anthony Hopkins (1984).

See also Bligh, William; breadfruit; Christian, Fletcher; mutiny; Pitcairn Island

Bouvet Island Uninhabited island in the South Atlantic, discovered and named in 1739 by the French navigator Jean-Baptiste-Charles Bouvet de Lozier while in the service of the Compagnie de Indes. Bouvet miscalculated the longitude of the island, and although Cook searched for it on his second voyage (1772–1775), he could not find it and decided that Bouvet had seen only an iceberg. The island was not found again until 1898, when a German expedition chanced upon it and determined its correct geographic coordinates. In 1927, a Norwegian team landed on and surveyed the island, and claimed it for Norway. Bouvetøya, as the Norwegians call it, is a single volcanic cone rising 3,000 feet out of the sea; it is Norway's only possession in the Southern Hemisphere. The southernmost manifestation of the Mid-Atlantic Ridge, it has the additional distinction of being the most isolated island on earth; there is no other land within 1,000 miles. **See also Mid-Atlantic Ridge**

Bowditch, Nathaniel (1773–1838) American navigator and mathematician, born in Salem, Massachusetts. Although he had no formal schooling, he taught himself Latin, French, and Spanish, as well as mathematics and astronomy. He first went to sea at the age of ten; on long voyages he studied navigation and corrected some eight thousand errors in Moore's *Practical Navigator*. His revisions were included in the edition of 1799. Bowditch's version was reissued as *The New American Practical Navigator* (later *The American Practical Navigator*) in 1802. It has been published by the U.S. Hydrographic Office ever since, and is still recognized as the basic text on celestial navigation.

See also Global Positioning System, sextant

bowhead whale (*Balaena mysticetus*) A heavy-bodied whale with a head that can account for a third of its 60-foot length, this species has the longest baleen plates of any whale, sometimes measured at 15 feet or more. Also known as whalebone, this baleen was used during the

bowhead whale

seventeenth, eighteenth, and nineteenth centuries in the manufacture of corset stays, skirt hoops, and buggy whips, and its oil fueled the lamps and tanneries of Europe. Dutch and British whalers hunted the bowhead (which they referred to as "the whale" or "the Mysticetus") to extinction in the eastern Arctic around Greenland and Spitsbergen. About seven thousand bowheads live in the Bering Sea and off the North Slope of Alaska, where they are hunted by the Inuit under a strict quota system established by the U.S. government.

See also baleen, Spitsbergen

bow riding Many species of dolphins are adept at riding the bow wave of fast-moving ships, often leading to greatly exaggerated estimates of their swimming speed. In fact, the dolphins take advantage of a pressure wave that the vessel pushes before it as it moves through the water, and are therefore able to match or even exceed the speed of the vessel. Those most frequently seen as bow-riders are the bottlenose dolphin (*Tursiops truncatus*) and the common dolphin (*Delphinus delphis*). There are records of common dolphins riding the bow wave of ships for 60 or 70 miles.

See also bottlenose dolphin, common dolphin

box jelly (*Chironex fleckeri*) Also known as the sea wasp, the box jelly is considered one of the deadliest animals in the sea. It is a cubomedusa (box-shaped jellyfish) with a fleshy arm at each "corner," and complex eyes. The stringlike tentacles are loaded with stinging cells (nematocysts) that contain venom so potent that it can kill an adult human in three minutes. The pain from the sting is said to be excruciating; one who experienced it likened it to having a red-hot poker applied to the skin. The "box" is about 8 inches across (about the size of a volleyball), and the tentacles can trail 20 feet below. Since 1880, some sixty-five people have been killed by jellyfish stings in Australian waters, but it was not until 1955 that *C. fleckeri* was identified as the culprit. Green turtles can eat the jellies with no ill effects, a process that is not understood. Adult box jellies aggregate in northern Australian river mouths, where they spawn and die. The resulting offspring—known as planulae—sink to the bottom and colonize the underside of stones before metamorphosing into a polyp and eventually into the "medusa" form of the deadly jellyfish. There are other jellies that are known as sea wasps (such as the bioluminescent and poisonous *Carybdea alata*, and *Chiropsalmus quadrigatus* from the Philippines and Malaysia), but the box jelly of Australia is the real killer.

See also jellyfish, nematocyst

brachiopoda A phylum of bivalve invertebrates that is divided into two classes, the Arcticulata and the Inarticulata. Both classes have asymmetrical shells (the ventral shell is larger than the dorsal), which are hinged with interlocking "teeth" in the former and with muscles in the latter. Brachiopods differ from clams and oysters (pelecypods) in that the two shells cover the top and bottom of the animal, while those of a clam cover the right and left sides. They are sometimes known as lampshells because of the resemblance to Roman oil lamps. Brachiopods are found mainly on the continental shelves, where they burrow into the mud or attach themselves to rocks or other shells. They feed by means of paired lophophores, spirally coiled crowns of hollow tentacles that extract minute food particles from seawater. They range in size from less than 1 inch to 6 inches, including a cordlike stalk in some species. The tongue-shaped shells may be white, pink, cream, brown, or gray. Of the three hundred–odd surviving species, the best-known is *Lingula*, an inarticulate brachiopod found in Hawaiian waters 40 to 130 feet deep. The fossil record is very extensive, and some thirty thousand species have been identified from the Cambrian Period onward. **See also clam, gastropoda, oyster**

Braer When her engines flooded and would not restart, the U.S.-owned, Liberian-registered supertanker *Braer* ran aground in Lerwick, Shetland Islands, on January 5, 1993, dumping 25 million gallons of Norwegian light crude onto the ocean's surface. The supertanker broke up on January 11 and 12, discharging the remainder of her 85,000-ton cargo—twice as much as the oil spilled from the 1989 wreck of the *Exxon Valdez*. Heavy storms and wave action dispersed much of the oil before it could wash ashore, and a major disaster was averted. Because a large quantity of the oil floated out to sea, it did not have the catastrophic effect on sea life that other spills have had, but its effect on fishes and invertebrates is unknown.

See also *Exxon Valdez*, oil spill, supertanker, *Torrey Canyon*

bramble shark (*Echinorhinus brucus*) Deepwater, bottom-dwelling shark that is unmistakable because of the large, thornlike dermal denticles scattered all over the body. Some of the denticles are joined together in clumps, their bases having a scalloped edge. This species, along with the closely related prickly dogfish or prickly shark, has two similarly sized dorsal fins, placed so far back that they are over the pelvic fins. They are usually dark brown or blackish, with lighter undersides, and can reach a length of 10 feet. Little is known about the habits of bramble sharks, but they are believed to be distributed in the North and South Atlantic, along the continental shelves at depths up to 3,000 feet. The prickly shark (*E. cookei*) is the Pacific version, found in Californian and Peruvian waters, as well as New Zealand, Palau, and Hawaii. It grows somewhat larger than the bramble shark (up to 13 feet), but the denticles are smaller and not fused together.

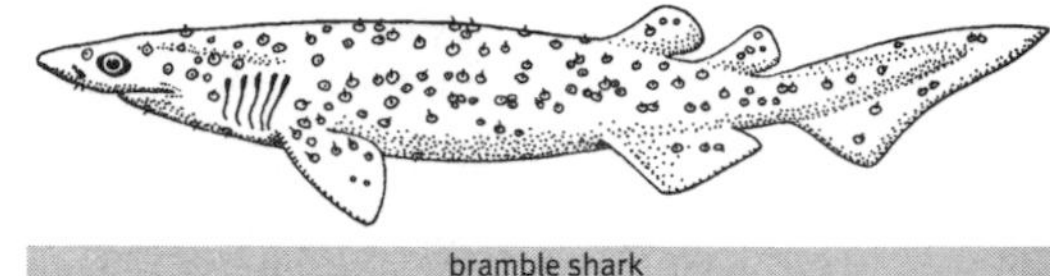

bramble shark

See also prickly dogfish

breadfruit (*Artocarpus communis*) A tree that grows to a height of 60 to 80 feet and has large, ovoid fruits that are 6 to 8 inches in diameter. The fruits are brownish green on the outside and have a white, fibrous pulp. Some types have seeds, while others do not. The seedless form is believed to have originated in the Malay Archipelago, but it has been cultivated throughout the tropical South Pacific. It has also been introduced into the West Indies, Mexico, and Central America. It is not eaten raw, but is roasted, baked, fried, boiled, or dried into flour. Cloth can be made from the fibrous inner bark, and the wood is used for canoes and furniture.

See also *Bounty*

Brendan (484–578) Irish monk who made a (possibly legendary) voyage in the Atlantic, held by some to be the first man to discover America. There are several manuscript copies in existence of the *Navigatio Sancti Brendani Abbatis* (*The Navigations of Saint Brendan the Abbot*) dating from the eleventh century. During the years 565–573, he is supposed to have made a voyage in a leather-covered rowing boat called a curragh, accompanied by seventeen monks. There are many adventures recounted in the *Navigatio* that place it squarely within the realm of fantasy, such as the one where he lands on what he believes is an island, builds a fire, and discovers that he and his crew are actually on the back of Jasconius the whale, or the one where the birds sang in Latin. On the other hand, the manuscript also contains a description of an iceberg, and islands that have been identified as the Faeroes and Iceland, so there might be some truth in it—but probably not the part where he is said to have crossed the Atlantic. (Irish monks colonized the Faeroes around 700, and Iceland by 800, so the stories are not all that far-fetched.) For many years "St. Brendan's Islands" (*Insulae Sancti Brendani*) appeared on maps, and no less an authority than Columbus reported seeing them. In 1977, Tim Severin built a 36-foot-long curragh, named it *Brendan*, and, with a crew of four, sailed it from Brandon Bay in Ireland to Newfoundland. **See also curragh, Hy-Brasil**

brigantine A two-masted vessel, square-rigged on the foremast and fore-and-aft rigged on the mainmast. The name comes from the fact that these ships were the favorite vessels of brigands, particularly in the Mediterranean. A "brig" was originally an abbreviation of brigantine, but it later came to denote a two-masted vessel that was square-rigged on both the foremast and the mainmast. **See also bark, ship**

bristlemouth

bristlemouth (*Cyclothone* spp.) The most numerous fishes in the world, bristlemouths get their common name from the tiny, brushlike teeth on the outer margins of their rounded jaws that trap minute planktonic organisms. There are perhaps ten species (the taxonomy is unresolved) of these 3-inch fishes, found in the deep waters of all the oceans of the world. Some are nearly transparent, others are dark-colored; the deeper the water in which they live, the darker the coloration. Schools of millions of bristlemouths migrate vertically through the water column to feed; these daily migrations are believed to account for the phenomenon known as the deep-scattering layer, where echoes have been picked up by fathometers, moving toward the surface in the evening and descending at dawn.
See also deep-scattering layer, lantern fishes

Britannic When the *Titanic* sank in 1912, work on the *Britannic* was suspended so that a double hull and more watertight bulkheads could be added, as well as oversized lifeboat davits and more lifeboats. These modifications, completed in 1914, made *Britannic*—whose name had hastily been changed from *Gigantic*—the largest of all the White Star liners. Instead of the Atlantic crossings that she was built for, however, *Britannic*'s first job was that of hospital ship, and with *Mauretania, Aquitania,* and her surviving sister *Olympic,* she was dispatched to the Mediterranean. She made five round-trip voyages from the Aegean to Southampton carrying the wounded and sick. On November 21, 1916, as she was steaming through the Kéa Channel in the Aegean, *Britannic* was hit by a tremendous explosion and sank in less than an hour. The cause of the explosion has never been determined, but it might have been a torpedo or, more likely, a mine. In 1976 the sunken hulk of *Britannic* was discovered by divers in 350 feet of water, intact except for the massive hole in her bow—the same injury that caused her sister ship *Titanic* to sink. **See also *Titanic***

British East India Company Chartered by Queen Elizabeth in 1600, the British East India Company was originally formed to sponsor shareholder voyages to the Indies and Japan, but by 1612 all voyages were made by the company itself. The original mission of the company was to break the Dutch monopoly on the spice trade with the East Indies, but it was also heavily involved in competition with the Dutch on the whaling grounds of Spitsbergen. The company's ships, called East Indiamen, were heavily armed to defend themselves from (and attack) ships of the rival Dutch East India Company, and to protect themselves from the pirates of the Malay states. After the Amboina Massacre in 1623, however, when the British forswore all intentions in the Spice Islands, the company concentrated on its activities in India, where they were enormously successful and established a presence in Bombay, Madras, and Bengal. By the time Robert Clive won the Battle of Plassey in 1757, the company had a monopoly on textile trading with the Mogul emperors, and was completely dominant in India. By 1770, they were essentially running the country, but the British government, recognizing that it ought to have more control, initiated the Regulating Act of 1773 and the India Act of 1784, which set up a governing board and gave the government more power. The company also maintained strict control of the shipping lanes in the Pacific, and even British sperm whalers had to petition the company for the right to take whales in the "South Seas." In 1813, trade was thrown open to the public, although the company retained its monopoly of trade with China until 1833, when this trade too was opened to all. The government of colonial India was left in the hands of the British East India Company until 1857, when the Indian Mutiny caused the British government to assume control. The company ceased to exist as a legal entity in 1873. **See also Dutch East India Company**

British storm petrel (*Hydrobates pelagicus*) A tiny petrel, dark all over except for a white area in the middle of the underwing and a white rump, this species is a habitual ship follower and can be recognized by its erratic, batlike flight as it flutters back and forth from one side of the ship to the other. It is smaller than Wilson's storm petrel, and its feet do not project beyond the tail when the bird is in flight. It breeds on islands in the eastern Atlantic and the Mediterranean, from Iceland and the Faeroes to the Canaries and Malta.
See also petrel, Wilson's storm petrel

British Virgin Islands Part of the island chain known collectively as the Virgin Islands in the Caribbean just east of Puerto Rico. The British colony consists of four large islands—Tortola, Anegada, Virgin Gorda, and Jost Van Dyke—and thirty-two smaller islets. The entire group was named by Columbus in 1493 for Saint Ursula and the eleven thousand martyred virgins, but after his discovery, he left the islands to the fierce Carib Indians that inhabited them. Dutch buccaneers held Tortola until English settlers took it over in 1666. This island—much the largest at 21 square miles—was annexed by the British in 1672 and used for the growing of sugar until the British abolition of slavery in 1833. In 1872, the colony became part of the British Leeward Islands; after 1956, it became the British Virgin Islands. The islands are devoted to the tourist industry, with many hotels, guest houses, and facilities for sailboats.

See also American Virgin Islands, Tortola, Virgin Gorda

brittle star (class Ophiuroidea) Among the most abundant inhabitants of the seafloor, brittle stars occur as far down as 20,000 feet in all the deep oceans of the world. They are very active, moving across the seafloor and occasionally swimming. Sometimes known as serpent stars because of their snakelike arms, they are very fragile and can break when handled. There may be as many as sixteen hundred species, ranging in size from the dwarf brittle star (*Axiognathus squamatus*), with an armspread of 2 inches, to species that can be 30 inches across. Some species are bioluminescent, but it is not clear what advantage this might confer.

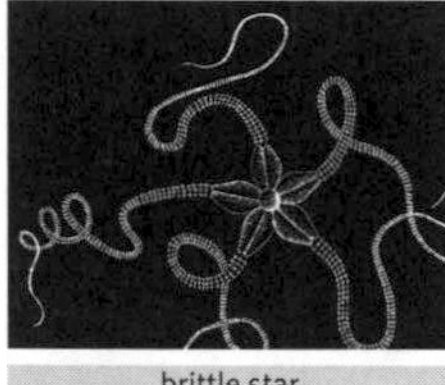
brittle star

See also basket star

broadbill swordfish

broadbill swordfish (*Xiphias gladius*) Found worldwide in temperate and tropical waters, the broadbill swordfish gets its common name from its smooth, flattened sword that is much longer and wider than the bill of any other billfish. The bill is used for defense and also to slash and debilitate its prey, which consists of squid, mackerel, bluefish, and many other mid- and deep-water species. There are many recorded instances of swordfishes attacking boats, and in 1967 the submersible *Alvin* was attacked by a swordfish that impaled itself and was subsequently brought to the surface and eaten by the crew. Swordfish are a highly sought food fish, and they are extensively captured by longline fishermen. They are also premier game fishes and are prized as trophies. The world-record swordfish weighed 1,182 pounds and was caught in 1953 off the coast of Chile. The single species is worldwide in distribution, and there are records from every ocean except the Arctic and Antarctic. A drastic reduction in swordfish populations as a result of longlining has engendered an American restaurant boycott of swordfish.

See also International Game Fish Association, longline fishing

bronze whaler shark (*Carcharhinus brachyurus*) Found in a patchy distribution in inshore temperate waters from the Mediterranean to Southern California to China and Japan, Australia and New Zealand, this species is characterized by its bronze or grayish-brown back and white undersides. It has narrow, finely serrated teeth, accounting for its common name in American waters, "narrowtooth shark." Bronze whalers reach a length of about 10 feet, and although they have a reputation in Australia for aggressiveness, it is mostly unfounded. They feed mostly on bottom fishes.

Brooke, James (1803–1868) The "white rajah" of Sarawak, Brooke was born in India, served in the British East India Company, and was wounded in the Anglo-Burmese War of 1825. After his father's death in 1835, he inherited a considerable fortune and planned to settle in Borneo. He arrived in the Sarawak port of Kuching in 1839 aboard his armed yacht *Royalist* and helped to put down a Dyak revolt against the Bruneian governor. His reward was the principality of Sarawak, and he was named rajah in 1841. The British named him consul general for Borneo, and for his services to the Crown, he was knighted in 1848. Brooke returned to England in 1857, leaving the government in the hands of his nephew Charles, who was succeeded by *his* son, Charles Vyner Brooke, who was able to make Sarawak a British colony. **See also Borneo, Brunei, Sarawak**

brotulid (family Brotulidae) Big-headed fishes with long, tapering tails, some brotulids have prominent eyes, and others have no eyes at all. While the brotulids superficially resemble the grenadiers, they are closely related to the shallow-water blennies; the grenadiers are related to the cods. Grenadiers frequently sport a chin barbel; brotulids do not. Brotulids are bottom dwellers, found from shallow waters to the greatest depths. The record holder for the world's deepest-living fish is a brotulid, *Abyssobrotula galatheae;* a 6-inch specimen was dredged up from 27,453 feet in the Puerto Rico Trench in 1970. **See also eelpout, grenadier, pearlfishes**

brown algae (Phaeophyta) The largest and structurally the most complex of all the seaweeds. The group includes kelp, gulfweed, and rockweed, which are highly differentiated by well-defined stipes, holdfasts, and blades. Even the color varies, ranging from yellow and olive green to brown and black, depending upon the relative amounts of xanthophyll (yellow) and carotene (red) pigments that mask the green chlorophyll. Floating in the eponymous Sargasso Sea (but also in the Sea of Japan and the Gulf of Thailand), the brown algae known as Sargassum weed provide living quarters and protection for many marine organisms. The bull or giant kelp (*Macrocystis pyrifera*) is the fastest-growing of all plants, increasing as much as 1 foot a day to lengths of 200 feet. Kelp "forests" are home to many species of invertebrates and fishes; California sea lions and sea otters often entwine themselves in the fronds. Brown algae are also a rich source of iodine, bromine, potassium, and potash, and kelps are harvested commercially off the Pacific coast of North America.

See also kelp, red algae

brown booby (*Sula leucogaster*) The commonest and most widespread of all the boobies, this species is a dark chocolate-brown color above with white underparts. (The specific name *leucogaster* means "white belly.") The maximum wingspan is about 59 inches. Brown boobies are found throughout the Pacific, Caribbean, tropical Atlantic, Indian Ocean, northern Australia, and the Great Barrier Reef. They fly in loose flocks and feed by making low-level plunge dives to capture fishes underwater.

See also blue-footed booby, masked booby, red-footed booby

Brunei An Islamic sultanate on Borneo, occupying all of the island in the sixteenth century but now reduced to 2,226 square miles in the north. In the early nineteenth century, the sultanate included Sabah, Sarawak, and Brunei, but in 1841, James Brooke, a British soldier, helped put down an insurrection and was proclaimed the rajah of Sarawak. In 1847, the sultan of Brunei signed a treaty with Britain for commercial relations and suppression of piracy. In 1877, Brunei was placed under British protection, and British residents (counselors) were introduced in 1906. In 1968, Sultan Hassanal Bolkiah succeeded his father, who had abdicated voluntarily. The country, now officially known as Negara Brunei Darussalam, became independent in 1984, and because of the enormous increase in revenues from oil and gas from 1973 onward, the country's 284,000 people have one of the highest per capita incomes of any country. The sultan, who is also the prime minister, is considered one of the richest men in the world. He lives in a 1,788-room palace and owns the Palace Hotel in New York, the Plaza Athénée in Paris, and the Dorchester in London. In August 1998, the sultan named his twenty-four-year-old son, Prince Al-Muhtadee Billah, as his successor and heir to his throne.

See also Borneo, Sarawak

Brunel, Isambard Kingdom (1806–1859) British engineer and ship designer who was also largely responsible for Britain's Great Western Railway. His first major ship was the *Great Western* (which was an adjunct to the railway), a wooden paddle steamer that was the first steamship to make regular crossings of the Atlantic. Then came the iron-hulled *Great Britain,* the largest ship of her time, which made her first Atlantic crossing in 1845. In 1858 he designed the *Great Eastern,* which was then, at a length of 680 feet, the largest ship ever built. Brunel died before *Great Eastern* was launched, but the vessel proved to be too large and too expensive to function as a passenger liner, so she was converted to a cable ship; she laid the first successful transatlantic cable in 1866. The *Great Britain* is now on display in Bristol, where she was built, after being towed back from the Falklands, where she was abandoned in 1886.

Bruny Island Located in the D'Entrecasteaux Channel of southeastern Tasmania, Bruny Island was first explored by Abel Tasman in 1642, and then by James Cook in 1773. It is named for Admiral Bruni Raymond D'Entrecasteaux, a French explorer who navigated its coastal waters in 1792. By 1804, the island was a base for whaling operations for the right whales of the Derwent River estuary. It is now a resort frequently visited by visitors from Hobart, 23 miles to the north.

See also Cook (James), D'Entrecasteaux, Tasman, Tasmania

Bruun, Anton (1901–1961) Danish zoologist and oceanographer, lamed by childhood polio, who yearned to be a sailor but instead became a scientist who went to sea. He sailed as assistant to biologist Johannes Schmidt aboard the research vessel *Dana* in 1928–1930. (Schmidt located the breeding grounds of eels in the Sargasso Sea, and Bruun never gave up his quest to find a giant eel, suggested by the 6-foot-long leptocephalus that had been collected by the *Dana* off the coast of South America.) Named curator of the Zoological Museum of the University of Copenhagen, Bruun published many papers on deep-sea fishes. His 1945 *Atlantide* expedition was the precursor of the 1950 *Galathea* expedition, where he led a group of zoologists on a 'round-the-world cruise in search of information on the creatures that inhabited the deepest parts of the oceans. He is responsible for the introduction of the word "hadal"—meaning "from Hades"—to refer to the fauna of the deep-sea trenches. The *Galathea* expedition trawled in the depths of the Pacific, the Atlantic, and the Caribbean, discovering many creatures never before seen by humans. In May 1952, in a dredge haul raised from 11,700 feet off Costa Rica, several single-shelled gastropods were found. These creatures were named *Neopilina*, and their only relatives are known from the fossil record of 400 million years ago. The *Galathea*'s deepest trawl—a collection of worms, echinoderms, mollusks, and crustaceans—came from 33,678 feet in the Mindanao Trench off the Philippines. During the 1957–1958 International Geophysical Year (IGY), the former presidential yacht *Williamsburg* was converted into a U.S. research vessel and named *Anton Bruun*.

See also *Galathea* expedition, leptocephalus, Schmidt

Bryde's whale

Bryde's whale (*Balaenoptera edeni*) One of the smallest of the rorquals (groove-throated whales), Bryde's (pronounced "Brewdah's") whale was named for Johann Bryde, the Norwegian consul to South Africa in 1913, when the species was first described. It reaches a maximum length of 50 feet and a weight of 30 tons. It closely resembles the sei whale, but where the sei has a single ridge on the top of the upper jaw (the rostrum), Bryde's whale has three ridges. Bryde's whale is the only rorqual that does not migrate to high polar latitudes, and is sometimes referred to as the tropical whale.

See also rorqual, sei whale

bryozoan (phylum Bryozoa) Commonly known as moss animals, bryozoans were thought to be plants because they formed lichenlike mats on underwater rocks. The colonies are composed of tiny animals, rarely 1/32 inch in length, with a delicate ring of hollow tentacles surrounding the mouth. The colony may be branching, creeping, bushy, leafy, tubular, or encrusting. Bryozoans are suspension feeders, trapping microscopic organisms in the tentacles and bringing them into the mouth. Each individual is enclosed in a chitinous or rubbery case, into which the tentacles can be withdrawn. There are about four thousand species of marine bryozoans (class Gymnolaemata) that can be found on seaweeds, driftwood, pilings, rocks, and the bottoms of ships. Some species dissolve their way into the shells of conchs or other heavy marine mollusks, creating a characteristic pitted appearance.

buccaneer Seamen of all nations that cruised the Spanish Main and the Pacific in the seventeenth century, raiding and plundering. They called themselves privateers, but they carried no authorizations from any government, and they were often nothing more than pirates. Inspired by the Elizabethan privateers, they were best known for their activities in the Caribbean after the English capture of Jamaica in 1655. Among the best known of the buccaneers were Sir Henry Morgan, who took Panama in 1671; Bartholomew Roberts; William Dampier; and Alexander Selkirk, who became the model for Defoe's Robinson Crusoe. The name "buccaneer" is derived from the Arawak Indian word *"buccan,"* a grill for cooking dried meat.

See also Dampier, L'Ollonois, Morgan, piracy, Port Royal, Roberts, Selkirk, Spanish Main

Bullard, Edward C. (1907–1980) British geophysicist, best known for his 1930s measurements of heat flow on the ocean floor. Long before the existence of hydrothermal vents was known, Bullard recognized the existence of a geothermal dynamo that was responsible for convection currents that rose beneath the oceans and sank beneath the continents. In 1964, using an early computer system, he confirmed Wegener's idea that the opposing Atlantic shores of South America and Africa match well, especially at the 1,000-meter (3,300-foot) depth contour, which is a better approximation of the edge of the continental block than the present shoreline. During the 1960s, Bullard was director of Cambridge University's department of Geodesy and Geophysics, where J. Tuzo Wilson, Frederick Vine, Drummond Matthews, and Harry Hess developed the theory of sea-floor spreading. He was knighted in 1953.

See also sea-floor spreading, Wegener

Bullen, Frank (1857–1915) British author who went to sea as a cabin boy aboard his uncle's ship *Arabella* and later signed aboard a whaler. Although the ship was actually the *Splendid* out of New Bedford, in his 1898 book, *The Cruise of the "Cachalot,"* he rechristened her *Cachalot* and disguised the names of all the crew members. Often read as a factual chronicle of a whaling voyage, *The Cruise of the "Cachalot"* is a mixture of fact and fiction, usually making a hero of its protagonist, a certain Frank Bullen. While employed as a clerk in the Meteorological Office in London, he wrote thirty-six books, among them *Idylls of the Sea, Our Heritage of the Sea, Deep-Sea Plunderings, The Log of a Sea Waif,* and *Denizens of the Deep.*

bull shark

bull shark (*Carcharhinus leucas*) A stocky, heavy-bodied carcharhinid shark that can reach a length of 10 feet and a weight of 500 pounds. It is usually a dirty gray color, with a short snout and small eyes. The upper teeth are broadly triangular and serrated, while the lowers have narrower cusps on a wide base. It penetrates easily into freshwater, and has been found far upstream in the Mississippi and the Amazon. The Lake Nicaragua shark, Zambezi River shark, Ganges River shark, and Java shark have all been shown to be bull sharks. The bull shark may be the most dangerous of all shark species, having been responsible for attacks on people in South Africa, Australia, India, North America, and elsewhere. It is now believed that bull sharks were responsible for the four fatal attacks along the New Jersey shore during a ten-day period in July 1916. **See also carcharhinid sharks, dusky shark, shark attack**

buoyancy The ability to float or rise in or on a liquid medium, as expressed by Archimedes's Principle: "A body immersed in a fluid is buoyed up by a force equal to the weight of the displaced fluid." The principle explains not only the ability of ships to float, but also the apparent loss of weight of objects underwater. Organisms living in water are subject to hydrostatic pressure, which increases by one atmosphere for every 10 meters (33 feet) of water depth. Since most animal tissues are equal to or denser than water, they have had to develop special adaptations to allow for equilibrium between body weight and buoyancy plus hydrodynamic lift. Many fishes have developed gas bladders that effectively reduce their overall density, allowing them to remain neutrally buoyant. In many cases, lipids or fats are less dense than water, and some fishes, such as coelacanths and orange roughys, have fat-filled swim bladders; the livers of sharks, which are rich in oil, also serve as buoyancy devices. Many fishes and sharks maintain buoyancy by generating lift with their fins, especially the pectorals, using a principle not unlike that which enables airplanes to fly. The chambered nautilus can float because of the gas-filled chambers in its shell, and many species of squids are ammoniacal, meaning that their tissues contain ammonium chloride, which is lighter than water and enables them to maintain neutral buoyancy. The nonammoniacal squids and most octopuses swim by means of fins or water jets that give them propulsion and lift. **See also gas bladder**

buoys Floating markers used to identify navigable channels or to indicate dangers to navigation, such as banks, shoals, rocks, mines, or cables. They came into use in European waters around the fifteenth or sixteenth century; Henry VIII introduced them to Britain in 1514. The first buoys consisted of floating timbers marked with a pole or flag (known as spar buoys); these were later replaced by metal ones. Some buoys are lighted, and others are equipped with bells or whistles, activated by the movement of the water. Radio buoys were introduced in 1939. A unified system of buoyage was established by the League of Nations in 1936, but World War II began before the agreement could be ratified, and it was not until 1980 that a system of shape, size, color, and markings was adopted by fifty maritime nations. It is not completely unified, however, since in Region A (the Americas, Japan, Korea, and the Philippines), a vessel is supposed to keep to the port side of a buoy ("red right returning"), while in Region B (Europe, Australia, New Zealand, the Persian Gulf, and most Asian countries), the system is reversed: it is the green buoy that is to be kept to the port side of a vessel.

Burano Island in the lagoon of Venice, founded in the fifth century by refugees fleeing Attila the Hun. For centuries it was subordinate to neighboring Torcello, but it gained importance when it became a center for lace making. The lace of Burano, known as *punto de Burano* or *punto in aria,* was considered the finest in Europe, but when the Republic of Venice began to decline, Burano lace making declined with it. In 1872, the Contessa Marcello took it upon herself to save the traditional industry and started a school to revive the almost lost art. By the twentieth century, five thousand lace makers were employed in Burano. With the introduction of machines, however, the meticulous art of making lace by hand became obsolete, and it will undoubtedly die out. Burano was integrated into the city of Venice in 1923, and it is once again a quiet suburb. It has no imposing churches, but it does have a museum of lace. Six miles from Venice, Burano can be easily reached by boat, from which visitors can view the surprising reds, blues, pinks, yellows, and greens of its two-story houses. **See also Murano, Torcello, Venice**

Burmeister's porpoise (*Phocoena spinipinnis*) Although it has been regularly described as black, this animal is, in fact, light brown. *Spinipinnis* means "spiny fin," and the name is derived from the denticles or spines on the leading edge of the unusually shaped dorsal fin, which is convex on the leading edge and concave on the trailing edge. Like all other phocoenids (true porpoises), Burmeister's porpoise has small, spade-shaped teeth. It is hard to spot in the water, and reaches a length of 6 feet. It is found in the coastal waters of eastern and western South America, from Uruguay around Cape Horn to Peru, where it is often trapped in fishermen's nets. **See also Dall porpoise, harbor porpoise, porpoise, spectacled porpoise**

Burmeister's porpoise

Bushnell, David (1742–1824) American inventor, best known for his submarine, the *Turtle.* During the War of Independence, when British warships had blockaded the Atlantic coast, George Washington was seeking a way to break the blockade, and also, if possible, to disable the British warships. He commissioned Bushnell to solve the problem, and the 1775 Yale College graduate developed an ovoid vessel that floated upright on its long axis; it was made of barrel staves and iron in the form of two tortoise shells, waterproofed with a thick coating of tar, and strengthened amidships by a baulk that also served as a seat for the pilot. (It actually looked more like an egg than a turtle.) An internal water tank that could be flooded and emptied by a foot pedal provided the means to raise and lower the little vessel in the water. Its propeller was powered by hand-turned screws, and the tight little boat could hold enough air to enable the pilot to remain below for about half an hour. Piloted by Ezra Lee, the *Turtle* attempted to attach a bomb to HMS *Eagle*, the flagship of the British fleet blockading New York harbor, but, because the submarine's screw device could not pierce the copper-sheathed hull of the British warship, the attempt was a failure. **See also submarine**

butterfish (*Peprilus triacanthus*) A small, bony food fish, with a thin, oval body and delicious oily flesh. Bluish or greenish with silvery highlights, they grow to be 1 foot long. Juveniles are often found swimming beneath large jellyfishes for protection from predators, but they are not immune to the stings from the tentacles, and they are sometimes killed. Butterfishes have small mouths and feed on minute planktonic organisms. Like other members of the family Stromateidae, they have an expanded, muscular esophagus that is equipped with ridges. They have been caught commercially since the early nineteenth century, and from 1920 to 1962 the annual domestic harvest averaged 3,500 metric tons. The arrival of the "distant water" foreign fleets increased the take substantially, but the species is no longer popular, and most butterfish caught nowadays are discarded. They are related to the longer-finned harvest fish (*P. alepidotus*), also from the North Atlantic, and the 10-inch California pompano (*Palometa simillima*), which is not a pompano at all.

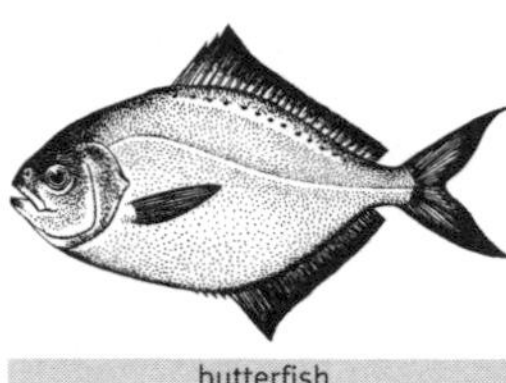
butterfish

butterfly fish (genus *Chaetodon*) Small, brightly colored inhabitants of coral reefs and adjacent waters around the world, butterfly fishes are somewhat smaller than their angelfish relatives and do not have the prominent spine on the gill covers. Angelfishes usually swim in a slower, more stately fashion. Butterfly fishes are named for their bright coloration and their active, flitting swimming movements. There may be more than 150 species, almost all vividly colored and many with a false eyespot near the base of the tail as well as a dark stripe that runs through the eye. These patterns may make it difficult for predators to differentiate between the head and the tail. Butterfly fishes feed by nibbling on coral polyps and other invertebrates. The species shown here is the raccoon butterfly fish, *C. lunula*, of the Indo-Pacific.

raccoon butterfly fish

See also angelfish

butterfly stingray (*Gymnura* spp.) Very broad, butterfly- or diamond-shaped rays, in which the body is half again as wide as it is long. Their tails are short, and they have no caudal fins. The 2-foot-wide smooth butterfly ray (*G. micrura*) occurs from Chesapeake Bay to the Gulf of Mexico, and south to Brazil. The spiny butterfly ray (*G. altavela*), named for the spines on its tail, grows larger, up to 6 feet across, and is found in the same area. Other species, found in Australia (*G. australis*) and South Africa (*G. natalensis*), can grow even larger. **See also stingray**

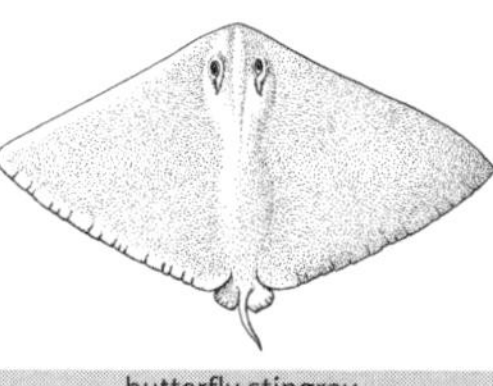
butterfly stingray

Bylot, Robert British navigator and Arctic explorer whose early life and death are unrecorded. Bylot was aboard the *Discovery* with Henry Hudson in 1610, and his demotion was one of the things that incited the mutiny that resulted in Hudson, his son John, and seven other sailors being put overboard in a dinghy and left behind in the bay that now bears Hudson's name. Almost single-handedly, Bylot piloted the *Discovery* home in 1611; he was absolved of all charges. Soon afterward, he joined Thomas Button's 1612–1613 search for Hudson (again in the hardworking *Discovery*), but they found no trace of him. Bylot led his own *Discovery* expedition in 1616 with William Baffin as pilot. They circumnavigated the icy vastness of Baffin Bay (a feat that would not be repeated until the nineteenth century), but they became convinced that there was no passage through Davis Strait to the west. On this voyage, Bylot and Baffin discovered the entrances to Jones and Lancaster Sounds and made accurate surveys of Hudson Strait and Southampton Island. Bylot gave his name to Bylot Island, just north of Pond Inlet on northern Baffin Island, as well as to Cape Bylot on Southampton Island. **See also Baffin, *Discovery,* Hudson**

Byrd, Richard E. (1888–1957) Born into an illustrious Virginia family, Richard Evelyn Byrd entered the U.S. Naval Academy in 1908 and received his commission four years later. During World War I he learned to fly, and in 1926 he and Floyd Bennett took off from Spitsbergen and claimed to be the first men to overfly the

North Pole. (Later investigations showed that they never got closer than 150 miles to the Pole.) In 1927, he tried to be the first to fly across the Atlantic, but he was beaten by Charles Lindbergh. (With three companions, Byrd made the third nonstop transatlantic flight.) In 1928–1930, he led a private expedition to the Antarctic, and from his base (which he called Little America) at the Bay of Whales, he and copilot Bernt Balchen made the first successful flight over the South Pole. During 1933–1935, he wintered alone in Antarctica at Bolling Advance Base, a weather station, and almost died from carbon monoxide poisoning from a faulty stove. His men, stationed 120 miles away at Little America, came to his rescue when they realized that his radio transmissions were becoming incoherent. He discovered and mapped the Rockefeller Mountains and Marie Byrd Land, which he named for his wife. His successes (and survival) inspired the U.S. interest in the Antarctic, and he led three official expeditions: in 1939–1941; again in 1946–1947, when Operation High Jump became the largest expedition in Antarctic history, involving 4,700 servicemen, fifty-one scientists, thirteen ships, and fifty helicopters; and Operation Deep Freeze in 1955–1956. Byrd was also a prolific author who celebrated his triumphs in books such as *Skyward* (1928), *Little America* (1930), *Discovery* (1935), *Alone* (1938), and numerous articles for the *National Geographic.*

See also Peary

Byron, John (1723–1786) British admiral known as Foul Weather Jack for his misfortunes with bad weather conditions. When he was a midshipman aboard the *Wager* in 1741, the ship was wrecked off the coast of Chile. In 1764, he made a fruitless search for Terra Australis Incognita in the *Dolphin,* commanded by Samuel Wallis (visiting the previously unknown Ellice Islands—now Tuvalu—en route), and in 1765, he discovered the Gilbert Islands. In 1779, he was sent to relieve the British naval forces in North America during the Revolutionary War. As he tried to leave England, a destructive gale slowed his fleet badly; he became involved in an indecisive action against the French off Grenada and therefore never fulfilled his mission. He was the grandfather of Lord Byron, the poet.

See also Gilbert Islands, Tuvalu, Wallis

C

cabezone (*Scorpaenichthys marmoratus*) Common name from the Spanish for "big-headed," the cabezone is the largest of the sculpins, reaching a length of 30 inches and a weight of 25 pounds. Found in the eastern Pacific from Alaska to Baja California, it is considered a desirable game fish, and the flesh is eaten, even though it is green. (A species of drum, *Larimus breviceps,* found in the Atlantic from the Caribbean to Brazil, is also commonly known as the cabezone.)

See also sculpin

cable Any large hemp or wire rope, but normally associated with the anchor of a ship. Originally, all cables were hemp, but by 1800 chain cables for anchors had been introduced. Also, a measure of distance at sea, 1000 fathoms or about 200 yards.

Cabot, John (1450–1498) An Italian navigator, born in Genoa, whose name in Italian was Giovanni Caboto (or perhaps Gaboto). In 1484 he went to England to seek support for his plan to search for a passage to China by sea, but before he sailed, he learned that Christopher Columbus had gone west and reached the Indies. In May 1497, he set sail from Bristol in the little ship *Matthew.* In June he arrived off the coast of Newfoundland, where he found the waters thick with codfish. He returned to England, where he was given a £10 reward by Henry VII and letters patent to assemble another voyage to the "New Land." With five ships and three hundred men he sailed from Bristol in 1498. He was never heard of again.

Cabot, Sebastian (1476–1557) Son of John Cabot, born in England, who may have accompanied his father to Newfoundland in 1497. He was named chief cartographer to Henry VIII in 1512. When troops were sent to support King Ferdinand II against the French, Cabot accompanied them. Due to his knowledge of North America he was commissioned a captain in the Spanish navy, and in 1518, he became chief cartographer and pilot major of Spain. In 1525, in the service of the Spanish, he led an expedition from Seville to find Cathay (China) and Cipangu (Japan), but on the way to the Straits of Magellan (which had been discovered in 1520) he put into La Plata in search of the gold and silver he had heard about. He did not find it and returned to Spain. For the failure of this expedition, he was banished to Africa, but he was pardoned two years later and went back to England, where he persuaded Edward VI to support his Company of Merchant Adventurers, which became the Muscovy Company a few years later, in 1555.

Cabral, Pedro Álvares de Gouveia (1467–1520) Portuguese navigator who left Lisbon for India with thirteen ships in 1500, following Bartolomeu Dias's 1488 successful rounding of the Cape of Good Hope and the return of Vasco da Gama from India in 1499. Cabral sailed southwest from the Cape Verde Islands and is credited with the first European landfall in Brazil, which he claimed for Portugal. He continued his voyage eastward, and rounded the Cape of Good Hope, landed at Mozambique and Malindi (as da Gama had done before him), and finally put into the port of Calicut. He thus became the first man to sail to India by way of South America and the Cape of Good Hope. After this epochal voyage, he retired to Lisbon and never sailed again.

Cadamosto, Alvise (1432–1477) Also Ca' da Mosto; a Venetian nobleman and trader, enlisted by Prince Henry the Navigator to explore the west coast of Africa. On his first voyage in 1455, in a ship provided by Henry, he visited Madeira and the Canary Islands and the coast in the vicinity of the Gambia River. The next year, in a caravel he fitted out himself, he repeated his voyage and reached Cabo Blanco off the Sahara. He was blown offshore in a storm and may have made the first sighting of the Cape Verde Islands, although this claim is disputed. He wrote detailed accounts of his voyages and the people he encountered, including a description of Cape Verde (the cape, not the islands) on the West African coast, in what is now Senegal, close to the city of Dakar. His journal, published in Venice in 1507, contains the first known description by a European of a country inhabited by black people, an account of the trade empire the Portuguese were building in Africa, and very early depictions of an elephant and a hippopotamus.

See also Canary Islands, Cape Verde Islands, Henry the Navigator, Madeira

cadborosaurus Named for Cadboro Bay in British Columbia, "Caddy" is a sea serpent that has been sighted by hundreds of people, but whose existence is verified by no hard evidence—other than an ambiguous photograph. Zoologists Paul LeBlond and Edward Bousfield have been campaigning long and hard for its acceptance as a real animal; they have written a book about it, in which they assigned it a scientific name (*Cadborosaurus willisi*), as if there was a specimen for other scientists to compare it to.

See also Loch Ness monster, sea serpent

cahow (*Pterodrama cahow*) Sometimes known as the Bermuda petrel, the cahow gets its unusual name from an interpretation of its hollow, howling call. It is very similar to the black-capped petrel (*P. hasitata*), which is found in the Caribbean, but the cahow has no white behind the dark cap. It was extremely plentiful when the first settlers arrived in Bermuda, but when they suffered a food shortage, they slaughtered the birds by the thousands. Cahows were believed to be extinct until a couple of dead specimens were found in the early twentieth century. Their nesting sites on tiny Nonsuch Island were not located until 1951. There are probably no more than a hundred cahows left, making them among the rarest birds in the world. **See also Bermuda, black-capped petrel, petrel**

cahow

caisson From the French *caisson,* meaning "large box or chest"; pronounced "cayson." It is an enclosed space below water level, with a means of flooding and evacuating water, that is used to gain access to underwater engineering projects in the building of bridges, tunnels, breakwaters, etc. Originally constructed of wood or reinforced concrete, caissons are now made of steel. They are also platforms or tanks that can be submerged by the admission of water and, once in position under a wreck, evacuated; the resulting buoyancy is used to lift heavy objects from the bottom. The best example of caissons (then also known as camels) used to refloat sunken ships was the raising of the German High Seas Fleet, which was scuttled in Scapa Flow, Scotland, in 1919. **See also Scapa Flow**

caisson disease Also known as decompression sickness or the bends, this is a painful—and sometimes fatal—condition in which nitrogen bubbles form in the blood when a diver or underwater worker returns to normal atmospheric pressure after being subjected to greater than normal pressure, as in a deep scuba dive or work in a submerged caisson. **See also bends, the**

calamari The Italian word for "squid," referring in English to small squid prepared for human consumption. The tubular body of the squid is eviscerated, then cut into rings and deep fried or grilled. After shrimp and tuna, squid are the third most lucrative catch from the sea. **See also squid**

California gray whale: See gray whale

California gull (*Larus californicus*) A white-headed gull with a gray back and white underparts, the California gull is found on the Pacific coast of North America, but its primary habitat is inland, around the lakes and rivers of the Great Plains, and even areas where there are no bodies of water. It resembles a somewhat smaller herring gull. In 1848, when the crops of Mormon settlers in Utah were threatened with destruction by "locusts" (actually shield-backed katydids, *Anabrus simplex*), California gulls came to their rescue by eating the insects. There is a monument to the California gull in Salt Lake City, and it is Utah's state bird.

See also gull, herring gull

California sea lion

California sea lion (*Zalophus californianus*) Although it is commonly seen performing in circuses and oceanariums as the "trained seal," this familiar, adaptable animal is a sea lion, with visible ear flaps and hind flippers that can be rotated forward for locomotion. Males are much larger and heavier than females, reaching a weight of 600 pounds and developing a prominent sagittal crest, or forehead bulge. As its name implies, the California sea lion lives in and around the waters of California and Mexico, but there are subspecies found in the Galápagos and Japan.

Canary Islands The Spanish/African equivalent of the Hawaiian Islands; seven large volcanic islands positioned over a "hot spot" off the coast of North Africa that created the islands as the seafloor spread in an easterly direction. The islands are considered a *comunidad autónoma* ("autonomous community") of Spain. The western group (Gran Canaria, Tenerife, La Palma, Gomera, and Hierro) consists of mountain peaks that rise directly from the ocean floor, while the eastern islands of Fuerteventura and Lanzarote are exposed parts of a single submarine ridge. The climate is subtropical, and the volcanic soils of the lowland areas support a wide variety of vegetation, including bananas, oranges, coffee, dates, sugarcane, and tobacco. On the slopes of the higher mountains, cereals, potatoes, and grapes are grown. The original inhabitants of the islands were a Berber people known as Guanches, but they were eliminated by the Spanish in the fifteenth century. Pliny the Elder referred to the large number of dogs (*canis* in Latin) on the islands, which are responsible for the name. The familiar yellow cage bird is descended from a small brown finch (*Serinus canarius*) that lives on the island, and received its name from the island, rather than vice versa. Various Genoese, Majorcan, Portuguese, and French navigators visited the islands during the thirteenth and fourteenth centuries, but in 1404 Juan de Bethéncourt was named king of the islands, by order of Henry III of Castile. The Spanish gained control of the islands in 1497, and when Columbus sailed westward, he replenished all of his expeditions in the Canaries. The islands were frequently raided by pirates and privateers; in 1595, Francis Drake failed to take La Palma, and Horatio Nelson was repulsed by the Spanish at Tenerife in 1797. The sweet wine known as canary was popular throughout Europe until a grape blight (phylloxera) hit the islands in 1853. General Francisco Franco used the islands in 1936 as the first base of Spain's Nationalist revolt. In addition to agriculture, the Canaries are popular vacation islands. They have a permanent population of 1.5 million.

See also Drake, Gran Canaria, Nelson, Tenerife

Cão, Diogo (active 1480–1486) Portuguese navigator who was the first European to sail south of the equator along the coast of West Africa. Sent by King João II of Portugal, Cão (sometimes spelled "Cam," but pronounced "koun" in both cases) carried with him stone pillars (*padrões*) inscribed in Latin, Portuguese, and Arabic, which he erected on conspicuous landmarks as he discovered them. In 1482 he set one up at the mouth of the Congo River and another at his southernmost landfall, Cape Cross, just north of Walvis Bay in what is now Namibia. Other *padrões* were set up at Monte Negro (also named Cape Santa Maria) and Shark Point. These pillars stood until the early twentieth century, but three of them were removed to museums, two in Lisbon and one in Germany, and only the one at Shark Point remains in place. Cão added another 1,450 miles of African coastline to the map, and sailed deep into the Congo. He was knighted upon his return to Portugal, and set out on another voyage, from which he never returned. It was rumored that he actually rounded the Cape of Good Hope, but we will never know: he disappears from the historical record after 1486. In 1487, the king sent Bartolomeu Dias to circumnavigate Africa, which he did not do; he did, however, round the Cape of Good Hope, becoming the first European to sail from the South Atlantic into the Indian Ocean and setting the stage for Vasco da Gama's successful voyage to India in 1497. **See also da Gama, Dias**

Cape Breton Island Large island immediately adjacent to Nova Scotia in Canada's Maritime Provinces, separated from the mainland by the 2-mile-wide Strait of Canso. The island is 110 miles long and 75 miles wide, occupying close to 4,000 square miles. It is deeply indented by Ann's Bay on the northeast, and the center of the island contains the Bras d'Or salt lakes. The Cape Breton Highlands dominate the wilder northern portion of the island, which slopes to flat plains in the south. It is believed that John Cabot landed here in 1497, and the Cabot Trail that encircles most of the highlands is named for him. It was a French possession from 1632 until 1763, but after the 1713 Peace of Utrecht, many Nova Scotians migrated there because they did not want to be ruled by the British. They renamed the island Île Royale and built a fortress at Louisbourg. When Canada was ceded to Britain in 1763, Cape Breton was made a part of Nova Scotia. It was briefly a separate colony with Sydney as its capital, but it was reunited with Nova Scotia in 1820. Extensive coalfields on the eastern part of the island led to the development of coal mining and steelmaking, and the towns of Sydney, Sydney Mines, Glace Bay, and Dominion grew up around the mines. Mining is still the island's major industry, but the 120,000 residents of Cape Breton Island also engage in lumbering, fishing, and the support of summer tourism. Since 1955, the island has been accessible by a causeway across the Strait of Canso.

See also Cabot (John)

Cape Horn The southernmost extremity of South America, actually located on Isla Hornos, Tierra del Fuego, Magellanes Province, Chile. It was not named for its shape, but rather for Hoorn, the birthplace of Dutch navigators Willem Schouten and Jakob Le Maire, who rounded it in 1616. In the days of square-rigged sailing ships, "rounding the horn" was considered the most dangerous part of a westerly voyage, because these ships were not designed for sailing to windward, and the winds roared out of the west, holding ships up for weeks or months and often destroying them completely. The opening of the Panama Canal in 1914 made this route an unnecessary risk.

See also Beagle Channel, Le Maire, Panama Canal, Schouten, Tierra del Fuego

capelin (*Mallotus villosus*) A species of smelt distributed in the Northern Hemisphere in both the North Atlantic and the North Pacific, south to Maine, the Straits of Juan de Fuca, Korea, and Norway. Slender, silvery little fishes, they reach a maximum length of about 7 inches. Females can lay anywhere from three thousand to fifty thousand eggs. Like grunions, they come ashore with the breakers to spawn on the beach at high

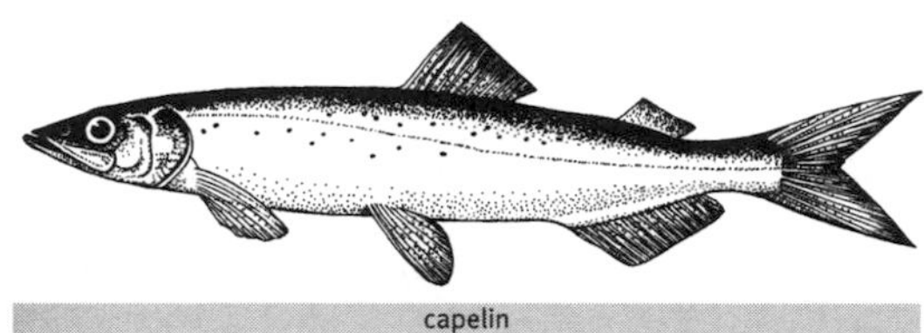

capelin

tide and are easily caught with dipnets and buckets. They are popular food fish, and are served fried, smoked, or salted. In Newfoundland, they are so numerous that they are used as fertilizer. Humpback whales also feed extensively on capelin in northern waters. **See also grunion, humpback whale, smelt**

Cape of Good Hope First rounded by Bartolomeu Dias, who named it Cabo da Boa Esperança in 1488, the Cape of Good Hope is not the southernmost tip of the African continent. That distinction belongs to Cape Agulhas, about 100 miles east of Good Hope. The cape is immediately south of Cape Town, at the end of a peninsula that forms the western boundary of Fish Hoek Bay. "Cape of Good Hope" is also the official name of the Cape Province, the largest of the four republics in the Republic of South Africa. **See also Dias**

Cape petrel (*Daption capensis*) Also known as the pintado petrel or Cape pigeon, this is probably the best-known seabird of the Southern Ocean, easily identifiable by the dark brown and white checkered markings on the back and the 34-inch wingspan. Devoted ship followers, they will stay with ships for days, often settling on the water in flocks.

See also fulmar, giant petrel

Cape petrel

Cape Verde Islands The Cape Verde ("Green Cape") Islands bear the same name as the most westerly cape of Africa. There are ten islands in the archipelago, divided into a Windward group (Santo Antão, São Vicente, Santa Luzia, São Nicolau, Boa Vista, and Sal) and a Leeward group (Maio, São Tiago, Fogo, and Brava). In addition, there are five islets: Raso, Branco, Grande, Luis Carneiro, and Cima. The largest port is Mindelo on the island of São Vicente. The date of the islands' discovery is disputed: Cadamosto claimed to have visited in 1456, and in 1460 the Portuguese navigators Diogo Gomes and António de Noli came upon São Tiago. In 1462 Portuguese settlers landed on São Tiago, founding Ribeira Grande, the most ancient city in the tropics. The islands served as an important station for the slave trade, but with the end of slavery, their prosperity vanished. In 1951, the status of the Cape Verdes was changed from that of a Portuguese colony to an overseas province; it's now the Cape Verde republic, an independent nation.

See also Canary Islands

Capri A single block of limestone almost 4 miles long and a maximum of 1.8 miles across, Capri is probably Italy's most famous small island. Emperor Augustus vacationed here, and Tiberius built twelve villas and eventually refused to return to the mainland. The two main settlements are Capri and Anacapri, inhabited since the tenth century. The island belonged to the Abbey of Montecassino and then to the republic of Amalfi before passing to the Kingdom of Naples. During the Napoleonic Wars, the island changed hands several times before it became part of the Kingdom of the Two Sicilies in 1813. It is across the bay from Naples; there are numerous ferries that make the hour-and-a-quarter trip daily, and hydrofoils do the crossing in twenty minutes. The permanent population of the island is around 7,500, but more than 1.5 million tourists visit the island every year, mostly during the months of July and August. They come to see the Blue Grotto, a cave enhanced by eerie blue light; the ruins of Tiberius's Villa Iovis; the Castello di Barbarossa and the Castiglione; and numerous other attractions, including many restaurants and beaches. The name of the island is believed to have been derived from *capreae,* which is Latin for "goats."

caravel A relatively small trading vessel of the Mediterranean (from the fourteenth to the seventeenth centuries), usually three-masted, averaging 75 to 85 feet in overall length. They were valued for their speed and their ease of sailing in contrary winds. Prior to the development of the galleon, they were the mainstay of Spanish shipping. Columbus's *Niña* and *Pinta* were caravels (the *Santa María* was a *nao*), as were the ships of Ferdinand Magellan's fleet when he circumnavigated the world in 1519–1522.

See also carrack, galleon, pinnace

carcharhinid sharks (family Carcharhinidae) One of the largest and best-known of all shark families; sometimes known as requiem sharks or whaler sharks. There are twelve genera, which include some of the most familiar of all sharks, including the bull, sandbar, dusky, Galapágos, blacktip reef, gray reef, spinner, tiger, blue, lemon, bronze whaler, and oceanic whitetip. They are characterized by a flattened snout, eyes with a nictitating membrane, and bladelike teeth, where the uppers are broadly triangular and serrated and the lowers are narrower and smooth-edged. The carcharhinids are voracious predators, feeding on fishes, other sharks, rays, mollusks, and crustaceans. Many of the larger species have been implicated in unprovoked attacks on swimmers and divers. They are also the most economically important shark family; many species are fished for food, oil, leather, fish meal, and many other uses. Longline fisheries have severely depleted some species, and many others are taken for their fins.

See also *individual species*, finning, shark attack, shark's fin soup

cardinal fishes (family Apogonidae) Small, brightly colored, warm-water fishes, primarily found in lagoons and around coral reefs around the world. Mostly nocturnal, they spend the day in caves and come out at night to feed. Cardinal fishes, which average 4 inches in length, are characterized by large heads and large mouths; the tail is usually forked. In most of the two hundred species, the males are mouth brooders, incubating the eggs in their mouths.

careen To heave a ship down on its side for the purpose of cleaning or repair. Often confused with "career," one definition of which is "to take a short gallop; to charge; to turn this way and that way in running (said of a horse)." In correct usage, a person can *career* down a hill, but he cannot *careen.*

Caribbean monk seal (*Monachus tropicalis*) First described by Columbus in 1494, this seal had been seen around various Caribbean islands such as the Antilles and the Bahamas, and even the Florida Keys. Since the discovery of these islands half a millennium ago, these animals were displaced and hunted so thoroughly that they are now believed to be extinct. The last documented sighting took place off Jamaica in 1952.

Caribbean monk seal

See also Hawaiian monk seal, Mediterranean monk seal

Caribbean Sea A suboceanic basin, covering approximately 1 million square miles. It is bounded on the south by Venezuela, Colombia, and Panama; to the west by all of Central America (Panama, Costa Rica, Nicaragua, Honduras, Guatemala, Belize, and the Yucatán Peninsula of Mexico); to the north by the Greater Antilles islands of Cuba, Jamaica, Hispaniola, and Puerto Rico; and to the east by the Lesser Antilles chain, composed of the arc that extends from the Virgin Islands to Trinidad, off the Venezuelan coast. The Caribbean's greatest depth is in the Cayman Trench, at 25,216 feet below sea level. Hurricanes are born in the Caribbean from June to November, but the biggest month is September. Since many Caribbean countries and islands offer a tropical climate and good weather, sailing, diving, and other popular recreations have led in recent years to a great increase in tourism.

Carib Indians American Indians who occupied the Lesser Antilles in the West Indies, persecuting and killing the inoffensive Arawaks, a century before Europeans arrived and finished the job. Expert navigators, the Caribs colonized various islands in sailing canoes. Extremely warlike and ferocious, they also indulged in cannibalism. Their society had little social structure, and warfare seems to have been their major preoccupation. In their raids on the Arawak, the Caribs killed the men and captured the women, which produced a most unusual linguistic differentiation: the men spoke Carib and the women spoke Arawak. The Spanish could not domesticate the fierce Caribs, so they slaughtered them, claiming they were cannibals and that they needed to be eliminated. Those few that remained in the West Indies were integrated into other societies, and the pure-blooded Indians eventually disappeared from the islands. Some still remain on the northern coast of South America. The Caribbean was named for them. **See also Arawak Indians**

Caroline Islands East of the Philippines and west of the Marianas, the Caroline Islands include the Yap, Truk, and Palau groups and the islands of Pohnpei and Kosrae, as well as many coral atolls. They were discovered by sixteenth-century Spanish navigators, who named the islands after their king, Charles II. In 1914, the islands were seized by Japan, but were occupied by U.S. forces during World War II, and in 1947 they became part of the United Nations Strategic Trust Territories. The islands are mostly volcanic, and the main exports are copra, fish products, and handicrafts. All of the group except Palau joined the Federated States of Micronesia in 1971.

See also Micronesia, Palau, Pohnpei, Truk, Yap

Carpentaria, Gulf of A large rectangular embayment in northern Australia of some 120,000 square miles bounded by Cape York on the east and Arnhem Land on the west. It was first explored by the Dutch explorer Willem Jansz in 1605, and in 1622, the *Pera* and the *Arnhem,* two ships sent by the Dutch East India Company to search for gold and spices, were blown across the gulf, but they found nothing but barren land and hostile natives. (Arnhem, the source of the name of the ship and the wide peninsula that forms the western boundary of the gulf, is a city in the Netherlands.) The gulf was named for Pieter Carpenter, a Dutchman who visited in 1628, but the misspelling of his name has not been explained. The southern and western coasts were investigated by Abel Tasman in 1644. Along the shores of the gulf are huge bauxite deposits, and manganese is mined on Groote Eylandt. Other islands in the gulf are the Wellesleys, in the south, off the town of Karumba, which is the center of a fishery for banana prawns (*Pe-*

naeus merguiensis), one of Australia's more important fisheries exports.

carpet sharks (family Parascyllidae) Small, elongated sharks found only in the western Pacific, similar to the catsharks (Scyliorhinidae) because of their catlike eyes. The three species of *Cirrhoscyllium* have barbels at the throat, a feature found in no other sharks. The most striking species is the necklace carpet shark (*Parascyllium variolatum*), with a dark, white-spotted collar around the gills and black and white splotches on the fins. Also found in the waters of southern Australia and Tasmania is the collared carpet shark (*P. collare*), a yellowish shark with several dusky saddles and scattered dark spots. Carpet sharks can grow to a length of 36 inches, but most are smaller. **See also catsharks**

carrack A larger type of trading vessel in use in northern and southern Europe from the fourteenth to the seventeenth centuries, developed as a compromise between the square rig of the northern European nations and the lateen rig of the Mediterranean. They are regarded as the immediate predecessors of the galleon, larger, beamier, and generally more robust than caravels. The carrack was designed primarily to fight medieval battles, that is, to resist enemy boarding parties, therefore the very high fore- and after-castles.

See also caravel, galleon, lateen sail

Carson, Rachel (1907–1964) Raised in rural Pennsylvania far from the sea, Rachel Carson became one of the most influential of all writers on the environs and inhabitants of the oceans. She taught at the University of Maryland and the Johns Hopkins University, later worked as a marine biologist for the U.S. Bureau of Fisheries, and was editor in chief of the U.S. Fish and Wildlife Service. She is recognized as one of America's foremost nature writers, having published such highly acclaimed books as *Under the Sea Wind, The Sea around Us,* which won the National Book Award for 1951, and *The Edge of the Sea.* Her *Silent Spring* (1962), which alerted the country to the hazards of pesticides, is regarded as one of the most important environmental exposés of all time.

Carteret, Philip (1738–1796) British naval officer who served aboard HMS *Dolphin* on John Byron's 1764–1766 voyage around the world. He was later given command of the *Swallow* and given instructions to complete another circumnavigation, also with the *Dolphin* (this time commanded by Samuel Wallis). The two ships were separated after passing through the Strait of Magellan, and Carteret sailed on alone. In 1766, he discovered (and named) Pitcairn Island, the Admiralty Islands, and New Ireland. Since New Ireland had never been claimed (as far as he knew), he took possession of it for England. When Bougainville arrived at the island two years later, he took possession too, even though he saw the plaque left there by Carteret.

See also Bougainville, New Ireland, Pitcairn Island, Wallis

Cartier, Jacques (1491–1557) French navigator who discovered the St. Lawrence River while searching for the Northwest Passage to China, laying the basis for subsequent French claims to Canada. In 1534 he sailed around Newfoundland; entered the Strait of Belle Isle; discovered Anticosti Island, the Magdalen Islands, and Prince Edward Island; and entered the mouth of the great river before sailing back to France. He returned to Canada in 1536 and sailed up the river as far as the Île d'Orléans, which he named for the French royal family. Anchoring there, he continued westward in longboats until he reached the Iroquois village of Hochelaga, the site of the present Montreal. On his third voyage in 1541, having been told by the Indians that a rich land ("Saguenay") lay to the west, he returned to conquer it by force of arms, but he failed to make contact with the seigneur de Roberval, the commander of the troops, who had been delayed in Europe. Although he had orders to go to Quebec, Cartier returned again to France. He made one last voyage in 1543, again in search of the rich land, but Roberval ordered him back to France. He returned to St. Malo, the place of his birth, and remained there until he died in 1557.

See also Anticosti Island; Newfoundland; St. Lawrence, Gulf of

cartilaginous fishes Fishes or fishlike vertebrates that have cartilage where teleosts have bone. Also known as Chondrichthyes, the group includes sharks, rays, skates, and chimaeras. Parts of their skeletons, especially the skull, spines, and vertebrae, are often strengthened by calcification, and while they may resemble bone, they are not. Cartilage disintegrates shortly after death, so cartilaginous fishes are rarely preserved as fossils. Only the teeth (which are composed of dentine) can fossilize, so often our only record of earlier species is the teeth, which are poor indicators of overall shape or lifestyle. The evolutionary history of cartilaginous fishes is therefore incomplete, and their relationships to other groups of fishes is unclear.

Caspian tern (*Sterna caspia*) The largest of the terns, with a 55-inch wingspan and a coral-red bill; its broad wings and unternlike soaring flight makes confusion with a gull possible. Its coloration is typical: light gray above, white beneath, with a black cap. Caspian terns breed in small colonies or singly on the coasts of North America, Eurasia, and Africa, and winter throughout the temperate zones of the Northern and Southern Hemispheres. They seem to prefer the company of ring-billed gulls (*Larus delawarensis*) to any other birds, and will often nest near them and rest with them on the same sandbar. **See also Arctic tern, tern**

Cassin's auklet (*Ptychoramphus aleuticus*) A small, plump alcid, dusky grayish brown above and lighter below, Cassin's auklet has a wingspan of about 10 inches. It is a gregarious bird, feeding on planktonic organisms far offshore in huge flocks. One of the most widespread of the Pacific alcids, it breeds from the Aleutians through the Gulf of Alaska and as far south as Baja California. **See also Alcidae, least auklet**

Castillo de Bellver Spanish supertanker that caught fire at midnight on August 6, 1983, 70 miles west of Cape Town. Seven hours later, she broke in half, and the stern section sank in water that was 1,500 feet deep. Rescue tugs managed to tow the burning forward section farther out, and it sank a week later. For the next five months, 78 million gallons of oil leaked from the wreck but the stern compartments still contain millions more gallons; if and when they rust through, the remainder will come bubbling to the surface. Even though this was the largest tanker spill in history, hardly any of the oil came ashore. The effects on pelagic fishes are unknown. **See also *Amoco Cadiz, Exxon Valdez*, oil spill, supertanker, *Torrey Canyon***

catamaran From the Tamil *katta,* "to tie," and *maram,* "wood"; a catamaran was originally a raft consisting of two or more logs or tree trunks lashed together and used as a surf boat in the East or West Indies. The term is also used to describe much larger rafts made from the trunks of balsa trees, growing on the western coast of South America. In contemporary nautical terminology, a catamaran is a twin-hulled sailing yacht, originally used for racing but now adapted for cruising purposes. Because of their low immersion area, catamarans can attain speeds considerably higher than those of conventionally hulled sailing boats. **See also trimaran**

catcher boat In whaling, the small, fast boats from which the whales are killed and then towed back to the factory ships for processing. Originally steam-powered, they later employed diesel engines. Catcher boats averaged between 150 and 200 feet in length and were characterized by the presence of the harpoon cannon in the bows and the catwalk that allowed the harpooner to get quickly from the bridge to the gun. The harpooner fired a heavy iron projectile into the body of the whale, with toggles that opened to keep it from pulling out. In the heyday of Antarctic whaling (1910–1935), there may have been as many as three hundred catcher boats prowling the Southern Ocean. **See also factory ship, Foyn, whaling**

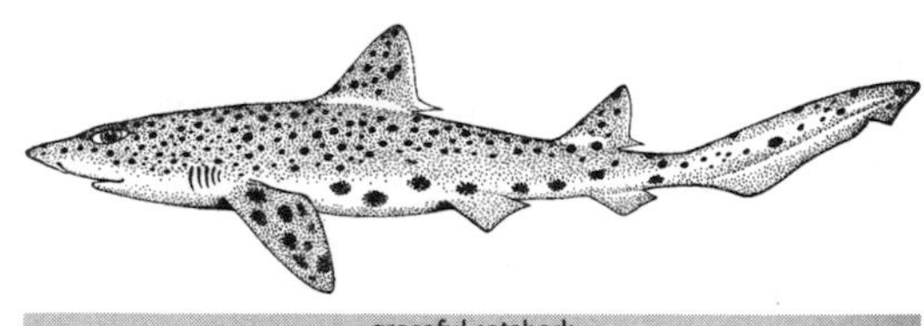

graceful catshark

catsharks (family Scyliorhinidae) The largest family of sharks with more than eighty species, catsharks are small bottom dwellers that are often marked with stripes, spots, or bars. In most species, there are two similarly sized dorsal fins and a long tail fin that does not curve up but continues in a straight line with the body. The origin of the name "catshark" is unknown, but it is believed to refer to the large, slitted eyes. Two common European spotted dogfishes (*Scyliorhinus canicula* and *S. stellaris*) are actually catsharks, a testimony to the confusion that can be caused by common names. The beautifully marked chain dogfish (*S. retifer*) is also a catshark. The genus *Apristurus* contains some twenty-five species, most poorly known but similar in general form, with a flattened, wedge-shaped head, pointed snout, and low dorsal fins positioned far to the rear. With their multi-cusped teeth they feed on small crustaceans and fishes. They average about 2 feet in length and are found in localized populations throughout the world's cooler waters. The swell sharks (*Cephaloscyllium* spp.) are classified with the catsharks, as is the peculiar little headshark (*Cephalurus cephalus*). The marbled catshark (*Galeus arae*) is a strikingly colored, 16-inch-long shark commonly found along the continental slopes from Georgia to Colombia. In South Africa, the shysharks (*Haploblepharus* spp.) are catsharks that get their common name (*skamoog* in Afrikaans) from their habit of curling up with their tail over their eyes when caught. **See also headshark, small-spotted catshark, swell shark**

Cavendish, Thomas (1560–1592) The third man to circumnavigate the world (after Magellan and Drake). After squandering a fortune inherited from his father, he decided to replicate Drake's 1580 circumnavigation of the world. With three ships, the *Desire,* the *Content,* and the *Hugh Gallant,* he sailed from London in 1586, reached the Pacific in 1587, and burned and pillaged his way up the coast of South America. Off Cabo San Lucas, the southern tip of the Baja California peninsula, he captured and burned the Spanish treasure galleon *Santa Ana,* which was heading from Manila to Acapulco, loaded the gold, pearls, and silks on his own ships, and returned to England via the Philippines, the Cape of Good Hope, and St. Helena. Estimated at between 100,000 and 200,000 gold pesos in value, the treasure was the richest payload ever collected from a Spanish ship. Shortly after his triumphant return to England, Cavendish bought four more ships and set out again to reach China. In the Strait of Magellan they were beset by winter storms, and their little fleet dispersed. The *Dainty,* with John Davis commanding, put in at Port Desire (now Puerto Deseado in Patagonia, which had been found and named by Cavendish on his first voyage), and made it safely back to England. In the galleon *Leicester,* Cavendish turned back into the South Atlantic. Having hung his pilot for insubordination, he failed to find either St. Helena or Ascension Island, and he died at sea.

Cayman Islands A British colony consisting of Grand Cayman (the largest), Cayman Brac, and Little Cayman, this group of islands in the Caribbean was first sighted by Columbus in 1503, on his third and last voyage to the New World. He named the islands "Las Tortugas" for the sea turtles that are found there, but the islands were later named "Cayman," after the Spanish word for alligator, perhaps a mistaken reference to the iguanas that inhabit these low coral islands. It was not until 1670 that the British laid claim to the islands, which, until 1959, were governed from Jamaica, 180 miles to the southeast. Tourism sustains the islands' economy, but George Town (the capital), with more than five hundred banks, serves as a base for a financial community that attracts many offshore investors because of favorable taxes and confidentiality laws.

See also Columbus, Jamaica

CCAMLR The Convention on the Conservation of Antarctic Marine Living Resources, an international treaty open to all nations of the world, whether or not they border on or have business in the Antarctic. CCAMLR (usually pronounced "camel-R") came into effect in April 1982, and is designed to function in place of any exclusive economic zones, since Antarctic waters (south of 60° south) are considered international. "Marine living resources" include fish, whales, seals, seabirds, and invertebrate life forms, such as squid and krill. Member nations are Australia, Austria, Belgium, Canada, the Czech Republic, Denmark, Finland, France, Germany, Greece, Hungary, Iceland, Ireland, Italy, Japan, Luxembourg, Mexico, the Netherlands, New Zealand, Norway, Poland, Portugal, South Korea, Spain, Sweden, Switzerland, Turkey, the United Kingdom, and the United States.

See also International Whaling Commission, whaling

Celebes Also known as Sulawesi, this large Indonesian island consists of four peninsulas separated by three gulfs, in a shape rather like that of an extremely irregular pinwheel. It is almost entirely mountainous, with many active volcanoes. The population of the sparsely populated eastern and southeastern peninsulas exist largely on subsistence agriculture. Celebes is the major source of copra for Indonesia, and corn, rice, cassava, yams, tobacco, and spices are also grown. The Dutch expelled the Portuguese and conquered the natives in the Macassar War (1666–1669). In 1950, the island became part of the nation of Indonesia. Endemic to the island is a dwarf buffalo known as the anoa, and a large wild pig known as the babirusa. The Celebes black ape (*Cynopithecus niger*) is a large, tree-dwelling, stump-tailed monkey that is also called the Celebes crested macaque. It is found only on Celebes and neighboring islands.

See also Indonesia

Central America Paddle-wheel steamer that set out from Panama in September 1857, ran into a monster hurricane, and sank off the Carolina coast with a shipment of gold from the San Francisco mint to New York banks and a shipload of forty-niners who were carrying their gold from California. In what was then the largest maritime disaster in American history, the 300-foot-long ship went down with 452 of the 600 passengers and crew. (The other 148 were rescued by passing ships.) On the bottom in 8,000 feet of water, the *Central America* seemed far beyond the reach of salvors (those who salvage ships or cargos from the sea) and treasure hunters. In 1989, however, after years of careful preparation, an engineer-inventor named Tommy Thompson and the Columbus American Discovery Group (named for Columbus, Ohio, where most of the money for the expedition was raised) not only located the wreck on the bottom, but managed to bring up 3 tons of gold. At 1857 prices (ninety cents an ounce for gold) the gold aboard the *Central America* was valued at about $1.2 million, but with gold selling at about $360 an ounce today, the treasure is worth an estimated $450 million. If the coins were sold individually, however—mint double eagle gold pieces are worth about $20,000 apiece—the value of the treasure might reach $1 billion.

See also Ballard, Manila galleon, marine archaeology

Cephalonia Known in Greek as Kefallinía, this is the largest of the Greek islands in the Ionian Sea. It is not as lush and beautiful as Corfu, and is therefore not the beneficiary of a tourist invasion like that island's. The Ionian islands differ significantly from the Aegean islands: they have a hotter climate (they do not experience the *meltemi,* the wind that sweeps the east coast of mainland Greece), and their history has been strongly influenced by Italy, their neighbor to the west. Successively, Cephalonia has been occupied by the Romans, the Byzantines, the Venetians, the Ottoman Turks, and the Venetians again. In August 1953, an earthquake of such magnitude struck the island that Mount Aenos was split in two and many of the island's villages were destroyed. Cephalonia—along with Zacynthus (Zákinthos)—is visited by breeding loggerhead turtles (*Caretta caretta*), but the increase in tourism and boats has disturbed the turtles and resulted in a serious decline of the turtle population. **See also Zacynthus**

cephalopod Literally, "head-foot," the term used to designate that class of marine invertebrates in which the arms are attached directly to the head. The class consists of octopuses, squids, cuttlefishes, nautiluses, and the extinct belemnites and ammonites. Cephalopods are classified in the phylum mollusca (clams, oysters, snails, etc.), but they differ from most other mollusks in being free-swimming and having only a vestigial, internal shell, known as the pen or gladius. (The nautilus is the only cephalopod with an external shell, and octopuses have no internal or external shell.) They have large brains, and are considered the most highly evolved of invertebrates. The tentacles—8 in octopuses, 10 in squids and cuttlefishes, 60–90 in nautiluses—surround a horny beak like that of a parrot. They move by taking water into the body cavity and expelling it through a funnel or siphon. (Octopuses also walk or crawl on the bottom.) With the exception of the primitive nautilus, cephalopods have three hearts: two (the branchial hearts) to pump blood to the gills, and one (the systemic heart) to supply blood to the rest of the circulatory system. The chambered nautilus has very primitive eyes, but the remaining cephalopods have highly developed, efficient visual systems. Many of them can change color rapidly and dramatically, and many species of squid have bioluminescent organs. Most octopuses and squids can produce a cloud of ink that serves as a smoke screen or forms in the general shape of the emitting animal, permitting it to jet away and escape a predator. The smallest cephalopods are the size of grains of rice; the largest is the giant squid, the largest of all invertebrates at up to 60 feet in total length.

See also chambered nautilus, cuttlefish, octopus, squid

cetacean From the Greek *ketos* and the Latin *cetus,* meaning "whale," the term includes all the marine mammals (whales, dolphins, and porpoises) that spend their entire lives in the water, breathe air through a single or double blowhole, and give birth to live young that they nurse underwater. (Manatees and dugongs have nostrils on the end of their muzzles and are classified as sirenians.) Cetaceans are further divided into two groups: the mysticetes, or baleen whales; and the odontocetes, or toothed whales. All the rorquals are mysticetes, as are the right whales and the bowhead, pygmy right, humpback, and gray whales. The sperm whales (including pygmy and dwarf sperm whales), all freshwater dolphins, pelagic dolphins, and porpoises are odontocetes. A person who studies cetaceans is known as a cetologist.

See also *individual species;* dolphin, porpoise, rorqual, whale

Challenger A 200-foot-long British steam corvette of 2,306 tons that was built in 1858 and sailed around the world from 1873 to 1876 on an international oceanographic cruise. The commanding officer was Captain George Nares, but the scientific team was under the direction of Charles Wyville Thomson (1830–1882). The *Challenger* expedition traversed 68,890 nautical miles, visiting the North and South Atlantic, the North and South Pacific, and the Arctic and the Antarctic, making 362 "samples" at specific locations, recording the exact depth, a sample of the bottom, the temperature at the bottom, and a collection of the bottom fauna. She returned to England with more than thirteen hundred plants and animals, 1,441 water samples, and hundreds of containers of seafloor material. Under the editorship of John Murray, the official fifty-volume *Report of the Scientific Results of the Voyage of the H.M.S. Challenger* was published between 1880 and 1895; among other important observations, it identified 4,417 new species and 715 new genera. **See also Murray, Thomson**

Challenger Deep The deepest point in the world's oceans. Located east of Guam in the deep cut known as the Mariana Trench, it is 35,800 feet deep. Although it would have made sense to name the oceans' greatest depth for the greatest oceanographic expedition of all time, it was in fact named for a successor to the original *Challenger,* the British research vessel *Challenger II.* In 1949, at the southern end of the Mariana Trench, oceanographers recorded an echo that corresponded to a depth of 5,900 fathoms, almost 1,000 feet deeper than the previously known greatest depth in the Philippine Trench. On January 23, 1960, with Jacques Piccard and Lieutenant Don Walsh aboard, the bathyscaph *Trieste* descended to the bottom of the Challenger Deep, setting a record that cannot be broken. **See also *Trieste,* Walsh**

chambered nautilus (*Nautilus pompilius*) The only cephalopod with an external shell, the chambered nautilus is the last survivor of a group of animals that flourished millions of years ago. The shell, which serves for protection and also as a flotation device, is composed of a series of chambers connected by a tube called the siphuncle, and the animal can regulate its buoyancy by adjusting the amount of gas in the chambers. The chambered nautilus never leaves its shell, but if it is re-

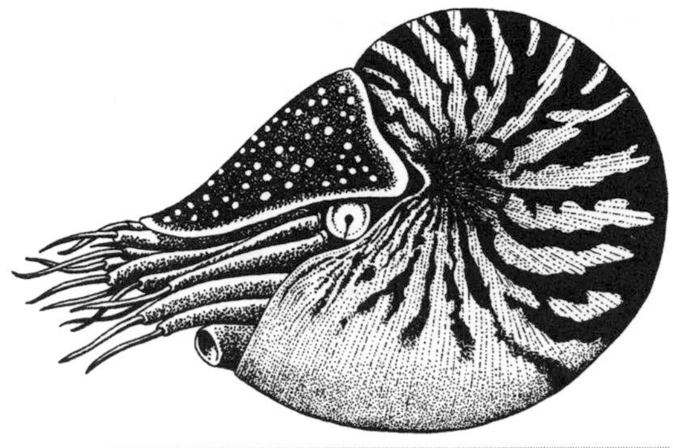

chambered nautilus

moved, the animal resembles a squid with an overabundance of tentacles. It can retract into the shell and close it off with a leathery hood. The nautilus propels itself slowly by taking water into the body and ejecting it through a funnel, which is not a closed tube as in other cephalopods. The eyes, which are pinholes open to the sea, are not very efficient, and the nautilus has to depend on other senses to find food and/or a mate. They may have as many as ninety tentacles, which are used for capturing prey and for sensing. There are perhaps six species of nautilus, all found in the Indo-Pacific region and differentiated by geography and minor anatomical variations. Nautiluses feed at night and make daily migrations down the face of a reef in search of food. They are believed to be mostly scavengers. The eggs are among the largest of any invertebrate's, more than 1 inch in diameter. Specimens collected off Palau have been raised in captivity, and they make extraordinarily interesting aquarium exhibits. The only threat to the chambered nautilus is man, who collects them for their shells, which are strikingly striped on the outside and iridescent on the inside. **See also ammonite, belemnite, cephalopod, octopus, squid**

Chancellor, Richard (d. 1556) British navigator who sailed with Sir Hugh Willoughby in 1553 on a voyage to discover the Northeast Passage. Willoughby was captain of the *Bona Esperanza;* Cornelius Durforth was captain and Stephen Borough was sailing master in the *Edward Bonaventure;* and Chancellor was captain of the *Bona Confidentia* and pilot general of the expedition. They sailed north toward Norway until a great storm separated Chancellor from the other two ships, and he went on to Archangel on the White Sea; he then continued to Moscow over 1,500 miles of frozen tundra and ice. (Willoughby, Durforth, and all their crews died.) In the Russian capital, Chancellor met with Tsar Ivan IV (Ivan the Terrible), and was given a letter granting the English favorable conditions for trade with Russia. This led to the formation of the Muscovy Company by Sebastian Cabot and others in 1555. Chancellor returned to Moscow on a trading mission in 1556 and died in a shipwreck off Aberdour Bay, Scotland, on his return voyage.

Channel Islands A group of four main islands and many smaller islets and reefs in the English Channel, 60 miles south of England. Guernsey, Jersey, Alderney, and Sark are much closer to France than they are to England—just west of the Cotentin Peninsula of Normandy. They are dependencies of Britain, but not part of the United Kingdom. Since the Norman Conquest of 1066, they have been ruled by the English Crown; even after Philip II of France confiscated the duchy of Normandy in 1204, they remained under the control of England. On many occasions, the French have attempted to reclaim the islands, but as recently as 1953, the International Court of Justice at The Hague ruled that they were under British jurisdiction. Both English and French are spoken, but French is declining. The Channel Islands were the only British territories to be occupied by the Germans during World War II. The islands differ in scenery, customs, and administrative procedures, but they are all popular tourist destinations. **See also Alderney; Guernsey; Jersey; Sark, Isle of**

Charcot, Jean-Baptiste (1867–1936) French polar explorer, trained as a doctor, who made his first voyage to the Antarctic in 1903–1905 in the *Français,* with the intention of rescuing the Nordenskjöld expedition, whose ship had been lost in the Weddell Sea. Finding that Nordenskjöld had already been rescued (by the Argentine ship *Uruguay*), Charcot spent two seasons exploring the Antarctic as far south as Alexander Island. Upon his return, his wife divorced him for desertion; when he remarried in 1907, he made his new wife promise never to oppose his expeditions. In 1910, he returned to the Antarctic in the *Pourquois Pas?* and explored Marguerite Bay. He found a previously unknown land to the southwest of Alexander Island that he named Charcot Land. After the war, he resumed his research and exploration in Greenland and Rockall. He was drowned in 1936 when the *Pourquois Pas?* was wrecked off the coast of Iceland.

Cheng Ho (1371–1433) Known as the Three-Jewel Eunuch, Cheng Ho (also Zheng He) was the greatest Chinese voyager of the fifteenth century. He was born a Muslim in Yunnan Province, captured when he was ten, and brought to the court of the newly established Ming dynasty, where he was castrated and sent into the army as an orderly. He became a confidant and adviser to the Ming emperor Yung-lo, who appointed him commander of the missions to the "Western Oceans." From 1405 to 1433, fleets of Chinese armed warships, some more than 400 feet long, made seven epic voyages through the seas of China and the Indian Ocean. They carried cargoes of silks, porcelains, and lacquerware to trade for ivory, pearls, and spices, but their main purpose was to impress local rulers with the riches of the Chinese empire. The first expedition, consisting of sixty-two ships and 27,000 men, set out in 1405 and visited Java and Sumatra, then Calicut and Ceylon. The second voyage (1407–1409) reached Siam and Malacca, and traveled to the Maldives and as far west as the sultanate of Hormuz at the entrance to the Persian Gulf. (On this voyage, Ceylon's military commander was taken prisoner because he refused to kowtow to the Chinese.) In 1415, Cheng Ho brought envoys of thirty states of Southeast Asia to pay homage to the emperor. From 1417 to 1419, Chinese squadrons visited the Ryukyus and Brunei; others traveled to Mombasa and Malindi on the coast of East Africa. The sixth expedition (1421–1422) traversed the full width of the Indian Ocean from Borneo to Zanzibar. The emperor Yung-lo died in 1424, but the seventh expedition (1431–1433) was the most ambitious of all, carrying forty thousand men to every port from Java to Mecca, intent upon showing the world how rich and powerful the Ming dynasty was. Cheng Ho died at sea in 1433 on the return voyage to Nanking. The fleet brought tributes from the Asian and Arab states, including horses, elephants, and a giraffe, but Yung-lo's successors had no interest in replicating his grandiose naval demonstrations. They forbade all overseas travel and disbanded the fleet, along with China's policies of outward expansion.

Chesapeake Bay The largest inlet on the Atlantic coast of the United States, Chesapeake Bay is 193 miles long and from 3 to 25 miles wide. The southern part of the bay is bordered by the state of Virginia and the northern part by Maryland. The first permanent British settlement in North America was in Chesapeake Bay at Jamestown, on a peninsula near the James River in Virginia. It was founded on May 14, 1607; the following year, Captain John Smith arrived and mapped the bay and its estuaries. Two major battles during the American Revolutionary War were fought between Britain and France in Chesapeake Bay in 1781, and during the War of 1812, the British invaded through the bay. The major rivers that empty into the bay are the Susquehanna, James, York, Rappahannock, Potomac, and Patunxet from the west, and the Wicomico, Nanticoke, Choptank, and Chester from the eastern shore. Baltimore, Maryland, is at the head of the bay; Annapolis is a little farther south, and Washington, D.C., is at the head of the Potomac River. Until the latter half of the twentieth century, Chesapeake Bay was a major source of fish and shellfish—particularly oysters and crabs—but residential and industrial development so polluted the bay that its productivity declined sharply, and various projects are under way in an attempt to rescue it.

Chichester, Francis (1901–1972) British deepwater sailor who is most renowned for his 'round-the-world solo voyage in *Gipsy Moth IV* in 1966–1967. In 1929, he flew a Gipsy Moth airplane from England to Australia, and in 1960 he won the first single-handed transatlantic yacht race in just over forty days. In 1966, he sailed alone from Plymouth, England, to Sydney, Australia, and back via Cape Horn in 274 days, celebrating his sixty-fifth birthday en route. Upon his return to England, he was knighted by Queen Elizabeth II, who used the same sword that Elizabeth I had used to knight Sir Francis Drake.

Chiloé At 3,241 square miles, Chiloé is the largest island along Chile's 2,800 miles of coastline. It is located about two-thirds of the way down, separated from the mainland by the Gulf of Corcovado. Immediately to the south of Chiloé, the coastline of Chile is broken up into an ragged archipelago of islands, fjords, and lakes that culminates in Tierra del Fuego. Chiloé is the only one of the coastal islands that has been settled, because its rainy climate encourages the growth of dense evergreen forests. The population is around 67,000; the inhabitants export timber and raise wheat and potatoes. The Spanish settled the island in 1567; it was the last stronghold of Spanish royalists when Chile asserted its independence in 1826. Early in 1835, Charles Darwin arrived at Chiloé in the *Beagle* and spent the better part of three weeks exploring the island. During that time, an enormous earthquake hit southern Chile, devastating the cities of Concepción and Valdivia immediately to the north of Chiloé. On May 22, 1960, the largest earthquake recorded in this century (magnitude 9.5 on the Richter Scale) occurred on the coast of southern Chile, killing more than five thousand people and generating lethal tsunamis that ran up on the shores of Hilo, Hawaii, and Hokkaido, Japan, 10,000 miles from the epicenter. The Chilean shoreline subsided 6 to 12 feet, leaving it open to invasion by the sea. Turbulent waves estimated at 12 to 15 feet high flooded the towns, destroyed boats and buildings, and then hissed out again, returning in a 26-foot-high wall of green water traveling at 125 mph. On Chiloé, the town of Ancud was completely destroyed. **See also Tierra del Fuego, tsunami**

chimaera (Holocephali) Sometimes called ratfishes, ghost sharks, or elephant fishes, chimaeras are cartilaginous fishes that are not exactly sharks and not exactly bony fishes (teleosts). They have large eyes and two dorsal fins, the first of which has a serrated, poison spine in front of it. While they have four pairs of gill openings like sharks, these are covered by an operculum as in the teleosts. Like sharks, they lay eggs encased in horny capsules. They also have claspers like the sharks, but the males have another clasper just in front of the eyes, the reproductive function of which is not

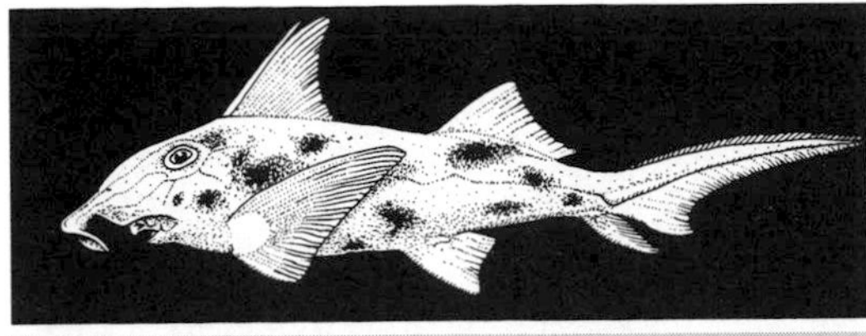
chimaera

known. There are some twenty-eight species, ranging in size from 2 to 6 feet. The three groups of chimaeras are differentiated by their noses: rounded, pointed, or plow-shaped. Most species are round- or short-nosed. Classified as *Chimaera* and *Hydrolagus,* they have a long, tapering tail, which accounts for the name "rat-fish." The long-nosed versions (*Rhinochimaera*) have a nose like that of a jet fighter, and the plow-nosed (*Callorhinchus*) have an appendage that hangs down and points back toward the mouth. Chimaeras are found in temperate to cold waters, from the shallows to the great depths. In Greek mythology the chimaera was a fire-breathing monster with the head of a lion, the body of a goat, and the tail of a serpent. **See also shark**

Chincoteague Island Made famous by Marguerite Henry's children's book about one of the local ponies, *Misty of Chincoteague,* this is Virginia's only resort island, connected to the mainland by a causeway and bridges. The island, 7 miles long and 1½ miles wide, is a national wildlife refuge and the gateway to Assateague Island National Seashore. Famous for its oyster and clam beds, Chincoteague has probably the world's only oyster museum, in the eponymous village that is also known for its carvers and other artisans. During the last week in July, volunteer firemen round up the Chincoteague ponies on nearby Assateague Island, believed to be descended from farm horses that escaped or were abandoned by their owners, and drive them across Chincoteague Sound back to Chincoteague, where an auction is held. Foals are sold for prices that can go as high as one thousand dollars.

See also Assateague Island

Chinese river dolphin (*Lipotes vexillifer*) One of a small group of freshwater dolphins, the *baiji* lives only in lakes and rivers of central China. It is grayish blue above and white below, with a long, upturned beak and eyes placed high on the head. Although efforts are being made to protect this animal, its propensity for becoming entangled in fishermen's nets has greatly reduced its numbers, and with the total estimated population numbering not more than three hundred animals, it is probably the most endangered of all cetaceans. **See also freshwater dolphins**

Chinese river dolphin

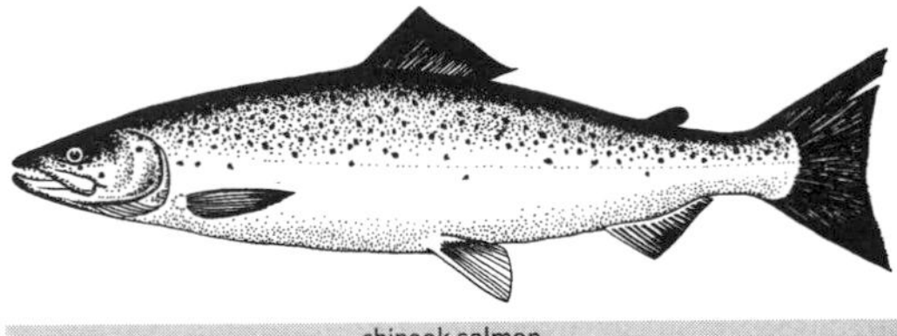
chinook salmon

chinook salmon (*Oncorhynchus tshawytscha*) Also known as the king salmon, the chinook is the largest of the Pacific salmons, known to reach a length of 5 feet and a weight of 126 pounds (the average weight is considerably less, about 20 to 45 pounds). Because of its size and previous abundance, the chinook used to be one of the most important food fishes in America. Like all Pacific salmon, the chinook spends its early years in freshwater rivers, descends to the sea, and eventually returns to breed and die in the place it was born. The first cannery on the west coast of North America canned .25 million pounds of chinook; catches rose rapidly until 1883, when 43 million pounds were canned. Overfishing, the destruction of the forests where the streams ran, and the construction of dams in the rivers that prevented the fish from returning to their breeding sites caused the chinook population to crash, and now the entire population is considered endangered. Most of the chinook salmon fisheries in Washington, Oregon, and Idaho have failed, and in 1996, the Alaska fishery was closed down by the U.S. government.

See also Atlantic salmon, chum salmon, coho salmon, Pacific salmon, pink salmon, sockeye salmon

chinstrap penguin (*Pygoscelis antarctica*) The derivation of this bird's common name is obvious; a narrow band of black feathers extends from ear to ear. Chinstraps belong to the same genus (*Pygoscelis*) as the Adélies, and their habits are similar. On South Atlantic islands and the Antarctic Peninsula, they build nests of stones, feathers, and bones, and both parents take turns incubating the eggs.

See also Adélie penguin

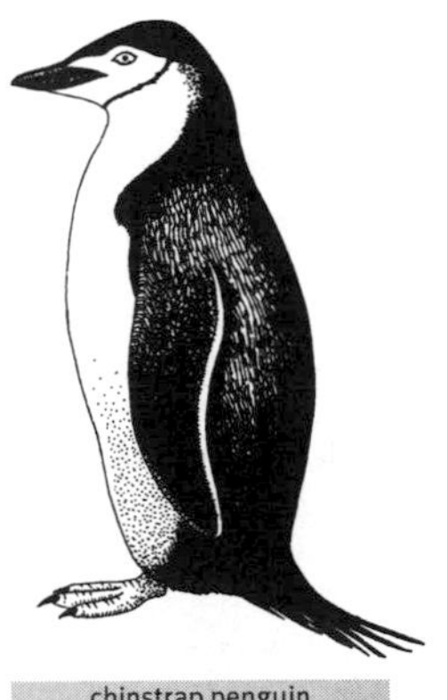

chinstrap penguin

chitons (class Polyplacophora) Primitive mollusks whose shell is composed of eight overlapping plates. They are believed to resemble the ancestors of snails, clams, and cephalopods, and the soft parts consist of a bilaterally symmetrical arrangement of mouth, radula, stomach, gonads, and digestive system. They have no tentacles or eyes. The body consists of a marginal mantle and a muscular, sole-shaped foot that can clamp so tightly to rocks or pilings that they cannot be dislodged. If they are taken by surprise, and a knife blade slipped beneath the foot before they can contract, they curl up like pill bugs. They are found on rocks in shallow waters around the world, where they graze on algae, bryozoans, and sponges. Most species are less than 3 inches long, but the gumboot chiton (*Cryptochiton stelleri*) of the west coast of North America can reach a length of 13.5 inches.

See also gastropoda, *Neopolina*

Choiseul Island in the Solomons, discovered and named by Louis-Antoine de Bougainville for Étienne-François de Choiseul (1719–1785), foreign minister of France when Bougainville made his 1766–1769 'round-the-world voyage. Choiseul is 83 miles long and covered with dense jungles and swamps; it rises to a height of 3,500 feet at Mount Maitabo. Japanese forces occupied the island in 1942, and although the Americans took other Solomon Island outposts such as Guadalcanal, New Georgia, and Bougainville, they could not oust the Japanese from Choiseul until the surrender in 1945. **See also Bougainville Island, Guadalcanal, New Georgia, Solomon Islands**

Christian, Fletcher (1764–1790?) Born at Moreland Close, a large manor house in Cumbria, Christian volunteered for naval service at the age of eighteen, served two years as a midshipman aboard the *Eurydice,* then made two voyages to the West Indies aboard the *Britannia,* Captain William Bligh commanding. In 1787, Bligh chose him to be master's mate aboard HMS *Bounty,* bound for Tahiti to pick up a cargo of breadfruit seedlings for transport to the West Indies. When they arrived in Tahiti in October 1788, they gathered the breadfruit plants and stayed for five months before setting sail for the West Indies. Three weeks out of Tahiti, the *Bounty* anchored off Nomuka in the Friendly Islands (Tonga), and there, for reasons that have never been successfully explained, Christian led a mutiny against Captain Bligh, putting him and eighteen loyalists over in a 23-foot-long open boat as the mutineers sailed away in the *Bounty.* When they landed at Tahiti again, they dropped off sixteen of the mutineers and picked up six Tahitian men and nineteen women. They sailed to lonely Pitcairn Island, where they burned the *Bounty* and built a settlement. During the eighteen years that intervened between their landing on Pitcairn and their discovery in 1808 by the Boston sealer *Topaz,* all the mutineers except Alexander Smith (who had renamed himself John Adams) had either died or been murdered. Among the surviving members of the colony were Fletcher Christian's sons, Friday Fletcher October Christian and Thursday October Christian. In 1855, the descendants of the *Bounty* mutineers, having outgrown Pitcairn, petitioned Queen Victoria for another home, and she offered them Norfolk Island, 1,000 miles east of Sydney and until 1854 the location of the most notorious of all Australian penal colonies. Many of the island's 2,000 inhabitants today are the descendants of the mutineers.

See also Bligh, *Bounty,* Norfolk Island, Pitcairn Island, Tahiti

Christmas Island Australian territory some 220 miles south of Java in the Indian Ocean. The population of the 52-square-mile island is 1,200, most of whom live in the island's only settlement, Flying Fish Cove. It was first sighted in 1615 by Captain Richard Rowe of the British East India Company. It was later named on Christmas day in 1642 by Captain William Mynns. The island was uninhabited until it was settled by the Scotsman George Clunies-Ross in 1827. Clunies-Ross obtained a ninety-nine-year lease to mine the island's rich phosphate deposits. The island became a British possession in 1857. With Singapore, Penang, Malacca, and the Cocos (Keeling) Islands, it became part of the Straits Settlements in 1900. In 1958 it was ceded to Australia, 850 miles to the southeast. It is one of the world's great seabird habitats, with eight species breeding there; some, such as the Abbot's booby, the Christmas Island frigate bird, and the golden bosun bird, breed nowhere else. It is best known, however, for the breeding of red crabs (*Gecarcoidea natalis*), hundreds of millions of which come ashore every year to mate and give birth to hundreds of millions more. A recent survey (December 1997) showed that approximately 40 percent of the island's coral had died, and the seabirds had almost completely deserted the island—perhaps due to the 1997–1998 El Niño, which was probably the largest and most influential on record. The island of Kiritimati in the Gilberts (now the Republic of Kiribati) was formerly known as Christmas Island, too.

See also Cocos (Keeling) Islands, El Niño, Kiribati

chromatophore A pigment cell in the skin of an animal or plant. (In plants, the color cells are also known as chloroplasts, because they contain chlorophyll.) In animals, they are most likely to contain melanin, a dark brown or black pigment, but they can also be red, blue, yellow, or any combination of colors. In animals that can change color voluntarily, particularly fishes and cephalopods, the chromatophores are connected to the nervous system and can be expanded or contracted at will. Many fishes can change color, often to match the background or to adopt a "breeding plumage," but cephalopods are the most striking color-change artists. Cuttlefishes, octopuses, and squid can flash different colors and patterns instantaneously, creating multi-colored stripes, spots, or blotches; some can make "washes" of color run across their bodies by activating the chromatophores in sequence. It is believed that some of the color change in cephalopods is used for communication; they "speak" in a language of color. This is especially interesting because cephalopods are color-blind. **See also cuttlefish, octopus, squid**

Chukchi Sea Part of the Arctic Ocean north of the Bering Sea between Siberia and Alaska. It covers some 225,000 square miles and has an average depth of 253 feet. Wrangel Island is in the western quadrant and the Beaufort Sea is to the east. It is navigable for ships only between the months of July and October, but seals, walruses, and whales are comfortable in the ice-filled waters all year round. California gray whales (*Eschrichtius robustus*) spend the summer here, feeding on bottom-dwelling crustaceans before making their annual migration to the lagoons of Baja California.

See also Beaufort Sea, Wrangel Island

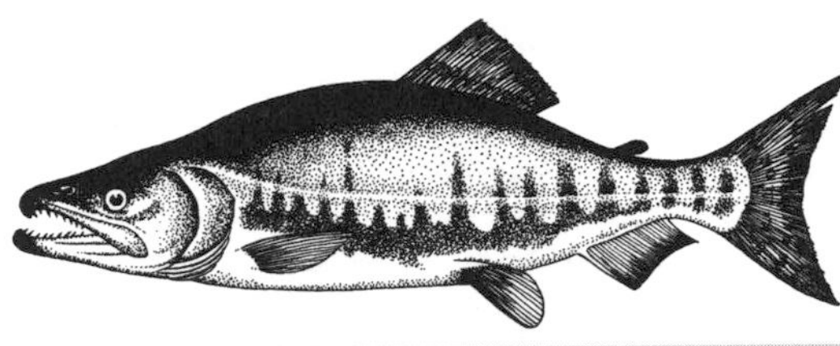

chum salmon

chum salmon (*Oncorhynchus keta*) Endemic to the Pacific and Arctic Oceans, the Bering Sea, and the Sea of Japan, and in North America from northwest Alaska to central California, chum salmon can weigh as much as 30 pounds. (The record is a 32-pounder caught in Alaska.) Breeding adults develop a jagged, blood-red marking that runs along the flanks; this is more pronounced in males. They are sometimes known as dog salmon, but it is not clear if this is a reference to their pronounced teeth or the fact that they used to be fed to sled dogs. The flesh is creamy white and not as popular as the pink flesh of other salmon species.

See also Atlantic salmon, chinook salmon, coho salmon, Pacific salmon, pink salmon, sockeye salmon

cigar shark (*Squaliolus laticaudus*) Adults of this species are mature at a length of 8 inches, making it the smallest of all sharks. (The largest fish in the world is also a shark, the 50-foot-long whale shark.) The derivation of its common name is obvious: it is the shape (and size) of a large cigar. It is the only shark with a spine on the first dorsal but not on the second. Its head is proportionally large, and although it has light organs on its undersides, bioluminescence has not been observed. It is found throughout the world in deep water, usually near landmasses, but it is not common anywhere. **See also cookie-cutter shark, pygmy shark, whale shark**

ciguatera A kind of poisoning caused by eating the cooked flesh of certain fishes. The symptoms, which may appear immediately or any time within thirty hours of eating the fish, are tingling about the lips, tongue, and throat; nausea; vomiting; abdominal cramps; and diarrhea. Dull muscle pains and aching increase until the victim is unable to walk. Teeth feel loose in their sockets, and temporary blindness may occur. About 7 percent of those infected die. There is no way of identifying ciguatera poisoning in a fish before it is prepared, and no way of predicting which species will transmit it. Fishes that are usually safe to eat can become ciguaterous in certain areas, and there are times when a fish known to be a potential carrier will be safe to eat. The fishes with the highest instance of ciguatera poisoning are the barracuda, various jacks, and some of the groupers. Although the etiology of ciguatera poisoning is unknown, it is now believed to have something to do with a toxic form of blue-green algae being passed up the food chain.

See also amberjack; barracuda, great

cirrate octopus Deepwater octopuses that have filaments (cirri; singular, cirrus) on the tentacles in addition to sucker disks. Because of their habitat, cirrate octopuses are poorly known, but those that have been seen have a pair of fins below the eyes and above the tentacles that are used for swimming. The fins are supported by a U- or V-shaped cartilaginous inner shell, and the body is usually gelatinous. The eyes of one species (*Cirrothauma murrayi*) are so degenerate that the species is functionally blind.

cirrate octopus

See also octopus, vampyroteuthis

Cladoselache A prehistoric, extinct shark that lived during the late Devonian period, about 350 million years ago. It reached a length of 6 feet and was built along the lines of the more recent sharks: torpedo-shaped, with a lunate tail and two dorsal fins, each of which had a strong spine before it. *Cladoselache*'s broad-based fins suggest it was not a fast swimmer, but it had powerful jaws and teeth and was obviously an ac-

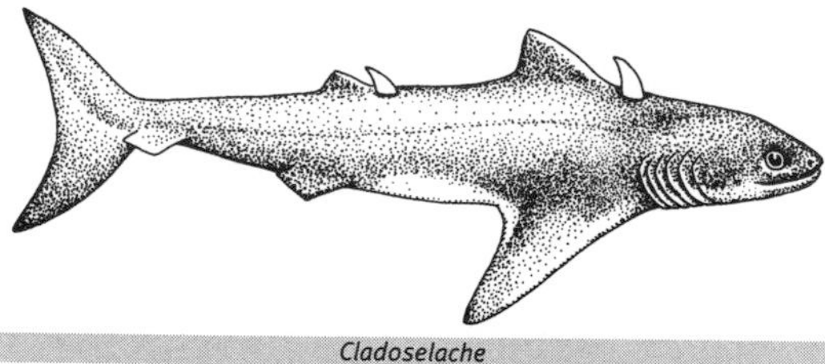

Cladoselache

tive predator. Most of the well-preserved fossils have been found in the Cleveland Shales in Ohio.

See also *Hybodus*, megalodon shark

clam Any one of twelve-thousand-odd bivalve mollusks that are flattened from side to side. The two valves are fastened to each other by an elastic, horny ligament at the umbo, an elevated knob that is the oldest part of the shell and enlarges as the animal grows. The smallest clam (a species of *Condylocardia*) is microscopic, but the largest (*Tridacna gigas*) can weigh more than 500 pounds. When undisturbed, clams lie partly buried in the sand, with the shells slightly agape, as moving water passes over the numerous cilia, providing food to the mouth as well as oxygen to the gills. Some clams, like the 1-foot-long Pacific geoduck (*Panopea generosa*) of the Pacific Northwest, have a long siphon through which they breathe while buried in the sand. Clams reproduce bisexually; females shed eggs into the water, where they are fertilized by males. They move (infrequently and slowly) by extending their foot, anchoring it in the sand, and then contracting the muscles. Many species of clam are edible, including the quahog (*Mercenaria mercenaria*); the geoduck and the soft-shell clam (*Mya arenaria*), also known as the longneck or steamer, is a common ingredient in soups and chowders. In Italian cuisine, the little clams known as *vongole* are frequently served with pasta. In Asia, different species are eaten, including the giant clam (*Tridacna gigas*), which is served (in small pieces) as sushi.

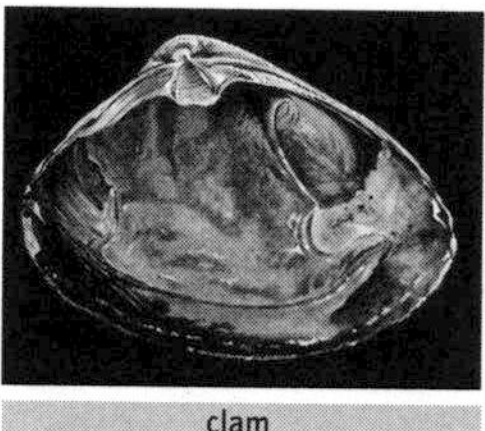

clam

See also geoduck, giant clam, mollusk, quahog

Clark, Eugenie (b. 1922) American zoologist, educator, and diver, born in New York City and educated at Hunter College (B.A.) and New York University (Ph.D.). She won a Fulbright fellowship in 1951 to study the fishes of the Red Sea; then she traveled to the South Pacific to study poisonous fishes. Upon her return in 1955, she was named director of the Cape Haze Marine Laboratory in Placida, Florida, which in 1960 would become the Mote Marine Laboratory in Sarasota. At the laboratory, Clark worked on groupers, blennies, and various other fishes, but she became fascinated with sharks and began to study them in earnest—to the point where the Mote Marine Lab became the world's most important facility for the study of sharks. Clark has traveled all over the world in pursuit of her subject, and has written hundreds of scientific papers and dozens of articles for magazines like *National Geographic, International Wildlife,* and *Natural History.* She is the author of the popular books *Lady with a Spear* (1953) and *The Lady and the Sharks* (1969). In 1967, she was appointed professor of zoology at the University of Maryland, and although she is now professor emerita, she continues to teach and travel. She has received seven fellowships, five scholarships, three honorary degrees, and six medals, including the Gold Medal of the Society of Woman Geographers.

See also Mote Marine Laboratory, shark

claspers Technically known as mixopterygia, claspers are modifications of the pelvic fins of male elasmobranchs (sharks, skates, and rays), which serve as intromittent organs for the transfer of sperm to the female. The size of the claspers is directly proportional to the age of the owner; they reach full size at sexual maturity. Normally trailing close to the animal's belly, the claspers are raised at right angles during copulation, and one is inserted into the genital opening of the female. In some species, the end of the clasper is modified into a hooklike structure that holds it in place during the mating process, which often includes the male grasping the female with his teeth. Semen is transferred into a groove in the clasper and pumped into the cloaca of the female by muscular contractions of a siphon sac. The sperm then enter the female's reproductive tract, where fertilization occurs.

See also shark

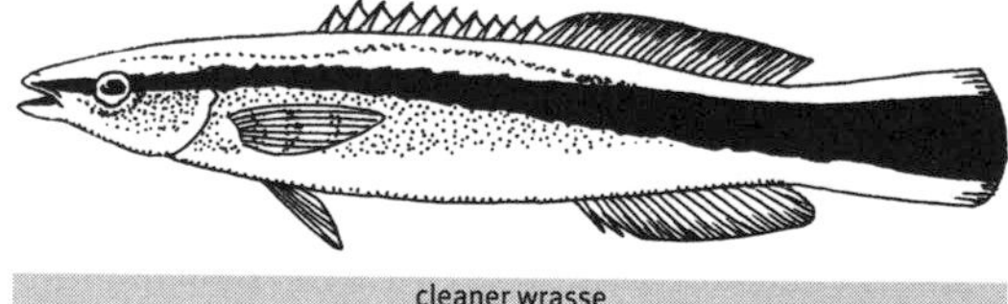

cleaner wrasse

cleaner fishes Fishes that feed on small parasitic organisms on other fishes by picking them off with tweezerlike teeth. The relationship of cleaner fishes and the fishes that they clean is symbiotic; both parties benefit from the association. Cleaners often establish "stations," where the larger fishes come to be cleaned. When the cleaners are at work, the "hosts" usually hover motionless with their fins fully displayed. Cleaners are usually small, brightly colored fishes that move fearlessly around the larger ones, often swimming into their mouths and out through the gill covers. Several species of wrasses of the genus *Labroides* are specialists in this procedure—they are known as cleaner wrasses—and make their living at it. Other fishes that regularly feed on ectoparasites are the Spanish hogfish (another wrasse), certain gobies, and some butterfly fishes and angels. Some fishes have developed color patterns that mimic those of the cleaners, but when they approach the "customers," instead of cleaning them, they take bites of the fish's flesh or fins. There are also several species of shrimp that pick parasites from fishes. **See also hogfish, mimicry, wrasse**

clipper ships A name originally used for the Baltimore clippers, fast schooners built in Virginia or Maryland that became famous during the War of 1812 as blockade runners. The name comes from the word "clip," meaning "speed," as in "to go at a good clip." Clippers were long and low, with a very sharply raked bow and an overhanging stern that reduced the area of hull in contact with the water, tall masts, and more sail than had ever been used before, all in the name of speed. The original clippers were built for the New York–China tea trade (hence the name "China clippers"), where the fastest delivery of the product was essential. The discovery of gold in California in 1848 and in Australia in 1850 made rapid delivery of goods to these places necessary, and in 1854 the *Champion of the Seas* made 464 miles in a day, a record never equaled by a sailing ship. Donald McKay of Boston designed and built some of the best of the clippers including *Flying Fish, Flying Cloud, Glory of the Seas, Sovereign of the Seas, Star of Empire,* and *Chariot of Fame.* Other famous clippers were the *Sea Witch, Houqua,* and *Rainbow.* The Tea Race of 1866 featured *Ariel, Taeping, Fiery Cross, Serica,* and *Taitsing,* and was decided—in favor of *Taeping* over *Ariel*—by a committee vote on which ship had actually docked first. The Suez Canal, which opened in 1869, marked the end of the clipper ship era, a gloriously romantic period that lasted less than thirty years. **See also *Cutty Sark*, McKay, Suez Canal**

Clipperton Island An uninhabited atoll in the eastern Pacific, about 800 miles southwest of Mexico. In the early eighteenth century, it was used as a base by (and is named for) John Clipperton, an English pirate, and was claimed by the French in 1855. Although Mexico claimed the island, King Victor Emmanuel II of Italy arbitrated the dispute and awarded it to France in 1930. (In 1914, the Mexicans landed a small garrison to protect the island, but they were evidently forgotten, and they nearly starved to death before being rescued by an passing American warship in 1917.) It is now an administrative part of French Polynesia, even though it is 1,100 miles from its nearest neighbor, Ua Huka in the Marquesas. **See also French Polynesia, Marquesas Islands**

clymene dolphin

clymene dolphin (*Stenella clymene*) A small dolphin, averaging about 6 feet in length and weighing approximately 165 pounds. For many years, the existence of a short-snouted spinner dolphin was suspected, but it was not until 1981 that the description of this species was published. The clymene dolphin closely resembles the longer-snouted spinner dolphin (*S. longirostris*), but it has a different color pattern, particularly on the head, and a much shorter snout. It has been identified primarily from the tropical waters of the Atlantic, but it may be more widely distributed. In Greek mythology, Clymene was a nymph, the daughter of Oceanus and Tethys. **See also spinner dolphin, spotter dolphin**

cnidarian Cnidarians (also known as coelenterates) are a group of about nine thousand invertebrates, including the jellyfishes, hydroids, corals, sea fans, and sea anemones. The name is derived from the Greek *cnidos,* which means "nettle," and refers to the stinging cells that are characteristic of these animals. Most corals have stinging cells, and most of the other cnidarians do. They are all "radiate" animals, built on a circular plan, either spherical, cuplike, or vaselike, with a ring of tentacles surrounding the mouth. Most pass through two stages—the polyp, or stalked phase, and the medusa, or jellyfish form—but there are some, like the corals, that remain polyps. Jellyfishes pass through a brief polyp stage, but the dominant, conspicuous stages of their lives are spent as medusae. Some cnidarians—like the siphonophores—are polymorphic, which means that they are composed of polyps of differing functions, colonially integrated into a single "animal." Corals also form colonies, not for the common good but rather because they are all descended from the growth and division of a single polyp. Coral polyps look like miniature sea anemones, with a ring of stinging tentacles around the mouth.
See also coral, hydroid, jellyfish, medusa, sea anemone, sea fan

cobia (*Rachycentron canadum*) A popular game fish throughout the world's tropical and temperate waters, the cobia—known as the prodigal son in South Africa and Australia—is an elongated, graceful fish that looks like a remora, to which it is not related. The largest

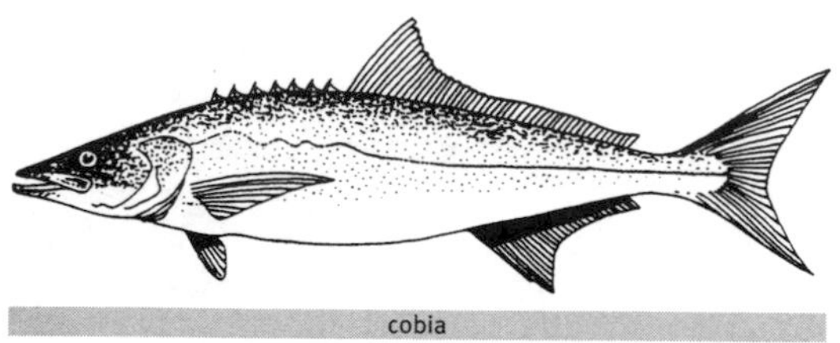
cobia

cobia on record, weighing 135 pounds, was caught in Shark Bay, Western Australia, in 1985. **See also remora**

cockle (order Eumelliabranchia) Common name for heart-shaped clams that are known for their ability to jump, which they accomplish by rapidly flexing the muscular foot beneath the shell. The shells are characterized by radiating ridges. Cockles are found around the world, from tidal waters to depths of 1,500 feet. In some species the muscle is too tough to be chewed, but in Britain, the European cockle (*Cerastoderma edule*) is a popular food item. They average about 2 inches across, but the great heart cockle (*Dinocardium robustum*) of the western North Atlantic can be 5 inches long. Because many of the shells are particularly graceful or colorful, cockles are very popular with shell collectors. **See also clam**

Cocos Island (1) On the Cocos Ridge—the same deep-sea ridge that thrusts up the Galápagos Islands some 400 miles farther south—there is a lonely speck of volcanic rock that takes its name from the Spanish word for coconut. It is 292 miles southwest of Costa Rica, to which it belongs. Only 10 square miles in area, Cocos is a steep-sided island, covered in dense vegetation, with two indentations in its fortresslike aspect, Chatham and Wafer Bays. The climate is cloudy and rainy, with more than 275 inches (22 feet) of rainfall per year. No one seems to have claimed credit for the island's discovery, but it appears on a map published in 1541. Because it was so isolated, it was a popular way station for pirates. From 1683 to 1685, with William Dampier and Lionel Wafer aboard *Bachelor's Delight,* Edward Davis used Cocos as a base for his privateering expeditions along the Pacific coast from Baja California to Guayaquil. It is said that he returned to Cocos to bury his treasure, and although it has never been found, it is not for want of effort. In 1984, German-born August Gissler, having heard tales of the buried treasure, convinced the Costa Rican government to allow him and others to settle on the island. They lived there for several years, but found nothing. Other treasure hunters continued to voyage to this remote island to search for gold, including British race-car driver Malcolm Campbell, who mounted an expedition in 1926, and Franklin D. Roosevelt, then president of the United States, who visited Cocos in 1935 and again in 1937 and 1940. To put an end to the digging and dynamiting, the Costa Rican government declared Cocos a national park in 1982, putting it off-limits to treasure hunters. Its submarine caves, clear waters, and diverse marine life have made it a treasure for divers, however, and people from all over the world come to swim with its hammerhead sharks, whale sharks, mantas, marlins, sailfish, sea turtles, and hundreds of other creatures. The boat trip from Puntarenas, Costa Rica, takes thirty-six hours, and it is a "live-aboard" expedition, for there are no accommodations on the island.

Cocos Island (2) Off the southern tip of Guam in the Marianas Islands is little Cocos Island, only 100 acres in size. On June 2, 1690, the Spanish treasure galleon *Nuestra Señora del Pilar de Zaragosa y Santiago,* en route from Acapulco to Manila, hit the reef at Cocos and sank in 30 to 85 feet of water. In 1991, a group of salvage divers called the Pilar Project began exploring the wreck and, so far, have recovered musket balls, lead hull sheeting, silver coins, pottery, and cannonballs. The Manila galleons, which carried treasure from China and Japan to Mexico (and thence across the land to Veracruz and then to Spain), often stopped in the Marianas for water. (A similar wreck occurred on September 30, 1638, when *Nuestra Señora de la Concepción,* the largest of all these treasure ships, sank off Saipan, 200 miles north of Guam.) **See also Guam, Manila galleon**

Cocos (Keeling) Islands Two separate island groups in the Indian Ocean, 800 miles southwest of Singapore. The Cocos Islands consist of twenty-six islets; Keeling is a single island, 15 miles to the north. They were discovered in 1609 by Captain William Keeling of the British East India Company, but not settled until 1826, when English adventurer Alexander Hare arrived with his Malay harem and slaves. The following year, the Scotsman John Clunies-Ross came with a boatload of Malays, and the British annexed the islands. In 1886, Queen Victoria granted the islands in perpetuity to the Clunies-Ross family, with the proviso that the Crown could use the islands for public purposes. In 1903, the islands were included in the Straits Settlements (along with Singapore, Penang, Malacca, and little Christmas Island), but when Singapore became independent and Penang and Malacca joined the Malay Union, the Cocos (Keeling) Islands (and Christmas Island) were placed under Australian administration. In 1977, Australia purchased the Clunies-Ross family's interests, except for the family estate on Home Island. The current population—mostly Malays and descendants of the Clunies-Ross family—numbers around 700 people, most of whom are engaged in the production of copra.

codfish (*Gadus morhua*) The cod is a fish of modest size (the largest known weighed 211 pounds), and omnivo-

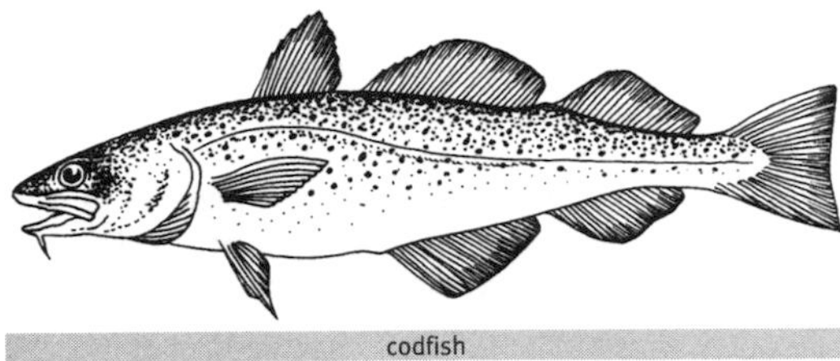

codfish

rous habits, found in the temperate waters from Greenland to North Carolina, including Hudson Strait; and from Novaya Zemlya to the northern reaches of the Bay of Biscay, including the North Sea and Iceland. Cod was sold fresh, salted, or dried (*baccalao*) and for more than three hundred years was one of the staples of the diet of Americans, eastern Canadians, and western Europeans. It was also the source of the vitamin-rich cod-liver oil. The cod was responsible for some of the earliest explorations of the coasts of New England and maritime Canada by people like John Cabot in 1497, and Basque and French fishermen shortly thereafter. The cod fishery was one of the primary industries of New England and Newfoundland until its complete collapse in the 1990s. There is no more cod fishery in New England and Canadian waters, because the greedy and short-sighted fishermen caught all the codfish.

See also Cabot (John), cod-liver oil, cod wars, Pacific cod

cod-liver oil Although the medicinal qualities of cod-liver oil were not fully realized until the twentieth century—it is especially rich in vitamins A and D—its virtues were known for hundreds of years to Lapps, Norwegians, and Icelanders. In eighteenth-century Europe, it was prescribed by physicians for the prevention and treatment of gout, rheumatism, and even tuberculosis. In later years—much to the discomfort and distaste of innumerable children—it became known as an all-purpose tonic and vitamin supplement.

cod wars From 1919 to 1951, Iceland's fishery limit was fixed at 3 miles, and any nation's boats could fish to within 3 miles of her shores. To protect what it considered its stocks of cod, Iceland increased the limit to 4 miles in 1952; to 12 miles in 1958; and in 1972, when most other countries maintained a 12-mile limit, Iceland unilaterally increased the limit around her shores to 50 miles. Ignoring the restrictions, British and West German trawlers persisted in fishing close to shore. Iceland's coast guard vessels had been equipped with trawl-wire cutters, which were employed to render useless the fishing gear of the offending boats. To prevent this interference, the British fishing boats tried to ram the coast guard cutters, and Britain also sent Royal Navy frigates to Icelandic waters to protect their boats, but no shots were fired. In October 1973, the British and Icelandic prime ministers met in London to resolve this dispute, and the British promised not to send their larger ships into Icelandic waters and not to fish in certain areas. But when Iceland adopted a 200-mile "exclusive economic zone" (EEZ), British fishing boats again intentionally violated Icelandic space, and the "cod war" began anew. British frigates accompanied fishing boats into Icelandic waters, and the Icelanders deployed coast guard ships, planes, and helicopters. Unable to resolve their differences by diplomatic means, Iceland broke off relations with Britain and threatened to close the American NATO base at Keflavík. In June 1976, the conflict was peacefully resolved, and again, Britain agreed to reduce her presence in Icelandic waters. **See also codfish, trawling**

coelacanth (*Latimeria chalumnae*) In 1938, fishermen from the South African city of East London hauled in a

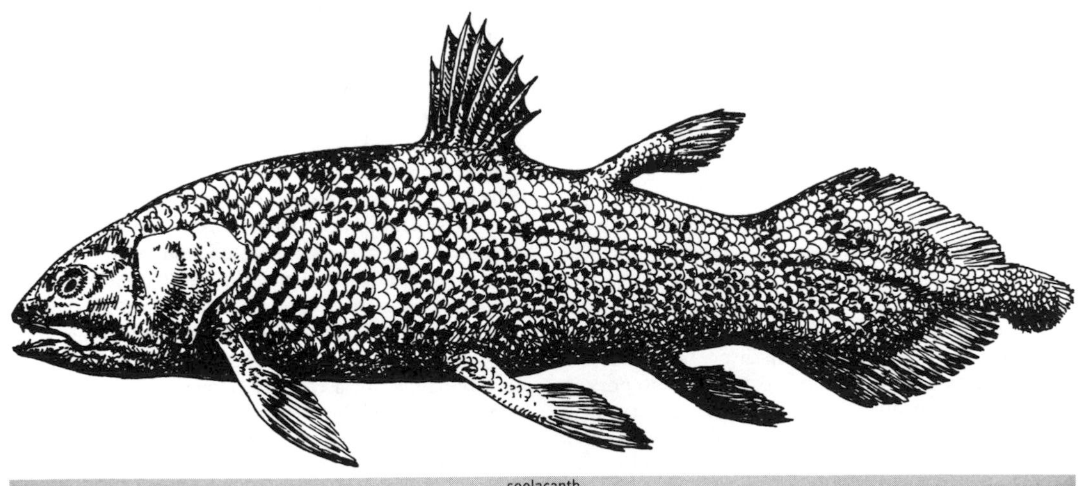

coelacanth

5-foot-long fish that was steely blue in color, with large bony scales and fins that appeared to be on leglike stalks. It was examined by Marjorie Courtenay-Latimer, who could not identify it, so she contacted J. L. B. Smith, a professor of chemistry at Rhodes University at Grahamstown and an amateur ichthyologist. Smith correctly verified it as a relative of a lobe-finned fish known as *Macropoma*, extinct for about 70 million years. He named it *Latimeria chalumnae*, after Courtenay-Latimer and the Chalumna River, near which it was found. No other specimen of the coelacanth ("*see*-la-canth") was seen until 1952, but since then, many more have been caught, usually in the vicinity of the Comoro Islands between Mozambique and the island of Madagascar. (The name "coelacanth"—which means "hollow spines" and refers to the first dorsal fin—was originally used in 1836 to describe the fossil species.) Local fishermen catch them unintentionally, usually while fishing for the oilfish (*Ruvettus pretiosus*), and what was once believed to be a stable population of about 650 animals is now thought to number no more than 300. This rare and zoologically significant creature is probably on the brink of extinction. Female coelacanths give birth to live young and did so long before the arrival of mammals. Coelacanths spend the day in lava caves and descend to around 2,000 feet to forage at night. In 1997, a coelacanth was spotted in a fish market on the Indonesian island of Celebes (now known as Sulawesi), and another specimen was hauled up in July 1998. Some 6,000 miles from East Africa, it appears that there is another, heretofore unexpected, population of coelacanths.

See also Comoro Islands, oilfish, Smith

coelenterate: See cnidarian

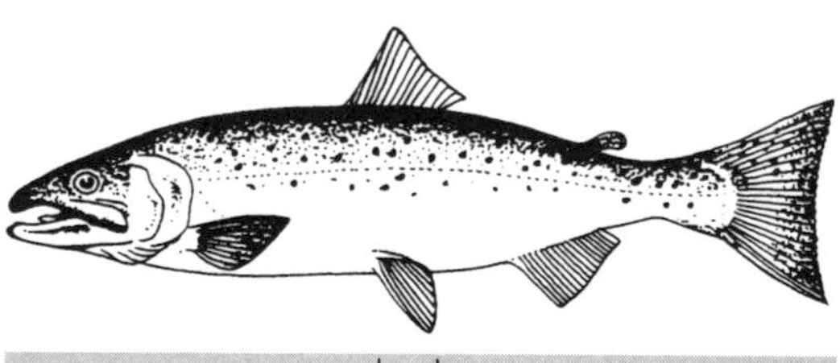

coho salmon

coho salmon (*Oncorhynchus kisutch*) Also known as silver salmon, the coho can reach a maximum weight of 33 pounds. Unlike many other Pacific species, male cohos do not undergo dramatic changes of color and shape as they prepare to spawn, but sometimes the males turn reddish while the females remain silvery blue. Of all the Pacific salmons, it is considered the most important sport fish. It is estimated that wild cohos are extinct in 55 percent of their former range in California, Oregon, and Washington. Originally found from Japan to Monterey Bay, the coho was introduced for sport fishermen into the Great Lakes and the waters of Maine, Maryland, Louisiana, Argentina, and Chile. In those regions where they do not spawn naturally, they do not breed and have to be restocked continuously.

See also Atlantic salmon, chinook salmon, chum salmon, Pacific salmon, pink salmon, sockeye salmon

Columbus, Christopher (1451–1506) Born in Genoa and christened Cristoforo Colombo, he was known as Cristobal Colón in Spanish and Christopher Columbus in English. After making several Mediterranean voyages and one to Iceland in 1477, he developed the idea of sailing west to reach the Indies. He moved to Lisbon and spent the next ten years trying to get support for his venture, finally obtaining the backing of Ferdinand and Isabella, joint sovereigns of Spain. In August 1492, with the ships *Niña, Pinta,* and *Santa María,* he left Spain and crossed the Atlantic, reaching land on October 12. Columbus believed he had arrived in Japan, but he was somewhere in the Bahamas, either on the island now known as San Salvador or on Watling's Island. He also encountered Cuba and another large island, which he named Hispaniola. In February 1493, he returned to Lisbon in triumph, with captive Arawak "Indians" to prove he had reached the Indies. The sovereigns of Spain appointed him "Admiral of the Ocean Sea," and by September he was ready to sail again, this time with seventeen vessels. On his second voyage he found and named Dominica, Guadeloupe, the Leeward Islands, the Virgin Islands, and Puerto Rico. At the settlement he called La Isabela on the eastern side of Hispaniola, he put his brother Diego in charge and spent the next five months cruising around Cuba, believing he had discovered a promontory of China. The admiral started the Indian slave trade by shipping five hundred captives home in 1495, and he returned home the same year. His third voyage commenced in 1498, and this time, after landing on Trinidad, he made what is considered the first European landfall in America when he landed on the Paria Peninsula of Venezuela. (Newfoundland, where Vinland was located and where John Cabot probably landed, is an island.) Turmoil at La Isabela caused the Spanish sovereigns to send Francisco de Bobadilla, who arrested the brothers Columbus and brought them home in chains in October 1500. After six weeks in prison, Columbus was released and outfitted for a fourth voyage, this time to explore the Gulf of Mexico in search of a passage to the Indian Ocean. In the flagship *La Capitana*, he explored the Caribbean, and on June 25, 1503, his fleet ran aground in Jamaica, where the expedition remained for a year. He returned to Spain in November 1504, and died less than two years later.

comb jelly (phylum Ctenophora) Radially symmetrical planktonic organisms that have differentiated tissues

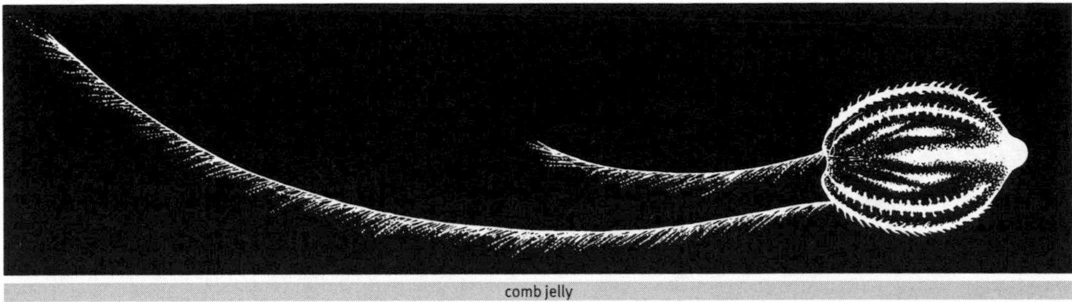

comb jelly

but no organ systems. There are about a hundred species of comb jellies (also known as ctenophores), most of them only 2 inches long, but some trail adhesive tentacles that may be 3 feet in length. Their tentacles, unlike those of the true jellyfishes (Cnidaria), have no stinging cells; they trap their prey, which consists of small organisms, with their sticky, feathery tentacles and bring it to the mouth. They have eight rows of *ctenes* (Greek for "combs"), which are used for locomotion, radially arranged on the spherical body surface. The transparent, jellylike body, usually globular or compressed, is divided into two hemispheres, with a mouth at the bottom and a balancing organ (statocyst) at the top. Underwater, comb jellies are bioluminescent, their combs flashing with bursts of multicolored light. Comb jellies are so fragile that they were hardly known at all until the advent of underwater viewing by scuba divers and submersibles. The most common species are the *Pleurobrachia,* otherwise known as sea gooseberries. Venus's girdle (*Cestum veneris*) is a flattened, ribbonlike ctenophore that can reach 1 meter in length. **See also jellyfish**

Commander Islands Known in Russian as Komandorskiye Ostrova, this group of four barren islands was named for Commander Vitus Bering, who died there after his ship *St. Peter* was wrecked in 1741. A survivor of this voyage was Georg Wilhelm Steller, the naturalist who first described the sea otter, the sea lion, the eagle, and the sea cow that bear his name. By 1769, only twenty-seven years after its discovery, Steller's sea cow had been completely eliminated, killed off for food and leather by Russian sealers who were stopping off at these islands. It probably reached a length of 30 feet and may have weighed as much as 4 tons. Its skin was thick, dark, and wrinkled, and it fed on seaweed that it crushed between the bony plates in its otherwise toothless jaws. Bering Island, the largest of the Commanders at 55 by 25 miles, has the largest community, Nikolskoye. Then comes Medny, about 35 by 5; and the two islets of Toporkov and Ary Rock. They are at the western end of the Aleutian chain, about 110 miles from the Kamchatka Peninsula, and because they belong to Russia, they are across the International Date Line from Attu, the westernmost of the Aleutians.

See also Bering, Steller, Steller's sea cow

Commerson's dolphin

Commerson's dolphin (*Cephalorhynchus commersoni*) A strongly marked, short-snouted dolphin with a characteristically rounded dorsal fin. It is found only in the Southern Ocean, particularly off the coasts of southern South America. These small (5-foot-long) dolphins have occasionally been captured for exhibition in Europe and North America. Only the males of the species have the saw-toothed serrations on the leading edge of the left pectoral fin; these unique structures may serve to enhance tactile stimulation during social contact, and are a gender-specific adaptation.

common dolphin (*Delphinus delphis*) In the vernacular name of this species can be found much of the confusion attendant upon the "porpoise-dolphin problem." Here is an animal known as the common dolphin (not to be confused with the dolphin fish, *Coryphaena*), but it is not the one that most people think of when they hear the word "dolphin." (That is the bottlenose, a huskier, grayish animal with a much shorter snout.) *Delphinus* is a slenderer, longer-snouted creature than *Tursiops,* and where the bottlenose is silver-gray in color, *Delphinus* (the name comes from the Greek for "womb," identifying these fish-shaped creatures as mammals) is among the most brightly colored of all dolphins, with a multicolored "hourglass" pattern and a complex arrangement of stripes around the eyes and mouth. With as many as 250 teeth in both jaws, common dolphins have more teeth than any other mam-

common dolphin

mal. This species is extremely gregarious, and schools of more than 300,000 have been observed. For many years, there was held to be only a single species of *Delphinus*, but recently, two other species have been recognized, the Cape dolphin (*D. capensis*) and the Arabian common dolphin (*D. tropicalis*).

See also bottlenose dolphin, tuna-porpoise problem

common octopus (*Octopus vulgaris*) Widely distributed throughout the shallow temperate and tropical seas of the world, *O. vulgaris* is the best known and most intensively studied of all cephalopods. They are usually mottled brown in color, but they can change color easily from almost white to dark red. They live in caves or crevices, and feed on crabs and lobsters, which they carry back to their den to consume. After the female lays her eggs, she guards and cleans them until they hatch, and then she dies. In captivity, common octopuses have displayed remarkable capabilities to escape from tanks and boxes, since they are able to squeeze their boneless bodies through paper-thin spaces. They are easily trainable, and it has been said that they have the intelligence and learning abilities of a house cat. In many areas of the world, particularly the Mediterranean, octopuses are considered a delicacy. Despite their reputation, most octopuses are harmless to people.

See also blue-ringed octopus, octopus

common tern (*Sterna hirundo*) Somewhat larger than the Arctic tern, but quite similar in coloration, with a black cap, and a red bill and feet. The common tern breeds around the world in the temperate zones of the Northern Hemisphere, from Scandinavia to the Canaries and throughout the Mediterranean in the eastern North Atlantic, and from Labrador to the Gulf of Mexico, including Bermuda and the West Indies, in the western. They also breed throughout the temperate re-

common octopus

gions of Eurasia. After breeding, they winter in such places as India and Japan, and Southern Hemisphere locations such as South Africa and Australia. They are common throughout the littoral areas of the temperate zones; this is the species of tern most likely to be seen around harbors and beaches. The flight of the tern is always a pleasure to watch, because they are so graceful and competent on the wing. They feed by diving from the air and snatching fish near the surface, and are therefore used by fishermen to locate schools of baitfish.

See also Arctic tern, tern

Comoro Islands An island republic at the northern end of the Mozambique Channel, consisting of the three main islands—Grand Comoro (on which the capital of Moroni is located), Anjouan, and Mohéli—and numerous islets and coral reefs. In 1841, the French annexed the islands; in 1958 they were granted autonomy within the French Union. In 1974, a referendum was held to decide on independence, and all but the island of Mayotte voted for it. (Mayotte, which is part of the Comoran archipelago, remained a French dependency.) Approximately .5 million people live in what is now known as the Federal Islamic Republic of the Comoros, subsisting primarily on agriculture and fishing. In 1938, fishermen from the South African city of East London hauled in a 5-foot-long fish that was steely blue in color, with large bony scales and fins that appeared to be on leglike stalks. It was identified as a living coelacanth ("*see*-la-canth"), a lobe-finned fish believed to have been extinct for about 70 million years. Since 1952, approximately two hundred specimens have been caught, almost all of them in the waters of the Comoro Islands. (In 1998, another population of coelacanths was discovered in Indonesian waters, off the island of Sulawesi.)

See also coelacanth, Madagascar

compass An instrument for determining direction. The mariner's compass consists of a magnetic needle freely suspended so that it aligns itself with the north/south magnetic field of the earth. The Chinese, having discovered the magnetic properties of lodestone (a naturally occurring iron ore) made the first compasses around the fifth century A.D., and by the eleventh century, they had developed the water compass, which had a fish-shaped magnetized needle floating in a basin of seawater. In Europe, tradition has it that the first compasses were developed in the Italian port of Amalfi around the end of the twelfth century. These early compasses consisted of a magnetized needle set on a chip of wood and floated in a bowl of water, but soon the chip and the water were eliminated, the needle was mounted on a brass pin, and the compass points were marked on the rim of a bowl. In a further development, the needle was affixed to a circular card on which the compass points with their subdivisions were marked, and the whole card would rotate, with the bowl becoming only a protective housing. The compass was mounted in a chest or binnacle at a convenient height, and set in the fore-and-aft line of the ship, usually with a lamp beside it at night. The introduction of iron ships necessitated a great deal of study to overcome the problem of the ship's magnetism. The gyroscopic compass consists of a rapidly spinning, electrically driven rotor, suspended in such a way that its axis automatically points along the geographical meridian. The gyrocompass is unaffected by magnetic influences, and came into wide use in warships and airplanes during World War II.

See also magnetic pole

conch (family Strombidae) A marine snail of the family Strombidae, which consists of some forty species, ranging in size from the 1-inch-long Indo-Pacific lipped strombus (*Strombus labiosus*) to the queen conch (*S. gigas*), which is much used for decoration because of the bright pink coloration of the aperture, which may be 12 inches across. Queen conchs (pronounced "conks") are harvested in the Caribbean and the muscular foot is eaten in stews and soups. Other well-known species are the West Indian fighting conch, the spider conch, and the rooster conch. Natives of Key West, the Bahamas, and some other West Indian islands are also known as conchs.

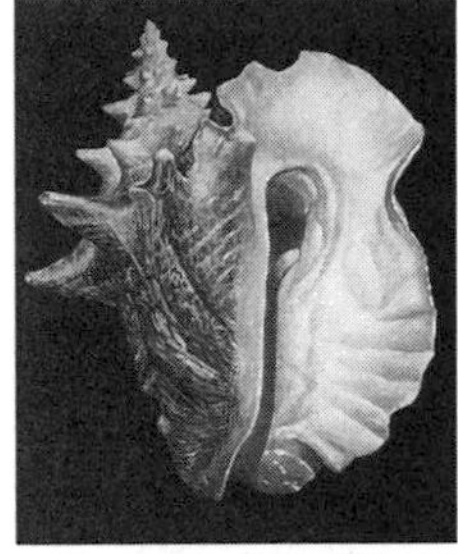
queen conch

cone snail (family Conidae) Mollusks that make tapered cylinders for shells, cone snails are among the most beautiful—and the most deadly—of all the gastropods. They are carnivorous and highly venomous and can shoot a tiny, harpoonlike dart from the proboscis into the victim, which may be a fish that comes too close or a human that handles the living snail carelessly, unaware of its proclivities. Some of the cones of the western South Pacific are particularly dangerous, though there are also hundreds of harmless species. For years, one of the shells most sought after by collectors was *Conus gloriamaris*, the "glory of the sea." Until its natural habitat was discovered off the east coast of New Guinea, the shell was so rare that it brought thousands of dollars at auctions.

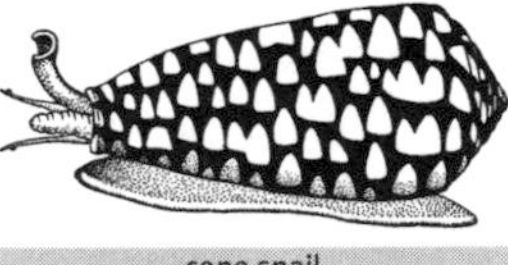
cone snail

See also gastropoda, mollusk, shell collecting

Coney Island Now a neighborhood in Brooklyn, Coney Island was originally known to the Dutch settlers as

Konijn Eiland (Rabbit Island), but the tidal creek that separated it from the mainland has been filled in, making it a peninsula. After the Civil War, railroad access made it a popular seaside resort—its 5-mile-long shoreline on the Atlantic Ocean faces south—and around the turn of the century, three amusement parks were opened: Steeplechase Park, Luna Park, and Dreamland. They proved to be enormously popular with New York City residents, and Coney Island was one of the city's most famous attractions, drawing as many as 1 million visitors in the summer. During the mid-twentieth century, the parks' popularity declined, and the food stands, carousels, roller coasters, and parachute jump were closed down. The New York Aquarium, located near the boardwalk and reachable by subway, is now the premier attraction of Coney Island, but its beaches are still popular with Brooklynites. **See also Manhattan**

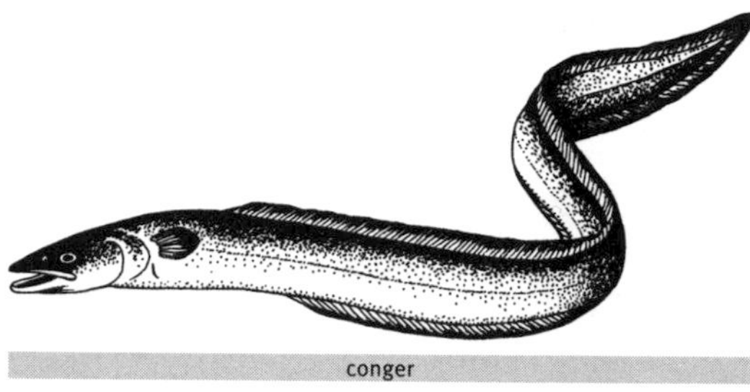

conger

conger (family Congridae) There are more than a hundred species of marine eels in the family Congridae, all of which are characterized by their scaleless skin, large heads, sharp teeth, and prominent pectoral fins. The European conger (*Conger conger*) is found from the Mediterranean to South Africa. It reaches a maximum length of 9 feet and can weigh 150 pounds. In Europe it is considered a game fish and a desirable food fish. The American conger (*C. oceanicus*) is a fierce game fish. Eels are true fishes and are in no way related to snakes. **See also eel, garden eel, moray, snipe eel**

Conrad, Joseph (1857–1924) Born in Poland and originally named Józef Teodor Konrad Korzeniowski, Conrad moved to Marseilles and went to sea on several French ships. In order to become a naturalized British citizen, he signed aboard the British ship *Mavis,* which took him to England in 1878. In 1879 he joined the crew of the full-rigged ship *Duke of Sutherland* for a voyage from London to Sydney and back. He then participated in a number of voyages to Southeast Asia, particularly Malaya, which deeply influenced his writing. He was second mate on the *Palestine,* bound for Bangkok in 1883, when her cargo of coal caught fire and she was abandoned off Java Head—a subject he would treat in his story "Youth," published in 1898. He sailed to Bombay aboard the *Narcissus;* later he would use the ship and the adventure in his *Nigger of the "Narcissus,"* which was written in English, like all his other novels and stories. In 1886 he received his first mate's certificate and his naturalization papers. He took ill on a voyage from Amsterdam to Semarang and, upon his recovery, spent several years sailing on the rivers of Malaya. In 1889, Conrad went to the Congo, where he was promised command of a river steamer under the Belgian colonial authority. The commission never materialized, but he used his African experiences to write *Heart of Darkness.* He settled in London in 1894, and under the name Joseph Conrad, he published his first novel, *Almayer's Folly,* in 1895. *An Outcast of the Islands* was issued in 1896 and *Lord Jim* in 1900. *Lord Jim, Heart of Darkness* (1904), *Nostromo* (1904), *The Secret Agent* (1907), *Under Western Eyes* (1911), and *Chance* (1913) are considered his greatest works. Although he did not speak English until he was twenty, Joseph Conrad's experiences served him so well that he is generally considered that language's finest writer about the sea, and one of its greatest prose stylists.

Constitution Generally regarded as the most famous ship in U.S. naval history, the USS *Constitution* was authorized by Congress in 1796 and launched in Boston the following year. In the war with Tripoli (1801–1805) she served as the flagship of Commodore Edward Preble, but it was in the War of 1812 that she achieved her greatest glory. In her victory over the British frigate *Guerrière* on August 19, 1812, under Captain Isaac Hull, so little British shot penetrated her sides that she was given the nickname "Old Ironsides." Three months later, under William Bainbridge, she destroyed *Java,* and in February 1815, she captured *Cyane* and *Levant.* After the war, she served in various capacities, but in 1828, she was declared unseaworthy and scheduled to be broken up. Oliver Wendell Holmes Sr. wrote "Old Ironsides," a nationally circulated poem that aroused such public sentiment that money was appropriated for her rebuilding. She was completely restored, and she was recommissioned in 1931 for a tour of ninety American cities. Now classified as a national historic landmark and based at the Charlestown Navy Yard in Boston, "Old Ironsides" celebrated her two hundredth birthday in July 1997 by heading out into Boston Harbor under full sail. **See also Tripolitan War**

continental drift The large-scale horizontal displacement of continents, relative to one another and to ocean basins, during one or more episodes of geological time. As early as 1620, Francis Bacon recognized the remarkable fit of the east coasts of North and South America and the west coasts of Europe and Africa, but it was not until 1912 that the German astronomer and geologist Alfred Wegener (1880–1930) proposed that the continents had once been joined—he called this single continent "Pangaea"—and had "drifted" to their present positions. Wegener was unable to identify the force that could move continents, but by the 1950s, geologists recognized that seafloor spreading was the mechanism that could (and did) cause the continents to slide over the lithospheric plates of the earth. The term "plate tectonics" was applied to this crustal movement, and analyzes the motions due to seafloor spreading and transform faulting. To date, however, the mechanism that governs the inner workings of the earth—and therefore causes the seafloor to spread, the mountains to rise, and the continents gradually to change their positions—has not been identified.

See also plate tectonics, seafloor spreading, Wegner

continental shelf and slope The margins of the continents below sea level that form the boundary of the ocean basins, consisting of a broad, shallow strip of land that extends from the coast to depths of about 300 to 600 feet, and sloping downward, generally at an angle of about 1°. At a much steeper angle, usually around 4° or 5° but occasionally much steeper, the continental slope leads from the shelf to the deep ocean floor. **See also submarine canyons, turbidity currents**

Cook, Frederick (1865–1940) American physician who began his career as an explorer when he served as surgeon on Robert E. Peary's expedition to northern Greenland in 1891. In 1897–1899 he performed similar duties for a Belgian expedition to the Antarctic. In 1906, he claimed to have scaled the then-unclimbed Mount McKinley in Alaska, but this claim turned out to be fraudulent. In 1907, he set out on an expedition to the Arctic. He returned on September 1, 1909, claiming to have reached the North Pole in April 1908. The accounts of his route did not gibe with those of the Eskimos who accompanied him, and he was denounced by an angry Peary, who claimed that he had reached the Pole in April 1909. After a sensational public controversy, the U.S. Congress voted Peary the winner, and Cook faded into oblivion. (Recent examination of the evidence has indicated that neither man made it and that both of them falsified their records.) Cook was later implicated in a fraudulent oil stock promotion scheme, and served time in prison from 1925 until he was paroled in 1929. He was given a pardon by President Franklin Roosevelt in 1940.

See also North Pole, Peary

Cook, James (1728–1779) Probably the greatest explorer in history. After seven years' experience with North Sea trading ships, Cook saw action in the Seven Years' War (1756–1763), spent five years charting the coast of Newfoundland, and then joined the Royal Navy, where he was given command of the *Endeavour* and ordered to convey gentlemen of the Royal Society to Tahiti to observe the transit of the planet Venus across the sun. Cook was then charged with locating Terra Incognita Australis, the southern continent that was then unknown but believed to exist. On his first voyage in 1769, he charted New Zealand (which had been discovered by Abel Tasman in 1642), and by sailing through the strait that now bears his name, he proved that South Island was not the northerly promontory of a southern continent. He sailed west, sighted the southeast coast of Australia, and anchored in Botany Bay (so named by Joseph Banks). After *Endeavour* was holed by a coral head in the Great Barrier Reef, Cook ingeniously refloated her and headed for Batavia in the East Indies, where his crew, previously healthy because of his precautions, was debilitated by dysentery and malaria. *Endeavour* returned to England on July 12, 1771, but Cook was unsatisfied with the results of the voyage; although he had reached 40° south, he knew not what lay farther south. Then followed his second voyage of discovery (1772–1775), this time in the *Resolution,* on which he became the first navigator to cross the Antarctic Circle, but still he sighted no southern continent. He then sailed to New Zealand, and after encountering a wall of ice at 71°10′ south (and proving to his satisfaction that there was no habitable continent there), he went to Tahiti and charted Easter Island, the Marquesas, the Friendly Islands (Tonga), the New Hebrides, New Caledonia, and Norfolk Island before heading for home. In November 1774, he rounded Cape Horn into the South Atlantic, where he claimed South Georgia and the South Sandwich Islands for England. *Resolution* arrived at Portsmouth in July 1775. Within a year he was off again, to round Cape Town and cross the Pacific to Tahiti, whence he was to head for the west coast of North America to search for the Northwest Passage. He discovered the Hawaiian Islands (which he named for the Earl of Sandwich) in January 1778, and sailed northeast until he reached Nootka Sound and the Aleutians. At the Bering Strait they met with an impenetrable wall of ice, so they came about and headed back to Hawaii for refitting. They anchored in Kealakekua Bay on the big island, where the returning Cook was welcomed like a god. After great celebrations, they left, but were forced to return two days later when the *Resolution* sprung her foremast. On January 14, 1779, when Cook went ashore again, he was surrounded and killed by natives who had become hostile.

See also Banks, Botany Bay, Great Barrier Reef, Hawaii, South Georgia, South Sandwich Islands

cookie-cutter shark (*Isistius brasiliensis*) Named for its habit of taking scooplike bites out of large prey, such as dolphins, billfishes, tuna, and even whales, the cookie-cutter is a small shark (maximum length 20 inches), with two very small dorsal fins placed far back on the body. Its teeth differ in the upper and lower jaws; the

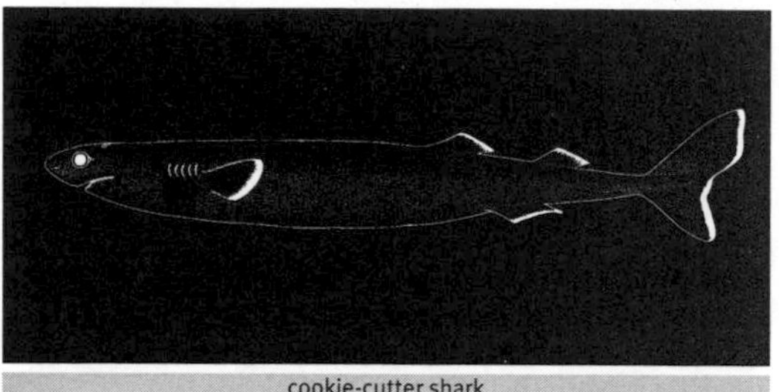
cookie-cutter shark

uppers are small and thornlike for getting a grip, while the lowers form a continuous sharp-edged band that removes the plug as the shark rotates on its long axis. (The lower teeth are shed and replaced as a unit, ensuring that the shark has a full cutting edge at all times.) Cookie-cutters are believed to school, and their undersides are bioluminescent, which may be useful for maintaining visual contact at depth. For years the pluglike bites taken from the rubber covering of submarines' sonar domes were a mystery until it was determined that cookie-cutters were responsible.

See also cigar shark, green dogfish

Cook Islands An island group in the southeast Pacific, about 2,100 miles northeast of New Zealand. The southern group comprises Rarotonga, Atiu, Takutea, Mauke, Mitiaro, Manu'ae, Mangaia, and Aitutaki; the northern Cooks are Manihiki, Penrhyn, Danger, Palmerston, Rakahanga, Suvarov, and Nassau. Captain Cook made three trips to the islands, in 1773, 1774, and 1777, and although he named them the Hervey Islands after Augustus John Hervey, the lord of the Admiralty, the islands were eventually named for Cook himself. Rarotonga was "discovered" in 1789, when the *Bounty* mutineers landed there, but they did not report their discovery. The islands were declared a British protectorate in 1888 and annexed to New Zealand in 1901. In 1965, they were granted self-government, with a prime minister, a cabinet, a legislative assembly, and a high commissioner appointed by New Zealand.

Cook Strait The passage between North and South Islands, New Zealand, about 14 miles wide at its narrowest point. In 1642, the Dutch navigator Abel Tasman entered the strait thinking it was a bay, and it was not until 1770 that Cook discovered its true nature. Both shores are lined with steep cliffs, and the passage from island to island is often difficult because of currents and winds.

copepod There are more than eight thousand known species of copepods, in 180 families and eight orders. These crustaceans are mostly minute components of the plankton, but some can get to be 1 foot long. They are characteristically ovoid (as shown by *Calanus finmarchicus*), with large antennae and a pair of projections that extend from the tip of the abdomen. Their legs are flattened like the blades of an oar—hence the name, which means "oar-footed"—but they swim by jerky movements of the antennae. The calanoid copepods (like *Calanus*) are among the most numerous animals in the world. Some are free-swimming, but others are parasitic, infesting other invertebrates, fish, and even whales. The largest of the copepods is the 13-inch-long *Pennella balaenopterae*, which is parasitic on fin whales. The Greenland shark (*Somniosus microcephalus*) is almost always found with a parasitic copepod on each eye.

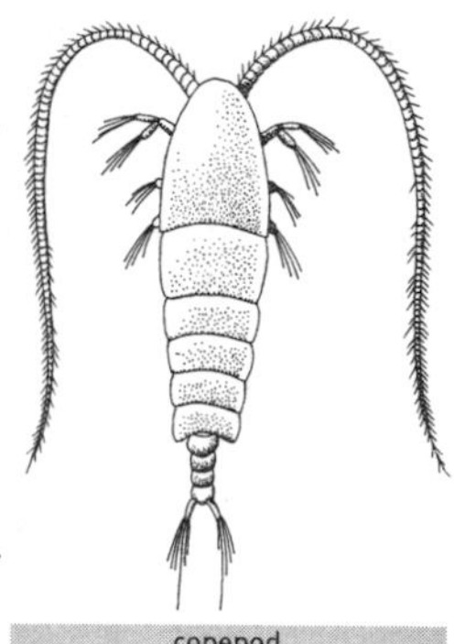
copepod

See also amphipod, fin whale, Greenland shark

coral Common name for a variety of small, sedentary marine organisms that are characterized by an external skeleton of a stonelike, horny, or leathery consistency. They are classified as cnidarians, with the jellyfishes, hydroids, sea fans, and sea anemones. There are true or stony corals (Madreporaria), black and thorny corals (Antipatharia), horny corals or gorgonians (Gorgonacea), and blue corals (Coenothecalia). The body of a coral animal consists of a polyp—a hollow, soft cylinder, attached at its base to some surface—with a mouth surrounded by tentacles. The tentacles are equipped with stinging cells (nematocysts) that paralyze the coral's prey. At night, they withdraw into the external skeleton, a cylindrical container known as a corallite or theca. Reproduction can be accomplished either by releasing eggs and sperm into the water or asexually by budding, where a fingerlike extension matures into a new polyp. Stony corals, the most familiar form, occur in various shapes, such as brain coral, mushroom coral, star coral, staghorn coral, and elkhorn coral, all named for their appearance. (The sprigs of whitened, branchlike material often found on beaches are the limestone casts of coral skeletons.) The success of corals as reef builders depends upon the coral's association with tiny, single-celled plants called zooxanthellae, which live in the polyp's tissues. Because they are plants, the zooxanthellae require sunlight for photosynthesis, the processing of the carbon dioxide produced by the polyp. Corals are not simply a cluster of

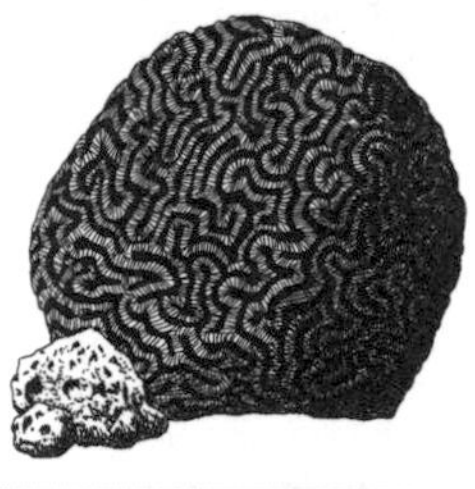
brain coral

animal colonies, but an association of animals and plants that is responsible for the creation of some of the most spectacular animal architecture in the world, such as the Great Barrier Reef. Although not a reef builder, the red coral of the Mediterranean (*Corallium rubrum*) is a popular material for jewelry, and the black coral of the Pacific (*Antipathes*), discovered in 1958, is even more exotic and even more expensive. Various creatures, including many fishes, worms, crabs, snails, and sea urchins, feed on coral polyps, but probably the most notorious coral predator is the crown-of-thorns starfish.

See also cnidarian, coral reef, crown-of-thorns starfish, gorgonians, Great Barrier Reef, soft corals

coralline algae A group of red algae or seaweeds (Rhodophyta) that are brittle and hard because they contain chalky deposits of calcium carbonate, giving them a texture somewhat like eggshells. The calcium carbonate is absorbed from the seawater to provide a strong supporting skeleton for the algae.

See also red algae

coral reef Limestone formations produced by living organisms, found in shallow, tropical marine waters. The reef is composed of the skeletons of dead corals that are bound together by their own limestone. Over thousands of years, coral growth, death, and cementing build a structure on which the living corals continue to live, but contributors to the reefs also include plants such as coralline algae, protozoans, mollusks, and tube-building worms. Reef-building corals do not grow at depths of more than 100 feet, or where the water temperature falls below 72°F. Most reefs occur within a band 30° north or south of the equator. Atolls are reefs surrounding a lagoon that has no central island in it, with passages through it to the sea. The way in which atolls form is not clearly understood. Because they are usually found in warm, tropical waters, coral reefs attract hosts of divers, who swim among the corals observing and photographing the colorful reef fishes and invertebrates. Many reef systems are in danger from coastal development, because pollutants have killed the living corals or because men have dynamited the reef to collect fishes, shells, or corals. The Great Barrier Reef off the east coast of Australia is the largest of all coral reefs; the second largest is off the Caribbean country of Belize. **See also atoll, coral, Great Barrier Reef**

Coral Sea Bordered on the west by Australia and New Guinea and defined in the east by Vanuatu (once known as the New Hebrides) and the Solomon Islands, the Coral Sea covers an area of about 2 million square miles. It is named for the formations of the Great Barrier Reef, which extends for 1,200 miles down Australia's northeast coast. Other islands in the Coral Sea are Lord Howe (which has the southernmost coral reef in the world), Norfolk, and the Louisiade Archipelago.

See also Great Barrier Reef, Lord Howe Island, Norfolk Island, Vanuatu

Coral Sea, Battle of the By April 1942, the Japanese had taken the Bismarck Islands, and from Rabaul, they were moving toward the Solomons and Port Moresby. After that, their next objective would be Australia. On May 5, when the Japanese landed at Tulagi in the Solomons, carrier-based planes from a U.S. task force commanded by Rear Admiral Frank Fletcher sank one Japanese destroyer and several other vessels. The Japanese fleet en route to Port Moresby included the giant carriers *Shokaku* and *Zuikaku,* the smaller carrier *Shoho,* nine cruisers, and fifteen destroyers. The American fleet included the carriers *Lexington* and *Yorktown,* five cruisers, and eleven destroyers. As the opposing fleets headed for Port Moresby, their planes took to the air. American dive-bombers sank a Japanese destroyer and three minesweepers; the Japanese sank an American destroyer. On May 7, when the Americans sank the carrier *Shoho,* the Japanese retreated. American reconnaissance planes located the adversary, and dive-bombers from the American carriers severely damaged the *Shokaku.* The *Lexington* was bombarded so heavily that she was abandoned and subsequent internal explosions sent her to the bottom. (The *Yorktown* survived but was sunk at the Battle of Midway in June.) While neither side can be said to have conclusively won the Battle of the Coral Sea, the Japanese lost so many planes that they had to turn back for Rabaul and could not continue to Port Moresby. It was the first naval battle in history where the opposing ships were never in gunshot range—in fact, the two fleets never got closer than 50 miles, and never actually saw each other. **See also Midway, Battle of**

Corfu Island in the Ionian Sea in northwestern Greece, the second largest (after Cephalonia) of the Ionian Islands. Although the fertile island produces olive oil, figs, wine, and lemons, it is now best known as an international resort. Corfu (also known as Kérkira) was probably the island of Scheria, the home of the Phoenicians in the *Odyssey.* It is believed to have been first settled around 730 B.C. by Corinthian colonists, and Thucydides records that the first naval battle was fought between Corfu and Corinth in 665 B.C. The island has been controlled by the Romans (229 B.C.), the Byzantines (A.D. 336), the Normans of Sicily (1080–1130), and the Venetians (1386–1797). Corfu was incorporated into the Napoleonic Empire in 1807, but became a British protectorate after the emperor's defeat at Waterloo. Along with the other Ionian Islands, Corfu was ceded to Greece in 1864. It was occupied by French forces during World War I and by Italians and Germans in World War II. Considered by many to be the most beautiful of all the Greek islands, Corfu is one of the most popular tourist destinations in the Mediterranean. **See also Cephalonia, Ionian Sea**

Coriolis effect Named for the French engineer and mathematician Gustave-Gaspard Coriolis (1792–1843), the Coriolis effect is the apparent deflection of an object moving within a system of rotating coordinates. On the earth, which rotates in an eastward direction, an object that moves along a north-south path (a longitudinal line) will undergo apparent deflection to the right in the Northern Hemisphere and to the left in the Southern. Because the velocity of the earth is essentially zero at the poles and attains its greatest velocity at the equator, a projectile fired northward from a cannon at the equator would land to the east of its due north path. This is because at the equator the projectile would be moving eastward faster than its target farther north. The Coriolis effect can be observed in the movement of winds and ocean currents. **See also current, tide**

cormorant (*Phalacrocorax* spp.) Large, long-necked birds that have four webbed toes pointing forward, putting them in the same order (Pelecaniformes) as the pelicans, tropic birds, frigate birds, gannets, and the (nonmarine) anhingas. There are about thirty species, distributed along temperate and tropical marine coasts around the world. In some places, they are known as shags. Plumage is generally dark with a metallic sheen, although some species have prominent white areas. They are fish eaters that dive from the surface to pursue their prey underwater. (This propensity has been exploited by certain humans, who put a ring around the neck of the bird so it cannot swallow the fish and send it down with a rope attached.) By the use of internal air sacs, cormorants can change their specific gravity, thereby adjusting their buoyancy, swimming at the surface like a duck or almost completely submerged with only the head showing. In flight they are recognizable by relatively short wings and their beeline flight pattern. Because their feathers are not waterproof, they have to dry them in the air; they are often seen perched with their wings spread. The spectacled cormorant, which lived on the Aleutian and Commander Islands in the North Pacific, is extinct.

common cormorant

See also blue-eyed shag, double-crested cormorant, flightless cormorant, olivaceous cormorant, shag, spectacled cormorant

Cornwallis Island One of the Parry Islands in Canada's Arctic Archipelago, Cornwallis was discovered by William Edward Parry in 1819, as he sailed west in search of the fabled Northwest Passage. The island was named for Sir William Cornwallis (1744–1819), a British admiral who was the younger brother of Charles Cornwallis, who surrendered to George Washington at Yorktown in 1781. The island is about 70 miles long and ranges from 30 to 60 miles wide; it consists of 2,700 square miles of snow and ice. It is the site of the Polaris mine, one of the most isolated mines in the world, which produces zinc and lead. Along with the other Parry Islands (Devon, Bathurst, and Melville) and most of the Northwest Territories, Cornwallis Island was turned over to the Eskimos in 1999, when it became part of the territory of Nunavut.

See also Bathurst Island (1) and (2), Devon Island, Melville Island (1) and (2), Nunavut

Coronel, Battle of Fought on November 1, 1914, between British and German cruiser forces off the coast of Chile. When German admiral Graf Maximilian von Spee was forced to leave Chinese waters because of British and German superiority, he took the heavy cruisers *Scharnhorst* and *Gneisenau,* along with the light cruisers *Nürnberg* and *Leipzig,* across the Pacific to South America, where he was joined by the light cruiser *Dresden.* He was met by British admiral Sir Christopher Cradock with the cruisers *Good Hope, Monmouth, Glasgow,* and *Otranto,* and the old battleship *Canopus.* Superior German gunnery and tactics resulted in the sinking of the *Good Hope* and the *Monmouth* with all hands; the other British ships escaped. When von Spee's squadron rounded Cape Horn and entered the South Atlantic, they were soundly defeated at the Battle of the Falkland Islands five weeks later. **See also Falkland Islands, Battle of the; Fisher; Spee**

Corregidor A 2-square-mile island at the mouth of Manila Bay that has been used for centuries as a protective fortress for the city of Manila. The Spanish fortified it in the eighteenth century, when it served as a registration point for incoming ships. (*Corregidor* means "magistrate" or "mayor" in Spanish.) After the Spanish-American War of 1898, it became a U.S. military installation, and an elaborate system of tunnels was created. The island was so heavily fortified that it became known as the Rock of the East, and it kept the Japanese from taking Manila for five months in early 1942, during which time it was subjected to such heavy bombardment that nearly the entire island was reduced to rubble. After the Japanese took Bataan on April 9, 1942, and General Douglas MacArthur had withdrawn, Corregidor became the last outpost of organized resistance in the Philippines. Under Lieutenant General Jonathan M. Wainwright, Filipino and American forces held off the Japanese for twenty-seven days, but they were forced to surrender on May 6, 1942. The island was retaken by American paratroopers and landing parties in March 1945. **See also Bataan, Philippines**

Corsica French island in the Mediterranean, 105 miles southeast of France and 7 miles north of Sardinia across the Strait of Bonifacio. It covers 3,367 square miles and has a population of about 250,000. After belonging to the Romans, the Vandals, the Byzantines, and the Lombards, the island was granted by the Franks to the Papacy in the eighth century. Genoa and Pisa battled over the island, but in 1768 it was ceded to France. (Napoleon, born in 1769 in the island's capital of Ajaccio, was therefore a French citizen.) In 1794, Corsica aligned itself with Britain, but the French (under Napoleon) recovered the island in 1796. During World War II, the island was occupied by German and Italian troops, but in 1943 the locals revolted and, with the help of a Free French task force, drove out the Axis soldiers. Corsica is largely agricultural, producing fruit, cork, wheat, wine, and cheese on its mountainous terrain. Although occasionally disturbed by political unrest, the island is now becoming a popular tourist attraction. **See also Sardinia**

Cortés, Hernán (1485–1547) Spanish conquistador who overthrew the Aztec empire and took Mexico for Spain. He went to the island of Hispaniola as a farmer in 1504, but six years later joined the expedition of Diego Velázquez to conquer Cuba. In 1518, after serving as mayor (*alcalde*) of Santiago de Cuba, he assembled his own force of eleven ships, six hundred men, and sixteen horses and sailed for mainland Mexico. He established his headquarters at Veracruz and burned his ships to prove that he was committed to conquest. Affiliating himself with the Tlaxcala nation, which was at war with the Aztecs, Cortés entered Tenochtitlán, the stronghold of Montezuma, who believed him to be the incarnation of the Aztec god Quetzalcoatl. He took Montezuma prisoner, but in 1520, he had to leave the city to fight off the efforts of another Spanish conquistador, Pánfilo Narváez, who intended to deprive him of his command. During his absence from Tenochtitlán, the Aztecs retook the city, but when Cortés returned, he found that they had rejected the authority of Montezuma, leaving him without a hostage, so he had to fight his way back out of the city. Reunited with the Tlaxcala, Cortés laid siege to the city, and by 1521 Tenochtitlán surrendered and the Aztec empire collapsed, leaving Spain in control of a huge territory from the Caribbean to the Pacific. In 1524, Cortés explored the jungles of Honduras, but while he was gone, his property in Mexico was seized by those he had left in charge. He returned to Spain in 1528, where King Charles V named him Marqués del Valle de Oaxaca, and in 1530, he returned to "New Spain," and settled at Cuernavaca, 30 miles south of Tenochtitlán (which became Mexico City). Eventually the greatest of the conquistadores returned to Spain. **See also Pizarro**

countershading In biology, the principle whereby the ventral (under) surface, particularly of fishes, is lighter than the dorsal (upper) surface, so that light from above (almost always the source of light in the ocean) forms a shadow on the underside, and equalizes the contrast between the fish and its environment. (Many birds and mammals are also lighter on the underside for the same reason.) Many fishes, e.g., lantern fishes and anglerfishes, take this solution one step further and are equipped with tiny light organs on their undersides, to equalize their silhouette in deeper, darker waters. **See also bioluminescence, mimicry, protective coloration**

Cousteau, Jacques-Yves (1910–1997) One of the pioneers of recreational diving and underwater photography, Cousteau popularized the underwater adventure as no person before him ever had. In 1943, with Émile Gagnan, a French engineer, he invented the Aqua-Lung, which made scuba diving possible for the masses. He also pioneered underwater cinematography and won the Academy Award for *World without Sun* in 1964. With support from the National Geographic Society, he published many articles in its magazine and wrote popular books such as *The Silent World* (1953) and *The Living Sea* (1963). By refitting a British minesweeper, he created *Calypso,* which became the platform for many of his adventures and films, on subjects that ranged from sharks and whales to underwater archaeology and Atlantis. His enormously popular television series was called *The Undersea World of Jacques Cousteau.* These extremely successful films spawned any number of tie-in books and videos, making Cousteau's name synonymous with oceanic education. The Cousteau Society, with headquarters in Chesapeake, Virginia, and in New York City, was founded in 1973 to promote the careful and intelligent management of ocean resources. **See also scuba**

cowfish (family Ostraciontidae) Also known as trunkfishes or boxfishes, these reef dwellers are encased in a

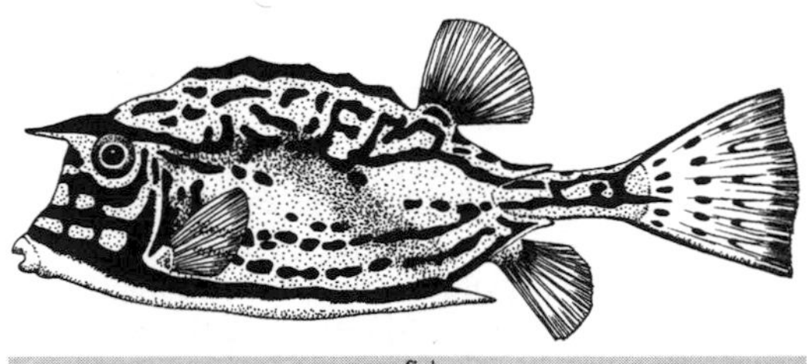

cowfish

hard, bony carapace that makes them almost impervious to attack by predators. In addition to this defense, many of the thirty-odd species can also discharge a poison (ostracitoxin) into the surrounding water that can kill other fishes that get too close. They are found throughout the world's warm, shallow waters, particularly around coral reefs. The largest species can reach 20 inches in length, but most are smaller.

See also filefish, puffer, triggerfish

cownose ray (*Rhinoptera bonasus*) An American stingray, found from New England to Brazil, but not in the Caribbean, the cownose gets its name from its blunt, upturned snout, which has an indentation that looks like the upper lip of a bovine. Brownish above and white on the underside, these rays with their 6-foot wingspan are renowned jumpers, and sometimes take to the air in groups. As with many stingrays, the spine at the base of the tail is poisonous. There are similar species in South America, the Mediterranean, and the western Pacific. **See also stingray**

cowrie (family Cypraeidae) The shells of cowries are easily identified by their glossy, polished surface and ovoid shape. Shy and seminocturnal, the animals spend their time under ledges, occasionally appearing at night. Cowries are often brilliantly colored (although the mantle is often more spectacularly colored than the shell) and mottled with spots, blotches, stripes, and intricate combinations of these. The snow-white species (*Ovula ovum*) is known as the egg cowrie, and the famed golden cowrie (*Cypraea aurantium*) is so rare that in Fiji only royalty were allowed to possess one. Several cowries are still considered the ne plus ultra of shell collecting, since only a few specimens exist in museums. (One of the world's rarest shells is *C. broderipii*, which is known from only three specimens, all of which came from the stomach contents of a deepwater Natal fish.) The unprepossessing money cowrie (*C. moneta*) is so common that quantities were used as currency and ornament in many African and Oceanic societies.

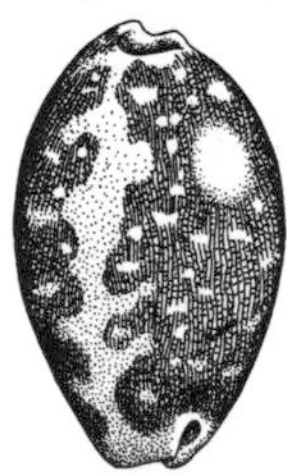

cowrie

See also gastropoda, shell collecting

Cozumel Mexico's largest island, located 12 miles off the eastern tip of the Yucatán Peninsula. Cozumel, which means "island of the swallows" in the Mayan language, is 31 miles long and averages 9 miles wide, with a total area of 189 square miles. The Mayans occupied the island before it was "discovered" by the Spanish explorer Juan de Grijalva in 1518. The conquistador Hernán Cortés visited the island during the next year, but he found no gold and moved on. There is little freshwater to be had on the island, so it was not colonized except by passing pirates. (Nowadays, water is brought in by tanker from the mainland.) During World War II, the United States built an air base for planes hunting U-boats in the North Atlantic (now reconstructed as Cozumel International Airport). It was not until the 1960s that the clear waters, coral reefs, and abundant marine life were made known to divers. It is now one of the premier dive spots in the Western Hemisphere, with a number of luxury hotels, many dive shops, and every amenity for tourists.

crab (order Brachyura) Crustaceans with a flattened shell (often toothed), and a very reduced abdomen that is held permanently flexed beneath the cephalothorax. (*Brachyura* means "short-tailed.") Like all decapods, crabs have five pairs of legs: four pairs of well-developed walking legs, and one pair adapted as large claws. Crabs are found throughout the world, at great depths in the oceans and in the shallowest tidepools. (Some species, like the robber crabs, hardly ever enter the water at all.) There is a species of blind crab (*Bythograea thermydron*) that lives in the vicinity of hydrothermal vents. After copulation and fertilization, the crab's eggs are held under the broad abdomen of the female. When they hatch, they go through several planktonic stages before assuming their final shape and settling on the bottom. There are sand crabs, rock crabs, lady crabs, walking crabs, mud crabs, fiddler crabs, toad crabs, sheep crabs, kelp crabs, ghost crabs, oyster crabs, box crabs, sponge crabs, porcelain crabs, spider crabs, and pea crabs. The Tasmanian giant crab, *Pseudocarcinus gigas,* can weigh as much as 30 pounds, and another (the Japanese spider crab, *Macrocheira kaempferi*) can measure 13 feet across its outstretched claws. Because they have no shell, hermit crabs are not classified as true crabs.

See also blue crab, fiddler crab, hermit crab, pea crab, robber crab, spider crab

crabeater seal (*Lobodon carcinophagus*) The name *carcinophagus* means "crabeater," but these Antarctic seals feed primarily on small euphausiid shrimp, the same krill that is eaten by baleen whales. Crabeaters have specially designed teeth that enable them to strain these small organisms from the water. Although the ac-

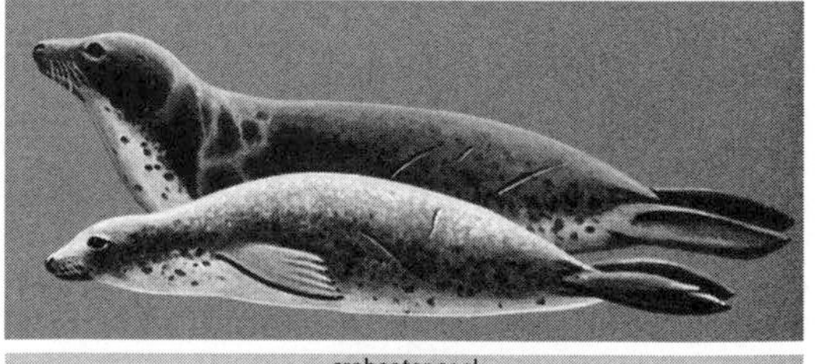

crabeater seal

tual estimates differ, there seems to be little doubt that the crabeater is the most numerous of all pinnipeds, and perhaps the most abundant large mammal in the world (after humans). Various studies have produced figures that range from 15 million to 75 million crabeaters. **See also euphausiids, krill**

crested penguins Members of the genus *Eudyptes,* characterized by a yellow eyebrow stripe that develops into drooping yellow plumes.

See erect-crested penguin, fiordland penguin, macaroni penguin, rockhopper penguin, royal penguin

Crete A Mediterranean island 152 miles long and 35 miles wide at its widest point, Crete forms the southern boundary of the Aegean Sea. It is an administrative division of Greece, with a population of some 500,000. The capital and largest city is Herakleion. First settled eight thousand years ago, Crete was home to the Minoan civilization, one of the most remarkable in ancient history. The Minoans built extensive palaces (Knossos, Phaistos, Zákros), filled with spectacular wall paintings, jewelry, and pottery that suggested a culture far more advanced than that of any other contemporaneous people. After an earthquake on Crete and the colossal eruption of the volcano on Thera (75 miles to the north) around 1500 B.C., the Minoan civilization mysteriously vanished. (Contrary to some theories, it seems this disappearance had nothing to do with the "lost continent" of Atlantis.) The Romans inhabited Crete by 67 B.C., and Byzantine Christianity was introduced around A.D. 400. Crete was sold to the Venetians in 1204, and after a bitter and protracted battle in 1669, the Ottoman Turks conquered the island. The Turks were ejected in 1898, and in 1913 Crete was officially allied with Greece. During World War II, advancing Axis armies forced the government of Greece, along with British and Greek troops, to retreat from mainland Greece to Crete. Launching the first airborne invasion in history, the Germans took the island as Allied forces were evacuated from the southern coast. Crete was occupied by Axis forces until the last German troops surrendered in May 1945. **See also Santoríni**

crinoids (class Crinoidea) Commonly known as feather stars and sea lilies, crinoids are echinoderms, like sand dollars, starfish, brittle stars, and sea urchins. Some varieties have stalks and others do not. The stalked ones (the sea lilies) are usually found at significant depths. The unstalked types, known as comatulids, are multiarmed starfishlike creatures that swim in currents or station themselves atop sponges, gorgonians, or corals and snare passing microorganisms with sticky mucous excreted from glands on the arms. (In fact, they have only five arms, but like those of the basket star, they branch and rebranch until a tangle of as many as two hundred arms is formed.) Unlike starfishes, feather stars position themselves with their mouth parts upward and their feathery arms waving in the current. The smallest feather stars are only 1 inch across, but *Heliometra glacialis,* from the Sea of Okhotsk, can be 3 feet in diameter. Until 1873, sea lilies were known only from fossils, but the *Challenger* expedition brought some up from the depths. Most of the eighty known species of sea lilies have a flowerlike body supported by a slender stalk, which is fastened to the bottom (or to undersea cables) with rootlike structures.

See also *Challenger,* echinoderms, starfish

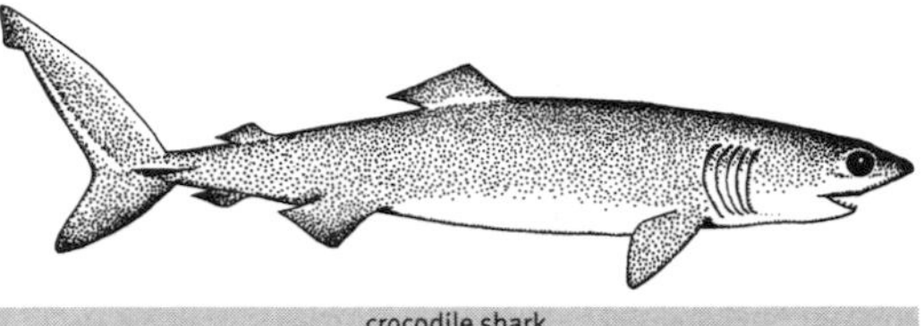

crocodile shark

crocodile shark (*Pseudocarcharias kamoharai*) The smallest of the lamnids (mackerel sharks), the crocodile shark was earlier classified with the odontaspids (sand tigers), but has recently been assigned its own family. Little is known about this 4-foot-long shark, but its large eyes suggest a deepwater habitat. It has been caught, usually accidentally, in tropical waters around the world.

See also mackerel shark, sand tiger shark

crown-of-thorns starfish (*Acanthaster planci*) A large, reddish, heavy-spined starfish that may be 24 inches across. Unlike the more traditional five-armed starfishes, *Acanthaster* has from twelve to nineteen arms. It feeds on coral polyps, and around 1963, it began to increase enormously on the Great Barrier Reef, decimating the coral formations. At first, shell collectors overharvesting the Pacific triton snail (*Charonia tritonis*), *Acanthaster*'s chief predator, were considered the reason for the increase, but when subsequent research suggested that these population explosions had evidently been taking place for eons, the shell collectors were exonerated. In addition to its nasty reputation as a destroyer of coral, the crown-of-thorns starfish has spines that are poisonous to the touch.

crown-of-thorns starfish

See also echinoderms, starfish

Crozet Islands Discovered in 1772 by Nicolas-Thomas Marion-Dufresne, this remote archipelago in the southern Indian Ocean is now (with St. Paul, Amsterdam, and the Kerguelen Islands) part of France's Terres Australes et Antarctiques Françaises. Named for Julien Crozet, second-in-command to Marion, the archipelago consists of several small islands, uninhabited by human beings but popular with penguins, elephant seals, and breeding albatrosses. Sealers arrived around 1800 and, as with the other subantarctic islands, began the systematic destruction of the elephant and fur seals. In 1873, the British oceanographic research vessel *Challenger* visited these lonely islands, along with Kerguelen, Marion, Macquarie, and Heard.

See also Amsterdam and Saint-Paul Islands, *Challenger,* Heard Island, Kerguelen Islands, Macquarie Island, Marion-Dufresne, Marion Island

Crozier, Francis R. M. (1796–1848) Irish-born British naval officer who entered the navy in 1810 and accompanied W. E. Parry on his three Arctic voyages between 1821 and 1825. After a voyage to the Davis Straits and Baffin Bay with James Clark Ross in 1836, he was named captain of the *Terror,* and accompanied Ross to the Antarctic in 1839. As captain of the *Terror* in 1845, he accompanied Franklin in the *Erebus* in the search for the Northwest Passage. In June, the ships were trapped in the ice off King William Island in the Canadian Arctic, and in June 1847, Franklin, who was fifty-one years old, died of natural causes. Under Crozier's leadership, the crew remained on the ice, hoping that the ships would break free, but after two years, when another twenty-three men had died (mostly of scurvy), Crozier decided to lead the survivors out. *Erebus* and *Terror* were deserted on April 25, 1848, and 105 men set out on foot for the Great Fish River to the south. Not one of them made it. Francis Crozier is remembered by a Cape Crozier on Marion Island in the Indian Ocean, and another one on King William Island, where *Erebus* and *Terror* were trapped.

See also Franklin, Parry, Ross (James Clark)

crustaceans Because there are no insects in the oceans, the arthropods (joint-legged animals) are well-represented by the crustaceans, which are distributed throughout the world's oceans, from the shoreline and the beach to the most profound depths. Probably sharing a common ancestor with the insects, crustaceans have segmented bodies and limbs; they breathe by means of gills or water intake through the skin; they have a chitinous exterior (sometimes but by no means always a shell); they have two pairs of antennae; like insects, they have an exoskeleton (body support is outside, not inside), but unlike insects, they never have wings. Many species are bioluminescent. There may be as many as forty thousand species in the order Crustacea, including the decapods (lobsters, crabs, shrimps, crayfish, prawns), euphausiids, copepods, amphipods, isopods, barnacles, and a vast array of less familiar forms. Amphipods, copepods, and isopods exist in vast profusion (in the upper layers of the sea, copepods may constitute 70 percent of the animal life), and serve as a major food source for most oceanic inhabitants. The euphausiids form the major proportion of the diet of the largest baleen whales, and some species of decapods, such as lobsters, crabs, and shrimp, serve as an important food source for *Homo sapiens.*

See also amphipod, barnacle, copepod, crab, euphausiids, isopod, lobster, shrimp

cryptozoology From the Latin *kryptos* for "hidden," this is the study of hidden—that is, unknown—animals. In recent times, cryptozoologists were delighted with the discoveries of the coelacanth and the megamouth shark, two sea creatures whose existence was completely unsuspected. They are still on the lookout for evidence to prove the existence of such things as sea serpents, 200-foot-long octopuses, 100-foot-long sharks, and of course, the Loch Ness monster.

See also coelacanth, Loch Ness monster, megamouth, sea serpent

Cuba The largest and westernmost of the West Indies, Cuba lies at the entrance to the Gulf of Mexico and forms the northwestern border of the Caribbean. It is separated from Haiti (the western half of the island of Hispaniola) by the Windward Passage, and it is about 90 miles south of Key West, Florida. The island is 745 miles long and 124 miles wide at its widest point; including the numerous smaller islands, it is about 42,800 square miles in area. At 850 square miles, the Isle of Pines is Cuba's largest outlying island. Cuba has been inhabited for at least three thousand years, first by settlers from South America, then by the Arawak Indians from the Antilles and the Bahamas. When Christopher Columbus arrived in 1492, the indigenous population was around 80,000, but most of the Arawaks died from diseases brought by the Europeans. By 1511, the Spanish had established a foothold on Cuba, and they used the island's numerous harbors as staging areas for their exploration of the Americas. The Spanish eliminated the local population and replaced it with slaves from Africa to work the sugar and tobacco plantations. (Slavery was abolished in Cuba in 1886.) The island remained in Spanish hands until 1895 when the poet-patriot José Martí led the fight that ended Spanish rule, thanks to U.S. intervention after the sinking of the battleship *Maine* in Havana harbor. In 1899, the treaty following the Spanish-American War made Cuba an independent republic. For the next six years, U.S. forces occupied the island during which yellow fever was surpressed there by Dr. Walter Reed. Sugar has always been the dominant crop in Cuba, and the fortunes of the country were tied to rising and falling prices. In 1933 Fulgencio Batista overthrew the provisional government of Gerardo Machado and remained in power as dictator until 1958, when Fidel Castro landed on the island and led a popular revolution that caused Batista to flee to the Dominican Republic on January 1, 1959. Castro aligned himself with the communism of the Soviet bloc and expropriated U.S. land holdings and banks, which led to the severing of diplomatic relations in 1961. An abortive invasion of the Bay of Pigs in April 1961 added to the friction between the two countries, and in October 1962, President John F. Kennedy demanded (and achieved) the removal of Soviet missiles from Cuban silos. Until the breakup of the Soviet Union in 1991, Cuba was philosophically, politically, and economically in the Soviet camp, but now that this country of 11 million people has to support itself, there is growing discontent with the Castro regime.

cunner (*Tautogolabrus adpserus*) A small wrasse found in the western North Atlantic, from Newfoundland to Chesapeake Bay. Also known as the sea perch or bergall, it schools around wharves and pilings and annoys fishermen by stealing their baits. It looks a bit like a small tautog, to which it is closely related.

See also tautog, wrasse

Curaçao The largest island in the Netherlands Antilles, located some 35 miles off the coast of Venezuela. (It is the "C" of the ABC Islands; the others are Aruba and Bonaire.) The capital is Willemstad; the population numbers about 170,000. The people claim descent from more than fifty different ethnic backgrounds, and the native language, known as Papiamentu, is based on Spanish, with a strong Dutch influence and contributions from Portuguese, English, French, various African languages, and Arawak Indian. Discovered in 1499 by Alonso de Ojeda (whose navigator was Amerigo Vespucci), the island was occupied by the Spanish until the Dutch took it over in 1634. The Dutch held it continuously except for the period during the Napoleonic Wars when France took control; since 1815, the island has been a Dutch possession. In 1918 the Royal Dutch Shell refinery was opened to process vast quantities of oil from Venezuela's Lake Maracaibo wells, and until it closed in 1989, it was one of the largest of its kind in the world, and the dominant element in Curaçao's economy. With the oil business shut down, the government is encouraging the development of tourism.

See also Aruba, Bonaire, Netherlands Antilles

curlew Any of a number of long-legged shorebirds, mostly brownish or grayish in color, with a long, downturned bill. All curlews breed on Northern Hemisphere tundras and migrate southward in winter, often crossing the equator. The largest of the genus *Numenius* is the long-billed curlew (*N. americanus*) of North America, with a body length of 25 inches. The bristle-thighed curlew (*N. tahitiensis*) migrates to Pacific islands from Hawaii to Tahiti, but its nesting grounds were unknown until 1948, when nests were discovered in the Canadian Arctic near the mouth of the Yukon River. The whimbrel (*N. phaeopus*), found throughout Europe, was known in America as the Hudsonian curlew, until it was reclassified as a North American version of the European species. The smallest of the curlews is (or was) the Eskimo curlew (*N. borealis*), which was extraordinarily plentiful in the United States until market gunners massacred them to the point where they are now believed to be extinct.

See also Eskimo curlew, shorebirds, whimbrel

curragh Also known as a coracle; a wood-framed, leather-covered small boat, used primarily by fishermen on the west coast of Ireland and the Aran Islands. The leather was treated with tallow or pitch to make it waterproof, and the boats were light enough to be carried on the backs of one or more men. St. Brendan, an Irish abbot of the sixth century, is supposed to have traveled in the Atlantic from Ireland to the Faeroes, Iceland, and perhaps even America, with seventeen monks. To duplicate this seemingly impossible feat, Tim Severin built a 36-foot-long curragh and, with a crew of four, sailed and rowed across the Atlantic. From May 18 to July 15, 1976, the *Brendan* sailed from Ireland to Iceland, by way of the Hebrides and the Faeroes. Then they regrouped, and the following year, from May 8 to June 26, sailed from Reykjavik to Newfoundland. Proving that it could be done, however, is not the same as saying that St. Brendan did it. **See also Brendan**

current The horizontal movement of the sea in a given direction. There are tidal currents, associated with the rise and fall of sea level, and nontidal currents, which are regular in their general flow and which are caused by the variable density of seawater, the Coriolis forces exerted by the rotating earth, and the frictional drag caused by steady winds that sets the water in motion. Atmospheric circulation over the ocean, where air flows are uninterrupted by landmasses, gives stability and permanence to such phenomena as trade winds, the most effective producer of surface drift movements. The Coriolis forces cause ocean currents to move clockwise in the Northern Hemisphere and counterclockwise in the Southern, creating cells known as gyres. The direction of a current is that point of the compass toward which it flows. In the north, the major currents are the Gulf Stream–North Atlantic–Norway Current and the Kuroshio–North Pacific Current; in the south the eastward-flowing circumpolar Antarctic Current carries a vast amount of water that produces the Peru and Benguela Currents, which draw their cold water from the Antarctic. The clockwise circulation of the Arctic Ocean is responsible for the movement of the ice pack and the southward distribution of icebergs from the Greenland Sea and Baffin Bay. In the tropics the great gyres flow as the Pacific North and South Equatorial Currents, but there is little exchange of water across the equator. In the Indian Ocean north of the equator, the dominant factor in the creation of currents is the monsoon winds, which causes an annual fluctuation of water movement from northeast to southwest. Pacific currents are less pronounced than those of the Atlantic and Indian Oceans; the North Equatorial Current becomes the Kuroshio as it passes Japan, and the California Current as it reaches the west coast of North America. Vertical circulation brings up deep waters and moves down surface waters (upwelling and downwelling); winds affect only the top 100 meters (330 feet), known as the Ekman Layer. Although many currents are steady and predictable, there are instances when major anomalies occur. When warm, humid circulation above the western Pacific is displaced eastward, the waters of the eastern Pacific are warmed, and the El Niño effect (also known as the Southern Oscillation) occurs, bringing drought to Australasia, storms to California, and death and devastation to the marine life along the west coast of South America.

See also Coriolis effect, El Niño, Gulf Stream, Kuroshio Current, tide

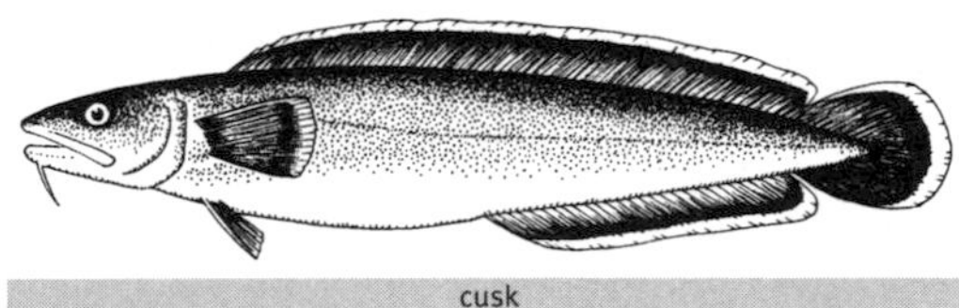
cusk

cusk (*Brosme brosme*) A relative of the cods, the cusk is easily identified by the barbel at the chin and the single dorsal fin that extends the length of the body. They vary in color, from reddish or greenish brown to yellowish gray, but the fins always have white edges. Cusks reach a maximum length of 40 inches and weight of 27 pounds. It is a deepwater fish from the western North Atlantic, ranging from Newfoundland to New Jersey. Like many other members of the cod family, cusks lay millions of eggs at a time, each with a yellowish oil globule that keeps it close to the surface. After the fry reach about 2 inches in length, they descend into the depths. The stock in the Gulf of Maine was fished intensively throughout the 1960s and 1970s, which so depleted it that the American fishery was suspended indefinitely. **See also codfish, haddock, hake, pollack**

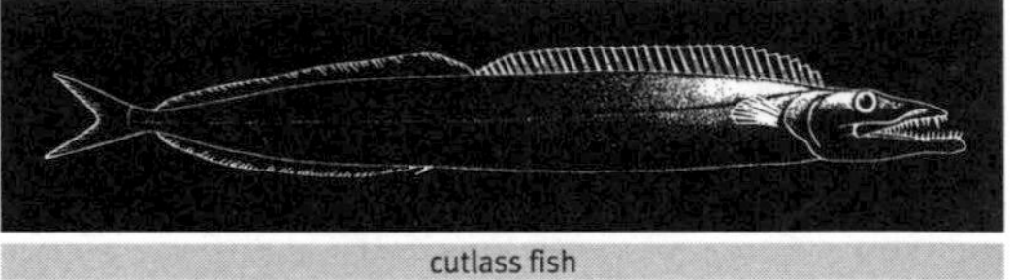
cutlass fish

cutlass fish (*Aphanopus carbo*) At a known length of 5 feet, with large eyes and a mouthful of teeth, the cutlass fish is a formidable predator at depths from 1,200 to 4,000 feet. Many rare and exotic deep-sea fishes have been collected from the stomach contents of *Aphanopus*. Shiny black in color, the cutlass fish is sometimes called the black scabbardfish, and in Madeira, Portuguese fishermen call it *espada preta*, or "black sword." Efforts are being made to develop a large-scale fishery for the *espada preta* in the Canary Islands and the Cape Verdes. **See also snake mackerel**

cuttlefish (family Sepiidae) Cephalopods with a shield-shaped, flattened body and fins that extend from the anterior to the posterior edge of the mantle, but are not connected at the posterior (tail) end. The body is strengthened by an internal shell (the pen or gladius) that lies just beneath the upper surface and is the common "cuttlebone" of bird cages. Cuttlefishes have eight arms and two retractile tentacles, used for the capture of prey. There may be as many as 180 species, ranging in size from a total length of 1¼ inches, to *Sepia latimanus* of the Indo-Pacific, which can reach a mantle length of

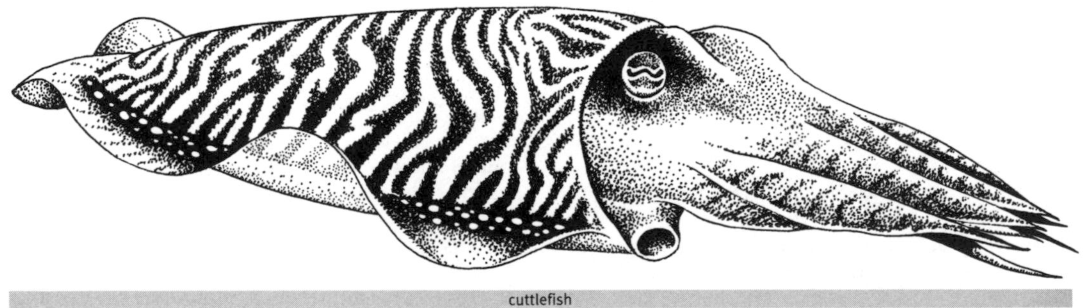
cuttlefish

2 feet and a total length of 5 feet. Cuttlefishes are good swimmers, but generally not as fast as squids, and they spend much more time resting on the bottom, often changing their color to match their surroundings. They are masters of color change, and probably have a greater range of color or patterns than any other animal on earth. In an instant, they can change from solid to stripes or from dark to light; they can also decorate themselves with moving waves of flickering color. The best-known species is the common cuttlefish (*S. officinalis*) of the eastern Atlantic, described twenty-two centuries ago by Aristotle and found from the English Channel to the Cape of Good Hope. From this species, people collected the ink sacs that provided the ink (and the color) known as sepia. Many species of the larger cuttlefishes are caught commercially.

See also cephalopod, octopus, squid

Cutty Sark Her name taken from a Robert Burns poem, *Cutty Sark* was one of the last of the British tea clippers, and the only surviving ship of her type. Commissioned by Captain Jock Willis as a challenge to the great *Thermopylae*, she was 212 feet long, with a beam of 36 feet. She was launched in 1869 and in 1870 took part in the annual tea race from China to London. *Cutty Sark* raced *Thermopylae* only once. In 1872 they were off South Africa, when *Cutty*, which had a 400-mile lead, lost her rudder in a gale. The opening of the Suez Canal in 1869, however, meant that steamships could use the much shorter canal route, and the fabled 'round-the-Horn races were finished. When the tea trade had been assigned to steamships, the sailing clippers had to look for other work, and *Cutty Sark* began regular voyages from Australia to England in the wool trade. She was sold to the Portuguese in 1895, but by 1922, she was back in England. In 1949, she was brought to a specially constructed dock at Greenwich, completely refitted and rerigged, and she was opened to the public in 1957. **See also clipper ships**

Cyclades An island group in the southern Aegean Sea, the name Cyclades comes from the Greek word *cyclus* for the roughly circular arrangement of the twenty-seven islands. Archaeological evidence shows that the Cyclades have been inhabited since 6000 B.C.; a Cycladic culture that flourished in the early Bronze Age around 3000 B.C. produced remarkably sophisticated statuary. The Minoans occupied Thera (modern Santoríni) a thousand years later, and the Myceneans arrived thereafter. After the Persian Wars, the islands became part of the Dorian League, and were later incorporated into the Athenian empire. By A.D. 1200, the islands were parceled out to various Venetian aristocrats, and occupied by the Ottoman Turks until Greek independence in the early nineteenth century. Long unknown except to their occupants, the Greek islands began attracting tourists in the 1970s, and nowadays, Delos, Mykonos, Ios, Paros, and Naxos are favored destinations for tourists from all over the world. With its spectacular caldera, splendid beaches, and Minoan relics, Santoríni, the southernmost of the Cyclades, is probably the most popular Greek island of all.

See also Delos, Mykonos, Santoríni

cyclone A low-pressure distribution surrounded by higher pressure, combined with the Coriolis effect, that causes the air to circulate in a counterclockwise direction in the Northern Hemisphere, and clockwise in the Southern. Near the surface of the earth, the frictional drag on air moving over land or water causes it to gradually spiral inward, and this inward movement causes rising currents near the center. When the winds of a tropical cyclone increase to severe intensity (more than 74 mph), it is known as a hurricane in the Atlantic and a typhoon in the Pacific. Often the terms are interchangeable, but "cyclone" is commonly used around the Bay of Bengal and the Indian Ocean.

See also hurricane, typhoon

Cyprus Island republic in the eastern Mediterranean, 40 miles south of Turkey and 60 miles west of Syria. Its capital is Nicosia, and other important cities are Famagusta, Larnaca, and Limmasol. The population of about 700,000 lives on an island that covers 3,578 square miles. Cyprus has been inhabited since Neolithic times, six thousand years ago, and has been occupied by Phoenicians, Assyrians, Egyptians, and Persians. Cyprus was conquered by Alexander the Great in 333 B.C., but after his death in 323 B.C. it fell under Egyptian rule. It was annexed by Rome in 58 B.C., and ruled by the Byzantines until 1191, when Richard I (Coeur de Lion) of England conquered it on his Third Crusade and bestowed it upon Guy de Lusignan, the dispossessed king of Jerusalem, who founded a feudal monarchy that lasted into the Middle Ages. Cyprus was annexed by Venice in 1489 and conquered by the Turks in 1571. In 1878, it was placed under British rule, then, in 1914, annexed outright by Britain. The Cypriots rebelled, leading to the departure of British administrators in 1954 and full independence in 1960. Archbishop Makarios, president since 1959, was overthrown in 1974 by the Cypriot National Guard. Ongoing conflicts between Greek and Turkish Cypriots have led to irregular violence, assassination of leaders, and intervention by the United Nations. Turkish forces invaded the island in 1974, and in 1983, Turkish Cypriots declared themselves independent and established the Turkish Republic of Northern Cyprus—recognized by no country but Turkey. The island is now divided into Greek and Turkish sectors, separated by a buffer zone supervised by UN troops.

D

da Gama, Vasco (1460–1524) Born in Sines, Portugal, da Gama was commissioned by Prince Henry the Navigator to reach India by sea. He departed from the Tagus River off Lisbon on July 8, 1497, in four ships, including his flagship, the caravel *São Gabriel,* and the *São Rafael,* which had his brother Paolo in command. They sailed south to the Cape Verde Islands, then to the southern tip of Africa, where they made the first contact with the Hottentot people. They rounded the Cape of Good Hope, and on Christmas Day, 1497, they passed the east coast of southern Africa, which they named Natal. From Malindi (in what is now Kenya) they headed across the Indian Ocean, and after twenty-three days out of sight of land, they reached Calicut in southwestern India. On September 9, 1498, da Gama was greeted with a tumultuous welcome when he arrived back in Portugal with a load of spices. He was named Admiral of the Ocean Seas, and in 1502, he took a fleet of fifteen ships back to India, where he led the retaliation against the Muslims who had massacred the Portuguese in India. He returned to Portugal in 1503. Not until 1524 did he again sail to India. He died in Cochin after only a few months as viceroy of Portuguese India.

Dall porpoise

Dall porpoise (*Phocoenoides dalli*) One of the fastest of the small cetaceans, the Dall porpoise dashes through the water approaching vessels under way, then zooms away, sending up a "rooster tail" of spray. It is found only in the North Pacific, from California up through Alaska and Siberia, and south to Japan. Its several color variations used to be considered separate species, but they are now all included as one. The Dall porpoise has the characteristic spade-shaped teeth of the Phocoenidae (true porpoises), but it also has "gum teeth," which project over the actual teeth and may help the porpoise to grasp small, slippery fishes. Thousands of Dall porpoises are killed every year in the offshore gill-net fisheries, and they are also hunted deliberately for food in Japanese waters. **See also bow riding, gill net**

Dampier, William (1652–1715) Apprenticed to a ship's captain at age eighteen, Dampier made a voyage to the Newfoundland fisheries, and when his apprenticeship was duly served, he joined an East Indiaman and sailed to Java. Upon his return to England, he enlisted in the Royal Navy and served aboard the *Royal Prince* at the Battle of Schoonveld in the Third Dutch War. After failing as manager of a sugar plantation in Jamaica, he joined a band of buccaneers led by Bartholomew Sharpe, which raided the town of Panama on the isthmus. From 1678 to 1691, he was engaged in piracy along the west coast of South America and in the South Seas, a voyage that would briefly maroon him on the Nicobar Islands, but ultimately take him around the world. His journal was published in 1697 as *A New Voyage round the World,* complete with maps and drawings of the exotic wildlife he had seen. Its success brought him to the attention of the British Admiralty, which gave him command of the HMS *Roebuck* and sent him on an exploring voyage to New Holland (Australia). He mapped much of the previously unexplored west coast, but lack of suitable harbors where he might take on water forced him to sail to Timor in the East Indies. On the return voyage to England, the *Roebuck*—which was in terrible condition even before she left England—foundered off Ascension Island, and Dampier and his men waited for two months before they were rescued and brought back to England. He was court-martialed and declared unfit for further service in the Royal Navy, but in 1703, he joined a privateering expedition to the Pacific as captain of the *St. George.* Again, Dampier's autocratic manner brought the expedition nothing but disappointment, and he returned to England in 1707. A year later, he began another privateering voyage, as navigator on Woodes Rogers's *Duke* and *Duchess.* This time, the voyage was a financial success, but Dampier died before he could get his share of the profits. In Western Australia there is a port named for him, as well as an archipelago and a promontory.

damselfishes (family Pomacentridae) Small, mostly brightly colored fishes that are found in warm, shallow tropical waters. Among the better-known species are the sergeant major (*Abudefduf saxatilis*), which is rather drab; the beau gregory (*Eupomacentrus leucostictus*);

and the largest of the damselfishes, the garibaldi (*Hypsypops rubicunda*), which can reach 1 foot in length. Also widespread is the genus *Dascyllus*, brightly colored little fishes of the Indo-Pacific regions. The anemone or clown fishes (*Amphiprion* spp.) are colorful damselfishes that live within the poisonous tentacles of sea anemones without getting stung. Some species are known as demoiselles.

damselfish

See also anemone fish, blue chromis, garibaldi, sergeant major

Dana, Richard Henry (1815–1882) American author, the son of a poet and essayist of the same name. Young Richard was enrolled at Harvard in 1833 when an attack of measles so injured his eyes that he was unable to continue his studies. He believed that a sea voyage would cure his eye troubles, so in 1834 he signed aboard the brig *Pilgrim* for a trading voyage around Cape Horn and up the western coast of North America. Upon his return to Boston (he had changed ships halfway through the voyage to escape the brutality of the *Pilgrim*'s captain), he returned to law school and used the notes of his voyage to write *Two Years before the Mast,* which exposed the oppression of American seamen. Published in 1840, the year Dana was admitted to the bar, it became one of the most influential and successful books of its time; it is now considered a masterpiece. Dana graduated from Harvard and became a well-known maritime lawyer; he was named minister to Great Britain in 1876, but he is best remembered for his writing.

Dardanelles Known as the Hellespont in ancient times, the Dardanelles is the narrow strait (¾ to 4 miles wide) that separates Gallipoli in Europe from Turkey in Asia, and connects the Mediterranean with the Sea of Marmara. In Greek mythology, Leander used to swim the Hellespont to meet his lover, the priestess Hero, until a storm put out the light she guided him with and he was drowned. (The British poet Lord Byron swam the Hellespont in 1810 in homage to Leander's romantic dedication.) In 480 B.C. the Persian army of Xerxes crossed the strait on a bridge of boats, and Alexander the Great did the same on his 334 B.C. excursion against Persia. The strait has always been of strategic importance because it is the gateway to Istanbul from the Mediterranean. In 1915, in the Dardanelles campaign, the Allies failed to take Gallipoli from determined Turkish troops, and most of the town was destroyed.

dartfishes (family Microdesmidae) The family Microdesmidae includes the wormfishes, firefishes, and dartfishes, all of which live in tropical waters of the Indo-Pacific and Caribbean regions. None of them gets to be more than 6 inches in length. The wormfishes are elongated, somewhat compressed fishes, with small eyes and a single dorsal fin that runs the length of the body. Firefishes (*Nemateleotris* spp.) are colorful creatures with a very high dorsal spine and a color pattern that fades from white at the head to dark blue or orange at the tail. They are popular with saltwater aquarists. Dartfishes are more gobylike, but most species do not have the goby's protuberant eyes.

Davis, John (1550–1605) English navigator, one of the greatest of Elizabethan explorers. In 1583, he set out to discover the Northwest Passage between Europe and the Far East. Failing to find it, he persuaded the English authorities to send him to explore the northern seas of Greenland. In 1585, in the bark *Sunshine,* with John Bruton in the *Moonshine,* Davis rounded (and named) Cape Farewell, the southern tip of Greenland, and coasted the west coast, which he accurately called the Land of Desolation. He then sailed north and located the strait that now bears his name between Greenland and Baffin Island, but this ice-choked passage was obviously not going to take him to the Orient. He outfitted the second voyage in the spring of 1586, again employing the *Sunshine* and the *Moonshine,* but adding two little pinnaces, the *North Star* and the *Mermaid.* Under Richard Pope, *Sunshine* and *North Star* were sent on a futile quest to force a passage between Greenland and Iceland, while Davis took the *Moonshine* and the *Mermaid* to Gilbert Sound (now Nuuk [Godthåb], the capital of Greenland), where they put on a show of English country dancing for the Eskimos and staged wrestling matches with them. Davis took the *Moonshine* across the strait that would be named for him, and rounding a point he named "Cape of God's Mercy" (now simply Cape Mercy), he passed Cumberland Sound and headed south to Labrador. (He missed the entrance to Hudson Strait, which would not have taken him to China but would have preceded Hudson's discovery by twenty-three years.) In the 1588 battle with the Spanish Armada, Davis commanded the *Black Dog* and sailed with Thomas Cavendish's 1591 expedition to seek a way through the Strait of Magellan. Aboard the *Desire,* Davis failed to locate a passage, and his crew was ravaged by scurvy on the return voyage. He sailed with Walter Raleigh to Cádiz and the Azores in 1596–1597, and voyaged to the East Indies as pilot of the Dutch ship *Lion* in 1598 and 1601. Eager to compete with the Dutch for gold and spices, the East India Company chartered its first official voyage in 1605, under the command of Sir Edward Michelbourne. As part of this expedition, Davis sailed for the Indies again, aboard the ship *Tiger.* At anchor off Sumatra, on December 27, 1605, he was killed by pirates.

dead man's fingers (*Alcyonium digitatum*) One of the soft corals (alcyonaceans) that have only a scattering of spicules to stiffen the body, dead man's fingers takes the form of a flesh-colored, orange, or white bloated, gelatinous mass, rising as much as 8 inches from the gravelly bottom. At night, the outline is broken by the emergence of numerous tiny polyps that form a white furriness over the surface. This species occurs in shallow waters on both sides of the North Atlantic, but in other parts of the world, alcyonaceans take on different forms and colors, like the organ-pipe coral (*Tubipora*), which has emerald-green polyps emerging from brick-red limestone tubes; or the blue coral, which excretes a skeleton that turns blue because of iron salts.

See also coral, soft corals

decapod From the Latin *deca*, meaning "ten," and *pod*, meaning "foot," the decapods are marine crustaceans that have a thorax fused with the head to form the cephalothorax, stalked eyes, a pair of mandibles, and five pairs of thoracic legs, hence the name.

See also crab, lobster, shrimp

Decatur, Stephen (1779–1820) American naval officer, born in Maryland. In 1804, during the Tripolitan wars against the Barbary pirates, he slipped the *Enterprise* into the harbor at Tripoli and burned the American frigate *Philadelphia*, which had been captured by the pirates. When the War of 1812 began, in command of the frigate *United States*, he captured the British ship *Macedonian* off Madeira. In January 1815, his ship *President* was engaged in a battle with the British *Endymion*, when Decatur was badly wounded and had to surrender. He was killed in a duel with James Barron, an American officer at whose court-martial he had been one of the judges. The cities of Decatur in Alabama, Georgia, and Illinois are named for him.

Deception Island One of the subantarctic South Shetland Islands, Deception was probably discovered by sealers around the turn of the nineteenth century. Because the fur seals (*Arctocephalus* spp.) of Deception Island had never seen humans, they were remarkably easy to kill, and by 1825, a quarter of a million of them had been slaughtered. When the first whalers arrived, around 1905, they found a magnificent harbor, and Whaler's Bay became the center of operations for the western Antarctic until around 1925, when the addition of the stern slipway to factory ships made it unnecessary for whalers to process their catch on shore. The island, which consists mostly of a caldera surrounded by black, cindery hills, is an active volcano, with steam issuing from vents on the beach, and pockets of hot water bubbling just offshore. (In 1842, when members of the U.S. Exploring Expedition arrived at Deception, they observed that the island appeared to be on fire.) When the volcano erupted in 1967, members of the Chilean scientific station had to be evacuated by helicopter. In 1969, five members of the British Antarctic Survey were awakened by strong earthquake tremors, and then the volcano erupted again. As the men shielded themselves from falling volcanic bombs, lava flowed down the hillside and carried the old whaling station into the bay. They, too, were lifted off by helicopter. The volcano erupted again in 1970, but there was no one on the island.

See also South Shetland Islands

deep-crested whale

deep-crested whale (*Mesoplodon bowdoini*) A dark-colored animal with a whitish beak and the usual scars and scratches, the deep-crested whale is a rare species in a rare genus. The pair of massive teeth in the lower jaws of the males often have a notch on the leading edge, but there is no adequate explanation for this. It is similar enough to Hubbs' beaked whale for some cetologists to believe that they may actually be the same species. *M. bowdoini* has been recorded (mostly from strandings) in the South Pacific and Indian Oceans, while Hubbs' beaked whale is a resident of the North Pacific waters of California, Washington, and Japan.

See also beaked whales, Hubbs' beaked whale, *Mesoplodon*

deep-scattering layer (DSL) A horizontal, mobile zone of living organisms that occurs below the surface in many oceans and is so called because it scatters or reflects sound waves. This phantom layer was not recognized until 1942, when U.S. Navy scientists, experimenting with underwater sound for locating submarines, detected an echo from 900 feet, where no bottom existed. The deep-scattering layer, or DSL, was later observed to move toward the surface in the evening and sink again at dawn, a most unlikely activity for the seafloor. Trawls pulled through the DSL have not resolved the mystery of its composition, but it is believed to consist of countless small organisms; lantern fishes and/or bristlemouths, considered the most numerous fish in the ocean; small squids; siphonophores; or a mixed or layered aggregation of all of them.

See also bristlemouth, lantern fishes, siphonophore

deep-sea anglerfish (*Ceratias holboelli*) Although most species of deep-sea anglers (Ceratidae) are fist-sized or smaller, *C. holboelli* is a giant, with the females reaching a maximum known length of 40 inches. The parasitic

deep-sea anglerfish

males, which are only a tenth the size of the females, attach themselves permanently to the females and serve only as a reproductive organ. Juveniles have functional eyes, but when a male attaches himself to a female for breeding, the eyes of both sexes degenerate, and adults are blind. Like most of the other anglers, *Ceratias* has a lure (the ilicium) projecting from its back behind the eyes that it can move forward or back by means of muscular contractions. The skin of mature females is covered with thornlike protrusions, and on the back are two bulblike structures, the caruncles, that luminesce like the tip of the ilicium. Anglers of this species are found throughout the deeper, colder waters of the world, and are sometimes eaten by sperm whales.
See also anglerfishes, sperm whale

Delos Located roughly in the center of the Cyclades, a circle of Greek islands, Delos is Greece's holy island, the mythical birthplace of Apollo and Artemis. At 1 square mile, it is the smallest of the Cyclades, but its historical importance is immense. After the defeat of the Persians at Salamis, the Athenians organized the Delian League with the seat of the treasury at the temple of Apollo on Delos. When Pericles removed the treasury to Athens in 454 B.C., Delos lost its political significance. It was occupied by the Romans in 146 B.C., but the island was invaded in 88 B.C. by Mithridates and its ten thousand inhabitants massacred. Nobody lives there now. The historical remains attract many visitors to this sacred site, but there are no tourist facilities, and unless you are an authorized archaeologist, it is illegal to stay overnight on the island. **See also Cyclades**

dense-beaked whale (*Mesoplodon densirostris*) The teeth in mature males of this species of beaked whale are so exaggerated that they protrude above the forehead like a pair of horns. These teeth can be more than 6 inches long, but the females have no teeth at all. The dense-beaked whale—sometimes known as Blainville's beaked whale—has been reported from scattered locations north and south of the equator, including Madeira, South Africa, Nova Scotia, Australia, Chile, Formosa, and New Jersey. **See also beaked whales, *Mesoplodon***

dense-beaked whale

D'Entrecasteaux, Antoine Raymond Joseph de Bruni (1737–1793) French naval officer and explorer in command of a fleet that explored New Caledonia, Australia, and Tasmania in 1785. After serving as governor of Mauritius from 1787 to 1789, he was appointed rear admiral in 1791 and sent to find Jean-François La Pérouse. Although he passed the site of the *Astrolabe* wreck in Vanikoro in the Solomon Islands, he did not notice any wreckage or survivors, and La Pérouse's fate remained a mystery until 1828, when Dumont d'Urville found the remains of the ships. D'Entrecasteaux died on the return voyage to France in 1793. **See Dumont d'Urville, La Pérouse**

D'Entrecasteaux Channel Inlet of the Tasman Sea between Bruny Island (also named for Antoine Bruni d'Entrecasteaux) and the southeast coast of mainland Tasmania; it eventually merges with the Derwent River estuary. Discovered in 1642 by Abel Tasman, it was explored by d'Entrecasteaux in 1792, and shown to be a channel rather than a bay. It is now the southern approach to the port of Hobart.

D'Entrecasteaux Islands Heavily forested island group off the eastern tip of the island of New Guinea in the southwest Pacific, these islands now belong to Papua New Guinea. During his failed search for La Pérouse in 1793, D'Entrecasteaux found and named the islands. In 1873 they were charted by Captain John Moresby in the *Basilisk*. **See also D'Entrecasteaux, New Guinea, Tasman**

dermal denticles Also known as placoid scales, dermal denticles are tiny, toothlike protrusions embedded in the skin of elasmobranchs. They are characteristic of all sharks and rays. They are actually modified teeth, since each one has an outer layer of enamel and a central pulp canal containing nerve cells and blood vessels. In most species, the denticles are aligned in one direction, pointing tailward, but in the basking shark, they are randomly arranged, making the skin rough in every direction. The very large, randomly distributed denticles in the bramble shark are responsible for its common name. It is the denticles that give sharkskin its abrasive character, not unlike sandpaper. When it was used to smooth wood surfaces or to provide a better grip on sword handles, it was known as shagreen. Until a process was developed for the removal of the denticles, sharkskin was commercially useless, but since then it has been marketed as a durable and attractive leather.
See also bramble shark, sharkskin

de Soto, Hernando (c. 1496–1542) Spanish explorer who accompanied Pedro Arias (Pedrarias) Dávila in 1516–

1520 to Central America and was named military commander of Nicaragua in 1523. He was in Peru with Pizarro in 1532, and was the first of the conquistadors to meet Atahualpa, the Inca chief that Pizarro later had executed. After the conquest of the Incas, de Soto returned to Spain but then decided to lead an expedition to Florida, in the largely unexplored southeastern region of North America. In 1538, with 1,000 men, 350 horses, and a pack of war dogs, which fed on Indians (dead or alive), he stopped briefly in Cuba, rounded the southern tip of Florida, and landed south of Tampa Bay on the gulf coast. He wandered through what would later become Florida, Alabama, Mississippi, Tennessee, and Arkansas, living off the land, which meant seizing food and enslaving others to carry it. Thousands of Native Americans were rounded up and snapped in iron collars; the women were raped and killed. In 1540, de Soto followed the Mississippi River, hoping it would lead him to rich cities, but he died on the way.

See also Pizarro

Devil's Island In French, Île du Diable, an Atlantic island off the northern coast of South America, officially part of French Guiana. The smallest of the three islands known as the Îles du Salut, Devil's Island is a narrow strip of land, less than 1 mile long and 400 yards wide, about 6 miles from the mainland. It was originally a leper colony until it was made a maximum-security prison in 1895. Its first and most famous prisoner was Alfred Dreyfus (1859–1935), the French army officer condemned for treason, who spent four years there before being remanded to France, retried, and pardoned. Devil's Island was later used to incarcerate political prisoners, especially deserters in World War I. The penal colony was abolished in 1938, and the island is now advertised as a resort.

Devon Island Located in the Northwest Territories of the Canadian Arctic, Devon Island lies across Lancaster Sound from the northern end of Baffin Island. It consists of some 21,000 square miles of snow and ice, and is largely uninhabited. Along with Melville, Bathurst, and Cornwallis Islands, Devon is one of the Parry Islands, named for their discoverer, Sir William Edward Parry (1790–1855). Parry named Devon Island for the native county of Matthew Liddon, his second-in-command. As of April 1, 1999, some 40 percent of Canada's northern wilderness was administratively separated from the Northwest Territories and amalgamated into a new territory, Nunavut, which is the Inuit word for "our land." Constituting one-fifth of all Canada, Nunavut consists of empty, ice-covered islands, and although it encompasses 770,000 square miles—an area larger than Mexico—the new territory has a total population of 26,000. (Mexico, with a climate somewhat more conducive to settlement, has 97 million people.) In one of the largest native-rights claims settlements ever paid, the Canadian government will pay to the inhabitants of Nunavut $1.148 billion in Canadian dollars ($840 million in U.S. dollars) over fourteen years. **See also Baffin Island, Bathurst Island (1), Cornwallis Island, Melville Island (2)**

Dewey, George (1837–1917) American admiral, hero of the Battle of Manila. An 1858 graduate of the U.S. Naval Academy, he served with distinction during the Civil War, commanding the steam gunboat *Agawam* during the North Atlantic blockade of the South. He later was appointed commodore of the U.S. Asiatic squadron. In Hong Kong in April 1898, when war was declared between the United States and Spain, he was ordered to head for the Philippines, where eight Spanish ships were lying in Manila Bay. He annihilated Admiral Montojo's fleet without losing a man, and joined forces with U.S. Army general Wesley Merritt to capture the city of Manila. He was promoted to the post of admiral of the navy and returned home to a hero's tumultuous welcome; he was so popular that a movement was started—by the same newspaper editors whose "yellow journalism" had done so much to bring about the war—to nominate him as the Democratic presidential candidate. He resisted all such pleas and retired to a quiet life. **See also Spanish-American War**

Dezhnev, Semyon Ivanov (1605–1673) When Tsar Peter the Great sent Vitus Bering to discover whether Asia was connected to America in 1725, neither man knew that the answer had been provided seventy-seven years earlier by a Cossack officer collecting fur tributes. In 1648, Semyon Dezhnev had taken six small boats down the Kolyma River on the northern coast of the Chukchi Peninsula, rounded the easternmost point, sailed through the narrow passage that separates the two continents, and ended up at the Anadyr River, which debouches into the Pacific Ocean. Dezhnev was barely literate, and his report to St. Petersburg remained unnoticed until 1736, when the German historian Gerhard Müller found it while researching historical materials on Siberia. Bering knew nothing of Dezhnev's accomplishment, and when he passed through the strait that now bears his name, he was not sure where he was; he was unable to confirm that the two continents were separated. As for Dezhnev, when he finally returned to Moscow, he was well paid for his services, and he returned to Yakutsk as an administrator. Cape Dezhnev, the eastern tip of Siberia, which is named for him, is only 55 miles from the Seward Peninsula, the westernmost part of North America.

See also Bering, Bering Sea and Strait

diamond squid (*Thysanoteuthis rhombus*) Because it resembles no other species, *T. rhombus* has been placed in

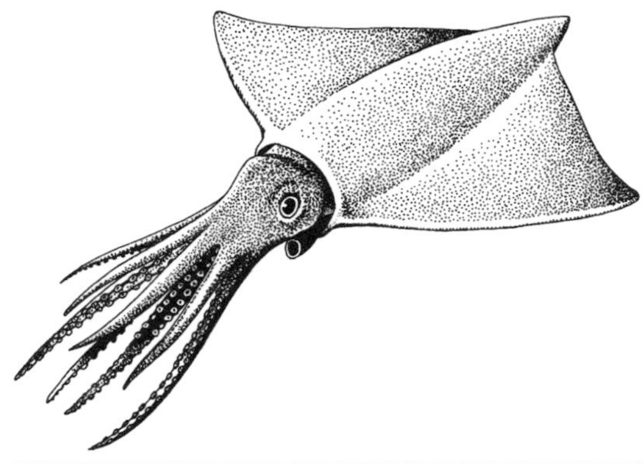

diamond squid

its own family, the Thysanoteuthidae. (*Thysanos,* meaning "fringe" or "tassel," probably refers to the broad fins.) It is a large, nektonic squid with a strong muscular mantle and very long rhomboidal fins that extend along the entire body, making it look like a stingray with a head and arms. It is found around the world in warmer waters, but it is not common anywhere. Mature animals are large and heavy, with a maximum known mantle length of 39 inches and a weight of 45 pounds. Japanese fishermen know it as the sleeved squid (*sode-ika*) or barrel squid (*taru-ika*). Like many squid species, *Thysanoteuthis* passes through growth stages in which the shape and proportions of the animal change dramatically. When hatched (at about ¼ inch in length) the diamond squid is a plump little creature with tiny terminal fins. As it matures, the mantle elongates and the head and arms become longer and narrower, but as the fins approach their full rhomboidal configuration, the arms become relatively shorter. Perhaps the strangest aspect of the life of this squid is its propensity for traveling in male-female pairs. Even when aggregations of these squids are trapped in nets, the pairs remain close together. No other squid species is known to do this.

See also giant squid, squid

Dias, Bartolomeu (c. 1450–1500) Portuguese explorer in the service of King João II; in 1488, the discoverer of the Cape of Good Hope. After Diogo Cão almost reached Walvis Bay in 1483, Dias was selected to circumnavigate the African continent with three ships, the *São Cristóvão,* the *São Pantaleão,* and an unnamed supply ship. In 1488, he reached the cape that now bears his name (off what is now Lüderitz in Namibia), where he erected a commemorative plaque, but when he set sail again, he was swept by gales to the southwest, away from the coast, for thirteen days. When he was able to turn eastward, he made a landfall at Mossel Bay, on the south coast of the Cape Colony, having rounded the southern tip of Africa without knowing it. He then coasted eastward, reaching the Great Fish River before his men, fearful that they would run out of food, forced him to abandon the quest and return home. Because of his experience, he named the southernmost point of Africa Cabo Tormentoso (Cape of Storms), but it was later changed to Cabo de Boa Esperança (Cape of Good Hope). Dias's voyage set the stage for Vasco da Gama's successful crossing to India in 1497, and when Pedro Álvares Cabral set out again in 1499 for India with a dozen ships, Bartolomeu Dias was given command of one of them. In May 1500, while rounding the very cape that he had named "Boa Esperança," Dias's ship was lost with all hands.

See also Cabral, Cão

diatoms One-celled planktonic organisms that live in lighted areas of the sea. They are encased in a glassy shell called a frustule, whose two halves (the epitheca and the hypotheca) fit together like a box with a lid. Among the most numerous organisms in the sea, they occur everywhere but are especially abundant in temperate and polar waters.

See also plankton

Diego Ramirez Islands Group of uninhabited islands about 60 miles south of Cape Horn in the Drake Passage. Now belonging to Chile, they were discovered by the brothers Bartolomé and Gonzalo Nodal in 1619. These Spanish navigators, who were sent to verify the reports that Jakob Le Maire and Willem Schouten had discovered a new route to the Pacific in 1616, passed successfully through the Le Maire Strait but were blown southward while trying to round the horn into the Pacific. They sighted a group of islands, charted their position, and named them for their pilot and cosmographer, Diego Ramirez.

dinoflagellates Single-celled organisms equipped with two whiplike organs (flagella; singular, flagellum) that enable them to swim. Dinoflagellates also have a light-sensitive organelle, the eyespot, that enables them to detect the light they need for photosynthesis. Many dinoflagellates are bioluminescent, causing the common phenomenon of ocean "phosphorescence." The dinoflagellates *Gonyaulax catenella* and *Ceratium horrida* are responsible for the neurotoxic occurrence known as the red tide.

See also bioluminescence, red tide

Diomede Islands Two rocky islands in the Bering Strait between Alaska and Siberia, 2½ miles apart, discovered by Vitus Bering in 1728. The larger island, Big Diomede, belongs to Russia and is uninhabited except for weather station personnel; Little Diomede is American, and inhabited by about a hundred Yupik Eskimos who rely on fishing for their subsistence. On a (rare) clear day, you can see Big Diomede from Little Diomede; the only place where Russia and the United States are visible to each other. Not only does the Russian-American border pass between the islands, the International Date Line does, too. **See also Bering Sea and Strait, International Date Line**

Discovery The name of several famous vessels in the history of British exploration. In 1602, under the command of Captain George Weymouth, the first *Discovery* explored what would later become known as Hudson Strait. Eight years later, Henry Hudson took the same ship through the strait and into the bay, where he failed to find an outlet to the west; he was abandoned there by his crew. In 1615, William Baffin took the *Discovery* through Hudson Strait, and in 1616, he passed through Davis Strait to the vast bay that now bears his name. Captain Charles Clerke commanded the small Yorkshire collier *Discovery* on James Cook's third and last voyage (1776–1779), and George Vancouver circumnavigated the world in another converted collier named *Discovery* from 1791 to 1795. In 1875 a converted whaler was renamed HMS *Discovery,* strengthened for traveling through ice, and taken (along with the *Alert*) by Captain George Nares as close to the North Pole as anyone had ever gotten. This *Discovery* was the prototype for the polar vessel commanded by Captain Robert Scott on his first Antarctic expedition from 1901 to 1904. Scott's *Discovery* was subsequently turned over to the Hudson's Bay Company, but was brought back into government service as a research vessel to carry out oceanographic surveys and investigations on the whaling grounds, from 1925 to 1927. (This research was published in a series, *Discovery Reports,* that is still ongoing.) The Australian explorer Sir Douglas Mawson used Scott's *Discovery* to survey Australian Antarctic territory, after which she was laid up and replaced by a new British research vessel that had been christened *Discovery 2.*

Disko Island Although the coasts of Greenland are heavily indented by fjords, Disko is the only large island that is separated from the mainland. On the west coast, in the Davis Strait, it covers some 3,312 square miles, and is 80 miles long. It was first explored by Eric the Red in 984, and in 1587, as the British explorer John Davis (for whom the strait is named) sailed past western Greenland, he landed on the island and his men entertained the Greenland Eskimos by playing music and dancing on the ice. In the eighteenth and nineteenth centuries, Disko Bay was a popular ground for British whalers hunting the bowhead—until the whales were eliminated. Greenlanders call the island Qeqertarsuaq, and its settlement is Godhavn. There are coal and iron deposits on the island, and offshore are some of the world's largest shrimp beds.

diving petrel (*Pelecanoides* spp.) Diving petrels fly low over the water with a swift buzzing flight, the wings almost constantly in motion; often they fly directly into waves, hence their name. They also "fly" underwater in pursuit of their prey, which consists of small crustaceans. In appearance and habits, they resemble the little auks of the north. Diving petrels are stocky, short-winged little birds (15-inch wingspan), dark above and white below. The four species (common, Georgian, Peruvian, Magellan) all inhabit the Southern Hemisphere, and although they all look very much alike, they can be differentiated by locality. Along with the albatrosses, shearwaters, and storm petrels, diving petrels are tubenoses. Their nostrils extend on top of the bill through two horny tubes, the function of which is not fully understood, but it is associated with salt secretion. **See also albatross, petrel, shearwater, storm petrel**

dodo (*Raphus cucullatus*) A very large flightless pigeon found only on the island of Mauritius in the Indian Ocean, extinct since 1681. The island, which is east of Madagascar, was discovered by the Portuguese navigator Alfonso Albuquerque in 1507. It took visiting sailors 174 years to eliminate the dodo from the face of the earth. Exactly why they killed the birds remains a question, since their flesh was described as tough and unpalatable, but the extinction was helped along by feral pigs that ate the eggs. The bird was larger than a turkey, weighing as much as 50 pounds and standing about 3 feet tall. It had a round body, a small head with a powerful hooked bill, stout yellow legs, stubby, useless wings, and a tuft of curly feathers high on its rear end. We know what it looked like because some bones and parts have been preserved, and there are numerous contemporary illustrations. Two close relatives of the Mauritius dodo are also extinct: the Réunion solitaire (*Raphus solitarius*) and the Rodrigues solitaire (*Pezophaps solitaria*). Mauritius, Réunion, and Rodrigues are known as the Mascarene Islands, named for Pedro de Mascarenhas, who visited the islands in 1512.

dodo

See also Mauritius, Réunion

dogfish A common name for a number of small sharks, usually of the family Squalidae (spiny dogfishes) and Triakidae (smooth hounds), but the name is also commonly used for a family of small deepwater sharks of the genus *Etmopterus,* such as *E. virens,* the green dogfish. (Because many species of *Etmopterus* have light organs, they are also known as lantern sharks.) Many other Squaliform sharks are also known (in English) as dogfishes, such as the black dogfish (*Centrocyllium fabricii*), the Portuguese dogfish (*Cen-*

troscymnus coelolepus), and the birdbeak dogfish (*Deania calcea*). That these sharks are all called dogfishes does not necessarily mean that they are related; it only means that the old French term *chien de mer* ("dog of the sea") was probably applied because of the tendency of some species to aggregate in packs. It was then applied more or less indiscriminately to various sharks, and translated into English with little regard for specific identification.

See also birdbeak dogfish, green dogfish, Portuguese dogfish, smooth hound sharks, spiny dogfish

Dogger Bank An extensive shoal in the North Sea, covering about 6,800 square miles, roughly equidistant between Denmark and the Northumberland coast of England. It is a huge moraine, 160 miles long and 60 miles wide, deposited during the last glaciation. Its depth below sea level varies; the shallowest portions may be no more than 30 feet deep, but most of it is between 60 and 120 feet deep. It is a particularly rich fishing ground for cod, plaice, haddock, and turbot, and was probably named for the *dogger,* a type of Dutch fishing trawler. In 1904 a Russian naval force en route to Japan fired on the Hull fishing fleet on the Dogger Bank, thinking they were Japanese torpedo boats, but an international incident was avoided. During World War I a major naval engagement was fought on the Dogger Bank, where the armored cruiser *Blücher* was sunk by the British and 954 Germans were killed. According to their exclusive economic zones, Britain, Germany, Denmark, and the Netherlands all have rights to fish the Dogger Bank.

doldrums The belt of calms that lies within the trade winds of the Northern and Southern Hemispheres. The doldrums are found to the north of the equator except in the western Pacific, where they are to the south. They are caused when a large amount of solar radiation creates intense heating of the land and ocean. The doldrums are noted for calms—periods when the winds disappear altogether, trapping sailing vessels for days or weeks. The term is also used to signify general depression; ships—and sailors—were said to be "in the doldrums" when the ships were lying motionless.

See also horse latitudes, trade winds

dolphin The common term used to refer to members of the family Delphinidae: aquatic mammals (cetaceans) with a single blowhole and numerous, canine-like pointed teeth. There are some forty-five species, sometimes separated into the Delphinidae (oceanic dolphins), Platanistidae (freshwater dolphins), and Phocoenidae (porpoises). They are found throughout the temperate and tropical waters of the world, often in large schools; only the killer whale (which is actually a large dolphin) is a regular inhabitant of high polar latitudes in both hemispheres. All are streamlined in shape, and some are capable of great speeds; bow riders get an assist from a moving ship's pressure wave. The largest of the dolphins is the aforementioned killer whale (*Orcinus orca*), which can reach a length of 30 feet and a weight of 10 tons; the smallest is the Franciscana (*Pontoporia blainvillei*), which is rarely more than 4½ feet in length, including the long beak. Pilot whales (*Globicephala*), false killer whales (*Pseudorca*), and pygmy killer whales (*Feresa*) are all dolphins. The best known of the dolphins is the bottlenose dolphin (*Tursiops truncatus*), famous for its appearance in oceanariums around the world, but the common dolphin (*Delphinus delphis*) is the one that interacted with people in ancient mythology. All dolphins identify their prey and its location by echolocation, a process where clicks are broadcast outward and the returning echoes analyzed. Some species, particularly spotters (*Stenella attenuata*) and spinners (*S. longirostris*), have been killed in huge numbers in past years because they became trapped in the purse seines of tuna fishermen in the eastern tropical Pacific.

See also *individual species;* bottlenose dolphin, common dolphin, echolocation, Franciscana, freshwater dolphins, killer whale, pilot whale, porpoise, tuna-porpoise problem

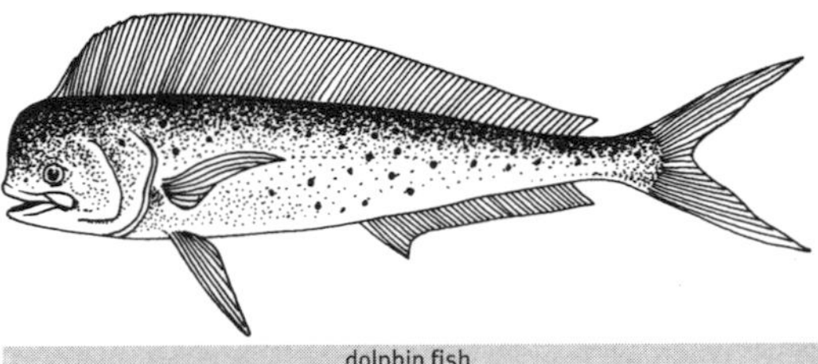

dolphin fish

dolphin fish (*Coryphaena hippurus*) Despite its confusing common name, *Coryphaena hippurus* is a proper fish and is in no way related to any marine mammals. Also known colloquially as dorado or mahimahi, it is a pelagic, migratory, schooling fish, found throughout the world's warm temperate and tropical waters. In life, dolphins are among the most colorful fish in the sea, but when they die, they turn a uniform yellowish gray. Males differ from females in that they are larger (maximum known length is 6 feet), and they have a high, vertical forehead, as compared with the sloping profile of the females. They are fast, powerful swimmers, and are particularly adept at catching flying fish—sometimes even in the air. They are extremely popular as food fishes, and their speed and acrobatics when hooked make them a favorite game fish, too.

See also flying fish

Dominica Island republic in the West Indies, between Guadeloupe and Martinique. Although Columbus sighted the island in 1493 (and named it for Sunday, *domingo*, the day he sighted it), attempts at colonization were thwarted by the Carib Indians, who had displaced the Arawaks who previously lived there, and who fended off European settlers. It was not until 1815 that the British occupied the island, almost eliminating the Caribs. (The remaining three thousand Indians have been relegated to a reservation on the eastern side of the island.) Dominica joined the West Indian Associated States in association with Britain in 1967 and became fully independent in 1978. Tourism is now the major industry, but hurricanes destroy the facilities with unfortunate regularity. Dominica is one of the poorest countries in the Caribbean.

See also Arawak Indians, Carib Indians, Guadeloupe, Martinique

Dominican Republic The eastern two-thirds of the Caribbean island of Hispaniola; the western third is occupied by Haiti. The countries share a tumultuous history. Columbus discovered the island in 1492 and established the settlement of La Isabela in 1496—the oldest European settlement in the New World—with his brother Bartholomew and his son Diego, respectively, as governor and viceroy. Later France claimed the western third of Hispaniola, which Spain ceded in 1697 (it is now Haiti). Spain surrendered the remainder to France in 1795. However, a slave uprising had begun in the French colony and the area now known as the Dominican Republic was conquered by forces led by former slave François-Dominique Toussaint-Louverture. Under General Charles Leclerc, the French defeated Toussaint and fended off the attacks led by Jean Jacques Dessalines, another ex-slave. In 1808, the people revolted again, and the revolutionaries captured the capital of Santo Domingo in 1809. With the aid of an English squadron, they evicted the French and Spanish rule was reestablished. But in 1821, the country declared independence, only to be reconquered in 1822 by the Haitians under Jean Pierre Boyer. The Haitians were evicted in 1844, and the Dominican Republic was established. Political instability forced the leaders to petition Spain for relief, and from 1861 to 1865, the country was a Spanish province again. The United States did not recognize the Spanish claim, and sent troops that remained there until 1934. Rafael Trujillo overthrew the established puppet government in 1930 and remained dictator until his assassination in 1961. Afterward, American presidents continued to send troops to maintain a semblance of stability and democracy, but the country is still plagued by political infighting and frequent coups. The population of this 18,000-square-mile country is about 8 million, of which some 2 million live in the capital, Santo Domingo. The country produces amber, coffee, cigars, and major-league baseball players. Because of the pleasant weather (except during the hurricane season), tourism is the Dominican Republic's major industry. In 1996, 2 million people visited the country. **See also Haiti, Toussaint-Louverture**

Dönitz, Karl (1891–1980) German grand admiral and last chancellor of the Third Reich. During World War I, Dönitz commanded the submarine *U-68*, which was sunk by the British ship *Queensland* in the Mediterranean in 1918. He was held as a POW in Britain; after being released because of ill health, he promptly went back to sea. After the war he campaigned for an increase in U-boat construction, in flagrant violation of the Treaty of Versailles, which forbade Germany from building warships. In 1935, as Germany was rearming, Dönitz was appointed to command the new U-boat division of the navy. When the war began, he introduced the "wolf pack" system, whereby the U-boats would stay in contact with one another and with his headquarters at Wilhelmshaven and converge upon Allied convoys, usually attacking at night. During the Battle of the Atlantic, German U-boats sank 2,828 merchant ships, while losing only 782 of their own vessels. It was not until May 1943 that the tide turned and the Allies had enough aircraft available to protect their convoys from wolf-pack predation. When Admiral Erich Rader resigned in 1943, Hitler appointed Dönitz commander-in-chief of the German navy, but the invasion of Europe by the Allies in 1944 was the beginning of the end of the war. Before Hitler committed suicide on April 30, 1945, he had appointed Dönitz his successor as chancellor. He was arrested in May, tried as a war criminal at Nuremberg, and sentenced to ten years in prison. He was released in 1956.

double-crested cormorant (*Phalacrocorax auritus*) The most common cormorant in North American waters, found from Newfoundland to the West Indies on the east coast and from Alaska to Baja California on the west. It is blackish all over, with a greenish gloss and no white markings. There is an orange patch at the base of the yellowish bill. Although cormorants consume a great quantity of fish, they do not interfere with commercial fishermen because they favor the "trash fish," such as sculpins, cunners, gunnels, and eels, that people do not want. **See also cormorant, shag**

dragonet (*Callionymus lyra*) A flattened, bottom-dwelling fish found in the Mediterranean and the North Atlantic. Dragonets sometimes bury themselves in sand or mud; they have eyes on the top of their head. During the breeding season, the males, which are always larger than the females, display by spreading their large, brightly colored fins and raising their elongated dorsal fin. If the female is interested, they both swim up from the bottom with their bodies close together, and the eggs

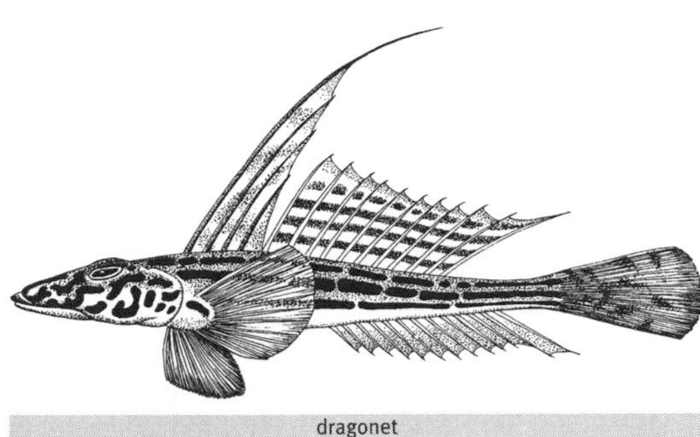

dragonet

are released and fertilized. There are several species, none more than 1 foot in length.

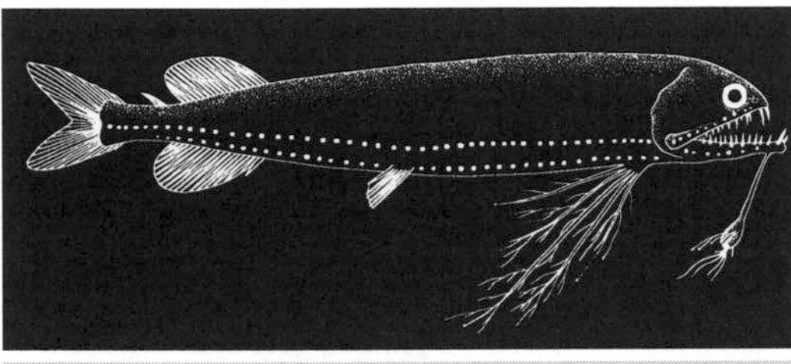

dragonfish

dragonfish (order Stomiiformes) A common name that loosely refers to various deep-sea fishes, also known as viperfishes, snaggletooths, or swallowers. They are characterized by rows of photophores on an elongated body, a disproportionately large head, and an impressive array of large, sharp teeth. Some species have additional light organs that function as lures, usually attached to barbels on the chin. They are almost always black, and they are all aggressive predators, ranging up to 20 inches in length. Because of their specialized jaw structure, some are capable of swallowing fishes larger than themselves, and one species, known as the rat-trap fish (*Malacosteus niger*), has a mouth with no floor, which enables it to snap open its jaws and engulf prey items with no water resistance. Male and female black dragonfish (*Idiacanthus fasciola*) begin life as slender little fishes with their eyes on long stalks, but only the females mature into powerful predators; the 2-inch-long males have no teeth, cannot feed, and function only to fertilize the females. The viperfish (*Chauliodus* spp.) has fangs so long that it cannot close its mouth, and an elongated first dorsal spine that serves as a lure.

See also bioluminescence, viperfish

Drake, Sir Francis (1543?–1596) The first Englishman to circumnavigate the world, Drake was apprenticed to a ship's captain at an early age and made voyages to Guinea and the West Indies. In 1567, Drake had command of the 50-ton *Judith* on a slave-trading expedition under the command of his cousin John Hawkins; their fleet was attacked by the Spanish off the coast of Mexico and all but two of the vessels destroyed. In 1573, Drake set out on his own marauding expeditions, capturing the town of Nombre de Dios on the Isthmus of Panama and bringing home to England 30 tons of silver captured from the Spanish. This brought him fame as well as wealth, and by 1577, with five ships, he embarked on what became a 'round-the-world privateering voyage. Four of the ships were lost or destroyed, and in the *Golden Hind,* Drake passed through the Strait of Magellan, coursed up the west coast of South America (where he plundered Valparaiso), and captured a Spanish treasure ship, complete with sailing charts. Seeking a passage to the Atlantic, he sailed as far north as the present state of Washington, sailed south to what is now San Francisco (claiming it for his sovereign, Queen Elizabeth I), and then turned westward, passing through the Moluccas, the Celebes, and Java before rounding the Cape of Good Hope and heading for England, landing on September 26, 1580. The Queen knighted him for his exploits in 1581. Continuing his authorized privateering, Drake then sacked Vigo in Spain and burned São Tiago in the Cape Verdes. Crossing the Atlantic, he took Santo Domingo and Cartagena, plundered the Spanish colony of St. Augustine in Florida, and rescued Sir Walter Raleigh's colony at Roanoke, Virginia. In 1587, as England prepared for war with Spain, Drake entered the harbor of Cádiz, destroyed thirty Spanish ships, and captured his richest prize, the Portuguese *San Felipe,* with goods valued at £114,000. Aboard the *Revenge,* he was vice admiral of the British fleet that defeated the Spanish Armada in 1588. After an unsuccessful raid on the West Indies in 1595, Drake died of dysentery.

Drake Passage Six hundred miles of open water from Cape Horn at the tip of southern South America to the South Shetland Islands, located 100 miles north of the Antarctic Peninsula. Considered one of the most difficult passages in the world—especially for sailing vessels—this region of the Southern Ocean is the only place on earth where winds can travel around the world uninterrupted by land. Those ships that did take the protected route through the Strait of Magellan often found themselves facing powerful westerly winds and mountainous seas that drove them back to the east, making a passage around the Horn especially treacherous. Francis Drake never actually sailed in the passage that was named for him; he passed through the Straits of Magellan on his way into the Pacific.

See also Cape Horn; Magellan, Strait of; South Shetland Islands

dreadnought Generic term for very large, heavily armored, and heavily armed battleships, first introduced into naval warfare by the British in 1906. In that year, they launched HMS *Dreadnought*, the precursor (and eponym) of many ships to come. By 1918, the British navy had forty-eight dreadnoughts; the Germans, twenty-six. The only dreadnought battle ever fought was the brief and inconclusive Battle of Jutland (May 31–June 1, 1916), where German and British commanders maneuvered their large vessels in complex strategies, firing with their huge guns, with little lasting effect. Large armored battleships dominated the world's navies for the next forty years. **See also battleship**

driftnet fishing First introduced by the Japanese in 1976, polypropylene driftnets can be 5 to 500 miles long, set below the surface to drift overnight. By 1983, there were as many as five hundred driftnet boats targeting squid in the central Pacific, and by 1987, the Japanese fleet had expanded to twelve hundred. Because they indiscriminately kill everything that blunders into them in addition to the object of the fishery, driftnets have been banned from commercial fisheries around the world. Despite the prohibition, which was passed by the United Nations in 1993, some Asian countries persist in deploying driftnets anyway. China, South Korea, and Taiwan set these nets, mostly for tuna and squid, in locations where they are not likely to be observed. The nets are supposed to be hauled in the day after they are set, but in some instances the nets break loose and are lost. These "walls of death" continue fishing, although no one will ever collect the catch. Hundreds of thousands of seabirds, sharks, dolphins, and whales have died in these insidious and illegal nets.

See also gill net, longline fishing, purse seining, trawling

dromon A type of large, oared vessel that operated in the Mediterranean between the ninth and the fifteenth centuries. Believed to be Byzantine in origin, the dromon was powered by many oars—often on two levels—and a single mast with a large square sail. Equipped with a lethal ram above the waterline, dromons were the ships in which the Byzantines defeated the Gothic invaders at Sena Gallica (Italy) in A.D. 551. They were also used to transport men and cargo. Most of the Christian armies in the Crusades traversed the Mediterranean in dromons. **See also galley**

ducking at the yardarm A form of naval punishment, in which a rope was slung under the arms of the miscreant, who was then hoisted up to the end of the yardarm and dropped violently into the sea. Sometimes the procedure was repeated, and sometimes an elaboration was employed, called keelhauling. Both ducking and keelhauling were replaced around the beginning of the eighteenth century by flogging.

See also flogging, keelhauling

dugong (*Dugong dugon*) Sirenians (the manatees and dugongs) spend their entire lives in the water and are unable to move on land. Like the manatees, dugongs are vegetarians, grazing on the bottom and then slowly surfacing to breathe. The mouth of the dugong turns more sharply downward than that of the manatees, and while manatees have a rounded, paddle-shaped tail, dugongs have tail flukes very much like those of a whale or dolphin. Dugongs have no fingernails. The largest known dugong measured 13 feet in length and weighed 1 ton. They are found throughout shallow waters of the Indo-Pacific region, where they are often hunted for their meat, oil, and blubber. In many areas, they are considered an endangered species.

dugong

See also manatee, sirenians, Steller's sea cow

Dumont d'Urville, Jules Sébastien César (1790–1842) French naval officer and explorer, who discovered a beautiful statue on the Greek island of Melos; it was acquired by the Louvre and became known as the Venus de Milo. A founder of the Paris Geographical Society, he set out to locate the lost Jean-François La Pérouse in a vessel he named *L'Astrolabe* after La Pérouse's missing ship. From 1826 to 1829 he explored the South Pacific, from Australia and New Zealand to New Guinea, the Celebes and Mauritius, and located the wreck of the original *Astrolabe* in the waters of Vanikoro, in the Santa Cruz group of the Solomon Islands. King Louis Philippe of France suggested that he try to surpass James Waddell's record for high southern latitudes, so in 1837, in the *Astrolabe* and the *Zelée*, Dumont d'Urville headed for the Antarctic. On his first voyage, he failed to penetrate the pack ice, but in 1840, he returned and landed on the continent at a place he named Terre Adélie, after his wife. (The name of the Adélie penguin has the same origin.)

See also Adélie penguin, La Pérouse

Dungeness crab (*Cancer magister*) An edible crab found in the waters of the Pacific coast of North America, from Alaska to lower California. At a carapace length of 7 to 9 inches, this is the largest and the most commercially important crab on the west coast of North America. Dungeness Point on Washington's Olympic Peninsula was named after a place of the same name on the south coast of England by Captain George Vancouver, who scouted the area in 1792. The *Cancer* crabs are commercially fished wherever they are found: *C. irroratus* (rock crab) and *C. borealis* (Jonah crab) on the east coast of the United States and *C. pagurus* (common or edible crab) in the British Isles. **See also crab**

Dunkirk Seaport on the northern coast of France (called Dunkerque by the French) on the Dover Straits.

The French privateer captain Jean Bart (1650–1702) was born and died at Dunkirk, and was famous for his knowledge of the coast and his victorious battles with the Dutch while in the service of Louis XIV from 1672 to 1678. From May 26 to June 4, 1940, after the German breakthrough into France, 198,000 members of the British Expeditionary Force and 140,000 French and Belgian troops were evacuated and brought across the Channel to England in hundreds of boats of every description. Dunkirk was liberated by the U.S. Army in 1945, but most of it was destroyed in the fighting. With Place Jean Bart as its center, the town was rebuilt, and is now one of France's most important ports. The population is around 70,000.

dusky dolphin

dusky dolphin (*Lagenorhynchus obscurus*) So similar to the Pacific white-sided dolphin (*L. obliquidens*) that many scientists believe they are the same species, the dusky dolphin is found only in the Southern Hemisphere. The dusky's dorsal fin is not so sharply hooked, and not so strongly two-toned. Like its Northern Hemisphere counterpart, the dusky forms large groups and is particularly acrobatic, leaping energetically out of the water and often somersaulting. Duskies are also inveterate bow riders and approach ships frequently. They are often seen in the company of right whales in Argentine waters. They are also found in the waters of the Cape of Good Hope, New Zealand, and many subantarctic islands, such as Campbell, Kerguelen, and the Falklands. In Antarctic waters farther south, the representative "lags" are the hourglass dolphin (*L. cruciger*) and Peale's dolphin (*L. australis*).

See also hourglass dolphin, Pacific white-sided dolphin, Peale's dolphin

dusky shark (*Carcharhinus obscurus*) True to its name, the dusky shark is a dirty gray above and white below. It is found around the world in inshore temperate waters; it reaches a maximum length of 11½ feet and a weight of 450 pounds. Unlike the bull shark (*C. leucas*)—with which it can easily be confused—the dusky does not frequent estuaries. Duskies can also be differentiated from sandbar sharks (*C. plumbeus*) because sandbars have a much higher dorsal fin. The dusky feeds on fishes of all kinds, as well as crabs, lobsters, octopuses, cuttlefish, and squid. There have been few reported attacks on people, but a shark of this size should be considered dangerous. In Australia, this species is commonly known as the black whaler.

See also bronze whaler shark, bull shark, carcharhinid sharks, sandbar shark

Dutch East India Company Founded by the Dutch in 1602 to protect their trade in the Indian Ocean and to assist in their war of independence from Spain (the Netherlands had belonged to Spain since 1547, when Habsburgs invaded and conquered the Low Countries). The company was granted a monopoly to negotiate with native rulers, to build forts, and to administer outposts, all in the interest of trade. In 1619, the company took Jakarta on the island of Java, renamed it Batavia, and made it the center of Dutch administration in the East Indies. Under the forceful administration of governor generals like Jan Pieterszoon Coen and Anthony van Diemen, the company prospered, but by the eighteenth century, it had changed from a predominantly commercial shipping enterprise to an agricultural organization more concerned with the production of tea, rubber, coffee, and spices. In 1652, the company established the colony at the Cape of Good Hope; it remained Dutch until it was taken by the British in 1814. In 1799, with the company heavily in debt and scandalously corrupt, the Dutch government revoked its charter and took over its debts and possessions. **See also British East India Company, Indonesia, Java**

dwarf sperm whale (*Kogia simus*) The Latin word *simus* means "snub-nosed," and it accurately describes this little whale. It is very similar to the pygmy sperm whale, but as far as is known, it grows no longer than 8½ feet. (The pygmy sperm whale grows to 13 feet.) The two species can be differentiated by the much taller, more dolphinlike dorsal fin of the dwarf sperm whale. Both species have an undershot lower jaw and tiny pythonlike teeth in both jaws, unlike the great sperm whale, which has teeth only in the lower jaw. *Kogia simus* is known primarily from animals stranded on scattered beaches around the temperate waters of the world. Both little sperm whales have mysterious "bracket marks" behind the head that look very much like gills. It is very difficult to imagine what evolutionary purpose such marks would serve. Why would a whale want to resemble a fish? **See also pygmy sperm whale, sperm whale**

dwarf sperm whale

E

eagle ray (family Myliobatidae) Open ocean rays that swim near the surface but find their food on the bottom. They feed on shellfishes and crustaceans, which they crush with their powerful flat teeth. Myliobatids have distinct heads, as contrasted with the typical rays, in which the pectoral fins blend into the often pointy nose. They have spines, but these are short and largely ineffectual. The spotted eagle ray (*Aetobatus narinari*), which may be 7 feet across the wings, may have as many as five spines at the base of its very long tail. Since they lack the rostal protrusion of the skates and stingrays, they have a pug- or snub-nosed appearance. (*A. narinari* is sometimes called the duck-billed ray.) A group with very broad wings is known as butterfly rays. Also in this group are bat rays (*Myliobatis californica*), the bullnose ray (*M. freminvillei*), and the cow-nosed ray (*Rhinoptera bonasus*). They often aggregate in huge schools over shellfish beds in shallow water.

See also manta, stingray

spotted eagle ray

Earle, Sylvia (b. 1936) A renowned undersea explorer, Sylvia Earle was trained at Florida State (B.S.) and Duke (M.S. and Ph.D.), and has been involved in both the private and public sectors. Trained as an algologist, she studies marine plants, and is a research associate in botany at the Smithsonian. She was chief scientist for the National Atmospheric and Oceanic Administration (NOAA) and a cofounder and director of Deep Ocean Engineering, designers of one-person submersibles and more than four hundred remotely operated vehicles (ROVs) that have been delivered to thirty-one nations and eleven navies since 1985. She has amassed more than six thousand hours underwater, and is explorer in residence for the National Geographic Society. She is the author of *Exploring the Deep Frontier* (1980; with Al Giddings) and *Sea Change* (1995).

East China Sea The China Sea, which washes the Asian mainland on the west, is usually subdivided into two parts, the East China Sea and the South China Sea. The East China Sea (*tung hai*—"eastern sea" in Chinese) is bounded on the east by Taiwan and the Ryukyu Islands. It is south of the Yellow Sea and connected to the South China Sea by the Taiwan Strait. Because the boundaries of this arm of the North Pacific Ocean are unclear, its area has been variously estimated at between 290,000 and 480,000 square miles. It is heavily fished for tuna, mackerel, anchovy, shrimp, and shellfish, and provides much of the animal protein consumed along the densely populated coast of Southeast Asia.

See also South China Sea, Taiwan, Yellow Sea

Easter Island Discovered on Easter Sunday, 1722, by Dutch navigator Jacob Roggeveen, Easter Island is one of the most remote islands in the world. The nearest island is Pitcairn, 1,200 miles away, and the closest major landmass is Chile, 2,200 miles to the east. (Easter Island inhabitants are citizens of Chile.) The earliest inhabitants were of Polynesian stock, but how or when they arrived on the island they called Rapa Nui is unknown. When the viceroy of Peru sent an expedition to Easter Island in 1770, there were perhaps 3,000 people on the island, but when Captain James Cook arrived four years later, disease or civil war had reduced the population to 700 men and 30 women. The most famous inhabitants of the island are the massive stone heads, locally known as *moai*, of which there are more than six hundred. Some are 50 feet high and weigh as much as 50 tons. (One unfinished statue, still attached to the rock, is 68 feet tall.) There are many theories about the fabrication and transportation of the *moai*, but no one really knows who made them, how, or the way in which they were moved from the quarries where they were carved and set upright in rows, staring out with gigantic, empty eyes. During his 1722 visit, Roggeveen saw no overturned statues, but when Cook arrived, some fifty years later, he noted that a large number of the *moai* had been overturned. In the 1860s, slave traders seeking cheap labor to mine the guano on Peruvian islands kidnapped a thousand Easter Islanders and put them to work. Most of them died, and those who returned to the island brought smallpox and other diseases, reducing the indigenous population to about 100. When the island was annexed by Chile in 1888, the population began to rise. It is now around 2,000.

East Siberian Sea Known as Vostochno-Sibirsokoye More in Russian, the East Siberian Sea is that part of the Arctic Ocean that lies west of the Laptev Sea and east of Wrangel Island. The Long Strait connects it with the Chukchi Sea. The East Siberian Sea covers some 361,000 square miles, and its greatest depth is 510 feet. Frozen most of the year, it is navigable only in August and September.

See also Kara Sea, Laptev Sea, Wrangel Island

ebb The flow of the tide as it recedes, from the ending of the period of slack water (when there is no apparent tidal movement) at high tide to the start of the period of slack water at low tide. It lasts approximately six hours.

echinoderms Echinoderms are the only major group of animals that are all confined to the marine environment. Most fulfill the requirements of the name; they have spiny skins (*echinos,* "spiny"; *dermis,* "skin"). There are five basic groups: sea lilies and feather stars (Crinoidea), starfishes (Asteroidea), brittle stars and basket stars (Ophiuroidea), sea urchins (Echinoidea), and sea cucumbers (Holothuroidea). They are all based on a five-pointed plan arranged around the axis of the mouth, known as pentamerous radial symmetry. Echinoderms have an endoskeleton that is composed of ossicles, crystals of calcite (calcium carbonate) that strengthen the body without adding much weight. These ossicles may be loosely joined, as in starfishes and brittle stars, or fused together to form a reinforced internal structure called a test, as in the sea urchins. (The test of the sand dollar—a kind of sea urchin—is probably the most familiar.) In the holothurians (sea cucumbers) the ossicles are substantially reduced, giving the animals their characteristic soft, fleshy appearance. Because of the absence of a head, there is no brain, and no aggregation of nerve organs in a particular part of the body, so they are sensitive to touch and waterborne chemicals. Echinoderms move by employing a water vascular system, consisting of water-filled tubes under pressure. The tube feet (*podia*) of sea cucumbers, sea stars, and sea urchins can be contracted or extended by water pressure from within, and moved by associated muscles, but the brittle and basket stars move by bending their arms. Some echinoderms, like the sea lilies and feather stars, feed almost passively, by trapping current-borne organisms in their waving arms, but others, particularly the starfishes, are active, aggressive predators. **See also basket star, brittle star, sea cucumber, sea urchin, starfish**

echolocation The process by which toothed whales (odontocetes), such as dolphins and sperm whales, can generate sounds in their heads that they can broadcast directionally and then read the returning echoes to identify prey species, predators, landscape features, boats, etc.

See also stranding

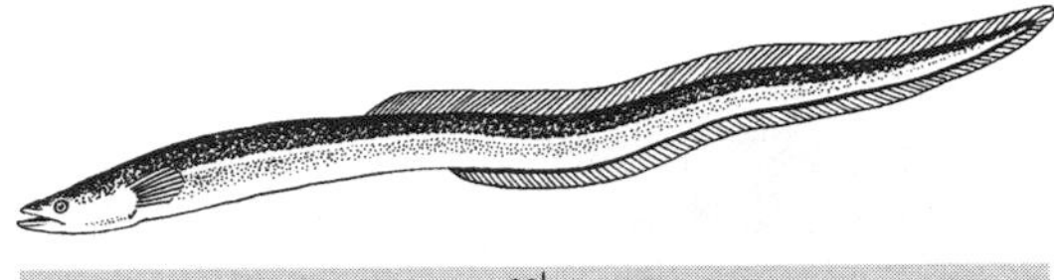

eel

eel (order Anguilliformes) Any of five hundred species of fishes, identifiable by their slender appearance, usually scaleless skin, and long dorsal and anal fins that are continuous around the tail tip. Snipe eels have elongated jaws, and snake eels have pointed tails that enable them to burrow backward into the sand. All eels pass through a transparent, leaflike form known as a leptocephalus. Common eels (*Anguilla anguilla*) are born in freshwater streams in Europe and North America, then migrate to the Sargasso Sea, where, after spawning, they die. The young eels (known as elvers) then return to either side of the North Atlantic, where they ascend the freshwater streams and rivers where they were born, and begin the process anew.

See also conger, leptocephalus, moray, snipe eel

eelpout (family Zoarcidae) Long, thin fishes, eelpouts resemble true eels in some ways, but they have larger heads, prominent gill openings, and thick, fleshy lips that give them a pouting countenance. They are drab-colored bottom dwellers that rarely get longer than 2 feet and that hide in rocky crevices, coming out to feed on crabs and mollusks. Eelpouts are found in all the world's cold oceans, from the Arctic to the Antarctic. The ocean pout (*Macrozoarces americanus*) is an Atlantic species, found from New Jersey to Greenland. While other species bear their young alive, the ocean pout is an egg layer, and both the male and the female take turns guarding the eggs. An eelpout named *Thermarces cerberus* was found in the vicinity of the hydrothermal vents of the Galápagos Rift Zone. (In Greek mythology, Cerberus was the three-headed dog that guarded the gates of hell.)

See also brotulid, grenadiers, hydrothermal vents, pearlfishes

eider (*Somateria mollissima*) Because it is the most widely distributed of the eiders, this species is also known as the common eider. A heavy-bodied seagoing duck with a sloping profile, it nests along the shores of Iceland, Scandinavia, and northern Britain (the Orkneys, Shetlands, and Hebrides), occasionally as far south as the Gulf of St. Lawrence. It winters throughout northern Europe, Canada, and northern New England. The drakes are boldly patterned in black and white; the ducks are a drab brown. Eiders are quite numerous; some estimates have placed the European population at 2 million or more. They aggregate in large flocks, rafting outside the surf line or flying low over the water. Like most other ducks, female eiders insulate their nests with down plucked from their breasts, but the down of eiders is particularly desirable: it is the famous "eiderdown" of commerce. It is used in quilts, parkas, pillows, and sleeping bags. (The down of eiders consists of fluffy feathers with hundreds of tiny filaments radiating from a central core, which traps pockets of air, providing maximum insulation.) In Iceland there is a relatively large industry devoted to the collection of eiderdown from the nests.

eider

See also king eider, sea ducks, spectacled eider, Steller's eider

Ekman, V. Walfrid (1874–1954) Swedish hydrographer who, among other things, worked with Fridjtof Nansen on the circulation of the oceans. When the *Fram* was trapped and drifting in the ice between 1893 and 1896, Nansen reported that she was drifting off the wind, some 20° to 40° to the right. Ekman built a glass tank, filled it with layers of colored liquid, and towed a model ship through it. He observed that waves below the surface, which he called internal or boundary waves, corresponded to a larger system of undersea currents, prevailing winds, and the rotation of the earth that combined to bring the deeper waters toward the surface. This movement of water, which became known as the Ekman spiral, explained the characteristic upwellings along the coasts of Morocco, Southwest Africa, California, and Peru. He also provided the data for the Ekman transport, which explained certain phenomena of wind-driven currents, and the Ekman layer, which explained oceanic interfaces occurring at certain depths. **See also Nansen**

elasmobranch Literally, "plate gills" or "strap gills," referring to the multiple nature of the external gill slits. In the class Chondrichthyes (cartilaginous fishes), which includes the sharks, rays, and chimaeras, the subclass Elasmobranchii consists of the sharks and rays. Chimaeras have soft gill covers, whereas the elasmobranchs have five to seven gill slits on each side. Most elasmobranchs have five gill slits, but the sixgill and sevengill sharks are self-explanatory, and one species of sawshark (*Pliotrema warreni*), and one stingray (*Hexatrygon bickelli*), have six gill slits. In the sharks these slits are on the side of the head, but in the skates, rays, angel sharks, and sawfishes, they are on the underside of the body. **See also chimaera, rays, sawfish, sawshark, shark, skates**

Elba Island off the west coast of Italy in the Tyrrhenian Sea, 6 miles from Piombino on the mainland. Elba was known to the Etruscans and the Romans for its iron ore, and there are still iron works at Portoferraio, its main town. It was ruled by Pisa, but it passed to Genoa in 1290, thence in 1399 to the dukes of Piombino, who ceded it to Cosimo di Medici in 1548. It was then ruled by Naples, and in 1802, it became the property of France. After Napoleon's disastrous retreat from Moscow in 1812, Paris fell to the allied forces of Prussia, Great Britain, and Sweden, and Napoleon abdicated, taking up residence on Elba on May 4, 1814. As sovereign of this newly designated independent principality, Napoleon remained there until February 26, 1815, when he returned to Paris for the "hundred days," during which he reassembled his armies and defeated the Prussians at Ligny on June 16, 1815. Two days later, he was defeated by Wellington at Waterloo, and he abdicated again, this time in favor of his son. He was not permitted to return to Elba, and was exiled by the British to the remote South Atlantic island of St. Helena, where he died in 1821. After its chaotic history, Elba has now become part of Tuscany. The permanent population, greatly enhanced by summer tourists, is around 27,000. **See also St. Helena**

electric ray (*Torpedo* spp.) Small rays, also known as torpedoes, known for their ability to produce electrical shocks. They are found worldwide in tropical and temperate waters, mostly shallow, but occasionally as deep as 3,000 feet. The electrical capability, used by the ray for defense and to capture prey, is generated by two organs of modified muscle tissue, one on each side of the head. In ancient Greece and Rome, the shocks of electric rays were used to cure gout, headache, and other maladies. **See also stingray**

Elephant Island Elephant Island, the northernmost of the South Shetlands, is a mountainous, ice-covered island with several prominent glaciers. About 24 miles long, it was named for the elephant seals that were hunted by sealers since the early nineteenth century. It was here in 1916 that Ernest Shackleton made his first recorded landing after the *Endurance* was trapped in the ice of the Weddell Sea. Leaving Frank Wild and twenty-one men behind, Shackleton set out in a longboat, the *James Caird,* for South Georgia, where he hiked over the mountains to the whaling station at Stromness. After four attempts to return to Elephant, Shackleton finally rescued his men.

See also Shackleton, South Shetland Islands

Elephant seals (*Mirounga* spp.) There are two species of elephant seals, the northern (*M. angustirostris*) and the southern (*M. leonina*). They are very similar (the southern is somewhat larger), and are differentiated primarily by distribution. The northern elephant seal is found only in the northeast Pacific, from Baja California to Alaska, and the southern, north of the pack ice in the Southern Ocean, on various subantarctic islands and on the Valdés Peninsula in southern Argentina. The males, by far the largest of the pinnipeds, can get to be 20 feet long, and weigh upward of 4 tons. Females are considerably smaller and do not reach more than 11 feet in length. Elephant seals get their common name from the large, inflatable proboscis of the males, through which they make a variety of gurgling and growling noises, especially during the breeding season. Males fight viciously for dominance over a harem of females. When not hauled out on their breeding grounds, elephant seals spend most of their time at sea, where they dive to prodigious depths—well over a mile—in search of the fish and squid on which they feed. Once brought to dangerously low levels by hunters, both species have staged remarkable comebacks. **See also sealing**

elephant seal

Ellef Ringnes An uninhabited island in the Northwest Territories of Arctic Canada across the Peary Channel to the west from Axel Heiberg Island, Ellef Ringnes is some 4,300 square miles of snow and ice. In 1898 Otto Sverdrup took Nansen's *Fram* north again, exploring the region around Ellesmere Island on foot and by sledge. He surveyed and named the islands of Axel Heiberg, Ellef Ringnes, and Amund Ringnes. (Heiberg and the brothers Ringnes, the joint founders of the Ringnes Brewery in Oslo, sponsored Sverdrup's expedition.) Now part of Canada, the three islands are known as the Sverdrup Islands. Since James Clark Ross located the magnetic North Pole on the Boothia Peninsula in 1831, it has drifted some 600 miles to the north, and is now located on Ellef Ringnes.

Ellesmere Island Part of the archipelago known as the Queen Elizabeth Islands (which includes Melville, Devon, Bathurst, and Cornwallis Islands), Ellesmere is Canada's second largest island, after Baffin, at 82,000 square miles. It is the northernmost island in the Arctic Archipelago, and Cape Aldrich is Canada's most northerly point. (It is only 30 miles farther south than Cape Morris Jesup in Greenland, the most northerly point of land in the Northern Hemisphere.) Ellesmere Island was first sighted by William Baffin in 1616, but it was not explored until Edward Inglefield, captain of the *Isabel* on a search for John Franklin, explored the island in 1852. It is named for Francis Egerton, first Earl of Ellesmere. In 1875–1876, George Nares in HMS *Alert* wintered at Cape Sheriden. A permanent ice cap covers the eastern half of the island, and it is largely uninhabited. In 1999, when Nunavut was separated from the Northwest Territories and put under control of the native peoples, all of the Queen Elizabeth Islands were included.

See also Baffin Island, Bathurst Island (1), Cornwallis Island, Devon Island, Melville Island

Ellis, Richard (b. 1938) Marine life illustrator and author. Born and raised in New York, Ellis has had a lifelong fascination with the sea and its inhabitants. After college and service in the U.S. Army (in Hawaii), Ellis began work as an exhibit designer at the Academy of Natural Sciences of Philadelphia. He then joined the staff of the American Museum of Natural History in New York, where, among other things, he designed the Hall of the Biology of Fishes and the life-sized blue whale model. His paintings of sharks, whales, and fishes hang in numerous galleries, museums, and private collections around the world. He is the author, and sometimes illustrator, of more than a hundred magazine articles and eleven books, including *The Book of Sharks* (1975), *The Book of Whales* (1980), *Dolphins and Porpoises* (1982), *Men and Whales* (1991), *Great White Shark* (1991; with John McCosker), *Physty: The True Story of a Young Whale's Rescue* (1993), *Monsters of the Sea* (1994), *Deep Atlantic* (1996), *Imagining Atlantis* (1998), *The Search for the Giant Squid* (1998), and this one.

Ellis Island Located in upper New York Bay, closer to New Jersey than New York, Ellis Island was originally named Oyster Island, but it was then named for Samuel Ellis, who acquired it before 1785. Purchased by the state of New York in 1808, it served as a fort and arsenal to defend the harbor, but it was designated as a federal immigration center in 1890. Between its opening in 1892 and the year 1924, some 12 million immigrants were processed here (71 percent of all immigrants to the United States), the largest mass migration in human history. The original wooden buildings were destroyed by fire in 1897 and replaced by a brick and limestone building in the style of the French Renaissance. As the number of immigrants increased to .5 million per year, the government added to the island by landfill taken from the construction of subway tunnels and Grand Central Station, and its original 3 acres were enlarged to 27½ acres. In 1924, when Congress curtailed mass immigration, the role of Ellis Island diminished, but it remained an extraordinarily important part of

American history. In 1965, it was designated as a national monument, but it was not until 1990 that it opened as a museum—the largest historic restoration in American history. The Statue of Liberty and Ellis Island National Park are reachable by ferry services from New York and New Jersey. In 1998, in response to a suit brought by the state of New Jersey, the U.S. Supreme Court ruled that 90 percent of the island—everything but the original three acres—belonged to New Jersey.

See also Liberty Island, Manhattan

El Niño A poorly understood meteorological phenomenon that gets its name (*El Niño* means "the Christ Child" in Spanish) from its frequent appearance around Christmastime. It consists of a weakening of atmospheric circulation, particularly the Southeast Trade Winds off the west coast of South America. (This is sometimes known as the Southern Oscillation.) The cold Peru Current, which usually flows eastward toward land, reverses and flows away from it. Sea surface temperatures can rise as much as 14°F. A halt of the upwelling of cold water along the shore either kills or drives the anchovies of that region into deeper water. Seabirds and larger fishes such as tuna that feed on anchovies begin to starve and die, too. Dead and decaying fish pollute the water, and the escaping gas is said to be strong enough to blacken the paint of passing ships. The anchovies, which supply a third of the world's supply of fish meal, are a major part of the economy of Peru. The arrival of El Niño is marked by a tremendous increase in rainfall on the west coast of South America, with flooding and massive soil erosion, but the environmental disruption can affect the weather around the world, causing droughts elsewhere, and maybe even cyclones and tornadoes. Completely unpredictable, destructive El Niño conditions occurred in 1891, 1925, 1941, 1958, 1972, and 1983. A major El Niño event took place in 1997–1998, causing, among other things, reduced rainfall in Indonesia, which resulted in major forest fires; flooding in southern Brazil and drought in the north; a drought in Hawaii; heavy rains in the Galápagos; flooding in Peru and Ecuador; major rainstorms and flooding, warmer waters, and unexpected fishes and squid off the coasts of California and Oregon; dying sea lions and elephant seals in California; and unprecedented South Pacific storms, such as super-typhoon Paka, which produced winds of 236 mph—the highest surface winds ever recorded.

emperor penguin (*Aptenodytes forsteri*) The largest of the penguins, the emperor can stand 4 feet high and weigh 80 pounds. Emperor penguins have extremely dense plumage and thick fat deposits, because they live and breed in the pack ice of the Antarctic, one of the most inhospitable environments on earth. In March, the adult birds leave the sea and trudge as many as 150 miles over the ice to reach their breeding colonies. Six weeks after their arrival, the female lays a single egg, which the male sits on for nine weeks, as blizzards howl, temperatures drop to –80°F, and the males huddle together for warmth. The female, who has been feeding at sea, now returns, and the male, who has been fasting for fifteen weeks, heads for the sea. When he returns, both parents take turns feeding the chick, making regular trips to the sea for food. When the chicks are about seven weeks old, they huddle together in a crèche; they are able to recognize their parents by their high-pitched squeals. After the chicks reach five months of age, their parents abandon them, and they make their way across the ice to the sea. Emperors are the deepest-diving penguins, capable of reaching depths of 1,700 feet. **See also Antarctica, king penguin, penguin**

emperor penguin

emperors (family Lethrinidae) Including some forty species of perciform fishes found only in the Indo-Pacific, emperors resemble sweetlips (Haemulidae) and snappers (Lutjanidae), to which they are related. They are found along the sandy fringes of coral reefs, where they forage for invertebrates. Most species are nocturnal and spend the day in caves or crevices in the reef. They are caught commercially in Australia and South Africa.

Endeavour Captain James Cook's flagship on his first exploring voyage, 1768–1771, HMS *Endeavour* was a bluff-bowed Whitby collier built in 1764, 97 feet in overall length, with a gross weight of 369 tons. With naturalists Joseph Banks and Daniel Solander aboard, Cook headed first for Tahiti, where they remained for three months to observe the transit of Venus. From there they sailed to New Zealand, where Cook established that the north and south islands were separate, and across the Tasman Sea to Australia, entering Botany Bay on January 27, 1770, a day now celebrated as Australia Day. While navigating the inner passages of the Great Barrier Reef in June, *Endeavour* was holed, but the leak was stopped by "fothering," drawing a sail impregnated with oakum under the ship's bottom. She was repaired at what is now Cooktown in Queensland, and the cruise continued around the world, to

Batavia, Cape Town, St. Helena, and finally England on July 12, 1771. **See also Cook (James), Great Barrier Reef**

Endurance A barkentine of 144 feet overall weighing 300 gross tons, chosen by Ernest Shackleton for his voyage to the Antarctic and attempt to reach the South Pole in August 1914. He and his crew sailed HMS *Endurance* to South Georgia, and then to the edge of the pack ice. They maneuvered through the ice for almost 1,000 miles before becoming stuck fast at 76° south, only 60 miles from the Antarctic continent, where they would have begun their overland journey. The ship drifted in the ice for nine months, and by October 1915 she was abandoned and her crew removed all their supplies and camped on the ice. On November 21, the battered ship sank; the expedition escaped to Elephant Island. With five others Shackleton began his epic rescue voyage, which involved sailing a small boat 800 miles across the Drake Passage back to South Georgia, and eventually returning to rescue the remaining crew members. The story of Shackleton's heroic exploits has been told in several books, many of them named *Endurance*. **See also Drake Passage, Elephant Island, Shackleton, South Georgia**

English Channel Known to the French as La Manche ("the sleeve"), the body of water that separates England and France is 560 miles long, and at its widest (between Lyme Bay and St. Malo), it is 150 miles across. The closest the two countries get to each other is the 21-mile stretch from Dover to Cape Gris-Nez. The channel is open to the Atlantic Ocean in the west; in the east, the Strait of Dover connects it to the North Sea. Islands in the channel are the Isle of Wight, which is part of England, and the Channel Islands (Guernsey, Jersey, Alderney, and Sark), which, although they are much closer to France, are dependencies of England. Principal channel ports in England are Dover, Plymouth, Southampton, and Portsmouth; their French counterparts are Calais, Cherbourg, and Le Havre. The entire battle of the Spanish Armada (1588) was fought in the English Channel except for the escape of the Spanish ships through the Strait of Dover. From May 26 to June 4, 1940, more than 300,000 Allied troops, cut off by the German advance through France, were evacuated from Dunkirk in hundreds of small boats. Freight and passenger ferries now connect England and the Continent at various points, and in 1994, a 31-mile-long train and automobile tunnel (officially the Eurotunnel or Channel Tunnel, but popularly known as the Chunnel) was completed between Folkestone, England, and Sangatte, France. The high-speed train trip takes about thirty-five minutes.

English sole (*Parophrys vetulus*) One of the right-eyed flounders, the English sole is found over soft bottoms to 300 fathoms in the Pacific from the Bering Sea to Baja California. At a maximum length of 22 inches, it ranks as one of the most important flatfishes caught by commercial trawlers.
See also flatfishes, flounder, halibut, plaice, winter flounder

Eniwetok Circular atoll in the Ralik (western) chain of the Marshall Islands in the central Pacific. Taken by Japan in 1942, it was retaken by American forces in 1944 and its anchorage turned into a naval base. After the war, the Marshalls came under the administration of the U.S. Navy, and in 1946, Eniwetok (along with Bikini Atoll) was designated a test site for atomic weapons. On November 1, 1954, the first hydrogen (thermonuclear) bomb, code-named "Ivy Mike," was exploded at Eniwetok. **See also Bikini, Marshall Islands**

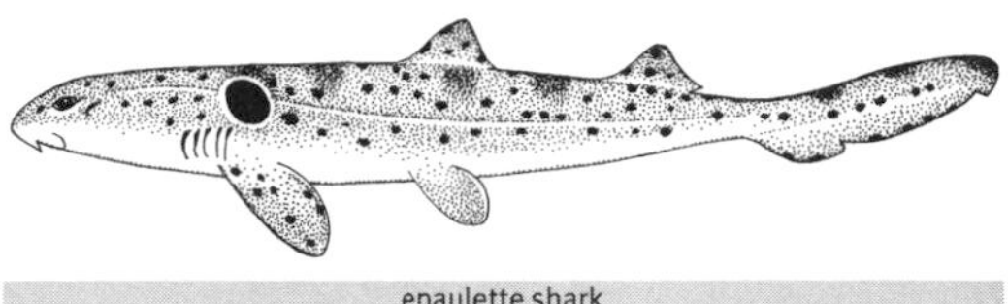

epaulette shark

epaulette shark (*Hemiscyllium ocellatum*) A relative of the bamboo sharks, the 3-foot-long epaulette shark gets its common name (and its specific name: *ocellatum* means "with spots") from the large black spot above each pectoral fin. It is found in the shallow waters of northern Australia and New Guinea.
See also bamboo shark

equator Imaginary great circle around the earth that is everywhere equidistant from the two poles, forming the base line from which latitude is measured. Designated as 0° latitude, it measures 24,902 miles around and passes through northern South America, central Africa, the islands of Sumatra, Borneo, and Celebes, the Gilbert Islands, and the Galápagos.

erect-crested penguin (*Eudyptes sclateri*) Found in the subantarctic waters of New Zealand and the neighboring islands (Bounty, Antipodes, Auckland Islands), the erect-crested penguin differs from the other crested penguins in having yellow crest feathers that it can raise upward; the crests of the other species droop. The population is estimated at 200,000 breeding pairs.
See also fiordland penguin, macaroni penguin, rockhopper penguin, royal penguin, Snares penguin

Ericsson, Leif: See Leif Ericsson

Eric the Red (c. 985) A Norse explorer named for the color of his hair, Eric the Red left Norway around A.D.

984 to escape trial for manslaughter and fled to Iceland. He left Iceland shortly thereafter, because he had heard of a new territory to the west, and when he sailed to investigate, he discovered a frozen land that he named Greenland, probably to entice settlers. In the summer of 985, with a group of hardy Icelanders, he rounded Cape Farewell, and landed at Eriksfjord, where he founded the first European settlement in Greenland, which he called Brattahild. He was the father of Leif Ericsson. **See also Leif Ericsson, Vikings**

Eskimo curlew (*Numenius borealis*) At a total length of 13 to 14 inches, this is the smallest of the curlews. Now so rare that it is considered extinct, the Eskimo curlew was once one of the most numerous birds in North America. As recently as the 1880s, they migrated in dense flocks that numbered in the millions from northern Hudson Bay and Labrador to Patagonia. Their return flight took them over the prairies and great plains of North America, where market hunters were waiting for them. The last confirmed sighting was in 1963, but conservationists and birders are still hopeful that more will be seen among flocks of the very similar—but larger—whimbrels or Hudsonian godwits.

Eskimo curlew

See also plovers, sandpiper, shorebirds

Essex Nantucket whaling ship that was sunk by a whale on November 20, 1820. Some 1,800 miles west of the Galápagos, first mate Owen Chase watched a large bull sperm whale charge the *Essex* and ram the ship with its head. The whale attacked the ship again and punched a great hole in her hull, causing her to take on water and sink. Three boats with twenty men aboard set out for the east coast of South America, 2,700 miles away. After a month at sea, they reached uninhabited Henderson Island, 300 miles west of Pitcairn. There was no food or water there, but three men decided to remain anyway. The others set out again, and by December, they began to die of starvation. The first victims were consigned to the sea, but by February 18 the survivors drew lots to determine who would have to die so the others would not starve to death themselves. Of the twenty men who left the *Essex*, five survived the ordeal. They had covered 4,500 miles and spent ninety-seven days in open whaleboats when they were rescued by the British frigate *Indian*. The story of the whale's sinking of the *Essex* was known to Herman Melville, and he used it in the writing of *Moby-Dick*.

See also *Moby-Dick*, Mocha Dick, sperm whale, whaling

euphausiids Shrimplike crustaceans, usually known as krill, that form the basis of many oceanic food chains. The best known is *Euphausia superba*, which reaches a length of 2½ inches, but *Thysanopoda cornuta* can be 4 inches long. In the North Atlantic, they are represented by *Meganyctiphanes*, *Nyctiphanes*, and *Thysanoessa*, which are eaten by various fishes and baleen whales. Euphausiids differ from shrimps in that their gills are exposed, whereas those of the shrimps are covered by the carapace. Most euphausiids are transparent or translucent, and, with the exception of some deepwater species, all are bioluminescent. **See also krill, shrimp**

Ewing, Maurice (1906–1974) Texas-born geophysicist who made fundamental contributions to the understanding of the formation of marine sediments and ocean basins. First using seismic methods in 1935, he studied the structure of the earth's crust in the Atlantic Basin and along the Mid-Atlantic Ridge, and proposed that earthquakes are associated with the rift zones that encircle the globe, and that seafloor spreading was the "engine" that caused the continents to drift. Ewing took the first deep-sea photographs in 1939. In 1949, he founded and was named the first director of Columbia University's Lamont Geological Observatory (now Lamont-Doherty Geological Observatory). He was the author of many important papers, popular articles, and books, including the *The Floors of the Oceans I: The North Atlantic* (1959). Ewing was the subject of William Wertenbaker's *The Floor of the Sea* (1974).

Exquemelin, Alexandre Oliver (c. 1645–c. 1707) The author of *The Buccaneers of America*, published first in Dutch in 1678 as *De Americaensche Zee-Roovers*, then in German (1679) as *Americanische Seeraüber*, and in Spanish (1681) as *Piratas de la America*. The first English edition appeared in 1684, as *Bucaniers of America*. Of Exquemelin, nothing is known but what he wrote himself, and it would appear to be true, as he wrote on the title page, that he, "of necessity, was present at all these acts of plunder." (His name is also spelled Exquemeling or Esquemelin.) Around 1666 he went to Tortuga, where he joined the buccaneers, served until 1674, and then returned to Europe. He may have been the "Oexmelin" who appears on Dutch rolls as a surgeon; he would have then written the original book in Dutch while living in Amsterdam. Whoever he was, he wrote a seventeenth-century best-seller, the principal source of our information about people like Henry Morgan, L'Ollonais, and other buccaneers.

See also buccaneer

Exxon Valdez Just after midnight on March 24, 1989, the 987-foot-long supertanker *Exxon Valdez* ran aground, dumping 11 million gallons of crude oil into

Alaska's Prince William Sound, the largest oil spill ever to occur in the United States. Joseph Hazelwood, captain of the *Exxon Valdez,* had handed control of the ship to third mate Gregory Cousins when the ship rammed Bligh Reef, tearing a gaping hole in the cargo tanks. Winds and shifting tides spread the oil over 10,000 square miles along the Alaskan Peninsula, qualifying it as one of the worst ecological disasters in history. The oil moved for 1,500 miles along the coastline of Alaska, contaminating portions of the Kenai Peninsula, lower Cook Inlet, and the Kodiak Islands. High winds blew the oil slick onto the shore, creating havoc with living creatures. The actual totals will never be known, but it has been estimated that at least 100,000 seabirds died, along with 5,000 sea otters; 150 bald eagles; hundreds of seals, sea lions, whales, dolphins, porpoises; and countless fishes. The devastation of the spawning grounds of pink salmon, black cod, and herring destroyed the livelihood of dozens of fishing communities. (Seven years after the spill, the Pacific herring population unexpectedly collapsed.) Exxon was completely unprepared to deal with such an event, and it failed to provide the necessary equipment or expertise to contain the spill or, when that failed, to clean it up. Under pressure from the U.S. government and Alaskan environmental groups, the oil company spent $1.9 billion to clean up the mess, and another $.5 billion on public relations campaigns. In addition, Exxon was ordered to pay $5 billion in punitive damages to the fishermen, native Alaskans, and landowners whose livelihood and property were affected by the spill. Captain Hazelwood was found guilty of negligence and fined five thousand dollars. Residual oil still persists on and under the surface of the beaches, and the cleanup continued into 1994.

See also *Amoco Cadiz, Braer, Castillo de Bellver,* oil spill, supertanker, *Torrey Canyon*

F

factory ship Any large fishing vessel that incorporates facilities for processing the catch before bringing it in. The term was first used for the "floating factories" used by Norwegian whalers in the Antarctic, but they were just large ships from which the whales were flensed (stripped of blubber). Later, the carcasses were brought to shore stations, such as Grytviken on South Georgia, but the invention in 1925 of the stern slipway completely changed the nature of whale processing. Now the 100-ton whales could be hauled right through the gaping hole in the stern and dragged onto the flensing deck. They would be chopped up and boiled into oil in huge pressure cookers aboard the ship, eliminating the umbilical that previously kept the ships close to the shore stations. The whales were still being killed with harpoons fired from cannons mounted on the bows of catcher boats, but now the catchers could roam all over the Southern Ocean, killing whales wherever they found them. The term "factory ship" was also used to refer to the giant fishing boats of the 1960s and '70s, usually owned by Germans, Russians, Poles, or Japanese, where the fish were not only caught, but also filleted, packaged, and frozen before being brought back to port. **See also Georges Bank, South Georgia, whaling**

Faeroe Islands Islands in the North Atlantic, between Iceland and the Shetlands. A self-governing part of Denmark, the group consists of eighteen inhabited islands and many uninhabited rocks and islets. The main islands are Streymoy (on which Tórshavn, the capital, is located), Eysturoy, Mykines, Suduroy, Sandoy, Kalsoy, Fugloy, and Svinoy. The current population of the 540-square-mile islands is around 47,000. The climate is mild and wet, and subject to frequent storms and fogs. First settled by Irish monks around A.D. 700, the islands were then occupied by Vikings and Christians sent by the king of Norway. Almost the entire population of the islands was wiped out by the Black Death in the fourteenth century, but Norwegians resettled the islands. Separated from Norway in 1709, the Faeroes became part of Denmark. Originally sheep farmers, the Faeroese have become fishermen and exporters of fish and fish products. The Faeroes did not join the European Economic Community with Denmark in 1972 because it would have opened their waters to other EEC members, but in 1977 they established a 200-mile exclusive fishing zone. The Faeroese insist upon their traditional right to slaughter pilot whales in a *grind*, which involves driving the whales into shallow waters and killing them with axes and spears.

fairy basslet (*Gramma loreto*) As their name implies, the basslets are little basses. These 3-inch fishes with a spectacular color scheme are found in the Caribbean, Bermuda, and the West Indies. They lives in caves and under ledges, and since they orient ventrally to their surroundings, they are often seen upside-down under ledges. Sometimes called the royal gramma, these little fishes are highly prized by saltwater aquarists.

fairy basslet

Falkland Islands British self-governing colony consisting of two main islands (East and West Falkland), and around two hundred smaller islets, lying about 300 miles northeast of the southern tip of South America. Based in Port Stanley, the capital and only town, the Falkland Islands government also administers the British territories of South Georgia and the South Sandwich Islands. The population, numbering some 2,100, is almost exclusively devoted to sheep raising, and there are an estimated 700,000 sheep on the islands. The Dutch navigator Sebald de Weerdt made the first documented sighting in 1600, and Louis-Antoine de Bougainville established the first settlement in 1764. The British claimed the islands in 1765, but after the Spanish bought out the French settlement, they claimed ownership of the islands. After the British withdrew in 1774, the islands (then known as the Malvinas) were officially recognized as belonging to Spain. When Argentina declared its independence from Spain in 1816, it proclaimed sovereignty over the islands, but in 1833, a British force evicted the Argentinean settlers, and in 1841, a British governor was appointed. The Falklands' status as a British colony was recognized in 1892. The long-running dispute over the Falkland Islands came to a head in April 1982, when Argentina launched a military invasion of the islands, also seizing South Georgia and the South Sandwich Islands. By the end of April, there were ten thousand Argentines occupying the islands. British prime minister Margaret Thatcher responded by sending a naval task force to retake the islands, and after steaming 8,000 miles from

England, they engaged the Argentines in a naval, air, and land war. The Argentine cruiser *General Belgrano* was sunk by a British submarine, and Argentine air attacks sank four British warships, but British forces successfully took the settlements of Darwin and Goose Green, and surrounded the Argentine-held capital of Port Stanley. The Argentine garrison surrendered on June 14, effectively ending the conflict. British sovereignty of the Falklands was reestablished, but the defeat so discredited the Argentine military junta that civilian rule was instated in Buenos Aires in 1983.

See also Bougainville, South Georgia, South Sandwich Islands

Falkland Islands, Battle of the After his victory at Coronel in Chile on November 1, 1914, Admiral Graf von Spee took the heavy cruisers *Scharnhorst* and *Gneisnau,* plus the three light cruisers *Nürnberg, Leipzig,* and *Dresden,* around the Horn into the South Atlantic, hoping to break through and return to Germany. The British Admiralty sent Admiral Frederick Doveton Sturdee with the battle cruisers *Invincible* and *Inflexible* and the light cruisers *Kent, Cornwall,* and *Glasgow* to intercept the Germans. Von Spee tried to make a run for it, but he was chased and caught by the British on December 8, 1914. They succeeded in sinking the *Scharnhorst,* with 765 men, including von Spee; the *Gneisnau* (850 dead); the *Nürnberg;* and the *Leipzig.* The *Dresden,* found hiding three months later in the Juan Fernández Islands, surrendered to the British, but was blown up and sunk by her crew.

See also Coronel, Battle of; Fisher; Spee

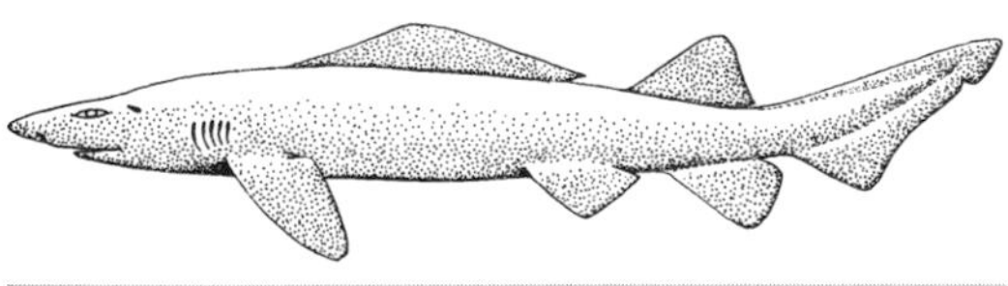
false catshark

false catshark (*Pseudotriakis microdon*) There are two species of false catsharks: *P. microdon* from the deep Atlantic, and *P. acrages* from Japan. As with true catsharks, the eyes are elongated and the base of the dorsal fin is exceptionally long, even longer than the tail fin. They are rare, deepwater sharks that are known to reach a length of 10 feet. From examination of the few specimens, it is believed that they are sluggish bottom dwellers that feed on crustaceans and demersal fishes, at depths of 1,000 to 5,000 feet. **See also catsharks**

false killer whale (*Pseudorca crassidens*) One of the largest of the dolphins, the false killer whale can reach a length of 18 feet. It is easily recognized by its all-black coloration, beakless profile, and strangely "humped" pectoral fins. A tropical and temperate water species, the false killer is probably best known for its occasional habit of stranding in large numbers. In 1946, an incredible 835 false killers stranded at Mar del Plata, Argentina, the largest recorded stranding for any cetacean. From other recorded strandings, it can be assumed that this species is fairly common, but methodical examination of the beached animals has failed to reveal the cause of these strandings. The animal gets its name from the similarity of the skull and teeth to those of the killer whale (*Orcinus orca*), but the false killer is smaller and slenderer and does not have the white markings or the high dorsal fin of the orca.

false killer whale

See also killer whale, pygmy killer whale, stranding

fangtooth (*Anoplogaster cornuta*) A deepwater fish that reaches a length of 4 inches, the fangtooth (also known as the ogrefish) is found from 300 to 3,000 feet down in the tropical and temperate waters of the world. The juveniles look so unlike the adults that they were originally classified as a different species. Fangtooths have no relatives (except other fangtooths); there is only one species in the genus. **See also deep-sea anglerfish**

fangtooth

Farallon Islands Located 28 miles west of San Francisco, the Farallon Islands (original spelling: Farallones) are a group of rocky outcroppings, largely inaccessible to humans but highly desirable real estate for seabirds and pinnipeds. They constitute the largest seabird colony in the contiguous United States; the only larger ones are in Alaska. Almost 30 percent of the breeding seabirds in California nest here. In a two-month period in 1854, the Farallon Egg Company collected and sold almost half a million eggs of the common murre (*Uria aalge*), eliminating this little bird from the islands completely. California sea lions (*Zalophus californianus*) and Steller's sea lions (*Eumetopias jubatus*) also breed on these islands, but their populations are in decline. The northern elephant seal (*Mirounga angustirostris*) was eliminated from the Farallons, but they recolonized the islands in the 1970s and are now breeding here again. The northern fur seal

(*Callorhinus ursinus*) used the Farallons for breeding in the nineteenth century, but Russian sealers wiped them out. Great white sharks (*Carcharodon carcharias*) haunt the Farallons to feed on pinnipeds—their favorite prey is juvenile elephant seals—and many useful studies of the shark-pinniped interactions have been conducted in these waters. Migrating gray whales pass the Farallons every year. The Farallon National Wildlife Refuge was established in 1909, and an area of some 1,235 square miles, from Bodega Head south to the islands, including Tomales Bay and Point Reyes, has now been declared a national marine sanctuary.

fathom The unit of measurement usually used for depths of the sea or lengths of ropes or cables. A fathom is 6 feet long.

featherduster worm (family Sabellidae) Also known as fanworms, featherdusters are classified with the segmented polychaetes that do not move around, Sedentaria, as opposed to the Errantia, which are mobile. Featherduster worms construct parchmentlike tubes that they attach to rocks or pilings, or wedge into crevices on open rocky bottoms. From the tube, which may be as much as 18 inches long and is often covered with sand, mud, or other debris, the worm extends a pair of armlike organs known as palps, each of which is equipped with a spray of feathery gill plumes, which also serve to trap food. The plumes are equipped with light-sensitive spots, and at the slightest hint of danger—even a shadow falling on the gills—the worm snaps back into the tube.

See also polychaetes, tube worms

Fernando de Noronha Archipelago of ten islands, 225 miles off Cape São Roque on the northeast coast of Brazil in the South Atlantic. The islands are part of a submerged volcanic range, not geologically part of the South American continent. Discovered in 1500 by Amerigo Vespucci, they were granted four years later to Fernando de Loronha (or Noronha), a Portuguese nobleman who had obtained the concession to harvest the red-dye-producing brazilwood tree. (The tree, *Caesalpinia echinata,* called in Portuguese *pau-brasil,* was supposedly responsible for the name of Brazil.) Noronha's islands were invaded by Dutch, British, and French adventurers, and the Portuguese built ten forts on the islands to defend them. In 1700, the islands reverted to the Portuguese Crown and became part of the state of Pernambuco. Charles Darwin visited the islands in the *Beagle* in 1832. Several of the forts were converted to prisons, and the islands served as Brazilian penal colonies from the mid-nineteenth century well into the twentieth. From 1957 to 1962, the United States maintained a guided-missile tracking station on Fernando de Noronha. Because of their fragile environment, the islands were declared a national park in 1988. They are the breeding grounds of numerous seabirds. Green turtles nest there, and immediately offshore, there is a resident school of spinner dolphins, known in Portuguese as *golfinhos rotador.* Reachable by plane from Recife, the islands have become a popular destination because of the wildlife of the national park, in addition to the snorkeling, scuba diving, and surfing they offer. Under the strict control of the Brazilian government, only 420 tourists may be on the island at any time. There are 1,200 permanent residents.

Fernando Po Now known as Bioko; an island tucked into the Gulf of Guinea, just off the coast of the country of Equatorial Guinea, of which it is a part. It covers some 783 square miles, consisting mostly of tropical rain forest, and supporting stands of timber and agricultural production. The island was discovered in 1472, by Fernão do Pó, one of the Portuguese navigators who was exploring the west coast of Africa. In 1778, Portugal ceded this and other islands (Annobón, Corisco, Elobey Grande, and Elobey Chico) to the Spanish colony at Río Muni, so the Spanish could have their own source of slaves. The Spanish settlers died of yellow fever, and the islands were abandoned. After some freed slaves settled on the island in 1844, the Spanish reclaimed it; they established a penal colony there in 1879. In 1900, the islands and Río Muni were grouped together as the colony of Spanish Guinea. In 1960, the colony was reorganized under Spanish rule, but a strong nationalist movement fought for independence, and the Spanish residents fled for their lives. In 1968, the country of Equatorial Guinea was formed, but it deteriorated almost immediately under the dictatorship of Francisco Macías Nguema. He was overthrown and executed in 1979, and the country is trying to reestablish itself economically and politically.

Ferro Also known as Hierro, this is the smallest and westernmost of the Canary Islands in the North Atlantic, off Morocco. Covering only 107 square miles, Ferro is almost completely mountainous and ringed with high cliffs, except at Valverde, the capital, where the forests slope down to the sea. There is no harbor, so landings are made at a short pier near the port of Estaca. Ferro's 7,000 inhabitants are mostly subsistence farmers. **See also Canary Islands**

fiddler crab (*Uca pugilator*) With one claw enormously enlarged, male fiddler crabs are easy to identify. The great claw of adult males may be 2 inches long, while the entire carapace can be less than 1½ inches across. Females have smaller claws of the same size, and they are identifiable by the presence of the large-clawed males. The ritualized mating behavior of the males involves elaborate posturing and making semaphorelike signals

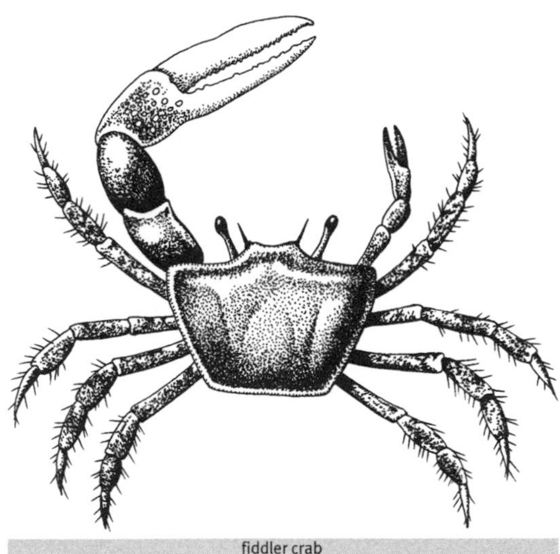
fiddler crab

with the great claw. Male and female fiddler crabs are great diggers and may make burrows that are 3 feet deep. Fiddler crabs live in colonies and are active by day. **See also crab**

Fiji Melanesian group consisting of more than three hundred islands, an independent country since 1970. The capital city of Suva is on Viti Levu, the largest island; the second largest is Vanua Levu. The remaining islands are scattered over 125,000 square miles of the South Pacific, including the Koro Sea. The population of Fiji is about 760,000, of which 50 percent are ethnic Fijians, 48 percent Indians, and the remainder Chinese, Europeans, and Australians. Abel Tasman encountered the islands in 1642, James Cook in 1774. Cannibalism was practiced in "Feejee" when the first European explorers arrived, and the islands were often referred to as the Cannibal Isles. In 1840, the Christian Missionary Society established the first European settlement on the island of Levuka, and by 1874, the islands had been annexed by Great Britain. British governors, opposed to using native Fijians as workers in the sugar fields, began to import Indians as "coolies," a practice that would eventually have serious repercussions. During World War II, although their islands were not threatened by the Japanese, Fijians fought alongside Allied soldiers at Guadalcanal and Bougainville. In 1970 Fiji was declared an independent country in the British Commonwealth. In 1987, elections seated a coalition government dominated by ethnic Indians. A coup led by army colonel Sitiveni Rabuka, a native Fijian, overthrew the civilian government. The Indian population of Fiji now controls the cities and the government, but Fijian law reserves 83 percent of the land for native Fijians. With luxurious new resorts springing up on previously uninhabited islands, the Fijian economy is prospering, but the century-old tensions between Fijians and Indians may ignite at any moment.

filefish (family Monacanthidae) Closely related to the triggerfishes, filefishes have the same spine-locking mechanism, but in this group, the spine is located directly above the eye. The common name comes not from this spine, but rather from the scales, which are roughly textured, like a file.
See also longnose filefish, triggerfish

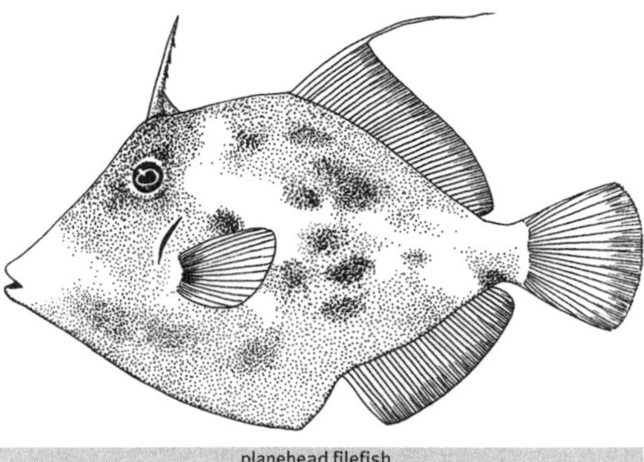
planehead filefish

finless porpoise (*Neophocoena phocoenoides*) One of the most unusual of all small cetaceans, this small (5 feet maximum length) beakless porpoise is easily identified by the total lack of dorsal fin. In place of a dorsal fin, *Neophocoena* has a series of tubercles on its back. The actual function of these bumps was only recently discovered: the babies ride on the backs of their mothers and the roughened skin of the parent helps them stay aboard. For years, the animal was known colloquially as the black finless porpoise, but on-site observations reveal it to be a light bluish gray. Some cetologists believe there are two species of finless porpoise—one from the waters of India and Pakistan, and the other in the Sea of Japan—but most studies still link the two forms. **See also porpoise**

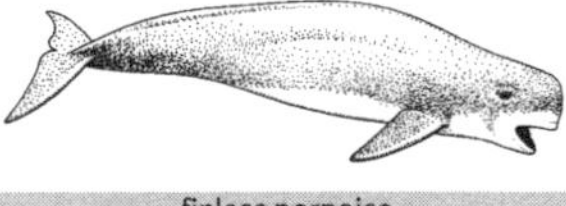
finless porpoise

finning The nefarious—and illegal—practice of catching sharks, cutting off their fins, and throwing the finless and helpless sharks (which cannot swim without their fins) back into the water to die. The fins are used to make shark's fin soup, an expensive delicacy in China, Singapore, Hong Kong, and other Asian countries. Many shark fisheries around the world—Mexico, for example—are in business largely to supply fins to this market. In some parts of the world, finning is so widespread that the sharks have become endangered.
See also shark's fin soup

fin whale

fin whale (*Balaenoptera physalus*) Second only to the blue whale in size, the fin whale can reach a maximum length of 80 feet and a weight of 50 tons. Finners are less robust than blues; they are also considerably faster and more graceful. The fin whale (also known as the razorback) gets its common name from its high, falcate dorsal fin, which is much larger than that of the blue whale. They are among the fastest whales, having been clocked at 25 mph. Their lower jaws are black on the left side and white on the right, making them the only consistently asymmetrically colored animals in the world. Moreover, the baleen plates on the forward part of the mouth match the coloration of the jaws: black on the left and white on the right. The reason—if there is one—for this curious coloration is unknown. They are found throughout the temperate and subpolar waters of the world, and are probably the most common of the large whales. After the whalers decimated the blue whale population of the Antarctic, they turned their harpoons on the fin whales and killed them in astonishing numbers. For example, in the period between 1946 and 1965, Antarctic whalers killed 417,787 fin whales—an average of 20,889 per year. During this period, fin whales were also being hunted in the North Atlantic and North Pacific, so these numbers are only part of the total. Although their numbers were severely depleted, they are now protected throughout the world. **See also blue whale, whaling**

Fiordland penguin (*Eudyptes pachyrhynchus*) Found only on South Island, New Zealand, Fiordland penguins belong to the genus *Eudyptes,* the crested penguins. Like the others, it is a chunky bird, characterized by a yellow eyebrow stripe that develops into silky plumes that droop down the sides of the nape. The species often has little white markings on the cheeks, parallel to the eyebrow stripe. They live in a thickly vegetated forest habitat and are therefore difficult to study, but it is estimated that there are between five thousand and ten thousand breeding pairs.
See also erect-crested penguin, macaroni penguin, rockhopper penguin, royal penguin, Snares penguin

fireship A small vessel, usually considered of no value, that was filled with combustibles and ventilated to insure rapid combustion. The fireship was sailed as close as possible to the enemy vessels and grappled alongside, then a fuse was lit to blow up the contents after the fireship's crew had escaped in a small boat that had been towed along for this purpose. The exploding or burning fireship was intended to damage enemy ships at anchor. Fireships were used successfully at the battle of the Spanish Armada (1588). Six Spanish ships were forced to raise their anchors to escape the fireships, and when they sailed from Calais into the open sea, they were attacked by the British. **See also Armada, Spanish**

First Fleet In 1787, under the command of Captain Arthur Phillip of the Royal Navy, a squadron of British vessels set sail for Australia with a cargo of 730 convicts and 550 officers, marines, their wives and children, and the ships' crews. The ships were *Alexander, Charlotte, Scarborough, Lady Anne, Friendship, Golden Grove, Fishburn, Borrowdale, Prince of Wales,* and, as the flagship, HMS *Sirius* with HMS *Supply* as her armed tender. Both male and female convicts were aboard many of the ships, which led to onboard conflicts but also produced the first white children born in the new colony. Their first landing in Australia was at Botany Bay (which had been named and suggested by Sir Joseph Banks), but Phillip found it too dry and inhospitable, so he sailed north for Port Jackson, which James Cook had named in 1770, but not entered. When Captain Phillip arrived with the fleet on January 26, 1788, he described it as "the finest harbour in the world," and there he founded the settlement that was to become Australia.

Fisher, John Arbuthnot (1841–1920) Probably Britain's greatest naval administrator, "Jacky" Fisher went to sea at the age of thirteen and rose quickly through the ranks until 1881, when he was made commander of the HMS *Inflexible,* then Britain's most powerful battleship. He was promoted to rear admiral in 1890, made third sea lord in 1892, and knighted in 1894. Appointed second sea lord in 1902 and first sea lord in 1904, he introduced innovative concepts of training and efficiency to a navy that had existed largely on tradition. By 1905, Germany was emerging as a naval military power. Fisher introduced the *Dreadnought,* the largest battleship ever built, with a battery of ten 12-inch guns, and also the fast and powerful *Invincible*-class battle cruisers. (*Dreadnought* was immediately copied by Germany, making their Kiel Canal obsolete.) In 1909 he was made Baron Fisher of Kilverstone, and retired from the navy. Winston Churchill, appointed first sea lord in 1911, continued many of his programs, including the adoption of oil instead of coal as the basis of the navy's power. As the war clouds gathered, Churchill recalled Fisher to the post of first sea lord, and Fisher oversaw the construction of a vast armada of ships of all sizes that were so necessary when the Germans launched their U-boat attacks in

1917. After the defeat of a British squadron by German admiral Graf von Spee's forces at the Battle of Coronel off Chile, Fisher sent the battle cruisers *Invincible* and *Inflexible* to the Falklands, where they destroyed von Spee's ships in the Battle of the Falkland Islands on December 8, 1914. He was the first great proponent of submarines for the Royal Navy at a time when their use was considered "ungentlemanly" and he clearly recognized the strategic value of this new system. Because he opposed Churchill's plan to concentrate on the Dardanelles and capture Constantinople, Fisher resigned in 1915, and although he tried to return, Admiral John Jellicoe, who was first sea lord in 1917, rejected his offer.

See also Coronel, Battle of; Falkland Islands, Battle of

FitzRoy, Robert (1802–1865) British naval officer, best remembered for the 'round-the-world voyage of the *Beagle* (1831–1836) with Charles Darwin as naturalist. FitzRoy was first appointed commander of the *Beagle* in 1829. He surveyed the Strait of Magellan and named the Beagle Channel. After the *Beagle*'s epic voyage, FitzRoy worked up the data, completed the charts, and wrote up his notes, which were published in two volumes, *Narrative of the Surveying Voyages of His Majesty's Ships Adventure and Beagle between the Years 1826 and 1836, Describing Their Examination of the Southern Shores of South America, and the Beagle's Circumnavigation of the Globe.* Elected to Parliament in 1841, FitzRoy lobbied unsuccessfully for improvement of conditions in the merchant navy. He was named governor of New Zealand in 1843. He was recalled in 1845, after he contended that the Maori had as much right to the land as the settlers. After his retirement, he devoted himself to meteorology, and devised a storm warning system that is the precursor of our daily weather forecast. It was rumored that an incorrect weather forecast caused him to commit suicide.

fjord Norwegian word, pronounced "fyord" (and sometimes spelled "fiord"), for a steep-sided, coastal inlet of the sea, characteristic of glaciated regions. Fjords were formed as glaciers moved toward the sea, carving out deep valleys; when the glaciers retreated or melted, the valleys were filled by the sea. Fjords are common along the coastlines of Norway, Alaska, British Columbia, and South Island, New Zealand. (An area of South Island is called Fiordland.)

flashlight fish (*Photoblepharon palpebratus*) Beneath each eye of these 4-inch-long fishes is an oval patch that contains a colony of luminous bacteria. Although the bacteria are always "on," the fish is equipped with an opaque, black eyelid below each eye that can be raised to shut off the light, enabling the fish to "blink" the lights on and off at will. (*Photoblepharon* means "eyelid light.") How the juvenile fishes become "infected" with the bioluminescent bacteria is a mystery, since it is obvious that they could not have been born with the genetic material for two completely different animals, and it has been demonstrated that the bacteria cannot live outside the host. Little is known about the lives of these deepwater fishes, but it is believed that the lights are used to see by, to communicate, to confuse predators, or to attract prey. They emerge from caves at night to feed, and return during the day. Found in the Comoro Islands, the Red Sea, the Philippines, and Indonesia, they have occasionally been exhibited in aquariums, where their flashing makes for a most unusual display.

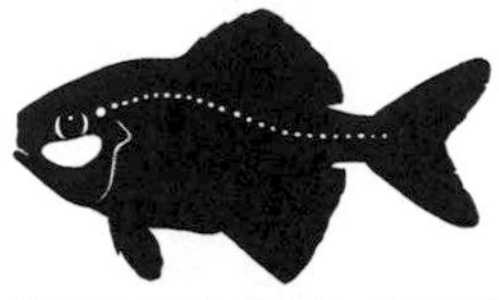
flashlight fish

See also bioluminescence, lantern fishes

flatback turtle (*Natator depressus*) The flatback turtle gets its common name from the smoothness of its shell, which shows little definition of the scales on the carapace. The flatback used to be known as *Chelonia depressus* (the same genus as the green turtle), but it has now been placed in its own genus. Its shell is about 36 inches long, and adults weigh about 175 pounds. Unlike most other sea turtles, flatbacks do not migrate; they spend their lives in and around Queensland, the Gulf of Carpentaria, and the Torres Straits, in shallow, soft-bottomed areas away from reefs. Flatbacks are carnivores, feeding on sea cucumbers, soft corals, and jellyfish. Like most of the sea turtles found in Australian waters, the flatback is hunted for food by Aborigines, and on the breeding islands, feral pigs dig up and eat the eggs, but it is still the only sea turtle species in the world that is not considered endangered.

See also green turtle

flatfishes (order Pleuronectiformes) A group of some six hundred species whose eyes migrate to one or the other side of the body, so that in the adults both eyes are on the same side of the head. As befits the name, these are flattened fishes that spend their lives lying on their blind side, with the eyes upward. If both eyes are on the right side of the fish, it is known as a right-eyed fish; if they are on the left, it is left-eyed. These circumstances are usually constant, but there are some species where it can vary with locality. Flatfishes are known for their ability to match their pigmented sides to the background; in most species, the blind side does not develop pigment and is whitish in color. The variety of flatfishes is enormous; they can range in size from less than a pound to 700 pounds. The group is found in every ocean, and includes some of the world's most important food fishes.

See also flounder, halibut, plaice, sole, tonguefish, turbot

flatworm (phylum Platyhelminthes) Probably the earliest bilaterally symmetrical animals, flatworms are also the first creatures with a head and even a rudimentary brain. There are free-living and parasitic types (tapeworms are flatworms, as are the blood flukes such as *Schistosoma*), but the best-known marine species are the polyclads, which are a couple of inches long and live in shallow waters around the world. Those of temperate waters are often drab, but their relatives of the coral reefs are often brightly colored. They look like little flying carpets as they undulate along the bottom. None of the polyclads are parasitic.

flensing The act of removing the blubber layer from a whale carcass. In the days of the eighteenth-century bowhead fishery in the Arctic, whales were tied alongside the ship and flensed at sea; Yankee sperm whalers would harpoon the whales and tow them back to the ship, where the flensers would work with long flensing knives from specially designed platforms, called stages, that hung from the side of the ship. When mechanized whaling was introduced in the early twentieth century, whales were flensed at shore stations, but the introduction of the stern slipway in 1925 made it possible to haul the whale aboard the factory ship and strip the blubber off on deck.

See also factory ship, sperm whaling, stern slipway

flightless cormorant (*Phalacrocorax harrisi*) Found only in the Galápagos Islands, the flightless cormorant has rudimentary wings and cannot fly at all. Like other cormorants, it is a strong underwater swimmer and captures fish in its strong beak, which is hooked at the tip. When it emerges from the water, it spreads its stubby wings to dry. It is believed that the absence of predators made flight unnecessary, and over time the bird lost the ability. The introduction of dogs, cats, and rats to the islands made this flightless bird extremely vulnerable to predation, and with only eight hundred pairs left on the islands of Fernandina and Isabela, it is considered an endangered species.

flightless cormorant

See also cormorant, Galápagos Islands

Flinders, Matthew (1774–1814) Royal Navy navigator who charted much of the coast of Australia and was the first to circumnavigate the island of Tasmania. In 1795, only seven years after the First Fleet arrived in Australia, Flinders sailed from England in the *Reliance* with George Bass, for whom the Bass Strait between the island of Tasmania (then known as Van Diemen's Land) and the mainland of Australia is named. In 1801, he sailed again for Australia, this time in the unseaworthy *Investigator,* and surveyed the entire southern coast, from Cape Leeuwin in the west, through the Great Australian Bight, to the Bass Strait. He charted the Gulf of St. Vincent (the future site of Adelaide) and Port Phillip (the future site of Melbourne), and from Port Jackson (Sydney) he headed north to the Great Barrier Reef and the Gulf of Carpentaria. By the time he returned to Port Jackson in 1803 the ship's company had been decimated by scurvy, and the *Investigator* was not fit to sail again. Flinders sailed for England as a passenger in 1803, but he was taken prisoner by the French and kept on the island of Mauritius until 1811. When he finally returned to England, he was in poor health; he died three years later.

flogging A form of punishment in the old days of the British and American navies, administered for more serious crimes and officially awarded only by a court-martial. The instrument of punishment was the infamous cat-o'-nine-tails, which consisted of nine lengths of cord, each with three knots, affixed to a larger rope that served as the handle. For lesser violations, officers could authorize a maximum of twelve lashes without benefit of court, but few captains abided by this rule, and one hundred, two hundred, or even five hundred lashes were sometimes given. The prisoner stood upright, his hands tied to some sort of grating, while the lashes were administered. For the most serious crimes, "flogging round the fleet" was the punishment: the offender was spread-eagled on a grating in a small boat, which visited every ship in the fleet, and the bosun's mate from each of them gave the prisoner a dozen lashes. Flogging was the punishment of choice in the penal colonies of Australia in the early nineteenth century, and offenders—those who tried to escape or refused orders—were sometimes given five hundred lashes, or even deliberately flogged to death.

Florida Keys A chain of limestone islands curving for about 180 miles in a southwesterly direction into the Gulf of Mexico from the southeastern tip of Florida. The Keys—from the Spanish *cayo,* meaning "small island"—are connected by the Overseas Highway (U.S. Interstate 1) from the mainland to Key West, the end of the chain. After the railroad that was supposed to connect the keys was destroyed by a hurricane in 1935, the highway, with its forty-two bridges (the longest over-water road in the world), was built in 1938. The keys extend from Virginia Key, south of Miami Beach, to Key Largo, at 21 miles in length the largest of the keys. Key Largo includes John Pennekamp Coral Reef State Park, the only underwater park in the United States. The large islands are Plantation Key, Islamorada, Upper

Matecumbe, Lower Matecumbe, Vaca, Marathon, Bahia Honda, Big Pine, Ramrod, Summerland, Sugarloaf, Boca Chica, and finally, Key West. The Florida Keys are known for their scenery, bird life, fishing, and diving, and there are two designated national marine sanctuaries here: Key Largo and Looe Key. Key West is a world-famous resort (and home to 25,000 residents), filled with restaurants, art galleries, hotels, shops of all sorts, and a historical roster of famous residents that includes John James Audubon, Harry Truman, Tennessee Williams, and Ernest Hemingway.

flotsam Any part of the wreckage of a ship or her cargo that is found floating on the surface of the sea. Under British admiralty law, such materials originally belonged to the Crown, but they are now considered derelict property and become the property of the finder. Articles on the bottom fall into different salvage categories. **See also jetsam, salvage**

flounder There are two basic plans for flounders: left-eyed and right-eyed. When they first hatch, flounders look like any other fishes, with a vertical body and one eye on each side of their heads. Within a few days, they begin to lean to one side, and the eye that is on the underside begins to migrate upward and across the head, so that when the fish flattens out, both eyes are on the top surface. Left-eye flounders (Bothidae) are those species where the right side is the bottom side, and both eyes are on the left side of the head. If you look at a drawing, the fish is almost always facing to the left. Otherwise, the gills would have to be shown on top of the head. The reverse is true for the Pleuronectidae, the right-eyed flounders. All flounders are dark above and light below, as befits a life spent hugging the bottom. Although they spend most of their time lying on the bottom, often buried except for the eyes, flounders are excellent swimmers, and when disturbed, they can swim efficiently with an undulating "magic carpet" movement. **See also English sole, halibut, plaice, sole**

Flying Dutchman One of the most famous of all legends of the sea, about a "ghost ship" under a Dutch captain named Vanderdecken, who, having cursed God for his failure to round the Cape of Good Hope, was doomed to sail forever as his crew cried out for help. The *Flying Dutchman* eternally haunts the waters off South Africa, and there is a superstition that any sailor who sees the ship will die. A similar legend concerns a German captain condemned to sail forever in the waters of the North Sea. The story has inspired many literary works, including Samuel Taylor Coleridge's "Rime of the Ancient Mariner," John Greenleaf Whittier's *Death Ship of Harpswell,* and Frederick Marryat's *The Phantom Ship,* as well as Richard Wagner's opera, *Der fliegende Holländer.*

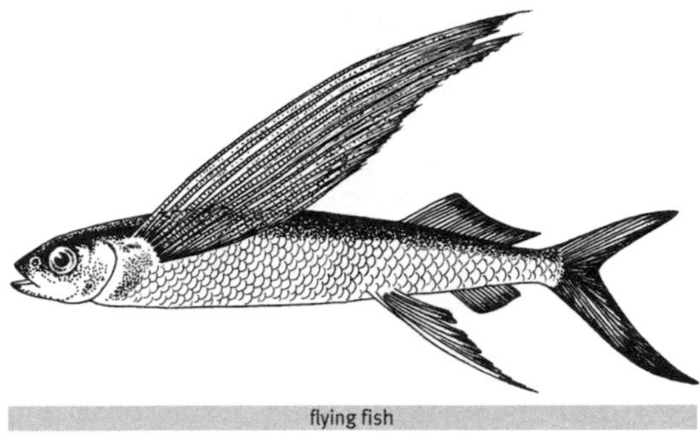

flying fish

flying fish (family Exocoetidae) Bursting from the water at speeds up to 40 mph, flying fishes spread their broad pectoral fins and glide for up to several hundred yards. As the lower, longer lobe of the tail fin leaves the water, it is vibrated to provide additional momentum. (When the pelvic fins are also exaggerated, the fishes are said to be four-winged.) The reason for this activity is not known, but they might be escaping from predators like the dolphin fish (*Coryphaena hippurus*), which are so fond of flying fish snacks that they sometimes launch themselves out of the water to catch them. The 18-inch California flying fish (*Cypselurus californicus*) is the largest; the others average about 1 foot in length. Various species of flying fishes can be seen airborne in the temperate and tropical waters of the world. They are related to the halfbeaks, needlefishes, and sauries.
See also dolphin fish, halfbeak, needlefish, saury

flying gurnard (*Dactyloptera volitans*) With its broad, winglike pectoral fins, this bottom-dwelling fish looks as if it ought to be able to fly, but its "wings" are probably used to startle predators, and there are no accounts of its voluntarily leaving the water. It has a single long dorsal spine that can project straight up from the back of the head, and its elongated pelvic fins are used for "walking" along the bottom. There are similar species in the temperate waters of the Eastern and Western Hemispheres. Flying gurnards can be confused with the sea robins (*Prionotus* spp.), which also have broad pectoral fins—not as large as those of the gurnard—and where the gurnard has a flat profile, the sea robin has a long tapering snout. **See also sea robin**

foraminifera Single-celled organisms that are encased in a multinucleated cytoplasmic body, enclosed within a shell known as a test. Foraminifera (nicknamed "forams") range in size from microscopic to 2 inches in diameter, and while they vary greatly in shape and number of chambers, the skeletons of all are made of calcium carbonate (therefore "calcareous"), pierced by innumerable tiny holes. (*Foramen* is "hole" in Latin.) They are an important component of the planktonic (floating) and benthic (bottom-dwelling) microfauna throughout the oceans of the world, and are found at all depths. Their steady drift toward the bottom creates oceanic snow, a never-ending fall of tiny particles that characterizes all deep oceans. On the bottom, the empty tests form the foraminiferal ooze (also known as calcareous ooze) that covers as much as 50 percent of the ocean floor. These tiny shells are the source of all the world's great deposits of limestone—such as the White Cliffs of Dover.

See also ooze, radiolarian, sediments

Forbes, Edward (1815–1854) British botanist, zoologist, and geologist who developed an early interest in mollusks and starfishes, and participated in dredging expeditions to the Irish Sea (1834) and the North Atlantic (1838) before serving as naturalist aboard HMS *Beacon* in the Mediterranean in 1841. After studying the invertebrates dredged up from the depths, he developed the idea that organisms of the ocean could be divided into eight layers, depending upon depth and the nature of the bottom. He theorized that life became less abundant the deeper one looked; by 300 fathoms (about 2,000 feet), he said, there was no life at all. He called this the "azoic zone," and many scientists believed Forbes's idea that the depths of the ocean were lifeless, despite evidence to the contrary. Shortly before his death at the age of thirty-nine, Forbes was appointed to the chair of natural history at the University of Edinburgh.

forceps fish (*Forcipiger flavissimus*) A common and widely distributed butterfly fish, the forceps fish gets its common name from its elongated snout. It has been seen by divers in the Red Sea, on the Great Barrier Reef, and off the coasts of Africa, lower California, Western Australia, and New Guinea. Its range overlaps that of the long-nosed butterfly fish, and they are often seen in mixed aggregations. They feed on small crustaceans and the tentacles of polychaete worms. As with the long-nosed butterfly fish, a black form is sometimes seen.

See also long-nosed butterfly fish

Foxe, Luke (1586–1636) British navigator and explorer who petitioned King Charles I to finance an expedition to search for the Northwest Passage in 1629. Calling himself "North-West Foxe" (which was also the title of his book, published in 1635), he set out from Bristol in the pinnace *Charles* in 1631 and reached Frobisher Bay, then traversed Hudson Strait and turned south into northern Hudson Bay until he reached Coates Island. In Hudson Bay he met Thomas James, another Englishman who was searching for the same passage, and the two dined together aboard James's ship, *Henrietta Maria.* (It was at this time that Foxe told James, who was carrying letters to the emperor of Japan, "You are out of the way for Japan, for this is not it.") Sailing along Baffin Island, Foxe realized he was not going to discover the Northwest Passage on a southwest shore, and the combination of impassable ice and incipient scurvy prompted him to turn back. Foxe named Roes Welcome Sound for Thomas Roes, one of his patrons; the westward continuation of Hudson Strait is named Foxe Channel, and the large body of water between Baffin Island and the Melville Peninsula is named Foxe Basin.

Foyn, Svend (1809–1894) Born in Vestfold, Norway, Svend Foyn is considered the godfather of modern industrial whaling. In 1863 he built his first whaling ship, *Spes et Fides* (*Hope and Faith*), a 94-foot-long steamship with no fewer than seven whaling cannons, to replace the earlier, inefficient shoulder guns. The harpoon cannon was a muzzle-loading device that shot a heavy iron harpoon with barbs that opened when the projectile pierced the whale's body. The harpoon shaft was fastened by a line to a series of springs belowdecks (called the accumulator), which prevented the whale from ever getting the line taut enough to break it. Because dead whales have a tendency to sink, Foyn pumped compressed air into them to keep them afloat. Foyn's new technology was terribly efficient; by 1880 there were twenty whaling companies killing blue and fin whales in Norway's waters. Foyn opened a whaling station at Vadsø on the Varanger Fjord, where he and his cohorts managed to kill almost all the available whales in the waters of Finnmark. In 1893 he dispatched the sealing ship *Antarctic* to investigate the possibility of taking his exploding harpoons and accumulators to the southern polar regions. By 1904, Norwegian whalers were hard at work in the Antarctic.

Fram Norwegian for "forward," *Fram* was the ship designed by Fridtjof Nansen and Scottish naval architect Colin Archer for Nansen's 1893 attempt to reach the North Pole by sea. *Fram* was a three-masted topsail schooner with an auxiliary engine, 127 feet in length and with a gross weight of 402 tons. Her 2-foot-thick hull, rounded so that the closing ice would push her up rather than squeeze her in, consisted of three layers of oak and greenheart. She was equipped with a complete workshop, so that anything from delicate tools to wooden shoes could be made on board, and a windmill was erected on deck to drive the dynamo, a source of (sporadic) electric light. With a crew of twelve men and thirty dogs, *Fram* set out from Bergen on August 24, 1893, and sailed along the coast of Siberia, getting trapped in the ice off the mouth of the Lena River in eastern Russia a month later. Nansen hoped that the drifting ice pack would take him north to the Pole, but instead the ship drifted southeast for two months before drifting north again and was immobilized and trapped at about 84° north. With Hjalmar Johansen,

Nansen left the *Fram* and set out by dogsled to try and reach the Pole. They got as far as Franz Josef Land, where they were forced to spend the winter starving and freezing in a stone hut until May 1896. They turned back and met the British explorer Frederick Jackson (who was going to Spitsbergen) in June. On August 13, 1896, after almost three years on the ice, the *Fram* burst free of the ice. She sailed triumphantly to Tromsö, where Nansen and his crew were reunited. In February 1909, the Norwegian Parliament authorized funds for Roald Amundsen to make another attempt at drifting over the North Pole in *Fram*, but when Amundsen heard that Peary had made it to the Pole in April 1909 (in fact, he didn't make it), Amundsen sailed to the Antarctic instead, and set up a base camp called Framheim ("home of *Fram*") at the edge of the Ross Ice Shelf. Leading an efficient dogsled expedition over 440 miles of ice, Amundsen became the first man to reach the South Pole, on December 14, 1911. The intrepid *Fram* is now on display in her own museum in Oslo.

See also Amundsen, Nansen, Peary

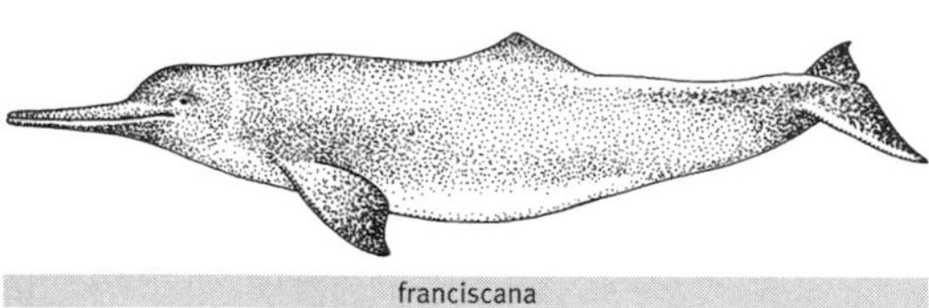

franciscana

franciscana (*Pontoporia blainvillei*) This little dolphin with the long name also has the longest beak in proportion to its body length of any dolphin. It may have as many as 240 teeth in its jaws. Maximum size is about 6 feet. It is thought to have been named for the Franciscan friars who wore habits of a grayish color that approximates the color of this dolphin. It is not strictly a freshwater species but is found in the shallow coastal waters of the Plata River off the coasts of Uruguay and Argentina. Many of these dolphins are accidentally caught in shark nets every year and drowned.

See also Amazon River dolphin, tucuxi

Franklin, John (1786–1847?) British naval officer and Arctic explorer. He entered the Royal Navy at the age of fourteen and was aboard the *Polyphemus* at the Battle of Copenhagen in 1801. He accompanied his cousin Matthew Flinders on an exploratory voyage to Australia from 1801 to 1803, and served at the battles of Trafalgar (1805) and New Orleans (1814). In 1818 he was chosen to command the *Trent* on Captain David Buchan's quest for the North Pole. From 1819 to 1822, he led an expedition that explored the western shore of Hudson Bay; on another overland expedition to the same region in 1825–1827, he explored much of the Mackenzie River delta. For his deeds he was knighted in 1829. He served as governor of Van Diemen's Land (Tasmania) from 1836 to 1843. In May 1845, accompanied by Captain Francis R. M. Crozier in the *Terror,* he sailed the *Erebus* in search of the Northwest Passage. British whalers reported sighting the ships at the entrance to Lancaster Sound off Baffin Island in July 1845; they were never heard from again. By 1847, when no word had been received, search parties were sent to locate the Franklin expedition. Sir John Richardson, who had traveled with Franklin in 1819, tried in 1848–1849 but found nothing. Nor did Sir John Ross, in 1850, turn up any evidence of the lost ships. In 1854, in the *Investigator,* Sir Robert McClure approached from the Bering Sea in the west, but his ship became trapped in the ice at Banks Island, and he and his men had to be rescued themselves. In 1852, the *Assistance,* commanded by Sir Edward Belcher, also failed to find any evidence of Franklin. In 1853, Eskimos reported to Dr. John Rae, of the Hudson's Bay Company, that the men had perished near King William Island, and Rae collected the ten-thousand-pound reward offered by the Admiralty for authentic information on the fate of the expedition. Enmeshed in the Crimean War, the British government refused to send out another search party, so in 1854, Lady Jane Franklin, Franklin's second wife, equipped the *Fox* and put Sir Francis McClintock in command. In 1859, participating in his third expedition in search of the missing explorers, McClintock found the remains of Franklin's men and remnants of the *Erebus* and the *Terror.* Thirteen years earlier, the ships had become trapped in the ice off King William Island, and twenty-three men died there. Led by Crozier, the remaining 105 men set out on foot, but none survived. Although Franklin never located the Northwest Passage, many of those who went to search for him surveyed and mapped much of the unexplored Canadian Arctic.

See also Crozier, Northwest Passage

Franz Josef Land Archipelago in the northeastern Barents Sea, north of Novaya Zemlya and 600 miles from the North Pole. There are more than a hundred islands, mostly covered with ice, with rocks and lichens occasionally visible. Polar bears prowl around the ice, and Arctic foxes trail them. There are also walruses, various seals, and numerous bird species in residence. In 1872, the Austrian explorers Karl Weyprecht and Julius von Payer, seeking the Northeast Passage in the *Admiral Tegetthof,* became trapped in the pack ice at Novaya Zemlya and drifted for a year before finally finding land in August 1873: they had accidentally discovered a previously unknown group of islands, which they named Franz Josef Land, after their emperor. After spending a year exploring the islands, they abandoned their ship and journeyed in a small boat for ninety-six days to Novaya Zemlya. Fridjtof Nansen tried to reach the Pole in 1893 by drifting in the specially reinforced *Fram*, but when he found himself drifting in the wrong direction, he left the ship with Hjalmar Johansen and attempted to reach his goal by sledge. They had to turn back, and they spent the winter of 1896 on Franz Josef Land in a stone hut with a walrus-skin roof. During his three-year expedition, Nansen became the first man to map this complicated collection of islands. In 1926 Franz Josef Land was annexed by the Soviet Union, and remote weather stations were erected there. They are now unoccupied. **See also *Fram*, Nansen**

Fraser's dolphin

Fraser's dolphin (*Lagenodelphis hosei*) This species was completely unknown until 1956, when Sir Francis Fraser of the British Museum discovered a skeleton in the collection that was known to have come only from Borneo. In 1971, however, specimens began turning up on beaches and in tuna nets, and this heretofore unknown species was identified as ranging throughout the temperate and tropical waters of the world. It was encountered in South Africa, Australia, Japan, Taiwan, and the Cocos Islands in the Pacific; it has recently been reported from the Caribbean. It is not common, but its range is extensive. It can be identified by its short snout, small dorsal fin, and very small flippers. Because it has features of both the lags (*Lagenorhynchus*) and the common dolphin (*Delphinus*), it was given the name *Lagenodelphis*. **See also tuna-porpoise problem**

French Polynesia South Pacific territory of France, consisting of 118 islands with a total land area of 1,359 square miles, spread out over an area the size of Western Europe. French Polynesia consists of five archipelagos. The Society Islands are divided into two parts: the Windward Group, which includes Tahiti, Moorea, and Tetiatora; and the Leeward Group, consisting of Huahine, Bora-Bora, Raiatea, and Maupiti. The Tuamotu Archipelago covers 600 square miles and includes eighty atolls, including Rangiroa (the second largest coral atoll in the world), Fakarava, Anaa, and Taaroa. To the south of the Tuamotus is the group known as the Gambier Islands. The Austral Archipelago consists of five islands to the south of Tahiti, one of which, Rapa, is 1,000 miles from Papeete, the capital of French Polynesia. The Marquesas, which include Nuku Hiva and Hiva Oa, are probably the best known after Tahiti, since Herman Melville jumped ship at Nuku Hiva in 1842 (and wrote *Typee* and *Omoo* about his experiences with the cannibals) and Paul Gauguin lived and painted on Hiva Oa. Clipperton Island, which is 3,400 miles northeast of Tahiti, is also part of French Polynesia.

See also Bora-Bora, Marquesas Islands, Moorea, Rangiroa, Tahiti, Tuamotu Archipelago

freshwater dolphins Technically known as the Platanistidae, this group of dolphins can be found in various freshwater rivers and lakes around the world. They all have reduced vision (one species has virtually no eyes at all) and highly developed acoustic senses, which enable them to hunt and navigate in the often murky waters in which they live. In addition, they have long beaks and numerous teeth. Because of their proximity to human habitation, many of the freshwater dolphins are seriously endangered.

See also Amazon River dolphin, Chinese river dolphin, Franciscana, Ganges River dolphin, Indus River dolphin

frigate A class of warship in all navies, usually fast and not heavily armed. In the eighteenth century, frigates were usually employed as lookouts, and later as long-range cruisers in search of enemy vessels. During World War II, the term was revived for a class of medium-speed antisubmarine vessels used in convoy work; it has become a generic term for smaller, all-purpose warships.

frigate bird (*Fregata* spp.) With the longest wingspan-to-body-weight ratio of any bird, the frigate bird is a truly spectacular flier. Their enormous wingspan (up to 7 feet), slim body, and forked tail make them among the most easily recognized of all seabirds. They are pirates and steal food from boobies, gulls, or tropic birds on the wing, often engaging in breathtaking displays of aerobatic maneuvers as they do so. They also are capable fishermen, swooping down to pluck their prey from the water with their long, slender, hooked bills. Frigate birds neither walk nor swim. They never land on the sea or on level ground, because the length of their wings would make it impossible to take off. They nest in large groups, often in mangrove stands. There are several species and subspecies, including the magnificent frigate bird (*F. magnificens*), the great frigate

great frigate bird

bird (*F. minor*), the lesser frigate bird (*F. ariel*), the Christmas frigate bird (*F. andrewsi*), and the Ascension frigate bird (*F. aquila*), but all are black above, with varying amounts of white on the undersides. Adult males have a red throat pouch that can be inflated to a vivid crimson balloon during the breeding season. Young birds have a white head, and females have a white chest. They can be seen soaring effortlessly in the sky throughout the tropics.

See also brown booby, masked booby, red-footed booby, tropic bird

frigate mackerel (*Auxis thazard*) A fairly large mackerel (1½ to 2 feet in length, and weighing up to 10 pounds), this schooling pelagic species is found in warm and temperate waters around the world. Because its tail is more lunate than forked, it resembles a tuna more than a mackerel. Frigate mackerels sometimes appear in abundance in certain areas, then disappear for years at a time. Their meat is dark and oily, and they are fished commercially only for the production of high-protein fish meal. **See also Atlantic mackerel**

frilled shark (*Chlamydoselachus anguineus*) Considered the most primitive of all living sharks, the frilled shark is eel-like in shape, with six gill slits on either side of its head; the first of these is continuous along the throat, which gives it a frilled appearance. The teeth, identical in both jaws, are unique for their three fanglike cusps with two smaller cusps between them. It is a deepwater species, found in the North and South Atlantic, South Africa, Japan, and California. The largest known specimen was 6 feet long.

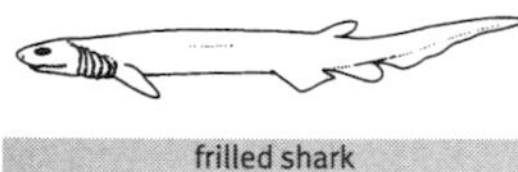
frilled shark

See also sevengill shark, sixgill shark

Frobisher, Martin (1535–1594) English navigator and explorer of northern Canada. In 1576, he set out in search of a northwest passage to the Orient with three ships, but two had to turn back, leaving only the *Gabriel* under Frobisher's command. He found Greenland, which had been identified as "Frisland" on the map he was using, and headed west. When he reached Baffin Island he entered Frobisher Bay (which he named to correspond with Magellan's eponymous strait in the south), and, believing he had found a route to the Pacific, he captured an Eskimo, gathered up a sample of what he believed to be gold-bearing ore, and headed home in triumph. Under the patronage of Queen Elizabeth, he returned to North America in the *Aid* in 1577, mined 200 tons of glittering "black rock," and delivered it to England. Even before the assayer's report was completed, his third expedition was under way, with 120 settlers, a prefabricated house, and orders to mine more "gold." It was actually iron pyrites ("fool's gold"), and upon Frobisher's return, Michael Lok, the principal backer of the expedition, was put in debtor's prison. In 1585, Frobisher was vice admiral of Drake's expedition to the West Indies, and three years later, in command of the *Triumph*, the largest ship in the British fleet, he played a major role in England's defeat of the Spanish Armada. He was mortally wounded in 1594 while fighting a Spanish force off the west coast of France. The town of Frobisher Bay has been renamed Iqaluit. In 1999, it became the capital of Nunavut, which occupies almost half of what was the Northwest Territories.

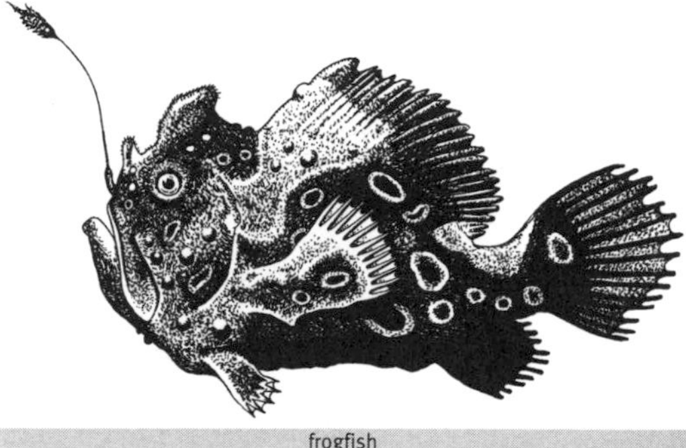
frogfish

frogfish (*Antennarius* spp.) Chubby, ugly little fishes that depend mostly on their camouflage and stealth to capture their prey, but some species have a well-developed lure above the mouth. Their unusual fins, shaped not unlike the feet of a land animal, enable them to "walk" across the substrate and clamber around on rocks and coral. In addition to swimming like other fishes—albeit slowly—frogfishes take water into their mouths, and expel it under pressure through the tiny, tubelike gill opening, "jet-propelling" themselves forward. They are usually covered with frilly or whiskery filaments that serve as disruptive coloration to conceal them. Some species carry their eggs on the side of their body; some dangle them from the dorsal spines; and others hold them in a pocket formed between the pectoral fin and the body. Most species are less than 12 inches long, but they are aggressive feeders and will attack anything they can swallow. The best-known frogfish is the sargassum fish (*Histrio histrio*), a perfectly camouflaged inhabitant of sargassum weed.

See also deep-sea anglerfish, goosefish, sargassum fish

fugu The generic name for puffer fishes in Japan. Because the liver, gonads, intestines, and, in some species, the skin contain a deadly nerve toxin, it is extremely risky to eat them. Nevertheless, they command high prices in Japanese fish markets, and some restaurants hire specially trained cooks to prepare them. If improperly prepared fugu is eaten, there is no antidote. Symptoms include tingling of the mouth, motor incoordination, and overall numbness, followed by excessive salivation, weakness, nausea, vomiting, convulsions, and death. One species of puffer in Japanese waters is known as *maki-maki,* the Japanese death puffer. **See also porcupine fish, sharpnose puffer**

fulmar (*Fulmarus* spp.) The fulmars belong to the family Procellariidae, which also includes typical petrels, prions, and shearwaters. (The albatrosses are closely related; they all belong to the order Procellariiformes.) Fulmars differ from gulls in that they have tubular nostrils on the upper mandible that gulls lack. The northern fulmar (*F. glacialis*) is the most common of the tubenoses in the northern latitudes, a bull-necked gray bird with darker wings, a stubby yellowish bill, and a white, round-looking head. It flies with its wings almost straight, not cranked like those of a gull. The Antarctic, or southern, fulmar (*F. glacialoides*) looks very much like its northern counterpart, but it is slightly smaller.

fulmar

See also albatross, prion, shearwater

Fulton, Robert (1765–1815) Born in Pennsylvania, Fulton was a talented painter who went to London to study art, but became interested in engineering and wrote a treatise on canal navigation. In 1797 he left for France, where he designed a submarine boat for the Directory (the government of Revolutionary France from 1795 to 1799), but it was rejected as being a dishonorable form of naval warfare. Napoleon had no such problem, and in 1800 he commissioned the *Nautilus.* Returning to America in 1801, Fulton built the *Clermont,* the first steamboat, which plied the Hudson from Albany to New York. His partner in this enterprise was Robert Livingston, who had been minister to France, and who had obtained a monopoly on steamboat travel on the Hudson River. (In 1808, Fulton married Livingston's niece, Harriet.) He returned to England in 1804 to promote his ideas for submersible boats that would carry explosives, but the victory at Trafalgar in 1805 more than adequately demonstrated British marine superiority without having to resort to underwater sneak attacks, and Fulton's ideas were rejected. In 1814, the *Demologos* was launched, the first steam warship, consisting of twin hulls with a paddle wheel between. Although Robert Fulton was not the first to build a steamboat, he was the first to make a commercial success out of them.

Fundy, Bay of Ninety-four-mile-deep inlet of the Atlantic Ocean between the Canadian provinces of New Brunswick to the north and west and Nova Scotia to the south and east. The bay covers some 3,600 square miles and is bordered by high rock formations and forests. At the entrance, where it is 32 miles wide, fast-running tides produce rises of 70 feet, the highest in the world. (In two inlets, Chigneteco Bay and Minas Basin, the tides can reach 46 and 53 feet, respectively.) At the mouth of the St. John River, the tidal surge can be 6 feet high at its crest and can rise 8 to 11 feet per hour. At Passamaquoddy Bay, on the Maine–New Brunswick border, the possibilities of using dams to harness the tremendous hydroelectric power of the tides have been investigated since the 1920s, but the immense costs and environmental concerns have impeded any such development.

fusiliers (family Caesionidae) Originally classified with the snappers (Lutjanidae), fusiliers have recently been assigned to their own family. They are slender, streamlined fishes found in schools, often in the region of outer reef drop-offs. Fusiliers are found only in Indo-Pacific waters, where they are fished for sport because their delicate flesh makes good eating. They sometimes swarm around divers in such numbers that the divers cannot see out of the school. **See also snappers**

G

gadfly petrel A generic term that refers to the petrels of the genus *Pterodroma,* which can be translated as "swift wing." There is no bird actually called "gadfly petrel," but the great-winged petrel (*P. macroptera*) of the Southern Ocean, the white-headed petrel (*P. lessonii*), the black-capped petrel (*P. hasitata*), the cahow (*P. cahow*), the soft-plumaged petrel (*P. mollis*), and another twenty-odd species are currently classified in the genus. The average span of the long, pointed wings is about 40 inches. Birders regard them as among the most difficult birds to identify at sea. They are usually dark above with varying amounts of white on the undersides; some have white bellies, others have white underwings, and so on, but the Mascarene petrel, only recently discovered, is wholly blackish brown.

See also black-capped petrel, cahow, petrel

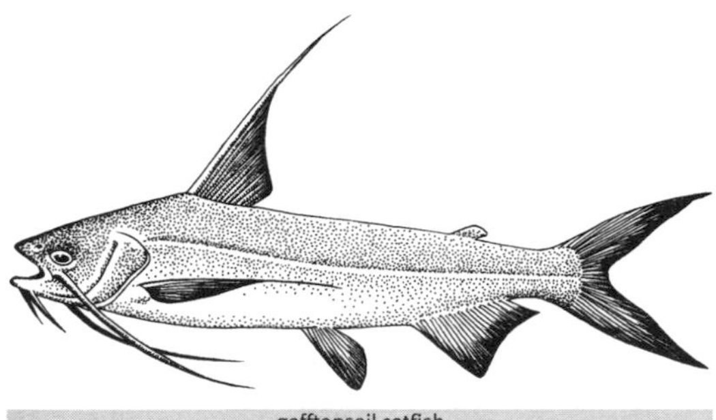

gafftopsail catfish

gafftopsail catfish (*Bagre marinus*) One of the few species of catfish that live in saltwater, the gafftopsail is found in the western North Atlantic from Cape Cod to Panama and Venezuela. Its maxillary barbels are unusually long, reaching all the way to the base of the anal fin, and the first ray of the topsail dorsal fin is modified into a long filament. It is steel blue above and yellowish gray below, and can reach a length of 2 feet. The male carries the eggs in his mouth until they hatch.

See also sea catfish

Galápagos Islands On the equator 600 miles off the coast of Ecuador, the Galápagos Archipelago consists of sixteen volcanic islands and numerous islets and rocks in the Pacific Ocean. In 1535, the Spanish bishop Tomás de Berlanga, blown off course on his way from Panama to Peru, landed and named the islands after the giant tortoises (*galápagos*) he found there. The Galápagos were subsequently visited by such luminaries as Charles Darwin aboard the *Beagle* in 1835, who stayed for thirty-five days and made many observations—particularly of the various species of finches—that were relevant to his *Origin of Species.* Herman Melville visited while aboard the whaler *Acushnet* in 1841 (he wrote a story about the islands, called "The Encantadas"—"The Enchanted Isles"—in 1854), and William Beebe wrote the popular *Galapagos: World's End* in 1932. The islands became part of Ecuador in 1832. Throughout the eighteenth and nineteenth centuries, pirates and whalers visited the islands, creating havoc with the wildlife, particularly the tortoises, which they took aboard their ships as a long-lasting source of food. Most of the islands have Spanish and English names, although the Spanish are more commonly used: Isabela (the largest island) is also Albemarle, Santa Cruz is Indefatigable, Fernandina is Narborough, San Cristóbal is Chatham, Floreana is Charles, Santiago is James, Española is Hood, Genovesa is Tower, etc. In addition to the 500-pound tortoises, the Galápagos are famous for their exotic wildlife, including land and marine iguanas, sea lions, the flightless cormorant (which exists nowhere else), albatrosses, boobies, frigate birds, gulls, terns, and all other manner of seabirds. There are penguins on Fernandina and Isabela, and flamingoes on Santiago. The tortoises are considered endangered, and many indigenous species are threatened by introduced animals such as pigs, goats, cats, and rats. The wildlife and the spectacular landscapes of the islands are the reason for the great increase in tourism in recent years, but all is not peaceful in Paradise. The El Niño event of 1982–1983 created anomalous weather and sea conditions that wreaked havoc with the wildlife and vegetation of the islands; the volcano on Fernandina erupted in January 1995; and local fishermen are destroying the islands' sea cucumbers, which they sell to Asian buyers, despite the Ecuadorian government's ban on such fishing.

Galápagos penguin

Galápagos penguin (*Spheniscus mendiculus*) The smallest of the spheniscid penguins (the others are the Humboldt, African, and Magellanic species), this is probably the most unusual of all penguins, since unlike its relatives, which are cold-weather, cold-water birds, it breeds on hot volcanic islands on the equator, but the waters are cooled by the Humboldt Current, flowing north from southern South America. It is a nonmigratory species and does not leave the waters of the Galápagos. To rid themselves of the heat of the equatorial sun, Galápagos penguins take to the water or stand on shore and pant. During the disastrous El Niño event of 1982–1983, the waters around the islands warmed up so much that the small fishes that make up the penguins' normal food supply were not available, and many penguins starved to death.

See also El Niño, Humboldt Current, Magellanic penguin

Galápagos shark (*Carcharhinus galapagensis*) A large gray shark (maximum size: 12 feet) with a broad, rounded snout and a high first dorsal fin. It is associated with island groups, including the Galápagos; Hawaii; Madagascar; various Caribbean, South Atlantic, and South Pacific islands; and Baja California. It can be confused with the dusky shark (*C. obscurus*) but its dorsal fin is proportionately higher. Galápagos sharks are aggressive and considered dangerous. Like the gray reef shark (*C. amblyrhynchos*) this species performs a threat display, consisting of hunching the back and dropping the pectorals just prior to attacking.

See also carcharhinid sharks, dusky shark, gray reef shark

Galápagos tortoise

Galápagos tortoise (*Geochelone elephantopus*) From its scientific name, which can be loosely translated as "elephant-sized land tortoise," one can get a good idea of the nature of this beast. The largest living tortoises, the Galápagos giants can weigh as much as 500 pounds and live for 150 years. (The only comparable tortoises are found on Aldabra Island, in the Indian Ocean.) When William Dampier visited the Galápagos in 1701, he noted that the slopes of the volcanoes were black with giant "land-turtles," and that they could sustain hundreds of men for months without any other food. Buccaneers and whalers visiting the islands realized that these tortoises could provide fresh meat while they were at sea; all they had to do was capture the slow-moving creatures and stow them on their backs in the hold. They couldn't right themselves, and they lived for months. It is estimated that visitors to the islands took or killed at least 100,000 tortoises. But this was only a small contribution to their demise: arriving seamen and whalers introduced animals such as hogs, cats, and rats, which ran wild and ate the eggs or the newborn tortoises. Because of their isolation from one another, at least ten subspecies of giant tortoise evolved on the islands; each island had its own particular type, differing in such details as shell height and breadth, and length of neck and legs. The island of Isabela has five subspecies, each one restricted to a different volcano.

***Galathea* expedition** Danish deep-sea expedition of 1950–1952 that sailed around the world collecting specimens and hydrographic data. The *Galathea* was a 266-foot-long converted naval vessel, built in New Zealand and equipped with the latest in trawling gear, some of it purchased from the Swedish *Albatross* expedition of 1947–1948. Under the direction of Anton Bruun, the voyage of discovery was specifically concerned with the deep sea, although many observations were made in shallow waters and ashore on various islands. Countless fishes, cephalopods, and other invertebrates were trawled up from the abyssal depths, many of them visible for the first time to science. (*Neopilina*, a chitonlike creature that was thought to have become extinct 350 million years ago, was rediscovered by Henning Lemche on this voyage.) The expedition, which began in Copenhagen in October 1950, rounded Africa into the Indian Ocean, and then visited the deep waters of India, Southeast Asia, Australia, New Zealand, and various islands of the Pacific, passed through the Panama Canal in May 1952, and arrived back in Copenhagen in June. The *Galathea* covered 63,700 miles, the equivalent of three times around the earth at the equator.

See also Bruun, *Neopilina*

galleass An intermediate stage in the development of the fighting sail, between the galley, with its banks of oars, and the galleon, which was powered only by sail. A galleass employed a single bank of fifty-two oars (twenty-six on a side), each one manned by four to seven rowers, but there were also three lateen-rigged masts. A protective deck covered the rowers and provided a fighting platform for the soldiers and room for the sailors to handle the rigging. Galleys depended upon manpower for propulsion and also for fighting, but the introduction of gunpowder in the fourteenth century marked the end of the "ram and grapple" method of naval warfare, and ships were now fitted with cannons. The dominance of the galleass was highlighted by the Battle of Lepanto (1571), in which the oared galleys of the combined Christian forces of Venice and Spain defeated those of the Turks in the Mediterranean off Greece. **See also galleon, galley**

galleon Somewhat smaller than the carrack that preceded it, the galleon was developed by Sir John Hawkins around 1570. Although they only displaced some 500 tons, galleons were designed to be extremely seaworthy, and they carried banks of squaresails for speed. The high forecastle had been reduced because it always caught the wind and made the ship less maneuverable, but the rear deck, which contained the officers' and captain's cabins, still towered high, giving the ship its characteristic high-pooped profile. A typical galleon was 125 feet long and 35 feet wide, and carried a complement of 340 sailors, forty gunners, and 120 soldiers. The captain and officers were housed in cabins, but the men had no quarters and slept wherever they could find space, usually on the gun deck. Armament consisted of sixteen large cannons (culverins) on the gun deck and fourteen smaller ones (demi-culverins and sakers) on the upper deck. Galleons were used first as warships, but their speed and (comparative) seaworthiness made them ideal for the Spanish transporting treasure from the New World. The famed Manila galleons, which carried Spanish silver and trade goods once a year between Acapulco and the Philippines, were considered the most desirable prize for pirates, privateers, and enemies of Spain. The *Santa Ana* was taken by Thomas Cavendish off Baja California in 1587, with a cargo of 122,000 gold pesos, and in 1710, Woodes Rogers took *Nuestra Señora de la Encarnación* with a cargo that he sold in England for £148,000. **See also carrack, galleass, Manila galleon**

galley Oared fighting ship, used from about 3000 B.C. by the Egyptians, Minoans, Greeks, Romans, and other ancient peoples; continued in use into the eighteenth century. Most versions had one tier of oars, but the Phoenicians introduced biremes (two tiers), and three tiers (triremes) were sometimes employed. In multibank galleys, the length of the oars differed according to their position, i.e., their distance from the water. For the lowest tier, an oar might be 8 feet long, but for the upper bank of a trireme, it might be 14 feet long. The number of rowers also varied, but in a typical Greek trireme, there would be thirty-one pairs on the upper bank, and twenty-seven pairs in the middle and lower banks. The Viking longships were small galleys, with ten oars on a side, capable of carrying fifty or sixty warriors. In addition to its fighting men, the weapon of the galley was the ram, an armored, pointed timber fixed to the prow. With the introduction of gunpowder, guns were mounted on a platform in the bows, but they could only fire directly ahead. Galleys were suitable for use in calm weather, and although they were sometimes equipped with sails, these were used primarily for making a passage and were always lowered before battle. The term "galley" is also used to denote a ship's kitchen. **See also bireme, trireme**

galley slave A prisoner condemned to pull an oar aboard a war galley, where he was chained to his station by a fetter around his ankle attached to an iron bar on deck. Because they had little or no chance of pardon, sometimes had to row for twenty hours at a stretch, and were whipped if they faltered, galley slaves often considered theirs a fate worse than death. In France and elsewhere in the eighteenth century, prisoners of war were used to man galleys, but they were considered undependable in battle.

Ganges River dolphin (*Platanista gangetica*) Known locally as the susu, this dolphin has a long narrow snout, wide squared-off flippers, and eyes that have so degenerated that they are almost nonfunctional. Despite this apparent handicap, the susu manages quite well, since it has acutely sensitive acoustic senses, which enable it to hunt its prey in the almost opaque waters in which it lives. Captive specimens have often been observed swimming on their side, making contact with the bottom of the tank with one flipper. Unlike its close relative of the Indus River system, the Ganges susu is not immediately endangered.
See also freshwater dolphins, Indus River dolphin

Ganges River dolphin

gannet (*Sula bassana*) As adults, northern gannets are large white birds, with a yellowish nape, black wing tips, and a yellow eye staring from a thin black mask. (Juveniles are brownish gray.) The body is torpedo-shaped, the wings long and pointed (maximum span: 6 feet); the tail is wedge-shaped and pointed, too. Gannets are powerful, purposeful flyers, and feed by plummeting into the sea with their wings folded from heights of up to 100 feet. They aggregate in huge breeding colonies on islands on both sides of the North Atlantic; among the most populous are St. Kilda (the largest colony, with fifty thousand breeding pairs), Ireland, the Gulf of St. Lawrence, Newfoundland, Labrador, and Iceland. The nest is a bulky affair built of seaweed and grass, and a single egg is laid. The parents take turns incubating the egg, and when the chick hatches, they take turns catching fish to feed it. Two other species, the Cape gannett (*S. capensis*) of South Africa and the Australasian gannet (*S. serrator*) of Australia and New Zealand, are similar in appearance and habits. Boobies are also closely related to gannets.

northern gannet

See also blue-footed booby, brown booby, masked booby, red-footed booby

garden eel (*subfamily Heterocongrinae*) Small, thin, conger eels (family Congridae) that live in worldwide tropical waters shallow enough that there is sufficient light for them to see their food. Their common name is derived from their habit of keeping their tails in their burrows and facing into the current to feed on passing plankton. They reach a length of about 40 inches, but a third of this is usually buried in the sand. At the approach of a predator, the entire colony withdraws into its burrows. They mate by entwining with a member of the opposite sex that they can reach without leaving their burrows. It is believed that garden eels are among the few vertebrates that remain permanently attached to the substrate. **See also conger, eel**

garibaldi (*Hypsypops rubicunda*) The largest of the damselfishes, sometimes reaching 1 foot in length, the garibaldi is found in the kelp forests of California. Juveniles look like a species of blue-spotted damselfish, but they mature into the brilliant orange-red adults. They are found at depths up to 40 feet, and their startling coloration makes them especially appealing to underwater photographers. The name "garibaldi" came from immigrant Italian fishermen in California who likened the coloring of the fish to the red shirts worn by Giuseppe Garibaldi's volunteers during the 1860–1867 battles for Italian unification. **See also damselfishes**

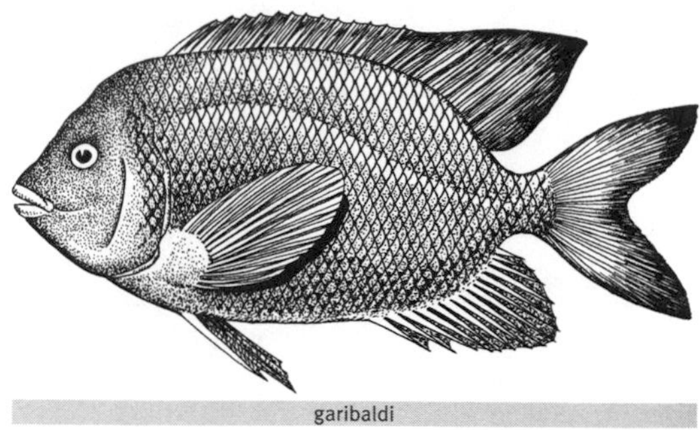

garibaldi

Garrett, George (1852–1902) British clergyman and submarine designer. In 1878, he built a small, teardrop-shaped, man-powered vessel that was submerged by admitting water to alter its displacement. Two years later, he designed and constructed the 45-foot-long *Resurgam* ("I shall arise"), a tapered double-ender that was powered by a steam engine that took up so much room that there was hardly any space for the three-man crew. On her shakedown voyage, *Resurgam* never got out of Liverpool Bay; in fact, she even got lost. She put into the harbor at Rhyl and continued her journey under tow. The hawser parted under the strain of wind and waves, and Britain's first practical submarine was wrecked on the north coast of Wales, miraculously without a loss of life. After the loss of the *Resurgam* in 1880, Garrett allied himself with Thorsten Nordenfelt, a Swedish arms maker and industrialist, who realized that the ideal platform for the firing of the torpedo was not a surface vessel but an underwater one. Under Nordenfelt's aegis, Garrett relocated to Sweden in 1882 and launched *Dykeri Pram En* ("Diving Barge One"), a 64-foot-long submarine boat that he piloted on the surface from Stockholm to Copenhagen, but he and his crew were felled by escaping gases and heat before they could conduct the underwater trials. Nordenfelt attempted to sell the vessel to anyone who would buy it, regardless of their political persuasions. In January 1886, he was commissioned to deliver two 110-foot submarine boats to the navy of the Ottoman Empire. At this time, Garrett was commissioned a *binbashi* in the Imperial Ottoman Navy, which enabled him to style himself "the Reverend Commander G. W. Garrett Pasha, B.A." The Turkish boats *Abdul Mejid* and *Abdul Hamid* were never used; they spent the next twenty years in a shed overlooking the Golden Horn. At the Barrow yards in England, he oversaw the construction of his last submarine, a 125-footer christened *Nordenfelt IV* that was sold to the Russians. Towed out to sea off the coast of Denmark, *Nordenfelt IV* immediately ran aground, causing the Russians to cancel their order. Although the Garrett-Nordenfelts were beginning to resemble modern submarines, with their elongated shape, low conning tower, and torpedo tubes, the propulsion problem continued to plague the designers. It appeared that they could get submarines to do almost everything under water, except move, dive, and maintain themselves at a level attitude.

See also submarine

gas bladder Also known as an air bladder and swim bladder, the gas bladder is the hydrostatic organ that regulates buoyancy in most bony fishes, enabling the fishes to equilibrate their internal pressure with that of the surrounding water. Because the density of a fish is nearly equal to that of water, it would be "transparent" to sound if not for its bone and gas bladder, both of which can act as sound conductors. Grunts, squirrelfishes, triggerfishes, drums, grenadiers, gurnards, and toadfishes are able to generate underwater sounds by making the bladder wall vibrate. In some species the initial filling of the bladder with air takes place when the young fish gulps a mouthful of air at the surface, but in deep-sea fishes that spend their whole lives at depth, gas is deposited in the swim bladder by passive diffusion from the blood. The deep-scattering layer (DSL), which moves up and down in the water column during a twenty-four-hour period, is tracked by sonar by listening to the reflections produced by millions of tiny gas bubbles: the gas bladders of lantern fishes and bristlemouths. Because of the reduction in pressure, the bladder of many fishes expands when the fish is brought rapidly to the surface. Lampreys, hagfishes, sharks, rays, and coelacanths do not have a swim bladder.

See also deep-scattering layer

gastropoda The univalve (single-shelled) subgroup of the phylum Mollusca, which incorporates the snails and slugs. Gastropod means "stomach-foot," and refers to the broad, tapered foot upon which the animals glide. Snails have hard, calcareous shells into which the head and foot can be drawn; slugs are snails without shells. There are perhaps fifty thousand species of gastropods, living in saltwater, in freshwater, and on land. Most gastropods are less than 1 inch long, but there are marine snails, such as the Australian trumpet (*Syrinx aruanus*) or the horse conch (*Pleuroploca gigantea*) from the western North Atlantic, which can exceed 2 feet in length. All snail shells are spirally curled—almost all of them a right-handed spiral—even those of the limpets (Fissurellidae), which begin as coiled shells before flattening out. Although their methods differ, most gastropods feed by grasping food substances with a tooth-studded tongue known as the radula. (The cone snails are carnivorous and feed on invertebrates and small fishes.) The gastropods display a dizzying variety of colors, shapes, and textures, from the smooth, polished ovals of the cowries to the architectural extravagance of the Muricidae and the intricate patterns of the cones.

See also abalone, conch, cone snail, cowrie, limpet, mollusk, nudibranch, shell collecting, Tyrian purple

gentoo penguin (*Pygoscelis papua*) After the emperor and king penguins, the gentoo is the third largest of the penguins. The total circumpolar population of gentoo penguins is about 350,000 breeding pairs. Often seen in the company of Adélies, they are identifiable by their pale orange feet, black-tipped orange-red bill, and the conspicuous white patch over each eye. Gentoos stand about 30 inches high and breed in loose colonies on land, with ritualized threat postures and courtship behavior. Their nests are scraped-out hollows lined with grass, stones, bones, and moss.

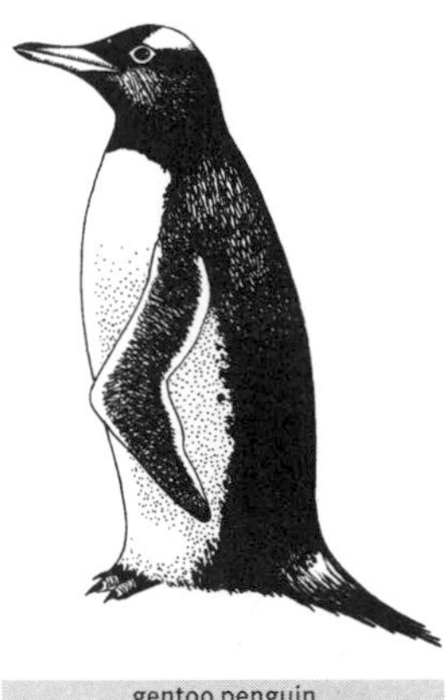

gentoo penguin

See also Adélie penguin, chinstrap penguin

geoduck (*Panope generosa*) At a maximum weight of 12 pounds, the geoduck is the largest burrowing bivalve in the world. The thin shell can be 8 inches long, and because the body and two siphons are much too large to be retracted, the two halves of the shell are always open. Occurring on mud flats from Southern California to British Columbia, geoducks make semipermanent burrows that can be 3 to 4 feet deep. The siphons are extended to the surface in order to draw in oxygenated water and nutrients. Digging for geoducks is popular in the state of Washington (where the word is pronounced "gooey-ducks"), but there is a limit of three per person per day.

See also clam

Georges Bank Twenty thousand square miles of turbulent shoal water off Nantucket, Georges Bank is—or was—one of the world's richest fishing grounds. Basque fishermen were probably the first Europeans to notice the abundance of cod in these waters, and when John Cabot arrived in 1497, he too was astonished by their numbers and caught the fish simply by lowering weighted baskets. Bartholomew Gosnold, who explored Nantucket and Martha's Vineyard in 1602, also named Cape Cod because the eponymous fish were so plentiful. The reason for the name Georges Bank is not known; it may have originally been called St. George's Shoal, or it may have been named for Sir Ferdinando Gorges (1566–1647), an early British proponent of fisheries in New England. (Although he never actually set foot in North America, Gorges received a grant for what eventually became the entire state of Maine.) Schooners from Gloucester and Marblehead fished for cod, herring, haddock, and halibut, and by the middle of the twentieth century, Georges Bank fishermen were taking about 10 percent of the total American catch. The herring fishery was so intense that it completely collapsed by 1976. Before the introduction of the 200-mile limit in that year, factory ships from East Germany, the Soviet Union, and Poland were sweeping American waters clean, nowhere more efficiently than Georges Bank. Since both the United States and Canada claimed jurisdiction over the Georges Bank, they took their dispute to the World Court in The Hague, which set up a boundary line in 1985 that neither country's boats could cross. In 1992, having run out of fish to catch, the Canadian government permanently shut down the cod fishery. Two years later, the United States did the same. A glimmer of hope for Georges Bank fishermen is the lowly hagfish (*Myxine glutinosus*), once regarded only as a "trash fish" to be discarded. A large fishery has grown up in New England for hagfish, which are shipped to Korea where the skins are processed and turned into "eelskin" wallets, belts, and other expensive leather goods.

See also Atlantic herring, codfish, hagfish, Nantucket

Gerlache, Adrien de (1866–1934) Naval officer who led the Belgian Antarctic expedition in the *Belgica* in 1897–1899. (On board were the Norwegian explorer Roald Amundsen, who would eventually be the first to reach the South Pole, and Dr. Frederick Cook, who would claim to have reached the North Pole.) Gerlache's ship was beset in the ice of the Bellingshausen Sea, probably because the scientists spent too much time collecting specimens. They remained for almost a year, thus becoming the first crew to winter over in the Antarctic. After his triumphant return to Belgium, Gerlache made a number of oceanographic cruises to the Arctic and the Persian Gulf. Gerlache Strait on the west side of the Antarctic Peninsula is named for him.

ghost pipefish (family Solenostomidae) According to one writer, the ghost pipefishes look like "pipefishes getting ready to fly." They are small, usually not more than 6 inches in length, with a body encased in segmented bony plates, a narrow head, an elongated snout, and enlarged pelvic fins. Some species are camouflaged by spines and spots, but the Indo-Pacific *Solenostomus cyanopterus* has only its yellowish-brown coloration to hide it in the sea grass. Unlike many of the sea horses and pipefishes, the males of which incubate the eggs, it is the females of the ghost pipefishes that do the job. **See also pipefish, sea horse**

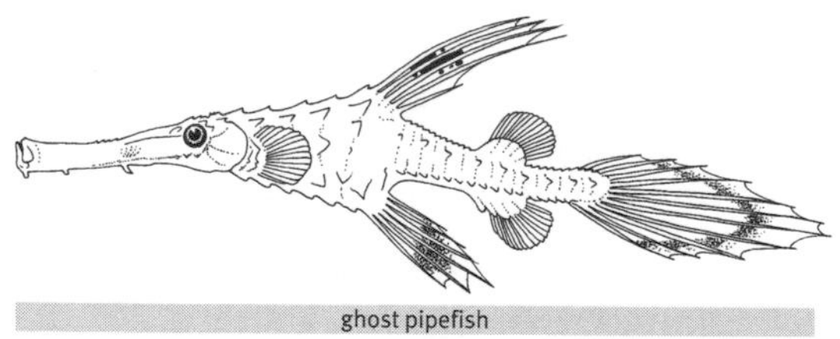
ghost pipefish

giant clam (*Tridacna gigas*) Found in the waters of the northern end of the Great Barrier Reef and islands to the north, *T. gigas* is the largest clam, and one of the largest of all mollusks. It has been measured at more than 4½ feet in length, and weighed at more than 500 pounds. There are many stories about pearl divers being trapped by the foot as the great shell closes, but these are undocumented. The largest known pearl, called the Pearl of Laotze, is a 14-pound convoluted object that looks a bit like a human brain, collected by pearl divers from a giant clam off the Philippines in 1934. Although the Great Barrier Reef is officially a preserve, unscrupulous fishermen raid the habitat of the giant clam to harvest clams for sushi.

See also clam, Great Barrier Reef, mollusk, pearl

giant octopus (*Octopus dofleini*) The largest of the octopuses, *O. dofleini* can reach a total armspan of 16 feet, although there are unsubstantiated reports of specimens reaching 20 and even 30 feet. The largest specimens have weighed upward of 100 pounds; again, there have been reports of individuals that weighed considerably more. The giant octopus is found in the northern Pacific, from Japan through the Aleutians, Alaska, and as far south as Oregon. Because of their size and the mythology surrounding them, they were long believed to be true terrors of the deep, and divers would often wrestle with them for sport or shoot them with spearguns. In fact, this species is harmless to everything but shellfish, and there are no reports of one ever attacking a human. They are fished commercially off Japan. **See also giant squid, octopus**

giant petrel (*Macronectes* spp.) There are two species of Southern Ocean giant petrel, the northern (*M. halli*) and the southern (*M. giganteus*), which are essentially similar, differentiated by the coloration of the tip of the bill and, to a lesser extent, by geography. The northern species (sometimes known as the northern giant fulmar) has no green on the bill, and breeds on islands north of the Antarctic Convergence, such as Marion, Crozet, Kerguelen, Macquarie, Chatham, and Auckland. The southern version, identifiable by a yellowish-green bill tip, breeds on South Georgia, South Sandwich, the South Orkneys, the South Shetlands, and the Antarctic Peninsula. The giant petrel, with a wingspan that can reach 80 inches, could be confused with a small albatross, but it has shorter wings and a heavier bill, and is a much less graceful flier. Unlike albatrosses, giant petrels feed on carrion, and are often seen tearing at bloody carcasses of whales, seals, and birds.

See also albatross, fulmar, shearwaters

giant sea bass (*Stereolepis gigas*) Also known as the California sea bass or California jewfish, this inhabitant of the warmer waters of the west coast of North America grows to a length of 7 feet and has been weighed at 563 pounds. It is believed that the larger specimens may be seventy-five years old. The giant sea bass of Florida, or jewfish (*Epinephelus itajara*), may reach 680 pounds, and *E. lanceolatus,* known as the giant groper in Australia, grows even larger, to a reported length of 10 feet and a weight of 800 pounds. Despite many exaggerated stories, there are no authenticated records of this species swallowing—or even attacking—a diver.

giant sea bass

See also grouper, jewfish

giant squid

giant squid (*Architeuthis* spp.) At a known length of 57 feet, the giant squid is the largest of the squids and the largest of all invertebrates. No one has ever seen a healthy, living *Architeuthis,* so almost everything we know has come from the examination of dead or dying specimens. Their eyes are the largest eyes in the animal kingdom, as much as 15 inches in diameter—the size of dinner plates. Despite wildly exaggerated reports about the size of sucker marks seen on the skin of sperm whales, the suckers of giant squid are about 1½ inches in diameter. They have no claws, but their suckers are ringed with tiny, chitinous teeth that allow them to get a better hold on their prey. We do not know how big they can get, what they eat, how fast they grow, where they live (we know more about where they die), or why they wash ashore. Giant squid have beached themselves in large numbers in various places, especially Newfoundland, Norway, and especially New Zealand, where in recent years more than thirty specimens have been caught in fishermen's nets deployed 2,000 to 5,000 feet down. **See also Humboldt squid, Pacific giant squid, sea serpent, squid**

Gibraltar Narrow peninsula (3 miles long) that is connected to the southern coast of the Iberian Peninsula by a 1-mile-long isthmus. The entire peninsula consists of a limestone and shale promontory known as the Rock of Gibraltar. Gibraltar comes from the Arabic *Jabal Tariq,* meaning "Mount Tariq"; it was named for Tariq ibn Ziyad, who captured it in A.D. 711. It overlooks the Strait of Gibraltar, the only entrance to the Mediterranean from the Atlantic. Wild olive and pine trees grow on the upper slopes of the rock, and in addition to rabbits and foxes, it is the home of the barbary ape, the only wild monkey in Europe. The ancients considered Gibraltar as one of the two Pillars of Hercules that guarded the entrance to the Mediterranean; the other is Mount Hacho (*Jabal Musa*) in Morocco, 10 miles across the strait. On Gibraltar there is a system of tunnels in the rock for vehicular traffic, and a car ferry crosses daily to Tangier in Morocco. The Muslims held Gibraltar until 1462; Queen Isabella of Spain annexed it to Spain in 1501. In 1704, during the War of the Spanish Succession (1701–1714), Gibraltar was taken by Admiral Sir George Rooke (1650–1709); it was formally ceded to Britain in 1713. It became a Crown Colony in 1830, and when the Suez Canal opened in 1869, Britain controlled the ingress and egress to the Mediterranean. During the world wars, Gibraltar served as a vital repair and assembly point for Allied convoys. A 1967 referendum offered the citizens of Gibraltar the opportunity to affiliate themselves with Spain rather than Britain; they voted for Britain, 12,138 votes to 44. Spain responded by closing the land border, which remained closed until 1980. After the last British battalion was removed in 1991, Spain suggested joint control of the rock, but Britain refused.

Giddings, Al (b. 1937) Now regarded as the foremost underwater filmmaker in the world, Giddings has worked in virtually every available medium and format, from super-8 to IMAX. He has designed many of the camera, lighting, and optical systems now being used by underwater cameramen from around the world. In Hollywood, he directed and shot the underwater sequences for many films, including *The Deep* (1977), *For Your Eyes Only* (1981), *Never Say Never Again* (1983), *The Abyss* (1989), and *Titanic* (1997), which he also coproduced. He has won Oscar nominations for his cinematography and numerous Emmys for his television documentaries, which have included specials on Cuba, the Galápagos Islands, the blue whale, sharks, hydrothermal vents, the North Pole, deep-sea volcanoes, humpback whales, seals and walruses, the *Andrea Doria,* and numerous other subjects. He has filmed in every ocean (and many lakes, rivers, and artificial environments) and has dived under the ice at the North Pole. He maintains an enormous library of still photographs and stock footage at his base in Pray, Montana. With Sylvia Earle he wrote *Exploring the Deep Frontier* (1980), and he was principal photographer for *Great White Shark* (1991), written by Richard Ellis and John McCosker.

giganturus

giganturus (*Giganturus chuni*) Because of its tubular eyes, directed straight forward, giganturus is sometimes known as the telescope fish. Half of its 10-inch length is tail. This species is found in deep water, and little is known of its natural history. When brought to the surface in trawls, it appears a burnished silver color, unlike so many other deep-sea fishes, which are black. A ferocious predator, giganturus has been found with a complete viperfish (*Chauliodus*) twice its own size in its stomach. The larval form is a big-headed, round-bodied little creature, so different from the adult that it was originally believed to be a completely different species. Long puzzled by its unusual eyes, scientists now believe that giganturus swims upright in a vertical position, using its upward-facing eyes to pick out prey species silhouetted against the lighter surface.

See also dragonfish, viperfish

Gilbert, Humphrey (c. 1539–1583) British soldier, navigator, first governor of Newfoundland, and half brother of Sir Walter Raleigh. In 1566, he wrote *A Discourse to Prove a Passage to the Northwest to Cathaia and the East Indies,* but instead of North America, Queen Elizabeth sent him to Ireland to put down an uprising. (He was knighted for his efforts.) By 1577 he had directed his attentions to North America and proposed a scheme to seize the codfishing fleets of Spain, Portugal, and France on the Newfoundland Banks. When the queen dismissed this plan, Gilbert outfitted his own expedition to North America, which never made it past Ireland. Finally, in June 1583, Gilbert sailed on a colonizing expedition from Plymouth with five ships, one of which—named *Raleigh* for its owner—turned back with disease, but the other four (*Delight, Golden Hind, Swallow,* and *Squirrel*) made it to Newfoundland, where they encountered a motley codfishing fleet composed of thirty-six English, Spanish, Portuguese, and French vessels. Gilbert entered the harbor (where St. John's is now located) and took possession of Newfoundland for England. His little fleet headed for home, but the *Delight* ran aground and sank. Gilbert chose to remain aboard the *Squirrel* for the return voyage, and in a great storm off the Azores, the ship went down with all hands. Edward Hayes, captain of the *Golden Hind,* was close enough during the storm to see that Gilbert was reading a book. When he returned to England, he reported that Gilbert's last words were, "We are as near heaven by sea as by land." Gilbert's *Discourse* was the inspiration for Martin Frobisher's 1576–1577 voyages.

See also Cabot, codfish, Frobisher, Raleigh

Gilbert, Perry W. (b. 1912) Probably the most influential shark biologist of his generation, Gilbert graduated from Dartmouth in 1934 and earned his Ph.D. at Cornell in 1940. After an instructorship at Dartmouth, he returned to Cornell and taught there until 1967, when he was named director of the Mote Marine Laboratory in Sarasota, Florida. In 1958 he was appointed chairman of the Shark Research Panel, sponsored by the Office of Naval Research and the American Institute of Biological Sciences, which created the original Shark Attack File. In addition to publishing more than a hundred scientific papers, Gilbert edited *Sharks and Survival* (1963) and *Sharks, Skates and Rays* (1967).

See also Shark Attack File

Gilbert Islands Sixteen islands that were administered by the British as the Gilbert and Ellice Islands since 1898, the Gilberts are now part of the Republic of Kiribati, which was established in 1979. First explored by British admiral "Foul Weather Jack" Byron in 1765, the islands were named for Thomas Gilbert, who explored them with John Marshall in 1788. Kiribati also includes the Line Islands (including Kiritimati, formerly Christmas Island), the Phoenix Islands, and Ocean (Banaba) Island, a producer of phosphates until it was mined out in 1981. The Gilbert islands of Tarawa, Butaritari, Abalang, Marakei, and Abemama were occupied by the Japanese in 1941 and liberated by U.S. forces in 1943. After the Ellice Islands gained independence as Tuvalu, the remaining islands incorporated as Kiribati. The government of Kiribati severed diplomatic relations with France in 1995 when the latter resumed nuclear testing in the Pacific. **See also Kiribati**

gill net Drifting nets that may be miles in length, gill nets may follow the fishing vessel or be anchored at or near the bottom. The mesh of the net is designed to allow the fish to pass its head through, and then, because it cannot back up, it is trapped there. The gill nets are then hauled aboard and the catch dumped out on deck. Unfortunately, nets cannot discriminate between the object of the fishery and any other creature, so everything else that gets trapped in the net—sharks, dolphins, seals—also dies.

See also drift-net fishing, longline fishing, purse seining, trawling

gills The respiratory organs of most aquatic animals, enabling them to breathe water instead of air. In fishes, water is taken in through the mouth and passed to the gill chambers at the rear of the mouth, where the gills themselves extract oxygen from the water and discharge carbon dioxide. (It is oxygen dissolved in water that is extracted, not the "O" in H_2O.) The gills consist of featherlike filaments with capillaries well supplied with blood, enclosed in a thin membrane. Their fine structure makes possible maximum contact of the blood with the water for the most efficient gas transport. The gills, which may be plated or tufted, are attached to a series of paired arches, which also include the gill rakers, comblike projections on the inner edges that trap particulate matter, and keep it from passing through the gill slits and direct it to the esophagus. In sharks, the gill slits, between five and seven on each side, are exposed, but in bony fishes they are covered by an operculum or gill cover. In crustaceans, the gills are covered by the carapace; in echinoderms they are branched appendages extending from various parts of the body. In mollusks, including clams, oysters, and cephalopods, the gills are internal and located inside the body cavity.

Gimbel, Peter (1928–1987) Educated at Yale, Peter Gimbel seemed destined to follow his father's footsteps into the merchandising business. (Gimbel's, New York's famous department store, was opened in 1909 by his great-grandfather.) After college, he went to work at a brokerage firm, but it was too tame for him, and he devoted the rest of his life to adventure and filmmaking. Gimbel parachuted into the Peruvian jungle in search of El Dorado. He was the first diver to photograph the wreck of the *Andrea Doria* in 1956 after her collision with the *Stockholm*. After making a short documentary film about diving with blue sharks off Montauk, Long Island, in 1965, he set out on a 'round-the-world expedition to film the great white shark in its natural habitat. Released to great acclaim in 1971, *Blue Water, White Death* is generally recognized as one of the best underwater documentaries ever made. In 1983 he returned to the *Andrea Doria*, made a film about the collision and the wreck, and brought up the purser's safe, which was reputed to contain a fortune in jewels and currency. When it was opened on national television, it was revealed to hold only a lot of very soggy bills stuck together. Gimbel laboriously separated the piles of money and traveler's checks and sold them as souvenirs of "the *Doria* Project."

ginkgo-toothed whale (*Mesoplodon ginkgodens*) This is one of the "newest" species of beaked whales, described for the first time in 1958. It was discovered offshore near Tokyo, and when scientists examined it they realized it was a type not previously seen. The male has two teeth in the lower jaw that resemble the leaves of the ginkgo tree, hence its peculiar name. This beaked whale is black in color, with the usual complement of scars and scratches. Since its original discovery, specimens have been reported from Taiwan, Sri Lanka, and Southern California.

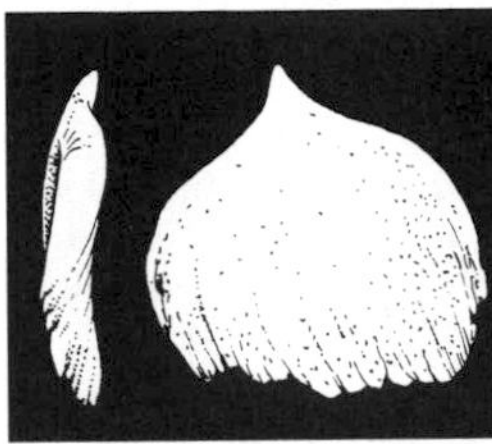

ginkgo-toothed whale (tooth)

See also beaked whales, *Mesoplodon*

glacier A moving mass of ice that survives from year to year, formed by the compacting of snow and set into motion by the pull of gravity. Glaciers are usually found at high altitudes and high latitudes, and they are found on every continent except Australia. Most of the Antarctic continent and Greenland are covered by a thick sheet of ice, and the glaciers at the edges of the ice sheet calve most of the world's icebergs. During the Ice Ages, the last of which ended about eleven thousand years ago, huge ice sheets spread over all of Antarctica, Greenland, and Canada, and covered large parts of Europe and North and South America, and small parts of Asia. Smaller glaciers are called mountain glaciers, and those that spread out at the foot of glaciated regions are called ice shelves. Many terrestrial formations owe their existence to glaciers, especially valleys and fjords. Although a glacier is said to flow, it actually moves as a solid rather than a liquid. The center of a glacier moves faster than the sides, and the surface moves more quickly than the bottom, because of friction. Glaciers are always in motion, either retreating or advancing. If the ice melts at the leading edge (known as the toe) faster than the glacier moves forward, it is said to be retreating. The largest glaciers are in Greenland and Antarctica, but Iceland's Vatnajökull Glacier, with an area of 5,000 square miles, is larger than all the glaciers on the European continent combined.

See also Antarctica, Glacier Bay, Greenland, Iceland

Glacier Bay Region of southeast Alaska; the northern area of the Alexander Archipelago. It was partially explored by George Vancouver in 1794; subsequent explorers marveled at the spectacular vistas of mountains and ice. From satellite imaging, we now know that most of the glaciers are retreating, but they still afford one of the world's most impressive views of tidewater glaciers—those that drive into tidal waters. The largest are named Brady, Margerie, Muir, McBride, Reid, Carroll, and Casement, but there are more than two hundred smaller glaciers in the park. Designated Glacier Bay National Monument in 1925, it was established as a national park in 1980. Most of the 300,000 annual visitors come aboard cruise ships and view the bears, mountain goats, whales, seals, and eagles from the comfort of the deck. At the northwestern corner of the park is Lituya Bay, the site of a 1958 earthquake that caused a landslide that raised the water level 1,700 feet.

See also Alexander Archipelago, Lituya Bay

glaucous gull (*Larus hyperboreus*) The only gull that approaches the great black-backed gull (*L. marinus*) in size, the glaucous is a pale gray bird with a white head and tail. Unlike most adult gulls, it has white wing tips. It has been called the only purely predatory gull; it captures adult auks and plovers on the wing, and raids nests for eggs and young. (Skuas are also predatory, but while they look like gulls, they are not.) It nests in the Arctic on cliffs and high ledges, often in the vicinity of guillemot or razorbill colonies.

See also great black-backed gull, gull, herring gull

Global Positioning System (GPS) An all-weather, twenty-four-hour-a-day radio-based satellite navigation system that enables users to accurately determine position, velocity, and time. Originally developed for military purposes, but now available for civilian use in a slightly degraded form, it is based on the U.S. Navy's Navigation Satellite Timing and Ranging (NAVSTAR) system, which consists of twenty-one satellites (plus three spares) operating in twelve-hour orbits at an altitude of 12,500 miles above the earth. The satellites are arranged in six orbits, each orbital plane inclined at 55° and equally spaced about the equator. The satellites continuously broadcast the time and their exact position, which are compared to the signal generated by a ground-based receiver. The resulting signal shift, along with the timing data, is used to determine the distance from the satellite, and thus the geodetic position of the point. Three satellites are required for two-dimensional (X and Y) positioning, and four for three-dimensional (X, Y, and Z) positioning. Three-dimensional positioning is required for aircraft, but ships at sea move primarily in a two-dimensional world. GPS units can be either attached to laptop computers or self-contained; they are small enough to hold in the palm of one's hand. For the first time, military, commercial, and privately owned vessels are able to determine their exact position on the sea. **See also astrolabe, backstaff, compass, quadrant, sextant**

Glomar Challenger Named for Global Marine, the company that built her, and the historic research vessel *Challenger,* the drill ship *Glomar Challenger* was launched in 1968. At 400 feet long with a 200-foot derrick amidships, she looked for all the world like an oceangoing oil rig. She was designed for extracting core samples from the bottom while maintaining her position by means of a sophisticated dynamic positioning system, which consisted of sonar beacons bounced off the ocean floor and relayed information to computers that activated the ship's stern and bow thrusters. From 1968 until her retirement in 1983, the *Glomar Challenger* plied the world's oceans, often working in water that was as much as 20,000 feet deep. The Joint Oceanographic Institutions Deep Earth Sampling program (JOIDES) was formed in 1964, and the *Glomar Challenger* was put to work four years later, employing scientists from almost every oceanographic and geological discipline who made major contributions to our understanding of seafloor spreading, continental drift, the structure of the earth's mantle, the earth's magnetic field, and the composition of ocean sediments. In her fifteen years of service, the ship made a total of ninety-three drilling voyages (known as legs), in all the major seas of the world, from the Arctic Ocean to the Ross Sea, from the North Atlantic to the South Pacific.

See also *Challenger*, continental drift, seafloor spreading, sediments

Glomar Explorer Originally developed to raise a Soviet Golf-class ballistic-missile submarine that had sunk to a depth of 13,000 feet in the North Pacific, *Hughes Glomar Explorer* was 618 feet in length, with a beam of 115 feet, and a displacement of 63,300 tons. In 1974, as part of the navy's high-secret Project Jennifer, she successfully raised the forward portion of the submarine with a giant claw. The ship was purchased by the U.S. Navy, the "Hughes" (for the company that had built her) was removed from the name, and she became the *Glomar Explorer.* She was leased to Global Marine for seafloor mining and is now employed as an oil-drilling ship.

goatfish (family Mullidae) Goatfishes are characterized by two widely separated dorsal fins, large scales, and two long chin barbels that are used for detecting food. There are six genera and about sixty species, most of which are less than 20 inches in length. Found worldwide in tropical and shallow temperate marine waters,

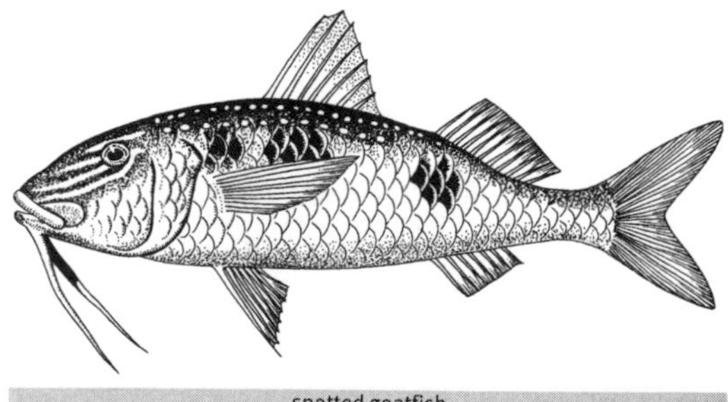

spotted goatfish

they are highly prized as food. Most species are brightly colored, as indicated by some of the common names: yellow-stripe goatfish, bicolor goatfish, blue-spotted goatfish, and the like. They are renowned for their rapid color changes. During the day, the spotted goatfish (*Pseudopeneus maculatus*) is white with a few large black spots; at night it becomes greenish yellow, horizontally striped in black. At ancient Roman banquets, a red mullet (*Mullus surmuletus*) was placed in a bowl so that the guests might marvel at the changing colors of the dying fish. The red mullet is a goatfish, but the common mullets (Mugildae) are different fishes altogether.

See also mullet

goblin shark

goblin shark (*Mitsukurina owstoni*) When this strange-looking shark was first discovered in 1898, it was believed to be the long-extinct *Scapanorhynchus,* and that was its original name. (It was named *M. owstoni* for Japanese ichthyologist Kakichi Mitsukuri and natural history dealer Alan Owston.) This 14-foot-long deep-water species has a very long tail and a bizarre extension that overhangs its snout and is responsible for its common name. The shark's use of this bladelike instrument is unknown; it may stir up sediment in the search for food, or it may contain sensory receptors like the head lobes of the hammerheads. The jaws of the goblin shark are greatly protrusible, which they would have to be if the shark is going to be able to eat. Goblin sharks are found in the deep waters off Portugal, France, South Africa, and Japan.

See also hammerhead shark

goby (family Gobiidae) An incredibly varied and colorful group, gobies are small bottom-dwelling fishes found in the shallow, warm, and temperate waters of the world. With some 220 genera and 1,600 species, it is the largest family of marine fishes in the world. These are elongated fishes, with two dorsal fins, an anal fin that mirrors the second dorsal, and two ventral fins that are fused together to form a cup-shaped disk. They usually have large eyes, which are often protruding. After an elaborate courtship, the females lay eggs, attach them to some sort of vegetation, and then leave the males to guard them. Many species live in burrows

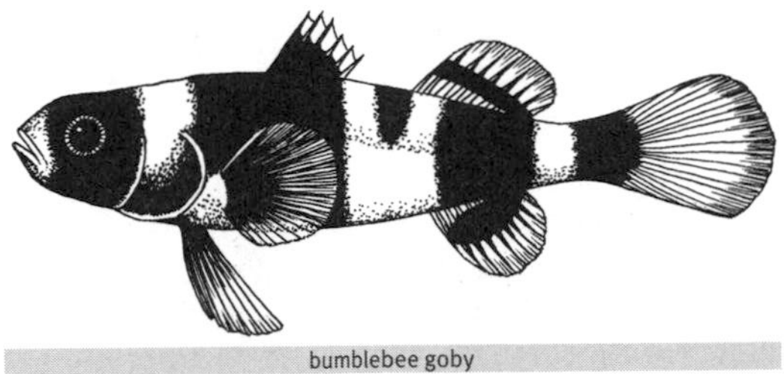

bumblebee goby

in sand or gravel, which they often share with other animals, particularly shrimps. Others live on branching corals, sea fans, or sponges. Several species, such as the neon goby (*Gobiosoma oceanops*), are parasite cleaners of larger reef fishes. The smallest of all vertebrates is the pygmy goby, *Pandaka pygmaea,* from the Philippines; at a maximum length of ½ inch, it is considerably smaller than its name on this page. The mudskipper (*Periophthalmus* spp.), a goby of the mangrove mudflats of the Indo-Pacific, spends more time out of the water than in it. The skin of this pop-eyed little fish contains blood vessels that can extract oxygen from the air, and it spends the days hunting flies and other insects while sitting in the mud, propped up on its strong pectoral fins.

golden plover (*Pluvialis dominica*) Like the passenger pigeon and the Eskimo curlew, the American golden plover was one of the most numerous birds in North America; its flocks darkened the skies of the Midwest as they migrated from the pampas of South America to Nova Scotia. The passenger pigeon and the Eskimo curlew were hunted to extinction, but the golden plover survived and its populations are recovering. A very similar species, *P. apricaria,* is found in Europe.

golden plover

See also Eskimo curlew, plovers

Gomes, Diogo (fl. 1440–1484) Sometimes "Gomez"; a Portuguese explorer sent by Prince Henry the Navigator to sail around Africa to reach India in 1455. His first expedition, which consisted of three caravels, reached Cape Verde (the cape, not the islands) on the West African coast, in what is now Senegal, close to the city of Dakar. At the Gambia River his men refused to sail any farther, believing they were at the edge of the ocean, so they sailed up the river before heading back. They met travelers from Timbuktu, but illness among the crew caused them to return to Portugal. Gomes sailed again in 1462 and claimed to have discovered the Cape Verde Islands, but they had actually been found six years earlier by Alvise Cadamosto, a Venetian in the service of Prince Henry. Gomes's account of his voyages, dictated twenty years after the fact to Martin Behaim, a German geographer, is not considered totally reliable.

Goodwin Sands A dangerous stretch of shifting sands and shoals, located at the southeast coast of England, where the North Sea meets the English Channel. It is said that the land now known as Goodwin Sands was originally 4,000 acres of dry land owned by the Earl of Godwine. William the Conqueror bestowed the property on the Abbey of St. Augustine at Canterbury, but in 1100, after the seawall had fallen into disrepair, the sea broke through and inundated the area. The unstable nature of the sands has made it impossible to erect a lighthouse there, so lightships and buoys have to serve to warn ships of the dangers. During a storm in 1703, thirteen British warships were anchored in The Downs, the anchorage just to the west of Goodwin Sands, and all were blown into the shallows and wrecked, one of the greatest disasters in British maritime history. In 1825, the *Baron of Renfrew,* a four-masted bark that was one of the largest ships ever built, crossed the Atlantic from Canada on her maiden voyage, and was wrecked on the Goodwin Sands. In 1836, thirty vessels were lost in a single storm.

goosebeak whale (*Ziphius cavirostris*) Although none of the beaked whales can accurately be described as "common," this species (also known as Cuvier's beaked whale) is probably the most frequently observed. The coloration varies from dark purplish brown to fawn, and the head and neck are lighter, often white. The forehead slopes to a short, poorly defined beak, and the flippers and dorsal fin are small. Like all other beaked whales, this species is usually covered with scars and

goosebeak whale

scratches, probably acquired during fights. It is found in all oceans of the world except the high Arctic and Antarctic, and relatively frequent strandings have been reported. **See also beaked whales**

goosefish (*Lophius americanus*) The largest of the anglerfishes, the goosefish can be 4 feet long and weigh 70 pounds. It lies on the bottom of the western Atlantic, from the Gulf of St. Lawrence to southern South America, waiting for prey to gobble up in its capacious, tooth-lined mouth. The width of the fish's huge, flattened head is equal to about two-thirds its length. It eats anything and everything, including all kinds of fishes and crabs, small sharks, and even seabirds. Goosefishes have almost armlike pectoral fins that they use to pull themselves along on the bottom. The goosefish is usually called monkfish when it appears on restaurant menus, and its white flesh is firm and tasty.

See also anglerfishes, frogfish

gorgonians Included are the sea fans, sea whips, sea plumes, and *Corallium,* the pink gorgonian from which the popular coral jewelry is made. Gorgonians do not secrete a limestone skeleton like the hard corals, but they are strengthened by gorgonin, a horny material. (*Corallium,* however, has no gorgonin; the hard, pink substance is the spicules fused together.) They are

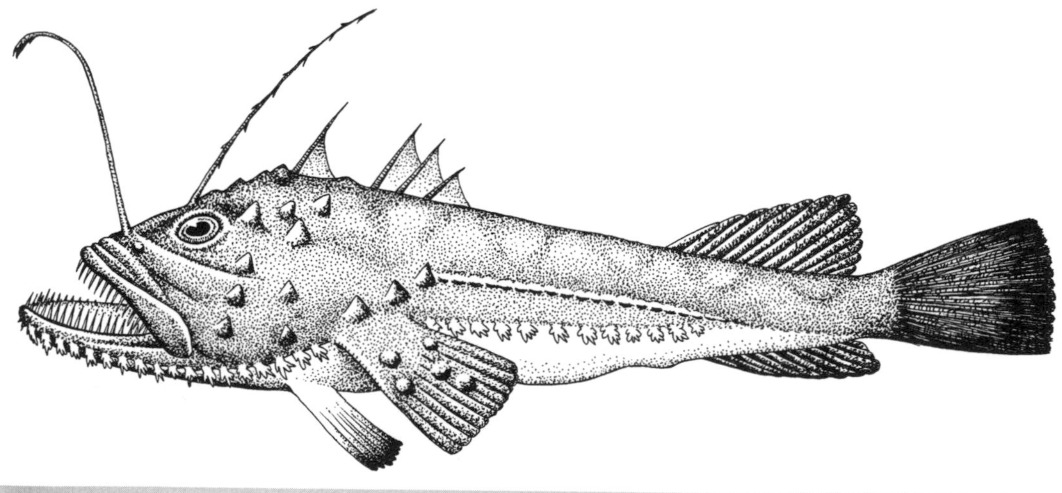

goosefish

plantlike in form, usually flattened to one plane, and they attach themselves to the bottom by means of a basal plate or *stolon*. Shallow-water gorgonians come in vivid shades of red, orange, yellow, and purple, and are plentiful around coral reefs, much to the delight of underwater photographers. Particularly impressive in the Caribbean are the Venus sea fan (*Gorgonia flabellum*) and the common sea fan (*G. ventalina*). Most gorgonians are found in shallow water, but there are also deep-water species, some of which are luminescent.

See also coral, soft corals

Gosnold, Bartholomew (1572–1607) English explorer, who left in the *Concord* for North America in 1602 to seek a location for a colony (and find gold, if possible). He navigated the coast from Maine to Narragansett Bay, naming Cape Cod (for the codfish) and Martha's Vineyard. He built a fort on Cuttyhunk, the westernmost of the Elizabeth Islands, but there was not much to recommend it for settlement, and he returned to England with a cargo of sassafras bark to be used for medical infusions. In 1606 he was put in command of the *Godspeed*, which carried some of the first settlers to Virginia. He protested against the choice of Jamestown for a settlement—with good reason, for he died there of malaria within a few months.

Gran Canaria After Tenerife and Fuerteventura, the third largest of the Canary Islands, off North Africa. Roughly circular, the island covers 592 square miles. The center consists of a 6,000-foot-high mountain range cut by numerous canyons, and ridges that extend to the shore. It is the most fertile of the islands, with bananas, tomatoes, and tobacco growing in the lowlands and native pine on the slopes of the mountains. The beaches of Las Canteras and Las Alcaravernas attract many tourists, as does Las Palmas, the island's largest city, with a population of 370,000. Named for its numerous palm trees, Las Palmas was founded in 1478. The port, the largest in the Canaries, is on the main shipping routes between Europe and America and is used for fuel bunkering. Las Palmas also has the most important airport in the islands. **See also Tenerife**

Grand Banks Extensive portion of the North American continental shelf, lying south and east of Newfoundland, known as a breeding ground for cod. Also known as the Newfoundland Banks, the Grand Banks extend for about 350 miles north to south and 450 miles east to west; they average about 180 feet in depth. Their richness was first discovered by John Cabot in 1497 when he lowered baskets overboard and brought them up full of fish. Cabot's discovery led to the seventeenth-century cod fishery, where vessels from Britain, France, Holland, and Spain crossed the Atlantic to take advantage of the incredible bounty. In addition to cod, there were also vast numbers of haddock, plaice, herring, and mackerel, which supported the great nineteenth-century fisheries of Maine and Massachusetts. By 1977, Canada had extended its economic zones to 200 miles off her coasts, which included most of the Grand Banks, but overfishing has greatly depleted the seemingly inexhaustible stocks of fish, and the once-dominant cod fishery has been indefinitely shut down by the Canadian government because the fish are gone.

See also Cabot (John), codfish, Georges Bank

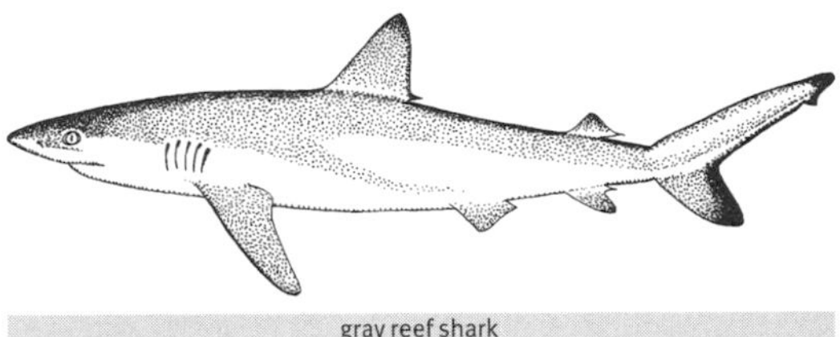

gray reef shark

gray reef shark (*Carcharhinus amblyrhynchos*) A medium-sized to large (up to 7½ feet) gray shark, characterized by a pronounced black band on the trailing edge of the tail fin, found throughout the islands of the Indo-Pacific region, including northern Australia. Common around coral reefs, it is often encountered in schools during the day. This species is known for its curiosity: it will often investigate novel events, even if no food stimuli are present. They are also attracted to spearfishermen, and will often contest ownership of a speared fish. The gray reef shark is known for its agonistic threat display, where it humps its back, drops its pectorals, and wags its head from side to side. If this does not scare off the intruder, the shark attacks.

See also carcharhinid sharks, Galápagos shark

gray whale

gray whale (*Eschrichtius robustus*) With its mottled grayish appearance and "knuckles" where a dorsal fin would be, the gray whale reaches an adult length of 40 to 50 feet, has an overhanging upper jaw, and is usually covered with barnacles and whale lice. (They are born black and acquire the barnacles and lice, which produce the scars and scratches.) Gray whales feed by scooping up mouthfuls of mud and tiny crustaceans from the bottom; after forcing the mud and water out through their white, sievelike baleen plates, they swallow the crustaceans. The world's only population of these whales lives off the coast of western North America, where they make a 13,000-mile round-trip from the Bering and Chukchi Seas to Baja California—the longest migration of any mammal. The gestation period for gray whales is a year; the young are conceived in the lagoons of Baja, then born there twelve months later. As recently as the seventeenth century, there were gray whales in the Atlantic, but they have all disappeared. Another population used to migrate from the Bering Sea down the eastern coast of Asia, but Japanese, Korean, and Chinese whalers eliminated them. Once thought to be nearly extinct because of heavy whaling pressure, the population has now stabilized at about 22,000, and is protected in Canadian, American, and Mexican waters. Russian whalers take about 175 gray whales annually for feeding captive minks, and the Makah Indians of Washington were recently granted permission to catch five gray whales per year in recognition of their traditional practices. Gray-whale watching, particularly off the coast of California and in the lagoons of Mexico, is enormously popular. The only "great" whales ever maintained in captivity were gray whales: "Gigi" was captured in Scammon's Lagoon in March 1971 and kept for a year at Sea World in San Diego before she was released. In January 1997, "J.J." was found sick and rolling in the surf near Point Loma, California; she was also taken to Sea World for rehabilitation and exhibition. When she was released in February 1998, she weighed 19,000 pounds and was 31 feet long. **See also whale, whaling**

great auk (*Pinguinis impennis*) A large flightless seabird that stood well over 2 feet tall; it has been extinct since the last two were killed by hunters in Iceland in 1844. On islands along the coast of Newfoundland and in Europe, they were hunted to extinction for their meat and eggs. They were originally called penguins: the name "penguin" was applied to the Southern Hemisphere birds because of their similarity to the great auk. Like penguins, they were strongly marked in black and white, and they stood upright with their little wings tucked into their sides. Also like penguins, they could not fly, and they dived for their food like their surviving relatives, the razorbills and the murres.

See also Alcidae family, auklet, murre, penguin, razorbill

great auk

Great Australian Bight The great indentation of Australia's southern coast, extending for 720 miles from Cape Pasley, Western Australia, to Cape Carnot, South Australia. The arid Nullarbor Plain overlooks the bight, and most of the shore consists of steep cliffs, some as high as 200 feet. First explored in 1627 by the Dutch navigator Pieter Nuyts, the coast was surveyed by Matthew Flinders in 1802. The Recherche and Nuyts Archipelagos are in the bight, as are the Investigator and Whidbey Islands.

Great Barrier Reef The largest coral reef in the world, ranging from 10 miles off the eastern Australia shore in the north to 100 miles in the south. At 1,250 miles, its length is comparable to the entire west coast of the continental United States. The largest structure ever built by living creatures, it actually consists of thousands of separate reefs, islets, and islands. At least 350 species of coral are responsible for its construction, each contributing its calcareous remains to the whole over a period of some 25 million years. The marine life of the warm, clear waters of the reef is plentiful and varied, consisting of anemones, worms, gastropods, crayfish, lobsters, prawns, crabs, cephalopods, and a great variety of fishes and seabirds. Among the better-known species are the crown-of-thorns starfish (*Acanthaster*), the giant clam (*Tridacna gigas*), the deadly box jelly, or sea wasps (*Chironex fleckeri*), and hundreds of endemic brightly colored fishes. Captain James Cook is believed to have been the first European to encounter the reef when his ship *Endeavour* ran aground on it in 1770. Although the entire reef has been declared a marine park, commercial development is taking place on many of the larger islands, such as Green Island, Heron Island, and Lizard Island, threatening the fragile balance of the reef's ecology.

great black-backed gull (*Larus marinus*) A heavy-bodied, broad-winged gull with a wingspan that can reach 66 inches, this is the largest of the gulls. It is a pugnacious, predatory bird that raids the nests of other birds and dominates other gulls. It breeds in North America from Labrador to New York, and also in Greenland, Iceland, Spitsbergen, the Faeroes, and the Baltic countries. Like most other gulls, juveniles are mottled brown, but breeding adult great black-backs are black above and white below, with a fierce expression and a barrel chest. Considerably smaller is the lesser black-backed gull (*L. fuscus*), with

great black-backed gull

its 50-inch wingspan and a much less intimidating demeanor. Its back is not as dark as its large relative's, but it is darker than that of the herring gull, the other large white-headed North Atlantic gull.

See also gull, herring gull

great circle The largest circle that can be inscribed on the surface of a sphere. The equator is the only line of latitude that is a great circle, but all lines of longitude are because they pass through both poles. In navigation, the shortest distance between any two points on the earth's surface lies along the great circle that passes through them both.

See also equator, latitude, longitude

great common cormorant (*Phalacrocorax carbo*) Along with the related shag (*P. aristotelis*), the common cormorant is found in European waters, but its range extends to North America, Asia, Africa, and the South Pacific as well. It is the largest of all cormorants, with a wingspan of almost 5 feet. Unlike the shag, which is dark metallic greenish black all over, this species (which tends toward bluish black) has a large white patch on the throat, and both males and females develop white patches on each thigh during the breeding season. It is a common sight in harbors, and shows a particular preference for resting on buoys and channel markers. At one time, this species was trained in Europe for fishing, but today that practice is continued only in China. **See also cormorant, double-crested cormorant, shag**

Great Eastern When she was launched in 1858, *Great Eastern* was the largest ship ever built, at an overall length of 692 feet and a displacement of 18,914 tons. Designed by Isambard Kingdom Brunel (1806–1859), she was equipped with two paddle engines, two screw engines, and six masts capable of spreading 6,500 square yards of sail. She could carry enough coal for a round-trip from England to Ceylon, and was designed to hold four thousand passengers or ten thousand troops. She never paid her way as a cargo and passenger ship, and in 1874 she was converted to a cable ship.

See also Brunel, *Great Western*, transatlantic cable

greater shearwater (*Puffinus gravis*) Until the 1920s, nobody knew where this common shearwater of the North Atlantic nested. The dark-backed, white-bellied bird with the dark cap appeared in great numbers off the eastern coast of North America and western Europe, but its nesting grounds were unknown until the nests were found on the remote South Atlantic islands of Tristan da Cunha, Inaccessible, and Nightingale. (Because of its nesting grounds, it is sometimes known as the Tristan great shearwater.) It is a larger, more torpedo-shaped bird than the Manx shearwater, with which it shares the North Atlantic habitat.

See also Manx shearwater, shearwater

Great Northern Expedition (1733–1742) The last of a series of expeditions planned by Tsar Peter the Great to map the northern sea route to the east. In the service of the tsar, the Danish explorer Vitus Bering crossed the North Pacific in the ships *St. Peter* and *St. Paul*, and landed on the coast of Alaska. Captain Alexei Chirikov also reached the Alexander Archipelago on the Alaskan coast, Semen Chelyuskin reached the remote coastal Siberian cape now named for him (the northernmost point in Asia), and the brothers Khariton and Dmitri Laptev charted the northern coast of Siberia from the Taimyr to the Kolyma Rivers. **See also Bering**

great skua (*Catharacta skua*) The great skua nests in Greenland, Iceland, the Faeroes, and the Shetland Islands, and ranges out to sea in the North Atlantic in pursuit of other seabirds to harass. It makes low-level attacks, forcing the other birds to drop their food, which the skua often catches on the wing before it hits the water. A mottled brown bird with conspicuous white wing patches, this is the largest and heaviest of the skuas, with a wingspan that can reach 5 feet. It is also the most aggressive of the species; both parents will dive-bomb any person foolish enough to enter the broad territory that they define when raising their chicks.

great skua

See also long-tailed skua, McCormick's skua, pomarine skua, skua

Great Western The first steamship to make regular scheduled crossings of the Atlantic, the *Great Western* was designed by Isambard Kingdom Brunel in 1838. She was 236 feet long, displaced 1,321 tons, and was built of wood trussed with iron. A fire during her shakedown cruise frightened potential transatlantic travelers, and on her maiden voyage she had only seven passengers. Her design showed that it was possible to carry enough coal to cross the Atlantic, a problem that had hampered previous steamships, and she enjoyed enormous popularity during the next eight years, carrying hundreds of passengers across the Atlantic at an average of fifteen days for the crossing. In 1845 Brunel designed the *Great Britain*, 100 feet longer and built of iron, which became the most popular ship afloat. In 1858 he launched the *Great Eastern*, which was the largest ship ever built until the White Star Line's *Oceanic* was launched in 1899. **See also Brunel, *Great Eastern***

great white shark

great white shark (*Carcharodon carcharias*) At a maximum known length of 21 feet and a weight of more than 3 tons, the great white is the largest shark that preys on warm-blooded animals. (The only larger sharks, the whale shark and the basking shark, feed on minute organisms.) White sharks—which are not white, but a dirty gray above and white only on the undersides—feed primarily on pinnipeds, but they also eat whales, dolphins, fish, squid, and, occasionally, human beings. They can be identified by their massive size, pointed snout, round black eyes, serrated triangular teeth, and frequent appearance in Hollywood movies. Great whites are found in most of the temperate waters of the world, but they are most common around South Australia, South Africa, and northern California. Because of the unprecedented human predation on these sharks, they are protected in all these areas. **See also basking shark, *Jaws*, megalodon, shark attack**

Greely, Adolphus Washington (1844–1935) American army officer and Arctic explorer, who was sent in 1881 to set up a series of meteorological stations. The ship *Proteus* dropped the twenty-two-man party off, intending to return with supplies the following year and then pick the men up in 1883. Without any previous Arctic experience, Greely and his party performed notable feats of exploration, charting the northwest coast of Greenland and crossing Ellesmere Island from east to west. Lieutenant James Lockwood reached the farthest north to date, 83°24′ north, on what was then named Lockwood Island. The *Proteus* did not appear (she had been wrecked by the ice), so the Greely party spent the winter at Lady Franklin Bay on Ellesmere Island, with minimal provisions. When Commander Wilfred Schley arrived in the *Thetis* in June 1884, all but Greely and six others had starved to death. Greely wrote *Three Years of Arctic Service* (1886), and was awarded the Congressional Medal of Honor in 1935.

Green algae (Chlorophyta) Because of the presence of chlorophyll, green algae are similar to terrestrial flowering plants, and this suggests that land plants evolved from the ancestors of the green algae. Among the most familiar green algae is *Codium,* the green branching seaweed that attaches itself to shells, rocks, and other fixed objects, and produces millions of zoospores that grow into new plants. Another common green alga is *Ulva,* known as sea lettuce, which grows well in polluted waters adjacent to urban areas. The presence of sewage or waste water encourages the growth of sea lettuce.

green crab

green crab (*Carcinus maenas*) A common British and European crab, this species gets its common name from its carapace, which is green with blackish spots. Males are yellowish on the underside, while females are orange-red. It is one of the swimming crabs (Portunidae), with its last pair of legs flattened but not paddle-shaped as they are in the blue crab. The carapace, which reaches a width of about 3 inches, is slightly wider than it is long. The specific name *maenas* means "frenzied," and refers to the fierce way this species fights to defend itself from predators, including people. (It is known as *crabe enragé* in French.) Also known as the shore crab, this aggressive, carnivorous species was introduced to North America in the early nineteenth century, whereupon it ravaged the softshell clam industry of New England and the Gulf of Maine. Since being accidentally transported to the California coast around 1989, it has begun moving into Oregon and Washington inshore waters, threatening the oyster, clam, mussel, and Dungeness crab fisheries.

See also clam, crab, Dungeness crab

green dogfish (*Etmopterus virens*) There are many species in the genus *Etmopterus,* but *E. virens* is probably the best known. It is a 1-foot-long little shark, with a pale yellow spot on the top of the head and a belly that iridesces bright yellowish green when the shark is alive. It is common in the deeper waters of the Gulf of Mexico, where it congregates in dense schools to feed on squid and octopuses.

See also cookie-cutter shark, dogfish

Greenland With the exception of the continent of Australia, Greenland is the largest island in the world. An autonomous part of the Kingdom of Denmark since 1953, it covers 840,000 square miles, most of which is covered by a permanent ice sheet that can be 10,000 feet thick. It lies mostly within the Arctic Circle, and its northernmost point is only 500 miles from the North Pole. Fewer than 60,000 live in Greenland, virtually all of them in settlements on the coast. The capital of Godthåb, known as Nuuk to the Greenlanders, has a population of about 11,000. The main industries of Greenland—which is called Kalaallit Nunaat in Greenlandic—are sealing, hunting, fishing, and mining. The island was first colonized in 985–986 by Eric the Red, who gave it its unlikely name in an effort to induce further settlement, but conditions were too harsh and the settlers perished or abandoned the island by the fourteenth century. It remained inhabited only by Eskimos until 1721, when Hans Egede, a Danish missionary, established a settlement there. In the nineteenth century, Dutch and British whalers plied the waters of east and west Greenland in pursuit of the bowhead whale (*Balaena mysticetus*), which they all but eliminated.

Greenland Sea Lying to the east of Greenland, between latitudes 60° north and 80° north, the Greenland Sea covers some 465,000 square miles. It is separated from the more moderate Norwegian Sea by an imaginary line drawn between northeast Iceland, Jan Mayen Island, Bear Island, and Spitsbergen. The Greenland Sea averages 4,750 feet in depth, and the deepest point is 16,000 feet down. It is linked to the North Atlantic via the Denmark Strait between Greenland and Iceland. Because of the large amount of dissolved nutrient salts, the Greenland Sea is inhabited by a profusion of lower life forms, which support large populations of invertebrates and such vertebrates as herring, redfish, halibut, and plaice, as well as seals, whales, and dolphins.

See also Bear Island, Iceland, Jan Mayen, Spitsbergen

Greenland shark (*Somniosus microcephalus*) At a known maximum length of 23 feet, this is one of the largest sharks and the largest of all deepwater fishes. A Greenland shark was filmed off Cape Hatteras at a depth of 7,436 feet. They are found in the colder waters of the North Atlantic, usually at depth, although they come closer to the surface in summer. They feed primarily on fishes, but these sharks have been found with reindeer in their stomachs; they also used to hang around whaling ships and docks for scraps. Most have a parasitic copepod covering each eye, and although it is not known what this does for the shark, there is speculation

Greenland shark

that the copepods are bioluminescent and attract prey species to the host, which then gobbles them up. Eskimos fish for these sharks through holes in the ice, and although the flesh is poisonous when fresh, it can be eaten if it is dried or boiled several times. A similar species, known as the Pacific sleeper shark (*S. pacificus*), is found from Japan to Puget Sound.

See also dogfish

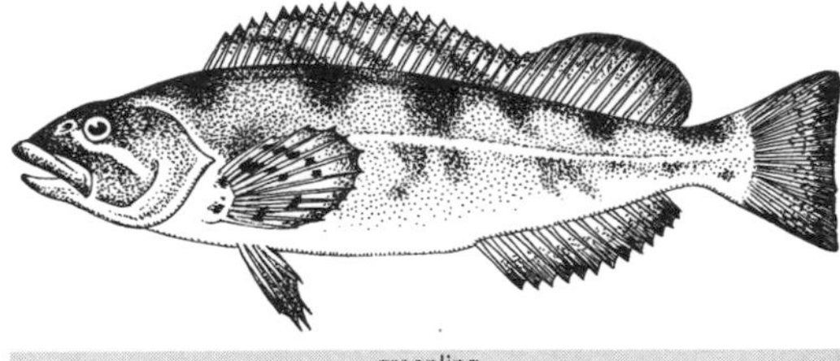

greenling

greenling (*Hexagrammos decagrammus*) Lingcod and greenlings, which make up the family Hexagrammidae, are found only off the Pacific coast of North America. The kelp greenling reaches 1½ feet in length, and the sexes sport different color schemes: males are a mottled blue; females tend toward reds and browns. The flesh is an unappetizing blue-green until it is cooked.

See also lingcod

Greenpeace Founded in 1969 to protest against nuclear detonations in the Aleutians, Greenpeace soon assumed the lead in the battle to save the whales from whalers. In 1975, from the ship *Phyllis Cormack,* Greenpeace dispatched inflatable boats to position themselves between Soviet whalers and the whales they were shooting. Although it failed to stop the whale killing, the attendant publicity soon raised the world's "whale consciousness," and the name Greenpeace became indelibly associated with whale preservation. The following year Greenpeace commandos tracked the whalers down again, and by 1977, they had sent guerrilla forces all over the world to interfere with commercial whaling: sinking some ships, chaining themselves to others, taking out prominent ads in newspapers. The Greenpeace *Rainbow Warrior* visited Iceland in 1978 and, using the same harassing tactics and publicity, made the world aware of Iceland's unwillingness to conform to International Whaling Commission rules. Greenpeace is not alone in the battle to save the whales, but because of its understanding of the power of the press, it is considered among the leaders of the preservation movement. They were major participants in the negotiations that eventually led to the moratorium on commercial whaling passed by the IWC in 1982.

See also pirate whaling, Watson

green turtle

green turtle (*Chelonia mydas*) The greenish fat of this species gives it its common name. Green turtles grow to about 40 inches in length and weigh about 300 pounds. They have a single claw on each foreflipper. Found in the Atlantic, Indian, and Pacific Oceans, green turtles nest on sandy beaches during the summer months. With their close relatives, the Indo-Pacific black turtles (*C. agassizii*), the adults are the only vegetarian sea turtles, feeding exclusively on sea grasses and algae. They are the turtle of choice for soup in the Caribbean, and the greenish color of the fat has given us "green turtle soup." Not surprisingly, their numbers have been greatly reduced by hunting, and they are considered endangered around the world. In the Persian Gulf, where man-made oil fires created an unprecedented environmental disaster in 1991, green turtles are threatened not only by oil spills, but by fishermen and fishing nets. Unless immediate steps are taken, the Persian Gulf green turtles may be on the way to extinction.

See also sea turtle

Grenada Sighted and named for the city in Spain by Columbus in 1498, Grenada was occupied by the Carib Indians, who had wiped out the Arawaks and who successfully drove off the Spanish. In 1650, the governor of Martinique bought the island and established a settlement at St. George's, and by 1672, it had become a French possession. British forces captured Grenada in 1762 and retained control until 1958, when the island joined the West Indian Federation. After the breakup of the federation in 1962, Grenada was a self-governing state in association with the United Kingdom, and in 1974, it became an independent nation. An internal coup by a group affiliated with Cuba brought an invasion by U.S. troops in 1983; the island was quickly occupied and the socialist government arrested and deported. A general election in 1984 reestablished democratic self-government. The island, approximately 133 square miles in area, is the southernmost of the Windward Islands and includes the dependency of the Grenadines to the northeast. The economy depends on agricultural exports (bananas, nutmeg, cocoa) and a burgeoning tourist industry.

grenadier (family Macrouridae) A number of deepwater fishes (perhaps 250 species, divided into thirty genera) characterized by the absence of a tail fin. The dorsal and anal fins taper to a sharp point, accounting for their other name, rat-tail. Most of the known species have large eyes, but given the darkness of their benthic habitat, it is not evident why these carrion feeders need acute vision. On the infrequent occasions when deep-sea fishes are observed, grenadiers almost always dominate the sightings, suggesting that they are very abundant near the bottom. They are rarely seen at depths of less than 600 feet, and they are often at depths of 2,500 feet or more.

See also cusk, eelpout, pearlfishes

grenadier

Grenville, Richard (1542–1591) British gentleman and naval hero, who passed the bar, was elected to Parliament, was knighted, and was made sheriff of Cornwall. In 1585 he commanded the seven ships that carried the first hundred settlers to the new colony at Roanoke Island (in present North Carolina), which had been established by his cousin, Sir Walter Raleigh. Grenville made another provisioning voyage in 1587, and he was not able to return to England until 1591, after the battle with the Spanish Armada. He was appointed second-in-command to Lord Thomas Howard in a fleet sent to capture Spanish treasure ships off the Azores. In the battle that ensued, Grenville's ship, the *Revenge*, became separated from the rest of the fleet, and he had to fight fifteen Spanish ships on his own. In the face of overwhelming odds, he fought valiantly, but he was mortally wounded and his ship captured—the first British war vessel captured by the Spanish since John Hawkins's *Jesus of Lubeck* was taken in Mexico in 1567. The action is immortalized in Tennyson's poem "The Revenge," published in 1880.

grey seal

grey seal (*Halichoerus grypus*) Because they occupy some of the same habitats, grey seals are sometimes mistaken for harbor seals. Greys are larger and have a much longer snout, which accounts for their Canadian name, "horsehead." The males are much larger than the females, reaching an overall length of 8 feet and a weight of 600 pounds, while the females reach only 6½ feet in length and are proportionately lighter. The 3-foot-long pups are born with a white coat (a lanugo), which they molt after three weeks. There are grey seal populations in Canada and various locations in northwestern Europe, but the largest concentration is in the British Isles, especially in the Hebrides (hence the spelling of "grey"). They have been the subject of much controversy between the fishermen who would kill them (believing they eat fish that rightfully belong to the fishermen) and the conservationists who would save them.

See also harbor seal, seals and sea lions

grouper (*Epinephelus* spp.) There are many species of groupers, ranging in size from the gigantic jewfish (*E. itajara*), at 800 pounds, to the red hind (*E. guttatus*), which weighs about 4 pounds. The Nassau grouper (*E. striatus*), as its common name implies, is common in the Bahamas and the Caribbean, and averages about 20 pounds in weight. Nassau groupers all start out as males, but as they mature, they develop into females. In Australia, the name is spelled "groper" and is applied to some fishes of the genus *Epinephelus*, which, to further confuse matters, are also referred to as rock cod. The potato cod (*E. tukula*) is a large, fearless grouper (or groper) that can weigh 200 pounds and often approaches divers, particularly along the Great Barrier Reef.

See also jewfish

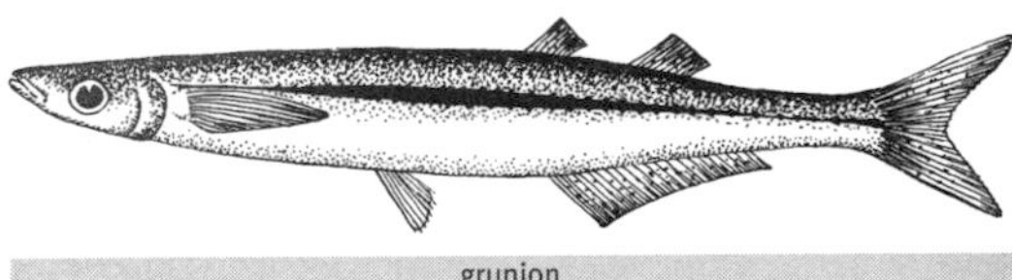
grunion

grunion (*Leuresthes tenuis*) One species in a large group of small fishes known as silversides (Atherinidae), the California grunion is famous for its spawning behavior, which is synchronized to the phases of the moon. Following the highest of the spring tides, after both the new and the full moon, millions of the 8-inch-long females come ashore at night to deposit their eggs, squirming tailfirst into the wet sand. The males arrive simultaneously to fertilize the eggs, which remain buried under the warm sand for two weeks; then they hatch and the young are washed out to sea by the next high tide. The process occurs every two weeks from March through September, at various beaches along the California coast, and each female may spawn from four to eight times during the season.

grunts (family Haemulidae) Several hundred species of grunts are found throughout the tropical and temperate waters of the world. The common name is derived from their ability to make grunting noises by grinding their pharyngeal teeth and amplifying the sounds through

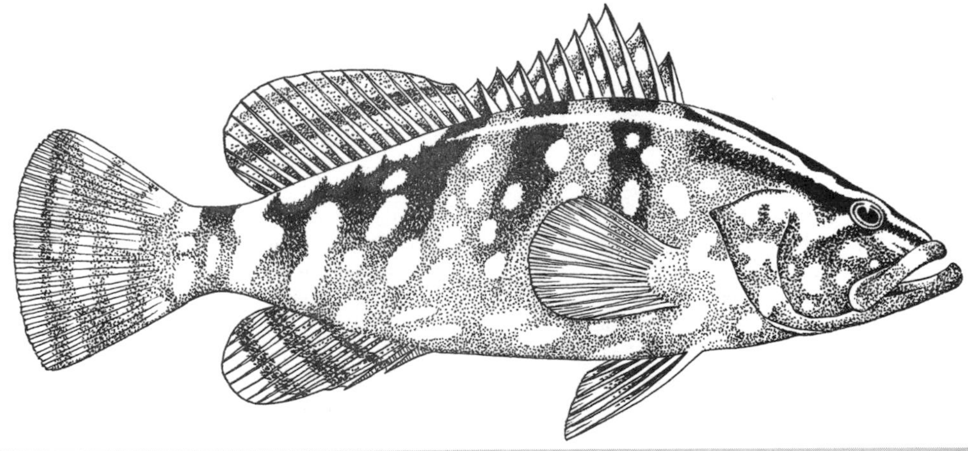
Nassau grouper

the swim bladder that acts as a resonator. They gather in large groups during the day and move out in the evening to forage for shrimps, worms, mollusks, and small fish. Of the twenty species in the Atlantic and the Caribbean, the best known are the white grunt (*Haemulon plumieri*), the French grunt (*H. flavolineatum*), the tomtate (*H. aurolineatum*), and the porkfish (*Anisotremus virginicus*), all of which share a blue-and-yellow-striped color scheme. Some species have a habit of approaching one another with wide open mouths, showing the bright red interior (*Haemulon* means "bloody gums"). The black margate (*A. surinamensis*) is a silvery-gray grunt with fleshy lips that is found from Florida and the Virgin Islands through the Gulf of Mexico and as far south as Brazil. In the Indo-Pacific, grunts are more commonly known as sweetlips, and several species there are more colorful than their Caribbean counterparts. The largest of the grunts is the giant sweetlips (*Plectorhinchus obscurum*), found from the Red Sea to Fiji, which can attain a length of 5 feet. **See also porkfish**

Guadalcanal At 2,510 square miles, Guadalcanal is the largest of the Solomon Islands, an independent country in the southwest Pacific. (The larger Solomon island of Bougainville is part of Papua New Guinea.) In 1568 the Spanish explorer Álvaro de Mendaña tried to settle there, but hostile cannibals drove his expedition away. Made a British protectorate in 1893, the island was occupied by Japanese forces during World War II. In August 1942, American marines landed, making the first large-scale invasion of a Japanese-held island. The Americans took the Japanese airstrip, but the outnumbered Japanese were firmly entrenched. Offshore, a major naval battle, the Battle of Guadalcanal, was also being waged, as Japanese cruisers, destroyers, battleships, and aircraft carriers engaged their American counterparts. Both sides suffered heavy losses, but in February 1943, after months of fierce fighting, the Americans conquered Guadalcanal. By the end of the year, the Japanese had lost Bougainville, their last stronghold in the Solomons.

See also Bougainville Island, Solomon Islands

Guadalupe Mexican island in the North Pacific, 180 miles west of Baja California and 180 miles southwest of San Diego. The island is about 20 miles long by 6 miles wide, and it has a 5,000-foot-high volcanic ridge running down its entire length. Nineteenth-century California whalers released goats onto the island, which destroyed almost all of the native vegetation; feral cats, also introduced by the whalers, completely eliminated the Guadalupe storm petrel (*Oceanodroma macrodactyla*), a bird that bred nowhere else. Goatherds, believing that the quelili (*Polyborus lutosus*), a large bird of prey, was taking the kids, systematically eradicated this bird too, the only premeditated extinction of any bird. (It was a species of caracara, known mostly as a carrion eater.) After the northern elephant seal (*Mirounga angustirostris*) was thought to be extinct because they had all been killed for their oil, the only remaining animals were found on Guadalupe Island in 1890. From this small herd, the elephant seal has recolonized its old breeding grounds on various islands and even the California mainland. The island is now unoccupied, except by the elephant seals and approximately fifty thousand goats.

Guadeloupe Located in the Leeward Island chain of the West Indies, Guadeloupe is an administrative region of France. Guadeloupe is composed of a butterfly-shaped pair of islands (Grande-Terre and Basse-Terre) separated by the Rivière Salée, a narrow channel. Other islands in the department are St. Barthélmy (St. Barts) and the northern half of St. Martin. Sighted by Columbus in 1493, the island was not enthusiastically colonized by the Spanish; it was abandoned in 1603. The French arrived in 1635, eliminated the Carib Indians, and imported African slaves to work on the sugar plantations. Britain and France contested ownership of the islands until 1815, when the French claim prevailed. The island's economy is heavily subsidized by the French government, and the majority of the tourists that sustain the island's major industry come from France. Like Montserrat, St. Vincent, and St. Lucia, Guadeloupe has a volcano called La Soufrière ("one that emits sulfur"), but—unlike that of Montserrat—it is inactive. **See also Leeward Islands (2)**

Guam Island in the western Pacific Ocean, an unincorporated territory of the United States. The 209-square-mile island is the most southerly and the most populous of the Mariana Islands, with a 1991 population of 141,000. Agana is the capital, and Apra is the site of a large U.S. Navy base. The inhabitants are chiefly Chamorro stock, with a mixture of Spanish, Filipino, and Micronesian blood. First explored by Magellan in 1521, Guam was a Spanish possession until 1898, when it was taken by the United States in the Spanish-American War. The Japanese captured the island in 1941, but it was regained by American forces in 1944. Except during the Japanese occupation, Guam was governed by a U.S. naval officer, advised by a local congress, until 1950, when jurisdiction was transferred to the U.S. Department of the Interior. Now a governor is elected every four years and a twenty-one-member legislature every two years. The major industries are tourism and providing goods and services for the U.S. bases. **See also Mariana Islands**

guano islands A term applied to several groups of islands, primarily off the west coast of South America, but also in the Caribbean and the South Pacific, where the droppings of various seabirds accumulate in such quantities that they can be "mined" and collected for fertilizer. When Alexander von Humboldt (1769–1859) explored the west coast of South America in 1802, he identified guano islands off the coast of Peru, and recognized that the nitrogen- and phosphorus-rich material would make an excellent fertilizer. (Humboldt is usually credited with being the "discoverer" of guano, but the Incas knew about it for centuries; "guano" comes from an Inca word.) By 1840, the Chincha Islands (off Pisco) had become a renowned source of guano, which was produced by the vast flocks of Peruvian boobies (*piqueros*) and cormorants that take advantage of the rich food sources brought up by the upwellings of the Humboldt Current. On some of the islands the guano was 300 feet thick, and it was considered such an important commodity that Congress passed a law giving U.S. citizens the right to collect it on islands belonging to other nations, offering to protect those rights by military force if necessary. Ordinary seamen refused to mine the dusty, stinking stuff, so convicts and coolies were used to perform work so odious that many of them threw themselves off the cliffs rather than do it. The guano so reeked of ammonia that when it was offloaded, the citizens of the ports took to the hills. In 1864, Spain seized the Chincha Islands from the recently independent Peru, but the United States invoked the Monroe Doctrine, and the combined navies of Chile and Peru drove the Spanish off. It is estimated that 20 million tons of guano were collected during the nineteenth century. The reckless abandon resulted in the destruction of nests and the slaughter of thousands of chicks and adult birds. In addition to the Chincha Islands, guano was mined on Peru's Lobos Islands and on Mona Island, off Puerto Rico. Several of the uninhabited Line Islands (now Kiribati) in the central Pacific, including Howland, Baker, Flint, Starbuck, and Malden, were annexed by the United States under the Guano Act of 1856, and their deposits collected until 1943. **See also Kiribati**

Guernsey One of the Channel Islands off the coast of Normandy that became part of England when William the Conqueror led the Normans across the English Channel in 1066. In 1204, the Channel Islands were put in the charge of a warden; by the fifteenth century a captain ruled, and later a governor. The Royal Court, a medieval throwback, administers the law of Guernsey, and points of law are referred to a bailiff. The Guernsey bailiwick includes Alderney and Sark, as well as the smaller islands of Herm, Brechou, and Jethou. Thirty square miles in area, the island is home to 55,000 people of mixed Norman and Breton descent. It is a popular destination for English and French tourists. Guernsey grows tomatoes, flowers, and grapes, and the original Guernsey dairy cattle originated here. Guernseys are larger than Jersey cattle, which are a solid fawn color; often marked with white, Guernseys have been exported around the world. **See also Alderney; Channel Islands; Jersey; Sark, Isle of**

guillemot: See murre

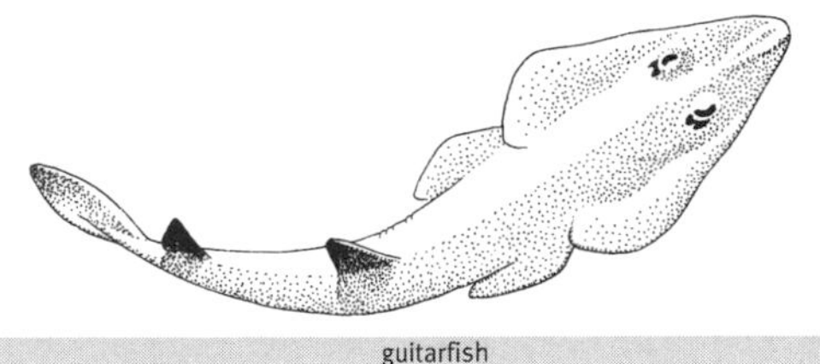

guitarfish

guitarfish (*Rhinobatus lentiginosus*) Because of the large, flattened pectoral fins and narrow, tapering hind parts, these rays look—to some people—like guitars. (In other countries, they are commonly called shovelnose rays, but they are also known as fiddler rays or banjo sharks.) They are classified with the rays because the mouth, nostrils, and gills are on the underside, while the sharks have eyes and gills on the sides of their heads. The largest is the giant guitarfish (*Rhynchobatus djiddensis*) of Australian waters, which can measure 10 feet from nose to tail. The spotted guitarfish (*Rhinobatus lentiginosus*), common throughout the shallow waters of the Atlantic, reaches a length of 3 feet.

See also rays, sawfish

Gulf of California The 750-mile long body of water, also known as the Sea of Cortez, that separates the Baja California peninsula from mainland Mexico. When Francisco de Ulloa saw it in 1539, he named it Mar Bermejo ("the vermilion sea"), because of a heavy bloom of red algae. The gulf is believed to have been formed by the pulling away of the peninsula from the mainland, since it lies over the San Andreas Fault system. The northern portion of the gulf, fed by the Colorado River, is shallow, seldom exceeding 600 feet in depth, but in the south there are several depressions, some as deep as 10,000 feet. It has a total surface area of some 62,000 square miles. In the gulf are the large islands of Angel de la Guarda, Tiburón, Carmen, San José, Espirítu Santo, and Cerralvo, as well as many smaller ones. The major ports are La Paz on the Baja side and Guaymas on the mainland. The gulf is rich in marine life, including a great variety of invertebrates, sharks, fishes, and sea lions, and a vast number of seabirds. The largest animals on earth, the blue whale and the fin whale, can be found in the gulf, as well as various dolphins and, occasionally, the gray whale, which breeds in the lagoons and bays on the Pacific side of the peninsula. Intensive fishing—legal and illegal—has greatly reduced the stocks of sharks, rays, fishes, and shrimp, and two species found nowhere but in the northern reaches of the gulf—the Gulf of California porpoise, or vaquita (*Phocoena sinus*), and the totoaba (*Cynoscion macdonaldi*)—are close to extinction. What was once one of the most productive fishing areas in the world has been reduced to a vestigial fishery, but the fishermen continue to fish, taking what little is left. In 1941, John Steinbeck (the author of *The Grapes of Wrath, Of Mice and Men,* and *Cannery Row,* among other famous novels) wrote *The Log from the Sea of Cortez,* an account of his experiences collecting invertebrates with biologist Ed Ricketts.

See also Cortés, gray whale, totoaba, vaquita

Gulf of Mexico A relatively shallow, oceanic basin, covering some 600,000 square miles, and bounded on the north and northwest by the United States (Texas, Louisiana, Mississippi, Alabama, and Florida); on the west, south, and southwest by Mexico; and on the east by Cuba. The Mississippi River and the Rio Grande drain into the gulf, whose greatest depth is 17,000 feet. Most of the U.S. shrimp catch comes from the gulf, and there are also fisheries for red snapper, flounder, mullet, oysters, and crabs. Oil and natural gas reserves are plentiful, and Mexico and the United States have erected oil rigs there. (The Mexican well Ixtoc I blew and caught fire in 1979. It burned for ten months and released 140 million gallons of oil into the gulf off Campeche.) Major ports on the Gulf of Mexico are Tampa and Pensacola, Florida; Galveston and Corpus Christi, Texas; and Tampico and Veracruz, Mexico.

Gulf of Tonkin Northwestern arm of the South China Sea, between Vietnam and China. The gulf is about 300 miles long, and 150 miles at its widest. The Red River flows into it, and Haiphong is its chief port. In August 1964, an alleged attack by Vietnamese gunboats on the U.S. destroyers *Maddox* and *C. Turner Joy* led to the passage of the Tonkin Gulf Resolution, which authorized the U.S. president to take all necessary measures to protect U.S. interests in South Vietnam. This led to the escalation of U.S. involvement in Vietnam; the resolution was used by presidents Lyndon Johnson and Richard Nixon to justify the Vietnam War. The "attacks" in the Gulf of Tonkin may never have occurred.

Gulf Stream Part of a clockwise-rotating gyre of currents in the North Atlantic, the Gulf Stream is only that portion of the system that extends from Cape Hatteras, North Carolina, to the Grand Banks of Newfoundland. It originates in the Caribbean, emerging through the Straits of Florida between the Keys and Cuba and being joined there by the confluent Antilles and North Equatorial Currents. It was first described in 1515 by the Spanish explorer Ponce de Léon and later studied by Benjamin Franklin and Matthew Fontaine Maury. It is not a simple ribbon of water flowing northward, but a complex system of constantly shifting currents and eddies that parallel the coast at speeds that average about 4 mph. It averages about 50 miles in width and is much bigger than any river; at its point of maximum flow off North Carolina, the volume of water is 70 million cubic meters per second, 3,500 times greater than the flow of the Mississippi. Bordered on the west by the Gulf Stream is the Sargasso Sea, a region of calm water, minimal winds, and no currents whatsoever. In the south the Gulf Stream is not particularly rich in fauna (there are mostly flying fishes, and migrating tuna and salmon), but when it meets the cold Labrador Current, it produces the thick fogs of Newfoundland and Labrador, and the turbulence and nutrients combine at the Grand Banks to create the most productive fishing grounds in the world. On July 16, 1969—the same day as the launching of *Apollo 11,* which put Buzz Aldrin and Neil Armstrong on the moon four days later—the submersible *Ben Franklin* was also launched. It drifted for 1,400 miles underwater in the Gulf Stream, emerging thirty-one days later in New York harbor.

See also Sargasso Sea

Gulf Stream beaked whale

Gulf Stream beaked whale (*Mesoplodon europaeus*) Although its primary habitat is the western North Atlantic and the Caribbean, this species was first described because of an animal found in the English Channel. Like most of its relatives, this *Mesoplodon* is known mostly from strandings, and from these accounts, the size of adults is estimated at 17 feet. In the males, the teeth are located about one-third of the way down the lower jaw; the females have no teeth at all.

See also beaked whales, *Mesoplodon*

Gulf War oil spill In August 1990, after an eight-year war with Iran, the Iraqi army invaded Kuwait, one of the world's most prosperous nations because of its enormous oil reserves. The UN Security Council condemned Iraq's actions and endorsed the use of force, if necessary, to achieve Iraq's withdrawal. In January 1991, Allied forces began Operation Desert Storm, a massive air and ground effort intended to liberate Kuwait. Six weeks later, the Iraqis were in full retreat. As they fled, they set fire to more than six hundred Kuwaiti oil wells, burning more than 6 million barrels of oil a day. (A barrel is 42 gallons, so that is 252 million gallons.) The fires, which burned for nine months, produced 3,400 metric tons of soot per day. Teams from around the world convened in Kuwait to put out the fires and stanch the flow of oil. Another 250 million gallons of oil flowed into the Persian Gulf, creating the largest oil spill in history. (For comparison, the tanker *Amoco Cadiz* spilled 68 million gallons of oil onto the coast of Brittany in 1978.) More than 400 miles of Saudi Arabia's coastline was soaked with oil, a situation that will not clear up for decades because of the sluggish circulation pattern of the gulf. Hundreds of miles of Kuwaiti desert remain uninhabitable because of oil lakes and soot. Millions of migratory birds died. The fishing industry was virtually wiped out. Sea turtles, whales, dolphins, and dugongs were also deleteriously affected. Among the factors contributing to the mysterious Gulf War Syndrome that affected many of the soldiers who fought there was the presence of heavy hydrocarbons and other chemicals in the air. (Nerve gas released by the Iraqis also played a part.) Five years after the war, the environmental crises precipitated by the Iraqi scorched-earth policy are still visible and largely unresolvable. **See also *Amoco Cadiz, Castillo de Bellver, Exxon Valdez,* Ixtoc blowout, oil spill, *Torrey Canyon***

gull (*Larus* spp.) Not all seabirds are gulls. This term is reserved for some forty-odd species of long-winged, solidly built, web-footed birds that are found almost everywhere except in the central Pacific. Adult gulls are normally gray and/or white, with variable head markings and bills and feet of varying colors. The juveniles are usually darker or brownish. Gulls are omnivorous and opportunistic feeders, sometimes scavenging, sometimes fishing, sometimes stealing eggs from other birds' nests. Gulls walk well, swim buoyantly, and fly superbly, meaning that they—and a few other seabirds—function competently on land, at sea, and in the air. They nest near the sea, but they can be found on the coast, at offshore islands, at all latitudes, and around many cities and garbage dumps. (The Mormons, looking for a place to settle in Utah in 1846, were led to Great Salt Lake by the presence of gulls.)

See also herring gull, kittiwake, skua, tern

gulper eel (order Lyomeri) Among the most bizarre fishes in the world are the pelican eel (*Eurypharynx pelecanoides*) and the black swallower (*Saccopharynx*). They are related to each other, but their relationship to other fishes is unclear. They are not eels (although they pass through a leptocephalus stage), and though they have luminous organs, they are not related to the anglers or the dragonfishes. The pelican eel gets its name from its pouchlike lower jaw, which it uses to surround its prey rather than attack it. The 2-foot-long *Eurypharynx* (also known as the umbrella-mouth gulper) is a midwater inhabitant, found at depths ranging from 600 to 9,000 feet. The mouth of the black swallower is large, but not as exaggerated as that of the pelican eel, and it has teeth, while *Eurypharynx* has none. *Saccopharynx* can reach a length of 6 feet. Both species have a luminous lure on the end of the tail, but it is difficult to imagine how the fish might use a lure so far from its mouth unless it swims in circles, chasing fishes that are chasing its tail.

See also bioluminescence, leptocephalus

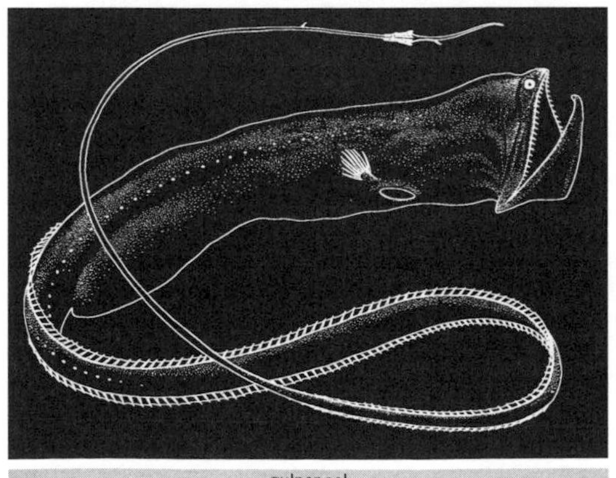
gulper eel

gunnel (*Pholis gunnellus*) A small fish of the eastern North Atlantic with a small head and a dorsal fin that runs almost the entire length of its 10-inch body. It is patterned with vertical stripes and a series of white-ringed black spots at the base of the dorsal fin. The gunnel is common in the inshore waters of the English Channel, the North Sea, and the western Baltic, and it is also common in the Pacific. It can be found under rocks or in tide pools at low tide. The gunnel secretes a coating of slime, which accounts for its other name, butterfish.

guyot Also known as a tablemount, a guyot is a flat-topped seamount rising from the seafloor, the top of which is more than 200 meters below the surface. The name, which is pronounced "gee-yo," with a hard "g," was coined in 1946 by oceanographer Harry Hess during his extensive sounding operations in the Pacific, in honor of Arnold Guyot, a nineteenth-century Swiss-American geologist. **See also seamount**

gyre The roughly circular path of water circulation in the open ocean, caused by the equatorial trade winds that drive the ocean surface waters to the west, and the westerly winds nearer the poles that drive the waters back toward the east. The North Atlantic gyre comprises the Gulf Stream, the North Atlantic Current, and the Canaries Current, and the return flow is the Equatorial Current. The speed of these oceanic currents is generally about 6 miles a day, but in the western margin of an ocean, where a narrower current is always found—such as the Gulf Stream or the Kuroshio Current—the speed may be 60 to 100 miles per day. In the Southern Hemisphere, the currents in the gyres are generally weaker, and the pattern is dominated by the great Antarctic Circumpolar Current, where the ocean flows around the globe, unimpeded by land.

See also Gulf Stream, Kuroshio Current

H

hadal zone The greatest depth of the oceans, from *Hades,* meaning "hell," applied to depths exceeding 6,000 meters (20,000 feet), which includes the bottom of the ocean's deepest trenches. At 20,000 feet, the pressure is more than 600 atmospheres; in the deepest trenches, it can reach 1,200 atmospheres. (An atmosphere equals air pressure at sea level, about 14.7 pounds per square inch.) The bottom is a reddish clay and radiolarian ooze. Hadal fauna is specialized and largely endemic, consisting primarily of amphipods, polychaetes, bivalves, and holothurians, with very few coelenterates and hardly any fishes. (The "fish" seen by the *Trieste* when she made her record dive to 35,800 feet in the Mariana Trench was actually a holothurian [sea cucumber].) **See also sea cucumber, *Trieste***

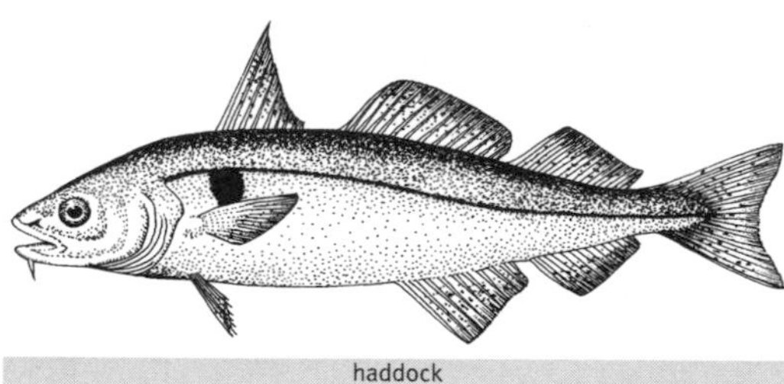

haddock

haddock (*Melanogrammus aeglefinus*) Found throughout the North Atlantic, haddock resemble cod (*Gadus morhua*) but can easily be identified by the black "thumbprint" behind the pectoral fin and the black lateral line. They grow to a length of 44 inches and can weigh 36 pounds. Smoked haddock is popular in Britain as "finnan haddie." Cod are fish eaters; haddock live near the bottom and feed on worms, mollusks, and echinoderms. In the western North Atlantic, there are two distinct stocks: the Gulf of Maine and the Georges Bank stock. In the early twentieth century, haddock were among the most commercially important food fishes in the Atlantic, but they were caught so heavily, especially by the Eastern European bottom trawlers in the 1950s, that the population was drastically reduced and the fishery has been suspended.

See also codfish, hake

hagfish (families Myxinidae and Eptatretidae) Primitive, eel-shaped vertebrates with a cartilaginous skeleton, several barbels on the end of the snout, and two pairs of toothlike rasps on the top of a tonguelike projection.

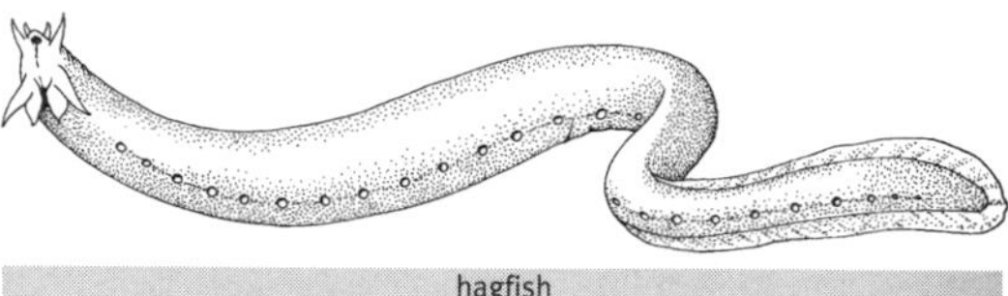

hagfish

Hagfishes have a single nostril and degenerated eyes that are buried under the skin. They have a tail fin, but no other paired fins, and no jaws or bones. They have a typical vertebrate heart and three auxiliary hearts: one to pump blood to the liver, one in the head, and a tiny one in the region of the tail. There are about forty species, ranging from 20 to 30 inches in length, found in all the cold, deep waters of the world. Unlike the lampreys, which attack living fishes, hagfishes are scavengers, remaining buried until some sort of carrion drifts down, then emerging to eat. From a series of glands along the side, hagfishes can emit gallons of nauseating, toxic slime, accounting for their common name, "slime eels." They can tie themselves in knots to provide leverage for pushing themselves into a dead fish, which they bore into and eat from the inside out. So-called eel-skin leather, recently popular for expensive belts, wallets, and briefcases, is actually the skin of hagfishes that have been collected in New England waters and shipped to Korea for processing. Once the skins have been removed and sent to China to be made into leather goods, the Koreans eat the hagfish meat, consuming nearly 5 million pounds each year.

See also Georges Bank, jawless fishes, lamprey

Hainan Island and province of southern China, in the South China Sea, immediately off the Liuzhou Peninsula. At 13,000 square miles, it is the second largest island off the China coast. (Taiwan is about 1,000 square miles larger.) For centuries, this semitropical, mountainous island was a backward and neglected region. With a population of about 6.5 million, it has now been designated a "special economic zone," which means that it can conduct foreign and domestic trade without the authorization of the central government in Beijing. Because of its climate, agricultural incentives are sought, and the Chinese government intends to encourage the construction of tourist facilities. It is also rich in minerals such as iron, tungsten, titanium, manganese, molybdenum, bauxite, copper, cobalt, and graphite. The island forms the eastern boundary of the Gulf of Tonkin; Vietnam is the western.

Haiti Independent country on the western third of the island of Hispaniola in the Greater Antilles. The country is mountainous, but about one-third of the land is arable. Haiti is one of the poorest, most densely populated countries in the world, with approximately 650 people per square mile. French is the official language, but about 90 percent of the people speak a Creole patois. AIDS in pandemic in Haiti, and many of the refugees who have fled their country to come to the United States in recent years have brought the disease with them. Based on beliefs of slaves from Dahomey (in West Africa), synthesized with elements of Christianity, voodoo is the national folk religion of Haiti. Its tenets include possession by evil spirits, exorcisms, and the resurrection of zombies. Before the Europeans, the island was inhabited by Arawaks (*Haiti* means "land of mountains" in Arawak), but the Spanish enslaved and killed them or gave them diseases for which they had no immunities. While establishing plantations on the eastern part of the island (now the Dominican Republic), the Spanish ignored the west, which became a haven for buccaneers, and was eventually ceded to France in 1697 as Saint-Domingue. It became France's most prosperous colony in the Caribbean, and one of the world's leading producers of coffee and sugar. In 1795, under Toussaint-Louverture, a former slave, the blacks took over the island, and Toussaint proclaimed himself governor. Napoleon sent General Charles LeClerc to put down the rebellion, and although Toussaint was captured, the Haitians did not capitulate, and the French troops were forced to leave. The rebels received support from U.S. president Thomas Jefferson, who feared that the French were planning to use Saint-Domingue as a base to invade Louisiana. In 1804, Haiti became the second independent nation in the Western Hemisphere. Henri Christophe proclaimed himself emperor in 1806, and there followed a long period of political and economic chaos, which culminated in the U.S. invasion of Port-au-Prince in 1915. The American presence in Haiti lasted until 1934, and the United States retained fiscal control of the country until 1947. In 1964, François ("Papa Doc") Duvalier named himself president for life and created a total dictatorship, patrolled by the Tontons Macoutes, his dreaded secret police. He was succeeded in 1971 by his son Jean-Claude, "Baby Doc," who oversaw another fifteen years of corruption and repression, until popular pressure forced him to flee the country in 1986. Since then, the country has gone through several failed attempts at democracy, each one overthrown by a military coup. In 1990, Jean-Bertrand Aristide was elected president. He was ousted by the army in 1991, but reinstated in 1995 after a UN multinational force was dispatched to intervene in Haiti's precarious political situation. **See also Arawak Indians, Dominican Republic, Toussaint-Louverture**

hake (*Merluccius bilinearis*) Several fishes commonly known as hake are found throughout the temperate waters of the world. They travel in large schools, usually just off the continental shelf, and down to a depth of 2,000 feet. Where cod (to which they are related) have three separate dorsal fins, hakes have the second and third joined to form a single fin. Compared to the firm flesh of the cod, that of the hake is soft; nevertheless, they are considered commercially important. The North Atlantic silver hake (*M. bilinearis*), also known as whiting, reaches a length of 36 inches and can weigh 20 pounds. As with many species in the North Atlantic, the arrival in the 1960s of the Eastern European ("distant water") fishing fleets greatly depleted the silver hake, and the population has not recovered. Also found in the western North Atlantic is the white hake (*Urophycis tenuis*), which is somewhat smaller and has a small chin barbel. The eastern Pacific hake is *M. productus,* caught off California, Oregon, Washington, and British Columbia. In New Zealand waters it is *M. australis,* and the stockfish (*M. capensis*) is one of the principal objects of the South African fishery.

See also codfish, haddock, pollack

Hakluyt, Richard (1552–1616) Born near London, Hakluyt (pronounced "Hacklit") became the first lecturer on geography at Christ Church, Oxford. Although he did little traveling himself, he was acquainted with most of the important sea captains, merchants, and sailors of sixteenth-century England, and he read several languages. In 1582, he published *Divers voyages touching the discouerie of America,* and in 1584 he prepared a report that set out the benefits to be obtained from the establishment of a colony in Virginia, titled *The Discourse on the Western Planting.* In 1583, he went to Paris as chaplain to the English ambassador and spent six years compiling his masterwork, *The principall Navigations, Voiages and Discoveries of the English nation within these 1500 Years,* a massive, chauvinistic summary of English exploration that appeared a year after the defeat of the Spanish Armada. When Hakluyt died in 1616, his work was taken over by Samuel Purchas, who published *Hakluytus Posthumus or Purchas his Pilgrimes,* in 1625. (The Hakluyt Society, founded in 1847, has published hundreds of volumes containing the texts and accounts of voyages and continues to do more.)

halfbeak (family Hemirhamphidae) One-foot-long slender fishes, related to the flying fishes, halfbeaks are

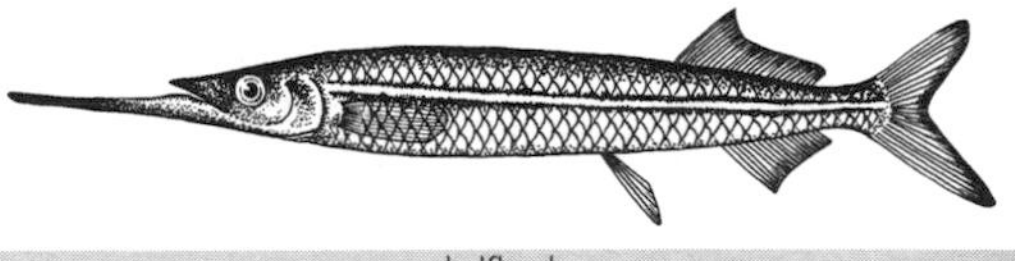

halfbeak

named for their greatly elongated lower jaw. There are some eighty species, found throughout the warm waters of the world, where they travel in schools close to the surface and are often seen skittering along the water, or even becoming airborne for short periods.

See also ballyhoo, flying fish

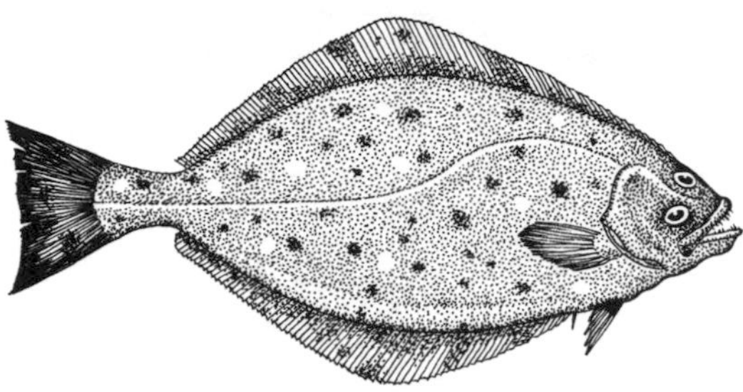

Atlantic halibut

halibut (*Hippoglossus hippoglossus*) Certainly the largest of the flatfishes, and one of the largest of all bony fishes, the Atlantic halibut can be 10 feet long and weigh as much as 720 pounds. It is found in the cold, deep waters of the North Atlantic, the Barents Sea, Iceland, and Greenland. The Greenland halibut (*Reinhardtius hippoglossoides*), sometimes known as turbot, is a North Atlantic species that has been the basis of a recent bitter rivalry between Canadian and Spanish fishing fleets. The California halibut, *H. stenolepis*, found along the west coast of North America all the way to Alaska, grows to a similar size. Both species are extremely important food fishes, but their numbers have declined in recent years. They feed on crustaceans on the bottom and fishes that they catch in midwater. Halibuts are classified as right-eye flounders, but they are often seen in a left-eyed configuration. **See also flounder, plaice, turbot**

Hall, Charles Francis (1821–1871) American Arctic explorer who in 1859 raised his own funds so he could join the search for Sir John Franklin, lost in the Arctic since 1845. He set out in the whaler *George Henry* to the north of Hudson Bay, where he became icebound and spent two years with the Eskimos on Baffin Island. (His book, *Arctic Researches, and Life among the Esquimaux*, is considered important to explorers and anthropologists.) He found no clues to the fate of Franklin, but he became the first explorer in three centuries to visit Frobisher Strait, where he discovered relics of Martin Frobisher's 1576 voyage. In 1864, he returned to the Arctic, and during a five-year stay with the Eskimos, he came upon traces of Franklin's party on the Boothia Peninsula and King William Island. In 1871, Congress granted him $50,000 for an expedition to sail to the North Pole, and he managed to take the *Polaris* to 82°11′ north, the farthest north anyone had gotten until that time. In his winter quarters, Hall fell ill; he died at Thank God Harbor, Greenland, on November 8, 1871. The *Polaris*'s captain tried to bring the ship home, but she was holed by an iceberg, and the crew drifted on a floe for 1,300 miles before they were rescued. **See also Franklin**

Halley, Edmund (1656–1742) English scientist who made a voyage in 1676 to St. Helena in the South Atlantic to observe the southern skies, catalogued 341 stars of the Southern Hemisphere, and studied trade winds and tides. His 1686 map of the world showed the prevailing winds over the oceans; it was the first meteorological chart ever published. Halley was also one of the pioneers of underwater exploration and salvage, inventing a "diving tub" that consisted of a large, tarred-wood, glass-windowed housing that trapped a large air bubble and was lowered to depths of up to 60 feet in the vicinity of a wreck. Three men could work in it for over an hour by breathing the air trapped in the diving bell and supplemented by air lowered in tarred and leaded casks. He also invented a diving suit. Halley financed the publication of Isaac Newton's *Principia Mathematica* in 1687 and, based on its principles, was able to calculate the orbit of the great comet of 1682 (now known as Halley's Comet), and predicted its return in 1758—and every seventy-six years thereafter. In 1698 he made the first primarily scientific sea voyage; in command of the war sloop *Paramour Pink*, he went to the South Atlantic to study the variations of the magnetic compass. He was appointed professor of geometry at Oxford in 1704, and was made Astronomer Royal in 1720.

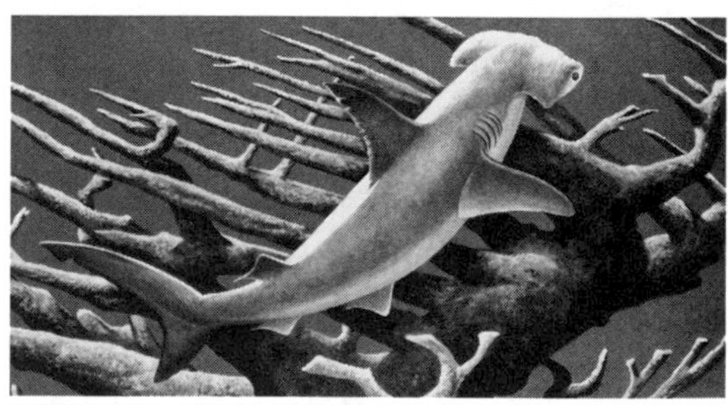

hammerhead shark

hammerhead shark (*Sphyrna* spp.) There are several species of hammerheads, all characterized by a flat, wide, shovel- or hammer-shaped head. The specific function of the "hammer" is not known, but it is equipped with many sensory organs, which the shark uses to detect prey—skates and rays are favorite food items—buried in the sand. They range in size from the 20-foot-long great hammerhead (*S. mokarran*), to the scalloped hammerhead (*S. corona*), which is less than 4 feet long. Hammerheads are found in temperate waters around the world, where they are often seen "finning"—swimming with their dorsal fins and tail fins above the surface. The dorsal fin of most species is proportionally taller than that of other sharks of comparable size. In the Sea of Cortez and other locations, huge schools of hammerheads have been observed, but the reason for these aggregations is not known.

See also blue shark, tiger shark

Hanseatic League A federation of north German towns and merchant communities (*Hanse* is German for "guild") established around the thirteenth century to protect their common trading interests. The first Hanseatic ports were Lübeck and Hamburg, followed by Lüneburg, Wismar, Rostock, Straslund, Tallin, and Danzig. Branches were formed in Novgorod, Bruges, Bergen, and even London. The league sought preferential treatment and, wherever possible, monopolies. They also wanted to protect their shipping from robbers and pirates. Principal trade items consisted of grain, timber, furs, tar, honey, and flax, which were sent to Flanders and England for cloth and manufactured goods and to Sweden for iron and copper ore. Herring caught in the North Sea was traded throughout Germany. In 1368, the league went to war with King Valdemar I of Denmark to prevent him from controlling trade in the Baltic, and by the Treaty of Stralsund (1370) the league was recognized as the dominant power. Over the following three centuries, however, the league lost ground to the growing power of the nation-states of the Baltic, Scandinavia, Russia, the Netherlands, and England. Its diet met for the last time in 1669.

harbor porpoise (*Phocoena phocoena*) "Porpoise" is derived from the Latin *porcus pisces,* or "pig fish," and even now the local names for this animal include "puffing pig" and "herring hog." *Phocoena* (pronounced "fo-*seen*-a") is the commonest cetacean in European waters, and is also found in the inshore waters of the western North Atlantic and the North Pacific. Commonly encountered around river mouths, it has gone upstream in various European rivers, including the Seine, the Thames, and the Danube. Among the smallest of cetaceans, the harbor porpoise rarely reaches 5 feet in length, and mature animals do not weigh more than 140 pounds. They are frequently entangled in fishermen's nets, and they have been heavily affected by organochlorine pesticides and heavy metals.

harbor porpoise

See also vaquita

harbor seal (*Phoca vitulina*) This small, stocky seal lives throughout the temperate and Arctic waters of the Northern Hemisphere, and has the widest distribution of any pinniped. Some populations are nonmigratory, breeding and feeding in the same area throughout the year, but others may migrate for hundreds of miles. Like all the other phocids (earless seals), the harbor seal uses its hind flippers for propulsion in the water, but on land, it hitches along using only its foreflippers, which are equipped with sturdy nails. Harbor seals eat almost anything they can catch, but their diet consists mostly of fish. They occasionally raid and ruin fishermen's nets, and they are killed for that reason, as well as for their meat and highly prized fur.

harbor seal

See also ringed seal, seals and sea lions

harlequin duck (*Histrionicus histrionicus*) Harlequins are small maritime ducks that spend their whole lives in and around rough, fast-moving water. They get their name from the breeding plumage of the drakes, which is dark gray with black-bordered daubs of white and chestnut flanks. The females are about the same size but brown in color with three white marks on each side of the head. At sea, they are often seen feeding (on small crustaceans, isopods, barnacles, and snails) in tumbling surf, then climbing out onto rocks to rest. They nest inland, near turbulent streams that replicate their oceanic habitat, where they can walk underwater, feeding on nymphs and insect larvae. Harlequins breed in Iceland, Greenland, Baffin Island, Labrador, northern Alaska, and Siberia, and winter as far south as Long Island, California, and Japan.

See also eider, old squaw, scoter, sea ducks

harpoon A spear with a barbed head and a line attached, used for catching and securing a whale. (When the whale was close enough, it was killed with a lance.) Many harpoons were designed with a toggle head that kept them from pulling out. In the early days of the whale fishery, harpoons were thrown by hand by a harpooner standing in the bow of a whaleboat, but more recently, harpoons have been equipped with explosive heads and are fired from a cannon. **See also harpoon cannon, whaling**

harpoon cannon Mounted on the bow of a catcher boat, the harpoon cannon was the innovation that enabled pelagic whalers to hunt and kill the larger, faster rorquals (blue, fin, and sei whales) that they couldn't kill with hand-thrown harpoons or shoulder guns. The first models, invented by the Norwegian whaleman Svend Foyn around 1865, were muzzle-loading, but this proved too dangerous to the harpooner, and by 1925, breech-loading models were in use. The projectile was a heavy iron shaft with a pointed, toggled head that was shot into the body of the whale. A grenade on the har-

poon head exploded inside the whale, killing it instantly—or at least mortally wounding it.
See also Foyn, whaling

harp seal (*Phoca groenlandica*) Newfoundland is not the only home of harp seals, which are so notoriously hunted by sealers. These 6-foot-long seals are also found throughout the Northern Hemisphere ice pack from Canada and Greenland to Siberia. They are born white and fluffy (it was these "whitecoats" that the Newfoundland sealers killed with clubs), but within about four years they develop the characteristic coloration of the adults. Harp seals are extremely aquatic and spend much of their lives in the water, where they display particularly effective diving skills. **See also sealing, sealskin**

harp seal

Harrison, John (1693–1776) British clockmaker who built the first seagoing chronometer of sufficient reliability to enable longitude to be calculated at sea. The Board of Longitude was set up in 1714, offering a prize of £20,000 to anyone who could design a chronometer that would be accurate within 30 miles after a six-week voyage to the West Indies. Harrison made four chronometers between 1735 and 1760, each more refined and more accurate than its predecessor. He easily qualified for the prize, but rivalries with astronomers and other clockmakers held up half of his reward, and he spent his last years bitterly fighting for his due. It was not until 1773 that the second half of the reward was paid, and then only through the intervention of King George III. Captain James Cook used one of Harrison's designs on his second voyage around the world (1772–1775) and found that it was the most accurate and useful chronometer he had ever experienced. Harrison's second chronometer is still running at the Royal Observatory in Greenwich. **See also longitude**

hatchetfish (*Argyropelecus hemigymnus*) Named for their deep, laterally compressed bodies, hatchetfishes are small (maximum length: 4 inches), silvery fishes with upward-pointing eyes and a large mouth that enables them to engulf their known prey of copepods and small fishes. They are mid- to deepwater inhabitants and are equipped with several rows of photophores on the ventral surface of the body. An unrelated freshwater species

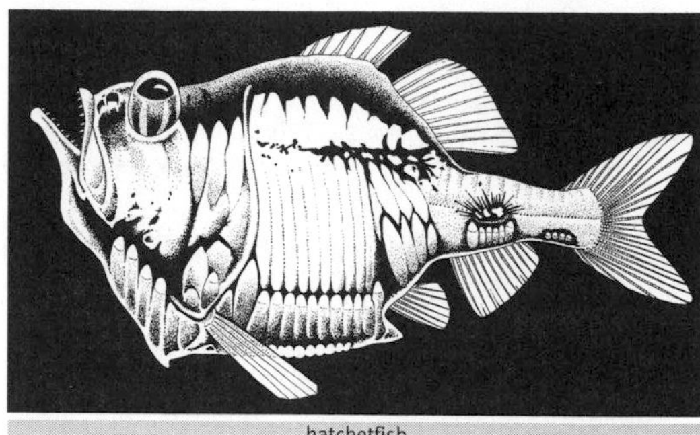
hatchetfish

(*Gasteropelecus sternicula*), also known as the hatchetfish, is popular with home aquarists, except for its habit of jumping out of the tank. **See also bioluminescence**

Hawaii Also known as the Big Island, Hawaii is the largest and the youngest of the Hawaiian Islands. Because the islands are moving in a westerly direction over a "hot spot" on the floor of the Pacific, each one in turn was formed by volcanic action. The whole island of Hawaii is composed of volcanic structures, including the volcanoes Mauna Loa, Mauna Kea, and Kilauea, the most active volcano in the world, with a huge bubbling lava lake inside its Halemaumau Crater. Because of its youth and this volcanic action, many of the island's beaches are black; they are made of cinders, not sand. Most of the rest of the island is wet and warm, with dense vegetation, as befits a tropical island, but Mauna Kea, at a height above sea level of 13,795 feet, is snowcapped in winter, making skiing possible. In January 1779, after a year of charting the islands he had discovered, James Cook returned and was welcomed as a god. When he departed, he encountered a storm and had to go back for repairs. This time, his welcome was not so warm, and under confused circumstances, he was killed by natives on the shores of Kealakekua Bay, on the western shore of the island of Hawaii. The volcanoes, along with the lush tropical scenery, including spectacular fern and bamboo forests, make Hawaii one of the islands' most popular tourist destinations, and many large new hotels have been built. Among the other important villages are Kailua-Kona (famous for big-game fishing) and Honaunau. In addition to tourism, which dominates the island's economy, Hawaii also has active sugar, cattle, and orchid industries. The 120,000 people who live on this island are sparsely scattered over 4,028 square miles, an area just slightly smaller than Connecticut. **See also Hawaiian Islands**

Hawaiian Islands First settled by Polynesian seafarers from other islands—probably the Marquesas—the Hawaiian Islands were the scene of a fully developed culture before James Cook arrived in 1778 and bestowed the name "Sandwich Islands" on them. The kingdom remained self-governing under rulers like King Kamehameha (1738–1819) and Queen Liliuokalani (1838–1917) until missionaries and businessmen decided that the islands could not be left to the natives. The introduction of diseases for which the natives had no immunities added to the conflict, as did the arrival of boisterous Yankee whalers into the port of Lahaina in the mid-nineteenth century. King Kamehameha III sought protection from the United States in 1851 to fend off claims from Britain and France, and in 1859 the United States was granted permission to build a naval base at Pearl Harbor. In 1893, Liliuokalani was deposed, and the Republic of Hawaii was declared, with Sanford R. Dole as president. Hawaii was annexed to the United States in 1898, became a U.S. territory in 1900, and was made the fiftieth state on August 21, 1959. Hawaii consists of eight volcanic islands, stretching for 1,500 miles in the central Pacific, 2,300 miles south-southwest of San Francisco. In order of size, the islands are Hawaii, Maui, Oahu, Kauai, Lanai, Molokai, Kaho' olawe, and Ni'ihau. It is unusual to find volcanoes far from the margins of tectonic plates, but the islands are sliding slowly over a "hot spot" on the ocean floor, which activates one island after another. Thus Kauai, the westernmost island, is the oldest, while Hawaii, at the eastern end of the chain, is currently over the plume, and is therefore the youngest and most active of the islands. Kilauea volcano on the southeast coast of Hawaii is currently erupting, and although it produces spectacular effects, it is not now considered threatening. (Extinct volcanoes are Haleakala on Maui and Diamond Head on Oahu.) The landscape is spectacular, with beautiful beaches, towering mountains, lush green valleys, and a plethora of tropical flowers, which are often made into leis. Although agriculture (mainly sugar and pineapples) and military installations (Schofield Barracks, Pearl Harbor, Kaneohe) were the mainstays of the economy in the past, tourism is now Hawaii's number one industry. Water sports, such as diving, surfing, sailing, and windsurfing, bring many visitors to the "Paradise of the Pacific," and there are many venues where the visitor can experience the characteristic Hawaiian history, dancing, crafts, and music. Honolulu, on the island of Oahu, is the state capital with a population of 377,000. With a total population of 1,108,000, Hawaii ranks fortieth among the states.

See also Hawaii, Kaho'olawe, Kauai, Lanai, Maui, Ni'ihau, Oahu, Pearl Harbor

Hawaiian monk seal (*Monachus schauinslandi*) Found primarily in the Leeward Islands, these animals sometimes stray eastward to the main Hawaiian group. They were heavily hunted by whalers and sealers in the nineteenth century, and disturbed again during World War II when U.S. forces occupied Laysan and Midway Islands. Recent surveys have estimated the population

Hawaiian monk seal

at about 1,400. They are fully protected, but they seem to be very susceptible to attacks by tiger sharks. Of all the monk seals, this species is the best known, because it has been studied in its native habitat and has been successfully maintained in captivity, particularly at the Waikiki Aquarium. **See also Caribbean monk seal, Mediterranean monk seal**

hawkfish (family Cirrhitidae) The name "hawkfish" comes not from any resemblance to a bird of prey, but rather to their unusual habit of perching on a coral branch or rock, and then dashing in to capture their unsuspecting prey. They are hardly ever seen swimming freely. Their dorsal spines are often tipped with fringes, known as cirri, and there are sometimes cirri aft of the nostrils. Many species have an unusual color pattern, in that the front half of the body is spotted, while the rear half is banded. The thirty-odd species are found mostly in the Indo-Pacific, but they also appear in the western Atlantic and off the coast of California.

See also longnose hawkfish

Hawkins, Sir John (1532–1595) English admiral who began his career as commander of hugely profitable expeditions from 1562 to 1565 to buy and capture slaves in West Africa and sell them in the West Indies. On an expedition to Mexico in 1567 (with his cousin Francis Drake in command of one of the ships), Hawkins's flagship, the *Jesus of Lubeck,* was captured by the Spanish, and only the vessels commanded by Drake and Hawkins himself returned home safely. In 1577 the queen appointed him treasurer and comptroller of the navy, and he is generally regarded as the architect of the Elizabethan navy, introducing many innovations in design and construction and upgrading the ships that Elizabeth had inherited from her father, Henry VIII. In the 1588 campaign against the Spanish Armada, Hawkins commanded the *Victory* and for his efforts was knighted during the battle. In 1595, he and Drake were sent to raid the Spanish settlements in the West Indies, but the expedition was less than successful, and both men died of dysentery. **See also Armada, Spanish; Drake; Hawkins (Sir Richard)**

Hawkins, Sir Richard (1562–1622) The only son of Sir John Hawkins, Richard Hawkins was also a renowned

British admiral. He commanded a small ship in Drake's expedition to the Indies in 1585 and was in charge of the *Swallow* at the battle of the Spanish Armada in 1588. In the 300-ton *Dainty,* he intended to circumnavigate the world in 1593; he rediscovered the Falkland Islands, passed through the Strait of Magellan, and attacked Valparaiso before he was forced to surrender to the Spanish off Peru in 1594. He was imprisoned until 1602, when he was ransomed for £3,000. When he returned to England he was knighted and elected to Parliament. In 1620, he sailed on an unsuccessful mission against the Barbary pirates; he died two years later. *The Observations of Richard Hawkins, Knight, in his Voyage into the South Sea,* published in 1622, is considered the classic work on Elizabethan naval conditions, customs, and history.

hawksbill turtle (*Eretmochelys imbricata*) Named for its narrow, sharply downcurved beak, the hawksbill is one of the smaller of the sea turtles; it reaches an average length of about 30 inches and a weight of about 120 pounds. It is found throughout the world's tropical and subtropical waters, and its favorite habitat is rocks and ledges in relatively shallow water. It feeds primarily on sponges, but it also consumes mollusks and sea urchins. Unlike many other sea turtles, the hawksbill does not nest in groups, but the females come ashore singly to dig their nests and lay their eggs. The beautiful, marbled shell of the young hawksbill has long been sought for the manufacture of decorative objects, and this quest for tortoiseshell has resulted in the worldwide depletion of the species. **See also sea turtle**

hawksbill turtle

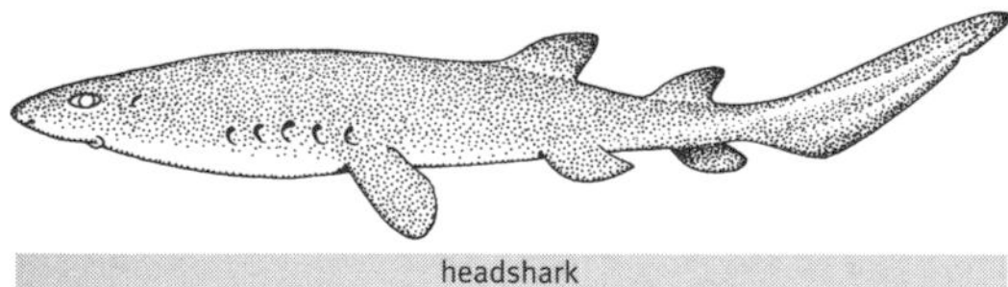
headshark

headshark (*Cephalurus cephalus*) A bizarre little cat shark that consists mostly of a huge, flattened head, hence its common name. (It is also known as the lollipop catshark.) A mature adult is only 11 inches long, which makes it one of the smallest of all sharks. It is found in and around the Gulf of California, and what little information we have about the species has come from specimens trawled up from depths greater than 900 feet. **See also catsharks, cigar shark, pygmy shark**

Heard Island Subantarctic island in the southern Indian Ocean, 2,500 miles southwest of Perth and 300 miles south of the Kerguelen Islands. It was discovered and named in 1853 by Captain James J. Heard, an American sealer on the bark *Oriental,* bound from Boston to Melbourne. As with other subantarctic islands, Heard served as a breeding area for southern elephant seals, and thus was visited regularly by those who would hunt the giant pinnipeds for their oil. In 1873, the British oceanographic research vessel *Challenger* visited this lonely island, along with Kerguelen, Marion, Macquarie, and the Crozets. The island, now an Australian territory intermittently visited by scientific researchers, is 27 miles long by 13 miles wide, and is dominated by the 9,000-foot-high dormant volcano known as Big Ben. Penguins of various species come ashore here, and the island has the greatest concentration of leopard seals in the world. The Indian Ocean Territory of Australia consists of Heard and the tiny MacDonald Islands, 25 miles to the west.

Heaviside's dolphin

Heaviside's dolphin (*Cephalorhynchus heavisidii*) The South African representative of the genus *Cephalorhynchus,* this little black-and-white dolphin closely resembles its Chilean counterpart, the black dolphin, except that the white ventral markings are considerably more pronounced. Its dorsal fin is low and rounded, and it rarely exceeds 5 feet in length. It was named for a certain Captain Haviside (the *e* was erroneously added later), who brought the first specimen from the Cape of Good Hope to England in 1827.

See also black dolphin, Commerson's dolphin, dolphin, Hector's dolphin

Hebrides A group of forty-odd diverse islands off the west coast of Scotland, subdivided into the Inner and Outer Hebrides, which are separated from each other by channels known as North Minch, Little Minch, and the Sea of the Hebrides. The Outer Hebrides, also known as the Western Isles, form a tight crescent consisting of Lewis with Harris, North Uist, Benbecula, South Uist, St. Kilda, and Barra, and are losing population—largely to Canada—because of the lack of employment opportunities. (Some of the islands, like St. Kilda, are now completely uninhabited.) The inner islands of Skye, Tiree, Rum, Eigg, Muck, Colonsay, Islay, Jura, and Mull are economically stronger because they are more accessible to mainland tourists. The climate is cool and wet, but the scenery is wild and beautiful, and wildlife abounds, particularly birds. Tourism is the major industry, but Hebrideans also engage in crofting (tenant farming), raising sheep and cattle, fishing, and weaving. (Harris tweed is woven on the island of Lewis with Harris.) The original Celtic inhabitants were converted to Christianity by Saint Columba in the sixth century, but the southern islands were conquered by the Vikings by the eighth century and held until 1266, when they were ruled by various Scottish chieftains. The Clan Macdonald ruled most of the Hebrides from 1346 until the sixteenth century, when the islands fell under control of the Scottish Crown.

See also Benbecula; Jura; Lewis with Harris, Isle of; Mull; Skye, Isle of

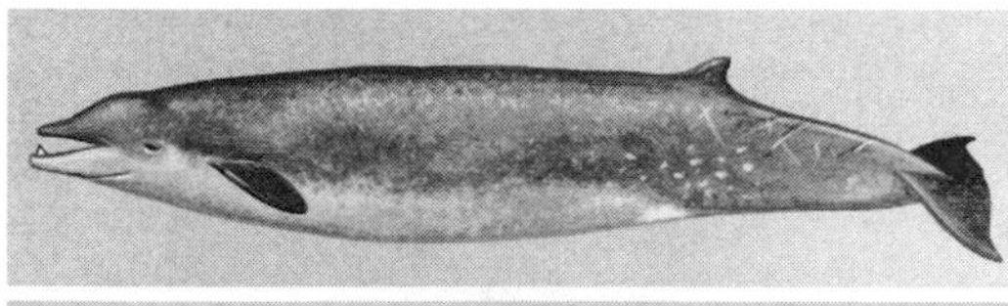
Hector's beaked whale

Hector's beaked whale (*Mesoplodon hectori*) Until 1977, there was no published information on the appearance of this species. It was thought to exist only in the Southern Ocean (off New Zealand, Tasmania, and Australia), but in 1975, a mother and calf and two other adults washed ashore in Southern California. From skeletal similarities to the Southern Hemisphere specimens, they were identified as *M. hectori*, which extended their range to another hemisphere. This is one of the smaller beaked whales, reaching a known length of 14 feet. **See also beaked whales**

Hector's dolphin (*Cephalorhynchus hectori*) Small dolphin found only in New Zealand waters. It is strongly marked in gray, black, and white, and has a fairly large, rounded dorsal fin. While other species of the genus *Cephalorhynchus* are shy and avoid ships, Hector's dolphin is a known bow rider and seems to seek out contact with vessels. It is threatened by incidental capture in gill nets. This species and Hector's beaked whale are named for Sir James Hector, a zoologist who was director of the Colonial Museum in Wellington, New Zealand. **See also black dolphin, Commerson's dolphin, dolphin, Heaviside's dolphin**

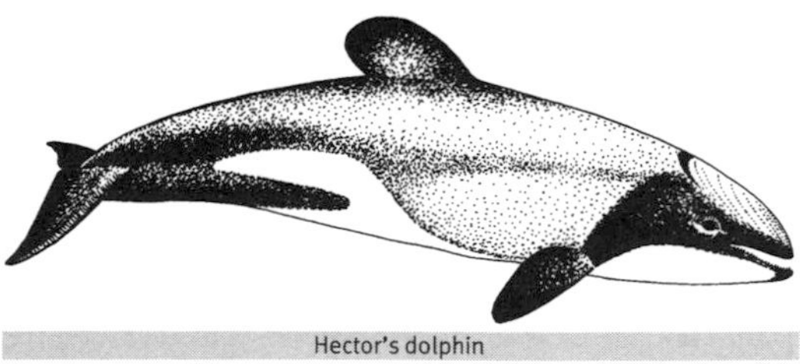
Hector's dolphin

Heemskerck, Jacob van (1567–1607) Dutch explorer who voyaged to the Arctic in 1596 with Willem Barents as his navigator; he discovered Bear Island and then Spitsbergen. Upon leaving Spitsbergen and rounding Novaya Zemlya, their ship was trapped in the ice, and the men spent the winter of 1596–1597 on the island in a hut made of driftwood. In June they set out in the ship's boats. Barents died after a week, but Heemskerck led the rest of the crew to safety. He spent most of the rest of his life in the East Indies and was appointed commander of the Dutch fleet in 1607. He was killed in a sea battle with Spanish ships off Gibraltar.

See also Barents, Bear Island, Spitsbergen

Heermann's gull (*Larus heermanni*) The only American gull with a white head and gray body, Heermann's gull is a Pacific species, found from Oregon to Mexico and Guatemala. Some 600,000 pairs of Heermann's gulls breed on Raza Island in the Sea of Cortez, its only breeding site. It was named for Adolphus Heermann, a nineteenth-century surgeon and naturalist.

See also California gull, gull

Heezen, Bruce (1923–1977) Oceanographer who was among the first to understand the nature of seafloor spreading and its profound effects on the formation of the surface of the earth. He was born in Iowa and was prepared to become a paleontologist until he met Maurice Ewing at Columbia University and turned his attention to the geology of the sea. Heezen and Marie Tharp produced the first map of the floor of the Atlantic in 1956, and eventually they mapped the seafloor of the entire planet. With Charles Hollister he was the author of *The Face of the Deep* (1971), an important illustrated discussion of the actual appearance of the ocean floor and the creatures that inhabit this previously unknown realm. In 1965, Heezen and Dragoslav Ninkovich published a paper showing that the ash found in the sediments of the eastern Mediterranean could be used to date the volcanic eruption on the island of Santoríni that destroyed most of the island and contributed to the downfall of the Minoan civilization on Crete. He died at work aboard the U.S. Navy's *NR-1*, a nuclear-powered deep submersible.

Heimaey Icelandic island in the Westmann group that was nearly subsumed by the eruption of the volcano Kirkjufell in 1973. The earth split open in January on the edge of the fishing village of Vestmannaeyjar, and fiery gouts of molten lava came spewing out. Houses burst into flames, and a wall of lava 100 feet high advanced on the town. The village was evacuated without mishap, but the lava threatened to close off the harbor, which is absolutely essential to Vestmannaeyjar, Iceland's most important fishing port. Firemen trained their hoses on the advancing wall of lava in an attempt to cool and solidify it before it closed off the harbor. Whether the hoses stopped the lava or the volcano stopped spontaneously, the harbor was not sealed off, and the residents of Heimaey were able to return to their village to clear away the mountains of cinders, which had destroyed some three hundred houses, including some that were completely buried. By 1975, most of the evacuated population had returned to the island and rebuilt their homes. **See also Iceland**

Helgoland Also Heligoland, an island in the North Sea off the coast of Schleswig-Holstein. It has been greatly reduced in area because of wave action and erosion, and now it covers only 150 acres, with a permanent population of some 2,000, most of whom are involved in raising cattle. Britain captured it from Denmark during the Napoleonic Wars (1803–1815), when it was used as a base for smugglers to break England's blockade of France. It was formally ceded to Britain in 1814, but traded in 1890 for the German colonies of Uganda and Zanzibar. Before World War I, it was fortified by Germany to protect the western approach to the Kiel Canal; the fortifications were torn down in 1919, only to be rebuilt in 1935, when Germany renounced the Treaty of Versailles. The island was evacuated in 1947 and the fortifications blown up in one of the largest nonatomic blasts ever recorded. It is a popular tourist resort and has an ornithological laboratory. In 1979 a German clergyman named Jürgen Spanuth published *Atlantis of the North*, in which he argued that Helgoland was actually Atlantis.

Helice City on the Gulf of Corinth on the north coast of the Peloponnesus in Greece that was engulfed by the sea after a massive earthquake in 373 B.C. There were no survivors. For the next five hundred years, the ruins of the sunken city were visible under the shallow waters, but they were eventually covered with silt brought down by adjacent rivers, and no traces can be seen. The remains of Helice are now believed to be under dry land, and since the precise location of the city is not known, archaeologists' attempts to find it have been unsuccessful. Since Plato was the originator of the story of Atlantis, and since this catastrophe occurred during his lifetime (428–347 B.C.), it is believed to have been one of the events that inspired him to write the story. It is also included in the works of the later Greek geographers Strabo and Pausanias. **See also Atlantis**

Hellespont: See Dardanelles

Henry the Navigator (1394–1460) Prince of Portugal who inspired and sponsored a succession of Portuguese explorers who made voyages of discovery among the Atlantic islands, down the west coast of Africa, around the Cape of Good Hope, and on the sea route to India. As governor of the Algarve, he lived at Sagres, where he established a school for the study of navigation, astronomy, and cartography. In 1434, one of his ships, commanded by Gil Eannes, first rounded Cape Bojador in West Africa; eight years later, Nuno Tristam brought the first slaves and the first gold dust from Africa to Portugal. He was also responsible for the voyages of Alvise Cadamosto and Diogo Gomes to the Cape Verde Islands and the coast of West Africa. Henry himself never went on a voyage of exploration.

See also Cadamosto, Cape of Good Hope, Gomes

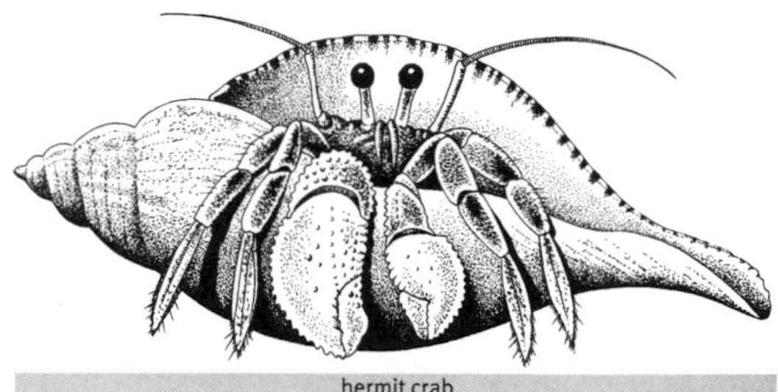

hermit crab

hermit crab (*Pagurus* spp.) A crab that makes its home in the shell of another animal by backing its unarmored abdomen into it. Hermit crabs normally occupy snail shells, but they may also use tusk shells, clam shells, corals, and, in some places, halves of coconut shells. They can leave their "home" at will, and do so when they have grown too large. Only the first two pairs of legs are fully developed; the hindmost pairs are used for purchase in the shell. The larger of the first pair of claws serves as a lid to block the shell opening when they draw inside. There are several species of hermit crabs; the largest, found on the Pacific coast of North America, measures more than 1 foot in length.

See also crab

herring: See Atlantic herring

herring gull (*Larus argentatus*) The most common gull of the North Atlantic region, the herring gull is a large white bird with a gray back, black wing tips, a yellow bill marked with a conspicuous red spot, and a wing span that can reach 56 inches. Juveniles, which are often as large as adults, are mottled brown and buff, with black beaks and legs. Herring gulls are scavengers, feeding on rubbish, waste food, other birds' eggs, and almost anything else that is edible. The great black-backed gull (*L. marinus*) is larger than the herring gull—it is the largest of the gulls—with a black back; the glaucous gull (*L. hyperboreus*), also larger than the herring gull, is almost white and found farther north. The ubiqui-

herring gull

tous herring gull ranges from Alaska to Greenland, and is found throughout Europe, Asia, and Africa. **See also black-headed gull, great black-backed gull, gull**

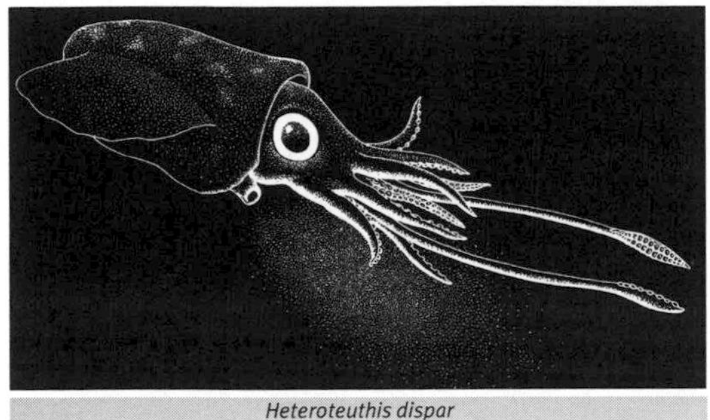

Heteroteuthis dispar

Heteroteuthis dispar This tiny (3 inches total length) squid employs a variation on the "ink-cloud" defense mechanism. Its ink is bioluminescent, like that of the tubeshoulder (*Searsia kofoedi*). **See also bioluminescence, squid, tubeshoulder**

Heyerdahl, Thor (b. 1914) Norwegian scientist-explorer who believed that the Polynesian islands could have been settled by voyagers from South America. To prove his theory, he and his associates constructed a raft made of Peruvian balsa (*Ochroma lagopus*), fastened only with materials that would have been available a thousand years ago. The raft, christened *Kon-Tiki* (after an Inca sun god), was launched from Callao, Peru, on April 28, 1947, and arrived at Rairoa, in the Tuamotus, 101 days later, having drifted with winds and currents for 4,800 miles. *Kon-Tiki*, the story of the voyage, was published in 1948. He also made a study of the great statues of Easter Island, which was published in 1958 as *Aku-Aku: The Secret of Easter Island*. When Heyerdahl attempted to cross the Atlantic in a papyrus raft to demonstrate that the ancient Egyptians could have been open-ocean sailors, the *Ra* disintegrated in heavy seas. Heyerdahl built another papyrus vessel, *Ra II*, and made a successful crossing from Morocco to Barbados in 1970. In 1977, he launched the *Tigris*, a raft made of reeds, and sailed from Iraq, down the Persian Gulf, to Pakistan and the Red Sea, to show that the ancient Sumerians might have settled southwest Asia by that route. In 1979, he published *Early Man and the Ocean*, a summary of his theories about cultural diffusion by sea. **See also Easter Island**

Hispaniola After Cuba, Hispaniola is the second largest island in the West Indies. Discovered and named Española by Columbus in 1492, it lies between Cuba and Puerto Rico. The French, who colonized the island in the seventeenth century, called it Saint-Domingue. The island, which covers some 29,000 square miles (a little smaller than South Carolina), is divided into two separate countries, Haiti in the west and the Dominican Republic in the east. **See also Dominican Republic, Haiti**

***Histioteuthis* spp.** The squid family Histioteuthidae includes several species characterized by a large number of light cells (photophores) all over the body and arms, accounting for one of their common names, jewel squids. Some are known as umbrella squids because of the substantial webbing between the arms. They all have dramatically different eyes, with the left eye much larger than the right, the smaller eye ringed by seventeen or eighteen small light organs around the periphery, which may be used as a flashlight. Histioteuthids are found around the world in temperate to cold waters, at depths between 1,500 and 5,000 feet. They are believed to be schooling species, and are a favorite food of sperm whales. **See also photophore, sperm whale, squid**

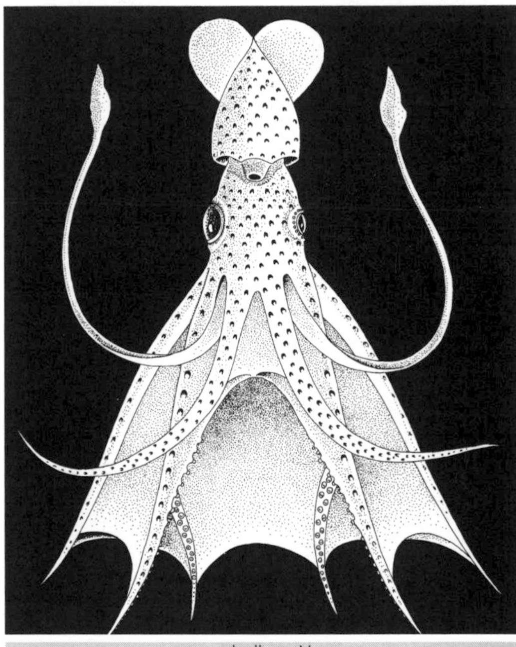

umbrella squid

Hjort, Johan (1869–1948) Norwegian oceanographer who began his career by studying sea squirts in local waters. In 1899, he persuaded the government to build the fisheries research vessel *Michael Sars*, which sailed throughout Norwegian waters from 1900 to 1909. In 1910, as expedition leader, Hjort took the *Michael Sars* to the Mediterranean, the Canaries, and the Azores for biological investigations, and he developed a new type of tow net that could be closed before being hauled to the surface. As a result of a 1910 voyage aboard the *Michael Sars*, he and John Murray wrote *The Depths of the Sea*, which was published in 1912. From 1924 to 1939, while president of the International Council for the Exploration of the Sea, he undertook several voyages to study the migration of commercially important food fishes. As a Norwegian, he was particularly interested in the relationship of whales and their environment; he played an important role in the formulation of international whaling policies. There is now a Norwegian fisheries research vessel named *Johan Hjort.* **See also Murray**

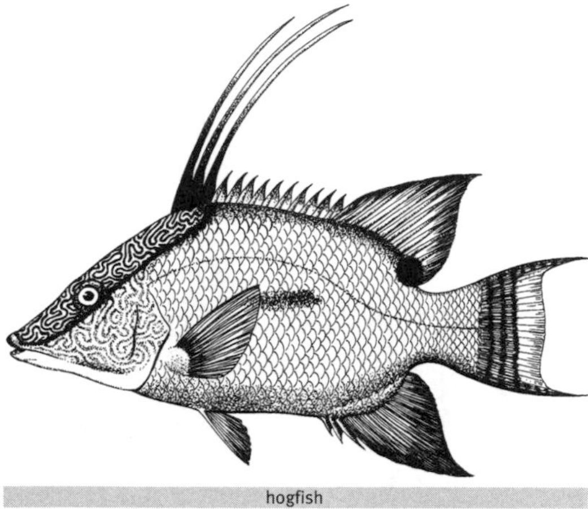

hogfish

hogfish (*Lachnolaimus maximus*) A wrasse found in the western North Atlantic, from the Carolinas south to Puerto Rico, the Gulf of Mexico, and the Caribbean. The hogfish lives in the vicinity of reefs, but it is more frequently encountered over open bottoms, especially in the area of gorgonians. It grows to a length of 3 feet and a weight of 25 pounds. Like its relatives, the Spanish hogfish (*Bodianus rufus*) and the spotfin hogfish (*B. pulchellus*), this species is easily approached and is often the target of spearfishermen. It is a popular food fish. **See also wrasse**

Holland, John P. (1840–1914) Irish American submarine designer, he was born in County Clare and emigrated to America in 1872. A schoolteacher in Paterson, New Jersey, Holland could not work for several months because of a broken leg, so he entertained himself by drawing up plans for a submersible boat, which he submitted to the Secretary of the Navy; however, they were rejected out of hand. Holland then sought private funding from the Fenian Society, a group of Irish nationalists dedicated to revolution in Ireland. It was the society's intention—but probably not Holland's—to deliver a squadron of submarines to British ports and release them to torpedo the Royal Navy at anchor. In 1878 they gave Holland $6,000 and he built *Holland I,* a 15-foot-long, one-man submarine that sank immediately on its unmanned launching in the Passaic River in New Jersey. In 1881 he completed the submarine that was fondly expected to accomplish all this mischief: *Holland II,* the legendary *Fenian Ram,* 31 feet long and weighing 19 tons. While the Fenians waited impatiently for him to launch his attack on England, Holland spent the next two years testing his invention in the Hudson River. Tired of waiting, the Fenian Society stole "their" submarine from Holland in 1883, and tried to tow it across Long Island Sound to New Haven, Connecticut, but they ran aground and abandoned the project. Holland then abandoned them. After several unsuccessful attempts to build another submersible boat, he won two successive competitions sponsored by the navy and delivered the *Holland* in 1898. Just over 50 feet in length, she used an electric motor for underwater propulsion and a gasoline engine at the surface. She was armed with bow torpedo tubes and dynamite guns pointed fore and aft. John Holland had designed the first successful submarine. **See also submarine**

Homer (9th–8th century B.C.) Almost nothing is known of the life of this Greek poet, who is believed to have been an Ionian, from the west coast of Asia Minor, not far from the Troy he wrote about. There is no evidence that he was blind, but the ancient Greeks thought that he was. Probably relying on oral traditions, he wrote the *Iliad* and then the *Odyssey,* epic poems that the Greeks believed were history, but that later scholars thought were romantic storytelling. Because the archaeologist Heinrich Schliemann believed that Homer's poems were fact, he dug for the fabled city of Troy where Homer said it was—and found it. **See also Odysseus**

Homer, Winslow (1836–1910) Born in Boston, Homer painted landscapes, genre paintings, and magazine illustrations of scenes from the Civil War, but is best known as a painter of the sea. During the summer of 1873, he sketched and painted in Gloucester, Massachusetts. In 1881, he sailed for England, where he took up residence at Tynemouth, a North Sea port. In 1883, he settled at Prout's Neck, Maine, where he was to paint some of his most memorable oils, including *Lost on the Grand Banks, Undertow, Eight Bells, The Herring Net, Cannon Rock,* and *The Northeaster.* He spent the summers visiting Florida, Bermuda, and the Bahamas, where he created the watercolors that have given him a permanent place among the world's greatest watercolorists. Here also he did the research for *The Gulf Stream,* the painting of a sailor aboard a dismasted ship with sharks circling ominously around, probably his most famous work. His reputation has expanded over time, and he is considered by many to be the greatest artist America has ever produced. **See also Turner**

hooded seal (*Cystophora cristata*) This seal is named for the inflatable sac that occurs only in mature males. This sac differs from the proboscis of elephant seals, which is inflated by muscular action; that of the hooded seal is

hooded seal

inflated with air to form the characteristic hood. These seals can also blow a bright red, balloonlike structure out of one nostril, accounting for their other name, "bladdernose seal." This behavior may be used to intimidate subordinate males, but it has also been observed when humans get too close. Hooded seals can reach a length of about 8 feet and a weight of 600 pounds.

Hormuz, Strait of Also known as Hormoz or Ormuz, the strait links the Persian Gulf with the Gulf of Oman, passing between Iran's Bandar Abbas on the north and Oman and the United Arab Emirates on the south. The strait, between 35 and 60 miles wide, is of great strategic and economic importance, because oil tankers coming to and from the Persian Gulf must pass through it. In the strait are the islands of Qishm, Henqām, and Hormuz. Hormuz was once an important center for the production of indigo, grain, and spices, but is now only sparsely populated. **See also Persian Gulf**

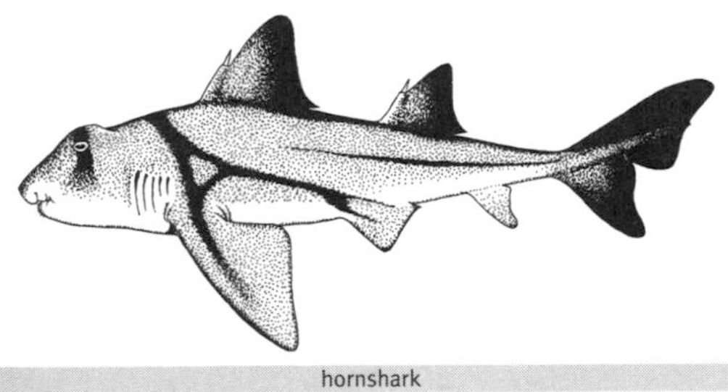
hornshark

hornshark (*Heterodontus* spp.) Small sharks, usually about 30 inches in length, with short, blunt snouts and prominent ridges above the eyes. The two dorsal fins are equipped with strong, sharp spines. Because of these spines, characteristic of some of the earliest extinct sharks, and the two types of teeth in the mouth (*Heterodontus* means "different teeth"), hornsharks are believed to be among the most primitive of living shark species. The egg cases of hornsharks are equipped with spiral flanges that the female shark screws into fissures in rock faces. These sluggish bottom dwellers are also known as bullhead sharks and Port Jackson sharks. (One Australian species is known as *H. portusjacksoni*, after the original name of Sydney.)
See also mermaid's purse

horse latitudes The belt of calms and light winds that borders the northern edge of the trade winds that blow steadily in a belt between 30° north and 30° south of the equator. The name may be derived from the practice of throwing overboard the "dead horse," a symbol of the period of time for which seamen had been paid in advance. This was usually about two months, which would have put them around 35° north or south. Another explanation for the name is that when ships were becalmed, the crew would throw the horses overboard to conserve water. **See also trade winds**

horseshoe crab (*Limulus polyphemus*) Not a crab at all, the horseshoe crab is more closely related to scorpions, spiders, and the extinct trilobites. It is often called a living fossil because it has remained unchanged for 200 million years. The body is divided into three parts that are hinged together: the broad frontal shell (the prosoma); a smaller middle section (the opisthosoma); and a long, sharp tail spine, called the telson.

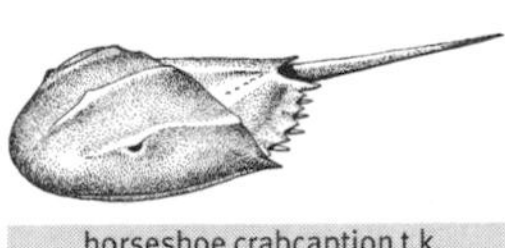
horseshoe crabcaption t.k.

Horseshoe crabs have six pairs of legs, of which the front two (chelicerae) are used for feeding. In spawning, which takes place on sandy beaches in spring and summer, the female, accompanied by one or more males, scoops out a depression in the sand and lays two hundred to three hundred eggs. **See also trilobite**

houndsharks: See smooth houndsharks

hourglass dolphin (*Lagenorhynchus cruciger*) A strikingly colored little dolphin (maximum length: 6 feet) found only in Antarctic waters, farther south than any other species of *Lagenorhynchus.* Because of the remoteness of its habitat, very little is known of this species. In addition to the prominent black-and-white markings that give this dolphin its common name, it also has a distinct dorsal fin, flattened back and sharply hooked. Like most of the "lags," hourglass dolphins are avid bow riders, and are often seen making long, high leaps as they approach the ship.
See also dusky dolphin, Peale's dolphin

hourglass dolphin

hovercraft A vehicle that rides on a cushion of air between it and the surface over which it is traveling. (Another name is air-cushion vehicle, or ACV.) There are two basic types of hovercraft: integrated, which uses the same power plant to lift and propel the craft, and nonintegrated, in which there are two distinct systems, one to force air underneath the craft and another to propel it forward. Both systems rely on a buildup of high air pressure under the craft to lift it off the surface, and the critical element in the hovercraft is the skirt or curtain that keeps the pressurized air underneath the vehicle. The principle was developed by John Thornycroft around 1875, but it was not until 1950 that British engineer Christopher Cockerell was able to produce a practicable design. In 1958 Cockerell persuaded the British government to underwrite the project, and in 1959, the hovercraft designated *SR.N1* crossed the English Channel. Their popularity rose in the 1960s, when they were being manufactured in Britain, the United States, Sweden, and France. In 1963 the "sidewall hovercraft" was introduced: a nonamphibious vehicle that has on each side a solid hull under which is an air chamber sealed by flexible skirts at the bow and stern. It does not rise completely out of the water, but only a small portion of the sidewalls remains in the water, reducing the drag considerably but not completely. Hovercraft can travel at more than 150 mph, but only over a smooth surface, since they cannot clear waves higher than 4 feet. Hovercraft never dominated shipping as their advocates hoped, but they are in regular use on the Øresund between Copenhagen and Malmø, Sweden; on the English Channel; in Hawaii and Alaska; and for various military purposes, such as transports and landing craft. There are also small recreational hovercraft. In the year 2000, a company called HoverQuest 2000+ is planning an attempt on the North Pole by hovercraft.

Howland Island Uninhabited island less than 1 square mile in area in the central Pacific, near the equator, about 1,600 miles southwest of Honolulu. It was initially sighted by American whalers in 1842, and named for the first man to see it. Along with Jarvis and nearby Baker Island, it was claimed by the United States under the Guano Act of 1856. With the decline of the guano industry, the islands were largely forgotten until they started being used as a stopoff for planes flying between Hawaii and Australia. American colonists were brought from Hawaii in 1935 to counter British claims, but the colony was abandoned at the outbreak of World War II. During her attempt to fly around the world in 1937, Amelia Earhart was on her way from New Guinea to Howland Island when she was lost in the Pacific. The island is now administered by the U.S. Department of the Interior. **See also guano islands, Kiribati**

Hubbs' beaked whale (*Mesoplodon carlhubbsi*) Named for the eminent American ichthyologist Carl Hubbs, this beaked whale is found only in the North Pacific. It is one of the "saber-toothed whales" that have large, prominent teeth that show outside the jaws. (The others are *M. bowdoini, M. stejnegeri, M. densirostris,* and *M. ginkgodens.*) It is characterized by a white patch just forward of the blowhole, which has been called a beanie. Adults are about 17 feet long.

Hubbs' beaked whale

See also beaked whales, *Mesoplodon*

Hudson, Henry (c. 1565–1611) English navigator and explorer. In 1607 Hudson made the first of two voyages for the Muscovy Company, seeking a route to China by way of the North Pole. He headed up the east coast of Greenland, but when he was blocked by the ice, he followed the edge of the polar ice to Spitsbergen. On his homeward passage, he discovered the island that came to be called Jan Mayen. On his second voyage, he again sought the Northeast Passage, but failed to find a way through the Barents Sea. In 1609 he made a voyage for the Dutch East India Company in the *Half Moon,* and after another unsuccessful go at the Barents Sea, he crossed the Atlantic and entered New York Bay, then sailed 150 miles up the river that now bears his name. He returned to England, and, forbidden by the English government to give his services to the Dutch, he sailed west again in 1610 in the *Discovery,* and passed through Hudson Strait into Hudson Bay. Unable to find the (nonexistent) passage to the Pacific from Hudson Bay, the ship sailed aimlessly until Hudson's crew mutinied and cast him, his son, and several crew members adrift in a small open boat. He was never heard from again.

See also Northeast Passage, Northwest Passage

Hudson Bay Covering some 316,000 square miles of central Canada, Hudson Bay is larger than France and Italy combined. Discovered by Henry Hudson in 1610, it is linked to the Atlantic Ocean by Hudson Strait, which passes between Baffin Island and Labrador. The extension in the south is known as James Bay. Its rich nutrients support large populations of crustaceans, which in turn feed various species of fishes, seals, seabirds, and, in the northern sector, walruses, dolphins, and killer whales. The settlement of Churchill, on the western shore, is visited by large numbers of polar bears every summer, causing a problem for the townspeople but also creating a major tourist attraction. The bay supports a large population of belugas. In 1670, the Hudson's Bay Company was incorporated in England, its charter authorizing it to find a northwest passage to Asia, to occupy the lands adjacent to the bay, and to carry on commerce. For the first two centuries of its existence, the company engaged in the lucrative fur-trading business, but by 1859 there were too many independent traders and its monopoly, held for almost forty years, was not renewed. The Hudson's Bay Company, with headquarters in Toronto, is still active in real estate, merchandising, and natural resources.

Humboldt, Alexander von (1769–1859) German explorer and scientist who achieved international fame as the advocate of the earth sciences. He is generally recognized as being the originator of ecology, the relationship of plants and animals and their environment. Born in Berlin, then the capital of Prussia, he developed an early interest in botany and mineralogy. In 1799, he embarked on a five-year scientific expedition to Central and South America with the French botanist Aimé Bonpland. They collected plant samples, made meteorological observations, and studied the magnetic field of the earth. In 1802 Humboldt took measurements of the current that would be named for him, showing the coldness of the flow in relation to the air and water around it. When he returned to Europe in 1804, he published a thirty-volume collection of his observations. He spent the last twenty-five years of his life writing *Kosmos,* a celebrated and comprehensive account of the structure of the universe as it was then known. **See also Humboldt Current, Humboldt penguin, Humboldt squid**

Humboldt Current Named for Alexander von Humboldt, and also known as the Peru Current, the Humboldt Current is a slow-moving, cold-water flow of the southeastern Pacific Ocean. It is about 500 miles wide. While most of the current passes through the Drake Passage around the tip of South America and into the Atlantic, a narrow stream turns north to parallel the continent as far as 4° south, where it turns west to join the Pacific South Equatorial Current. It is a cold current except during El Niño events, when it warms up and wreaks havoc with the weather systems. The cold flow is intensified by the upwelling of deeper water, which brings nutrients closer to the surface, allowing for a rich plankton growth and the proliferation of fish and squid that feed on the plankton and on each other. The waters of Peru, Chile, and Ecuador are therefore among the world's most productive fishing grounds, for such small fishes as anchovies and for larger fishes like tuna. On the islands of the west coast of South America, the birds that feed on the smaller fishes deposit vast quantities of guano that is collected for fertilizer. **See also Humboldt**

Humboldt penguin (*Spheniscus humboldti*) The Humboldt penguin is the least known of all the members of the genus *Spheniscus,* which includes the Magellanic, Galápagos, and African penguins. Humboldt penguins live almost exclusively in the long, narrow band of coastal water where the Humboldt Current comes into contact with the coasts of Peru and Chile. In northern Chile, the range of the Humboldt penguin overlaps that of the Magellanic penguin, and the two species sometimes interbreed. Overfishing of the Humboldt Current has caused a substantial decrease in the number of penguins, because not only has their food source been depleted, but the penguins are eaten by fishermen and also used for bait.

Humboldt penguin

See also African penguin, Galápagos penguin, Magellanic penguin

Humboldt squid (*Dosidicus gigas*) Also known as the jumbo squid, *Dosidicus* reaches a length of 10 feet (including tentacles) and a weight of 300 pounds. It is a fast, powerful hunter, inhabiting the waters from the west coast of South America north to Southern California. In 1940, members of a *National Geographic* big-game-fishing expedition to the Humboldt Current off Peru found themselves hooking these cannibalistic squids that came up their lines gnashing their beaks and squirting ink at them. In 1991, a team of filmmakers

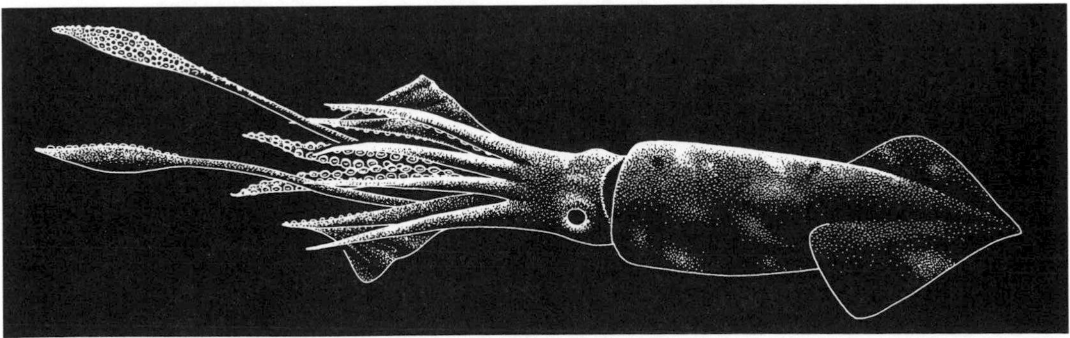

Humboldt squid

humpback whale

in the Sea of Cortez were attacked as they tried to film the Humboldt squid. There is a Japanese fishery for *Dosidicus* off the coast of Peru, and another in the Sea of Cortez. **See also giant squid**

humpback whale (*Megaptera novaeangliae*) Easily recognized by its long flippers and lumpy dorsal fin, the humpback is one of the best known of the baleen whales, but it is rare nonetheless. Because they are slow swimmers and breed in inshore waters, humpbacks were among the first species to be decimated when commercial whalers moved into a new area. From an estimated preexploitation population of perhaps 250,000, there are probably no more than 15,000 left in the world. Southern Hemisphere humpbacks come north from their Antarctic feeding grounds to calve in South African, South American, Australian, and New Zealand waters; the northern populations that breed in Hawaii and the Caribbean feed in Alaska and New England, respectively. Humpbacks are the object of whale watchers in New England, Hawaiian, Alaskan, and Japanese waters. Their spectacular feeding behavior, where many whales corral small fishes underwater and then lunge open-mouthed to the surface to engulf them, has been the subject of many films. Humpbacks are the whales that "sing"; their eerie, repetitive vocalizations have been recorded and analyzed since the early 1970s, when Roger and Katy Payne made *Songs of the Humpback Whale,* the best-selling animal recording of all time. (The humpbacks in Australia sing completely different songs than those in New England.) It is believed that only the males sing, but the mechanism by which they produce these whoops, yawps, moans, and gurgles has not been identified. Although they were long considered "gentle giants," it has recently been observed that male humpbacks fight viciously for females, butting each other with their barnacle-encrusted heads and slashing with their flippers.

See also whale watching, whaling

humuhumunukunukuapua'a (*Rhinecanthus aculeatus*) Pronounced "hoomoo-hoomoo-nookoo-nookoo-a-poo-a'a," this is a very long Hawaiian name for a very small triggerfish. Also known as the reef triggerfish, it reaches a maximum size of about 10 inches. It is found not only in Hawaii, but throughout the reefs of the Indo-Pacific region. **See also triggerfish**

Hurley, Frank (1885–1962) James Francis Hurley was born in Sydney. He ran away from home at age thirteen, worked at various jobs, and bought his first camera in 1905. He talked his way aboard Douglas Mawson's 1911 Antarctic expedition and was part of the team that set the sledging record of 41 miles in a day. He was the official photographer of Shackleton's *Endurance* expedition. When the ship sank on November 21, 1915, he could save only 120 of the more than 500 glass-plate negatives. He had to abandon his professional equipment and took some of his most impressive pictures with a Vest Pocket Kodak and three rolls of film. After the *Endurance* sank, Hurley continued to photograph the expedition and, eventually, its rescue. He remained on Elephant Island as Shackleton sailed the *James Caird* to South Georgia. (Hurley's remarkable photographs and movies were exhibited in 1999 at the American Museum of Natural History in New York, along with the *James Caird.*) He was an official Australian photographer in both world wars and undertook film ventures in New Guinea and central Australia. Between 1929 and 1931, he joined Mawson again on the joint British, Australian, and New Zealand Antarctic Research Expedition (BANZARE). Hurley remained a professional photographer until his death in Sydney. **See also *Endurance*, Mawson, Douglas, Shackleton**

hurricane Severe tropical storm of the Northern Hemisphere with winds that spiral inward counterclockwise. (In the Southern Hemisphere, where the storm is known as a typhoon, the winds spiral clockwise.) In the eye, there are low barometric pressures and light winds. It usually takes four to eight days for a hurricane to develop, and it can last for a week to ten days. They start as tropical depressions, then evolve into tropical storms, and when the wind speed reaches 71 mph, they are classified as hurricanes. These winds always cause very heavy seas and torrential rains, often reducing visibility to zero. The most disastrous hurricane on record hit Galveston, Texas, in 1900, killing six thousand people and causing $20 million worth of damage, but typhoons, which can occur in the western Pacific and the Bay of Bengal, have killed far greater numbers of people and destroyed much more property.
See also Beaufort Scale, cyclone, storm surge, typhoon

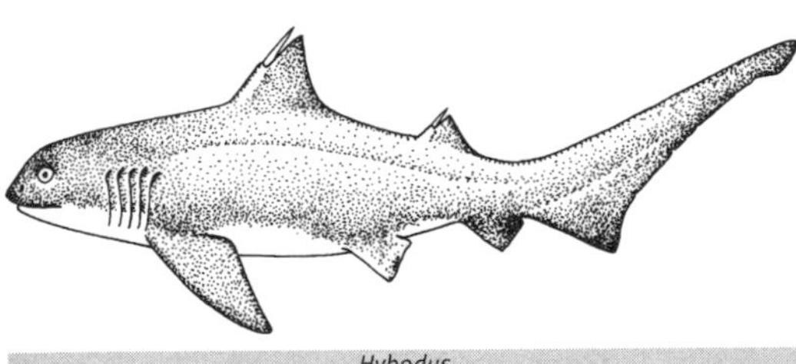

Hybodus

Hybodus A widespread, ancestral shark species of the late Permian to late Cretaceous periods, from 286 million years to 100 million years ago, *Hybodus* looked not unlike some recent sharks. It reached a length of 6½ feet, and had a markedly heterocercal tail, where the upper lobe is much longer than the lower. Like the hornsharks and dogfishes of today, it had a spine in front of each of its dorsal fins.
See also *Cladoselache*, dogfish, hornshark, shark

Hy-Brasil Legendary (and imaginary) island always said to be located off the west coast of Ireland. It began to appear on maps in 1325, usually as a round island, but sometimes bisected, like the two halves of a walnut. It was said that it was close enough to be sighted from shore, but it was always obscured by a fog bank. John Cabot searched for it on the 1497 voyage from which he never returned. It is believed to have been placed on maps by early cartographers, then copied regularly until 1835, when it disappeared forever. It was christened long before the South American country of Brazil, which is named for the red dyewood that was found there. **See also Atlantis, Cabot (John)**

hydroid A class of cnidarian (creatures with tentacles, nematocysts, and radial symmetry, also known as coelenterates), commonly called sea plumes and sea fans because of their branching patterns. The plumelike colonies can be 6 feet tall, but most are much smaller, and are manifested as white, pink, violet, or brown tufts that adhere to rocks, pilings, seaweeds, and many gastropod shells. They are often mistaken for seaweeds because of their plantlike appearance, but they are animals, and they feed voraciously on minute crustaceans, eggs, larvae, and even small fishes. Some hydroids can inflict painful stings if handled, but they should not be confused with stinging corals, which used to be grouped with the hydroids but differ in that they secrete massive limestone skeletons.
See also cnidarian, stinging corals

hydrothermal vents First discovered by scientists aboard the research submersible *Alvin* in 1977 at a depth of 8,000 feet in the Galápagos Rift Zone, hydrothermal vents are cracks in the seafloor at the juncture of two tectonic plates. Volcanic activity beneath the plates releases hot gases and dissolved minerals into the ocean, and heats the water to temperatures of nearly 700°F. At these vent sites—subsequently discovered along many other mid-ocean ridges—minerals are spewed into the water in clouds known as black smokers, which eventually dissolve and disperse into the water column. In the vicinity of these vents, completely unknown fauna were discovered, living not on oxygen, as every other known life-form does, but on hydrogen sulfide, a substance poisonous to most living creatures. These chemosynthetic (as opposed to those that are photosynthetic, i.e., able to process sunlight) life-forms include 8-foot-long tube worms with red, feathery plumes but no mouth and no gut; football-sized, snow-white clams with blood-red innards; ghost-white crabs; yellow mussels; floating "dandelions" that are related to the jellyfishes; and eyeless shrimp with light-detecting organs on their backs. They are nourished by symbiotic bacteria living inside them that are able to process the sulfides into a usable form. These creatures of the hydrothermal vents flourish in a pitch-black, superheated, sulfide-rich environment without any connection whatever with sunlight. They are as far removed from life as we previously understood it as life on another planet.
See also *Alvin*, rift shrimp, tube worms

I

Ibiza After Majorca and Minorca, Ibiza (also known as Iviza) is the third largest of the Balearic Islands off the east coast of Spain. It is the closest of the islands to the Spanish mainland; it is less than 100 miles southeast of Valencia in the western Mediterranean. Ibiza covers 221 square miles and has a permanent population of some 50,000 people. In ancient times, the island was inhabited by Phoenicians and Carthaginians, and was ruled by the Moors from the ninth to the twelfth centuries. In 1235, Catalan crusaders conquered the island, and it has been Catalan and Catholic ever since. Its mild climate and fine beaches have made it a popular tourist spot. In the 1960s it became known as a hangout for hippies and dropouts. It is now one of Europe's best-known destinations for 'round-the-clock parties, particularly in the capital city of Ibiza Town, also known as Eivissa. Only 2½ miles south of Ibiza is the tiny unspoiled island of Formentera, which also attracts vacationing Europeans, especially Britons, Germans, and Spaniards. **See also Majorca, Minorca**

Ibn Battuta (1304–c. 1368) On what started out to be a simple Muslim pilgrimage to Mecca and Medina, Ibn Battuta traveled a total of 75,000 miles from 1325 to 1354, making him the medieval world's greatest explorer. On land by foot and camel, he crossed North Africa from his birthplace in Tangier to Mecca, then voyaged out of the Red Sea and down the coast of East Africa as far as Mombasa and Kilwa (now Tanzania); he then returned to the Persian Gulf and crossed the Arabian Peninsula back to Mecca. He then wandered through Anatolia, Samarkand, and the steppes of central Asia, arriving in India in 1333. He spent seven years there, becoming a judge in Delhi. When Muhammad ibn Tugluk, the sultan of Delhi, appointed him ambassador to China, he was off again. In Calicut, a typhoon swept away the ships containing his gifts for the Chinese, so he went to Cathay on his own. He sailed down the Malabar coast, stopped briefly in the Maldives (where he was married and divorced six times), and then traveled by junk through Southeast Asia, passing through the Straits of Malacca before finally ending up in China. He returned to Morocco by way of Sumatra, India, Arabia, Persia, and Damascus, where he saw the Black Death kill 2,500 in a single day. After an absence of twenty-four years, Ibn Battuta had come home to Morocco. The sultan, Abu Aiman, provided him with a secretary who transcribed the records of his journeys, a process that took another thirty years. These records—three times longer than those of Marco Polo—are still considered the most reliable descriptions of the customs and geography of a period when most people spent their lives within sight of the house in which they were born.

ice Water in its solid form is ice, but it can also appear as hail, frost, and snowflakes. In glaciers it moves like a liquid; it flows downhill like a very slow river. Pure water freezes at 32°F. The presence of salts lowers the freezing point. When saltwater freezes, the dissolved salts are rejected, and even sea ice is nearly pure water. The temperature of ocean water must fall between 28°F and 29°F before it freezes, but the ice so formed will not melt until the temperature has risen to 32°F. Water is one of the three substances (the others are the elements antimony and bismuth) that expands when it freezes. If water contracted like almost everything else, it would sink to the bottom as it froze and stay there. It would not melt in the spring, because it would be on the bottom, where the sun's heat would not reach it. Soon all the oceans, lakes, and rivers in the cooler parts of the world would become solid ice. The tropics would become unbearably hot, and most of the rest of the world would be solid ice. **See also glacier, iceberg, sea ice**

iceberg A large piece of ice that has broken off a freshwater glacier. Pack ice is frozen saltwater that may break up into floating pieces. Pack ice is usually only a couple of years old, but glacier-born icebergs may have existed for thousands of years. Icebergs in the Northern Hemisphere come primarily from Disko Bay, Greenland; those in the Southern come mainly from the Antarctic ice sheet. Some fifteen thousand icebergs calve from Arctic glaciers each year, but very few of them get as far south as Newfoundland. (In 1907 and 1926, a combination of winds and currents carried icebergs as far south as Bermuda.) Most southern icebergs are concentrated south of the Antarctic Convergence at about 60° south. Arctic icebergs vary in size; those about the size of a piano are called growlers; up to the size of a small house, they are called bergy bits. Larger than that, they are full-fledged bergs. Those in the Antarctic can be enormous. Tabular bergs of 5 miles in length are not unusual, nor is a height of 150 feet above water. (The bulk of an iceberg, of course, is under water.) One of the largest Antarctic icebergs ever measured was 200 miles in length. **See also glacier**

icebreaker A ship with a reinforced bow and forefoot and with powerful engines, designed to force a way through pack ice in extreme northern latitudes. (Pack ice is less of a problem in the Antarctic.) Because so much of Russia is ice-choked during winter (and some parts all year round), the Soviets designed and built most of the world's icebreakers; the first of them were powered by steam, and later ones by nuclear energy. Russian icebreakers clear a shipping lane in the "northern sea route" that connects the Atlantic and the Pacific across the northern coastline of Siberia. In 1969 the *Manhattan,* an American tanker refitted as an icebreaker, plowed through the ice-filled Canadian Arctic from Greenland to Alaska, and then returned with a cargo of oil, thus making the first commercial transit of the long-sought Northwest Passage.

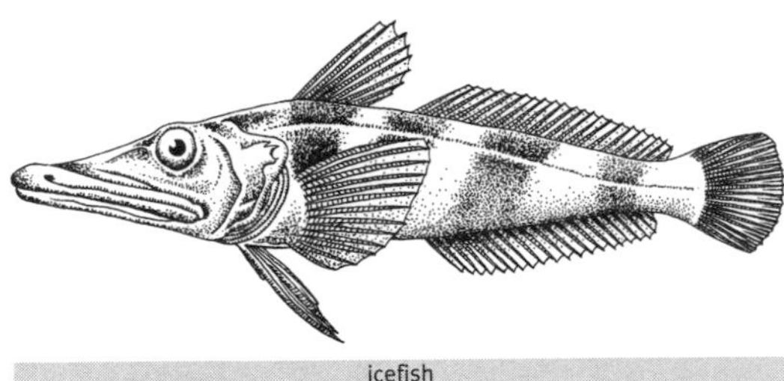

icefish

icefish (*Chaenocephalus* spp.) Icefishes have no scales, which is not so unusual, but they also lack hemoglobin, the substance that binds and transports oxygen molecules in the red blood cells, which is very unusual indeed. Their blood is colorless, and the gill rakers, pink or red in other fishes, are cream-colored. Instead of red blood cells, icefishes have a much larger quantity of blood than one would expect for a fish of this size (about 27 inches in length), accounting for as much as 9 percent of their body weight, as compared to 2 to 3 percent in most other fishes. Their blood also contains certain glycoproteins that prevent the blood from freezing at lower temperatures; these have been referred to as blood antifreeze. The metabolic rate of icefishes is also lower than that of comparably sized fishes that do not live in such harsh conditions. Active predators, they are equipped with a large mouthful of sharp teeth and are sometimes known as crocodile fish.

ice formation The first stage of seawater ice formation is "frazil," when platelets of ice float to the surface and form "slush" or "sludge." Circular floes with slightly raised edges are called "pancake" ice; "young" ice is unhummocked and about 1 foot thick. "Fast" ice is held to the shore and is often covered with snow. An ice "floe" is a free-floating area of ice whose complete extent is visible, and an ice "field" is an expanse that has no visible exterior borders. Larger bergs break up into bergy bits and growlers, and when swells and collisions with other ice formations result in fragmented bits of icy detritus, the result is "brash" ice. **See also sea ice**

Iceland An island country west of Norway and southeast of Greenland, Iceland covers some 39,000 square miles, a little less than Cuba. The population is about 260,000, of which 100,000 live in Reykjavik, the capital. In the southeast quadrant of the largely unoccupied interior is Vatnajökull, the largest glacier in Europe. It is believed that Iceland's first inhabitants were eighth-century Irish monks who sought isolation. Later came Norwegians escaping the political strife rampant in their homeland. Eric the Red, who grew up in Iceland as the son of a Norwegian exile, colonized Greenland in 982; his son Leif is popularly held to be the first European to explore the coast of North America. From the end of the fourteenth century until 1874, Iceland was under Danish rule; it became an independent state within the Kingdom of Denmark in 1918. When Denmark was occupied by Germany during World War II, Iceland requested independence, which was granted on June 17, 1944. Because it sits immediately on the Mid-Atlantic Ridge, Iceland is one of the most violently volcanic areas on earth. The island was formed by volcanic action, and if the two hundred active volcanoes, hot springs, geysers, and fumaroles were not evidence enough, there is always the offshore island of Surtsey, which burst from the sea in a cataclysm of steam and fire in 1963, and the town of Vestmannaeyjar on the island of Heimaey, which was half buried in flowing lava by the 1973 eruption of the Kirkjufell volcano, causing the evacuation of the town's 5,000 residents. Iceland's volcanic nature is also responsible for its power: geothermal heat and the waters from hot springs provide 90 percent of Reykjavik's heat and industrial energy. Iceland's economy depends on the sea, and fishing (for cod, hake, herring, and capelin) accounts for more than 80 percent of its export income. During the 1970s, Iceland and Britain were involved in the "cod wars," armed confrontations in Icelandic waters over fishing rights. Iceland's insistence on its right to kill whales despite restrictions and quotas caused it to resign from the International Whaling Commission in 1990.

ice worm (Hesionidae) Polychaete worms that were discovered in 1997 on a frozen methane outcrop 1,600 feet down in the Gulf of Mexico. Although the animal was not unusual, the habitat was. Methane hydrate, a kind of ice that forms when gas leaking from seafloor sediments combines with water molecules, was not considered conducive to life. The eyeless pink bristleworms, about 2 inches long, are believed to live on bacteria that feed on the hydrocarbons within the methane hydrate. They evidently make burrows into the frozen methane.

Like the hydrothermal vents, the methane outcrops are a completely unexpected life-supporting environment.

See also hydrothermal vents, polychaetes

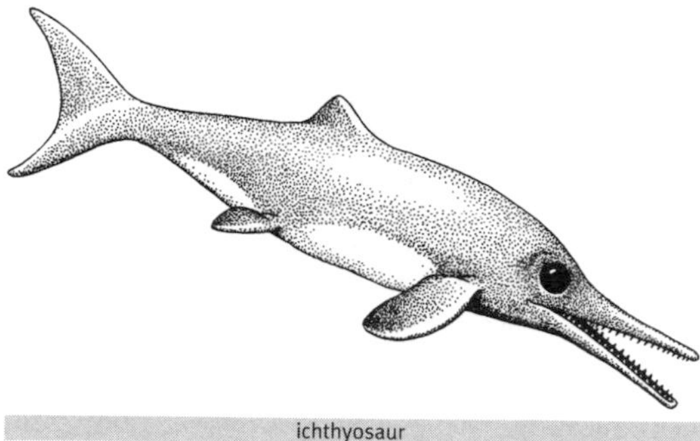

ichthyosaur

ichthyosaur An extinct marine reptile that flourished during the Mesozoic Era, between 65 million and 245 million years ago. There are several species of ichthyosaurs. The best known is *Ichthyosaurus*, which was dolphinlike in shape and habits, with a long, toothy snout, no visible neck, large eyes, and nostrils positioned far back on top of the skull. But while dolphins have only forelimbs, the ichthyosaurs also had small hind flippers, corresponding to the hind legs of terrestrial lizards. In the fossil beds of southern Germany, specimens have been found where not only bones were preserved, but also the outline of the body, which shows that *Ichthyosaurus* had a dorsal fin. They reached a length of about 10 feet (the size of a large bottlenose dolphin) and could probably swim quite fast. Most species of ichthyosaurs were smaller, but there were some true giants, such as *Shoniasaurus*, which reached a length of 45 feet, and an unnamed monster from British Columbia that was 75 feet long. From the fossil evidence, it is clear that ichthyosaurs gave birth to live young at sea. Why the ichthyosaurs died out is a mystery, but they might have been replaced by sharks.

See also mosasaur, plesiosaurs

Inca tern (*Larosterna inca*) A large, dark, atypical tern found only on the Pacific coasts of South America, from Ecuador to central Chile. It is an all-over blue-gray bird with a massive red bill and a handlebar mustache of white plumes that curve downward from the base of the bill. A gregarious bird, the Inca tern often gathers in the thousands. **See also tern**

Indianapolis U.S. cruiser, active during the war in the Pacific, including the invasions of Iwo Jima and Okinawa in March 1945. In July, she was used to carry elements of the atomic bombs that would be dropped on Hiroshima and Nagasaki from San Francisco to Tinian (Marshall Islands). On July 30, after leaving Tinian, the *Indianapolis* was torpedoed by a Japanese submarine and sank in twelve minutes, unable to send a distress signal. When the ship did not show up at Leyte in the Philippines, patrol aircraft were sent out, but they did not locate the survivors until August 2, and were not able to complete the rescue until August 8. Of her complement of 1,200 men, some 850 escaped the sinking, but they were left floating in shark-infested waters, and many who did not drown were killed by sharks. Of the 850 who did not go down with the ship, only 318 were rescued.

Indian Ocean The Indian Ocean, including the Red Sea and the Persian Gulf, is the third largest body of water in the world, representing approximately 20 percent of the world's ocean area. It covers some 28.4 million square miles, and its volume is approximately 70 million cubic miles. It is defined by Africa on the west, India and Indonesia on the north and northeast, and Australia on the east. To the south is Wilkes Land, on the Antarctic continent. The greatest depth of the Indian Ocean is the Java Trench, maximum depth 23,344 feet, where the Australian Plate is believed to be subducting beneath the Eurasian Plate. Immediately north of this trench is—or was—the island of Krakatau, which blew up in 1883 in one of the largest explosions in recorded history. The large islands of Madagascar and Sri Lanka are structurally part of the adjacent continents, and there are comparatively few islands in the Indian Ocean: the Seychelles, Maldives, Laccadives, Comoros, Mauritius, and Réunion, and, deep in the south, Kerguelen, Heard, and the Crozets. Vasco da Gama sailed around Africa in 1497 to cross the Indian Ocean and reach the shore of western India. Others who made extensive contributions to European knowledge of the Indian Ocean were Abel Tasman and James Cook. Though the floor of this ocean is rich in manganese nodules, the technology for mining these minerals has not been developed. The Persian Gulf, on the other hand, is the largest oil-producing area in the world. Twice a year, water drawn from the Indian Ocean is drawn into a northerly monsoon system that drops heavy rainfall onto the Indian subcontinent and Southeast Asia.

Indonesia The fifth most populous country in the world with more than 200 million people, Indonesia is composed of more than thirteen thousand islands that stretch for almost 3,000 miles along the equator from the tip of Sumatra, west of the Malaysian mainland, toward Australia. Previously the Dutch East Indies, Indonesia now includes the Greater Sunda Islands (Java and Sumatra); the Lesser Sunda Islands (Bali, Komodo, Flores, Sumba, Lombok, and the western part of Timor); the Moluccas (Ambon, Ceram, Halmahera); central and south Borneo (Sabah and Sarawak); Celebes (known as Sulawesi); and Irian Jaya, the western half of the island of New Guinea. Abundant rainfall and rich volcanic soils permit a rich agricultural yield, but crude oil is Indonesia's most valuable resource. It is also the world's largest producer of rubber. About 87 percent of the population is Muslim; 9 percent are Christian, and 2 percent are Hindu. More than three hundred languages are spoken, but Bahasa Indonesia, a pure form of Malay, has been adopted as the national tongue. By the eighth century A.D., traders from India had established colonies on Sumatra and Java, but eventually they were overpowered and displaced by Arab traders, who introduced Islam to the islands. In 1511, the Portuguese landed at Molucca; the English arrived in 1596, and the Dutch in 1600. After a series of Anglo-Dutch conflicts (1610–1623), the Dutch emerged as the dominant power in the "East Indies," and remained in control until the twentieth century. During World War II, the Japanese took many of the islands, but after the surrender, Sukarno (1901–1970) proclaimed Indonesia's independence from the Dutch and began the long process of consolidating the country. He made himself president for life in 1963, and withdrew Indonesia from the United Nations in 1965 in opposition to the seating of the Federation of Malaysia. In the same year, he was overthrown by General Suharto, who ousted the Communists and had hundreds of thousands of them executed. In 1975, when the Portuguese half of the island of Timor seceded, Indonesian troops invaded the island and began a war that killed tens of thousands of people and ended in 1999.

Indo-Pacific beaked whale (*Indopacetus pacificus*) Very little is known about many of the beaked whales, but *Indopacetus pacificus* is the least known of all. Most of the other species have stranded in greater or lesser numbers around the world, but this is the rarest of the beaked whales and, therefore, one of the rarest large animals in the world. The only evidence of its existence are two skulls, one that washed ashore in Queensland, Australia, in 1882, and another that was discovered in Somalia, on the east coast of Africa, in 1955. The Queensland skull is almost 4 feet in length, so we can assume that this is one of the larger of the beaked whales. In 1999, several cetologists published a paper suggesting that the Indo-Pacific beaked whale might not be as poorly known as had been believed. They correlated many sightings of heretofore unidentifiable beaked whales with a single species, which they tentatively named the tropical bottlenose whale.

See also beaked whales, bottlenose whale

Indo-Pacific humpback dolphin (*Sousa chinensis*) Although the humpback dolphin is an inshore species, found from South Africa throughout the Indo-Pacific region as far east as the South China Sea, it is very little known. It is even unclear if there is a single species

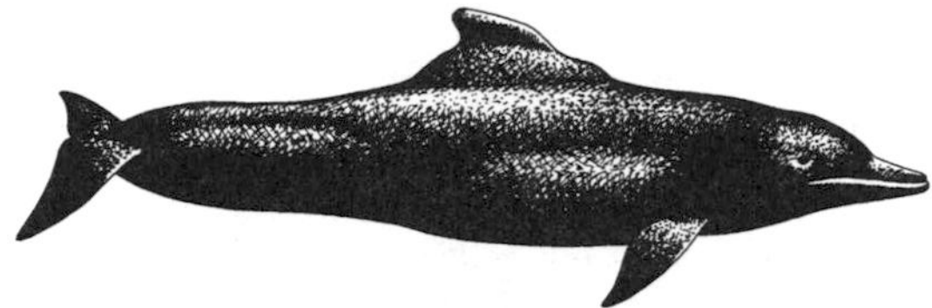

Indo-Pacific humpback dolphin

throughout this vast range, or as many as three. Some scientists assign full species recognition to the white dolphin (*S. borneensis*), the lead-colored dolphin (*S. plumbea*), and the freckled dolphin (*S. lentiginosa*), while others claim they are all color morphs of *S. chinensis.* However many species there are, they are good-sized dolphins, reaching a length of 9 feet and a weight of 600 pounds. As with the Atlantic (West African) humpback dolphin, the dorsal fin of the Indo-Pacific species sits on a pronounced ridge on the back, which is particularly noticeable when the animal arches its back to dive. Some populations have little or no ridge, which makes it easy to confuse them with bottlenose dolphins, which share most of their range.

See also Atlantic humpback dolphin

Indus River dolphin

Indus River dolphin (*Platanista minor*) This species shares many of the characteristics of its eastern relative, the Ganges River dolphin, but there are minor skeletal differences, as well as the geographical intrusion of the entire Indian subcontinent. Early observers believed that these dolphins grubbed in the mud for food, but they are excellent echolocators and can easily find and catch swimming fishes, despite the almost complete absence of visual apparatus. Both Indian freshwater dolphins have suffered major population losses as a result of hunting and the building of dams in their native rivers, but the numbers of *P. minor* have been reduced to less than three hundred animals, making it one of the most endangered cetaceans in the world.

See also freshwater dolphins, Ganges River dolphin

International Council for the Exploration of the Sea (ICES) Established in Copenhagen in 1902, ICES is the oldest intergovernmental organization in the world concerned with marine life and fisheries. Each year, ICES holds more than a hundred meetings of its various working groups, study groups, workshops, and committees. Since the 1970s, a major area of the council's work as an intergovernmental marine science organization is to provide information and advice to member country governments and international regulatory commissions (including the European Commission) for the protection of the marine environment and for fisheries conservation. Membership has increased from the original eight countries in 1902 to the present nineteen, which come from both sides of the Atlantic and include all European coastal states except the Mediterranean countries from Italy eastward. In many instances emphasis is placed on the influence of changes in hydrography (e.g., temperature and salinity) and current flow on the distribution, abundance, and population dynamics of finfish and shellfish stocks. These investigations are also relevant to marine pollution studies because physical oceanographic conditions affect the distribution and transport of contaminants in the marine environment. In support of these activities, the ICES secretariat in Denmark maintains three data banks: the Oceanographic data bank, the Fisheries data bank, and the Environmental (marine contaminants) data bank. The North Pacific Marine Science Organization (PICES) was created in 1992 from the ICES model.

international date line A line that runs mainly along the 180th meridian of longitude, but with adjustments to avoid bisecting certain islands or island groups that lie along that longitude, including the Aleutians in the Northern Hemisphere, and Fiji, Tonga, the Kermadec Islands, and New Zealand in the south. If a traveler sets out around the world in an easterly direction from Greenwich (0 degrees longitude), he advances his clock by one hour for every 15 degrees of longitude, so by the time he has reached the 180th meridian, he has put his clock forward by one hour twelve times. If he were to continue east until he reached Greenwich again, he would have advanced his clock twenty-four hours, and would be one day ahead of everyone else in Greenwich. To correct this anomaly, he changes the date upon crossing the international date line, subtracting a day when he crosses it eastward and adding a day when he crosses it westward. When you cross in an easterly direction, it is the same calendar day for two days in succession; crossing westward eliminates one day completely. The international date line passes between Big Diomede and Little Diomede Island in the Bering Strait; when it is Saturday on Big Diomede, it is Sunday on Little Diomede, 2½ miles away.

International Game Fish Association (IGFA) An organization based in Dania, Florida, that was established in 1939 to establish ethical international angling regulations and to serve as the processing center for data on world-record catches. In order to be recognized in the IGFA record book, the angler must submit data on the length and weight of the fish, where and when caught, bait and equipment used, and a photograph of the fish to verify the species. The IGFA Executive Committee then rules on the submission and decides whether or not to accept it. There are categories for saltwater and freshwater fishes, and for every line class, from 2-pound test to "all-tackle," which includes everything from 2-pound to 130-pound test line. Fly-fishing is a separate category. The IGFA publishes an annual volume that includes all world records.

International Whaling Commission (IWC) Originally chartered in 1949 as an organization whose stated purpose was to insure that there were enough whales so that the whaling industry could continue killing them, the IWC oversaw the most massive destruction of whales in history. With a membership drawn only from the whaling nations, the IWC set rules and catch quotas that annihilated one population of whales after another. When the Antarctic blue whales were too scarce to hunt, the IWC authorized the commensurate slaughter of the fin whales, and then the sei whales. During the late 1960s, with the blessing of the IWC, Soviet and Japanese factory ships were processing some 25,000 sperm whales per year, more than the entire New England whaling fleet had taken in fifty years. By the early 1970s, this needless slaughter had been brought to the world's attention, and various conservation organizations led the movement to "save the whales." (There was never an actual organization named Save the Whales.) Although they struggled mightily to stay in business, the whaling nations quit one by one, first Britain, then the Netherlands, and finally Japan and Russia. In 1982, the IWC passed a moratorium, effectively putting an end to commercial whaling. Japanese whalers still continue to kill a small number of whales every year under the rubric of "research whaling," and Norway authorizes the killing of some five hundred minke whales annually because they never agreed to the moratorium when it was passed. The International Whaling Commission has now become an organization that supervises hardly any whaling, and its ranks have been filled with conservationists as well as whalers.

See also sperm whaling, whaling

invertebrate Any animal that lacks a vertebral column or backbone, as contrasted with vertebrates, which have some sort of cartilaginous or bony internal skeleton. Invertebrates account for more than 90 percent of all living animals, and most invertebrates are insects. Invertebrates have no special shape; they share no common characteristics except the lack of a backbone. They vary in size from microscopic protozoans to the giant squid. There are very few marine insects, but invertebrates are incredibly abundant and varied in the sea. They include protozoans, sponges, corals, worms, sea anemones, jellyfishes, barnacles, clams, oysters, scallops, sea urchins, sea cucumbers, snails, nudibranchs, octopuses, squids, crabs, lobsters, and shrimps.

Ionian Sea Part of the Mediterranean between Greece and southern Italy, connected to the Adriatic to the north by the Strait of Otranto. One can sail through the Gulf of Corinth, between the Greek mainland and the Peloponnese, to get to the Aegean. Corfu and the other Ionian Islands (Paxoi, Cephalonia, Zacynthus, Ithaca, Levkas, and Kythera) are in the eastern quadrant. The Battle of Lepanto, where the Turks were defeated by the combined forces of Spain, Venice, Genoa, and the Papal States, was fought just off the western entrance to the Gulf of Corinth in 1571.

See also Cephalonia, Zacynthus

Ipnops murrayi The first specimen of *I. murrayi* was dredged up from 11,400 feet by the *Challenger* expedition of 1872–1876. Related to the tripod fishes, this 6-inch fish has no eyes but, instead, a continuous flat, cornealike organ that covers the upper surface of its flattened snout and is thought to be light-sensitive. This species is hermaphroditic, meaning that adults have both male and female organs, so in their lightless environment, if they meet another *Ipnops,* they can perform the role of the male or the female, whichever is required. **See also tripod fish**

Ipnops murrayi

Ireland An island nation separated from Great Britain by the Irish Sea, Ireland is 302 miles long from north to south and about 171 miles wide, covering some 27,000 square miles. The island consists of a low central plain, almost entirely ringed by coastal highlands. The climate is wet and mild, and the better part of the country is green all year round. Saint Patrick arrived in Ireland in the fifth century and Christianized the entire country. Subsequently, after a long history of conquest by Celts, Norsemen, Normans, English, and Scots, the island now has a remarkably heterogeneous Roman Catholic population, numbering around 3.5 million. In 1801, by the Act of Union, England and Ireland were joined as the United Kingdom of Great Britain and Ireland. In 1846 a catastrophic famine resulted in the death or emigration of half the population. By the time World War I began, the Irish were battling for home rule. The Easter Rebellion of 1916 failed, but continued agitation resulted in the establishment of the Irish Free State in 1922. The six northern counties remained part of the United Kingdom, and today, the problems of Northern Ireland—"the Troubles"—have not been resolved. During World War II, Ireland remained neutral. Eamon De Valera (1882–1975), who led the Irish in their fight for independence from Britain and served as prime minister from 1932 onward, was defeated in the 1948 elections, and in 1949, Ireland withdrew from the Commonwealth. Throughout the 1960s, Ireland was torn by internal politics, with the Irish Republican Army (IRA) wreaking havoc as it tried to bring Northern Ireland into the republic and oust the British soldiers. In 1972, British prime minister Edward Heath suspended the constitution and parliament of Northern Ireland. Since this predominantly Protestant region has a large Catholic minority, the struggle is also being fought on religious grounds. Under Sean Lemass, prime minister from 1959 to 1966, Ireland's stagnant economy was revived, bringing social and cultural changes to what had been one of the poorest and most backward countries in Europe. In 1987, Ireland joined the European Community (since 1993 the European Union), but it still lags behind other countries in economic progress. Mary Robinson, Ireland's first woman president, was elected in 1991, and in 1995, a referendum allowing divorce, previously forbidden, under certain conditions was narrowly passed. Ireland's major industry is tourism, with thousands of visitors coming to the countryside and the urban centers of Dublin, Limerick, Cork, and Galway and buying woolens, crystal, linen, and laces. Ireland's most significant export is probably her literature, represented by such literary giants as Jonathan Swift, Edmund Burke, Oscar Wilde, William Butler Yeats, James Joyce, George Bernard Shaw, and Samuel Beckett.

Irian Jaya Indonesian province occupying the western half of the island of New Guinea. (The eastern half is the independent country of Papua New Guinea.) The province is inhabited by hundreds of tribes of Papuans, each with its own language and customs. The Dutch claim to the western half of the island was recognized by Britain and Germany in treaties of 1885 and 1895, but even after Indonesian independence in 1949, the Dutch retained control of western New Guinea. In 1962, Indonesian troops landed on the island to take it by force, and in 1963, the Netherlands agreed to interim administration by the United Nations, followed by transfer to Indonesia—providing the people agreed. In 1969 a plebiscite was held, and the province officially became Irian Jaya—*Irian* is Indonesian for the island of New Guinea, and *jaya* means "glorious" or "victorious."

See also Indonesia, New Guinea

ironclad General term referring to ships whose wooden hulls were covered with iron plates as a means of defense against artillery shells. In the age of wooden warships, the ships were not usually destroyed or sunk by shellfire but were reduced to a shambles by shells or shot that tore their masts and sails away, caused them to burn, and led to the surrender of the crew. When cannons were fired directly into the hulls of opposing ships, some sort of protection was needed. Ironclads were introduced around the middle of the nineteenth century, particularly after the Battle of Sinope in 1853, when the wooden-hulled Turkish fleet was completely demolished by the guns of the Russians. Experimental wrought-iron hulls proved to be unsatisfactory, because this material became brittle at low sea temperatures and lacked resistance to the direct impact of shells. Built in 1859, the French frigate *Gloire* was the first true ironclad warship; she led to the introduction of ironclad warships into the navies of England and Russia. In 1862, during the U.S. Civil War, the Union navy sent the ironclad gunship *Monitor* into battle against the Confederacy's *Merrimack* off Hampton Roads, Virginia, an inconclusive battle that essentially led to the employment of armor for most warships and the eventual development of the iron hull.

See also dreadnought, *Merrimack*, *Monitor*

Irrawaddy River dolphin (*Orcaella brevirostris*) *Orcaella* means "little orca," a name probably derived from the beakless profile, not from its ferocity. This little grayish-blue dolphin is never found far out to sea, but is restricted to riverine and estuarine habitats in southeast Asia and Australasia. Irrawaddy River dolphins have been observed spitting mouthfuls of water as they surface, but the purpose of this peculiar behavior is not known. These animals have been exhibited and trained in an Indonesian oceanarium, and in 1979, the first *Orcaella* calf was born in captivity.

See also freshwater dolphins

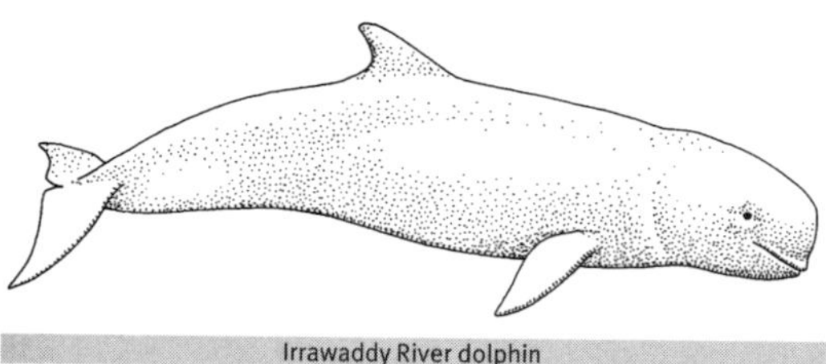

Irrawaddy River dolphin

Ismay, Thomas H. (1837–1899) British shipowner, one of the founders of the White Star Line. In 1868, he realized that square-rigged clipper ships were obsolete and that the future of shipping lay in steam. He established the Oceanic Steam Navigation Company, and in 1870 launched the *Oceanic*, the ship that was to set the standard for luxury travel across the Atlantic. Ismay's son Bruce was aboard the *Titanic* on her maiden (and only) voyage in 1912, and unlike 1,489 of the other passengers, he managed to get into a lifeboat.

See also *Titanic*, White Star Line

isopod (order Isopoda) "Isopod" means "similar feet," even though their legs are rarely the same length. Although it is far from universally applicable, a good rule for differentiating isopods from amphipods is that isopds are flattened from top to bottom, whereas amphipods are flattened from side to side. Moreover, in the isopods, the back is arched and the thorax is divided into six or seven segments. These crustaceans usually have six pairs of legs, but some have only two. On land they are the familiar wood lice, pill bugs, and sow bugs, but in the ocean, isopods reach the full breadth of their incredible diversity. Some are parasitic on fishes and other crustaceans; the gribble (*Limnoria lignorum*) is a tiny isopod that lives in submerged wood, destructively chewing its way through piers and pilings throughout the world. The genus *Sphaeroma* is named for its habit of rolling itself up into a ball when threatened. Other species are eyeless but have elongated legs and feelers that enable them to find food in the abyss. Found at depths of 4,000 feet or more, the giant isopod (*Bathynomus giganteus*) is the largest of its kind, and also, at a length of 14 inches, one of the largest of all crustaceans.

See also amphipod, copepod, crustaceans

ivory gull (*Pagophila eburnea*) True to its name, the ivory gull is the only gull that is pure white. It is found in the high Arctic, in the vicinity of ice-filled water, except during the breeding season, when it nests on cliffs or on flat islands in northern Canada, northern Greenland, Spitsbergen, Franz Josef Land, and Novaya Zemlya. Because of the difficulty of access to the nesting sites, little is known about the biology of the ivory gull.

ivory gull

See also Ross's gull

Iwo Jima One of three islands in the Volcano group, south of the Bonin Islands and north of the Marianas. (The other two are Kita-jima and Minami-jima.) Dominating the scrubby, barren island is Mount Suribachi, an extinct volcano 546 feet high, but there is a large flat area on which the Japanese built a military airfield during World War II. In late 1944, American bombers needed a base closer to Tokyo than Saipan, which was 1,500 miles away, so the decision was made to take Iwo Jima and use the airfield for bombing raids. American landing forces found almost 21,000 Japanese troops dug into underground tunnels and bunkers, but the marines landed on February 15, 1945, and after fierce fighting (six thousand marines were killed, as well as almost all the Japanese defenders), the American flag was raised on Mount Suribachi. The bombing of Tokyo continued from Iwo Jima and other bases in the Marianas, and led to the attack on Okinawa, a Japanese base even closer to Tokyo.

See also Bonin Islands, Mariana Islands, Okinawa

Ixtoc blowout After the intentional discharge of 250 million gallons of oil into the Persian Gulf by Iraq as its troops retreated during the 1991 Gulf War, the Ixtoc blowout is the second largest oil "spill" in history. On June 3, 1979, in the Gulf of Mexico, off Ciudad del Carmen, the Ixtoc well site blew and caught fire. Petroleum Mexicanos (Pemex) first estimated that the fire was burning 400,000 gallons of oil per day, but in fact the fire was only burning natural gas, and the oil was flowing out on the surface, forming a slick that grew to 100 miles long and 50 miles wide—and heading west, toward the mainland. Attempts to cap the well failed. Pemex drilled two wells close to Ixtoc in an effort to relieve the pressure, but the fire kept burning. The slick reached the Mexican coastline, destroying the shrimp fisheries there. In a much diluted form, it also reached the coasts of Texas, Louisiana, and Florida. Pemex injected 100,000 tennis-ball-sized steel and lead spheres in a heavy solution into the well, which cut the outflow to 300,000 gallons a day. Then they tried to lower a 310-ton steel "sombrero" over the well. Although this decreased the flow, the fire was still burning ten months after the blowout. On March 23, 1980, famed oil-well firefighter Red Adair dropped the third gigantic cement plug into the well from the adjoining two wells, and the burning gusher was sealed off. The cleanup cost was estimated at close to $400 million. The already-endangered Kemp's ridley sea turtle lays its eggs only on the beach at Rancho Nuevo in Mexico, just south of Brownsville, Texas, and because the hatchlings would have to swim through the oil in July and August, conservationists airlifted ten thousand baby turtles to an oil-free region of the gulf. **See also oil spill**

J

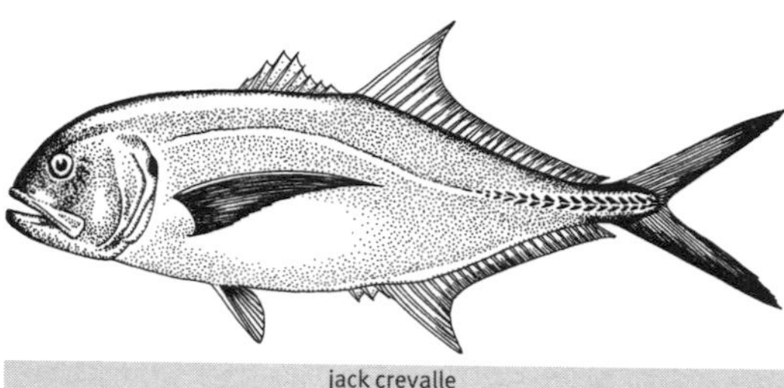
jack crevalle

jack crevalle (*Caranx hippos*) There are more than two hundred species of Carangids, or jacks, and they are found in all the oceans of the world. They are all fast, powerful predators, and many of them are fished commercially. The jack crevalle (also known as crevalle jack, cavally, and horse cravalle) is a common schooling species of the western North Atlantic, found from Nova Scotia to Uruguay. Jacks often feed by herding the prey fish into a tight mass and then dashing in to feed. They are a favorite among light-tackle fishermen, who take them by fly-fishing, spinning, trolling, or surf casting. The record jack crevalle, caught off Angola in 1992, weighed 57 pounds.

See also amberjack, permit, pilot fish, pompano, rainbow runner

jacks (family Carangidae) A large family (about two hundred species) of fast, open-water fishes, usually characterized by a teardrop shape, a caudal keel, a forked tail, and the absence of finlets. They occur around the world in tropical and temperate seas, schooling when young but often swimming alone as adults. The amberjack (*Seriola dumerili*) of the California coast is a popular game fish that can reach a weight of 150 pounds; another well-known jack is the yellowtail (*S. dorsalis*). The sleek rainbow runner (*Elagatis bipinnulata*) is a jack, and so is the 1-foot-long pilot fish (*Naucrates ductor*). Deeper-bodied jacks are the permits, palometas, and pompanos, and in the roosterfish (*Nematistius pectoralis*) the dorsal spines are almost as long as the fish itself. The horse mackerel (*Trachurus trachurus*) and the jack mackerel (*T. symmetricus*) are jacks, not mackerels.

See also amberjack, jack crevalle, permit, pilot fish, pompano, rainbow runner, roosterfish, yellowtail

jaeger: See skua

Jamaica Island in the West Indies, 90 miles south of Cuba and 100 miles west of Haiti. At 4,411 square miles, it is somewhat smaller than Connecticut. It was inhabited by Arawak Indians when Columbus arrived in 1494 but they died from diseases introduced by the Europeans. (Columbus named the island St. Iago, but the Arawak name was Xamayaca.) The island remained under Spanish rule until 1655, when the British claimed it. Port Royal was the de facto capital because buccaneers like Henry Morgan made it their base for plundering the Caribbean, but the city was destroyed by a massive earthquake and tsunami in 1692. Slaves were brought from Africa to work the sugar plantations, but the emancipation of the slaves and a drop in sugar prices brought about conditions that led to an uprising in 1865. The change in Jamaica's status from possession to colony improved conditions somewhat, as did the introduction of a new crop: bananas. Jamaica is also known for the production of coffee, ginger, citrus fruits, cocoa, and tobacco. Bauxite, the ore from which aluminum is made, is Jamaica's most important export. In 1953, Jamaica gained internal autonomy. Five years later, it led in organizing the West Indian Federation. Sir Alexander Bustamente led the campaign for withdrawal from the federation, and after a national referendum, Jamaica became independent in 1962. The capital of Jamaica is Kingston, but Montego Bay, Ocho Rios, and Negril are known for their world-class tourist facilities. The Rastafarian movement, a religion based on the divinity of Haile Selassie (the emperor of Ethiopia, who died in 1975; his name before he was crowned was Ras Tafari), began in Jamaica in the 1950s and still flourishes there. Many of the political tenets of Rastafarianism are proclaimed in reggae music, made popular by such singers as Bob Marley.

James, Thomas (1593–1635) British navigator and explorer who set out to seek the Northwest Passage in 1631, almost simultaneously with Luke Foxe. Although it had been suggested that they combine their expeditions, James left from Bristol two days before Foxe sailed from the Thames. He met with the self-styled North-West Foxe in Hudson Bay, and the two dined together aboard James's ship, the *Henrietta Maria.* James then traveled into the southern reaches of Hudson Bay, giving the name of his ship to the cape that he passed on the way into what is now James Bay, the southern extension of Hudson Bay. With their ship locked in the ice, James and his crew spent the winter of 1631–1632 on Charlton Island, subsisting on the ship's dwindling stores. Many men died of starvation and scurvy, and it was not until spring that the ice broke up enough to allow them to escape and sail back to England. Upon his return, James had to report that there was no passage from southern Hudson Bay to the South Sea. In 1635 he published *The Dangerous Voyage of Captain Thomas James.* **See also Baffin, Bylot, Foxe**

Jan Mayen Island in the Arctic Ocean, due north of Scotland; about 300 miles north of Iceland and the same distance east of Greenland. The island is 25 miles long and 9 miles wide, and covers 145 square miles. It is the peak of a submarine ridge that culminates in the forbidding, 7,470-foot-high Beerenberg volcano, which last erupted in 1732 and is Norway's only active volcano. The island was discovered in 1607 by Henry Hudson as he was searching for the northeast passage to Asia. By 1614, after Dutch sea captain Jan Jacobsz May claimed the island, whalers discovered the rich whaling grounds in the vicinity and contested with the British for the bowhead whales there. The Dutch tried to establish a shore-whaling station on Jan Mayen, but there were no available harbors, so in 1622, they settled Smeerenburg ("blubber town") on Spitsbergen. The island remained uninhabited until 1882, when the Norwegians built a radio station and meteorological observatory. Norway annexed the island in 1929, and during World War II, the United States maintained a weather station there. A NATO airstrip and radio station were erected on Jan Mayen in 1959, and other than station personnel and weather-station staff, no one lives on the Norwegian-owned island.

Japan Island nation in the western Pacific, consisting of four main islands: from north to south, Hokkaido, Honshu, Shikoku, and Kyushu. Numerous smaller island groups are also part of Japan, among them the Bonins, the Ryukyus (including Okinawa), and the Volcano Islands (including Iwo Jima). With adjacent Yokohama, Tokyo is the most populous city in the world, with a population approaching 30 million. The population of the entire country, with a total area of 145,000 square miles (about the same as Montana), is 125 million. The main islands are mountainous; the tallest and most famous mountain is Mount Fuji, 12,389 feet high. Only an eighth of the total land is arable, which means that much of the food must be imported or fished from the sea. From the fifth century onward, Japan was ruled by an emperor, but the real power lay with the shogun. The first European contact occurred in the mid-sixteenth century, when Portuguese and Spanish traders and Jesuit missionaries arrived, followed by the English and the Dutch. (Japan was the fabled "Cipango" that the Europeans were seeking.) Trade with Westerners was forbidden until 1853, when Commodore Matthew Perry sailed into Tokyo Bay. Japan quickly made the transition from a feudal state to an industrialized one, conscripting an army, creating a navy and a merchant fleet, and expanding its territory and influence. It defeated Russia in the Russo-Japanese War (1904–1905), acquired Formosa (Taiwan) and the Pescadores Islands, and annexed Korea in 1910. During World War I, Japan seized Germany's Pacific Islands and controlled large areas of China. In 1937, Japan invaded China, and in 1941, it attacked Pearl Harbor in Hawaii, setting off the world war in the Pacific. After a crushing defeat, which included atomic bombs dropped on Hiroshima and Nagasaki (the only nuclear weapons ever used in war), Japan surrendered on September 2, 1945. It was forced to cede Manchuria to China; the Pacific islands to the United States; and Sakhalin and the Kuril Islands to the Soviet Union. Since then Japan has rebuilt its economy to the point that it is the second greatest economic power in the world, after the United States. Even though virtually all the raw materials have to be imported, Japan is one of world's leading producers of motor vehicles and steel, and a primary exporter of electrical and electronic appliances. It is also prominent in fishing and fishing technology to feed its people, but when other countries established economic zones to keep foreign fleets out, Japan was forced to concentrate on its own coastal and inland waters.

Japan, Sea of Arm of the North Pacific Ocean that is bounded on the east by the islands of Japan and the Russian island of Sakhalin, and by Russia and Korea on the west. It covers a surface area of approximately 377,600 square miles, and it is a fairly deep basin, averaging around 5,000 feet in depth. Known as Nihon-kai in Japanese, and Yaponskoye More in Russian, the Sea of Japan is separated from the East China Sea to the south by the Tsushima Strait, and from the Sea of Okhotsk to the north by the La Perouse Strait. This semienclosed body of water contributes to the mild climate of Japan because of the steady evaporation of its relatively warm waters. Vladivostok, eastern Russia's only year-round, ice-free port, is on the Sea of Japan, just north of the Korean border.

Japanese flying squid (*Todarodes pacificus*) Also known as the common squid or the Japanese red squid, *T. pacificus* is the object of the world's largest directed squid fishery. It is an oceanic species that reaches a maximum total length of about 30 inches, including arms. In the 1950s, the fishing was done on a small scale, but it has been improved into a multimillion-dollar industry, with larger boats and mechanized fishing devices. In Japanese, Siberian, and Alaskan waters, hundreds of millions of tons of these squids are caught every year for human consumption. They are either dried, frozen, or delivered fresh to the markets. **See also squid fishing**

Java Indonesian island south of Borneo. Although it is only the fifth largest of Indonesia's thirteen thousand islands, Java contains two-thirds of the country's population. Indeed, it is one of the most densely populated areas in the world, with more than 100 million people living in an area the size of Alabama. Jakarta (formerly known as Batavia), the capital of Indonesia, is located on Java; it has a population of 8.25 million. Much of the central highlands of the island are planted in teak, accounting for much of Indonesia's timber production. After 1619, the Dutch East India Company's settlement at Batavia became the trading and shipping center for the East Indies. The disastrous Allied defeat in the Battle of the Java Sea in February 1942 left the island wide open for Japanese occupation. When the Japanese left the island in 1945, there was much fighting between Dutch and Indonesian forces, but the creation of the nation of Indonesia in 1950 drove the Dutch out.

See also Borneo, Indonesia

jawfishes (Opistognathidae) Small, robust, blenny-like fishes, with large heads, large mouths, and large eyes. The dorsal fin is continuous, and the tail is either rounded or pointed. They live in burrows of their own construction, and many species spend the day hovering a few inches above the opening of their burrows, ready to dash in at the first sign of danger, often tail first. There are about ninety species, found throughout the shallow Atlantic and Pacific Oceans. **See also blenny**

jawless fishes Officially classified in the class Agnatha, order Myxiniformes, the jawless fishes include the hagfishes (Myxinidae) and the lampreys (Petromyzonidae). Considered the most primitive of the true vertebrates, they are eel-like in form, with a cartilaginous or fibrous skeleton that has no bones; they have no paired fins, no scales, and no jaws or teeth. Lampreys possess eyes while hagfishes have only vestigial optic organs under the skin. Lampreys feed by attaching themselves to living fishes by means of a sucker-disk mouth, while hagfishes—also known as slime eels because they can emit a large quantity of noxious mucous—feed mostly on carrion by rasping their way into the body of a dead fish and eating it from the inside out.

See also hagfish, lamprey

Jaws A 1974 novel by Peter Benchley that Steven Spielberg made into a very scary, enormously successful movie. It is the story of a huge great white shark that terrorizes the beaches of the fictional town of Amity on Long Island, New York, eating people and sinking boats, until it is hunted and dispatched by an intrepid shark hunter named Quint. Because the movie was such a bonanza, there were three sequels, none of which lived up to the original. The theory that a "rogue shark" would become fond of human flesh and become a man-eater is pure fiction, but many people thought it was fact, and the novel and the original movie set off a shark mania where people were afraid to go swimming and many sharks were unnecessarily slaughtered.

See also great white shark, shark attack

Jellicoe, John R. (1859–1935) British admiral of the fleet who entered the Royal Navy in 1872. During the Boxer Rebellion (1900) he was shot in the lung but survived. He was promoted to rear admiral in 1907. As second sea lord, he was placed in command of Britain's Grand Fleet, in Scapa Flow, and flew his flag from the super-dreadnought *Iron Duke.* Jellicoe commanded the British fleet against the German High Seas Fleet at the Battle of Jutland, a large, complicated, but ultimately indecisive engagement that was fought off Denmark on May 31, 1916. Both sides lost many ships, but Britain probably fared better, because Germany never threatened in the North Sea again. After America entered the war in 1917, Jellicoe introduced the convoy system, which protected Atlantic shipping from German U-boats. Conflicts with Prime Minister David Lloyd George led to his abrupt dismissal at the end of 1917, and he was sent on missions to India, Australia, New Zealand, and Canada, to advise these colonies on how they might contribute to Britain's war effort. One result of this mission was the construction of a large naval base at Singapore, to service Britain's Pacific fleet. For his service as governor-general of New Zealand from 1920 to 1924, he was given an earldom.

See also Jutland, Battle of; Scapa Flow

jellyfish Common name for the invertebrate animals of the class Schyphozoa (phylum Cnidaria), characterized by a transparent or translucent gelatinous body. There are some two hundred species, ranging in size from 3/50 inch to 6½ feet in diameter. They live as polyps for a brief time, but they spend most of their lives as medusae: a bell-like organ above a cluster of tentacles. The bodies of jellyfish are about 99 percent water, and when they wash up on shore they dry out quickly and die. They usually drift with the currents, but they can move through muscular contractions of their bells. Beneath the bell is a more or less tubular projection, with a mouth at the free end. The upper surface of the bell is armed with nematocysts (stinging cells), as are the tentacles. Jellyfish have light-sensitive organs (eyespots) and balance organs at the base of the tentacles. They feed on fishes and other animals that come close enough

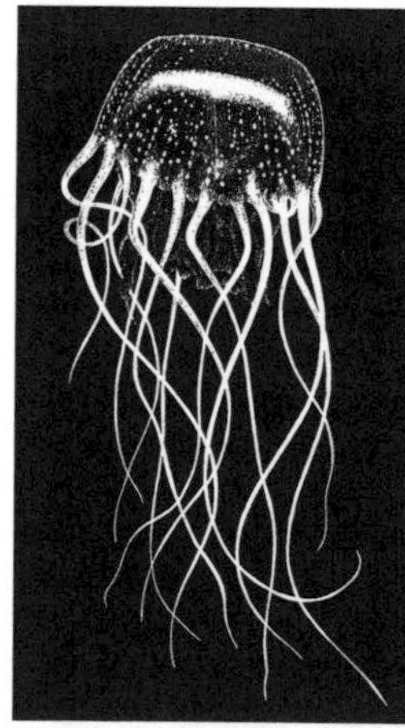
jellyfish

to the tentacles to be stung and paralyzed. The largest is *Cyanea capillata,* the lion's mane jellyfish of the cold waters of the Northern Hemisphere, whose bell gets to be more than 6 feet across. There are warm-water, cold-water, deepwater, and shallow-water jellyfish. Some are bioluminescent, some have tentacles 100 feet long, and there are some that rest on the ocean floor.

See also box jelly, cnidarian, nematocyst

Jenkins's Ear, War of A conflict that preceded the War of the Austrian Succession (1740–1748). It was precipitated in 1738 when a British merchant captain, Robert Jenkins, displayed what he said was his own ear to a committee of the House of Commons and claimed that it had been cut off by Spanish coast guards when his ship *Rebecca* was boarded in the West Indies in 1731. England and Spain were already on the verge of war, and indignation over this seven-year-old incident was so high that war was declared in 1739. The War of the Austrian Succession broke out when the Austrian archduchess Maria Theresa succeeded her father, Holy Roman Emperor Charles VI, as the ruler of the Habsburg empire. The succession was disputed by Charles Albert, elector of Bavaria, and also Philip V of Spain and Augustus III of Poland, who had claims of their own. Complicated alliances brought France, England, Saxony, Holland, and Austria into the war, which raged throughout Europe, and even appeared in North America as a part of the French and Indian Wars, with France and England fighting for dominance in the colonies.

Jenny Haniver An artificial "sea monster," made by doctoring the carcass of a skate or ray, and then drying it in such a position that the nostrils (on the underside of the head) look like eyes, the fins look like wings, and the claspers (if it is a male) look like legs. From the Middle Ages onward, these creations were used to verify the existence of mermaids and dragons, but the origin of the name is a mystery. A later device consisted of attaching the skeleton of a monkey's torso to that of a fish and claiming it was the skeleton of an actual mermaid. In 1842, P. T. Barnum exhibited one of these composites, the Feejee Mermaid, to thousands of people at his museum in New York, and we can only assume that a large number of them thought it was real.

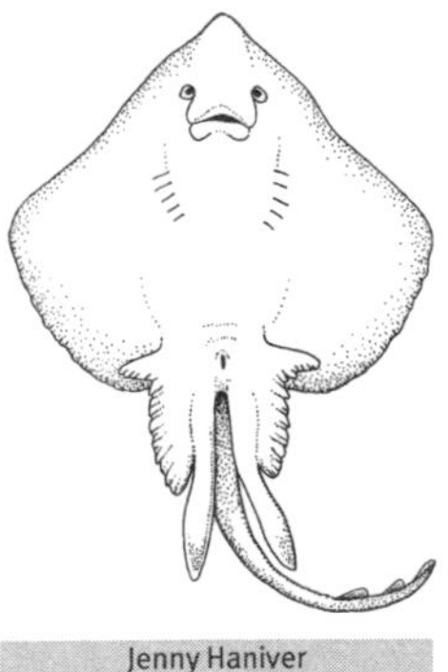

Jenny Haniver

See also mermaid

Jersey The largest of the Channel Islands, 12 miles west of the Cotentin Peninsula of Normandy, but a dependency of the British Crown. Known to the Romans as Caesarea, the island has been occupied since the Bronze Age. Like the other Channel Islands, Jersey has been the subject of a long-running struggle for ownership between the French and the British, but it is now governed by an elected assembly, with the proceedings conducted in French. The 45-square-mile island is a large plateau, with deep valleys sloping from north to south; more than half of the island's 82,000 people live in the capital of St. Helier. Its mild climate and fertile soil make it ideal for growing vegetables (tomatoes, potatoes, broccoli). With the other Channel Islands, it is a popular destination for British and French tourists. Around the fifteenth century, weavers in Jersey and Guernsey began knitting wool shirts for fishermen that became popular throughout Europe. They are still called jerseys in the United Kingdom, Australia, and New Zealand, but in America they are called sweaters. The fawn-colored Jersey dairy cow was developed here, and since 1789, it has been the only breed allowed on the island. Naturalist Gerald Durrell (1925–1995) founded the Jersey Zoological Park in Trinity Parish expressly for the purpose of breeding and displaying animals in danger of extinction.

See also Channel Islands, Guernsey

jetsam Any goods or equipment that have been intentionally thrown overboard from a ship at sea, differentiated from flotsam, which refers to goods or equipment (or the wreckage of the ship itself) that have been lost accidentally or have been found floating because of a shipwreck. Like flotsam, jetsam belongs to the finder.

See also flotsam, salvage

jewfish (*Epinephalus itajara*) A huge sea bass that is found in the western Atlantic from Florida to Brazil (including the Gulf of Mexico), and in the eastern Pacific from Costa Rica to Peru. It is usually found in water no more than 60 feet deep, although it occasionally wanders into deeper waters. It can be differentiated from the giant sea bass (*Stereolepis gigas*) because it has more soft rays than spines in the dorsal fin. The jewfish also has a rounded tail fin, while the giant sea bass has a tail that is concave on its terminal margin. The record jewfish of the International Game Fish Association (IGFA) weighed 680 pounds; it was caught in Florida in 1961. "Jewfish" is thought to be derived from anti-Semitic characterizations of this large, ugly fish. The name is being dropped from current usage.

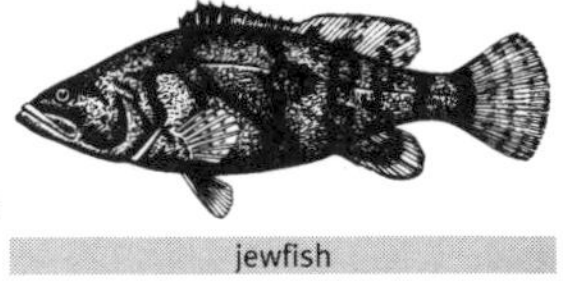

jewfish

See also giant sea bass, Queensland grouper

John Dory (*Zeus faber*) There are several species of John Dory; the European *Z. faber,* the American *Zenopsis*

John Dory

ocellata, and similar species found in Japan and the Indo-Pacific region. They are all midwater fishes that are flattened like pancakes, and have a large black spot, encircled by a lighter ring, in the center of the body. They are 20-inch-long, solitary fishes that can be found close to the surface or at depths of 600 feet. The mouth is turned down, and the jaws are greatly distensible. They are slow swimmers, but they feed by approaching their prey slowly, then swinging their jaws forward to gobble up small fishes. Some people find the John Dory a delicious eating fish, but others regard it as completely unpalatable and fit only for the manufacture of fish meal. **See also orange roughy, oreo**

jolly roger The name popularly given to the pirate flag, a white skull and crossbones on a black field. There is no evidence that such a flag was ever flown by pirates, who, if they flew any flag, would probably have displayed a plain black one. **See also buccaneer, privateer**

Jonah A minor prophet, whose story is told in Jonah 1–4 in the Old Testament. While he was en route to Nineveh to preach against sin, a great storm overtook his ship, and the sailors, suspecting him to be responsible, threw him overboard. He was swallowed by a "great fish" that vomited him up on land three days later. The creature that swallowed Jonah has been variously identified as a whale, a shark, or any number of large fishes. His name has survived as a person who brings bad luck to a ship, and more generally, a person who brings misfortune to those with whom he comes in contact.

Jones, John Paul (1747–1792) One of America's greatest naval heroes, Jones was born John Paul, but added "Jones" after he killed the ringleader of a mutinous crew in self-defense (the victim was leading a mutiny) and wanted to avoid trial. During the War of Independence, he was commissioned in the Continental navy and given command of the *Ranger.* After successfully taking the battle to France (where he captured the British warship *Drake*), he was given command of the *Bonhomme Richard,* and at Flamborough Head, on September 23, 1779, in one of the most memorable battles in naval history, he defeated the *Serapis.* (It was during this battle that Jones is supposed to have said, "Sir, I have not yet begun to fight.") With his own ship sinking, Jones boarded the burning *Serapis* and brought her safely to the Texel in the Netherlands. Jones was a hero in France and spent a considerable amount of time there, trying to collect the prize money that was due him. He was given the rank of rear admiral in the Russian navy and was asked by Catherine the Great to lead a Russian squadron into battle with the Turks. In 1792, under the urging of Thomas Jefferson, President George Washington appointed him American consul in Algeria, but he died before he learned of the appointment. He was buried in France. In 1905 his remains were removed to Annapolis and enshrined in a crypt at the U.S. Naval Academy.

Juan de Fuca Strait Named for a Greek whose real name was Apostolos Valerianos, who sailed in the service of Spain in the sixteenth century, the Juan de Fuca Strait is a 100-mile-long narrow passage between the Olympic Peninsula of Washington and Vancouver Island, British Columbia. (Part of the U.S.-Canadian boundary lies in midchannel.) The strait is the site of Victoria, B.C., and Port Angeles, Washington, and is used as an access from the Pacific to the cities of Vancouver and Seattle.

Juan Fernández Islands Four hundred miles west of Valparaiso, Chile, these three small islands were discovered around 1563 by Juan Fernández, a Spanish navigator who lived there for some years and stocked them with goats and pigs. The islands are volcanic peaks rising from the Juan Fernández submarine ridge. They include Isla Santa Clara, Isla Más Afuera (island farther out), sometimes called Isla Alejandro Selkirk, and Isla Más a Tierra (island nearer land), also known as Robinson Crusoe Island. In 1704, British seaman Alexander Selkirk quarreled with his captain and asked to be put ashore. He remained there for four years before being rescued by Woodes Rogers, a privateer. His adventures are the basis for Daniel Defoe's *Robinson Crusoe,* published in 1719. After the Battle of the Falklands in 1914, the German light cruiser *Dresden* took refuge in Cumberland Bay in the Juan Fernández Islands, and was blown up by her crew to avoid capture by the British.

junk (1) A sailing vessel common to Far Eastern seas, used by the Chinese and Javanese. Flat-bottomed, with a high stern and two or three masts carrying lugsails (four-sided sails), often made of batting stiffened with horizontal battens. The name is derived from the Portuguese *junco,* which in turn comes from the Javanese *djong,* a ship. Most junks nowadays are small trading vessels, but in the fifteenth century, Chinese voyagers in enormous junks, some up to 400 feet in length, plied the seas of China and the Indian Ocean, as far afield as Ceylon, the Philippines, and Mombasa.

See also Cheng Ho

junk (2) In the nose of a sperm whale, there is a huge reservoir of oil that the Yankee whalers called the case, and below that there is a spongy, oil-impregnated section that was known as the junk. Because the oil in the case was liquid, it could be bailed out with a bucket, but the junk had to be squeezed or pressed to get the oil out. **See also sperm whale, sperm whaling**

Jura Island in the Inner Hebrides of Scotland, north of its immediate neighbor, Islay. The name is derived from the Norwegian *dyr øe,* which means "deer island," and refers to the population of about 4,000 red deer. (The human population is about 1,000.) Many of the crofts have been abandoned, and the only village is Craighouse. Where Islay is relatively flat, Jura has a dramatic landscape of cone-shaped peaks—the "Paps of Jura"—and windswept moors. It is a popular island for hikers. In 1949, George Orwell wrote *1984* at Barnhill on Jura. **See also Hebrides**

Jutland, Battle of Known in Germany as the Battle of the Skagerrak, Jutland was the greatest naval engagement of World War I and the last sea battle to be fought without airplanes. It took place some 60 miles north of Jutland (the continental portion of Denmark) on May 31, 1916, and arose from a German plan to station U-boats off the entrances of the main British bases, and torpedo them. The German fleets (commanded by Vice Admiral Franz von Hipper and Vice Admiral Reinhold Scheer) and the British (commanded by Vice Admiral David Beatty and Sir John Jellicoe) were deployed in force and engaged in battle on a previously unimagined scale. Although regarded as a contest of naval tacticians, the missed and confused signals, bad weather, and wholly inadequate preparations for night fighting made the Battle of Jutland a chaotic convocation of armored ships lobbing shells at one another, often without an idea of who they were shooting at or where anyone else was at any given time. The British lost the battlecruisers *Queen Mary, Invincible,* and *Indefatigable,* the armored cruisers *Defence, Warrior,* and *Black Prince,* and eight destroyers; the Germans suffered equally devastating losses to their navy and their pride. Thousands of men on both sides died. On June 1, when it was over and the surviving ships limped into their home ports, both adversaries believed they had won, the Germans because they had wreaked such havoc on the British fleet, the British because the Germans never again challenged their control of the North Sea.

See also dreadnought, Jellicoe

K

Kaho'olawe The smallest of the main Hawaiian Islands, Kaho'olawe is but 12 miles across and occupies 45 square miles. Only 7 miles from Maui, it is the last visible part of a sunken volcanic caldera that reaches 1,483 feet above sea level at Puu Moaulanui, its highest point. Before James Cook arrived in 1778, the island was used by fishermen and as a training area for ocean navigators. It has been uninhabited since 1953, when the U.S. Navy moved the few natives of the island to Molokai and Maui so they could use Kaho'olawe for bombing and gunnery practice. Native Hawaiians regard the island as a sacred place, and in 1976 and 1977, a group called Hui Alaloa moved out to the island at great risk. In 1990, President George Bush ordered a halt to the bombing of Kaho'olawe, but Hawaiian nationalists want the island returned to their control. The navy maintains that there are so many unexploded shells on the island that it cannot be safely returned to civilian use.

See also Hawaiian Islands

Kaiko Remote-controlled deep-diving vehicle that replicated the world's deepest dive, originally made by the manned bathyscaphe *Trieste* in 1960. Owned and operated by the Japanese Marine Science and Technology Center (JAMSTEC), *Kaiko* (Japanese for "trench") descended to 35,798 feet in the Mariana Trench in March 1995. Since this South Pacific trench is the deepest part of the ocean, future claimants to the record can only equal the descents of *Trieste* and *Kaiko*.

See also *Trieste*

Kamchatka A peninsula in far eastern Siberia, between the Sea of Okhotsk and the Pacific Ocean. It is about 750 miles long and 300 miles across at its widest point. At 140,000 square miles, it covers approximately the same area as California. Two ranges of mountains, many of which are active volcanoes, extend along its length. The weather in Kamchatka is severe, with long cold winters and damp, cool summers. The main city is Petropavlovsk (St. Peter and St. Paul), named for the ships that Commander Vitus Bering built there before setting out on his 1733–1741 voyage of Arctic exploration.

See also Bering; Kuril Islands; Okhotsk, Sea of

Kane, Elisha Kent (1820–1857) American doctor of medicine and Arctic explorer. After graduating from the University of Pennsylvania in 1843, Kane was appointed surgeon to the American missions to China, western India, Egypt, and Europe. In 1850, he was named senior medical officer aboard the *Advance* as part of the 1850 Grinnell expedition to search for John Franklin, who had been lost in the Arctic since 1845. (American merchant Henry Grinnell sponsored the ships *Advance* and *Rescue* as part of the massive effort to find Franklin.) Despite chronic rheumatism and a cardiac condition, Kane dedicated himself to travel and exploration, and in 1852, he returned to the Arctic in the *Advance* to continue the search for Franklin. His ship was stopped by the ice off western Greenland, but he explored and named Grinnell Land on Ellesmere Island, and proceeded southward to Baffin Bay. There the *Advance* was trapped in the ice for twenty-one months and had to be abandoned as the men began to show the symptoms of scurvy. After a ten-week trip of 1,300 miles, Kane brought them safely to Danish settlements in Greenland, with the loss of only one man. He named Rensselaer Harbor for his grandmother and Kane Basin for himself. Upon his return to America, he was awarded a Congressional Gold Medal. Shortly after the publication of his two-volume *Arctic Explorations*, Kane died at the age of thirty-seven.

See also Ellesmere Island, Franklin, Greenland

Kangaroo Island Island in the Gulf of St. Vincent, South Australia, 45 miles from Adelaide, and separated from the mainland by the 9-mile-wide Backstairs Passage. The island is 90 miles long and 34 miles wide. It was first described by Matthew Flinders, who came ashore in 1802 and shot many kangaroos for food. It has a permanent population of some 2,000, most of whom live in Kingscote. Kangaroo Island also supports a population of more than 1 million sheep. Flinders Chase National Park occupies the western portion of the island, providing a safe haven for sea lions, kangaroos, wallabies, koalas, platypuses, emus, and more than two hundred species of birds. It is a popular holiday destination for South Australians. **See also Flinders**

Kara Sea Karaskoye More in Russian, the Kara Sea is a shallow section of the Arctic Ocean, defined on the west by Novaya Zemlya and Franz Josef Land and on the east by the islands of Severnaya Zemlya. It covers some 340,000 square miles, and because it lies on the Siberian Shelf, its average depth is only 417 feet. Like the Laptev Sea to the east, it is frozen for most of the year and navigable only by icebreakers in August and

September. The impassability of the northern Siberian coast—the longest east-west coastline in the world—is the reason that Russia has to have ice-free ports. On her western boundary, St. Petersburg opens to the Gulf of Finland, and Murmansk enables Russian shipping to enter the Barents Sea and sail around northern Norway; on the Pacific side, the only Russian port is Vladivostok. **See also Barents Sea, Laptev Sea, Novaya Zemlya**

Karlsefni, Thorfinn: See Thorfinn Karlsefni

Kauai During his third voyage, in January 1778, Captain James Cook landed at Waimea on the south coast of Kauai, thus becoming the first European to set foot on the Hawaiian Islands. He did not have time to explore very much before he headed north (he was looking for the Northwest Passage), but he named the islands that he saw after his patron, Lord Sandwich. Kauai is the westernmost, and therefore the oldest, of the Hawaiian Islands. It is the most verdant, and it has therefore been nicknamed the Garden Isle. It is nearly round, covering 553 square miles, but because of the Na Pali cliffs on the western side, there is no road that encircles the island. Inland of these cliffs is Mount Waialeale, reputed to be the wettest spot on earth, with an annual rainfall of 500 inches (41 feet). Here also is Alaka'i, a dense, tropical jungle filled with birds and plants found nowhere else on earth, and also Hawaii's answer to the Grand Canyon, Waimea Canyon, which is best viewed from a helicopter. There are many hotels on Kauai, and the island is famous for its beaches. (Much of the movie *South Pacific* was shot here.) The major settlements are at Lihue (where the airport is located), Princeville (with a 1,000-acre resort, shopping center, golf course, and condominiums), Kilauea, Waimea, and Hanalei. The population of the island is around 50,000. In September 1992—as *Jurassic Park* was being filmed there—Hurricane Iniki devastated large portions of the island.
See also Hawaiian Islands

kayak An Eskimo word for a light, covered-over, canoe-type boat that is used for fishing. In common use in northern waters from Greenland to Alaska, it is made by covering a wooden frame with sealskin, leaving a hole in the center of the top covering for the kayaker to sit in. It is usually propelled by a double-ended paddle.

keel The lowest and principal timber of a wooden ship, or the lowest continuous line of plates of a steel or iron vessel, which extends the entire length of the vessel, and to which the stem, sternpost, and ribs or timbers are attached. In ocean-racing yachts, there has been much modification of the traditional centerline ballast keel, including such modifications as twin keels ("ballast keels"), a cut-out version (the "fin keel"), and the addition of a weighted bulb to the bottom of the keel, to give it greater stability without adding that much resistance. The term "keel" is also used to describe the sharp-edged parts of whales, dolphins, and certain fishes; it can be used for the sharp underside or for the horizontal elements at the base of the tail of tunas, billfishes, and some sharks.

keelhauling A naval punishment, said to have been invented by the Dutch but introduced to other navies around the fifteenth and sixteenth centuries. A rope was rigged from yardarm to yardarm, passing under the bottom of the ship, and the delinquent was attached to it, sometimes with weights attached to his legs. He was hoisted up on one yardarm, dropped into the sea, hauled underneath the ship, and hoisted up on the other yardarm. This process was often repeated. If the sailor came into contact with barnacles on the underside of the ship, he could be seriously injured or even killed. Keelhauling passed out of fashion around the beginning of the eighteenth century, to be replaced by the even more dangerous flogging with a cat-o'-nine-tails. **See also ducking at the yardarm, flogging**

kelp Large seaweeds found in colder seas, classified as Laminariales or brown algae. *Laminaria*, abundant along the west coast of North America and also in the British Isles, is a source of commercial iodine; it produces acetic acid (used in the production of vinegar, films, textiles, and solvents) when it ferments. The bull kelp (*Macrocystis pyrifera*) is the largest known kelp and the fastest growing of all plants. It can grow 1 foot a day, to lengths of 200 feet. It has a large, rootlike holdfast that anchors it to the bottom, a hollow stem (called a stipe in seaweeds), and a long, branching stalk with blades that are kept afloat by air bladders known as pneumatocysts. Kelp "forests" are home to many species of invertebrates and fishes; California sea lions and sea otters often entwine themselves in the fronds when not diving for sea urchins or abalones.
See also abalone, sea urchin, seaweed

kelp gull (*Larus dominicanus*) The only gull in the Antarctic, it closely resembles the black-backed gulls of the north, with its white head, black back, and yellow bill with a red spot. The kelp gull breeds in southern South America, South Africa, and southern Australia, and also on the South Shetlands and South Georgia. Its northern counterparts are omnivores, eating anything and everything, including refuse, but the kelp gull subsists mainly on limpets. The mollusks are swallowed whole, and the shells are later regurgitated intact.

Around human habitation, however, the gulls are not above scavenging in garbage dumps.

See also greater black-backed gull, gull, herring gull

Kemp's ridley turtle (*Lepidochelys kempii*) The smallest and most endangered of all sea turtles, Kemp's ridleys rarely weigh more than 100 pounds. They can be found in the open waters of the Gulf of Mexico, and along the gulf coasts of Mexico, Texas, Louisiana, and Florida. They nest only on the beaches of Rancho Nuevo, Mexico, and in the past forty years, the number of nesting females has dropped from tens of thousands to about four hundred. The beaches are now patrolled by Mexican marines, but still the eggs are stolen by people or eaten by pigs and dogs. In addition, many juvenile turtles are drowned every year in shrimpers' nets in the Gulf of Mexico. The introduction of a turtle excluding device (TED) into the nets has saved many turtles, but many scientists fear that the numbers, estimated at only thirteen hundred to fifteen hundred animals, are already too low for the species to survive. The olive ridley (*L. olivacea*) is widespread in the Pacific, the Indian Ocean, and parts of the Atlantic, but the ranges of the two types do not overlap. They may actually be a single species. **See also sea turtle**

Kerguelen Islands An archipelago in the southern Indian Ocean, consisting of the island of Kerguelen (also known as Desolation Island) and nearly three hundred small islets, which together cover about 2,700 square miles. Kerguelen is about 100 miles in length, and consists of glaciers and rocky, inhospitable terrain. It was discovered by Yves-Joseph de Kerguélen-Trémarec in 1772 and later explored by Captain James Cook. The archipelago was frequently visited by elephant seals, and therefore by sealers, one of whom was Captain Joseph J. Fuller of the schooner *Pilot's Bride,* who was shipwrecked for a year (1881–1882). The islands are now part of Terres Australes et Antarctiques Françaises, the French Antarctic territories, which also includes St. Paul Island, New Amsterdam Island, and the Crozet Islands.

See also Saint-Paul Islands, Cook (James), Crozet Islands

Kerguélen-Trémarec, Yves-Joseph de (1734–1797) French explorer, the discoverer of the Kerguelen Islands in the southern Indian Ocean. After service in the Seven Years' War, he was given his own command and chosen to lead an expedition in search of Terra Australis Incognita. Sailing from Mauritius in 1771, he discovered the barren, desolate islands that now bear his name but was unable to land because of fog and ice; he returned to France. He named his discovery "La France Australe" and claimed it was part of the Antarctic continent. He sailed for the south again in 1793, but scurvy and quarrels among his officers forced him to turn back. He was court-martialed and sentenced to twenty years in prison. He served only four, but during the French Revolution, he spent another eight months in prison. He died planning more voyages of discovery.

Kidd, William (c. 1645–1701) "Captain Kidd," one of history's most notorious pirates, began life as a merchant seaman. In 1694 he was commissioned by the Earl of Bellomont, governor of New York, as a privateer to defend English shipping in the Red Sea and the Indian Ocean, but disease, mutiny, and the failure to take prizes apparently caused him to turn pirate. In 1698 he captured the Armenian *Quedagh Merchant* in the West Indies and returned in triumph to Anguilla, where, although he claimed the prizes that he took were lawful, he was arrested for piracy and the murder of one of his men. He was taken to London, tried, convicted, and hanged. With the exception of a small trove that was found after his death on Gardiner's Island off Long Island, "Captain Kidd's treasure," usually said to be buried in the West Indies, has not been found. Edgar Allen Poe's "The Gold Bug" is but one of the many stories of Kidd's buried treasure.

See also buccaneer, piracy, privateer

Kiel Canal Artificial waterway constructed between 1887 and 1895 through the Schleswig-Holstein Peninsula in north-central Germany, to create a passage from Kiel on the Baltic to the mouth of the Elbe River and the North Sea. Originally named the Kaiser Wilhelm Canal, it was conceived by the kaiser to facilitate the movement of Germany's fleet and was designed to accept the largest battleships of the day. When the British launched the *Dreadnought* in 1906, it meant that all other navies had to follow suit, so the canal was no longer wide enough. Work to widen it was completed on the eve of World War I. After the war, the canal was internationalized by the League of Nations, but Hitler repudiated the treaty in 1936 and seized the canal. U-boats were built at Krupp's Germania Shipyard at Kiel, the industrial center at the canal's eastern terminus, and it was one of the prime targets for Allied bombers during the war.

killdeer (*Charadrius vociferus*) A small plover whose characteristic cry, "killdee-ee, kill-dee," is responsible for its common name. Unlike most plovers, killdeers are not exclusively inhabitants of beach areas, but can also be found in upland sites, often far from water. They breed from New England and the Canadian Maritimes south to the Caribbean and Mexico, and winter as far south as Venezuela and northwestern

killdeer

Peru. Many birds do a "broken-wing" act when their nest is threatened, but the killdeer is the master of the ruse. When danger approaches, the incubating parent scurries away and pretends to brood elsewhere, but if the threat to the nest or chicks appears imminent, the bird fakes injury by flopping on the ground, dragging one wing as if it were broken. This usually attracts the intruder, and the bird continues to feign injury as it moves away from the nest, staying just out of range of the predator. **See also plovers, shorebirds, turnstone**

killer whale

killer whale (*Orcinus orca*) The most widely distributed of all cetaceans, the killer whale is found throughout the oceans of the world, from the poles to the tropics in both hemispheres. Males can reach a maximum length of 30 feet, females 26, which makes them by far the largest of the dolphins (Delphinidae). They are easily distinguished at sea by their jet black and white coloration and the high, usually vertical, dorsal fin of the bulls. Females and juveniles have much smaller, more sharply curved dorsal fins. They live all their lives in tightly knit family groups. Killer whales are among the oceans' apex predators, feeding in packs on everything from sharks, fish, and squid to seals, sea lions, dolphins, and even the great whales. They are fast, powerful, and intelligent, but they are surprisingly docile in captivity. There are no records of killer whales in the wild attacking people. **See also dolphin**

king eider (*Somateria spectabilis*) King eiders nest along the Arctic Ocean coasts of both hemispheres and winter farther south. Of all the eiders, this species is the most likely to stray south of its natural range, and it has been reported from California, Georgia, and the Great Lakes. In the north, the king eider is often seen associating with the common eider. Although the black-and-white drakes are similarly colored, the two species can be differentiated during the breeding season by the king eider's large orange "shield" at the base of the bill and the common eider's black mask. In all eiders the females are mottled brown, but the female king eider is a somewhat richer color. **See also eider, sea ducks, spectacled eider, Steller's eider**

king eider

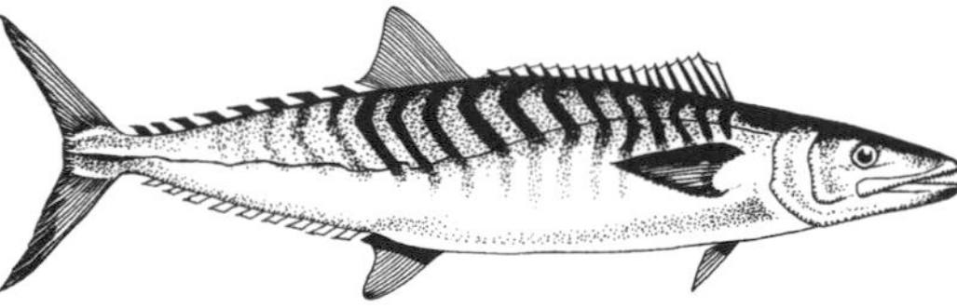

king mackerel

king mackerel (*Scomberomorus cavalla*) A coastal species found in the western Atlantic from Maine to Rio de Janeiro, including the Gulf of Mexico. It is the largest of the mackerels, growing to 5 feet in length and weighing up to 100 pounds. The narrow-barred mackerel (*S. commerson*) of Australian waters grows almost as large. **See also Atlantic mackerel, Spanish mackerel**

king penguin (*Aptenodyptes patagonicus*) Second only to the emperor penguin in size, the king is one of the

king penguin

heaviest birds in the world. It can reach a standing height of 3 feet and a weight of 30 pounds. It is a stately bird with silvery gray upper parts, a black head and chin, and a bright golden ear patch. An estimated 1 million pairs breed on Antarctic and subantarctic islands. The king penguin is an accomplished diver; it has been recorded diving to 1,056 feet. For breeding, they aggregate in large colonies, where there is much calling and ritualized courtship behavior. Five weeks after the chicks are born they form crèches, and these huge chicks in brown, fluffy suits may be the silliest-looking birds in the world. In the early sealing days, king penguins were killed and boiled for oil, but now their only predators are leopard seals and killer whales.

See also emperor penguin

Kiribati The Republic of Kiribati is composed of thirty-three islands, twenty of which are inhabited, scattered across 2,400 miles of the Pacific Ocean. The islands are all coral atolls; none is higher than 25 feet above sea level. Spanish explorers of the sixteenth century may have sighted some of the islands, but in 1765 John Byron discovered Nikunau, and in 1788 Thomas Gilbert saw Tarawa and John Marshall sighted Aranuka. American sperm whalers came upon the rest of the islands during the nineteenth century, by which time the main group had been designated as the Gilbert Islands. The British made the Gilbert and Ellice Islands a protectorate in 1892, and annexed Ocean Island (Banaba) in 1900. In 1916, the islands were made a Crown Colony. Japanese troops invaded the Gilberts in 1942, and some of the fiercest fighting of the war took place on Tarawa when U.S. Marines recaptured the island. In 1977, the colony became self-governing, and in 1979, it became the independent nation of Kiribati, which includes the Line Islands (including Kiritimati, formerly Christmas Island), the Gilberts (Tarawa and others), the Phoenix Islands, and Banaba Island, a producer of guano until it was mined out in 1981. Most of the population of 75,000 is concentrated in the Gilbert Islands, and more than one-third of those live on Tarawa. The government of Kiribati severed diplomatic relations with France in 1995 when that country resumed nuclear testing in the Pacific.

See also Gilbert Islands, Tarawa

Kiritimati Once known as Christmas Island, this is the largest all-coral island in the world. It is located in the central Pacific, about 1,000 miles south of the island of Hawaii. Now part of the Republic of Kiribati (formerly the Gilbert and Ellice Islands), along with Palmyra, Baker, Howland, Jarvis, and Johnston, the "island" consists of many small coral outcrops, covering some 222 square miles and enclosing one of the largest atolls in the Pacific. The population is about 1,700, many of whom are involved in the production of copra, the island's main export. The island was sighted and named by James Cook on Christmas Eve, 1777; claimed by the United States in the Guano Act of 1856; annexed by Britain in 1888; and included in the Gilbert and Ellice colony in 1919. The British evacuated the atoll and conducted nuclear tests there in 1957; the Americans did likewise in 1962. Half of the island has now been leased to the Japanese government, which plans to build satellite launch facilities and a hotel.

See also guano islands, Kiribati, Palmyra Atoll

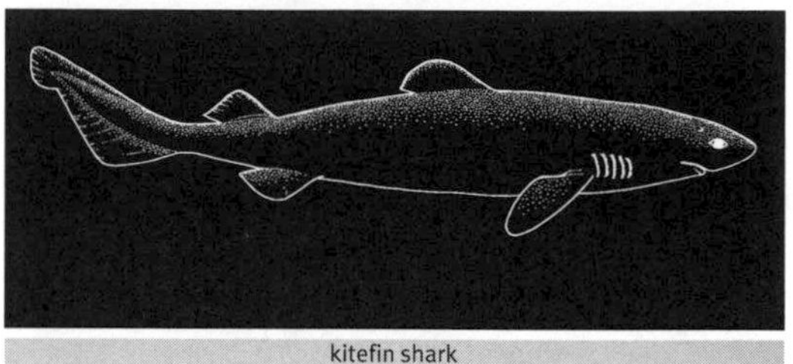

kitefin shark

kitefin shark (*Dalatias licha*) A small black dogfish (maximum length: 5½ feet) that gets its common name from its dorsal fins, which are narrower at the base than at the top. Sparsely distributed throughout the world's tropical and temperate waters, the kitefin is a large-prey predator with huge teeth and heavy, powerful jaws. It is sometimes caught in deepwater trawls in the eastern Atlantic off Spain and Portugal, but otherwise it has no commercial value. **See also dogfish**

kittiwake (*Rissa tridactyla*) A medium-sized gull of the Arctic, the North Pacific, and the North Atlantic, identifiable by its black wing tips, black legs, and yellow bill. The common name comes from its mewing call, which sounds something like "kitti-wake." Kittiwakes nest in large numbers on narrow cliff ledges. A similar species, the red-legged kittiwake (*Larus brevirostris*) is found around the Bering Sea.

kittiwake

knorr The Viking longship, employed either in raiding or transporting cargo. They were double-ended, clinker-built (with overlapping planks) of oak, with a single squaresail and auxiliary oars. As shown in some contemporary depictions—such as the Bayeux Tapestry, which depicts the conquest of England by William of Normandy in 1066—they had high stem and stern posts, often decorated with a dragon figurehead. Several Norwegian longships that were used for burials have been found, including the Oseberg, Gokstad, and Tune ships, all of which were found around Oslo Fjord and excavated, restored, and exhibited in the Viking Ship Museum in Oslo. The Oseberg ship, which was probably a private sailing ship rather than a warship, is 70 feet long and 16 feet in the beam amidships. The warships, none of which have been preserved, are believed to have been as much as 110 feet long and able to carry as many as 120 armed warriors. **See also Vikings**

knot (1) An intertwining of the parts of one of more ropes or cords, usually for the purpose of fastening them together or to something else.

knot (2) A measure of nautical speed. One knot equals one nautical mile (6,080 feet) per hour. Originally a log attached to a rope with evenly spaced knots on it (47 feet 3 inches apart) was heaved overboard to measure the speed of the ship. By counting the number of knots let out in a specific time period—timed by a 28-second sandglass—a sailor could calculate the speed in knots (never "knots per hour").

knot (3) A large, short-billed sandpiper (*Calidris canutus*) that travels in dense flocks and breeds from Arctic Canada and Greenland to California and Massachusetts, and winters in Europe, South America, Africa, Asia, and Australasia.

Kodiak Island Alaska's largest island, separated from the mainland by the Shelikof Strait. The mountainous, wooded island is 100 miles long and 60 miles wide; it covers 3,588 square miles, approximately the area of Puerto Rico. Because of the warm Kuroshio Current, the weather is unexpectedly warm and moist. Three-quarters of Kodiak (2,836 square miles) is devoted to a wildlife refuge, and is the habitat of the huge Kodiak brown bear, a species of grizzly that is among the largest of all land predators, rivaled only by the polar bear. Discovered in 1763 by Russian fur traders, the island was first settled in 1784 by Grigory Shelikof at Three Saints Bay. The Russians used the island as a base for seal and sea otter hunting until 1873, when the United States purchased Alaska. The 1912 eruption of Novorupta Volcano near Mount Katmai, across the strait to the west, covered the island with ash and interfered with its agricultural development. The Good Friday earthquake of 1964 caused widespread destruction on the island, and when the *Exxon Valdez* ran aground in Prince William Sound in 1989, the oil spread as far as the shores of Kodiak. The 13,000 inhabitants—6,000 of whom live in the city of Kodiak—have recovered from these disasters, and are engaged in fishing (particularly for salmon), agriculture, and dairy farming.
See also *Exxon Valdez*

Komodo Island of the Lesser Sunda group of Indonesia. A hilly island, covering about 200 square miles with only one village, which is also known as Komodo. It is best known as the home of the Komodo monitor lizard (*Varanus komodoensis*), also known as the Komodo dragon, the largest lizard in the world, which reaches 10 feet in length and weighs up to 300 pounds. (Salvador's monitor, *V. salvadori*, may grow longer, but it has an exceptionally long tail.) They are found primarily on the island of Komodo, but there are small populations on Rintja, Padar, and Flores. (The lizards may swim from island to island.) Although they are said to prey on deer, pigs, and occasionally humans, Komodo monitors are primarily carrion eaters. The Komodo dragon was discovered and named in the early twentieth century and has been kept in captivity in various zoos.
See also Indonesia

***Kon-Tiki:* See Heyerdahl, Thor**

Kotzebue, Otto von (1797–1846) The son of the German playwright August von Kotzebue (1761–1819), Otto von Kotzebue was a navigator in the service of the Russians who made three circumnavigations of the earth. The first was with Adam Ivan Krusenstern in 1803. In the brig *Rurik*, with a crew of only twenty-seven men, he rounded Cape Horn and visited Easter Island and many of the islands in Polynesia. Sailing north, he charted much of the coast of Alaska; he discovered and named Kotzebue Sound. Unable to find a passage through the Canadian Archipelago to the Atlantic, he returned to Russia in 1818. He made two more 'round-the-world voyages, collecting plants and geological specimens, and published extensive accounts of his journeys and discoveries.

Krakatau Volcanic island between Sumatra and Java in Indonesia that was almost completely destroyed by an eruption on August 27, 1883. After three months of rumbling earthquakes, the island blew up in a series of thunderous blasts that were heard 3,000 miles away. The explosion created a vast hole in the ocean, and what remained of the mountain collapsed into the abyss, causing the sea to rush in. As the steam came into contact with the red-hot magma, it exploded with catastrophic violence, and 4 cubic miles of rock and ash were hurled as much as 50 miles high. Since it was uninhabited, nobody on Krakatau was killed, but giant tsunamis (some reaching 135 feet high) radiated in all directions, flooding the coasts of the neighboring islands and killing more than 36,000 people. Ash in the stratosphere generated lurid sunsets and lowered temperatures around the world for a year. A Hollywood movie, *Krakatoa, East of Java*, was made in 1969, but Krakatau is actually *west* of Java. When the movie was made, the volcano was popularly known as Krakatoa, but the spelling has now been revised to conform to the Indonesian pronunciation.

kraken A sea monster of enormous size, said to inhabit the waters off Norway and Sweden and, for some reason, often seen by clergymen. In 1555, Bishop Olaus Magnus of Sweden said that the monster was ½ mile long. The name is said to be derived from a Norwegian word meaning "uprooted tree," because of the similarity between the multiple arms of the creature and the roots of a tree, but this is not so. "Kraken" was introduced by Bishop Erik Ludvigsen Pontoppidan of Bergen, in *The Natural History of Norway,* and he used it to refer specifically to a sea monster. Later descriptions referred to a snakelike creature (the "sea-serpent") or giant worms, which increased the possibility that what they had seen was in fact a giant squid, a creature unknown to science until 1853. In 1830 Alfred, Lord Tennyson wrote a poem called "The Kraken," in which he said, "There hath he lain for ages and will lie / Battening upon huge sea-worms in his sleep." **See also giant squid, sea serpent**

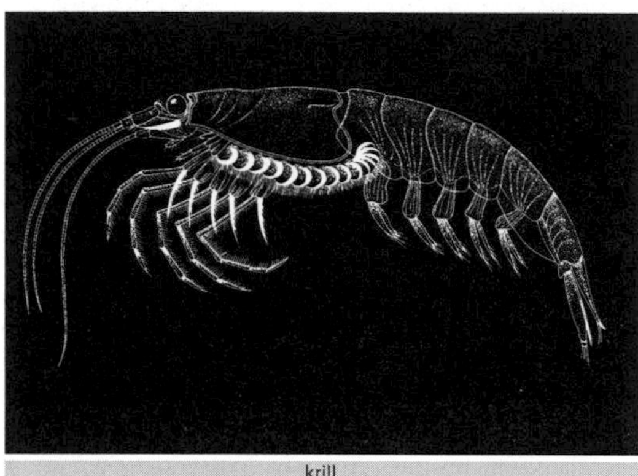

krill

krill A Norwegian word that means "whale food," krill is the base of the food chain in the Antarctic. The term is usually applied to *Euphausia superba,* a shrimplike crustacean that reaches a maximum length of 2½ inches and aggregates in huge rafts that may be miles in diameter. It is these aggregations that draw the feeding blue, fin, and sei whales to Antarctic waters in the summer, but various species of seals, penguins, and seabirds also feed on these abundant creatures. It is estimated that the standing crop of krill is in the neighborhood of 750 million metric tons, which makes it attractive to Japanese and Russian fishing fleets.

See also blue whale, euphausiids, fin whale, sei whale

Krusenstern, Adam Ivan (1770–1846) Russian admiral and hydrographer who served in the British navy from 1793 to 1797. Interested in a direct route from Russia to China via Cape Horn, he wrote to Tsar Alexander I and was commissioned to make such a voyage. In 1803, in the *Nadezhda* and the *Neva,* accompanied by Otto von Kotzebue and Yuri Lisianski, Krusenstern sailed from Kronstadt (at the head of the Gulf of Finland, near St. Petersburg) to Kamchatka and around the Horn, via the Sandwich (Hawaiian) Islands. After exploratory voyages in the North Pacific, the expedition returned to Kronstadt by way of the Cape of Good Hope. As leader of the expedition, Krusenstern became the first Russian to circumnavigate the globe, and he was showered with honors and appointed admiral. He published an atlas of the Pacific Ocean in 1827. Cape Krusenstern in Alaska, on Kotzebue Sound, north of the Bering Strait, is named for him.

Kuril Islands A chain of fifty-six islands extending for approximately 750 miles from the tip of the Kamchatka Peninsula to Hokkaido, the northernmost of the main Japanese islands. To the west of the Kurils (sometimes spelled "Kuriles") is the Sea of Okhotsk; to the east, the Pacific Ocean. As part of the Pacific "Ring of Fire," the islands have nearly a hundred volcanoes, thirty-eight of which are considered active. Parallel to the islands is the Kuril-Kamchatka Trench, listed among the oceans' deepest at 34,586 feet. Before World War II, the islands belonged to Japan, but they were ceded to the Soviet Union in 1945, and the Japanese population was replaced by Russians. There has been no peace treaty between Japan and the former Soviet Union, because the issue of the "Northern Territories" is still unresolved. Japan still claims the southernmost of the Kuril Islands—Etorofu, Kunashir, Shikotan, and Habomai.

Kuroshio Current A strong, beltlike surface current of the eastern Pacific, flowing in a northeasterly direction from the Philippines to the coast of Japan and strongest from May to August. It is known as the Black Current (*kuro-shiro* means "black stream" in Japanese) because it appears a deeper blue than the surrounding waters. Hydrographers have theorized that the general circulation of the ocean is maintained by the stresses of prevailing winds over the ocean, which in this case leads to an intensification of the current closest to its western boundaries. Like the Gulf Stream in eastern North America, the warm waters of the Kuroshio have a marked effect on the climate of the southern Japanese islands and Taiwan. The Kuroshio is one of the greatest oceanic currents in the world, and has been known to Japanese navigators and fishermen since ancient times. Its existence was first published in the west by the mapmaker Varenius in 1750, and it was described by von Krusenstern in 1804 as the "Japan Current," a name still sometimes used. **See also current**

Kwajalein A coral island in the central Pacific, in the Ralik chain of the Republic of the Marshall Islands, Kwajalein consists of some ninety-seven islets and the main atoll, which is the largest coral lagoon in the world, covering almost 1,100 square miles. (The second largest is Rangiroa in French Polynesia.) At its longest, the atoll is 75 miles, from Kwajalein Island at the eastern end of the lagoon to Ebadon at the western. The large Japanese naval base located on Kwajalein during World War II was taken by American troops in 1944. In 1959, Kwajalein was designated the target area for the Western Missile Test Range in California, and there are now U.S. missile installations on the island. The Marshallese who work at these bases do not live on Kwajalein, but commute from Ebeye Island in the lagoon. With Majuro, "Kwaj" is the most populous of the islands in the Marshalls, with approximately 13,500 Marshallese living on fourteen islands within the atoll, and about 12,500 of these on Ebeye. The battle for Kwajalein (February 1944) left many ships sunk in the lagoon, and these have made the island a popular destination for divers. **See also Marshall Islands**

L

Labrador duck (*Camptorhynchus labradorius*) The only species of waterfowl to become extinct in North America, the Labrador duck was a black bird with a white head and white markings, sometimes known as the skunk duck. It inhabited eastern Canada and the United States, and may not have lived in Labrador at all. Although it was hunted, its flesh tasted strongly of fish, and it was never a popular food item. It does not seem to have ever been plentiful and was recorded as rare by 1844. Although the actual cause is not known, it seems likely that hunting of the remaining population was partially responsible for its demise. It had a peculiar bill with numerous lamellae (filter plates), and it may somehow have been deprived of its food source—whatever that was. The last one was shot on Long Island in 1875. **See also eider, sea ducks**

Labrador duck

ladyfish

ladyfish (*Elops saurus*) A much smaller cousin of the tarpon, the ladyfish (also called the ten-pounder) is also a fighter when hooked. It leaps, tail-walks, and somersaults in an attempt to throw the hook. Ladyfishes average about 5 pounds in weight, but the machete (*E. affinius*), a somewhat larger version found in the eastern Pacific, has been caught at 23 pounds.
See also bonefish, tarpon

Lafitte, Jean (1782–1826?) Notorious privateer and smuggler who became an American hero during the Battle of New Orleans in the War of 1812. With his brother Pierre he operated a depot for smugglers and slavers on the islands of Barataria Bay, south of New Orleans. He also held a privateer's commission from the Colombian port of Cartagena, then in revolt against Spain, and plundered Spanish shipping. The British wanted to capture Barataria Bay in 1814, and they promised him $30,000 for his allegiance, but when he reported that to Governor W. C. C. Claiborne, Claiborne attacked him instead. Lafitte then offered his forces to the beleaguered General Andrew Jackson to help defend New Orleans in return for a full pardon. Jackson accepted, and the Baratarians distinguished themselves in battle. They were pardoned by President James Madison, but after the war, Lafitte went back to his old privateering ways and began attacking American shipping as well as Spanish. After burning his own colony of Campeche (where Galveston, Texas, now stands), he is believed to have continued his piracy on the Spanish Main.

lagoon From the Spanish *laguna*, meaning "pool"; a lagoon is a shallow waterway, separated from the open sea by either sandbars, barrier islands, coral reefs, or some combination of the three. If it abuts a sandbar or barrier island, the lagoon is long and narrow; if enclosed by a coral atoll, it is roughly circular.
See also atoll, barrier island, barrier reef

Lake, Simon (1866–1945) American submarine designer. His first effort, the *Argonaut Jr.*, was built in 1894, a 14-foot-long wooden box lined with canvas and shaped like a flatiron, with wheels on the bottom. His idea was that submarines could be driven across the bottom (by the use of an interior hand crank) in order to send divers out to cut cables, destroy mines, and communicate enemy ship movements back to the submarine. His second attempt, known simply as *Argonaut*, was twice as long as its predecessor and a much more ambitious undertaking. He equipped this version with a double hull, which enabled her to float comfortably on the surface when her ballast tanks were empty; when they were filled, she descended on an even keel. *Argonaut* was also equipped with a modest air-intake tube, which enabled her crew to breathe as long as the boat did not dive too deeply. To interest the U.S. Navy in his inventions, Lake then built *Argonaut I*, a 36-foot iron vessel, and in 1898 sailed her 1,000 miles from Norfolk, Virginia, to New York under her own power, marking the first time a submarine had operated extensively in the open sea. Lake's third boat, *Protector*, built in 1902, was the first submarine to mount a gun on its foredeck, and also the first to have a periscope, which Lake called an omniscope. When he failed to convince the U.S. government of the virtues of submarines, he sold *Protector* and five more of the same design to the Russians, who were then at war with the Japanese. Neither Lake's nor any other submarines saw service in the Russo-Japanese War, but Lake continued to sell his vessels to European powers, including the Austrians. A joint enterprise with Krupp was initiated in Germany, but Lake's patents were not protected and he received no royalties. He returned to America to build the *Seal* for the U.S. Navy; at 161 feet long and displacing 400 tons, it was the largest submarine up to that time. Before the turn of the century, Lake had designed the first cargo submarine, only to see his plans appropriated by the Germans, who built the *Bremen* and the *Deutschland,* the world's first successful submarine freighters. In 1932 Lake manufactured *Explorer,* a 10-ton vessel powered by an electrical cable attached to a surface ship. **See also Garrett, Holland, submarine**

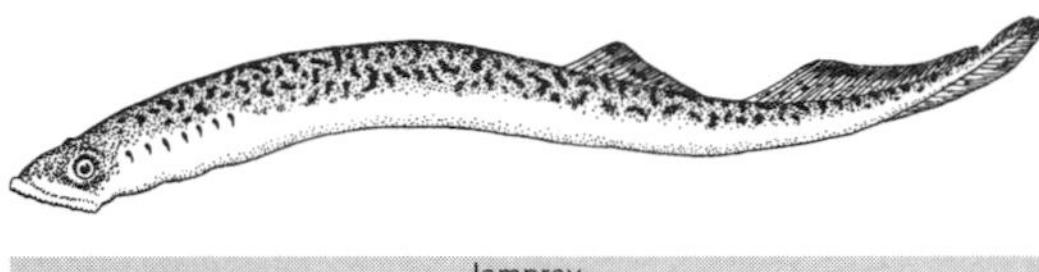

lamprey

lamprey (family Petromyzonidae) Lampreys are primitive fishlike vertebrates that are placed with the hagfishes in the class Agnatha, the jawless fishes. They have well-developed eyes (hagfishes have none), a single nostril on top of the head, and seven gill openings on each side. They have no bones, no jaws, and no paired fins. Parasitic predators on fishes, they are equipped with a mouth that consists of a round, horny-toothed sucker disk. Lamprey eggs hatch into burrowing larvae, called ammocoetes, that are completely unlike the adults they will become. Before the sucker disk develops, the juveniles feed on suspended food particles that they capture on mucous-covered cilia in the mouth. Of the twenty-two-odd species of lampreys, some spend their entire lives in freshwater, but in the early nineteenth century, when a canal was built to allow ships to sail from Lake Ontario into Lake Erie, sea lampreys (*Petromyzon marinus*) invaded all the Great Lakes and reduced the resident populations of trout to nearly zero. Attempts to restock the lakes have failed, and the lampreys have established themselves as permanent landlocked residents. **See also hagfish, jawless fishes**

lamp shells (phylum Brachiopoda) Brachiopods—whose common name refers to their resemblance to a Roman oil lamp—are truly creatures of the past, for they dominated the late Paleozoic seas some 250 million years ago. More than 30,000 fossil species of brachiopods have been identified; only about 250 are alive today. Today's lamp shells average about 1½ inches in length, but some of their extinct ancestors were 1 foot across. Whereas the shells of bivalve mollusks are hinged in a left/right arrangement, those of the brachiopods are arranged as an upper and a lower, with the lower shell being the larger. They attach themselves to rocks by means of a short, muscular stalk (the pedicel) that emerges through a hole in the lower shell and that can be extended or contracted to raise and lower the animal. There are two classes of brachiopods: the articulata and the inarticulata, which are differentiated by the variations in the hinge arrangement of the shells. They feed by straining minute organisms from the water and are found throughout the world's oceans. **See also gastropoda**

Lanai One of the Hawaiian Islands, across the Auau Channel from Maui. In the mid-nineteenth century a group of Mormon settlers founded a colony on Lanai, but it failed within a decade. For years afterward it was used for grazing cattle until, in 1922, James Dole bought all 140 square miles of it for a pineapple plantation. After statehood (1959), management of the island was taken over by Castle & Cook, the island conglomerate that had bought Dole. Not surprisingly, Lanai is known as the pineapple island, although pineapple growing has decreased drastically because of cheaper imports from Asian countries. David Murdock, a California developer, bought the Castle & Cook property in 1987, plowed under most of the fields, and erected two luxury hotels and a golf course. Lanai is one of the least developed of the Hawaiian Islands. It has a permanent population of only 2,500, most of whom live in Lanai City. **See also Hawaiian Islands**

lancelet (*Amphioxus*) Named for their bladelike shape, lancelets are small, semitransparent, primitive animals that lack a cranium, brain, eyes, or other sense organs but have a stiffening rod known as a notochord that sets them apart from the invertebrates. In addition to the notochord, they also have several typically vertebrate features, such as chevron-shaped muscle blocks (myotomes), a digestive tract, and a vascular system. The animals' sperm structure and innervation of muscles more closely resemble those of echinoderms, so lancelets are not classified as true vertebrates. Less than 2 inches long, *Amphioxus* (sometimes known as *Branchiostoma*) has a buccal opening at one end and a small tail fin at the other. Lancelets breathe by means of numerous gill-like openings in the pharynx (pharyngeal slits), and they feed by straining particles through the tentaclelike cirri that are located where other creatures have jaws. Although they can swim for short distances, they spend most of their time buried in the sand with the anterior tentacles exposed. The 2-inch-long Caribbean lancelet (*B. caribaeum*), is found buried in sandy bottoms from the Chesapeake Bay south to the West Indies.

lancet fish (*Alepisaurus ferox*) The lancet fish is a swift, powerful predator of the depths that can reach a length

lancet fish

of 7 feet. It is light-bodied, however, and a 7-foot-long lancet fish weighs only 10 pounds. It is equipped with a mouthful of fearful fangs and, like many deepwater fishes, an emerald-green eye. In life, it is iridescent blue, green, and bronze, but the color fades when it is brought to the surface. Like the sailfish, the fastest fish of surface waters, the lancet fish is equipped with a greatly exaggerated dorsal fin, but this appendage cannot add much to its speed and is probably laid back when the animal races through the water. Lancet fishes are hermaphroditic, meaning they have both male and female sex organs. Many deep-sea fishes are known only from specimens found in lancet fish stomachs; the only two juvenile giant squid ever examined were acquired this way. **See also cutlass fish, sailfish, snake mackerel**

La Niña Where the El Niño phenomenon is marked by weakening of the trade winds and the warming of the sea-surface temperatures in the eastern tropical Pacific, La Niña (Spanish for a female child) is its cold counterpart, where surface temperatures fall below normal. It produces warm waters in the southeastern United States, colder winters from the Pacific Northwest to the Great Lakes, and unsettled winters in the Northeast and Mid-Atlantic regions. El Niño and La Niña are extreme phases of a naturally occurring climate cycle known as the El Niño/Southern Oscillation (ENSO). Some meteorologists believe that there are only two states, El Niño and non–El Niño, and that one or the other is always present to a greater or lesser degree. La Niña events usually—but not always—follow El Niños, at three- to five-year intervals, and last for as long as two years. A major La Niña event occurred between June 1988 and February 1989. **See also El Niño**

lantern fishes (family Myctophidae) With the bristlemouths, the lantern fishes are among the most numerous fishes in the world, but they are divided into many more species, perhaps as many as 250. The most populous genera are *Diaphus, Lampanyctus, Myctophum,* and *Lampadena.* They are differentiated on the basis of the arrangement of the fifty to eighty light organs (photophores) on the head, belly, and flanks, which emit a bright blue light, the color of an electric spark. The function of these bioluminescent organs is not self-evident; they may serve to camouflage the fish in the water column by eliminating its shadow (counter-illumination), or to provide a visual cohesion to the school, but a fish that lights up is also calling a potential predator's attention to itself. (Also like the bristlemouths, lantern fishes are believed to be a component of the deep-scattering layer.)

See also bioluminescence, bristlemouth, deep-scattering layer, photophore

lantern fish

lantern sharks (*Etmopterus* spp.) Small deep-sea sharks, rarely exceeding 24 inches in length, that are characterized by spinous dorsal fins, large eyes, and bioluminescent capabilities. There are seventeen recognized species distributed throughout all the world's oceans. Some of them are quite common (such as the velvet belly, *E. spinax,* and the green dogfish, *E. virens*), while others are known from only a few specimens. The teeth in the lantern shark's upper jaw are cusped and sharply pointed, while those in the lower are broad and fused into a single razorlike cutting band. Although they have not been observed in action in the depths, lantern sharks are believed to be schooling sharks that attack large prey in numbers, employing the bioluminescent organs to maintain the visual integrity of the schools. Many lantern sharks (and a few other shark species) have a pale yellowish spot on the top of the head, known as the pineal window, with which they are able to read the ambient light levels and adjust their bioluminescence. The blackbelly lantern shark (*E. lucifer*) is found close to landmasses throughout the Southern Hemisphere, while the great lantern shark (*E. princeps*) is the largest species, growing to 30 inches in the waters of the North Atlantic.

See also bioluminescence, green dogfish, velvet belly shark

La Pérouse, Jean-François (1741–1788?) French navigator who explored much of the Pacific in *La Boussole* (which means "compass"), accompanied by *L'Astrolabe* ("sextant"). Sailing from France in 1785, he rounded Cape Horn and visited Easter Island and the Sandwich Islands (now Hawaii) in his search for the Northwest Passage. Heading north, he put in at Alaska, and then sailed south along the North American coast as far as Monterey, California. He then crossed the Pacific to the east coast of Asia, visiting Macau and Manila in 1787.

La Pérouse then sailed north through the Sea of Japan and found the strait that is named for him, between Sakhalin Island and Hokkaido, Japan. He took his ships to Samoa, where eleven men of *L'Astrolabe* were murdered, thence to Tonga and Botany Bay in Australia, where the first settlement of convicts from England had just arrived. After La Pérouse left Botany Bay, his ships were wrecked on Vanikoro, in the Santa Cruz group of the Solomon Islands. Admiral Bruni D'Entrecasteaux was sent to look for him in 1791, but he passed Vanikoro without finding any signs of *L'Astrolabe*. The wreckage was discovered in 1828 by Dumont d'Urville, sailing in a ship that had been named *Astrolabe* for the lost vessel of La Pérouse.

Laptev Sea A division of the Arctic Ocean on the northern Siberian coast, between the Taymyr Peninsula and the New Siberian Islands, immediately to the east of the Kara Sea. It covers about 276,000 square miles; its average depth is 1,896 feet. This shallow sea is frozen for most of the year and navigable only during August and September; even then, icebreakers are required. It was originally called the Nordenskjöld Sea after the Swedish explorer Adolf Nordenskjöld, but the Russians renamed it after Dmitri and Khariton Laptev, two members of Bering's second Arctic expedition in 1742.

See also Arctic Ocean, Bering, Kara Sea

Larsen, Carl Anton (1860–1924) Norwegian whaler, held to be the man responsible for large-scale whaling in the Antarctic. During two Antarctic voyages as a sealing captain (1892–1894), he realized that the large rorquals (blue, fin, and sei whales) and the plentiful humpbacks represented an unprecedented bonanza for whalers, but not if they had to make an annual voyage from Norway. While sailing with Otto Nordenskjöld aboard the sealer *Antarctic* in 1903, Larsen explored the South Shetlands and South Georgia, but their ship was trapped in the ice and they spent the winter eating penguins. They were rescued by Argentines and taken to Buenos Aires, where Larsen raised the money for a joint Norwegian-Argentine whaling company, Compañia Argentina de Pesca. They returned to South Georgia and built the first whaling station at Grytviken. Larsen remained active in whaling until his death aboard the Norwegian factory ship *Sir James Clark Ross* in 1924. The Larsen Ice Shelf in the Ross Sea is named for him.

lateen sail Its name derived from the word "Latin," meaning "of the Mediterranean," this was an innovation thought to have been introduced to the eastern Mediterranean around the second century A.D., by Arabs of the Persian Gulf. It was a marked improvement in navigation. The square sail used previously permitted sailing before the wind only, but with lateen rigging—a long sloping yard slung from the top of the forward-leaning mast—the ship could take the wind on either side. The ability to tack greatly increased the maneuverability of the sailing ship. It is still used on the Mediterranean felucca and the Arabian dhow.

lateral line A system of sense organs that runs along the side of most bony fishes and sharks, from the gill openings to the tail. It is composed of mucus-filled canals under the skin that contain sensory cells known as neuromasts; each one has a fine hairlike projection above the epithelium, encased in a gelatinous cup. The cells are sensitive to disturbances near the fish and to low-frequency sounds, and the system is used to detect the presence of other sea creatures, obstructions, and directional changes in the flow of water. It can therefore be described as either a long-distance touch organ or a low-frequency hearing system. The lateral-line system is also used to coordinate the movements of schooling fishes. In many species of deep-sea fishes, such as anglers and gulper eels, there are no lateral-line canals, but the sensory organs are carried on the ends of skin papillae and are directly exposed to the water. In some species, such as those with reduced or nonexistent eyes, the lateral line is the primary sensory organ.

See also ampullae of Lorenzini

latitude From the Latin *latitudo,* "breadth," latitude is used to identify a point on the surface of the earth measured by the angular distance from the equator, which is 0° of latitude. Lines of latitude are parallel to the equator. **See also equator, longitude**

Lauria, Ruggiero de (d. 1305) Italian admiral (known in English as Roger Lauria) in the service of Aragon and Sicily, considered the most remarkable naval commander of his age. He won important naval victories over the Angevin kings in the war that followed the Sicilian Vespers, an uprising and massacre of the ruling French, which led to the separation of Sicily from Naples in 1282. He was named grand admiral by Peter III of Aragon, and for the next twenty years, he was undefeated at sea. Where his opposition relied on grappling and boarding, de Lauria's crews used the ram and the crossbow; they developed gigantic crossbows that were used as guns would be later. He overcame the French at Malta in 1283, and a year later captured the enemy fleet's commander, the future King Charles II of Naples. De Lauria subsequently defeated the French fleet off Catalonia, and conquered the island of Majorca. He also served Peter's successor, James II of Aragon, who had become king of Sicily in 1285, but when James ceded Sicily to the pope, who bestowed it on Charles II of Naples, de Lauria fought with the Angevins against the Sicilians. After the Peace of Caltabellotta in 1302, de Lauria retired to his estates in Spain and died peacefully two years later.

lava gull (*Larus fulginosus*) Found only in the Galápagos, the lava gull is a small (wingspan: 40 inches) gray gull with a sooty-brown head and white crescents above and below the eyes. It is a coastal scavenger, often seen in the vicinity of fishing camps or other settlements. It rarely alights on water. The lava gull is among the rarest of all seabirds; there are only 300 to 400 on the islands. The only other gull that breeds in the Galápagos is the much larger swallow-tailed gull (*L. furcatus*), easily identified by its forked tail.

See also Galápagos Islands, gull, swallow-tailed gull

Law of the Sea After several preliminary (and unproductive) conferences, beginning in Geneva in 1958, the Law of the Sea treaty was passed at Montego Bay, Jamaica, on December 10, 1982. It was intended to codify international law regarding territorial waters, ocean dumping, fishing and fish stocks, endangered species, sea-lanes, and ocean resources. According to the treaty, each nation's territorial waters extend 12 miles beyond its coasts, but foreign commercial vessels are permitted "innocent passage" through the 12-mile zone. In addition, every sovereign nation has exclusive rights to the fish and other marine life in waters up to 200 miles from shore; in those cases where countries are separated by less than 400 miles, an agreement must be negotiated. The oil, gas, and mineral resources of the continental shelf of any nation belong to that nation, up to 200 miles from shore. According to the treaty, minerals on the ocean floor are "the common heritage of mankind," but because they believed that unrestricted seabed mining would inhibit free enterprise, the United States, the United Kingdom, West Germany, Israel, Italy, and several other industrialized nations did not sign the original treaty. After another twelve years of negotiations, the United States ratified the treaty on July 29, 1994, and the treaty took effect in November. As of 1997, 116 nations were party to the United Nations Convention on the Law of the Sea (UNCLOS).

See also mare clausum, salvage

Laysan albatross (*Diomedea immutabilis*) This dark-backed albatross with a white head and white underparts breeds in the North Pacific on the islands of the northwest Hawaiian chain (Laysan, Midway, Lisianski, and Pearl and Hermes Reef) and has recently been reported nesting on Kauai. When airstrips were built on Laysan and Midway, albatrosses roamed freely over the landing strips, and there were frequent bird-plane collisions, which caused all sorts of problems for the planes (not to mention the birds) and led to the name "gooney birds." Black-footed albatrosses (*D. nigripes*) breed on many of the same islands, but they nest in bare open sites, while the Laysan prefers some scrub or grass cover. The wingspan averages around 80 inches.

See also albatross, black-footed albatross, waved albatross

Laysan Island Like Midway and Lisianski Islands, Laysan Island is the peak of a submerged volcano at the northwestern end of the Hawaiian chain in the North Pacific. It is not known how many times the islands were visited beforehand; the first official observation was made in 1857 by Captain John Paty, aboard the schooner *Manu-o-ka-wai,* when he annexed the islands

leafy sea dragon

to Hawaii. The islands were declared a U.S. possession in 1867 and made a naval reservation in 1903. Guano collectors introduced rabbits and guinea pigs; they ate every plant, turning the places into deserts, and three native species of birds became extinct. The vegetation has come back, and many birds now breed on the islands, including shearwaters, terns, boobies, tropicbirds, an extremely rare duck known as the Laysan teal, and the Laysan albatross.

Leach's storm petrel (*Oceanodroma leucorhoa*) With its deeply forked tail and white rump patch, this long-winged little bird is easily identifiable as it swoops, darts, and suddenly changes direction. Its 19-inch wingspan makes it larger than the British or Wilson's storm petrel. It breeds in the North Atlantic (Nova Scotia, Maine, Massachusetts, Faeroes, the Shetlands, St. Kilda, etc.) and in the North Pacific (Japan, Aleutians, Alaska), and then disperses southward, appearing in the equatorial Pacific and the South Atlantic, as far as the Cape of Good Hope. Because it is so widespread, this bird may be confused with many other species. **See also British storm petrel, petrel, Wilson's storm petrel**

leafy sea dragon (*Phycodurus eques*) This 1-foot-long Australian sea horse is abundantly festooned with branched appendages that camouflage it in the weedy bottoms or kelp forests in which it lives. There are several species of these "leafy" sea horses, all found in Australasian waters, including the ribboned sea dragon (*Haliichthys taeniophora*) and the weedy sea dragon (*Phyllopteryx taeniolatus*), which has only a few appendages. **See also pipefish, sea horse**

least auklet (*Aethia pusilla*) One of the most numerous birds in the North Pacific, the least auklet occurs in vast numbers; some colonies may be 1 million strong. There are believed to be 6 million least auklets in Alaska. It is also the smallest alcid, with only a 6-inch wingspan. It is sooty black above and white below, with a red bill and white irises. Also found in the North Pacific are two close relatives, the crested auklet (*Aethia cristatella*), which has a forward-drooping feather crest, and the whiskered auklet (*A. pygmaea*). The rhinoceros auklet (*Cerorhinca monocerata*) is a somewhat larger bird, also found in the North Pacific, that has a "horn" that erupts at the base of the bill during the breeding season.

leatherback turtle (*Dermochelys coriacea*) The largest of the sea turtles, and also the heaviest living reptile in the world, the leatherback can be 8½ feet long and weigh more than 2,000 pounds. It is the deepest-diving sea turtle, capable of reaching depths of 3,000 feet; only the elephant seal and sperm whale can dive deeper.

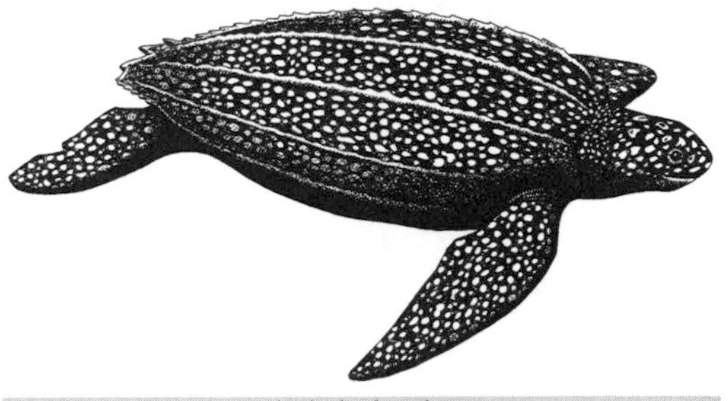
leatherback turtle

Leatherbacks can regulate their body temperature, an ability unique among reptiles. Where all other sea turtles have a shell or carapace, the leatherback has smooth, leathery skin that covers seven prominent bony ridges on the back and five on the underside. Its mouth and throat are equipped with backward-pointing spines that enable it to eat its favorite food, the slippery jellyfish. (Unfortunately, the turtles sometimes mistake floating plastic bags for their usual prey, then choke to death.) Females gather on the beaches of Indonesia, New Guinea, Central America, the Guianas, and the Pacific coast of Mexico to lay their eggs. Marked turtles have been found as far as 3,000 miles from their nesting beaches. **See also sea turtles**

Leeward Islands (1) Sometimes referred to as the Northwest Hawaiian Islands, the Leewards are a chain of islands stretching about 1,300 miles to the west of Kauai, the westernmost of the main Hawaiian islands. Officially part of Honolulu but administered by the U.S. Navy, the group consists of Nihoa, Necker, French Frigate Shoals, Gardner Pinnacles, Maro Reef, Laysan, Lisianski, Pearl and Hermes Reef, Midway, and Kure Atoll. Some 2,000 navy personnel are stationed on Midway (which is not considered part of Hawaii), and there is a tracking station on Tern Island in French Frigate Shoals, but otherwise the Leewards are uninhabited by humans. They abound in bird life, however, and millions of terns, boobies, frigate birds, plovers, curlews, and Laysan albatrosses are in constant motion above and around the islands. The islands are also the last refuge of the endangered Hawaiian monk seal. Eight of the islands have been designated as the Hawaiian Islands National Wildlife Refuge. **See also Laysan Island, Midway Island**

Leeward Islands (2) The northern group of the Lesser Antilles in the Caribbean, extending from Puerto Rico to the Windward Islands. The designation is a geographical rather than a political one, since the islands are under different jurisdictions. The northernmost group consists of the American and British Virgin Islands, then the jointly French- and Dutch-owned island of St. Martin, St. Barthélemy (France), St.

Eustatius (Dutch), St. Kitts and Nevis (independent), Montserrat (United Kingdom), Antigua and Barbuda (independent), and Guadeloupe (France). The islands, many of which were first sighted by Columbus in 1493, are volcanic in origin and mostly characterized by tropical vegetation, abundant rainfall, and a pleasant climate, which makes almost all of them popular tourist destinations. A notable exception is Montserrat, where the volcanic Soufrière Hills began to erupt in 1995; they have been in a state of constant agitation ever since, blanketing most of the island in ash and causing the residents to flee for their lives.

Leif Ericsson (active c. 1000) Generally accepted as the first European to set foot in North America. The son of Eric the Red (therefore "Leif Erics-son"), he was born in Iceland and spent his youth in Greenland. Around 999, after converting to Christianity, he was commissioned by King Olaf I of Norway to carry the faith to Greenland. Blown off course to the south, he discovered an unknown land that he named Vinland for the wild grapes growing there. No one knows exactly where on the North American continent he landed, but many historians believe it was at L'Anse aux Meadows, on the northern coast of Newfoundland. (According to Icelandic sagas, a trader named Bjarni Herjolfsson may have been the discoverer of America, but he is believed only to have sighted the coast of what is now Labrador and never landed there.) **See also Eric the Red**

Le Maire, Jakob (1585–1616) Dutch navigator who sailed in 1616 with Willem Cornelisz Schouten in the *Eendracht* and the *Hoorn,* in a voyage to discover a way around the tip of South America that was not controlled by the powerful Dutch East India Company. They were seeking the riches from the southern continent, Terra Australis Incognita. On the coast of Patagonia, the *Hoorn* caught fire and was destroyed, so both crews piled onto the *Eendracht* for the remainder of the voyage. They found a channel between Staten Island (Isla de los Estados) and Tierra del Fuego, which they named the Le Maire Strait. When they rounded South America, they named the tip Cape Hoorn after their home port, but it soon became known as Cape Horn. They reached a previously undiscovered group that they christened the Hoorn Islands (now, like the cape, known as Horn), and became the first Europeans to visit the Tonga Islands. When they arrived at Jakarta, the governor-general refused to believe they had not passed through the Strait of Magellan, confiscated their ship, and sent them back to Holland. Le Maire died on the way home at the age of thirty-one. The Le Maire Strait (Estrecho de Le Maire) between the tip of Tierra del Fuego and Staten Island is named for him, but the picturesque Lemaire Channel off the Antarctic Peninsula was named (by Adrien de Gerlache) for Charles Lemaire, a Belgian explorer in the Congo.

See also Magellan, Strait of; Schouten; Tierra del Fuego, Tonga

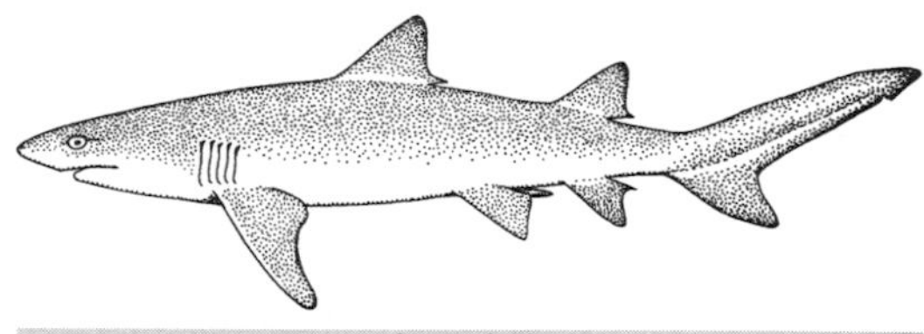

lemon shark

lemon shark (*Negaprion brevirostris*) A carcharhinid shark that does well in captivity, the lemon shark is often seen in aquariums and oceanariums. It gets its name from a slightly yellowish cast to its otherwise brownish coloration. Lemon sharks can be easily differentiated from other shark species—especially bull sharks, which they closely resemble—by the two dorsal fins of approximately the same size. Lemon sharks are found in the western North Atlantic, from New Jersey to Brazil, and in the eastern Pacific from the Sea of Cortez to Ecuador. They reach a maximum length of 10 feet and are potentially dangerous to swimmers.

See also bull shark, carcharhinid sharks

leopard seal

leopard seal (*Hydrurga leptonyx*) Named for its spots as well as its predaceous habits, the leopard seal is found throughout the Antarctic and surrounding island groups. Leopard seals are active carnivores, feeding on krill (which they strain from the water in the manner of a crabeater seal), fish, squid, other seals, and penguins—particularly Adélies—which they devour by grasping the birds in their powerful jaws and shaking them out of their skin. Unlike many other seals and sea lions, leopard seals are solitary, and it is rare to encounter more than one at a time. Unusual for pinnipeds, female leopard seals are larger than males. The only animal powerful enough to prey on a leopard seal is a killer whale. **See also Adélie penguin, crabeater seal, killer whale**

leopard shark (*Triakis semifasciata*) A small (3 to 5 feet in length) shark found in the shallow waters of the Pa-

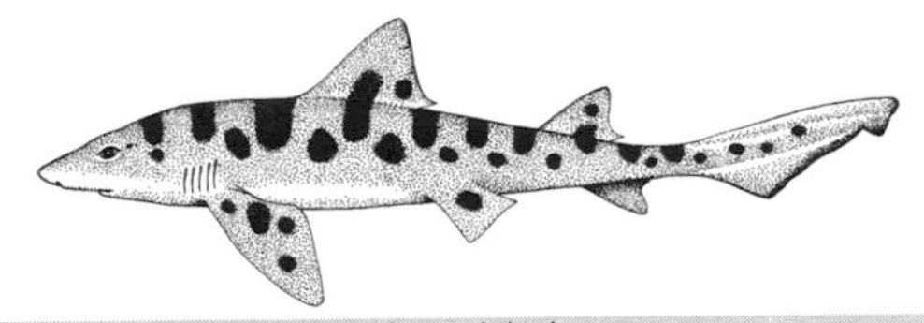

leopard shark

cific coast of the United States. It is yellowish gray and prominently marked with transverse black bands on its back and spots on its sides. It is generally considered harmless to man. **See also white-tipped reef shark**

Lepanto, Battle of After the Ottoman Turks took Constantinople in 1453, they were determined to control the entire Mediterranean. On May 25, 1571, to stop the Turkish expansion, an alliance was signed between the Republic of Venice, the Kingdom of Spain, and the Dominions of the Pope. Don John, the twenty-five-year-old bastard son of Emperor Charles V of Austria (and the half brother of Philip II of Spain), was appointed commander-in-chief of the fleet that was to engage the Turks. The Spanish, led by the Marquis de Santa Cruz, contributed eighty galleys, twenty-two other vessels, and 21,000 fighting men. Don John had sixty-four galleys; the Genoan Giovanni Andrea Doria (nephew of the famous admiral) commanded fifty-four; and the Venetians sent more than one hundred vessels, most of which were galleasses armed with cannons. The commander of the Turkish fleet was Ali Pasha; his second-in-command was Uluch Ali, a Calabrian fisherman who had been captured and enslaved by Turkish corsairs. On October 7, 1571, at Lepanto (now Návpaktos), in what is now the Gulf of Corinth, four squadrons of the Christian league met three hundred vessels of the Turkish fleet in the largest naval battle fought up to that time, and the last battle contested by ships rowed by slaves. In fierce fighting and, eventually, hand-to-hand combat, the Christian league defeated the Turks. The battle was of little strategic importance because Venice surrendered Cyprus to the Turks in 1573, but it was immortalized in the contemporary paintings of Titian, Tintoretto, and Veronese. Wounded while serving aboard *La Marquesa*, one of Andrea Doria's galleys, was Miguel de Cervantes, who would achieve greater fame than anyone else at Lepanto: in 1605, he would write *Don Quixote.* **See also galleass, galley**

leptocephalus From the Greek *leptos,* meaning "slender," and *cephalos,* meaning "head," leptocephalus refers to a small-headed, transparent, ribbonlike larval phase that eels, gulpers, and—surprisingly—tarpons, ladyfishes, and bonefishes pass through before metamorphosing into their more familiar shapes. Leptocephali are so different from the adult forms that until the middle of the nineteenth century they were believed to be a distinct group of fishes. The metamorphosis is not merely growth, but rather a profound process whereby the small head grows, the skin thickens to form the mouth and snout, the larval teeth are resorbed, and the digestive tract, which has been confined to a narrow strip on the ventral surface of the body, begins a radical reconstruction. (No leptocephalus has ever been found with food of any kind in its gut.) During metamorphosis, the animal becomes rounder in cross-section and it often *decreases* in length before it begins the normal growth process. The longest leptocephalus ever found was 6 feet long. It was caught at a depth of 1,000 feet off the coast of South Africa in 1930, and led to some wild speculation about 100-foot-long eels. Although the species of *Leptocephalus giganteus* has not been identified, it is likely that it would have matured into some kind of eel that is less than 6 feet long. **See also bonefish, eel, gulper eel, ladyfish, tarpon**

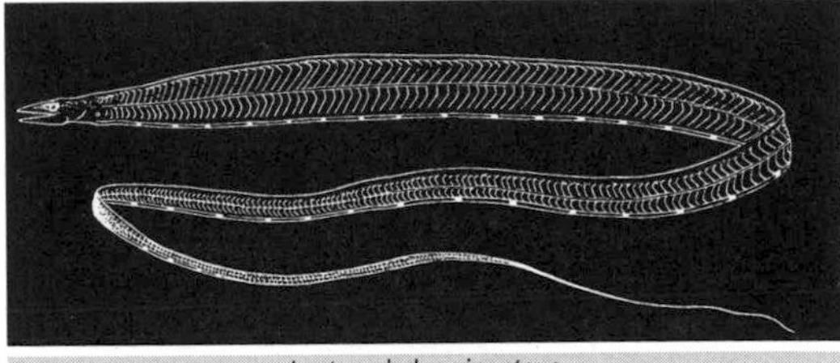

Leptocephalus giganteus

Lesbos Sometimes known as Mitilíni after its capital city, Lesbos is the third largest island in Greece after Crete and Euboea. Located about 10 miles off the coast of Turkey, the island is marked by two deep gulfs, Kallonis and Geras, and supports a population of about 88,000 people. In the sixth century B.C., the island was a center of artistic and philosophical achievement; it was home to the musician Terpander, the poet Arian, and Lesbos's most famous citizen, the lyric poet Sappho. Although little of her poetry has been preserved, Sappho wrote mostly for women, and the term "lesbian" is derived from her home island. Like many of the Greek islands, it was invaded many times: in 527 B.C. by the Persians, by the Athenians in 479 B.C., and in 70 B.C. by Julius Caesar. Subsequent waves of Byzantines, Venetians, Genoese, and Turks left their marks on the island before it joined the Greek kingdom in 1911.

Lesseps, Ferdinand de (1805–1894) French diplomat and engineer who directed the building of the Suez Canal. He served at Lisbon, Tunis, and Alexandria before being posted to Cairo in 1833. He was then stationed in Rotterdam, Malaga, Barcelona, and Madrid, where he was appointed consul general in 1848. After retiring from the diplomatic service, he conceived the idea of a canal across the isthmus of Suez, and was invited by the viceroy, or khedive, of Egypt to direct the building of the Suez Canal. Under de Lesseps's direction, the canal was built under the aegis of a joint French-Egyptian stock company. Construction began in 1859; it was officially opened ten years later. When the new khedive of Egypt experienced financial difficulties in 1875, he sold his 44 percent share to the British government at the instigation of British prime minister Benjamin Disraeli. In 1878, de Lesseps assumed the presidency of a French company dedicated to building a canal across the isthmus of Panama; work began in 1881. The project went bankrupt seven years later, and after an official inquiry into the massive failure, the French government sentenced de Lesseps to prison for misappropriation of funds. The sentence was commuted by an appeals court. **See also Panama Canal, Suez Canal**

letter of marque A commission issued in Britain by the lord high admiral and by equivalent authorities in other countries, licensing the commander of a privately owned vessel to cruise in search of enemy merchant vessels and seize them. Ships equipped with letters of marque were known as privateers, and they often made little distinction between friendly and unfriendly shipping, acting more like pirates than authorized men of war. Privateering was outlawed by the Declaration of Paris of 1859, but this treaty was not signed by the United States, and privateering persisted in American waters through the U.S. Civil War.

See also piracy, privateer

leviathan In the Old Testament, an evil aquatic monster with shining eyes and a scaly body, usually perceived as a whale, a crocodile, or a dragon. The English political theorist and philosopher Thomas Hobbes (1588–1679) wrote a treatise he called *Leviathan,* in which he wrote that the life of man is "poor, nasty, brutish, and short." More recently, the term is used to refer to whales or, by extension, a very large ship. The original name of Brunel's *Great Eastern* (when it was launched in 1858, the largest ship ever built) was *Leviathan.* In 1901, a 14,000-ton British armored cruiser was christened *Leviathan,* and when the Americans took the German liner *Vaterland* before they entered World War I, they renamed her *Leviathan.*

Lewis with Harris, Isle of Separated from the Scottish mainland by the 24-mile-wide North Minch, Lewis with Harris is a single island in the Outer Hebrides, although it is often thought of as two islands. The larger and more northerly part of the island is known as Lewis; Harris is the smaller, southerly portion. Small tenant farmers known as crofters raise meager crops, but the best land is reserved for sheep and cattle, and although fishing was important in the past, it has declined in recent years. Weaving the woolen cloth known as Harris tweed is the major cottage industry of Lewis with Harris, and its production and sale contributes substantially to the tourist industry. In 1919, Lord Leverhulme purchased Harris and spent more than £500,000 in a futile attempt to develop the island. The population of the island is approximately 23,000 people, many of whom speak Gaelic and adhere to the old cultural mores. **See also Hebrides**

Leyte Gulf, Battle of Important sea battle of World War II, which took place October 23–26, 1944; it crippled the Japanese fleet and enabled the United States to retake the Philippines. It was a last desperate attempt by the Japanese to prevail in the Pacific, for they knew that if the Philippines were lost, the war was lost. The plan was that, under Vice Admiral Jisaburo Ozawa, four aircraft carriers and a dozen other ships would come down from the north and draw off the main American forces. Two battleship forces would penetrate the central Philippines and converge on American invasion ships in Leyte Gulf, on the eastern perimeter of the central Philippines. The more powerful of these forces, under the command of Vice Admiral Takeo Kurita, consisted of five battleships, among them the *Yamato* and the *Musashi,* the largest battleships in the world; twelve heavy cruisers, and nineteen destroyers. The American Seventh Fleet consisted of 738 vessels (most of which were amphibious landing craft) under Vice Admiral Thomas C. Kinkaid; it was supported by the Third Fleet under Admiral William Halsey. American submarines found and sank two Japanese heavy cruisers west of Palawan on October 23, 1944, and there followed an almost continuous air and sea conflict. By October 24, three major battles were being fought, almost simultaneously. At the Surigao Strait, battleships and cruisers of the American Seventh Fleet destroyed the battleships *Fuso* and *Yamashiro* and forced the remaining Japanese ships to withdraw. Meanwhile, off the island of Samar, the Japanese attack force passed through the unguarded San Bernardino Strait, inflicted heavy damage on the American Essex-class carriers (often by kamikaze suicide pilots), and sank several destroyers. Just as it appeared that the Japanese would triumph, Admiral Kurita unexpectedly ordered his forces to break off the action and retreated. The third major engagement took place at Cape Engaño, where Halsey's Third Fleet was chasing Kurita's retreating task force. Poor communications between Kinkaid and Halsey almost lost the battle for the Americans, but when the smoke cleared, it was apparent that the Japanese fleet had effectively ceased to exist. In the various engagements of Leyte, the Japanese lost twenty-six ships; the Americans six. The path was cleared for General MacArthur to lead the triumphant American return to the Philippines.

Liberty Island Ten-square-acre island in New York harbor, just south of Ellis Island. Originally Bedloe's Island, it was renamed Liberty Island in 1956 because it is the site of the Statue of Liberty. When a group of Frenchmen decided to present a monument to the United States on the occasion of America's one hundredth birthday, they commissioned sculptor Frédéric-Auguste Bartholdi to design it. As Bartholdi sailed into New York harbor in 1871, he passed Bedloe's Island, which at that time was the home of star-shaped Fort Wood. At that moment, he decided to design a statue of a woman with a torch in her hand, which would stand on a star-shaped base. He titled the statue *Liberty Enlightening the World.* Construction began in 1875, but it was not completed in time for the centennial; the hand holding the torch was the only part that was shipped by 1876. It was exhibited in Philadelphia and later in New York City. The statue, based upon a likeness of Bartholdi's mother, was made of copper plate, supported by an armature designed by Gustave Eiffel. It stands 151 feet high on top of an 89-foot pedestal, atop a 65-foot-high base. The statue and Ellis Island now make up the Statue of Liberty National Monument, accessible by ferry from New York and New Jersey.

See also Ellis Island

liberty ships Mass-produced, prefabricated merchant vessels with all-welded hulls made in U.S. shipyards between 1941 and 1945 to replace tonnage sunk by Axis submarines during World War II. Working around the clock, American builders created 2,770 liberty ships, which were outfitted as cargo vessels, aircraft transports, and tankers. The record time for the construction of a single ship was 80½ hours—just over three days. The entire project was developed and directed by Henry J. Kaiser (1882–1967), an American industrialist who had also been the force behind the building of Hoover, Bonneville, and Grand Coulee Dams and the San Francisco–Oakland Bridge.

lighthouse A building or other structure erected to display a light as a warning of danger or as an aid to navigation. The forerunners of lighthouses were beacon fires kindled on hilltops. One of the earliest man-made lighthouses was the Pharos of Alexandria, built during the reign of Ptolemy II (283–247 B.C.) and said to be 600 feet high. In the first century A.D., the Romans erected lighthouses at Dover in Britain and in Italy at Ostia, Ravenna, and Messina. These and those that followed were lit by wood or coal fires in braziers, although oil lamps were sometimes used. In 1584, a magnificent lighthouse was begun on the island of Cordouan in the estuary of the Gironde River near Bordeaux. It was 100 feet high, with a richly decorated interior. By the time the lighthouse was completed in 1611, the island was completely submerged, making it the first lighthouse at sea. In 1699, Henry Winstanley designed the Eddystone Light off Plymouth, England, but it was swept away in a storm four years later, taking Winstanley with it. Other Eddystone lights followed; the current one was built in 1882. The first lighthouse on the North American continent was built in 1716 on the island of Little Brewster off Boston. Oil and acetylene lamps were used in the beginning of the twentieth century, and some were designed so that they self-extinguished in daylight. At Helgoland in Germany in 1913, an arc lamp produced 38 million candlepower, the most powerful lighthouse in the world. Electric filament lamps, now the standard illuminant, were introduced during the 1920s. As early as 1777, mirrors were used to concentrate the light into a beam, and the revolving beam was invented in Sweden in 1781. In 1828, Augustin Fresnel of France developed the refractive lens that now bears his name, a central bull's-eye with concentric prismatic rings that greatly increased the candlepower of the light source. For the most part, lighthouse keepers have now been replaced by automated systems.

lightship Where fixed structures were impracticable, lightships were used to establish fixed marks. They are usually nonnavigable vessels, moored over a shoal or bank, with a lighthouse structure amidships. The first lightship was at Nore Sands in the Thames estuary, but others quickly followed. Many are now equipped with sound equipment as well as lights. Their very nature makes them prime candidates for automation, and many have now been replaced by buoys.

limpet (class Gastropoda) Small, flattened snails found on littoral rocks. They are mostly herbivorous, scraping algae from rocks and seaweeds, but the keyhole limpet (*Diodora aspera*) feeds primarily on bryozoans. They can trap water beneath their shells, so they do not have to remain submerged in order to breathe. Most species are about 2 inches long, but the giant keyhole limpet (*Megathura crenulata*) of Southern and Baja California reaches a length of 5 inches. **See also bryozoan**

Lind, James (1716–1794) British naval surgeon, considered the father of naval hygiene, who discovered that scurvy could be prevented by ingesting the juice of limes, lemons, or oranges. Although Lind published his *Treatise of the Scurvy* in 1753, his recommendations were not officially adopted until 1795. Limes were used by the British navy (hence the name "limeys" for British sailors) because they were grown in the British West Indies, but lemons, grown in Mediterranean countries, are a much more effective antiscorbutic.

See also scurvy

lingcod (*Ophiodon elongatus*) These colorful relatives of the greenling (*Hexagrammos decagrammus*) are

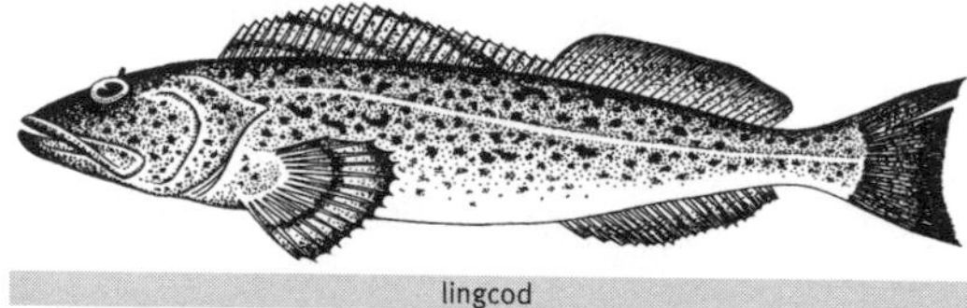
lingcod

found in the Pacific from Alaska to Baja California, where they are caught commercially and by sport fishermen. The green color of the flesh vanishes when the fish is cooked. Female lingcod can reach a length of five feet, males only half that. **See also greenling, Pacific cod**

Linschoten, Jan Huyghen van (1563–1611) Dutch explorer who sailed for India in 1583 as bookkeeper to the Portuguese archbishop of Goa, and remained there until 1590. On the voyage home, his ship was wrecked in the Azores. He spent two years there and wrote a history of Sir Richard Grenville's valiant battle in the *Revenge*, which took place off the island of Flores in 1591. He also wrote two books about the people and customs of India. In 1594 he accompanied Willem Barents from Holland on a voyage to search for the Northeast Passage to India. While Barents only got as far as Novaya Zemlya, Linschoten reached the Kara Sea and reported open water to the north. Believing he had found a northern route to China and India, he returned to spread the news. The following year seven ships attempted the passage, but they were blocked by ice. Linschoten wrote *Itinerario* (1596), a chronicle and atlas of his voyages, which, among other things, pointed out the scattered nature of the Portuguese outposts in India. His writings encouraged other nations to establish colonies there, and led to the formation of the British and Dutch East India Companies, established in 1600 and 1602 respectively, which were dedicated to the exploration and exploitation of the Indies.

lionfish (*Pterois volitans*) Because of its spectacular appearance, this member of the scorpionfish family has acquired numerous popular names, including turkeyfish, firefish, and zebrafish. Found throughout the reefs of the Indo-Pacific, *P. volitans* (the name means "winged flyer") is beautiful and deadly. The dorsal spines of this 15-inch-long fish are poisonous, although the venom is not as powerful as that of the stonefish, another scorpionfish. Lionfishes swim slowly around reefs, often in pairs, unafraid of anything, including divers, and will aggressively point their spines at a potential predator. Lionfishes are rarely found in the stomach contents of any fish, no matter how large. **See also scorpionfish, stonefish**

lionfish

lion's mane jellyfish (*Cyanea capillata*) The world's largest jellyfish, with a bell that can reach 8 feet across and tentacles that can be 200 feet long; known sometimes as the sea blubber. The umbrella has eight primary lobes and up to thirty-two lappets, and the tentacles are arranged in eight clusters. The yellowish or reddish tentacles hang down in great clumps, the gaps between them revealing the purplish mouth lobes. As with all jellyfish, the tentacles are equipped with nematocysts, and *Cyanea* is a powerful stinger. (One of the Sherlock Holmes stories, "The Adventure of the Lion's Mane," is about a fatal case of *Cyanea* poisoning.) It is found throughout the North Atlantic; similar species are found in the North Pacific and the Southern Hemisphere. **See also jellyfish**

Lipari Islands: See Aeolian Islands

Lisianski, Yuri (1773–1837) Russian explorer who accompanied Adam Krusenstern on his 'round-the-world voyage at the behest of Tsar Alexander I. The ships *Nadezhda* and *Neva* explored the North Pacific from 1803 to 1806. Their outward voyage was by way of Cape Horn, and they returned by rounding the Cape of Good Hope, thus becoming the first Russian circumnavigators. Lisianski Island, at the far western end of the Leeward Hawaiian chain, is named for him.

Lissa Now belonging to Croatia and known as Vis, Lissa is an island off the Dalmatian coast in the Adriatic Sea. In 1811, ten Franco-Venetian ships under the command of Commodore Dubordieu attempted to take the island, which was defended by four British frigates under the command of William Hoste. The British soundly defeated Dubordieu, and Hoste became a hero. In 1866 the waters of Lissa were again the scene of a major naval battle, this one between the Austrians under Admiral Wilhelm von Tegetthof and the Italians under Admiral Carlo di Persano. In the first battle between armored ships, the Italian fleet of twelve ironclads and fourteen wooden vessels attempted to take the Austrian-held island, but the Austrians brought up seven ironclads and seven wooden ships, and Tegetthof won the day by ramming the Italians. Because of the Austrian success at Lissa, the ram was a prominent feature on warships for the next forty years.

little auk (*Alle alle*) Also known as the dovekie, the little auk has a small bill, a chubby little body, and minute, whirring wings. This 8-inch-tall bird is black above and white below, and resembles the Southern Hemisphere diving petrels, to which it is not related. Little auks breed on most Arctic coasts, often nesting in huge, noisy colonies. Severe storms sometimes blow large numbers of these birds far inland; these "wrecks" have taken the birds as far as the Great Lakes of North America. **See also diving petrel, great auk, guillemot**

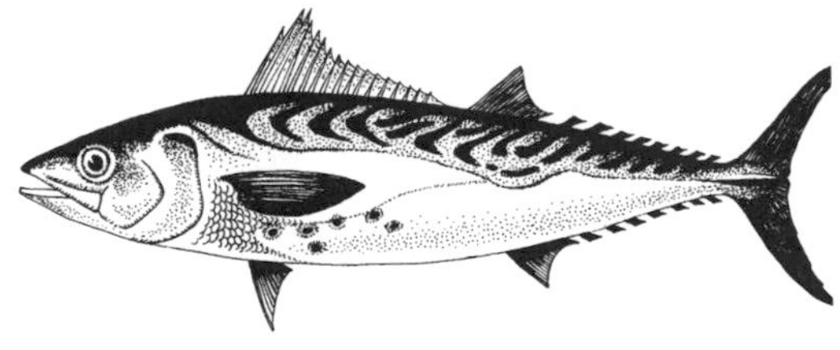

little tunny

little tunny (*Euthynnus alleteratus*) An inhabitant of the warmer waters of the western Atlantic from New England to Brazil and, in the east, from Great Britain and the Mediterranean to South Africa, the little tunny is a smallish tuna that aggregates in very large schools. It reaches a maximum length of 2½ feet and can be differentiated from similar-sized tunas, such as the skipjack (*Katsuwonus pelamis*) and the Atlantic bonito (*Sarda sarda*), by the presence of six black, fingerprint-like spots between the pectoral and ventral fins. It is a popular game fish, but because its flesh is dark, it does not have the commercial importance of some other tunas. **See also bluefin tuna, bonito, skipjack**

Lituya Bay Deep bay in southeast Alaska, now located within the borders of Glacier Bay National Park. It was the scene of a violent earthquake on July 10, 1958. The earthquake (magnitude 8.3 on the Richter Scale) caused a great piece of the hillside—estimated at 90 million tons—at the head of the bay to fall, creating a wave 1,740 feet high, the largest wave ever recorded. That the water reached this unbelievable height was determined from the examination of trees on the slope that had been knocked down by the wave, which was technically known as a swash. Three salmon trollers were anchored at the mouth of the bay: the *Summore* sank immediately; the *Badger* was carried over the trees and sank in the ocean, but her crew was rescued; and the *Edrie*, with Howard Ulrich and his seven-year-old son aboard, snapped her anchor chain and floated to safety.

lizardfish (family Synodontidae) Shallow-water, cigar-shaped fishes with big mouths, lizardfishes are named for their general reptilian appearance—especially the head—and their habit of waiting motionless on the

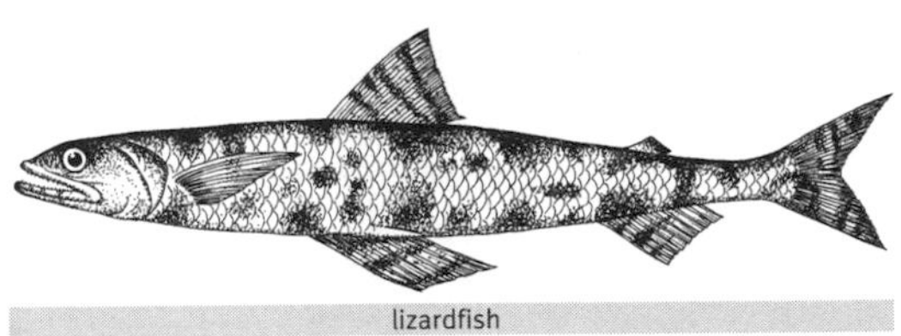

lizardfish

bottom, propped up on their ventral fins, until prey swims by, whereupon they dart rapidly upward to snag it. Most of the forty-odd species are less than a foot in length and have a blotched color pattern. They are found around the world.
See also Bombay duck, lancet fish

lobster (*Homarus americanus*) The familiar lobster is a carnivorous scavenger found only off the east coast of North America, from Labrador to North Carolina, from subtidal shallow waters to the edge of the continental shelf. They are most abundant in deeper water, down to 1,200 feet. The American lobster (also known as the northern lobster) is dark green or brownish green above and yellowish underneath. They are characterized by large pinching claws, which the spiny

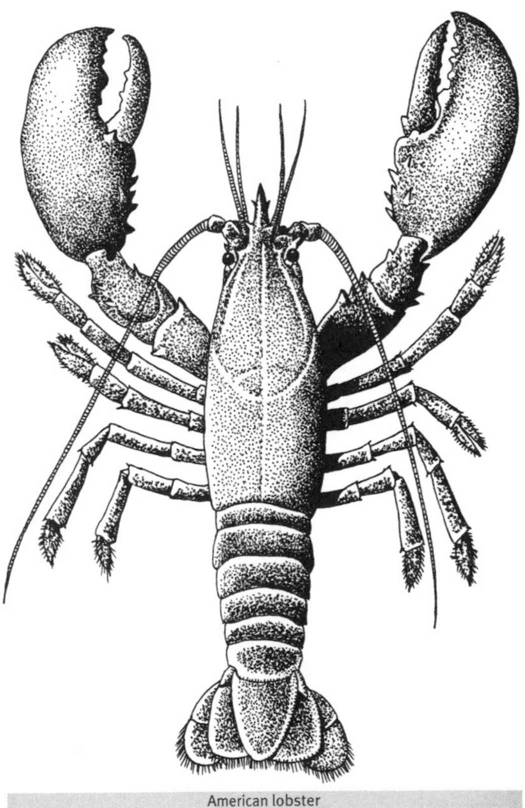

American lobster

lobster (*Palinurus*) does not have. There have been blue lobsters taken, and white ones too, but the familiar bright red color is the result of cooking. Lobsters served in restaurants usually weigh 1½ to 3 pounds, but there are records of individuals weighing more than 40 pounds. Lobsters are usually fished in relatively shallow waters, 5 to 55 fathoms deep, and are caught in "pots," traps (usually made of wooden slats) that are baited with dead fish and sit on the bottom, with their location marked on the surface by buoys. The spiny lobsters found in other parts of the world lack the American lobster's large crushing claws.

See also spiny lobster

lobster krill (*Pleuroncodes planipes*) Neither lobster nor krill, *P. planipes* is a species of swimming crab found principally in the Sea of Cortez, but also on the Pacific coast of the Baja Peninsula, sometimes as far north as San Francisco. The body and claws are bright reddish orange, hence the name "lobster krill," a reference to the color of a cooked lobster. These 2-inch crabs aggregate in enormous numbers, sometimes coloring the sea red for many square miles, and provide sustenance for seabirds, whales, and yellowfin tuna. (They are also known as tuna crabs.) There are instances when billions of them have washed ashore, piling up on the beach in windrows 3 feet deep and stretching for miles. Another species, *P. monodon,* is found off the coasts of Ecuador, Peru, and Chile. **See also crab, krill**

Loch Ness monster Loch Ness in the Scottish Highlands is not actually connected to the sea, but there are those who believe that "Nessie" has been so hard to spot over the years because it can enter the loch at will and disappear back into the ocean. From the first reports in the tenth century A.D., there have been regular descriptions of a monster of some sort in Loch Ness. The sightings have been identified as dinosaurs, giant eels, sharks, seals, slugs, snakes, logs, and anything else from pilot whales to elephants, but so far, there has been no hard evidence to support any of them. The most famous photograph, taken by a supposedly respectable London surgeon in 1934, was revealed in 1993 to have been a hoax involving a foot-long toy submarine with a plastic-wood head and neck.

See also plesiosaurs, sea serpents

Lofoten Islands With the Vesterålens, two contiguous archipelagoes that stretch for about 110 miles off the coast of northern Norway, just south of the city of Tromsø. Although they are well north of the Arctic Circle, the North Atlantic Drift gives these islands a temperate climate. The larger islands include Gimsøya, Flakstadøya, Vaerøya, Vestvågøya, Austvågøya, and Moskenesøya. The Vesterålen Islands include Hinnøya (Norway's largest island), Andøya, and Langøya. (*Øya* is "island" in Norwegian.) The population numbers around 66,000. The shoals on the east coast of the islands have long been considered one of the richest cod-fishing grounds in the world, and in March and April, thousands of Norwegian men come to the Lofotens to land and process the catch. The Maelström, also known in Norwegian as Moskenstraumen, is a 5-mile-wide channel and strong tidal current of the Norwegian Sea that flows between the Lofoten islands of Moskenesøya and Mosken. In Edgar Allan Poe's 1841 story "A Descent into the Maelstrom" and Jules Verne's 1870 novel *Twenty Thousand Leagues under the Sea,* this current is exaggerated into a deadly vortex that swallows up ships and men.

log Any device for measuring the speed of a vessel through the water, or the distance she has sailed in a given time. It is also a shortened version of "logbook," a daily record of information about a ship's position, speed, crew, and other pertinent data. The earliest logs were simple wooden boards attached to a line and trailed behind a vessel under way. Assuming that the board would remain stationary in the water as the ship moved forward, the speed of the ship could be ascertained from the length of line that ran out in a specified period of time. With the introduction of the nautical mile in the fifteenth century, the line was marked by knots, spaced uniformly at a distance proportional to segments of a mile. At that time, it was believed that a nautical mile—calculated as the distance on the earth's surface subtended by one minute of latitude at the earth's center—measured 5,000 feet, but calculations by Richard Norwood in 1637 established the nautical mile at 6,120 feet. In fact, the nautical mile is 6,080 feet.

See also knot

loggerhead turtle

loggerhead turtle (*Caretta caretta*) A loggerhead is a ball of iron on the end of a long handle used in the melting of tar or pitch, or, by extension, an unusually large head. (It was also a wooden bitt or post in a whaleboat, around which the harpoon line is wrapped as it runs out after striking a whale.) The loggerhead turtle has an oversized head, hence its name. It reaches a length of 38 inches and can weigh between 200 and 400 pounds. It feeds on crabs, mollusks, and shrimps, cracking their shells with its powerful jaws. The loggerhead can also be distinguished from the other sea turtles by the presence of two claws on each of its foreflippers. Many female loggerheads come ashore on Florida beaches to lay their eggs during the late spring and summer. It is believed that the baby turtles that make it to the sea drift in ocean currents to areas of seaweed, such as the Sargasso Sea, but no one knows where they spend their first year, which is known to biologists as the lost year.

See also sea turtle

L'Ollonois, François (1630–1671) French pirate who was born Jacques-Jean David Nau but was called L'Ollonois because he was born at Sables d'Olonne. Expelled from the colony of Santo Domingo for buccaneering, he set up on Tortuga and began a career of unmatched bloodthirstiness. He was reputed to have torn out and eaten the hearts of his victims. He sacked Maracaibo in 1667, but he was shipwrecked in Panama and lynched by Indians.

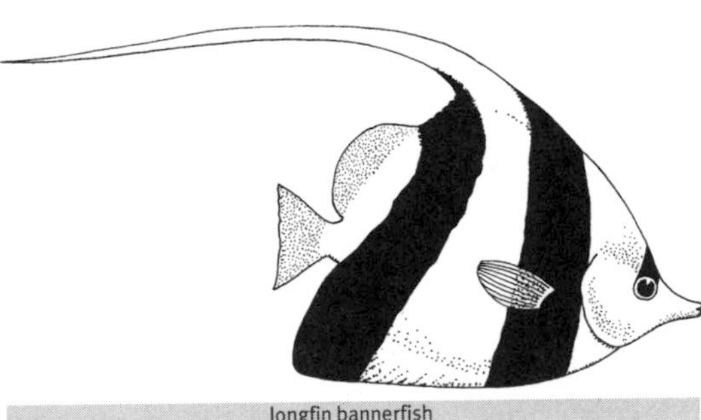

longfin bannerfish

longfin bannerfish (*Heniochus acuminatus*) Common on coral reefs from East Africa to the Persian Gulf, Hawaii, and Australasia, this 10-inch (counting the dorsal filament) butterfly fish is often seen in pairs. Because of its high dorsal fin and black, yellow, and white coloring, it is sometimes confused with the moorish idol, but the two are completely unrelated. **See also moorish idol**

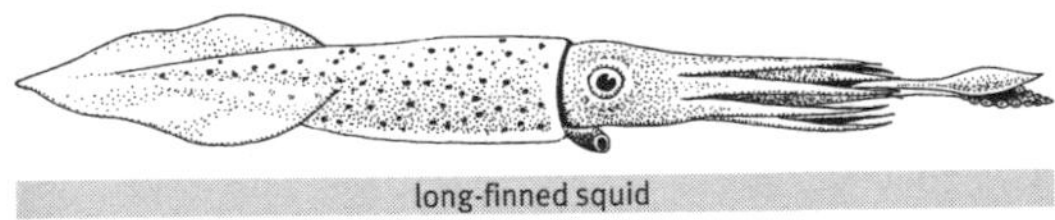

long-finned squid

long-finned squid (*Loligo pealei*) A pelagic, schooling squid found in the western Atlantic from Newfoundland to Venezuela, this species reaches a total length of about 30 inches, including arms. It is named for its long, pointed tail fins. Originally taken as a bycatch in the finfish and shrimp trawl fisheries, *L. pealei* is now the object of an international directed fishery. This is one of the species popularly marketed as calamari.

See also calamari, squid, squid fishing

Long Island The largest American island outside of Alaska and Hawaii, Long Island is 118 miles long and 23 miles wide at its widest point. It covers 1,401 square miles and is bordered by Long Island Sound on the north, the Atlantic Ocean on the south, and New York City on the west. (Two of the city's five boroughs, Brooklyn and Queens, are on Long Island.) The population of the island is close to 7 million. At its eastern end, the island is split into two peninsulas separated by Peconic Bay; in the bay are Shelter Island and Gardiners Island. The terminus of the north fork is Orient Point; of the south, Montauk. The Dutch and British established farming, fishing, and whaling communities on the island in the seventeenth century, and during the American Revolution, the Battle of Long Island (August 22, 1776) was fought here. It was not until the introduction of railroads, highways, bridges, and tunnels in the nineteenth and twentieth centuries that the island became so densely populated. After World War II, commercial fishing dropped off, farms were replaced by housing (Levittown, one of America's first mass-produced residential communities, is in Nassau County, on Long Island), and harbors, yacht basins, parks, and golf courses were built. Hundreds of thousands of Long Islanders now commute to work in New York City every day, and the highways and railroads are the busiest in the country. The south shore is a popular recreational area, and in addition to its many beaches, it contains Robert Moses and Jones Beach State Parks. Two of New York City's major airports, La Guardia and JFK, are on western Long Island, and the Hamptons—East, South, and West—are well-known, affluent residential and weekend destinations.

longitude The distance east or west on the earth's surface, measured by the angle that a particular place makes with a standard meridian. By international agreement, the meridian that passes through the site of the original Royal Greenwich Observatory outside of London is designated the prime meridian; all points along this line drawn from pole to pole are at 0° longitude. All other points on the earth are at longitudes ranging from 0° to 180° east and 0° to 180° west longitude. (Except where it is moved to account for populated areas, the international date line is exactly opposite the prime meridian, at 180°.) Lines of latitude are parallels, remaining parallel to each other as they circle the globe; the meridians of longitude are never parallel, but they converge at the poles. Measuring latitude was always easy, but determining the exact longitude of a particular place was a problem for most of recorded history. In order to accomplish this with precision, one had to know not only the time aboard ship, but also the time at, say, a home port at the very same moment. This required chronometers that would be accurate within fractions of a second over long periods of time and in all weather conditions. In 1714, the British parliament passed the Longitude Act, which offered a prize of £20,000 to anyone who could build a clock that would pass their rigorous tests. The prize was not awarded until 1773, when a reluctant Board of Longitude was forced to admit that the British clockmaker John Harrison (1693–1776) had constructed an instrument that fulfilled the requirements.

See also Harrison, latitude

longline fishing One of the most efficient—and destructive—of all fishing techniques. Longlining involves a single line that may be 100 miles long (the distance from New York to Philadelphia), supported along its length by floats. The lines are hung with thousands of hooks baited with live or frozen baitfish and deployed in an area where a particular species is being sought. Dangling from the longlines are about two thousand hooks on "branch lines" that can be adjusted to fish at depths ranging from 180 to 500 feet. Longlining accounts for about 30 percent of the world catch, including most of the billfishes taken commercially. The target species takes the bait, but so does every other kind of fish in the area. If, for example, yellowfin tuna is the object of the fishery, then sharks, billfishes, and any other fishes that are caught are often discarded, because it is too much trouble to separate them. In 1995, an estimated 1,500 shy albatrosses (*Diomeda cauta*), out of a breeding population of 8,000, were caught in longlines and killed. Another 45,000 albatrosses are killed every year; 8,000 of them are wanderers. Probably the most heinous use of longlines is in subantarctic waters, where fishers for the Patagonian toothfish (*Dissostichus eleginoides*) are scouring the waters for their target species and killing in the process hundreds of thousands of other fishes, whales, seals, and dolphins and as many as 150,000 seabirds annually.

See also albatross drift-net fishing, purse seining

long-nosed butterfly fish (*Forcipiger longirostris*) Although similar in coloration and range to the forceps fish (*F. flavissimus*), the long-nosed butterfly fish has a snout that is almost twice as long. Including the snout, this fish measures about 7 inches. It is found in the shallows from East Africa to Polynesia, and an all-black color phase has been seen in Hawaii and, less frequently, on the Great Barrier Reef.

See also butterfly fish, forceps fish

longnose filefish (*Oxymonacanthus longirostris*) The scientific name of this fish does an almost perfect job of describing it: *Oxymonacanthus* means "sharp single spine," and *longirostris* means "long snout." The only thing missing from the name is a description of the color scheme: it is an emerald-green fish with a pattern of orange spots. (It is sometimes called the orange-spotted filefish, even though a look at the snout would lead one to believe that it was really an orange fish with a green latticework pattern.) It reaches a length of 4 inches and is found throughout the Indo-Pacific.

See also filefish, triggerfish

longnose hawkfish (*Oxycirrhites typus*) Most of the hawkfishes have snub noses, but the longnose hawkfish has an elongated snout that it uses to pick small organisms from cracks and crevices in the reef. (A similar adaptation is found in the unrelated long-nosed butterfly fish.) This species is often seen perched head down in coral branches, at depths of between 100 and 300 feet, ready to pounce on small crustaceans.

See also hawkfish, long-nosed butterfly fish

long-tailed skua (*Stercorarius longicaudus*) The smallest of the skuas, about the size of a kittiwake, with only a 36-inch wingspan, the long-tailed skua is unmistakable because of the two greatly extended central tail feathers, which may equal the length of the body. It is a dark-backed bird, with dark wing tips, no white wing patches (except in immature birds), lighter underparts, a black cap, and a yellowish face. At sea, it can be differentiated from the two other *Stercorarius* skuas, Arctic and pomarine, by its proportionally longer wings and smaller head—and the tail feathers. Like others of its genus, it breeds north of the Arctic Circle in the circumpolar Arctic tundra regions and migrates south to regions like South Africa and southern South America. Except for the more northerly breeding areas, it has virtually the same range as the Arctic skua. Long-tailed skuas are pirates, chasing other birds and forcing them to disgorge their stomach contents, but they also prey on lemmings.

long-tailed skua

See also Arctic skua, pomarine skua, skua

lookdown (*Selene vomer*) A species of small jack, found along the Atlantic and Pacific coasts of North and South America, from Cape Cod to Brazil and from Baja California to Peru. Its imperious look comes from the

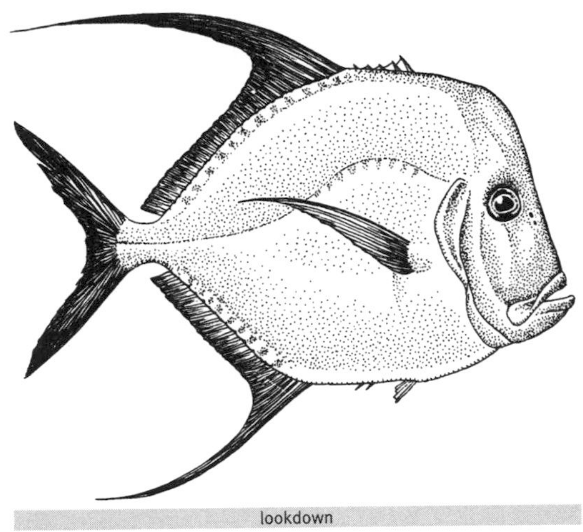
lookdown

high placement of the eye and the turned-down, disapproving mouth. Its exceptionally thin body is beautifully iridescent. Average length: 8 inches.

See also jack crevalle, pompano

loran Low-frequency radio navigation system used by airplanes and ships to determine their geographical position. Developed by the United States during World War II, the name is derived from **lo**ng-**ra**nge **n**avigation. It consists of two signals, transmitted from separated ground stations. A pulse from the master station triggers the slave station into transmitting a signal after a set time delay, and the navigator, using a loran chart, locates his line of position. By taking a similar reading from another pair of stations, a definite geographic fix is obtained. Loran has largely been replaced by the satellite-generated Global Positioning System (GPS).

See also Global Positioning System

Lord Howe Island A tiny island (total area 5 square miles; population 300) in the southwest Pacific, some 436 miles northeast of Sydney, Australia. (It is part of the state of New South Wales.) It is heavily wooded, and its Mount Gower rises to a height of 2,840 feet. Its main income is derived from tourists, some of whom come to dive the southernmost coral reef in the world. It was discovered in 1788 by Lieutenant Henry Lidgbird Ball of the British navy and named for Lord Richard Howe (1726–1799), the first lord of the Admiralty. Ball's Pyramid, a 1,300-foot-high rocky spire that rises out of the ocean 12 miles south of the Lord Howe Island, was named for Lieutenant Ball.

louvar (*Luvarus imperialis*) A 6-foot-long, silvery-pink fish with bright red fins, the louvar evidently lives in deep temperate and tropical waters; it has been recorded from New Zealand and South Africa. It has minute scales and tiny teeth and feeds on jellyfishes and other planktonic animals. It is usually classified with the billfishes, but instead of a beak, it has a bulging forehead like that of a pilot whale. The vent is located much farther forward than in most other fishes; it is below the base of the pectoral fins. Louvars produce an enormous number of eggs: one stranded female had ovaries that weighed 4 pounds and contained an estimated 47.5 million eggs.

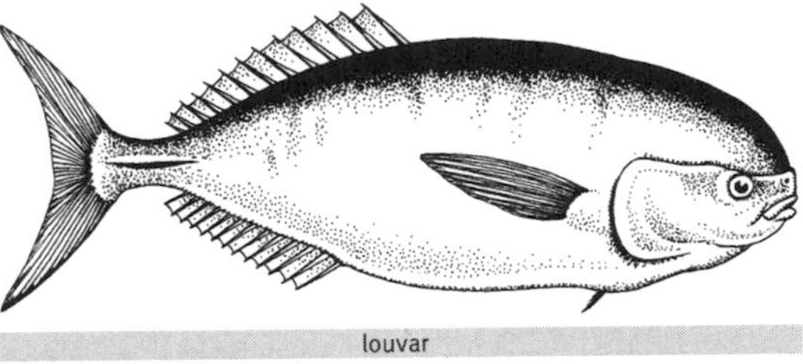
louvar

lumpfish (*Cyclopterus lumpus*) Named for its lumps, bumps, and ridges, the lumpfish also has a sucker on its ventral surface, which it uses to cling to rocks or pilings. (It is also known as lumpsucker.) Lumpfishes vary in color from green to brown to reddish and can reach a length of 24 inches. After eggs are laid, the male guards them and fans water through the egg mass to keep it oxygenated. Found on both sides of the North Atlantic, Europeans trawl lumpfishes for the roe, which is marketed as "lumpfish caviar."

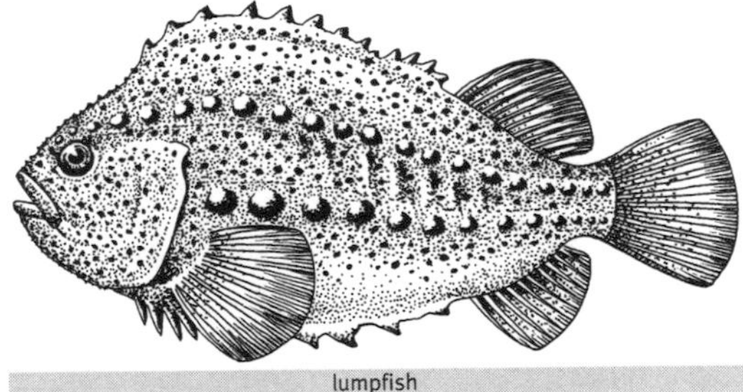
lumpfish

Lusitania Launched in 1906, *Lusitania* was the pride of the Cunard Line. When she made her first Atlantic crossing, in 1907, 200,000 people witnessed her departure. At an overall length of 785 feet, she was 5 feet shorter than her sister ship *Mauretania,* but they displaced the same 40,000 tons. Both ships were "floating palaces" that offered the height of luxury at speeds that won them the coveted Blue Riband, identifying the fastest ships in the transatlantic sweepstakes. On May 1, 1915, when she left New York, her passengers had been warned that a state of war existed between Britain and Germany, but they believed that the Germans would not attack an "American ship"—that is, one that had departed from the United States—and besides, they believed that "Lucy" could outrun any submarine. They were wrong on both counts. Off the Old Head of Kinsale on the Irish coast, the German submarine *U-20* was waiting for British vessels, and on May 7, in a heavy fog, when the *Lusitania* was only 20 miles from the coast, she was hit by a single torpedo. As the ship took on water and listed heavily, lifeboats were lowered, but for some unknown reason they shattered when they hit the water. Numerous ships picked up *Lusitania*'s distress signals, but with no lifeboats, the passengers had to remain on board the sinking ship. When she finally went down, twenty minutes after the hit, she took 1,198 passengers and crew members with her. (Some 764 were saved, including the captain.) Many Americans who had thought the United States ought to remain neutral in the war between Britain and Germany changed their minds after this "act of aggression," and President Wilson sent a note to Berlin, insisting that neutral shipping should be left in peace. It was later learned that the *Lusitania* was indeed carrying munitions.

See also *Mauretania*

macaroni penguin (*Eudyptes chrysolophus*) The macaroni gets its name from the "macaronis," fashionable British dandies of the 1770s, around the time that Captain James Cook discovered South Georgia, where these penguins are found. (This is also the source of the line from "Yankee Doodle," who "stuck a feather in his hat and called it macaroni.") The largest of the crested penguins, the macaroni can stand 27 inches high. It can be differentiated from the other species by the downward-pointing angle of the black coloration in the neck. As with the others, conspicuous golden-yellow plumes project from behind the eye and droop behind the head. On their breeding grounds on subantarctic islands and the Antarctic Peninsula, the total population has been estimated at 11 million breeding pairs. Some ornithologists consider the royal penguin a subspecies of the macaroni.
See also Antarctic Peninsula, crested penguins

MacInnis, Joseph B. (b. 1937) Trained as a physician and acknowledged as an expert in deep-sea diving, Dr. Joseph MacInnis has spent the last thirty years studying human performance in high-risk environments, especially the high Arctic. He was the first person to dive and film beneath the North Pole, and he shot some of the first films of Arctic whales, including the bowhead, beluga, and narwhal. In medicine, his work with decompression made possible dives by himself and others that set new records for depth and duration. He designed and built the first Arctic dive station (called Sub-Igloo) that allowed humans to explore below the polar ice cap. He led the team that discovered the world's northernmost shipwreck, HMS *Breadalbane*, under the ice in Canada's Northwest Passage. Dr. MacInnis was a consultant to the *Titanic* discovery team and the first Canadian to explore the world's most famous shipwreck. In 1991, he was coleader of the seven-million-dollar expedition to film *Titanic* in the IMAX format. Recently, Dr. MacInnis has been studying environmental and science-education issues related to the Great Lakes and the St. Lawrence River on behalf of the International Joint Commission. He has written five books on the ocean and has contributed articles for *Scientific American* and *National Geographic*. He now serves as Chair, Canada Trust Friends of the Environment, and has been honored with three honorary doctorates, the Queen's Anniversary medal, the Admiral's Medal, and his country's highest honor, the Order of Canada. **See also *Titanic***

mackerel: See Atlantic mackerel

mackerel shark Another name for the group of powerful, fast-swimming sharks that includes the makos (*Isusrus oxyrhincus* and *I. paucus*), the porbeagle and salmon sharks (*Lamna ditropis* and *L. nasus*), and the great white shark (*Carcharodon carcharias*). Mackerel sharks (also known as lamnids) are characterized by their streamlined shape, lunate tails with almost equal-sized lobes, keels on the caudal peduncle, and sharply pointed snouts.
See also great white shark, mako shark, porbeagle

Macquarie Island Volcanic island in the Southern Ocean, 600 miles south of New Zealand. It belongs to Australia and is considered a dependency of Tasmania. Macquarie is 21 miles long, but only 2 miles wide; it consists of an undulating plateau marked with several lakes. No trees or shrubs grow on this subantarctic island; the vegetation is primarily tussock grass. Macquarie was discovered in 1810 by an Australian sealer named Frederick Hasselburg (or Hasselborough or Hazelburgh), who named it after Lachlan Macquarie, governor of New South Wales at the time. Within a decade of their discovery, the plentiful fur seals were completely wiped out. King, gentoo, and rockhopper penguins breed on Macquarie Island, and it is the only known breeding ground of the royal penguin. Many species of petrels nest on the island, and several kinds of albatrosses. Southern elephant seals mate and deliver their pups here, and fur seals have returned in small numbers. **See also penguin, Tasmania**

Macropoma The lobe-finned fish, extinct for 70 million years, that is the close relative of the coelacanth (*Latimeria chalumnae*) that was found for the first time in 1938 off South Africa. *Macropoma* was less than 2 feet long, to *Latimeria*'s 5, but according to the fossil record, it had the same deep body, three-lobed tail, and fins on short, thick stalks. The fossils were found in Czechoslovakia and England. **See also coelacanth**

Madagascar The fourth largest island in the world (after Greenland, New Guinea, and Borneo), with an area of 226,444 square miles, a little smaller than Texas. The population numbers about 7 million, and the capital is Tananarive. Although the island is separated from the African continent only by the 500-mile-wide Mozambique Channel, the people are related not to

Africans but to Indonesians, and have virtually no commercial relations with the East African countries. The original settlers are believed to have come from Indonesia, and the Malagasy language spoken on Madagascar is derived from Indonesian. The economy of Madagascar is overwhelmingly agricultural, with the principal crops being rice, cassava, millet, sugarcane, coffee, tobacco, cloves, and vanilla. The island consists of a narrow coastal strip in the east, a large central plateau, and a huge plain in the west. Madagascar was once heavily forested, but woodcutters have denuded much of the island, and many of the endemic wildlife species are seriously threatened by loss of habitat. Three-quarters of the known species of lemurs are found only on Madagascar; the tenrecs, primitive insectivorous mammals that resemble long-nosed hedgehogs, are endemic only to the island. In addition, there are many reptiles (particularly snakes and chameleons) and insects (particularly butterflies) that are poorly known; many species remain to be discovered in the densely wooded and still unexplored interior. In the late seventeenth and early eighteenth centuries, Madagascar was a "pirate island," its numerous harbors serving as bases for cutthroats because it commanded the trade routes between the Cape of Good Hope and Europe, and also the Indian Ocean.

Madeira Portuguese island in the North Atlantic, 620 miles southwest of Lisbon, 340 miles west of Morocco, and to the north of the Canary Islands. Madeira is an autonomous region of Portugal, incorporating the island of Porto Santo and the uninhabited Desertas and Selvagens. The 260,000 Madeirans have their own parliament and government. About half the population lives in the capital city of Funchal. (Porto Santo, 26 miles to the northeast, has a population of 4,700.) Like the Canaries and the Cape Verdes, these islands are the peaks of mountains that originate on the floor of the deep ocean. Madeira is 34 miles long and 14 miles wide, a green, luxuriant island with many rare and unique plant species. The mountainous interior is uninhabited and uncultivated, given over to wild trees and vegetation. (*Madeira* means "wood" or "timber" in Portuguese.) It is possible that the Phoenicians first visited Madeira, but it is known that Genoese adventurers mapped the islands in 1351. João Gonçalves Zarco visited Madeira in 1450, and Prince Henry the Navigator of Portugal sent the first colonists there. They burned down much of the forest to clear the islands for cultivation, and established what is believed to have been the world's first sugarcane plantation. Around 1480, Christopher Columbus, whose wife was the daughter of the governor of Porto Santo, moved to the island. Prince Henry had planted grape vines from Cyprus and Crete, and by the seventeenth century, Madeira wine (aged in oak casks and fortified with brandy) had become a popular export. The islands still export sugar, wine, and bananas, but they grow almost every kind of fruit and vegetable, including sweet potatoes, oranges, lemons, guavas, mangoes, figs, and pineapples. Tourism is Madeira's most important industry, and many Madeirans are engaged in creating embroidery and wicker- and woodworking handicrafts to sell to the visitors.

See also Azores, Canary Islands, Cape Verde Islands

Maelström Commonly used to denote a large whirlpool, the word actually refers to a 5-mile-wide channel and strong tidal current of the Norwegian Sea in the Lofoten Islands. The strongest tidal current in the world, it flows between the islands of Moskenesøya and Mosken; its Norwegian name is Moskenstraumen. In Edgar Allan Poe's 1841 story "A Descent into the Maelstrom" and Jules Verne's 1870 novel *Twenty Thousand Leagues under the Sea,* this current is exaggerated into a deadly vortex that swallows up ships and men.

See also Lofoten Islands

Magdalen Islands Known in French, the primary language of the islands, as Îles de la Madeleine, this group of nine islands and numerous islets and rocks lies in the Gulf of St. Lawrence between Newfoundland and Prince Edward Island. The islands are Alright, Amherst, Brion, Coffin, East, Entry, Grindstone, Grosse, and Wolf; the capital is Havre-Aubert on Amherst. Now part of the province of Quebec, the islands are occupied by about 13,000 people, who are primarily engaged in fishing and fish processing. The islands are accessible by plane from Montreal and various other locations.

Magellan, Ferdinand (1480–1521) Commander of the first voyage to circumnavigate the globe. In 1505, Magellan volunteered to accompany Francisco de Almeida, the first Portuguese viceroy to Asia, to India. In 1509 he joined Diogo Lopes de Sequeira on a voyage from Cochin to the Spice Islands, where they encountered Malay pirates; he was appointed captain for his bravery in this conflict. In 1511, he joined the expedition of Viceroy Afonso de Albuquerque to capture Malacca. In 1512, he returned to Portugal. Having been accused of trading with the enemy, he renounced his nationality and proceeded to Spain in 1517. There he convinced King Charles V that he could reach the Spice Islands by sailing westward, and he was given command of a five-ship fleet led by the *Trinidad.* Crossing the Atlantic in September 1519, his first landfall was Tenerife in the Canary Islands, then Pernambuco in what later became Brazil. He sailed south along the coast until he reached a passage to the west and made a thirty-eight-day journey through the long and difficult strait that now bears his name. Emerging into the Pacific (which he named), he sailed through the islands for ninety-eight days, his men beset by thirst, hunger, and scurvy, until he landed at what is now believed to have been Guam. From there he proceeded to Cebu in the Philippines, where he was killed on April 27, 1521, in a battle with the inhabitants of nearby Maetan Island. Juan Sebastian del Cano took command of the *Victoria* and sailed home by way of the Cape of Good Hope, arriving on September 8, 1522. Only 17 of the original 270 seamen made it back to Spain. Included among them was Antonio Pigafetta, who wrote an account of the voyage, *The First Voyage Round the World by Magellan, translated from Pigafetta.*

Magellan, Strait of The Strait of Magellan, which is 330 miles long and between 2 and 15 miles wide, separates the South American mainland from the islands of Tierra del Fuego. It was discovered and first navigated by Ferdinand Magellan in 1520 as he passed into the Pacific Ocean. Before the Panama Canal was built, it was the preferred passage for ships passing from the Atlantic to the Pacific, because the waters of Cape Horn are almost always beset by powerful and dangerous winds. For most of its length, it passes through Chile, and the only city on the strait is Punta Arenas.

See also Cape Horn, Tierra del Fuego

Magellanic penguin (*Spheniscus magellanicus*) This species is the only member of the genus *Spheniscus* with two black bands across its chest, one of which loops down toward the feet while the other comes forward from the back to form a collar. (The coloration could also be described as consisting of two *white* bands, one of which comes up and over the eye while the other comes over the chest and down to the feet.) Colonies of these penguins are found all around the southern tip of South America, from the Patagonian coast of Argentina to the coast of Chile. They are also found in the Falkland Islands. They fish in groups and have been seen hundreds of miles offshore. They lay their eggs in burrows that they dig out of sand or shingle, or sometimes, as in the Falklands, in grassy turf; if the burrows are flooded, the chicks can drown. Magellanic penguins are a favorite prey species for sea lions and giant petrels; in recent years, oil spills have caused the death of thousands of them.

Magellanic penguin

See also African penguin, Falkland Islands, Humboldt penguin

magnetic pole The points on opposite sides of the earth where the magnetic intensity is greatest. Unlike the geographical North Pole, which is always located at 90° north, the north magnetic pole wanders according to the vagaries of terrestrial magnetism. The magnetic North Pole was first located in 1831 by James Clark Ross at the west side of the Boothia Peninsula in the Canadian Arctic, at 79° north and 110° west. A compass needle points not to the geographical North Pole but to the magnetic pole, so navigators must take into account this variation to obtain a true bearing. (When a compass needle points to magnetic north, its opposite end will not necessarily point to magnetic south.) The magnetic South Pole (72°25′ south and 155°16′ east, on the Antarctic continent) was reached in 1901 by the Australian explorer Sir Douglas Mawson.

See also compass, North Pole, South Pole

Mahan, Alfred Thayer (1840–1914) American admiral and naval historian. He entered the U.S. Naval Academy in 1857, after two years at Columbia College in New York, and graduated in 1859. After several assignments at sea, Mahan was appointed to the newly formed Naval War College in Newport, Rhode Island, and was named president of the college in 1886. Out of his lectures at the college, he developed *The Influence of Sea Power upon History, 1660–1783*, which was published in 1890 and is considered the most influential book of its time on naval strategy. His book was enthusiastically reviewed by Theodore Roosevelt, who became assistant secretary of the navy under President McKinley in 1897 and was an advocate of naval expansion. Mahan propounded the idea that a nation with an active fleet, no matter how large or small, had a decided advantage over a power whose fleet was in harbor, and this thinking affected many of the naval endeavors of World War I. He was president of the American Historical Society in 1902 and was appointed rear admiral in 1906.

Maine U.S. battleship, launched in 1890, 319 feet overall, with a complement of 354. Because there were riots in Cuba in late 1897, President William McKinley dispatched the USS *Maine* to Havana to protect American citizens. She arrived on January 25, 1898, and anchored in the harbor. On February 15, at 9:40 p.m., she blew up with a tremendous roar, broke apart, and sank. Of the ship's company, 268 men died and many others were injured. Although the cause of the explosion was unknown, certain newspaper editors in the United States who wanted the country to go to war with Spain printed lurid headlines accusing the Spanish of sinking the battleship. Their battle cry was "Remember the *Maine*!" The United States issued a resolution declaring Cuba's right to independence, demanded Spain's withdrawal, and authorized America's use of force. On April 24, the Spanish-American War commenced. Colonel

Theodore Roosevelt led his Rough Riders up San Juan Hill in Cuba; Admiral Dewey's squadron sailed to the Philippines, another Spanish territory, and captured Manila; and the entire Spanish fleet was destroyed in the Cuban harbor of Santiago. The Treaty of Paris (December 10, 1898) provided for Spain's withdrawal from Cuba, and ceded Guam, Puerto Rico, and the Philippines to the United States. It is now believed that the Spanish had nothing whatever to do with the sinking of the *Maine,* and that the explosion was caused by a coal fire that was burning close to the ship's reserve powder magazine. **See also Spanish-American War**

Majorca Also spelled "Mallorca," the largest of the Balearic Islands off the east coast of Spain in the Mediterranean. (The other Balearics are Minorca and Ibiza.) It is a popular resort, especially with the British, but tourists come from all over Europe. The Balearic Islands were occupied by Iberians, Phoenicians, Greeks, Carthaginians, Romans, and Byzantines; the Moors, who came in the eighth century, established an independent kingdom that became a haven for corsairs and pirates in the Mediterranean. The kingdom of Majorca rejoined Aragón in 1349. In 1708, it was captured by the British, who held it until 1802. In 1833, the Balearics were established as the Spanish province of Baleares. Frédéric Chopin and his mistress Aurore Dudevant (the novelist George Sand) moved to a villa on Majorca in 1838, but when he became ill, rumors of tuberculosis forced them to live in an abandoned monastery. They left for France the following year.

See also Ibiza, Minorca

mako shark (*Isurus oxyrhinchus*) The mako is a smaller, slenderer, faster version of the great white, to which it is closely related. Like its larger, more notorious cousin, the mako is a "warm-blooded shark," with special blood vessels that can heat the muscles for greater swimming efficiency. One of the fastest fish in the ocean, the mako is capable of great leaps out of the water. Where the teeth of the white shark are triangular, those of the mako are jaggedly curved. Makos are brilliant blue above and white below. They are eagerly sought as game fish, and the world's record—for a mako caught off New Zealand—is 1,115 pounds. *Mako* is a Maori word. **See also warm-blooded fishes**

mako shark

Maldive Islands Previously a dependency of Ceylon (now Sri Lanka), the Maldives are occupied by a mixed populace who speak Divehi, an Indo-European language. Stretching 500 miles north to south in the northern Indian Ocean, the Maldives consist of about twenty atolls and more than a thousand coral islets, which are the exposed tops of a submarine ridge. (The word "atoll" is derived from the Divehi word *atholhu.*) Two hundred of the islands are inhabited by a total of about 260,000 people. They raise coconuts, breadfruit, mangoes, papayas, plantains, pumpkins, and sweet potatoes; until recently, the chief industry was fishing. The first settlers probably arrived in the northernmost islands around 500 B.C. from Ceylon. The Portuguese occupied the island of Male from 1558 to 1573. In the seventeenth century the Maldives were under the protection of the Dutch rulers of Ceylon. The islands were an important source of the "money cowrie" (*Cypraea moneta*), which was used as currency throughout the Middle East and India. The Maldives were a British protectorate from 1887 until Sri Lanka became independent in 1948. The islands remained under the control of various sultans and sultanas until 1965, when they became an independent country. The Republic of the Maldives was proclaimed in 1968, with the capital at Male. Improved transportation has made the Maldives much more accessible to tourists, and fish-watching in the clear waters of the various lagoons has become enormously popular. There are now more than seventy-five hotels and guest houses in the Maldives. Some enterprising Maldivians have taken *dhonis,* the traditional fishing boats, and converted them to luxury houseboats. **See also Sri Lanka**

Malta Island in the Mediterranean, 60 miles south of Sicily. Malta is also a republic that includes the islands of Gozo and Comino, as well as some uninhabited rocks, all of which are sometimes collectively known as the Maltese Islands. Malta covers 95 square miles and has no rivers or lakes, few trees, and hardly any natural resources. Because of its location in the Mediterranean, however, in the Sicilian Channel between Sicily and Tunisia, it has always been considered of great strategic value. It was possessed successively by the Phoenicians, Greeks, Carthaginians, Romans, and Arabs. Saint Paul is supposed to have been shipwrecked on the island in A.D. 60, whereupon he converted the Maltese to Christianity. The Normans of Sicily occupied the island in 1090, and in 1530, the Habsburg king Charles V granted Malta to the Knights Hospitalers, a religious military order, also known as the Knights of Malta. The knights withstood a siege by the Turks in 1565 and held on until 1798, when the island was surrendered to Napoleon. Two years later, the British ousted the French, and for most of the nineteenth century Malta was ruled by a military governor appointed by the British. Because of its excellent deepwater harbor, Valletta was the home of the British Mediterranean Fleet for a century and half. It was subjected to heavy bombing by the Germans and the Italians during World War II, but the island was not invaded. Malta was a Crown Colony until 1964, when it became independent, choosing to remain in the British Commonwealth. The constitution was revised in 1974 to make Malta a republic. The population of Malta is 357,000, and the capital is Valletta, named for Jean de la Vallette, the grandmaster of the Knights of Malta who withstood the Ottoman siege of 1565. The people of Malta are of mixed Italian, Arab, and British descent, and they speak a dialect of Arabic written in the Latin alphabet, which includes such place names as Marsaxlokk, Tarxien, and Qormi. Tourism is Malta's most important industry, and most of the tourists are from Britain. **See also Mediterranean Sea, Sicily**

Man, Isle of Island in the Irish Sea, about 80 miles northwest of Liverpool. The Isle of Man covers 221 square miles, with a climate so mild that tropical plants can be grown without protection, and has a population of 70,000. It is a popular tourist destination, and the site of the T.T. (Tourist Trophy) motorcycle races. Occupied by Irish monks in the fifth century (the name of the island is probably a corruption of the Latin *mona,* which means "monk") and by the Vikings in the ninth century, the island was a dependency of Norway until 1266, when it was sold to Scotland. From the fourteenth to the eighteenth centuries, it belonged to the family of Sir John Stanley; then it passed to the dukes of Atholl in 1736. To eradicate the rampant smuggling, Parliament purchased the island from the Atholl family in 1765 for £70,000. The Isle of Man is not officially part of the United Kingdom, but it is a dependency of the Crown for purposes of defense and international affairs. It is traditionally governed by the open-air parliament known as the Tynwald, the world's oldest legislative body, with a thousand-year unbroken history. Gaelic was spoken on the island until the nineteenth century, and it is sometimes known as Ellan Vannin. A British lieutenant governor lives in Douglas, the capital, which has a population of 20,000. The Calf of Man, a separate islet off the southwest tip, is a bird sanctuary administered by the Manx National Trust. The tailless breed of cat known as the Manx did not originate on the island, but was probably brought by eighteenth-century seafarers from Asia.

manatee (*Trichecus manatus*) The North American manatee is found from the coastal waters of the southeastern United States, through the Caribbean, and to the northern coast of South America. It is a fat, fully aquatic mammal, reaching a length of 10 feet and a weight of more than 500 pounds. Manatees eat water plants, and they are such efficient feeders that they have been used to clear clogged waterways of plants, especially water hyacinths. In Florida waters, they are endangered by traffic on the canals, where fast-moving motorboats run over these lethargic animals, which are too slow to avoid the lethal propellers. A closely related species (*T. inunguis*) is found in the Amazon River and its tributaries; it is differentiated from its more northerly relative by the absence of nails on its flippers. A third species (*T. senegalensis*) is found off the coast of West Africa. **See also dugong, Steller's sea cow**

manatee

manganese nodules First collected by the *Challenger* expedition of 1873–1876, manganese nodules are usually found in vast quantities on flat, deepwater plains. Almost always round, although far from perfectly so, they are usually the size of grapes or potatoes, but they can be as large as grapefruit or cannonballs. How they are formed is a complete mystery, but examination of the nodules shows that they are composed of microscopic onionlike layers of manganese oxide and iron hydroxide, with in-between layers of nickel, copper, cobalt, and zinc. We do not know how long it takes for one to accrete, why they are round, or why they are not buried in the sediment. They are found in the North and South Pacific and the Indian Ocean, where they sometimes cover 90 percent of the surface of the ocean floor. Manganese nodules represent the largest mineral deposit on earth, but the technology to extract them from the sea has not been developed. A small portion of them could solve many of the world's mineral shortages. It has been estimated that the manganese cobbles contain enough titanium to fulfill the world's needs for 2 million years; enough manganese for 400,000 years; enough cobalt for 200,000 years, nickel for 150,000 years, and aluminum for 20,000.

Manhattan Like Singapore and Venice, Manhattan is an island that is completely urbanized. Its only greenswards are its parks; at 843 acres, Central Park is the largest. Measuring 13 miles from north to south and

2 miles wide, the island is surrounded by the Harlem River on the northeast and north, the East River on the east, Upper New York Bay to the south, and the Hudson River to the west. At 22⅗ square miles Manhattan is the smallest of the five boroughs that make up New York City, but it is the oldest, the most densely populated, and the most important. It was the first part of New York to be settled by Europeans when the Dutch West India Company established the first outpost at the southern tip of the island in 1624. Two years later, Peter Minuit purchased the entire island for trinkets and cloth valued at sixty guilders (about twenty-four dollars) from the Manhattan Indians. The island was incorporated as the city of New Amsterdam in 1653; it was taken by the British in 1664 and renamed New York City. The city was the first capital of the United States, and on April 30, 1789, George Washington was inaugurated as the country's first president there. In 1898, Greater New York was formed, consolidating Manhattan with the other boroughs of Brooklyn, Queens, the Bronx, and Richmond (Staten Island). Manhattan is one of the world's most important commercial, cultural, and financial centers; only Tokyo and Hong Kong can compare for such intensely concentrated activity in a small area. Although it is often deemed synonymous with New York City, Manhattan is only a small portion. While the city has a population of 7.5 million (it is the most populous city in the United States), the permanent population of Manhattan island is around 1.5 million. With the daily arrival of commuters from the other boroughs, as well as neighboring New Jersey and Connecticut, however, the population increases tenfold. **See also Long Island, Staten Island**

manila A type of cordage used extensively aboard ships before the introduction of artificial fibers. Manila is obtained from the manila hemp plant (*Musa textilis*), grown mainly in the Philippines, where it is known as *abacá*. It is not related to true hemp but it is closely related to the banana (*M. sapientum*). The long fibers of the plant are woven into twine, matting, and rope that is especially desirable for marine cordage because it is impervious to saltwater and does not have to be tarred. Along with manila paper, which is used for wrapping paper and envelopes, manila hemp gets its name from the capital city of the Philippines.

Manila galleon Treasure ships sent from Manila, in the Philippines, to Spain, or from Acapulco to Manila. After the Portuguese explorer Ferdinand Magellan reached the Philippines in 1521, the Spanish established colonies there in 1565 and began a most profitable trading pattern, primarily with China. (Japan closed her doors to the West in 1638.) The route from Manila to Acapulco was pioneered by Miguel López de Legazpi (who would later become the first royal governor of the Philippines), a Basque sailor in the service of Spain, who crossed the Pacific in 1565. Thereafter, galleons sailed westward with their holds filled with coins of Mexican silver and returned with cargoes of silks and spices; porcelains; furniture; gold, ivory, and jade artifacts; lacquered screens; sandalwood boxes; and richly brocaded fabrics. When the goods arrived at Acapulco, they were carried by mule across Mexico to Veracruz, thence across the Atlantic to Spain. After Sir Francis Drake captured a galleon in 1571, the treasure ships became the targets of pirates, privateers, and admirals of nations hostile to Spain. Several galleons bound for Mexico were taken by the British, including the *Santa Ana*, which was looted and burned by Thomas Cavendish off Baja California in 1587 and had a cargo of 122,000 gold pesos and so much other treasure that he could not carry it. In 1743, Commodore George Anson captured the *Nuestra Señora de Covadonga* with more than a million silver pesos and 36,000 ounces of bullion. More galleons were lost to wind and weather than to privateers, and some simply vanished without a trace. It was the eastward route, from Manila to Acapulco, that saw most of the disasters: *Nuestra Señora del Pilar de Zaragosa y Santiago* hit a reef off Guam in 1690 and sank; *Nuestra Señora de la Concepión* sailed from Manila in 1638, was broached off the island of Saipan in the Marianas, and went to the bottom. For all the hazards, some 95 percent of the galleons completed their voyages, and the exotic wealth of Asia flowed to Spain and Western Europe while the Chinese reaped a fortune in Spanish silver. **See also galleon**

manta (family Mobulidae) The largest species of the largest of all rays are the Atlantic manta (*Manta birostris*) and the Pacific manta (*M. hamiltoni*), both of which can have wingspans of 20 feet and weigh more than 1½ tons. They swim—usually near the surface, where they feed on plankton—by flapping their powerful pectoral wings. On each side of the head are the hornlike cephalic fins, which are used to guide small food items into the wide mouth, and which account for its other name, devilfish. Mantas are gentle, harmless creatures, and tales of their towing boats or attacking swimmers are nonsense. Indeed, they often allow swimmers to hang on to the cephalic fins and hitch a ride as they swoop gracefully through the water. Of all the skates and rays, mantas are truly creatures of the surface—and beyond. A spectacular sight is a one-ton manta, black on the dorsal sur-

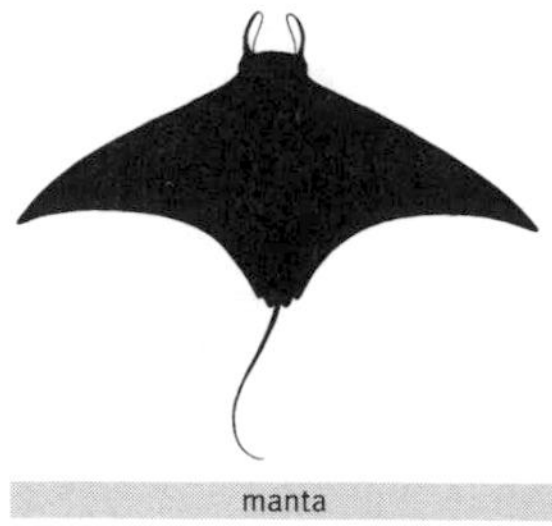
manta

face and white on the ventral, launching itself completely out of the water and reentering with a thunderous splash. The reason for these prodigious leaps is unknown. **See also eagle ray, stingray**

mantis shrimp (*Squilla mantis*) One naturalist wrote that "of all the crustaceans, the most intelligent may be the mantis shrimp." Not a true shrimp, but a member of the order Stomatopoda ("mouth-feet"), it gets its common name because the second pair of limbs are shaped like the grasping forelimbs of the praying mantis, an insect. Found in the North Atlantic and the Mediterranean, *S. mantis* is among the largest of the mantis shrimps, reaching a foot in length. The 250-odd species are found throughout the oceans, from shallow waters to 1,300 feet, where they burrow in sandy bottoms. Some species are striped; others have a tail appendage that resembles a sea anemone. Mantis shrimps have remarkable eyesight (their eyes have three pupils), and they can see color. Humans cannot see ultraviolet light, but mantis shrimps can, since they have at least four types of photoreceptors that are located in cells of the eye, known as R8 cells. Even in the larval stage, mantis shrimps are aggressive predators, and the carnivorous, pugnacious adults attack anything, even creatures larger than themselves. If two are placed in an aquarium, one will invariably kill and eat the other. By rubbing its tail fans against their covering, mantis shrimps make a loud rasping noise. One species, *S. raphidea,* is known as the killer prawn in Australia, and in 1999, the new species *Erugosquilla grahami,* measuring a foot or more in length, was discovered right off Sydney Harbor. **See also shrimp**

Manx shearwater (*Puffinus puffinus*) Dark gray above and white beneath, the Manx shearwater breeds on islands in the North Atlantic, such as the Westmann Islands of Iceland, the Faeroes, Shetlands, Hebrides, Scilly Isles, Azores, Ireland, and Madeira. It can be seen in the Atlantic, swooping low with a couple of rapid wingbeats followed by a glide over the waves, rising and falling and alternately showing the black upper side and white underparts. They winter as far south as Brazil and Argentina in the South Atlantic. Subspecies found in the Mediterranean are brown, whereas the nominate species is black.

See also shearwater, short-tailed shearwater, sooty shearwater

Maori Among the earliest Polynesian navigators were the "moa hunters," so known because they hunted this giant, now extinct bird; they are believed to have arrived in New Zealand around A.D. 800. Maori legends say that two hundred years later, the navigator Kupe came from Hawaiki to an island covered in fleecy clouds. He named it Aotearoa—"Land of the Long White Cloud"—which is still the Maori name for their homeland. The Maori (pronounced "*mow*-ree") culture of New Zealand was one of tribes, subtribes, and clans, with high chiefs, nobles, priests, and slaves. Men of the highest rank were most elaborately tattooed; women were tattooed only on the chin. Theirs was a society based on oral traditions, in which genealogy was paramount in the delineation of status, and they subsisted on agriculture, hunting, and fishing. The first contact between Europeans and Maoris ended in disaster: when Abel Tasman arrived in the present-day Golden Bay in Cook Strait in 1642, four of his men were killed by the natives. Captain James Cook visited the islands in 1769, and his men shot several Maoris as a result of each side's confusion about the other's intentions. When European whalers came around 1820, pursuing the right whales in the protected bays of New Zealand, they were abetted by Maori boatmen, who were recognized as competent and fearless harpooners. Many of the whalemen remained on the islands, often taking Maori wives. Unfortunately, they also introduced two things that would have a disastrous effect on the Maori for the rest of the century: guns and alcohol. When not fired up by drink to kill the *Pakeha*s (white men), the Maori carried on vicious tribal warfare against each other, resulting in a drastic population crash from 100,000 Maoris to fewer than 30,000. Partly to end the fighting, the British annexed the islands in 1840 by the Treaty of Waitangi, which the Maori chieftains who signed it may not have understood. Instead of settling matters, the British annexation led to the Maori Wars of 1860–1865, after which the native lands were appropriated by the *Pakeha*s. Subsequent Maori-*Pakeha* relations have been both good and bad. The Maori, who now make up some 10 percent of New Zealand's population, are increasing, and there is today a greater acknowledgment of their contributions to New Zealand's history. In 1995, however, an offer by the government to establish a $1-billion (New Zealand dollars) fund to compensate the Maori for their loss of land was summarily rejected, and militant Maoris protested, often violently. There are Maoris who believe that the only satisfactory solution is the return of New Zealand—all of it—to its original inhabitants. **See also New Zealand**

Marajó Brazilian island in the Amazon River Delta, Ilha de Marajó in Portuguese (pronounced "mara-*zho*"). The island, 183 miles long and 124 miles wide, covers 15,750 square miles, an area larger than Switzerland. The largest river-sea island in the world, it splits the outflow of the Amazon River into two distinct paths: the Amazon to the north of the island, and the Pará to the south. Cattle graze on the western part of the island, but the east is flooded annually during the rainy season, and the terrain is suitable only for water buffaloes. (The water buffaloes are descended from a herd that swam ashore after the ship that was carrying them from Southeast Asia to French Guyana sank off the shore of Marajó in the eighteenth century.) There are ranches (*fazendas*), beaches, and twelve towns and villages on the island; the largest is Breves, with a population of about 70,000. Soure, across the Pará River from the city of Belém, is reachable by ferry. Approximately 200,000 people live on this island, engaged in farming, ranching, and tourism. In 1990, American archaeologist Anna Roosevelt discovered the remains of a major mound-building culture that flourished on the lowland floodplain of Marajó from approximately 4000 B.C. to A.D. 1300. Because of the flooding, it is difficult to excavate on the island, but a large amount of sophisticated pottery has been collected, indicating an advanced culture, similar to those of the pre-Columbian Andes. (Some archaeologists believe that Marajó was simply an outpost of a Peruvian civilization.) The Arawak Indians, who were displaced from various Caribbean islands by the fierce, cannibalistic Caribs, established communities on Marajó before dispersing into the jungles of Amazonas.

marbled murrelet (*Brachyramphus marmoratus*) A little-known alcid of the North Pacific, the marbled murrelet is a 10-inch-long bird that occurs in vast numbers off Alaska and in smaller numbers in Japan and California. Despite the enormous population, the nesting sites remain unknown. They are brown-barred above and white below, with a short neck and a sharply pointed little bill resembling that of a wren. Kittlitz's murrelet (*B. brevirostris*) has a much shorter bill (hence *brevirostris*), more heavily marbled upper parts, and whiter underparts. Like those of the marbled murrelet, the breeding grounds of Kittlitz's murrelet are unknown.

See also Alcidae, least auklet

mare clausum Literally, "closed sea." In international law, a term used in connection with the claim made by certain maritime nations to exclusive ownership of areas of the open sea, in the same fashion that they owned land. In the fifteenth and sixteenth centuries, Spain claimed entire oceans, and Britain claimed dominion over her surrounding seas in the seventeenth century. In 1609, the Dutch jurist Hugo de Groot (known as Grotius) wrote *Mare Liberum,* in which he advocated the free use of the sea by all nations, particularly as it applied to Dutch boats fishing in British waters. In a 1635 response, John Selden, an Englishman, wrote *Mare Clausum,* in which he maintained that marine waters contiguous to a country were under the exclusive dominion of that country. In a compromise, it was agreed that dominion of the sea should be restricted to the range from which it could be protected by a cannon from shore, i.e., 3 miles. This defined the "3-mile limit," which was later extended to 12, and now to 200, which brings it close to the original doctrine of *Mare Clausum.*

See also Law of the Sea

Mariana Islands Sixteen islands in the western Pacific, of which six are inhabited; the most important are Guam, Saipan (the capital), Rota, and Tinian. The islands were first explored by Ferdinand Magellan in 1521, who named them the Ladrones (thieves); they were renamed in 1688 by Spanish Jesuits to honor Mariana of Austria, then regent of Spain. They remained a Spanish possession until 1898, when they were sold to Germany—except for Guam, which was ceded to the United States. For not siding with Germany during World War I, Japan was awarded all of the German islands in the Pacific north of the equator, including the Carolines, the Marshalls, and the Marianas. In 1944, in preparation for the invasion of Japan, U.S. forces took the islands. In 1947, the group was incorporated into the U.S. Trust Territory of the Pacific Islands, and in 1975, they voted for commonwealth status. Three years later, they became self-governing under U.S. military protection. Guam, the most populous of the Marianas, is a U.S. territory.

Mariana Trench Deep-sea trench in the southwestern Pacific, east of the Mariana Islands and part of the system of trenches that trends northward in the western Pacific as far as the Aleutians. At a maximum known depth of 35,800 feet, the part of the trench known as the Challenger Deep is the deepest point in the ocean. It was measured by (and named for) the British research vessel *Challenger II* in 1949. On January 23, 1960, with Jacques Piccard and U.S. Navy lieutenant Don Walsh aboard, the bathyscaph *Trieste* landed on the bottom there, thus setting an unbreakable depth record. In March 1995, the Japanese unmanned submersible *Kaiko* repeated the feat.

mariculture The saltwater version of aquaculture: farming the sea for plant and animal crops in more or less controlled environments. "Fish farming" probably began in ancient Egypt and biblical times, and was practiced around 500 B.C. in China, but until the 1960s, modern commercial fish culture was mainly restricted to raising rainbow trout. The increased demand for seafood, coupled with the decline of some important food fishes in the wild, has led to a great increase in mariculture. So far, the most successful species have been shrimp, prawns, salmon, and oysters, largely because they demand high prices in fish markets and restaurants, which can justify the high costs of raising them. (Pearl oysters are also farmed, especially in Japan.) Marine algaes, such as seaweed, are raised in such areas as Japan and Korea, where they are important items in the diet. The Japanese are experimenting with raising squid and octopuses, but the successful cultivation of lobsters has so far evaded them. Generally, prey species are easier to raise, because predators—like tuna and swordfish—require the cultivation of prey species, which in turn require the smaller, planktonic items upon which they feed. Ocean Thermal Energy Conversion (OTEC), which is the technology that converts heat energy stored in the ocean to electric power, brings to the surface cold, deep seawater, which is rich in nutrients and has been shown to be an ideal medium for the cultivation of various species, including salmon, oysters, giant clams, and abalone.

marine archaeology The systematic study of historic material found on or beneath the ocean floor. Ever since the first ships sank, men have been interested in bringing up what was on them, so the first underwater investigations were probably in search of sunken treasure, like that of William Phips in 1687, who found the wreck of the Spanish galleon *Nuestra Señora de la Concepción,* which sank off Hispaniola in 1641. The first archaeological investigation occurred in 1959, when Peter Throckmorton, a scuba-diving historian, located a Bronze Age wreck off southwest Turkey in the Mediterranean. A team led by George Bass of the University of Pennsylvania excavated the wreck in 1960, and during the next four years, on a seventh-century Byzantine wreck located 120 feet down near Bodrum, Turkey, introduced the concept of mapping by photogrammetry and other methods that allowed precise positions of artifacts to be properly recorded. In 1956, Swedish archaeologists raised the *Vasa,* a 200-foot-long, 64-gun wooden warship that went down in Stockholm harbor on August 10, 1628, just after she was launched. Since then, ships of ancient Greek and Roman, medieval, and later times have been found, and the field of marine archaeology has expanded greatly. In 1972, George Bass founded the Institute of Nautical Archaeology, now affiliated with Texas A&M University, which has investigated numerous wrecks and excavated the seventeenth-century town of Port Royal, Jamaica, which slid into the Caribbean during an earthquake in 1692. Using submersibles, submarines, and robotic cameras, Robert Ballard had located the *Titanic,* the *Lusitania,* and the *Bismarck,* and is now continuing to search for historic shipwrecks using the nuclear research submarine, *NR-1.*

See also Ballard, Bass (George, b. 1932), *Bismarck, Central America, Lusitania,* Port Royal, *Titanic, Vasa*

marine iguana (*Amblyrhynchus cristatus*) The only lizard in the world whose range is limited to the intertidal zone, the marine iguana is further restricted to the Galápagos Islands. Marine iguanas are the only lizards whose habitat is mostly aquatic, and they have blunt snouts to enable them to graze underwater on the algae that makes up most of their diet. They swim by propelling themselves with their tails, and they can dive to depths of 35 feet and remain submerged for nearly an hour at a time. On land—which is mostly lava—they bask in the sun in large groups. Reaching a maximum length of 5 feet (including tail), the lizards are sooty black, except during the breeding season, when they acquire bright red or green patches of color. The Galápagos land iguana (*Conolophus*) is a completely different reptile that spends all of its life on land, dining on plants, particularly cacti. **See also Galapágos Islands**

marine iguana

marine reptiles The only living marine reptiles are sea turtles, sea snakes, and the marine iguana, but in ancient seas, all kinds of aquatic reptiles, large and small, roamed the depths. (The saltwater crocodile lives mostly in freshwater habitats but can travel for great distances in the sea.) They are not dinosaurs because they are not descended from the first dinosaurs; they represent a separate group of vertebrates. The earliest known marine reptiles are the nothosaurs, which had slender bodies, long necks and tails, and webbed feet. They flourished in the Triassic period, about 255 million years ago. They were related to the plesiosaurs, which ranged in length from 8 to 50 feet and looked like seagoing dinosaurs (or what the Loch Ness monster is supposed to look like), but were true lizards. Other forms of marine lizards were the ichthyosaurs ("fish lizards"), dolphin-shaped lizards with a lunate tail fin, that lived from the early Jurassic (200 million years ago) to the Late Cretaceous (85 million years ago). Mosasaurs (related to today's monitor lizards, including the Komodo dragon) inhabited all the world's oceans during the Late Cretaceous, but the finest fossils have been found in Kansas. The best known of the mosasaurs is *Tylosaurus*, a 30- to 40-foot-long predatory lizard. Some marine reptiles came ashore to lay their eggs (like the sea turtles), but the ichthyosaurs and probably the mosasaurs gave birth to live young in the water, most likely in the shallows.

See also ichthyosaur, marine iguana, plesiosaurs, saltwater crocodile, sea snakes, sea turtle

Marion-Dufresne, Nicolas-Thomas (1729–1772) French naval officer who discovered the Prince Edward Islands and the Crozet Islands in the southern Indian Ocean between southern Africa and the Antarctic. Marion-Dufresne, who was sailing from Mauritius to the Pacific in 1772, mistakenly believed that the Crozets (which he named for his lieutenant, Julien Crozet) were offshore islands of the southern continent. While refitting his ships in New Zealand in March 1772, Marion-Dufresne and several members of his crew were killed and eaten by Maoris. In 1776, when Captain James Cook rediscovered these islands, he named them for Marion and Prince Edward, one of the sons of King George III.

Marion Island Isolated subantarctic island in the southern Indian Ocean, some 1,200 miles southeast of Cape Town, part of the Prince Edward Islands. The islands, which are only 12 miles apart, were discovered in 1772 by Nicolas-Thomas Marion-Dufresne, a French navigator who was searching for the Antarctic continent—and believed he had found its outlying islands. Four years later, the islands were rediscovered by James Cook, who gave them their present names. Marion is the larger of the two islands, covering about 100 square miles and dominated by the snow-capped President Swart Peak, 3,800 feet high. The coastline is characterized by steep cliffs, and grassy heights where wandering and sooty albatrosses nest. On the island's two sandy beaches, rockhopper penguins, southern elephant seals, and Antarctic fur seals come ashore.

Markham, Sir Clements (1830–1916) British geographer, explorer, and writer, who accompanied Sir Edward Belcher in the unsuccessful 1850–1851 search for Sir John Franklin, who had been lost in the Arctic since 1845. Markham retired from the British navy in 1852 and the following year joined the civil service. He is credited with introducing the Peruvian quinine-bearing tree (cinchona) into India from Peru and the Brazilian rubber tree into Malaysia. A prodigious author, Markham wrote more than fifty books, including important biographies of explorers like Christopher Columbus and John Davis. As honorary secretary of the Royal Geographical Society, he was responsible for the choice of Robert Falcon Scott as leader of the 1901–1904 expedition to the Antarctic.

See also Franklin, Scott

Marquesas Islands A group of volcanic islands in the South Pacific, some 700 miles north of Tahiti. The largest is Nuku Hiva, the second largest Hiva Oa. The islands are spectacularly beautiful, with lush forests and towering green mountains. The southern cluster was discovered in 1595 by Álvaro de Mendaña, whose navigator was Pedro Fernández de Quirós. Mendaña was searching for the Solomon Islands, which had been discovered and named by Pedro Sarmiento de Gamboa some thirty years earlier, but Mendaña was way off course, and could not find them. Instead, he discovered the Marquesas, which he named for the Marqués de Mendoza, the viceroy of Lima. (The southern cluster is now known as the Mendaña group.) The northern group was discovered in 1791 by the American navigator Joseph Ingraham, and naval officer David Porter claimed it for America in 1813, while searching the Pacific for British vessels to engage during the War of 1812. The American Congress never recognized the claim, and the French took possession of Nuku Hiva in 1842, as part of what would eventually be known as French Polynesia. Of all the Pacific islands, the Marquesas have been among the hardest hit by European diseases: from an eighteenth-century population of some 21,000, there are now only 5,000 Marquesans left. Young Herman Melville landed on Nuku Hiva when he jumped ship from the New Bedford whaler *Acushnet* in 1842; he told the story of his experiences with the cannibals in his first and second novels, *Typee* and *Omoo.* The French artist Paul Gauguin (1848–1903) lived and is buried on Hiva Oa.

See also Mendaña, Solomon Islands

Marshall Islands An independent country as of 1991, with a total population of some 60,000, the Marshall Islands consist of two parallel chains in eastern Micronesia: the Ratak ("Sunrise") group to the east, and the Ralik ("Sunset") group to the west. The twenty-nine separate atolls and innumerable islets are spread out over an area of 780,000 square miles, an area larger than Alaska. The capital is Dalap-Uliga-Darrit, three islands on Majuro connected by landfills. Most of the islands, including Eniwetok, Bikini, Kwajalein, Rongelap, and Majuro, are atolls that sit atop submerged volcanoes. They were discovered in 1529 by the Spanish navigator Álvaro de Saavedra, and later exploited by British naval officer John Marshall, who named the islands for himself. Russian explorers Adam Ivan Krusenstern and Otto von Kotzebue mapped the Marshalls in the early nineteenth century. The Spanish sold the islands to Germany after the Spanish-American War (1898), but Japan was awarded them after World War I because it refused to cooperate with Germany. During World War II, the Americans captured the Marshalls after heavy fighting, particularly on Eniwetok and Kwajalein. In

1954, American nuclear testing took place at Bikini and Eniwetok, the radioactivity rendering Bikini uninhabitable. (Eniwetok was always uninhabited.) The atolls of Rongelap, Rongerik, Utirik, and Ailinginae were covered with radioactive ash, and the residents were evacuated. Many of them suffered from radiation-induced diseases and gave birth to deformed or stillborn babies. In 1983, the U.S. government gave the Marshall Islanders $183.7 million for damages caused by the tests.

See also Bikini, Eniwetok, Kotzebue, Krusenstern, Kwajalein, Rongelap

Martha's Vineyard Island off the coast of southeastern Massachusetts, separated from Cape Cod by Vineyard and Nantucket Sounds. The island, about 100 square miles in area, was named by Bartholomew Gosnold in 1602 for the wild grapes. In 1641, Thomas Mayhew bought Martha's Vineyard, along with Nantucket and the Elizabeth Islands, for forty pounds and established the first community at Edgartown, where the settlers engaged in farming, brick making, salt production, and fishing. In the eighteenth century, Martha's Vineyard became an important whaling center, but never as important as smaller Nantucket, 10 miles to the east. By the middle of the nineteenth century, Martha's Vineyard's beaches and scenic attractions had made it into a popular summer resort. It is divided into the towns of Aquinnah (Gay Head), Chilmark, Edgartown, Oak Bluffs, Tisbury, and West Tisbury, each of which has its own particular character. **See also Nantucket**

Martinique Island in the Windward group in the Caribbean, with approximately 350,000 people living on 425 square miles. Martinique was visited by Columbus in 1502, but was otherwise ignored by Europeans until the French established a settlement there in 1635. They eliminated the indigenous Arawak and Carib Indians, and then imported slaves from Africa to work the sugar plantations. By the eighteenth century, Martinique was one of France's most valuable colonies because of its sugar production. Now administered as an overseas department of France, the island still produces sugar and rum, but it depends on tourism for most of its income. In 1902, Mount Pelée erupted, sending a broiling cloud of gas and ash rolling down the mountainside, completely obliterating the city of St. Pierre and killing every one of the city's thirty thousand inhabitants except for Auguste Ciparis, who was in an underground jail. **See also Mount Pelée**

Mary Celeste The brig *Mary Celeste,* Captain Benjamin Briggs commanding, sailed from New York on November 5, 1872, with a cargo of seventeen hundred barrels of alcohol, bound for Genoa. On December 4, she was sighted about halfway between the Azores and Portugal by Captain David Morehouse of another New York brig, the *Dei Gratia.* When calls failed to elicit any signs of human occupation on the *Mary Celeste,* Captain Morehouse boarded the ship to investigate. There was not a living soul aboard. The ship's sextant, chronometer, and register were missing, but the ship's logbook was there, with the last entry dated November 25. There was food enough for six months, but no cooked food on the table. It appeared that the entire ship's company—including Captain Briggs's wife and their two-year-old daughter—had departed in a great hurry. Not knowing what to make of the abandoned ship, Morehouse put a crew aboard her, and they sailed in tandem to Gibraltar. After much deliberation, a salvage fee of seventeen hundred pounds was awarded to the *Dei Gratia*'s captain and crew, but the mystery has never been solved. Every kind of solution has been proffered, from mutiny to abandoning ship for fear that the fumes from the alcohol were about to explode, and even a collision with a giant squid, but if the *Mary Celeste*'s captain and crew left the ship alive, they took the reason to their graves.

masked booby (*Sula dactylatra*) Very similar to the gannet, the masked booby is found in warmer waters, breeding in the western Pacific to the Coral Sea, western Mexico and the Galápagos, and throughout the Caribbean and the southern Indian Ocean. The range of this species overlaps that of the red-footed, blue-footed, and brown boobies. Like the gannets, the adults are white with a black mask and black wing tips, but where the gannets have a yellowish head and a bluish beak, the masked booby has a yellow bill and an all-white head. The masked booby is also known as the blue-faced booby. **See also blue-footed booby, gannet, red-footed booby**

Matsu: See Quemoy

Maui Second largest of the Hawaiian islands with an area of 728 square miles, Maui is named for the Polynesian god who fished the islands out of the sea with a large hook. It is nicknamed the "valley isle" for the narrow, valleylike isthmus that joins its two volcanic peaks. The larger is the dormant volcano on east Maui known as Haleakala ("House of the Sun"), 10,000 feet high and so prodigious a producer of lava that its flows joined it to what is now the smaller portion of the island to the northwest. West Maui is rainier, has better beaches, and is the location of the huge tourist complex at Ka'anapali. The island was spotted by James Cook in 1778, but he could find no suitable landing site and sailed on to Kauai. In 1786 the French explorer Jean-François de La Pérouse became the first European to land on Maui.

After conquering Maui in 1790, King Kamehameha made Lahaina his capital; in the 1840s, this port on the northwestern shore of the island became one of the world's most important centers for sperm whalers. They offloaded their valuable cargoes there and often wreaked havoc on the city. Lahaina is now the scene of a whaling museum and the home port of a fleet of whale-watching boats that take tourists out to see humpback whales in the offshore waters. Hawaiian statehood in 1959 coincided with the introduction of jet travel to the islands, and almost overnight what had been a sleepy South Pacific island devoted mostly to growing pineapples and sugarcane became a mecca for tourists. The permanent population is around 91,000, but another 2 million visit the island every year, making it the most popular tourist destination in the islands. Even though there are now hotels, condominium complexes, shopping centers, and overcrowded highways, Maui still has some spectacular natural wonders, including the Iao Valley, and some of the best beaches in the islands. **See also Hawaiian Islands**

Mauretania Launched in 1908, the Cunard liner *Mauretania* was a sister ship to the *Lusitania,* which had been christened a year earlier. Splendid ships like these, as well as the White Star Line's *Olympic* and *Titanic,* marked the apex of the age of the great transatlantic liners. These gigantic ships were fitted out in the height of early-twentieth-century luxury: marble walls, brocade curtains, and strategically placed trees gave the feeling of somewhat decadent floating villas. They were exactly what the rich and self-indulgent citizens of Europe and America wanted. The main staircase of the *Mauretania* was paneled in French walnut; the writing room was pure Louis XVI; the café was modeled after the Orangery at Hampton Court. *Mauretania* was the fastest ship afloat; she took the Blue Riband (the award for the fastest crossing) in 1907 and held it for twenty-two years. During the war, along with *Aquitania, Olympic,* and *Britannic, Mauretania* served as a hospital ship in the Mediterranean. As larger and faster ships were built, the gleam of those in the *Mauretania* class diminished. By the time the *Queen Mary* (80,000 tons) made her maiden voyage in 1936, the *Mauretania* (40,000 tons) had already been broken up for scrap.

See also *Lusitania*

Mauritius Island republic 500 miles east of Madagascar in the southwest Indian Ocean; one of the Mascarene Islands. (The others are Rodrigues, a dependency of Mauritius, and Réunion, a department of France.) A mountainous island surrounded by coral reefs, Mauritius covers 787 square miles. It was settled in 1598 by the Dutch, who named it for Prince Maurice of Nassau. When the French came in 1715, they called it Île de France. The British conquered the island in 1810, restored the Dutch name, and held it until Mauritius declared its independence in 1968. The capital is Port Louis; there are 1,140,000 people. With the eradication of malaria in 1960, the population grew so quickly that Mauritius became one of the most densely populated countries in the world. When the Dutch first arrived, they encountered huge flightless birds that they called *walghvogels,* or "nauseating birds," because they tasted so bad when they were cooked. Nevertheless, in less than a hundred years, they had killed all of these 50-pound pigeons anyway, and with the exception of some specimens and some bones that were brought to Europe, no one ever saw a living dodo again.

See also dodo, Madagascar, Réunion

Maury, Matthew Fontaine (1806–1873) American hydrographer and one of the founders of modern oceanography. Maury entered the U.S. Navy in 1825 as a midshipman, circumnavigating the globe in the *Vincennes.* Having prepared for a career in the navy, he was lamed in a stagecoach accident in 1836 and deemed unfit for active duty. In 1842 he was placed in charge of the navy's Office of Charts and Instruments, which evolved into the U.S. Naval Observatory and Hydrographic Office. To gather data on winds and currents, he examined the logs of hundreds of ships, and his analyses, published as *Wind and Current Charts* and *Sailing Directions,* produced shorter and faster routes. He created charts of the Atlantic, Pacific, and Indian Oceans, the first ever universally available. His *Physical Geography of the Sea,* published in 1855, was the first modern oceanographic text. An advocate of deep-sea soundings, he served as adviser to Cyrus Field in the laying of the first transatlantic telegraph cable. Because he was born in Virginia, he threw in with the South in the Civil War and became head of coastal, harbor, and river defenses for the Confederate Navy. He tried to establish a Confederate colony in Mexico, but when that failed, and the war ended badly for the South, he emigrated to England. Upon his return to the United States in 1868, he was made a professor of meteorology at Virginia Military Institute, a position he held until his death.

Mawson, Douglas (1882–1958) Australian geologist who became his country's most renowned Antarctic explorer. He accompanied Shackleton's *Nimrod* expedition (1907–1909), and was one of the party to climb Mount Erebus and reach the magnetic South Pole. Between 1911 and 1914, he led the Australian Antarctic Expedition, which was intended to map the Antarctic continent directly south of Australia. Unfortunately, he established his camp at Commonwealth Bay, held to be the windiest place on earth, and the expedition's work was severely limited by howling winds and gales. Accompanied by B. R. S. Ninnis and Xavier Mertz, Mawson was exploring the region east of Cape Denison, when Ninnis disappeared into a crevasse with most of the party's food, its dog food, extra clothing, and the tent. Mertz died on the 300-mile trek back to camp, and Mawson made it back alone, starving and close to death, after sawing the sledge in half with his pocket knife. He arrived in time to see his ship *Aurora* sailing away. Luckily, six men had been left behind to await the survivors. Because of a storm, the ship could not return to pick them up, so they were forced to spend another winter in a hut, surviving blizzards and even a hurricane. Mawson (who was knighted in 1914) returned to the Antarctic in 1929 as the leader of the joint British, Australian, and New Zealand Antarctic Research Expedition (BANZARE). The Australian base in the Antarctic is named for him, as are the Mawson Escarpment, Mawson Peak, Mawson Peninsula, and Mawson Point on Macquarie Island. **See also Shackleton**

Mayflower The ship that carried the first contingent of Pilgrims from England to Massachusetts in 1620. No details of the *Mayflower* have been passed down, but she is believed to have been a three-masted wine ship of 90 feet overall, displacing about 180 tons. On her scheduled voyage to the Virginia colony (founded thirteen years earlier), she was to have been accompanied by the smaller *Speedwell,* which had come from Holland with other settlers. They were supposed to rendezvous at Southampton and cross the ocean together, but the *Speedwell* was leaking so badly that she turned back, transferring some of her passengers to the *Mayflower,* under Captain Christopher Jones. The *Mayflower* crossed the ocean with 101 people—31 of them children—a crew of 34, and 2 dogs. It was beset by storms, and instead of reaching Virginia after a sixty-five-day voyage, the ship fetched up at today's Provincetown. It dropped anchor on November 11, 1620, and Myles Standish explored the coast, returning after two days with his assessment that the area was unfit for settlement. He then ventured to Wellfleet, which also seemed inhospitable (the Indians shot arrows at them), and finally crossed Cape Cod Bay to the location marked on John Smith's 1614 map as "Plymouth."

McClintock, Francis Leopold (1819–1907) British explorer who first sailed to the Arctic with James Clark Ross in the *Enterprise* in 1848–1849, in a search for Sir John Franklin. In 1852, while commander of the sloop *Intrepid* in an Arctic expedition under Edward Belcher, he made a sledge journey of 1,210 miles in 105 days from Beechey Island to the west coast of Prince Patrick Island, establishing himself as the greatest sledger in Arctic history. After the Admiralty gave up on finding Franklin, his wife, Lady Jane Franklin, outfitted the *Fox* and appointed McClintock commander. In 1857, twelve years after the disappearance of Franklin's ships *Erebus* and *Terror,* McClintock found Francis Crozier's last notes on King William Island, and was able—finally—to establish the fate of Franklin's expedition. In 1859 McClintock published *The Voyage of the Fox,* in which he discussed his significant discoveries. He subsequently served in the Mediterranean and the West Indies. **See also Crozier, Franklin, Ross (James Clark)**

McClure, Robert John (1807–1873) British Arctic explorer who discovered the Northwest Passage in 1850. In the search for John Franklin, McClure took the *Investigator* around the tip of South America accompanied by Richard Collinson in the *Enterprise,* but the two ships were separated off Cape Horn and proceeded independently. McClure stopped briefly in Honolulu, passed through the Bering Strait, and entered the Beaufort Sea from the west. He found no sign of the Franklin expedition, but, passing through what is now known as McClure Strait, between Banks and Melville Islands, he realized that the clear water that he saw to the east meant that he could reach Lancaster Sound and thus make the Northwest Passage. The ice closed around his ship, however, and not only did he not make the fabled passage, he was forced to spend two years trapped in the ice, and the *Investigator* was crushed. Suffering from the cold and scurvy, McClure and his men were rescued by Henry Kellett in the *Resolute* and brought back to England in 1854.
See also Franklin, McClintock, Northwest Passage

McCormick's skua (*Catharacta maccormicki*) The Antarctic and Southern Ocean version of the great skua (*C. skua*), McCormick's is a slightly smaller bird, somewhat lighter in color, but no less aggressive to other birds, as well as to people, who enter its nesting territory. Like the other skuas, it is a predator and kleptoparasite, feeding on fishes that it forces other birds to surrender and on eggs, chicks, and carrion. Whether *maccormicki* is a valid species is the subject of ongoing ornithological debate; there are those who believe that it is and others who think it is the same bird as the great (northern) skua, relocated to the south. *C. maccormicki,* sometimes known as the south polar skua, has been known to wander north across the equator, appearing in India, Japan, and Canada. To further confuse matters, there are several races of the south polar skua, such as the slightly smaller Chilean skua (*C. chilensis*), which inhabits both coasts of South America, and the Antarctic skua (*C. antarctica*), which is even smaller and breeds throughout the Antarctic. (Some ornithologists have designated these forms as full species.) All southern skuas are polymorphic—that is, individuals can range in color from all-over dark brown to those with undersides that are pale grayish-brown. **See also great skua, long-tailed skua, pomarine skua, skua**

McKay, Donald (1810–1880) Born in Nova Scotia, Donald McKay became the greatest builder of sailing ships in history. After his apprenticeship in New York and Newburyport, Massachusetts, he moved to Boston and opened his own shop. There he produced the swift, sleek clippers that are considered among the most graceful ships ever to sail the sea. The first to emerge from McKay's enterprise was the *Stag Hound,* in 1850 the largest merchant vessel afloat. Then came the *Flying Cloud,* which in 1851 set the record of eighty-nine days, twenty-one hours, for a 'round-the-Horn voyage from New York to San Francisco. (Before the 1849 California Gold Rush, two hundred days was considered a respectable time for the 15,000-mile voyage.) McKay also built the *Flying Fish, Glory of the Seas, Sovereign of the Seas, Lightning,* and *James Baines,* and the sister ships *Star of Empire* and *Chariot of Fame*—all clippers that were as famous for their beauty as they were for their speed. **See also clipper ships**

Mediterranean monk seal (*Monachus monachus*) There are probably no more than fourteen hundred Mediterranean monk seals alive today, making them among the rarest of all pinnipeds. Small populations are reported from the Canary Islands, Madeira, and Morocco and along the European and African coasts of the Mediterranean. Adults may be 9 feet long and weigh 600 pounds. In 1997, a morbillivirus killed 270 of these highly endangered seals on the western Saharan coast of Africa. The Caribbean monk seal, a close relative, is extinct.

Mediterranean monk seal

See also Caribbean monk seal, Hawaiian monk seal

Mediterranean Sea Almost completely enclosed by Europe and Asia Minor to the north and Africa to the south, the Mediterranean's 1.45 million square miles are open on the west to the Atlantic through the Strait of Gibraltar and on the east to the Black Sea through the Bosporus. Since it opened in 1869, the Suez Canal has connected the eastern Mediterranean to the Red Sea. Included in the Mediterranean are the Alboran, Balearic, Tyrrhenian, Ionian, Adriatic, and Aegean "Seas," all of which are merely convenient geographical distinctions; all of them comprise a single body of water. The Mediterranean is 2,400 miles from the Strait of Gibraltar to the shore of Lebanon, and the longest north-south distance, from Trieste to the shore of Libya, is about 1,000 miles. Because of its restricted access through the Straits of Gibraltar, the Mediterranean is virtually tideless, and the ebb and flow changes the water level of the entire sea only 1 or 2 inches. The largest rivers that flow into the sea are the Po, the Rhone, the Ebro, and the Nile. The largest islands are Corsica, Sardinia, Sicily, Crete, Cyprus, and Rhodes, but there are hundreds of others, including the Balearics, Malta, and the Greek islands. Clockwise starting from Spain in the far west, the countries bordering the Mediterranean are France, Monaco, Italy, Croatia, Slovenia, Bosnia, Albania, Greece, Turkey, Syria, Lebanon, Israel, Egypt, Libya, Tunisia, Algeria, and Morocco. Of the islands, only Cyprus is a sovereign nation. All the countries that border the Mediterranean use the sea for recreation, fishing, and, unfortunately, waste disposal. Among the more important fish species caught in the Mediterranean are such bottom fishes as flounder, sole, turbot, whiting, congers, croakers, sea bream, and hake, and pelagic (migratory) species such as sardines, anchovies, bluefin tuna, bonitos, and mackerel. Various sharks and rays are also harvested, and the largest great white shark on record, a 23-footer, was caught off Malta in 1989. Corals, sponges, and seaweed are also harvested in the Mediterranean, but almost all the sea's marine resources are depleted because of overexploitation. From the Egyptians whose ships first plied its waters to the Phoenician and Greek navigators who explored it and made the first maps, to the Renaissance sea powers such as Spain, Portugal, Venice, Genoa, and the Ottoman Empire, to the pirates of the Barbary Coast, the Mediterranean has contained the history of Western civilization.

medusa In Greek mythology, Medusa was a woman who was once beautiful but as a punishment had her hair turned into a nest of writhing serpents. The term is sometimes used in reference to the jellyfishes, an obvious reference to the tentacles. It also applies to the form that occurs in certain stages of cnidarians where they assume an unattached form in an umbrellalike shape with tentacles around the edges. The alternative shape is the polyp, with the tentacles upward, usually attached to a firm support. Many cnidarians pass through alternating polyp and medusa generations, which is confusing to all but cnidarian taxonomists.

See also cnidarian, jellyfish

***Medusa,* wreck of the** On July 2, 1816, the French frigate *Medusa* ran aground on a sandbank off the coast of what is now Mauritania in West Africa. Of the ship's 400 passengers, 250 managed to board the four lifeboats, leaving the remaining 149 men (and 1 woman) to clamber aboard a raft, hastily built from the ship's timbers and masts, that measured 65 by 23 feet. The raft was supposed to have been towed to shore by the lifeboats, but it was too heavy and unseaworthy, so it was left floating as the boats sailed away. Storms and overcrowding took their toll, and many of the raft's passengers fell (or were pushed) into the Atlantic. The bag of soaked biscuits, two casks of water, and six barrels of wine were not nearly enough for the remaining occupants, and fierce

giant squid
(above)

walrus
(left)

RICHARD ELLIS

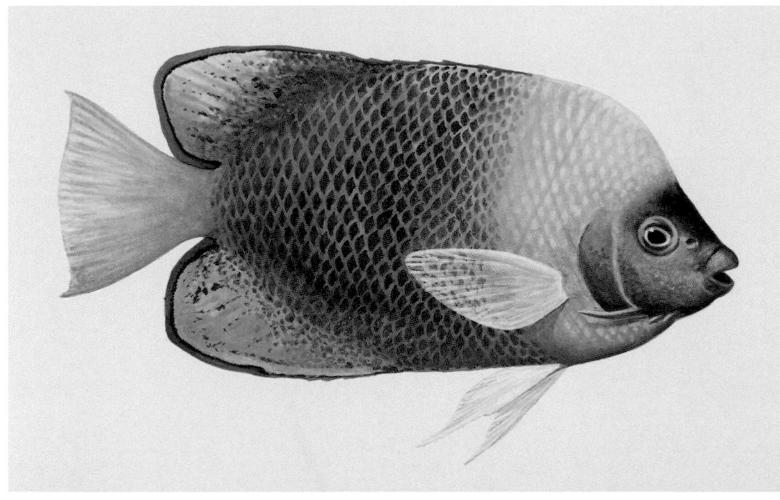

common dolphin
(opposite)

clarion angelfish
(left)

sailfish
(below)

striped marlin
(above)

Murex shell
(right)

great white shark
(opposite)

RICHARD ELLIS

sperm whales
(opposite)

wahoo
(left)

belugas
(below)

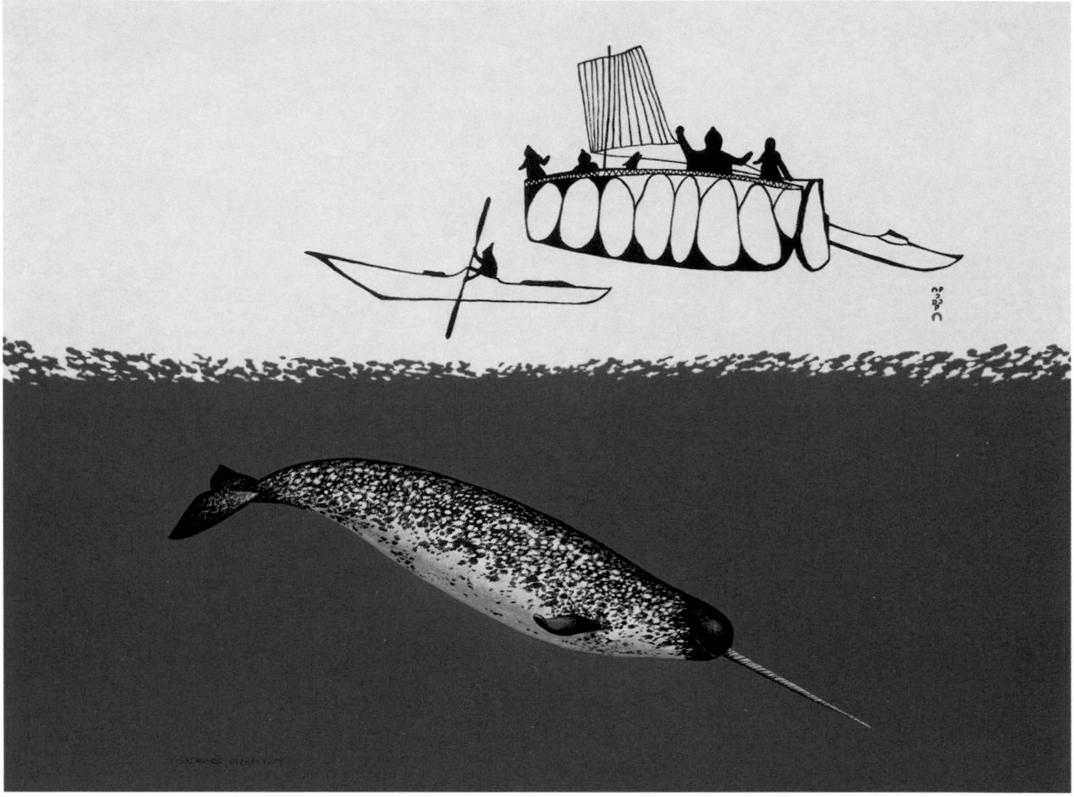

narwhal
(above)

harbor seal
(right)

megalodon

fighting broke out, resulting in the death of another sixty people. By July 7, there were only thirty left alive, and as hunger overcame the survivors, they began hacking pieces off their dead comrades to eat. These men continued to fight over the one remaining barrel of wine, and the stronger of them threw the weaker—including the only woman—overboard. The raft's fifteen survivors floated around until July 17, when they were rescued by the *Argus*, another French warship. The French painter Théodore Géricault (1791–1824) painted a monumental canvas of the raft—it was entitled *Shipwreck Scene*, but has since come to be known as *The Raft of the Medusa*—which was the sensation of the Paris Salon of 1819.

megalodon (*Carcharodon megalodon*) A 50-foot-long ancestor of the great white shark, this terrifying creature has been extinct for at least several hundred thousand years. Where the teeth of recent sharks are all white, because they are made of dentin, those of megalodon—which may be 6 inches in height as compared with the 2½-inch-high teeth of the white shark—are all fossilized, and are gray, dark brown, or black. A shark this size could swallow a horse. It is not known why this powerful predator became extinct.

See also great white shark

megamouth (*Megachasma pelagios*) The 1976 discovery of the first megamouth shark off the Hawaiian island of Oahu was one of the most remarkable zoological finds of the century. Completely unexpected because nothing like it had ever been seen, this shark, tangled in the parachute sea anchor of a U.S. Navy research vessel, was 14½ feet long and weighed 1,653 pounds. A male, it had big rubbery lips, and the examination of its stomach contents showed it to be a plankton feeder. It was originally classified in its own family, genus, and species, but it has now been shown to be related to the Lamnidae or mackerel sharks. Since the original specimen, ten more have been collected, usually dead, but in 1990, a healthy specimen was caught in a gill net by a California fisherman, then filmed, tagged, and released. Other specimens have come from Japan, Australia, Brazil, and the Philippines, indicating a wide—if unsuspected—distribution. When a female washed ashore in 1994 in Fukuoka, Japan, the carcass was examined by a veritable faculty of biologists, resulting in the 1997 publication of the definitive *Biology of the Megamouth Shark*.

See also basking shark, whale shark

megamouth

Melanesia With Polynesia and Micronesia, one of the traditional ethnographic groupings of South Pacific island cultures, including those of New Guinea, the Admiralty Islands, the Bismarck and Louisiade Archipelagoes, the Solomon Islands, New Caledonia, Vanuatu (formerly the New Hebrides), Fiji, and various smaller islands. Melanesia is not a political entity, but the term is applied both to the Papuan people of the New Guinea regions and to another group, the Austronesians, who are believed to have originated in the Philippines and Indonesia. The cultures of the various groups differ, but are based on agriculture and trading. During the twentieth century, profound changes occurred in these islands as a result of European, Japanese, and American influences, particularly the war in the Pacific, much of which was fought on Melanesian islands. **See also Admiralty Islands, Bismarck Archipelago, Fiji, New Caledonia, New Guinea, Solomon Islands, Vanuatu**

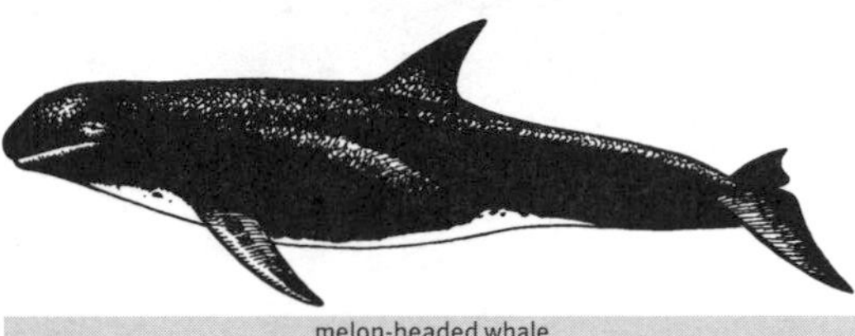

melon-headed whale

melon-headed whale (*Peponocephala electra*) Originally classified in the genus *Lagenorhynchus* (as *L. electra*), this rare delphinid was known only from skeletal material until 1963, when a live specimen was caught in Japanese waters. In 1965, some five hundred more appeared in Japan's Sugura Bay, and a careful examination of the animals led to the establishment of a new genus, *Peponocephala,* which means "melon-head." It is now known to be distributed worldwide in tropical and subtropical waters, where it has been observed in fast-swimming herds numbering up to fifteen hundred animals. Although this species closely resembles the pygmy killer whale, it is solid black where the pygmy killer has a dark "cape," and *Peponocephala* has smaller teeth, suggesting that it feeds on smaller prey. Both species reach a length of about 8½ feet. Despite its common name, this species is a dolphin, as are the pygmy killer *whale,* the false killer *whale,* the killer *whale,* and the pilot *whales.*

See also false killer whale, killer whale, pilot whale, pygmy killer whale

Melville, Herman (1819–1891) American author, born in New York City, who shipped out at the age of eighteen aboard the New Bedford whaler *Acushnet.* He jumped ship in the Marquesas in 1842 and found temporary refuge among the cannibalistic natives before escaping on the Australian whaler *Lucy Ann* headed for Tahiti. From there he enlisted as a seaman on the frigate *United States;* he returned to Boston in 1844. He began writing novels based on his experiences, the first of which was *Typee* (1846), then *Omoo* (1847), followed by *Mardi* (1849), *Redburn* (1849), and *White-Jacket* (1850). In 1850, he moved to Massachusetts, where he began a friendship with Nathaniel Hawthorne. It was here that he wrote the book for which he will always be remembered, *Moby-Dick* (1851). The story of Ahab and the white whale was not a critical or commercial success during his lifetime, and although he continued to write—*Pierre, or the Ambiguities* (1852), *Israel Potter* (1855), and *The Piazza Tales* (1856)—he could not support his family and so took a job as a customs inspector. He died unrecognized and almost unremembered until the 1920s, when literary critics began to realize his importance in American literature. (*Billy Budd,* written shortly before his death, was not published until 1924.)

See also Marquesas Islands, *Moby-Dick*

Melville Island (1) Located in the Timor Sea, off the coast of Arnhem Land, Northern Territory, Melville is Australia's second largest island, after Tasmania; it measures 80 by 55 miles and covers about 2,240 square miles. The island was first sighted by the Dutch explorer Abel Tasman in 1644, and in 1818 Captain Phillip Parker King named it after Robert Saunders Dundas, Viscount Melville, first lord of the Admiralty. At that time, the island was occupied by the Tiwi, a tribe of Aborigines (who called the island Yermalner), and it still is. Under the 1976 Aboriginal Land Rights Act, ownership of the island passed from the Australian government to the Tiwi Land Council.

Melville Island (2) Also named for Robert Dundas, second Viscount Melville, this Melville Island is the westernmost of the Parry Islands in Canada's Northwest Territory. The Canadian island is some 200 miles long and 130 miles wide and is punctuated by the deep Hecla and Griper Bay, which was named for the two ships in William Parry's 1818 expedition. Inhabited only by musk oxen, wolves, and polar bears, it became part of the native-administered territory of Nunavut in 1999.

Mendaña, Álvaro de (1541–1596) Spanish navigator (full name: Álvaro de Mendaña de Neira) who sailed westward from Peru in 1567 with Pedro Sarmiento de Gamboa in search of the southern continent. They encountered a chain of islands that they believed to be the islands of King Solomon, and named them accordingly. Constant skirmishing with the natives eventually drove the explorers off, and they returned to Peru. Mendaña could not arrange to get back to the islands for almost three decades, but in 1595, with Pedro Fernández de Quirós as his navigator, he set sail again. He also brought his wife, Doña Isabel, and her three brothers. Mendaña could not find the Solomon Islands, but he did discover the Marquesas, which he named for the Marqués de Mendoza, the viceroy of Lima. After a difficult time with the Marquesans, several of whom were shot, the expedition then sailed to an island Mendaña named Santa Cruz, where he tried to establish a colony. (The name would eventually be applied to the group of islands.) Dissension and illness plagued the settlement, and before he died, Mendaña chose his wife as commander of the expedition. Fortunately for the starving survivors of this disastrous enterprise, Pedro Fernández de Quirós brought them back to the Philippines, refitted the leaking ships, and returned to Peru.

See also Marquesas Islands, Quirós, Solomon Islands

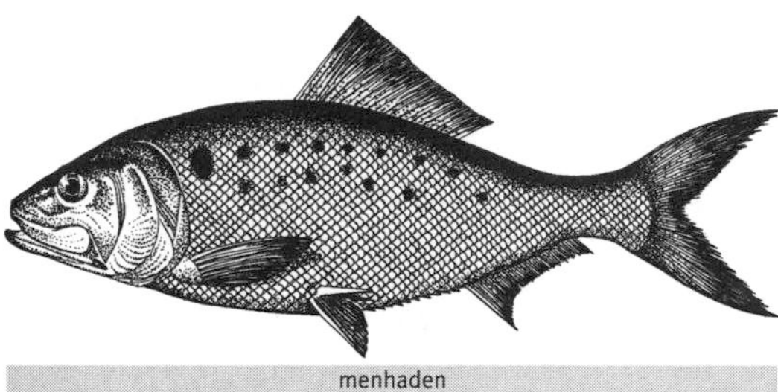

menhaden

menhaden (*Brevoortia tyrannus*) Also called mossbunker and pogy, this 1-foot-long member of the herring family (Clupeidae) is found in the continental waters of North America, from Nova Scotia to Florida. As in the other herrings, the belly scales of menhaden form a sharp keel. It is easy to identify this species by the large black spot just behind the gill cover. Menhaden are schooling fishes and swim near the surface. They are too bony and oily for a commercial fishery, but large numbers are collected for oil and the manufacture of fish meal.

See also alewife, Atlantic herring, shad

Mercator, Gerardus (1512–1594) Flemish geographer; he used the Latinized form of his name, which was Gerhard Kremer. His first maps, based on those of Ptolemy, showed two adjacent hemispheres. In 1554 he was given a chair in oceanography at the university at Duisberg (Germany), where he forsook Ptolemy and developed his own projection, with parallels and meridians at right angles. In 1569, he published his first work embodying these new principles, an eighteen-sheet world map. This was followed by more maps and finally a world atlas, produced in 1594, the year of his death. His projection did not become widely used until 1640, when French seafarers adopted it. For most sixteenth- and seventeenth-century European navigators, it did not matter that Mercator's map became more and more distorted as they approached the higher latitudes, deteriorating completely at the poles, which, in fact, cannot be shown on the Mercator projection.

mermaid A mythical creature that is supposed to be half woman and half fish. (If it is a man, it is known as a merman.) Although mermaid stories can be traced back to Babylonian times, the modern form was invented by the Danish storyteller Hans Christian Andersen (1805–1875) with "The Little Mermaid." (The statue of *Den Lille Havefrue* in Copenhagen is that country's leading tourist attraction and the most photographed statue in the world.) In most recent mermaid stories, including the films *Splash* (1984) and Disney's *The Little Mermaid* (1989), the heroine cannot talk and is rescued by a handsome prince. There are many affiliations between mermaids and manatees, but it is difficult to see how anyone could confuse a plump, pig-eyed, bewhiskered aquatic mammal with a beautiful, long-haired siren. One of P. T. Barnum's most notorious attractions was the "Feejee Mermaid," a dried-up combination of monkey skull and fish tail that he convinced thousands was the real thing. For centuries, people have been drying skates or rays in such a way that they seem to resemble winged fishes with human or mammalian heads. These dried rays were known as Jenny Hanivers, but the origin of the name is a mystery.

See also Jenny Haniver, manatee

mermaid's purse Often found in beach wrack, the leathery capsule with tendrils at each corner is actually the egg case of the skate. The tendrils serve to fasten it to plants or rocks before it hatches, but if it breaks loose and comes ashore, the embryo inside will die.

See also skates

Merrimack Originally a U.S. frigate that had been scuttled by Union forces when they left the Norfolk, Virginia, navy yard, the *Merrimack* was raised by the Confederates in 1861 and rebuilt into a slow-moving ironclad vessel armed with three 9-inch guns, two rifled guns fore and aft, and a cast-iron ram below the waterline. (Although she was rechristened CSS *Virginia,* she is almost always known as *Merrimack.*) The first warship to completely dispense with rigging of any kind, she was 275 feet long with a 51-foot beam. Wooden warships of the Union navy were helpless before the ironclad *Merrimack,* and on her maiden voyage out of Hampton Roads on March 8, 1862, she rammed and sank the *Cumberland,* set the *Congress* afire, and ran the battleship *Minnesota* aground. She sailed from Norfolk the next day, prepared to finish off the Union ships, but she found the *Monitor* waiting for her in Hampton Roads. Immediately the two ironclads began pounding away at each other, but the shells bounced harmlessly off their iron sheathing. Although the first battle of the ironclads ended in a draw, it was obvious that they represented the future of warships, and the revolving turret of the *Monitor* was an important innovation. When the Confederates evacuated Norfolk in May 1862, the *Merrimack* was scuttled—for the second time—in the James River. **See also ironclad, *Monitor***

Mesonychoteuthis hamiltoni A very large species of squid known only from the stomach contents of Antarctic sperm whales. *Mesonychoteuthis,* meaning "middle-hooked squid," refers to the location of the double row of hooks on the middle of each arm, between the basal and terminal ringed suckers. It is the only large squid with hooks on the tentacles as well as on the arms. Its tentacles are comparatively short and thick, and its tail fins are broad, muscular, and heart-shaped. The body of *Mesonychoteuthis* can be as large or larger than that of *Architeuthis,* but its tentacles are much shorter. *Mesonychoteuthis* is believed to have a circumpolar Antarctic distribution, and may be the predominant species in Antarctic waters. It is a cranchid squid, characterized by the fusion of the mantle to the head at one ventral and two dorsal points and by photophores on the ventral surface of the prominent, sometimes protruding, eyes.

See also giant squid, sperm whale, squid

Mesoplodon The largest genus of beaked whales; the name, meaning "middle tooth," refers to the location (in the males only) of the only two teeth in the lower jaw. Mesoplodonts are spindle-shaped, tapering noticeably at both ends. The blowhole is crescent-shaped, with the horns facing forward, and there are two throat grooves. The dorsal fin is usually small and located far back, and the tail lacks the median notch present in most other cetaceans. **See also Bahamonde's beaked whale, deep-crested whale, dense-beaked whale, ginkgo-toothed whale, Gulfstream beaked whale, Hector's beaked whale, Hubbs' beaked whale, Peruvian beaked whale, scamperdown whale, Sowerby's beaked whale, Stejneger's beaked whale, strap-toothed whale**

Micronesia With Melanesia and Polynesia, one of the three main divisions of Oceania, in the western Pacific Ocean north of the equator. Although there is a self-governing island group called the Federated States of Micronesia (Kosrae, Pohnpei, Truk, and Yap), "Micronesia"—from the Greek for "small islands"—applies to a loose collection of islands that share a common culture and history. They stretch for 3,000 miles and include the Carolines (Palau, Yap, Truk, Pohnpei, Kosrae); the Marshalls (Bikini, Eniwetok, Kwajalein, Majuro); the Marianas (Guam, Saipan, Tinian, Rota); the Gilberts (Tarawa, Butaritari); and the isolated island of Nauru. Many of the islands have mutually unintelligible Malayo-Polynesian languages.

See also Caroline Islands, French Polynesia, Gilbert Islands, Mariana Islands, Marshall Islands, Melanesia

Mid-Atlantic Ridge That part of the globe-circling seam of underwater mountain ranges that bisects the Atlantic Ocean, beginning north of Jan Mayen and Iceland at the Nansen Cordillera and running almost to the Antarctic continent. Most of the ridge is deep under the Atlantic, but occasional elements break the surface, such as Iceland, the Azores, St. Paul's Rocks, Ascension, Tristan da Cunha, and Bouvet Island. All along their length, the mountains are split by a volcanic rift that is constantly spewing lava and creating new ocean floor. It is this that is responsible for seafloor spreading, the movement of the tectonic plates that move the continents. The *Challenger* expedition of 1873–1876 first mapped the ridge, but modern research began in 1947, when oceanographers Maurice Ewing, Bruce Heezen, and Marie Tharp of Columbia University began taking soundings from the vessel *Atlantis* and produced the first accurate maps of the floor of the Atlantic Ocean. In 1973–1974, scientists of Project FAMOUS (French-American Mid-Ocean Undersea Study) became the first to view the wonders of the mid-Atlantic seafloor. Using three submersibles—*Alvin* of Woods Hole; the French *Archimede,* a 69-foot-long bathyscaph that had been to 32,000 feet in the Pacific; and the newly built *Cyana,* at a length of 20 feet, a little smaller than *Alvin*—they made a total of forty-seven dives to depths of approximately 3,000 meters, and carried the first human beings to the floor of the Atlantic rift valley. Although hydrothermal vent animals were first discovered in the Pacific in 1977, they are now known to inhabit many sites along the Mid-Atlantic Ridge.

See also *Alvin,* Ascension Island, Azores, Bouvet Island, Ewing, Heezen, hydrothermal vents, Iceland, Project FAMOUS, seafloor spreading, Tristan da Cunha

middle passage The second of three legs of the slaving voyages of the sixteenth to nineteenth centuries. The first leg was from England to the west coast of Africa, with a cargo of rum, brass goods, and firearms, which would be exchanged for slaves who were then shipped on the middle passage to the West Indies or the southern states of the United States. The money from the sale of slaves was used to buy cotton, sugar, or tobacco, to be carried back to England along the third leg of the triangle. The entire circuit was also known as the triangular trade. **See also slave trade**

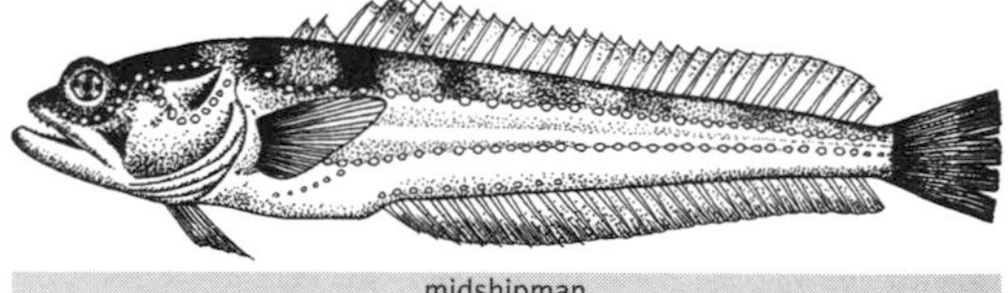

midshipman

midshipman (*Porichthys notatus*) The midshipman is named for the bright photophores on its sides that glow like a midshipman's brass buttons. It is notable for the number of these light organs, which may number eight hundred and which occur all over the fish's body. Unlike most other bioluminescent fishes, the midshipman is not a creature of the depths; it is found in shallow water. In some areas it is known as the singing midshipman, because it makes an audible humming noise by vibrating its air bladder. Like its relatives the toadfishes, the midshipman has a very large head for its body, forward-facing eyes, and a thick-lipped, downturned mouth. There are several similar species found in the western coastal Pacific from Alaska to Mexico and in the eastern Atlantic. Maximum size: 15 inches.

See also bioluminescence, toadfish

Midway, Battle of In May 1942, after their failure to take Port Moresby, the Japanese were preparing to attack the Aleutians to draw the Americans north and then capture Midway. Under the leadership of Admiral Isoroku Yamamoto aboard the battleship *Yamato,* the Japanese fleet consisted of eight other battleships, four carriers, cruisers, and destroyers, and two thousand men in landing craft. The Americans had already broken the Japanese code, so they knew of Admiral Yamamoto's plans, and they deployed the carriers *Enterprise* and *Hornet,* under the command of Rear Admiral Raymond Spruance, to wait for the Japanese. On June 5, American dive-bombers sank the carriers *Kaga, Soryu,* and *Akagi* and disabled the *Hiryu.* Planes from the *Hiryu* followed the Americans back to the *Yorktown* and so disabled her that she was abandoned and later sunk by torpedoes. The Battle of Midway conclusively demonstrated that bombers from aircraft carriers could defeat a superior surface force. It cost the Americans 150 planes and 307 lives, the Japanese 253 planes and 3,500 lives. The Japanese attack on the Aleutians ended in the occupation of two islands, Attu and Kiska. In May 1998, Robert Ballard, the discoverer of the *Titanic,* located the *Yorktown,* which went to the bottom of the Pacific during the Battle of Midway.

See also Coral Sea, Battle of the

Midway Island Except for tiny Kure (Ocean) Island farther to the west, Midway is the western terminus of the Leeward chain, stretching 1,000 miles northwest of Hawaii. The Midway Islands consist of Midway, Sand, and Eastern Islands. It was discovered by Captain N. C. Brooks in the Hawaiian bark *Gambia* in 1859 (who named it after himself, though the name did not stick) and he claimed it for the United States. (Its current name was applied because it lies midway between California and China.) It was formally annexed in 1867. Theodore Roosevelt designated it a naval reservation in 1903. In that year, Midway was chosen as the relay station for the transpacific cable, which involved the construction of concrete buildings and water towers and the importing of 150 tons of soil for gardens. An airstrip was built on the island as a stop on the San Francisco–Manila mail run, and in 1940, the United States established an air and submarine base on Midway. In June 1942, the island was shelled by Japanese ships, and the attempt to land on the island brought about the Battle of Midway, which was decisively won by American planes. It was a turning point of the war in the Pacific. The island, the only one of the Leewards not considered part of Hawaii, is now a military installation and a national wildlife reserve, established to protect the abundant bird life. **See also Hawaiian Islands, Leeward Islands, Midway, Battle of**

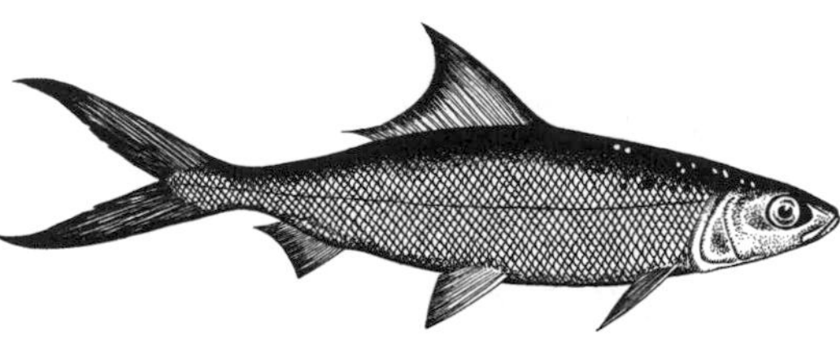

milkfish

milkfish (*Chanos chanos*) Found off the coasts of Pacific South America, lower California, and the western Pacific and Southeast Asia, milkfishes are considered one of the world's most important food fishes. They are now being raised on extensive fish farms in the Philippines, where the females lay millions of eggs and the fry are gathered into pens and fed on plant matter. Because they grow rapidly and will tolerate crowding, milkfish can be raised in large numbers. They reach a length of 4 feet and a weight of 30 pounds and are usually found in shallow coastal waters. In some areas, they venture into shallow waters where the water reaches 90°F, too hot for a human to stand in comfortably.

See also mariculture

mimic octopus In what is probably the most extreme example of mimicry, or protective coloration, or both (or something completely different), the mimic octopus, found in Indonesian and Philippine waters, can assume the shape and coloration of almost any animal it sees. There are photographs of this creature in the shape of a jellyfish (with its arms flared out and the tips tucked under); a sea snake (where all but two of its arms are balled up, and the visible ones stretched out and striped like the snake); a mantis shrimp (where the octopus lies buried in the sand, exposing only a couple of arms, which it makes look like the shrimp); a flounder (where it stretches all eight arms behind it, flattens out on the bottom, and adopts the coloration of a flatfish); a stingray; and several others. The "reason" for this unusual behavior is unknown, and the species has not yet been named.

See also mimicry, octopus, protective coloration

mimicry In biology, the advantageous resemblance of one species to another, often unrelated. (The imitation by an animal of its environment is usually referred to as protective coloration; certain flatfishes and many cephalopods can imitate the background on which they find themselves by altering their skin color or texture, and there are many fishes, such as stonefishes, scorpionfishes, and frogfishes, that blend into their backgrounds without having to affect any modifications.) In the evolutionary context, however, certain animals have somehow evolved to replicate the appearance of other species. For a prey species, there are advantages to resembling a predator; its own predators might leave it alone if it looks like a more threatening species. As an example, consider the small (nonpoisonous) filefish *Paraluteres proinurus,* which so closely resembles the poisonous black-saddled puffer, *Canthigaster valentini,* that even ichthyologists have difficulty differentiating them. It is even more problematic to identify the evolutionary mechanism that determines their similarity. It is often stated that certain creatures assume the appearance of others to "avoid being eaten," but this suggests some sort of evolutionary or genetic determinism. If evolution is an ongoing process, without an identifiable goal or end product, will the two species continue to evolve so that they no longer resemble each other? The juvenile cobia (*Rachycentron canadum*) bears a striking resemblance to the remora, *Echineis naucrates;* it lacks only the sucker disk. But where the remora is a harmless hitchhiker, the cobia is an aggressive predator. What mechanism accounts for this uncanny resemblance?

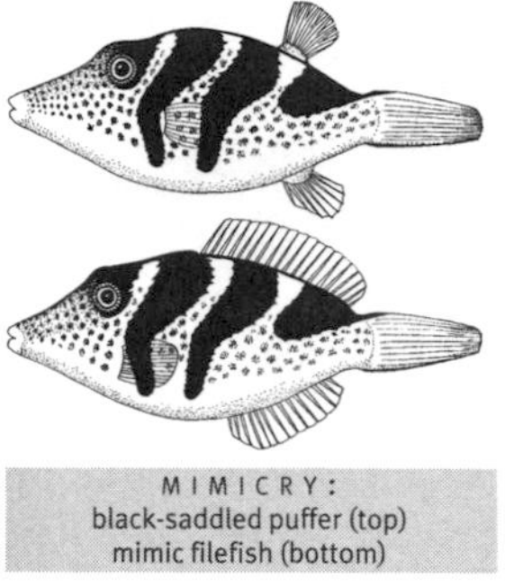

MIMICRY: black-saddled puffer (top) mimic filefish (bottom)

See also bioluminescence, countershading, protective coloration

minke whale (*Balaenoptera acutorostrata*) At a maximum length of 30 feet and a weight of 10 tons, the minke is the smallest of the rorquals ("grooved whales"), but compared to anything but other whales, it is a very large animal indeed. Known as the little piked whale until around 1870, this species supposedly gets its name (which is pronounced "minky") from a Norwegian whaler named Meineke who supposedly mistook a minke whale for a baby blue whale. There are northern and southern minkes: the Northern Hemisphere type is characterized by a broad white stripe on the flipper, a feature often lacking in the southern version. Along with the larger rorquals (the blue, fin, and sei whales), southern minkes come to the Antarctic to feed in the summer. Only when the larger species became too scarce to hunt economically did the whalers go after the minkes. Even though commercial whaling has ended, Antarctic minkes are still hunted by the Japanese for "research," and the Norwegians and Icelanders continue to take a couple of hundred every year. The minke is the only Antarctic whale whose numbers have increased; as the larger species were removed from the food chain, the minkes proliferated to fill the spaces.

minke whale

See also blue whale, Bryde's whale, fin whale, rorqual, sei whale

Minorca Also known as Menorca; one of the Balearic Islands, located in the western Mediterranean some 220 miles from Spain, to which it belongs. Second largest of the Balearics (after Majorca), Minorca covers 270 square miles and is low with a hilly center. The 58,000 inhabitants grow cereals, wine grapes, olives, and flax, and support a tourist industry that is particularly popular with the English. The island was occupied by the British during the War of the Spanish Succession (1701–1714) and held by the British until 1756, the beginning of the Seven Years' War. It was taken by the French at this time, but in 1763 the Treaty of Paris restored it to Britain. French and Spanish troops seized it during the American Revolution, but in 1798, during the French Revolution, the British recaptured it. The Peace of Amiens (1802) awarded Minorca to Spain. **See also Ibiza, Majorca**

Moby-Dick Herman Melville's massive, mysterious novel is generally considered the consummate achievement of American literature: "The Great American Novel." Told by the narrator Ishmael, it is on the surface the story of Ahab, a mad whaling captain who obsessively pursues the white sperm whale that took off his leg, but even though it contains the most detailed and important descriptions of the nineteenth-century Yankee whale fishery, it is much more than a whaling yarn. As the whaleship *Pequod* voyages around the world, Ishmael tells of the mates Starbuck, Flask, and Stubb; the harpooners Queequeg, Dagoo, and Tashtego; and the rest of the motley crew drawn into Ahab's ultimately fatal obsession. It has been called an elegy to democracy, a tract on the nature of religion, an investigation of man's relationship to the natural world, and a conflict between the eternal forces of good and evil. Unappreciated when it was first published in 1851, *Moby-Dick* has come to be recognized as one of the greatest novels ever written in English. It has been translated into uncounted languages, and has been the subject of any number of movies, including *The Sea Beast* (1926); *Moby Dick* (1930), John Huston's 1956 version, with Gregory Peck as Ahab; and a 1998 made-for-television movie that was four hours long. **See also Melville**

Mocha Dick A real white sperm whale believed to be one of the important inspirations for Melville's *Moby-Dick.* Around 1810, this whale (named not for his color, but for the Chilean island of Mocha) began a malicious rampage, sinking everything from lumber ships to whaling ships. According to a 1839 magazine article written by Jeremiah Reynolds, Mocha Dick was killed around 1820, but there are other reports of his continuing to attack ships until 1842.

See also *Moby-Dick,* Reynolds, sperm whale

Mohole Project Named for the Mohorovicic Discontinuity (which in turn was named for Andrija Mohorovicic, the Hungarian geologist who discovered it), the 1957 Mohole Project was an attempt to drill a hole into the boundary between the earth's crust and the mantle. The first phase, backed by the U.S. National Science Foundation and the National Academy of Sciences, drilled a hole off the coast of California under 2 miles of water, but inadequate technology, frightful cost overruns, and congressional infighting doomed the project, and it was abandoned in 1966. The lessons of the Mohole Project were employed in later deep-sea drilling projects, such as the multiple voyages of the *Glomar Challenger.* **See also *Glomar Challenger***

mojarra, yellowfin (*Gerres cinereus*) A small, silvery fish with a greatly protrudible mouth, the mojarra inhabits reef areas, mangrove swamps, and tidal creeks of all tropical seas, but they are most common in the Americas and the Caribbean. They feed on worms, mollusks, and other invertebrates that are found in the sand. Mojarras—the name comes from the Spanish for a kind of small knife—reach a length of about 1 foot.

mollusk There are close to 100,000 species of mollusks, divided into seven classes: after the insects, they are the most numerous, varied, and widely distributed of all animal groups. Although the name comes from the Greek *mollis,* which means "soft," most mollusks have shells. The class Monoplacophora was believed to be extinct until the surviving genus *Neopilina* was discovered in the eastern Pacific. The class Polyplacophora includes the chitons, which are elongated, flattened creatures with segmented shell plates. The Aplacophora are wormlike, deepwater mollusks. The Gastropoda are enormously diverse snails and slugs, and the Scaphopoda are tusk shells, elongated, tapered animals that bury themselves vertically in the sand. Bivalves (Bivalvia) are bilaterally symmetrical with two shells hinged together; the Cephalopoda are the squids, octopuses, cuttlefishes, and nautiluses. (The shells of most cephalopods are internal, taking the form of the "gladius" or "pen." The chambered nautilus, a primitive cephalopod, has an external shell.) They come in an incredible variety of sizes and shapes, from microscopic snails to 400-pound clams and 60-foot-long giant squids. They live on the surface or on the bottom, in midwater or burrowed into sediment or wood. Some are parasites that live in the bodies of other animals. Some mollusks, like clams, oysters, squids, and octopuses, form the basis of major fisheries.

See also chambered nautilus, chitons, clam, cuttlefish, gastropoda, giant clam, giant squid, *Neopilina,* octopus, oyster, squid

mollymawk A term used by sailors to refer to the smaller, dark-backed albatrosses, as contrasted with the larger, white-backed varieties, such as the wanderer and the royal albatross. The coloration of these smaller albatrosses, such as the black-browed and the yellow-nosed, is not unlike that of a large black-backed gull. The name comes from the Dutch *mallemowk,* which means "foolish gull." Most albatrosses approach ships closely, and sailors have traditonally regarded them as stupid; in some parts of the world they are called gooney birds. **See also albatross**

Moluccas Island group along the equator, located in the Banda and Molucca Seas, between Celebes (Sulawesi) and Irian Jaya. There are around a thousand islands, covering about 28,000 square miles. The largest are Halmahera, Ceram, Buru, and Ambon, the capital of Indonesia's Maluku Province. Earthquakes are frequent in these mountainous islands, and there are many active volcanoes. The original Spice Islands, the Moluccas first attracted Chinese, Arab, and Indian merchants, who came for the nutmeg and cloves, and in the sixteenth century, the Portuguese and Dutch fought fiercely to control them. By 1667, the Dutch had completed their conquest, and through their East India Company, they established vastly profitable plantations. When the spice trade diminished by the end of the eighteenth century, the Moluccas became an economic backwater, and the "Dutch East Indies" were incorporated into the Republic of Indonesia in 1949. The population of these islands is around 1.7 million, most of whom are engaged in fishing, and growing coconuts, timber, and spices.

See also Ambon, Celebes, Dutch East India Company, Indonesia, Irian Jaya

Monitor When the Union navy learned that the Confederates had raised a wooden warship and armored her with iron plates, they commissioned Swedish-American engineer John Ericsson to design a worthy competitor, and he came up with the *Monitor.* At an overall length of 172 feet, she was almost 100 feet shorter than the *Merrimack,* but she rode much lower in the water, which made her harder to hit, and she sported the first revolving turret, which meant that she did not have to maneuver to bring her two heavy guns into play. At Hampton Roads on March 8, 1862, the *Merrimack* devastated three Union ships, but when she returned the next day to finish them off, she found the Yankee "cheese box on a raft" waiting for her. With Union and Confederate supporters watching from nearby ships and shore stations, the two ships began firing at each other. Most of the shells were ineffective against the ironclads, but one shot hit the *Monitor*'s turret and blinded Lieutenant John Worden, the Union commander. The *Merrimack* was low on ammunition and the tide was falling, so she returned to Norfolk. This initial contest between ironclad warships had ended in a stalemate. In 1862, the *Monitor* foundered and sank in a storm off Cape Hatteras. Her hull was located by a Duke University oceanographic expedition in 1973, and although she is still under 225 feet of water, she was placed in the National Register of Historic Places, and was designated the first National Marine Sanctuary in 1975. **See also *Merrimack***

monkfish: See goosefish

monsoon From the Arabic *mausim,* "season," a wind that changes direction with the seasons, particularly in India and Southeast Asia. These winds blow from the southwest during one half of the year, and from the northeast during the other. The dry, or winter, monsoons of Asia are the result of high-pressure cells that develop over Siberia; the wet (summer) monsoons are caused by low pressure that develops in southern Asia as the landmass warms. When moisture-laden air over the oceans is drawn into these low-pressure areas, it cools as it ascends mountain barriers, producing heavy rainfall. (The temperature difference between the ocean and the land is called differential heating.) In the western part of the Arabian Sea and the northern China Sea the monsoon winds can reach gale force, but for the most part it is not the wind but the rain that causes destruction. The most powerful monsoons are over Asia—the South Asian monsoon over India, Pakistan, and Bangladesh, and the East Asian monsoon over China and Japan.

See also cyclone, hurricane, typhoon

Montserrat British Crown Colony in the Leeward Island group of the West Indies. Montserrat was discovered by Columbus in 1493; after changing hands several times, it was ceded to Britain in 1783. In 1966 the citizens of Montserrat rejected self-government, choosing to remain a colony of Britain. The 38-square-mile island had a population of some 12,000, most of whom were employed in the tourist industry. In July 1995, the volcano Soufrière Hills began to erupt, the first such occurrence in recorded history. (In the early twentieth century volcanoes on the nearby islands of St. Kitts, St. Lucia, and Guadeloupe erupted; the 1902 eruption of Mount Pelée on Martinique, 150 miles from Montserrat, killed thirty thousand people.) As ash fell on the capital of Plymouth, the city's six thousand people were evacuated to the north end of the island. A new dome formed on the mountain, and several lava spires grew and then collapsed. Throughout 1996, ash clouds towered over the island, sometimes 40,000 feet high; earthquakes shook the land, and avalanches of hot gas and ash (pyroclastic flows) rolled down the mountain, igniting trees and buildings in their path. By the summer of 1997, ash had covered most of the island, seven villages had been destroyed, the evacuated capital of Plymouth was aflame, and nineteen people were confirmed killed. In August, the British government ordered the complete evacuation of the island, but some residents refused to leave. By mid-1998, the volcano had subsided somewhat, but there were still pyroclastic flows, ash clouds, and thunderous eruptions. The island continues to be inhabited, but is considered dangerous. **See also Mount Pelée**

moonfish: See opah

moon jellyfish (*Aurelia aurita*) The common jellyfish of the North Atlantic, which appears offshore and washes up on beaches from Greenland to the West Indies. Although it can be 10 inches across, most are smaller. Numerous, short tentacles fringe the bell; periodically the jellyfishes sweep the tentacles around the rim of the bell, gathering the food items that have accumulated on the animal's upper surface. Moon jellyfishes are often seen in great shoals from the deck of a ship, far from shore. Although they have nematocysts, the tentacles of *Aurelia* are not particularly venomous.

See also cnidarian, jellyfish

moon snail (*Lunatia heros*) A gastropod with a globular shell, commonly found on North Atlantic beaches from Labrador to North Carolina. These snails, which can be more than 4 inches long, are responsible for the "sand collars" often found on beaches, which are actually the eggs of the snail, laid in a ribbon composed of sand and mucus.

See also gastropoda

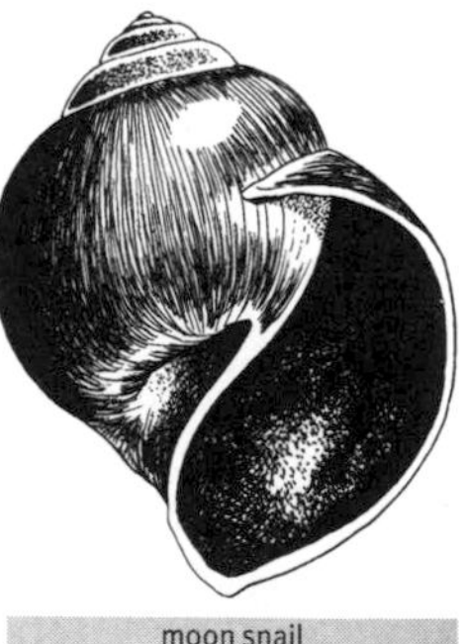

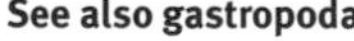
moon snail

Moorea Located 12 miles northwest of Papeete, Moorea is visible from its sister island, Tahiti, across the Sea of the Moon. It has spectacular, jagged peaks, enhanced by long white beaches; the location was chosen for the 1984 film version of *The Bounty*, based on the famous mutiny. (Evidently, the producers thought it looked more like Tahiti than Tahiti did.) British explorer Samuel Wallis passed the island in 1767, but he did not land there. When James Cook sailed the *Endeavour* from England to Tahiti in 1769 to observe the transit of the planet Venus, he did the actual sightings from Point Venus on Tahiti, but he put his naturalists ashore on Moorea. (He named the group the Society Islands after the Royal Society, the sponsors of his expedition.) In 1792, Moorea was conquered by King Pomare I, using arms he had obtained from the *Bounty* mutineers. The first missionaries arrived shortly thereafter and translated the Bible into Tahitian. Through some tricky political maneuvers, Tahiti, Moorea, and the rest of the Society Islands were first "protected" and then annexed by France in 1847. Some of the villages were described in Herman Melville's second South Seas novel, *Omoo*, published in that year. The arrival of Europeans (and European diseases) reduced the population of these islands from about 150,000 to 8,000. Moorea, which covers some 51 square miles, is now a popular tourist island with numerous resorts (including a Club Med) and pineapple, coconut, and vanilla plantations, but it does not have the busy, frantic pace of Tahiti. **See also Tahiti**

moorish idol (*Zanclus cornutus*) Found at depths up to 500 feet in Indo-Pacific waters, the moorish idol is usually seen singly or in pairs. Its body is very compressed, and it can reach a length, nose to tail, of 10 inches. Although the arrangement of black, white, and yellow is quite different, this relative of the surgeonfishes is often confused with the longfin bannerfish (*Heniochus acuminatus*), which is much more common. Even though it is included in the Acanthuridae (the surgeonfishes), the moorish idol lacks the characteristic spine at the base of the tail. **See also longfin bannerfish, surgeonfishes**

moorish idol

moray (family Muraenidae) A large family (perhaps as many as eighty species) of laterally flattened eels that are found mostly in tropical and subtropical waters around the world. They have no pectoral fins, and the dorsal fin usually runs the entire length of the body. The gill opening is a small round aperture behind the head. Like other eels, morays pass through a transparent, ribbonlike leptocephalus phase before assuming their adult shape. They feed mostly on fishes that they snare with their sharp canine teeth as they pass by the moray's habitat in a hole in a coral reef or rock wall. They look more frightening than they actually are, and most accidents come from unsuspecting divers poking their hands into a hole that contains a moray, or trying to manhandle the fish. (The bite of a moray is not poisonous.) Morays come in an enormous variety of color schemes, ranging from solid colors to stripes and polka dots. The pygmy moray (*Anarchius yoshiae*) of North America only reaches a maximum length of 8 inches, while the green moray (*Gymnothorax funebris*) is said to reach a maximum length of 10 feet.

See also eel, leptocephalus

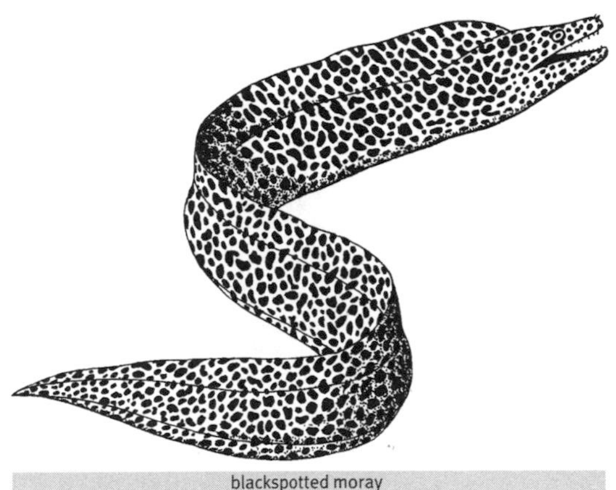

blackspotted moray

Moresby, John (1830–1922) The son of Admiral Sir Fairfax Moresby (1786–1877), who was admiral of the fleet in 1870, Captain John Moresby explored New Guinea, including the harbor at Port Moresby (which he named for his father and which is now the capital of Papua New Guinea), during an 1871–1874 voyage in the *Basilisk*. On this voyage he also charted the D'Entrecasteaux Islands off the eastern tip of New Guinea. He served in the Crimean War, and was captain in charge at Bermuda until his retirement in 1881.

Morgan, Henry (1635–1688) A Welshman who went to the West Indies as a youth and eventually joined a band of buccaneers. Upon the death of the pirate leader Edward Mansfield, Morgan assumed control and operated as a privateer, with a commission from the British admiralty. He captured Puerto Principe (now Camagüey) in Cuba in 1668, seized Maracaibo (Venezuela) in 1669, and ravaged the Cuban and American coasts. His methods were brutal but effective, and after he captured and burned Panama in 1671, one of the Spanish empire's major cities, he was caught and sent to England to be tried for piracy. When he was captured, England was at peace with Spain, but when hostilities threatened again, he was knighted by Charles II and sent to Jamaica in 1673 as lieutenant governor. He died a rich landowner in Jamaica. **See also buccaneer**

Morison, Samuel Eliot (1887–1976) American naval historian; he took his Ph.D. at Harvard in 1912 and began teaching there in 1915. He served in the U.S. Army during World War I and was a delegate to the Paris Peace Conference in 1919. Upon his return to academic life, Morison began to devote himself to research, particularly of naval exploration, and when the United States entered the war in 1941, President Franklin Roosevelt appointed him official historian of the U.S. Navy. He was commissioned as a lieutenant commander so he could serve afloat in combat areas and thus gain first-hand knowledge of naval operations during the war. From 1947 to 1962, he published the fifteen-volume *History of U.S. Naval Operations in World War II,* one of the most comprehensive histories ever written. Morison won the Pulitzer Prize twice, the first time for his 1942 biography of Christopher Columbus, the second for his 1959 biography of John Paul Jones. He also wrote the two-volume *European Discovery of America,* with the first volume (*The Northern Voyages*) published in 1971, and the second (*The Southern Voyages*) appearing in 1974. Retired from Harvard in 1955, Morison was the recipient of the Presidential Medal of Freedom, the gold medal of the American Academy of Arts and Letters, and the Francis Parkman medal of the Society of American Historians.

Moroteuthis robusta: **See Pacific giant squid**

mosasaur Extinct marine reptiles that flourished in the seas of the Cretaceous period, from 90 to 65 million years ago. The first mosasaur was found in 1780 in the Netherlands, but fossils have since been found in Africa, Israel, the Antarctic, and the most productive areas of all, Kansas, North Dakota, and South Dakota. Mosasaurs ranged in size from 12 to 50 feet, and with the exception of *Globidens,* a shellfish-eater, they were all powerful swimmers and apex predators, chasing down cephalopods, fish, and other marine reptiles. In addition to the large serrated teeth in the jaws, mosasaurs (like snakes) had teeth on the palate, which helped them grip their often slippery prey and also helped to "walk" it back into the throat. A prominent hinge in the middle of the lower jawbones also suggests that mosasaurs are related to snakes, which also have the intramandibular hinge. Mosasaurs were air-breathers; they had four flippers and a powerful, laterally flattened tail that they used for swimming. Like ichthyosaurs—to which they were not related—mosasaurs gave birth to live young in the water. From the examination of the shoulder bones of *Plioplatecarpus,* it has been suggested that this species "flew" through the water. The best known of the forty-odd species are *Clidastes, Platecarpus, Mosasaurus, Tylosaurus,* and *Plioplatecarpus.* Along with the nonavian dinosaurs, the mosasaurs all became extinct 65 million years ago, but the cause of this mass extinction is a mystery.

See also ichthyosaur, plesiosaurs

Mote Marine Laboratory Founded by William and Alfred Vanderbilt in 1955 as the Cape Haze Marine Laboratory at Placida, Florida, the Mote Marine Lab was renamed in 1962 when it was relocated to Sarasota. The first director was Eugenie Clark, who served from the opening of the lab to 1965, when she was succeeded by interim directors Sylvia Earle (1966) and Charles Breder (1967). Much important research has been done at the MML, especially during the tenure (1967–1978) of Perry Gilbert, one of the world's foremost shark researchers. The lab moved to much larger quarters in 1978 and now has a museum, an aquarium, and a marine mammal rehabilitation center, in addition to its full-scale research facilities that employ more than a hundred people. **See also Clark, Earle, Gilbert**

Mount Pelée Volcano overlooking the city of St. Pierre on the island of Martinique in the Windward Islands of the West Indies. In February 1902, the inhabitants of the city they called "the Paris of the West Indies" noticed an odor of sulfur, and in March and April, puffs of steam started to appear at the summit of the roughly circular cone. By May 2, Pelée continued to rumble, and ash blanketed areas close to the foot of the mountain. Flash lighting accompanied torrential rains, and the crater began to fill with water. On May 5, the walls of the crater gave way, and an avalanche of hot water, boiling mud, and jungle debris rolled down the mountainside, swept over a sugar mill, and killed twenty people. At about eight o'clock in the morning on May 8, Mount Pelée exploded. An immense crack appeared in the side of the mountain, and two huge black clouds shot out of the summit; one rose straight up to expand and darken the sky, and the other, a pyroclastic flow, roared down the mountain at more than 100 mph, killing, burning, or asphyxiating everything in its path. No lava poured out of the mountain, but it has been estimated that the temperature of the pyroclastic flow was between 1,300°F and 1,800°F, hot enough to soften glass and cause the wooden decks of the ships in the harbor to burst into flames. Of the thirty thousand residents of St. Pierre, only one survived: Auguste Ciparis, who was serving time in an underground dungeon. In October, a shaft of solidified lava began to rise from the crater, forced upward by pressure in the mountain. The spine, which came to be known as the Tower of Pelée, was 500 feet thick at the base, and often grew at a rate of 50 feet a day until it reached 1,020 feet, twice as high as the Washington Monument. In September 1903, the obelisk collapsed into a mass of rubble.

See also Martinique, Montserrat

Mozambique Channel The 1,000-mile-long passage between the East African country of Mozambique and the island of Madagascar. At the northern end of the channel are the Comoro Islands, and at the southern, the islands of Bassas da India and Europa. On the coast of Mozambique are the port cities of Maputo and Moçambique; the Zambezi River empties into the channel.

See also Comoro Islands, Madagascar

Mull At 351 square miles, Mull is the largest island of the Inner Hebrides, separated from the Scottish mainland by the Sound of Mull and the Firth of Lorne. The island is mountainous, and its coastline deeply indented. It is known for its gardens and farms, and Tobermory, its main town, is a popular summer resort. The population of the island is about 2,600 people. The Spanish galleon *San Juan de Sicilia,* fleeing from the British after the defeat of the Armada, was blown up in Tobermory Bay in 1588. In 1829, while on his "Grand Tour," the German composer Felix Mendelssohn-Bartholdy (1809–1847) visited the island of Mull and also the nearby island of Staffa, where the famous columnar basaltic formation known as Fingal's Cave is located. Mendelssohn's overture "The Hebrides, or Fingal's Cave," evoking the crashing of the seas and the vast cave, is considered one of his most important compositions. **See also Hebrides**

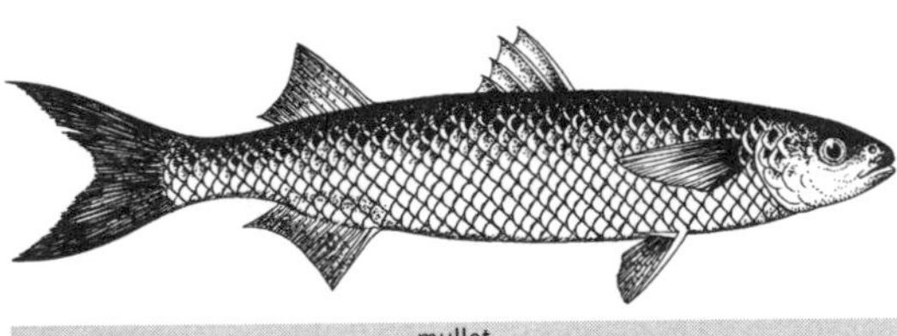

mullet

mullet (family Mugilidae) There are some ninety-five species of mullets, separated into thirteen genera. Found worldwide in warm seas and often up freshwater streams, mullet are spindle-shaped fishes with two widely separated dorsal fins and a small, triangular mouth designed for eating detritus from the sea bottom. Most species have a special pharyngeal apparatus for processing this material, and muscular stomachs and a long gut to facilitate digestion. The striped mullet (*Mugil cephalus*), which grows to a length of 3 feet, is common on both sides of the Atlantic Ocean, in every kind of water, from salt to brackish and fresh. Commercial fishermen catch large numbers of white mullet (*M. curema*) in Florida waters. Mullet are energetic jumpers, leaping to escape nets and predators (bottlenose dolphins are big mullet eaters), and fish jumping at twilight are likely to be mullet.

See also bottlenose dolphin

Munk, Jens (1579–1628) Danish explorer whose parents sailed to and settled in Brazil in 1591. Munk was back in Europe by 1601, employed by Baltic shipowners, and after becoming an owner in his own right, he set out to discover the Northeast Passage to the Far East. In 1609, with two ships, he rounded Novaya Zemlya, but one of them was wrecked and the other put into Archangel. Upon his return to Denmark, he was able to convince King Christian IV that he could find a route to the east, and in 1610 he was given command of two more ships, the *Angelibrand* and the *Rytteren.* This time they were stopped by the ice before they reached Novaya Zemlya. Having failed to find a northeast passage, he tried for the northwest, and in 1619, he sailed into Hudson Bay with the *Enhjørningen* (Unicorn) and the *Lamprenen* (Lamprey). He anchored at the mouth of the Churchill River, planning to search for the passage the following summer. By February, however, scurvy was affecting almost everybody, and by June, only Munk and two others remained alive. Munk managed to bring the *Lamprenen* back to Denmark with a crew of two. Although the king wanted him to return to recover the abandoned *Enhjørningen,* the expedition never took place.

Murano The largest of the islands in the lagoon of Venice, Murano has been inhabited since Roman times. The island occupies some 1,134 acres. Until the tenth century, it was a prosperous, self-governing trading center. In 1291, as a precaution against fire, all the glass furnaces in Venice were moved to Murano, and glassmaking quickly became the predominant occupation of the island. Indeed, from the fifteenth to seventeenth centuries, the island was the principal supplier of glass to all of Europe. During this time, it also became a popular resort for Venetian noblemen, who built country houses there. At its height, Murano had 30,000 inhabitants. (It now has about 7,000.) Like Venice itself, Murano was laced with canals, but by the nineteenth century, they had been filled in for residential areas. Glassmaking continues on the island, but on a much reduced scale. There is a Museum of Glass Art in the Palazzo Giustinian, but the island's most important building is the basilica of Santa Maria e San Donato, whose foundation goes back to the seventh century. It has been rebuilt several times, but it is still a fine example of Romanesque architecture, with a marvelous treatment of the exterior of the apse and a mosaic floor similar to that of St. Mark's. **See also Burano, Venice**

murex snail (family Muricidae) Murex snails, characterized by their spiny shells, are popular with shell collectors because of their astonishing shapes. Many of them have comblike spines; others have dramatically curved protrusions that look like the antlers of a fantastic deer.

murex snail

There are perhaps as many as a thousand varieties in the temperate and tropical waters of the world, ranging in size from a ½ to 8 inches. The apple murex (*Murex pomum*) occurs from North Carolina to Florida; other species are found in Florida and Mexico. Murexes were collected in and around the ancient Phoenician city of Tyre, and the snails themselves (particularly the Mediterranean species *M. brandaris*) produced the famous purple dye Tyrian purple. This color, also called royal purple, was so valuable that only emperors could wear clothing dyed with it.

See also Tyrian purple

Murray, Sir John (1841–1914) Born in Canada of Scottish parents, Murray was educated at the University of Edinburgh but had little patience with regular courses, sat for no examinations, and designed his own curriculum for the study of natural history. In 1868 he spent seven months aboard an Arctic whaler, and was a last-minute replacement aboard the *Challenger*'s 1872–1876 oceanographic research expedition. He was putatively in charge of bird specimens, but it was in the area of publications that he demonstrated his greatest worth. Upon the death of Wyville Thomson, the leader of the expedition, Murray took over the editorship of the *Challenger* reports—finally the *Report on the Scientific Results of the Voyage of HMS Challenger*—supervised the production of all fifty volumes, and actually wrote seven of them. From 1882 to 1894 he directed a biological investigation of Scottish waters, and in 1906, he surveyed the depths of the Scottish lochs. In 1910, with Norwegian oceanographer Johan Hjort, he went on a four-month cruise aboard the research vessel *Michael Sars,* and together they wrote *The Depths of the Ocean,* which was published in 1912. The following year Murray wrote *The Ocean: A General Account of the Science of the Sea,* a concise summary of oceanographic knowledge at that time.

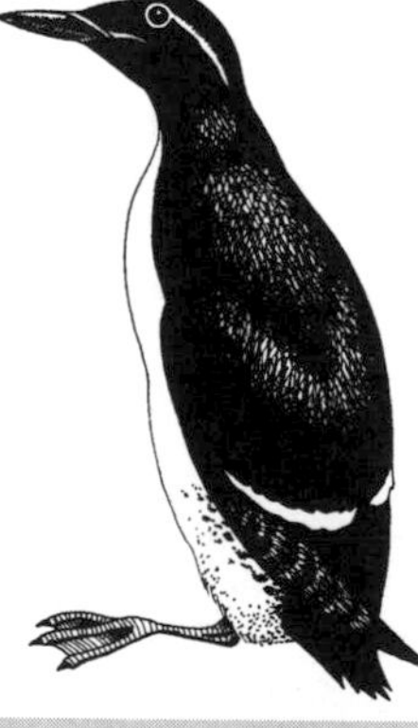
common murre

murre (*Uria aalge*) Also known as the black guillemot, this 17-inch-long bird with a 28-inch wingspan is a resident of the high Arctic. (Guillemots and murres are the same birds.) It breeds in the North Pacific from northern Japan to the coast of northern California, and in the North Atlantic from Labrador and Greenland to Scotland, Norway, and the Baltic. With its black head and back-and-white undersides, it resembles the razorbill (which lives in the same areas in the North Atlantic, and may even nest on the same cliffs), but the bill of the razorbill is much thinner. Other guillemots are Brünnich's guillemot (*U. lomvia*), whose range overlaps both the razorbill and the common murre; the pigeon guillemot (*Cepphus columba*) of the northern Pacific, which has wing bars like those of a pigeon; another black guillemot (*C. grylle*), recognizable by its all-black breeding plumage; and the spectacled guillemot (*C. carbo*), with its conspicuous white patch around the eye.

See also razorbill

Mururoa One of two atolls (the other is Fangataufa) in the southern Tuamotus Archipelago in French Polynesia. Some 750 miles north of Tahiti, Mururoa is the larger of the two, with an indigenous population of some 500 people. France conducted nuclear tests here for more than thirty years, using everything from balloons to barges. The most recent tests, conducted from 1995 to 1996, consisted of 137 nuclear detonations set off in the underwater volcanic cones of the coral atolls. Opponents claimed that these explosions would destroy the atolls and introduce radioactive elements into the water that will remain active for centuries. Despite riots by the Tahitians and protests from the United Nations, the International Atomic Energy Agency, Greenpeace, and almost every other body concerned with the safety of the planet, France persisted with the testing. (In 1985, French agents blew up the Greenpeace vessel *Rainbow Warrior* in Auckland harbor, killing one of the activists who was planning to protest the Mururoa nuclear tests.) In 1996, succumbing to intense worldwide pressure, France dismantled the test facility at Mururoa and terminated its nuclear testing in French Polynesia.

See also French Polynesia, Greenpeace

Muscovy Company Also called the Russia Company; the first joint-stock trading company, formed in 1555 by Sebastian Cabot after Richard Chancellor returned from Moscow with a letter granting favored trading status to the English. (Of the three ships that had sailed in 1553 seeking the Northeast Passage to China and the Indies, only Chancellor's *Edward Bonaventure* returned.) The British traded guns and cloth for Russian furs, timber, oil, and fish. Between 1562 and 1579, the company financed overland expeditions to Persia, and in 1576, Queen Elizabeth awarded the Muscovy Company an exclusive license to hunt whales "within any seas whatever." Although Tsar Alexis ended the company's privileges in 1649 and it lost the English monopoly of the Russian trade in 1698, the Muscovy Company profited from the Anglo-Russian trade of the eighteenth century and survived until 1917.

See also Cabot (Sebastian), Chancellor

mussel (*Mytilus* spp.) Marine and freshwater bivalve mollusks that move by means of a muscular foot and breathe by filtering water through extendable tubes known as siphons. Mussels are widespread, especially in cooler waters, where they form dense clusters. They are able to anchor themselves in place by extruding strong threads, known as the byssus. (Clams do not anchor themselves.) The blue mussel (*M. edulis*), which grows to a length of 3 inches, is found on both sides of the Atlantic on pilings and rocks, and is a favorite food item in Europe and America. Because it can be grown in controlled conditions, it is farmed commercially. The horse mussels (*M. modiolus, M. capax, M. rectus*), found on both coasts of North America, are much larger, reaching an overall length of 10 inches. Two species of mussel, *Calyptogena* and *Bathymodiolus,* occur in the vicinity of the hydrothermal vents; like the other vent animals, they can survive in heated water that is suffused with hydrogen sulfide. Freshwater zebra mussels are becoming a major environmental problem, as they form colonies in the pipes of water processing and heating plants.

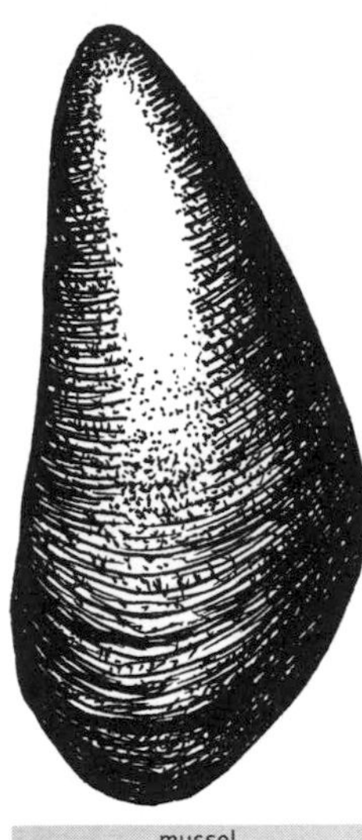

mussel

See also clam, hydrothermal vents, zebra mussel

mutiny Concerted disobedience, either by force or by refusal to obey orders, in military or naval service, or by sailors aboard commercial vessels. Mutiny may range from a combined refusal to obey orders to an active and armed revolt. During the days of sail, the penalty for mutiny in all navies was hanging from the yardarm; when an entire ship's company mutinied, the ringleaders were hanged. Probably the most famous of all mutinies was that of Fletcher Christian aboard the *Bounty* in 1789, when he and his men took the ship and put Captain William Bligh and eighteen men loyal to him over in a longboat. (Christian took the *Bounty* to Pitcairn Island, and Bligh piloted his small boat from Tonga to Timor, a distance of some 3,600 miles.) A mutiny with more lasting effects was that of the British navy at Spithead in 1797, when all the ships of the Channel fleet mutinied to protest the pay and conditions of naval service. It is not known who started this affair—which was so peaceful that it was more like a strike than a mutiny—but the Admiralty agreed to almost all the demands of the mutineers, and the Spithead mutiny ended with no lives lost. Not so with a sympathetic mutiny at the Nore, which was led by Richard Parker. Even when he learned that the Admiralty had acceded to all the mutineers' demands at Spithead, Parker did not call off the Nore mutiny, and he was captured and hanged. In 1905, during the years of unrest in Russia that led up to the Russian revolution, the crew of the battleship *Potemkin* mutinied and killed the captain and several officers. In October 1918, after the first Allied demands for surrender, the German navy mutinied at Kiel, refusing to take their ships into the North Sea for a decisive battle. The naval mutiny led to armed insurrections in Hamburg and Bremen, hastening the armistice.

See also Bligh, *Bounty,* Christian, Parker, *Potemkin*

Mykonos Although it is rocky, treeless, and lacking in ancient sites, the island of Mykonos in the Aegean is one of the most glamorous—and expensive—of the Greek Islands. Because of its beautiful beaches and sophisticated hotels and restaurants, Mykonos has long been a destination for jet-setters from all over the world. It is also the undisputed gay capital of Greece. The island is 10 miles long and 7 miles wide, and its two highest peaks—both named Profitis Ilias—are less than 2,000 feet high. The capital of the island is also named Mykonos.

See also Cyclades

N

Nansen, Fridtjof (1861–1930) Norwegian explorer and scientist who made the first crossing of the Greenland ice sheet in 1888–1889. He sailed in 1893 in the reinforced wooden schooner *Fram* ("Forward") to investigate the drift of polar sea ice and try to reach the North Pole. *Fram* was frozen into the ice and drifted for three years, as the crew took meteorological measurements and astronomical observations to determine the direction of their movement. In 1894, when they were some 400 miles from the Pole, Nansen and Hjalmar Johansen left the *Fram* with two kayaks, three sledges, and twenty-eight dogs, and before they were forced to turn back, they managed to get to 86° north, farther than anyone else up to that time. They headed for Franz Josef Land and arrived there three months later, having barely survived the cold and polar bear attacks. Nansen and Johansen spent the winter of 1895–1896 in a hut that they built of stones and walrus skins, and were found in June by the British explorer Frederick Jackson. By then, the *Fram* had escaped from the ice north of Spitsbergen, and had sailed back to Tromsö. After World War I, Nansen dedicated himself to famine relief and the repatriation of prisoners. In 1921, the League of Nations appointed him high commissioner for refugees, and the following year he was awarded the Nobel Peace Prize for his humanitarian efforts. **See also *Fram***

Nantucket Discovered in 1602 by Bartholomew Gosnold (who also discovered and named Martha's Vineyard and Cape Cod), this island 25 miles off the southern coast of Massachusetts was purchased from the Plymouth Colony by Thomas Mayhew in 1641 for forty pounds and two beaver hats, and administered as part of New York. It was ceded to Massachusetts in 1692 and given its present name, which means "faraway land" in the language of the Wampanoag Indians. American sperm whaling is said to have begun in 1712 on Nantucket, when Christopher Hussey, searching for right whales, was blown out to sea, where he found and killed the first cachalot. Shortly thereafter, Nantucket became the epicenter of Yankee whaling activities (in Melville's novel *Moby-Dick*, the *Pequod* is a Nantucket whaler), superseded by New Bedford only when whaling ships became too big to enter Nantucket's shallow, sandbar-protected harbor. Nantucket never regained its maritime prominence, and in the twentieth century, it became a popular summer resort and tourist destination for thousands of people. **See also Gosnold, Martha's Vineyard, whaling**

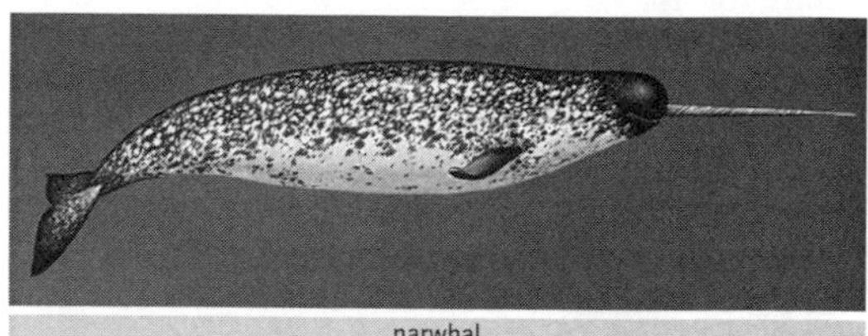

narwhal

narwhal (*Monodon monoceros*) Like its relative the beluga, the narwhal is found only in the Arctic. Unlike the beluga—and unlike any other animal on earth—the male narwhal has a greatly elongated "tusk" protruding from its upper jaw. This tooth, which can reach a length of 8 feet, is believed to be at least partially responsible for the myth of the unicorn: when these ivory spirals were brought to western Europe by early Arctic travelers, it was probably necessary to invent an animal to go along with them, since it seemed unlikely that the original owner could possibly have been a dumpy, spotted little whale. What the narwhal actually uses the tooth for is not known with certainty, but males probably employ it to assert dominance, and possibly even for sparring. Eskimos in Greenland and Canada still hunt narwhals for their blubber and their tusks. Narwhals can grow to a length of 16 feet—not including the tusk. **See also beluga**

national marine sanctuary Under Title III of the Marine Protection, Research, and Sanctuaries Act of 1972, the U.S. Congress is authorized to designate certain areas as national marine sanctuaries, and supervise the conservational, recreational, ecological, and educational uses of these areas by the National Oceanic and Atmospheric Administration (NOAA). The first national marine sanctuary, established in 1975, was the wreck of the ironclad *Monitor*, which was sunk in a storm in 1862 off Cape Hatteras, North Carolina. Other marine sanctuaries are the Channel Islands (California); Cordell Bank (California); Fagatele Bank (American Samoa); Florida Keys (Florida); Flower Garden (Texas); Gray's Reef (Georgia); Gulf of the Farallons (California); Hawaiian Islands Humpback Whale Sanctuary (Hawaii); Monterey Bay (California); Olympic Coast (Washington); and Stellwagen Bank (Massachusetts).

Nauru A small, coral island in the western Pacific, with a land area of 8.1 square miles, located north of the Solomon Islands and south of the equator. First inhabited by Polynesian and Melanesian explorers, it was sighted by British navigator John Fearn in 1798, but not really known to Europeans until whaling ships began to visit in the 1830s. The introduction of firearms and alcohol disrupted the peaceful existence of the island's 1,400 inhabitants, and by 1880, their number had been reduced to 900. In 1886 Nauru was allocated to Germany for the mining of phosphates. Australian forces took the island in 1914, and the League of Nations established a joint phosphate commission to be administered by Australia, Great Britain, and New Zealand. The island was occupied by the Japanese in 1942, and most of the able-bodied Nauruans were taken to the Caroline Islands to work as laborers. The survivors returned to Nauru in 1946, and the island became a UN Trust Territory. Despite a plan to settle the islanders on an island off Queensland, they voted instead for independence in 1964. In 1968, Nauru became the world's smallest republic, funded almost entirely by phosphate exports of some 1.75 million tons, giving the 8,000 Nauruans one of the highest per capita incomes in the world. In 1993, the Nauruans won a claim against the phosphate trust countries for $72 million, to compensate for their failure to fulfill their "trust obligations" and for the ecological damage done to the tiny country. By 1999, however, because of poor investments, the trust had dwindled so seriously that the Nauruans were forced to appeal to Australia for financial aid.

See also Solomon Islands

Nautilus Probably in honor of the chambered nautilus (*Nautilus pompilius*), the cephalopod with the gracefully curled, striped shell, this is a very popular name for submarines. It was the name of Captain Nemo's vessel in *Twenty Thousand Leagues under the Sea*, and also of Robert Fulton's prototypical submarine, built in 1800. Two men named Ash and Campbell launched a *Nautilus* in the Thames in 1888, but it sank and was released from the mud only when the six passengers ran back and forth in unison. In 1931, Sir Hubert Wilkins tried to take a refitted American submarine under the North Pole, but his *Nautilus* lost its rudder and did not make the epic voyage. There was a World War II submarine christened *Nautilus*, and in 1955, the USS *Nautilus*, the world's first nuclear-powered submarine, became the first vessel to sail under the North Pole.

Navarino Bay, Battle of On October 20, 1827, a combined British, French, and Russian fleet, under Admiral Sir Edward Codrington, defeated an Egyptian-Turkish fleet in Navarino Bay (now Pylos), off the southwestern Peloponnese. The Egyptian-Turkish ships, commanded by Ibrahim Pasha, were anchored in a semicircle in Navarino Bay, and to prevent them from escaping, Codrington, who had commanded the *Orion* at Trafalgar, ordered his fleet inside the semicircle. During this maneuver, one of the Turkish ships opened fire; it was returned by the British, and soon the battle was joined. For four hours the opposing navies fired broadsides at each other, and when the smoke had cleared, the Egyptians and Turks had been soundly defeated. They lost a total of thirty-four ships and numerous smaller vessels, and suffered more than 4,000 casualties, while Codrington's forces had 182 killed, 489 wounded, and no lost ships. Codrington had been under orders not to engage the enemy, but in London he was cleared of the charge of disobeying orders. The Turkish defeat was so complete that they began to evacuate Greece, leading to the creation of the independent Kingdom of Greece in 1832.

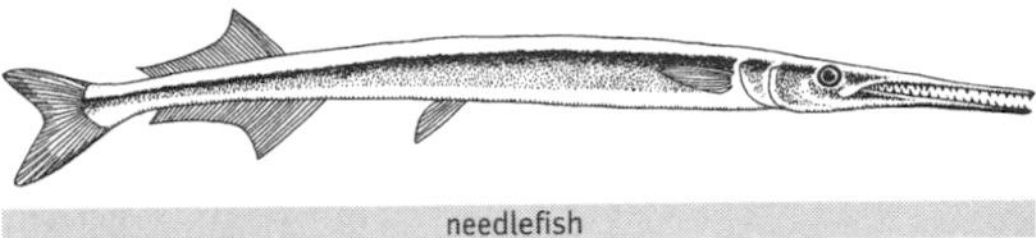

needlefish

needlefish (*Strongylura marina*) Named for its elongated, pointed shape, the needlefish is found throughout the western Atlantic, from Brazil to Massachusetts. Unlike the halfbeaks, to which it is closely related, the needlefish has two elongated jaws. They leap from the water when pursued or excited, and have a habit of leaping toward lights at night, presenting a serious hazard to night fishermen. In some instances fishermen have been stabbed by a flying needlefish. The houndfish (*Tylosaurus crocodilus*) of the Caribbean is a needlefish that can reach 5 feet in length.

See also ballyhoo, halfbeak

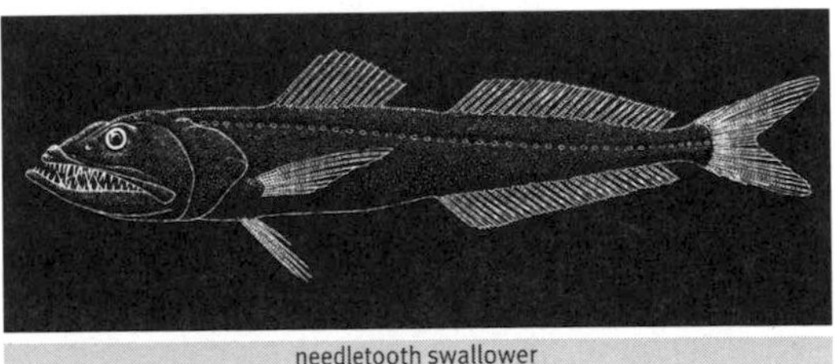

needletooth swallower

needletooth swallower (*Kali normani*) Its generic name is derived from Kali, the Hindu goddess of destruction, who is usually depicted as a blood-smeared hag with bared teeth and a protruding tongue, but this 6-inch-long deep-sea fish shares only the bared teeth. Both *Kali* and the "black swallower" (*Chiasmodon niger*) are known for their ability to swallow fishes much larger than themselves.

See also black swallower

Nelson, Horatio (1758–1805) England's most famous and beloved naval hero was born into modest circumstances in the Norfolk village of Burnham. At the age of twelve he went to sea, and eight years later, he was appointed captain of HMS *Badger*. After the American

War of Independence, he returned to England for five years, living on half pay with his wife, Frances Nisbet, and his stepson. In the land battle to secure Corsica from the French, he was blinded in his right eye. In 1793, when the war with France began, he was given command of the *Agamemnon* and then the *Captain*. In 1797, he distinguished himself at the Battle of Cape St. Vincent (Spain), where four prizes were taken as a result of his initiative. For this he was knighted and appointed rear admiral. In an unsuccessful attempt to take Tenerife in the Canaries, Nelson was so seriously wounded that his right arm had to be amputated. In HMS *Vanguard,* he found the French at Aboukir, in the Mediterranean off the mouth of the Nile, and on August 1, 1798, he nearly annihilated the opposing fleet. Wounded in the temple, Nelson was taken to Naples, where he was cared for by the British ambassador, Sir William Hamilton, and his wife, Emma. Nelson was placed in charge of a fleet defending the Two Sicilies, but the French conquered Naples in 1799 and established the Parthenopean Republic. When Nelson was ordered to Minorca, he refused to go, insisting that his presence was needed in Naples, but it was more likely that he did not want to leave Lady Hamilton, who by then had become his mistress. He returned to England in 1800, divorced his wife in 1801, and saw Lady Hamilton bear him a daughter named Horatia. Nelson was then named second-in-command to Admiral Sir Hyde Parker, defending the neutrality of the Baltic powers. He defeated the Danes at the Battle of Copenhagen in 1801, ignoring Parker's orders to break off the action by putting his telescope to his blind eye and then claiming he could not see the signal. Returning to England, Nelson lived at Merton, his country estate in Surrey, with Sir William and Lady Hamilton as his permanent guests. Sir William died in 1803. Nelson was given command of the Mediterranean fleet from the flagship HMS *Victory.* His assignment was to keep the French commander, Admiral Villeneuve, in Toulon, but the French fleet escaped and sailed to the West Indies, where they joined with the remaining French forces. As they returned, Nelson caught up with them on October 21, 1805, at Trafalgar, off the southwestern coast of Spain, and fought one of the last great battles of the age of sail. Twenty-seven British ships of the line defeated thirty-three French and Spanish, but in the moment of victory, Nelson was shot and killed. He is commemorated by the Nelson Column in Trafalgar Square, London. **See also Trafalgar, Battle of**

nematocyst All cnidarians (corals, hydroids, jellyfishes, anemones, the Portuguese man-of-war, etc.) have cnidocytes, stinging cells that are located in the tentacles and contain threadlike nematocysts, covered by a lidlike device known as the operculum. The microscopic nematocyst consists of a coiled hollow tube, tipped on the end with a miniature harpoon that pierces the flesh and injects the poison. The tentacles are covered with stinging cells, so all it takes is a passing contact with the animal to trigger the nematocysts. After the victim has been stung, the thread is not retracted; a new one is generated in the cnidocyte. The venom of the Portuguese man-of-war is highly toxic and can cause serious injury, but the nematocysts of the Australian box jelly (*Chironex fleckeri*) contain venom so potent that it can kill an adult human in three minutes. Sea turtles seem to be able to eat jellyfishes with no ill effects, a process that is not understood. **See also box jelly, coral, hydroid, jellyfish, Portuguese man-of-war**

***Neopilina* (class Monoplacophora)** In May 1952, in a dredge haul raised from 11,700 feet off Costa Rica by the Danish research vessel *Galathea,* several gastropods were found. Unlike chitons, whose name "polyplacophora" explains their multiplate shells, these had a single shell, and were therefore classified with the monoplacophora, creatures that were only known from the fossil record of 400 million years ago. Henning Lemche, who examined them when they were brought aboard, named them *Neopilina galatheae.* Since Lemche originally described them in 1957, another species was found at 19,000 feet in the Peru-Chile Trench. The cap-shaped shell is about an inch long, and the body consists of a repetition of organs: there are eight pairs of retractor muscles attaching the animal to the shell, five or six pairs of gills, and six or seven pairs of kidneys. Not only are the monoplacophorans unexpected "living fossils," but they are believed to show the anatomical characteristics of all molluscan ancestors.
See also chitons, limpet, mollusk

Neptune: See Poseidon

Netherlands Antilles Two groups of Caribbean islands belonging to the Netherlands; one is the ABC Islands (Aruba, Bonaire, and Curaçao), off the coast of Venezuela, and the other, some 500 miles to the northeast, consists of Saba, St. Eustatius, and the southern half of the island of St. Martin (which the Dutch call Sint Maarten) in the Leeward Islands. Although the islands were visited by Amerigo Vespucci in 1499, they were not settled by the Spanish until 1527. The Dutch captured the islands in 1634 and have maintained control ever since, except for the years 1800–1803 and 1807–1816, when the British occupied them.
See also Aruba, Bonaire, Curaçao

New Bedford Massachusetts city on the Acushnet River and Buzzards Bay; the home of the original Yankee whaling fleet. Founded in 1640, New Bedford was a haven for American privateers before the American

Revolution, which caused the British to burn the town in 1778. Nantucket was the first major whaling port, but because heavily loaded whaling ships could not pass over the sandbar in Nantucket's harbor, it was superseded by New Bedford around 1830. During its heyday, from about 1840 to 1860, New Bedford was the richest municipality per capita in America, and almost every aspect of its burgeoning economy was connected with whaling. In 1857, New Bedford whaleships brought home six million dollars' worth of oil and bone. In addition to the spermaceti oil that made the owners millionaires, the fishery provided income for shipwrights, chandlers, carpenters, coopers, rope makers, and blacksmiths, and ready markets for farmers and greengrocers. The coming of the Civil War, the discovery of gold in California, and, most significantly, the discovery of petroleum in Pennsylvania in 1859 marked the beginning of the end of sperm whaling, and with it, New Bedford's prosperity. By the late nineteenth century, New Bedford merchants had turned to textile manufacturing, which industry sustained the city until the 1920s, when it became the home port for some of New England's largest commercial fishing fleets. The textile mills are gone, the fishing fleets in decline—they are running out of fish—and New Bedford (current population: 95,000) is struggling to reinvent itself. Its most important facility is the whaling museum.

New Britain With an area of approximately 14,000 square miles, New Britain is the largest island in the Bismarck Archipelago, which is a part of Papua New Guinea. The island is mountainous, with several active volcanoes, two of which erupted in 1994. William Dampier discovered and named the island during his voyage in 1700, and it became part of German New Guinea in 1884. The Australians invaded New Pommern, as the Germans called it, in September 1914, bringing about the surrender of the whole colony of German New Guinea. After the war, the island was officially mandated to Australia by the League of Nations. In World War II, the harbor at Rabaul was the primary Japanese air and naval base for the projected invasion of Australia. During the battles for the Solomon Islands (Guadalcanal, Russell Islands, New Georgia, Bougainville), American forces concentrated on the destruction of the Japanese facilities at Rabaul.

New Caledonia French Overseas Territory in the South Pacific, 1,000 miles northeast of Sydney, and immediately southwest of Vanuatu. Nouvelle-Calédonie et Dépendances includes New Caledonia (also known as Grand Terre), the Isle of Pines, the Loyalty Islands, and the Huon, Chesterfield, and Bélep groups. The cigar-shaped island, 248 miles long and only 30 miles wide, is rich in minerals—it is the third biggest producer of nickel in the world—particularly iron, manganese, cobalt, silver, and gold. A large proportion of the island is heavily forested, but most of the kauri pines have been harvested. James Cook sighted and named the islands in 1774 (Caledonia is a Roman name for northern Britain, roughly the area covered by modern Scotland). The first European settlers were missionaries. After a group of French surveyors were killed and eaten by the islanders (the Kanaks), the French annexed the islands in 1853. American troops used the islands as a military base in 1942, but there was no fighting. The capital is Nouméa on the main island, which contains 90 percent of New Caledonia's total population of 185,000. The restaurants, architecture, customs, and language have a decidedly French accent, and it is one of the few cities in the South Pacific with a white majority. In 1985, the French National Assembly passed a bill that paved the way for eventual autonomy, but violence erupted when some factions demanded immediate independence and others wanted to retain their ties to France. In 1988, France resumed administrative control, promising a referendum that would enable the people to determine their political fate.

Newfoundland After Baffin, Victoria, and Ellesmere (all in the Northwest Territories), Newfoundland, at 42,000 square miles, is the fourth largest island in Canada. Now a Canadian province, it was an independent country until 1948. Around the year 999, Vikings led by Leif Ericsson landed at a place that they named Vinland for the wild grapes growing there. The probable spot, now called L'Anse au Meadows, was the first European landing in North America. After John Cabot's 1497 voyage, other explorers and codfishermen came to this new-found land, including the French, who also claimed it. In the nineteenth century, the fur trade and the fishing industry drew many European settlers, particularly from Ireland, and in 1866, the port of Heart's Content became the western terminus of the transatlantic cable. Negotiations to join Canada continued unsuccessfully until 1948, when Newfoundland (with Labrador, from which it is separated by the Strait of Belle Isle), under the leadership of Joey Smallwood, became the tenth province. The capital city is St. John's. The stony soil is largely unfit for agriculture, and codfishing was Newfoundland's primary industry for four hundred years. Whaling, seal hunting, and iron mining (in Labrador) have also been successful industries in Newfoundland, but for various reasons, each of them has failed, and the island province is now in dire financial straits; unemployment is rampant among the 540,000 Newfoundlanders. Seal hunting has been reintroduced in an attempt to stave off total economic ruin.

New Georgia Island group in the Solomons, named for the largest island, but also including Vella Lavella, Kolombangara, Rendova, Vangunu, and hundreds of tiny islets. The group is separated from Choiseul and Guadalcanal by "the Slot," which was what American forces called New Georgia Sound. During the American efforts to take the Solomons in the summer of 1943, some of the most intense fighting of the war took place on and around New Georgia. The U.S. light cruiser *Helena* was sunk, and several Japanese destroyers went down in the surrounding waters. On August 2, 1943, in Blackett Strait, which separates Kolombangara from New Georgia, *PT-109* was rammed by a Japanese destroyer, and a young navy lieutenant named John Fitzgerald Kennedy swam to safety, dragging an injured shipmate. **See also Solomon Islands**

New Guinea With an area of 342,000 square miles, New Guinea is the second largest island in the world, after Greenland. It is located in the southwest Pacific, separated from Australia to the southwest by the Arafura Sea and the Coral Sea to the southeast. The Cape York Peninsula of Queensland reaches to within 95 miles of New Guinea across the Torres Strait. Approximately 1,500 miles long and 400 miles wide at the center, the island is largely tropical, with the Owen Stanley and Bismarck mountain ranges forming the highlands. The highest point is Djaja Peak, at 16,503 feet. Rivers such as the Fly, Sepik, Mamberamo, and Purari flow to the sea through swampy lowlands. The people are of Melanesian ancestry, and although more than eight hundred indigenous languages are spoken, pidgin is the lingua franca. In 1527, the Portuguese explorer Jorge de Meneses named it Ilhas dos Papuas ("Island of the Fuzzy-Haired"), and the Spaniard Inigo Ortiz de Retes called it New Guinea because to him the people resembled those of the Guinea coast of Africa. The Dutch claimed the western portion of the island in 1824, and Germany took possession of it all in 1884. At the outbreak of World War I, Australian troops occupied the island, and it became an Australian mandate of the League of Nations in 1921. The densely populated highland valleys were "discovered" by Western anthropologists only in the 1930s. Between 1942 and 1945, the Japanese occupied the lowlands, but after the war, the territory was administered by Australia as a UN trusteeship. The independent country now known as Papua New Guinea (PNG), which occupies the eastern half of the island (and includes the islands of the Bismarck Archipelago, the Trobriand Islands, Samarai Island, Woodlark Island, the D'Entrecasteaux Islands, the Louisiade Archipelago, and the Northern Solomon Islands of Buka and Bougainville), was established in 1975. In the capital of Port Moresby, there is a governor general, a prime minister, and a cabinet. The western half of the island is the Indonesian state of Irian Jaya.

New Hebrides: See Vanuatu

New Ireland Volcanic island in the Bismarck Archipelago, which is part of Papua New Guinea. A long, thin island that is 220 miles long but only 30 miles wide at its widest point, it is the second largest island in the archipelago, after New Britain. It was first sighted in 1616 by the Dutch navigator Jakob Le Maire, who believed it was contiguous with the island of New Britain. Actually it is separated from that island by the 20-mile-wide St. George's Channel, which was discovered and named in 1767 by the Englishman Philip Carteret. (Carteret also named the island New Hibernia, which became New Ireland.) From 1888 to 1914, New Ireland was part of the German protectorate, during which period it was known as New Mecklenburg. New Ireland carvers produce *malanggan,* extraordinarily sophisticated sculptures in the form of masks, shields, and other objects, intricately detailed and colored in shades of terra cotta, black, and white.

New Providence Island in the Bahamas, on which Nassau, the capital city, is located. New Providence was settled around 1650 by colonists from Bermuda, and it soon became the most populous of the Bahamian islands. In 1670, the Bahamas were granted to the Carolina colony, but the proprietors paid little attention to their acquisition, and the ungoverned island soon became a haven for pirates. By 1710, there were perhaps five hundred pirate "families" on the island, and the town of Nassau was filled with grog shops and prostitutes. Various pirate leaders used the harbor as a staging area for their raids throughout the Caribbean and along the east coast of the American colonies. To clean up the colony, the exprivateer Woodes Rogers was appointed royal governor of the Bahamas in 1717; he offered pardons to those pirates who would surrender. Most of them did (except for Charles Vane and Blackbeard), and by 1725, the colony was back in the fold. **See also Bahamas**

New Zealand An island nation in the South Pacific Ocean, 1,000 miles southeast of Australia. In addition to the two main islands (North and South Islands), which are separated by the Cook Strait, the country also includes the Chatham Islands, the Aucklands, the Antipodes, Stewart Island, Kermadec, Campbell, Snares, and Solander Island. North Island is the site of Wellington (the capital), as well as the cities of Auckland and Hamilton; South Island's largest city is Christchurch. In 1642, Abel Tasman was the first European to set foot on South Island, but the islands had been inhabited since at least A.D. 1000 by the Maoris, a people of Polynesian descent. When Captain James Cook arrived in the *Endeavour* in 1769, he circumnavigated and mapped the islands of New Zealand. There were marine mammals aplenty in New Zealand waters, and the first white settlers were sealers and whalers. The first contacts between the Europeans and the Maoris were peaceful, but eventually hostilities broke out, culminating in the massacre of the crew of the brig *Boyd* in 1809 and their consumption by the Maoris. Attempts to govern New Zealand from across the Tasman Sea proved futile, and England planned to declare it a British colony. In Parliament, Edward Gibbon Wakefield was advocating the colonization of the islands, and in 1840, Captain William Hobson in HMS *Rattlesnake* sailed into Waitangi harbor and presented the Maoris with the Treaty of Waitangi, which opened up the islands for British occupation but was only vaguely explained to the Maoris who signed it. In 1860, a series of sporadic "land wars" broke out between the Maoris and the *Pakehas* (the Maori name for white people). They did not officially end until 1872. New Zealand was granted dominion status in 1906, and the monarch of England is the titular head of state, with an appointed governor general and an elected prime minister and parliament to govern the country. There are no large land mammals endemic to New Zealand, but the 3 million people are outnumbered by approximately 70 million sheep.

New Zealand sea lion (*Phocarctos hookeri*) Also known as Hooker's sea lion, this species is found mostly on the subantarctic islands of New Zealand, but it is sometimes seen on South Island as well. Males are much larger than females and can be 10 feet long and weigh 800 pounds. In addition to the size disparity, bulls are dark brown, while the females are silvery gray. Many of the sea lion species of the Southern Hemisphere were hunted to the brink of extinction by the end of the nineteenth century. **See also sealing**

New Zealand sea lion

Ni'ihau Hawaii's "forbidden isle," 17 miles off the west coast of Kauai. It was sold by King Kamehameha IV in 1863 to Elizabeth Sinclair of Scotland, and it has remained in her family ever since. Her descendants, the Robinson family, refuse to allow visitors, and only pure-blooded Hawaiians may live there. English is taught in the school, but Hawaiian is the primary language. The residents of Ni'ihau voted against statehood in 1959. Although it is officially part of Kauai County, there are no services, no county roads, no garbage collection, no county employees. The 73-square-mile island is dry and unproductive because it is in the rain shadow of Kauai, and the Robinson family raises mostly sheep and cattle. Ni'ihau shell leis, made by the islanders from tiny shells that wash ashore, are highly valued and are the island's only export. **See also Hawaiian Islands**

Nimitz, Chester (1885–1966) American admiral, born in Fredericksburg Texas, and a 1905 graduate of Annapolis. During World War I he was chief of staff to the commander of U.S. submarines, and then he served as executive officer on a battleship. In December 1941, after the Japanese attack on Pearl Harbor, he replaced Admiral Husband E. Kimmel as commander of the U.S. Pacific Fleet. Headquartered in Hawaii, he masterminded the Battle of Midway (May 1942), where a superior force of Japanese warships was soundly defeated by American forces. Then followed the campaigns of Guadalcanal and the Gilbert and Marshall Islands, and the capture of New Guinea and the Marianas, all of which required massive naval support. The Battle of Leyte Gulf (October 1944) was one of the decisive sea battles of World War II; it crippled the Japanese fleet and enabled the United States to retake the Philippines. Nimitz was named fleet admiral in December 1944; on the battleship *Missouri* in Tokyo Bay, in August 1945, he accepted the formal surrender of Japan. His last posting was chief of naval operations. **See also Guadalcanal; Leyte Gulf, Battle of; Midway, Battle of**

Niue A tiny country 350 miles south of Samoa in the South Pacific, consisting of a single uplifted coral island, 11 miles wide. When James Cook arrived in 1774, he was repulsed by natives throwing stones and spears, so he departed and named the island Savage Island. When the missionary John Williams arrived, he was also thrown out, by Niueans with red-painted teeth. In 1900 Niue was made a British protectorate, and the following year it was annexed by New Zealand. Niueans now manage their own affairs with New Zealand's political and economic assistance. A fringing reef surrounds the island except on the east, where the surf breaks directly against the cliffs. The clear waters and coral reefs of Niue provide excellent diving; the island itself is rich in tropical butterflies, and the smells of frangipani and bougainvillea fill the air. Most visitors to Niue are divers from New Zealand, although tourist facilities are limited to a couple of hotels and guest houses. The population of the island is around 2,300, but another 14,000 Niueans have emigrated to New Zealand, where all Niueans hold citizenship.

Noah Early shipbuilder and voyager. Shortly after the world was created, God felt that there was too much violence and wickedness, so he decided to drown everybody except Noah, his wife, and their three sons, Ham, Shem, and Japheth, and their wives. He commanded Noah to build a three-story ark out of gopher wood, 300 cubits overall, with a beam of 50 cubits, and load on it two of every living thing. It rained for forty days and forty nights, so that the entire earth was covered in water, and everything that wasn't on the ark was drowned. After 150 days, the waters began to recede, and the ark came to rest on Mount Ararat. Noah sent out a dove and when the bird returned with an olive leaf in its mouth, he knew that the earth was dry enough for disembarkation. All the animals were offloaded, and the three sons of Noah and their nameless wives became the ancestors of everybody on earth today, by virtue of a lot of begatting. Noah, who was 600 years old at the time of the flood, lived another 350 years.

noddy (*Anous* spp.) Noddies are terns of a different color. Where most of the others (*Sterna* spp.) are white or gray with a black cap, noddies are dark grayish-brown with a white cap. Where the bills of terns are usually red, those of noddies are black. Where most terns plunge-dive, disappearing into the water for a moment, noddies feed by hovering above the surface and then swooping down to pick up a fish. Noddies are known for their habit of settling on the rigging of ships at sea, or on any other floating object that is large enough. The brown noddy (*A. stolidus,* also known as the common noddy) is the largest, with a wingspan of 34 inches. It breeds widely in the Pacific, Atlantic, and Indian Oceans, and is a common sight in many island groups, from the Caribbean to the Galápagos and the Great Barrier Reef. The range of the lesser noddy (*A. tenuirostris*), a somewhat smaller bird with a similar coloration, overlaps that of the brown noddy in the Indian Ocean. The black noddy (*A. minutus*) is smaller still, and darker. **See also tern**

white-capped noddy

Nordenskjöld, Otto (1869–1928) Swedish geographer and polar explorer, the nephew of Nils Adolf Erik Nordenskjöld, a Finnish-born geologist and explorer who made the first successful traverse of the Northeast Passage in 1879. After voyages to Patagonia and Greenland, Otto Nordenskjöld commanded a 1902 expedition to the Antarctic, where his ship (the *Antarctic,* under the command of Carl Anton Larsen) was crushed by the ice; Nordenskjöld and five crew members spent the winter on Snow Hill Island. (They also crossed over to Seymour Island, where they found the bones of an extinct, 5-foot-tall penguin.) Captain Larsen had sledged to Paulet Island, and his men also spent the winter in a stone hut, eating penguins. None of the explorers knew that their ship had sunk, and by a completely fortuitous accident, they all convened at Snow Hill Island. The Argentine ship *Uruguay* rescued the men in November 1903 and brought them back to Buenos Aires. There Larsen, who had seen the abundance of great whales in Antarctic waters, raised the money for the Compania Argentina de Pesca, the first Antarctic whaling venture.

Norfolk Island Discovered in 1774 by Captain James Cook, who named it for the Duke of Norfolk, this island was settled as a penal colony in 1788, immediately after the First Fleet arrived in Australia with its cargo of convicts. Lieutenant Governor Philip Gidley King selected the men and women who would be Norfolk's first colonists, as much to keep the French from occupying the island as to create a settlement. The original Norfolk Island prisons were abandoned in 1814 but reopened in 1826 to function as the harshest penal settlement in Australia, described as "punishment short of death." Conditions were so bleak that many convicts preferred hanging. In 1855, the prison was closed, and the last of the convicts removed to Tasmania. In 1856, the descendants of the *Bounty* mutineers arrived from Pitcairn Island, settled on Norfolk, and became the ancestors of many of the island's 2,000 inhabitants. More than ten thousand tourists per year come to Norfolk to see the island and the partially restored remains of the penal colony. The 34-square-mile island is 1,000 miles east of Sydney, and is a territory of Australia. The indigenous Norfolk Island pine (*Araucaria heterophylla*) was used extensively for shipbuilding, especially for masts, since its straight trunk can grow to 200 feet in height, but most such trees have long since been cut down.

Normandie When she first sailed in 1935—a year before the *Queen Mary*—the French liner *Normandie* was considered the most beautiful ship ever designed. With her soaring spaces, opulent suites, ballrooms, theaters, crystal fountains and chandeliers, and of course, four-star cuisine and nonpareil wine cellar, she was considered the best of France—a seagoing Versailles. Her maiden voyage was a triumph; the corps de ballet of the Paris Opera performed on her last night in port, and when she reached New York, thirty thousand spectators cheered as the "Marseillaise" was played. Along with the French liners *Paris, Ile de France, Lafayette,* and *Champlain,* the *Normandie* was the toast of the Atlantic run. But the *gloire* was not to last. In 1938, the *Lafayette* burned to the waterline in Le Havre; in 1939, the same fate befell the *Paris.* On August 28, 1939, as she completed her 140th transatlantic crossing, the *Normandie* arrived in New York. Four days later, Germany marched into Poland. *Normandie* sat in New York harbor until Pearl Harbor, when the American government seized the dormant liner to turn her into a much-needed troop transport. Early in 1942, a crew of eight hundred workmen removed her expensive fittings and prepared her for war. As a worker in the grand salon was cutting a iron pillar with an acetylene torch, sparks ignited a pile of kapok life jackets that had been stored nearby, and in minutes, the ship was in flames. The fire burned for eleven minutes before anyone thought to call the New York Fire Department, and when the firefighters finally arrived, all they could do was hose thousands of gallons of water into the burning ship. Even when the fire was doused, the water caused the ship to list sharply to starboard, and no amount of human effort could keep her from capsizing. In June 1944, she was finally removed from her resting place in the mud of the Hudson River and cut up for scrap.

Normandy landings On June 6, 1944 (D-Day), British, American, and Canadian troops under the command of General Dwight Eisenhower began the invasion of France (code-named "Operation Overlord") by dropping airborne troops behind enemy lines and approaching the coast of Normandy with four thousand troop transports, eight hundred warships, and innumerable small craft. Before dawn, American airborne divisions had captured the town of Saint-Mère-Église, while British commando units took key bridges and knocked out German communications. The largest invasion fleet in history carried 73,000 American and 83,000 British and Canadian troops, while ten thousand planes dropped bombs on German positions. The American troops came ashore at the beaches they called Utah and Omaha, while the British landed at Gold, Juno, and Sword to the east. (Of all the landings, the one at Omaha was the bloodiest, encountering the most German resistance.) The Allied troops were stopped short of Caen, but they prevented the German commander, Field Marshal von Runstedt, from bringing up his reserves, and by June 18, they had captured the German communications center at St.-Lô, cutting off the German force under Field Marshal Erwin Rommel. Then the U.S. Third Army, under General George S. Patton, broke through the German left flank at Avranches and raced to Brittany to outflank Paris. By July 9, the British had taken Caen, and the major part of the German Seventh Army was wiped out within a month. By August 23, the Allies had taken northern France.

Norris, Ken (1924–1998) Probably the best-known—certainly the best-loved—professor of marine science and cetological researcher of this century, Ken Norris was born in Los Angeles. He received bachelor's and master's degrees from UCLA and a Ph.D. from the Scripps Institute of Oceanography at La Jolla. In 1951, he was founding curator of Marineland of the Pacific at Palos Verdes, California, and was among the first to confirm that dolphins use sound to navigate underwater. He also recognized that they use echolocation to find and identify objects in the water, and that they communicate with bursts of sound. As the first curator of Sea Life Park in Hawaii, which opened in 1963, he became one of the first to work with spinner, spotter, and rough-toothed dolphins, in addition to bottlenoses. In 1972, he was named director the Center for Coastal Studies at the University of California at Santa Cruz, and remained there for the rest of his professional life. He was scientific adviser to the Marine Mammal Commission, and helped write the Marine Mammal Protection Act of 1972. In 1983, with Bertel Møhl, he published the first description of the "sonic blasts" that toothed whales and dolphins use to stun their prey prior to capturing it. In addition to numerous scientific papers, he is the author of *The Porpoise Watcher* (1974), *Dolphin Societies* (1991; with Karen Pryor), and *Dolphin Days* (1991), and the editor of *Whales, Dolphins, and Porpoises* (1966). **See also echolocation**

Northeast Passage Before the extent of the ice along the northern coastline of Russia was known—even before the coastline itself was known—travelers attempted to find a way around the north of Europe to Asia. The English were among the first to try: Sir Hugh Willoughby and Richard Chancellor in 1553, Stephen Burrough in 1556, James Bassendine in 1568, and Henry Hudson in 1607 and 1608. Between 1594 and 1597, the Dutch also sponsored explorers, the best known of whom was Willem Barents, who reached Novaya Zemlya in 1596. Most of these early expeditions happened on new discoveries (Spitsbergen, Jan Mayen Island, etc.), but none of them made it to the Bering Sea in the Far East. The Russians tried; the Austrians tried (they failed, but discovered Franz Josef Land); and from 1733 to 1743, the Russians mounted "the Great Northern Expedition," conceived to map the entire northern coast of Russia. (As part of this enormous project, Vitus Bering discovered Alaska and the Bering Sea in 1741.) In 1875–1876, the Swedish explorer Nils Adolf Erik Nordenskjöld took the *Vega* through the ice and achieved the goal of three centuries.

See also Northwest Passage

northern bottlenose whale (*Hyperoodon ampullatus*) Males and females of this species are so distinct from each other that for years scientists believed they were different species. Males are much larger than females, lighter in color (sometimes almost white), and they have an overhanging bulbous forehead that the females lack. Found only in the North Atlantic, this beaked whale can reach a length of 30 feet. It is believed to be among the deepest-diving of all cetaceans, and has been recorded to remain submerged for two hours in its search for squid. In the past, the species was hunted by Scots, Danes, Icelanders, Canadians, and Norwegians, but it is now considered too rare to be hunted commercially. (Small cetaceans are not protected by the quota regulations of the International Whaling Commission [IWC].)

See also beaked whales, International Whaling Commission

northern fur seal (*Callorhinus ursinus*) Also known as the Pribilof fur seal, this eared seal breeds on the chain of islands that extends from Alaska to Kamchatka, and has been known to venture as far south as Santa Barbara, California, in the eastern North Pacific, and Honshu, Japan, in the western. The large males (8 feet long and weighing 600 pounds) arrive first on the breeding grounds, and then the females arrive to form into harems. Once hunted mercilessly for their luxurious pelts, the northern fur seal was "harvested" in the Pribilofs under the North Pacific Fur Seal Convention, an agreement between the United States, Japan, Canada, and Russia. This treaty has now run out, and no nation is legally involved in a seal hunt. As with the Steller's sea lion, with which it shares much of its habitat, the population of northern fur seals has declined seriously, perhaps because of food limitations.

northern fur seal

See also Pribilof Islands, Steller's sea lion

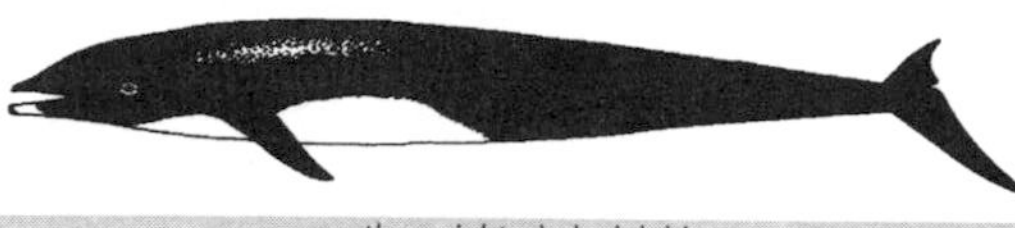
northern right whale dolphin

northern right whale dolphin (*Lissodelphis borealis*) This slim and graceful dolphin gets its common name from its lack of dorsal fin, virtually the only characteristic it shares with the ponderous right whale. It is perhaps the slimmest of all dolphins; a 10-foot-long specimen weighs only about 180 pounds. Unlike its southern relative, *L. borealis* is mostly black, except for a white hourglass pattern on the ventral surface and white-tipped upper and lower jaws. It is found in the North Pacific from Southern California to Alaska, Siberia, and Japan, but is difficult to see because of its unobtrusive silhouette and its low-angled leaps from the water.

See also southern right whale dolphin

North Pole The northern end of the earth's axis, located about 500 miles north of the northern tip of Greenland. Unlike the South Pole, which is located on land, the North Pole is in the Arctic Ocean and, depending on conditions, can be on floating ice or on open water. The water at the Pole is 13,410 feet deep. It does not coincide with the magnetic North Pole, to which magnetic compasses point, which is located at about 76°20′ north. The North Pole has been eagerly sought by explorers, and many of them have exaggerated their accomplishments in order to claim the prize. For his 1909 attempt, Robert E. Peary was long recognized as the first, but it is now believed that he faked his diaries and did not make it. The same is true of Peary's contemporary, Frederick Cook. In 1926, Roald Amundsen, Lincoln Ellsworth, and Umberto Nobile passed over the Pole in a dirigible, and Admiral Richard Byrd flew over it in an airplane in 1929. The American nuclear submarine *Nautilus* passed under the Pole in 1958. In 1967, Ralph Plaisted of Duluth, Minnesota, led a snowmobile expedition that got within 384 miles of the Pole before the weather forced them to quit. The following year, Plaisted mounted another attempt, and reached the Pole on April 18, 1968. (The achievement was documented by a U.S. Air Force plane that flew overhead and verified the position.) There are those, however, who do not regard a mechanized conquest of the Pole as valid, and therefore they award the honor to Wally Herbert, an English cartographer and explorer, who made the trip by dogsled across the surface of the Arctic Ocean and arrived at the Pole on April 6, 1969.

See also Amundsen, Cook (Frederick), magnetic pole, Peary

North Sea An arm of the North Atlantic between Norway and the British Isles, connecting the Skagerrak (the channel between Norway and Denmark) and the English Channel, between England on the north, and France, Belgium, the Netherlands, and Germany to the south. Although the 220,000-square-mile North Sea is generally shallow, the bottom configuration varies considerably, from the shallow Dogger Bank, off northern England, which may be as little as 50 feet below sea level, to the 1,830-foot-deep Norwegian Trench off the southern coast of Norway. Off the coast of Kent, where the North Sea meets the English Channel, are the treacherous sandbanks known as the Goodwin Sands. Since the Middle Ages, the North Sea has been the scene of important British, Norwegian, Dutch, and German fisheries for herring and cod. Herring are still caught in commercial quantities, but the cod populations are depressed. In the 1970s, large petroleum deposits were discovered in the North Sea, and oil and natural gas became economic bonanzas for Britain, Norway, and the Netherlands.

See also Dogger Bank, Goodwin Sands

Northwest Passage Until the nuclear-powered icebreaker *Manhattan* made a continuous east-west transit in 1969, a passage through the island-strewn, ice-choked waters of northern Canada has been the

Holy Grail of mariners. One of the first voyagers in search of a westward route from Europe to the Orient was Christopher Columbus, who found the Bahamas instead. John Cabot, dispatched by King Henry VII of England in 1497, encountered Newfoundland on his first voyage, and never returned from his second. His son Sebastian (who may have accompanied his father on the first voyage) sailed much farther north, suggesting that he believed there was a way through. Giovanni da Verrazano headed west for King Francis I of France in 1525, and although he did not find a passage to the east (no one suspected there was a whole continent and another ocean in the way), he did explore the east coast of America, from Virginia to New York. Then came Jacques Cartier, whose second voyage in 1535 took him up *La grande rivière* (the St. Lawrence) as far as Montreal. Martin Frobisher was sent in 1576, and Humphrey Gilbert was lost at sea in the *Squirrel* in 1583. John Davis in the *Sunshine* and the *Moonshine* made a gallant effort in 1585; and Henry Hudson entered the bay that bears his name in 1611, but found no westward egress and died there. In 1615, William Baffin in the aptly named *Discovery* could not get past the gigantic island that was named for him. Subsequently, navigators like Luke Foxe and Thomas James (1635) tried and failed. There were many eighteenth-century attempts to make an overland crossing, and in 1778, James Cook took the *Resolution* and the *Discovery* through the Bering Strait, hoping to be able to sail eastward, but even the greatest navigator in history could find no passage. Whaling captain William Scoresby reported in 1817 that the pack ice between Greenland and Spitsbergen had disappeared, and that this might be the moment to search for an open sea lane to the Pacific. The following year, the Admiralty offered a prize of twenty thousand pounds to the first ship to reach the Pacific. In 1818 John Ross took the reinforced *Isabella* and *Alexander* deep into Lancaster Sound, but he could make no further westward progress. William Parry almost made it in 1819, and Frederick Beechey (1824) tried to find a way through from the west. None of them was successful, but their attempts and those of their predecessors are immortalized in the names of the straits, islands, and bays that they encountered. The worst tragedy in the search for the Northwest Passage occurred in 1845, when Sir John Franklin and 138 men aboard the *Erebus* and *Terror* vanished. After Franklin's disappearance, many expeditions were dedicated to finding him and his men instead of the elusive passage. The remains of Franklin and his men were discovered in 1859, nine years after they died, on King William Island. In 1850, Robert McClure sailed through the Straits of Magellan along the west coasts of North and South America and up to the Bering Strait, and penetrated the Prince of Wales and McClure Straits, proving that the passage was possible, but his ship *Investigator* was trapped in the ice at Mercy Bay and had to be abandoned. In 1904, Roald Amundsen completed a three-year, east-to-west voyage in the 47-ton *Gjøa*, having spent two years of the trip locked in the drifting ice.

See also Amundsen, Baffin, Cabot (John), Cabot (Sebastian), Cook (James), Davis, Drake, Franklin, Gilbert, Hudson, McClure, Parry, Ross (John), Russ, Scoresby, Verrazano

Novaya Zemlya A group of islands in European Russia that separate the Barents and the Kara Seas in the Arctic Ocean. Novaya Zemlya ("new land" in Russian) consists of two large islands, Severny ("northern") and Yuzhny ("southern"), plus several smaller islands, aligned in a SSW-NNE direction. A continuation of the Ural Mountain system, the islands are mountainous and covered with ice for most of the year. Even when early explorers managed to get through the ice, they were often stopped by the unexpected appearance of land, always an impediment to sailing ships.

nudibranch (order Nudibranchia) Named for their "naked gills," nudibranchs are snails without shells. They are sometimes known as sea slugs, a most unfitting name for animals that are often spectacularly colored; "butterflies of the sea" would be a better name. Some have retractile gills; others breathe through the skin. The three-thousand-odd species are found in all seas, from the surface to the depths, but they are most common in shallow tropical waters. Some nudibranchs crawl on the bottom; others swim with graceful undulating movements. **See also gastropoda, sea hare**

nudibranch7

Nunavut As of April 1, 1999, some 40 percent of Canada's northern wilderness was administratively separated from the Northwest Territories and amalgamated into the new territory named Nunavut, the Inuit word for "our land." Constituting one-fifth of all Canada, Nunavut consists of empty, ice-covered islands and is divided by the Arctic Circle. Two of the world's largest islands, Baffin (the fifth largest) and Ellesmere (ninth), are included. Although it encompasses 770,000 square miles, an area larger than Mexico, Nunavut has a population of only 26,000. In one of the largest native-rights claims settlements ever paid, the Canadian government will give to the inhabitants of Nunavut \$1.148 billion Canadian (\$840 million in U.S. dollars) over fourteen years.

Nunivak Island Located off the west coast of Alaska in the Bering Sea, Nunivak is Alaska's second largest island (after Kodiak). It is 55 miles long and 40 miles wide, and is separated from the Alaskan mainland by Etolin Strait. Encompassing 1,700 tundra-covered square miles, the island is mostly a wildlife refuge; it is home to musk oxen, innumerable shorebirds, and reindeer, which were introduced as a food source for the native Yupik Eskimos. The human inhabitants of Nunivak number only a couple of hundred and are well known for traditional handicrafts, especially masks and soapstone sculptures.

nurse shark (*Ginglymostoma cirratum*) With its long upper tail lobe, two large dorsal fins, and long barbels at the front of each nostril, this is an easy shark to recognize. It is found in the subtropical waters of the eastern and western Atlantic, usually resting on the bottom under ledges, or in caves in shallow water. (There are similar species in Australia and South Africa.) Unlike most sharks, which have to keep moving in order to pass oxygenated water over their gills, the nurse shark can remain motionless, with its gills pumping water through. Because it appears sluggish, divers often attempt to ride it or pull its tail, but it is a shark and can be dangerous if provoked. It reaches a maximum length of 14 feet, but smaller ones are often seen in oceanarium tanks. **See also wobbegong**

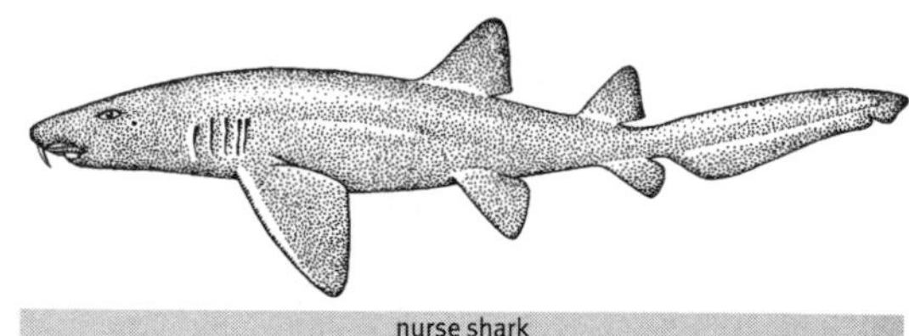
nurse shark

O

Oahu Third largest, after Hawaii and Maui, and most populous of the Hawaiian Islands, Oahu covers 607 square miles and has a population of 836,000 of whom 377,000 live in Honolulu, the state capital. The island is roughly rhomboidal in shape, with two mountain ranges, the Waianae and the Koolau, traversing its center, in a northwest-southeast configuration. The south shore is indented by several deep harbors, one of which contains the naval base at Pearl Harbor, and another, the port of Honolulu (the name means "protected bay" in Hawaiian). The city, now a busy metropolis with high-rise office towers, was the royal capital of Hawaii until American businessmen abolished the royalty in 1893. In front of the Iolani Palace stands the famous statue of King Kamehameha, which is festooned with leis every July 11, Kamehameha Day. The state capital building, the University of Hawaii, and the Bishop Museum, which celebrates Hawaii's history and natural history, are also in Honolulu, as is Waikiki, one of the world's most famous beaches, framed by the volcanic cone known as Diamond Head. Before statehood and jet travel made it so accessible, Waikiki was a tranquil little resort, but it is now dominated by high-rise hotels, many of them built and owned by Japanese investors to accommodate the legions of tourists from Japan and everywhere else. On the north shore are Waimea Bay and Sunset Beach, where only the most daring professional surfers attempt the steep and powerful waves. The mountainous center of Oahu is occupied by pineapple and sugar plantations, and also by the town of Wahiawa and Schofield Barracks, the U.S. Army installation that was bombed along with Pearl Harbor on December 7, 1941, as the Japanese began their war on the United States. Other tourist destinations on Oahu are Sea Life Park at Makapu'u, the Polynesian Cultural Center at La'ie, and Waimea Falls.

See also Hawaiian Islands

oakum Tarred hemp or manila fibers picked from old ropes that are used for caulking the seams of wooden ships in order to make them watertight. Oakum was rammed into the seams with a caulking iron and a heavy hammer, and then held in position as hot pitch was poured along the seams. The unpicking of oakum, a tedious job, was often used as a punishment aboard ship.

See also manila

oarfish (*Regalecus glesne*) The longest of all the bony fishes, the oarfish—sometimes known as king-of-the-herrings—has been measured at a length of 26 feet. (Two species of sharks—the whale shark and the basking shark, which have cartilaginous skeletons—grow longer.) Since it is known mostly from stranded specimens, very little is known of its biology. It has a bright red "crest" that is thought to be responsible for some sea serpent sightings.

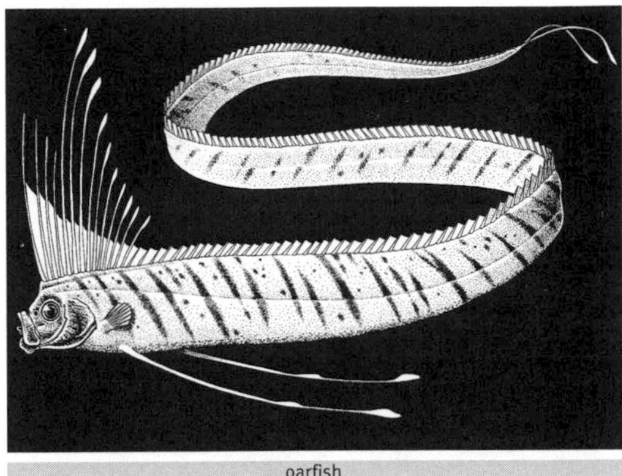
oarfish

See also bony fishes, ribbonfish, sea serpent

oceanic whitetip shark (*Carcharhinus longimanus*) A large pelagic shark, identifiable by its long pectoral fins and tall, rounded dorsal fin, which is usually tipped with white. Oceanic whitetips may reach a length of 13 feet, but most are in the 10-foot range. They are found offshore in all the world's tropical and temperate waters, and although they have been rarely implicated in attacks on swimmers or divers, they are attracted to shipwrecks and downed airplanes and, in these situations, have been known to eat humans.

oceanic whitetip shark

See also carcharhinid sharks, shark, shark attack

ocean sunfish (*Mola mola*) A huge relative of the triggerfishes and puffers, the ocean sunfish is one of the largest of all the bony fishes. With its terminally placed dorsal and anal fins and its almost nonexistent tail, it appears to be all head. They have been measured at 10 feet long and weighed at two tons. Molas feed on jellyfish and salps, but their flesh is not edible by people.

They spend most of their time floating at or near the surface as though dead. *Mola* is the Latin word for "millstone."

See also opah, triggerfish

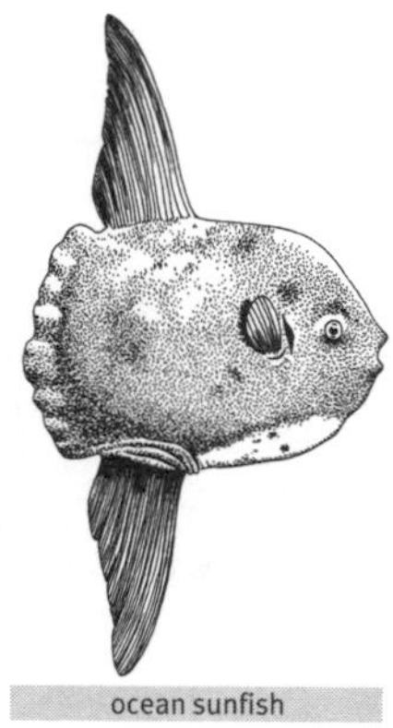

ocean sunfish

Oceanus In early Greek mythology, Oceanus is more of an element than a personality; in Homer he is the river that encircles the world and in which the stars and sun rise and set. He is the son of Coelus (the sky) and Terra (the earth), and with his wife, Tethys, he fathered the sea and river gods, known as the Oceanides. He was rarely depicted in Greek sculpture or pottery, but the Romans represented him as an old man with a flowing beard, sitting on the waves of the sea.

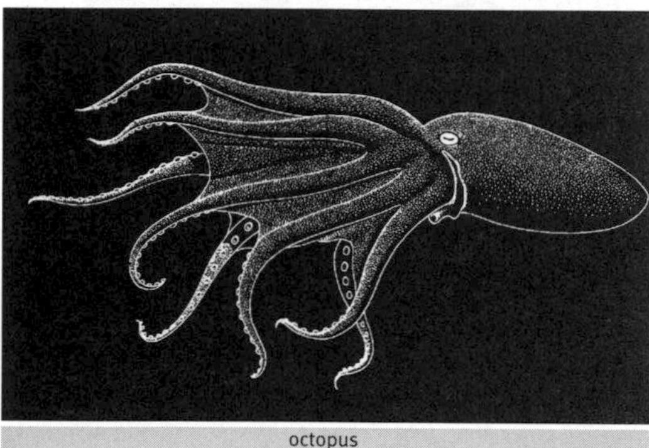

octopus

octopus Eight-armed cephalopods that range in size from 2 inches to 20 feet, measured from outstretched arm tip to outstretched arm tip. Octopuses (the correct plural) have a saclike body that is positioned atop the cluster of arms. On the underside and in the center of this cluster is the sharp, horny beak, like that of a parrot. Except for one species that is blind, their eyes are large and highly effectual. They locomote by walking along the bottom or by swimming, usually head-first with the arms trailing. Most octopuses are shallow-water dwellers, but there are some deepwater species that are softer-bodied. The tentacles are equipped with suction cups that can attach themselves to any surface, including their prey. "Cirrate" octopuses are those deepwater octopuses that have two fins above the eyes and filaments (cirri) on the tentacles as well as to suckers. Most octopuses feed on crustaceans. They usually live in holes or crevices in rocky terrain, and bring their prey back to their den to eat it.

See also blue-ringed octopus, cirrate octopus, common octopus, giant octopus, mimic octopus

Odysseus The son of Laertes and Anticleia, Odysseus (also known as Ulysses) is probably the most famous hero in Greek mythology. The story of his ten-year journey is told in Homer's *Odyssey*. After the Trojan War and the capture of Troy with the wooden horse, he set sail for Ithaca, but he was forced to cross unknown seas and brave dangers that included Scylla and Charybdis, the enchantress Circe, and the Cyclops. After nine years on the island of Calypso, Odysseus finally arrived at Ithaca, where he was recognized only by his dog and a nurse. With the help of Athena and his son Telemachus, he slew the suitors of his wife, Penelope, and reclaimed his throne.

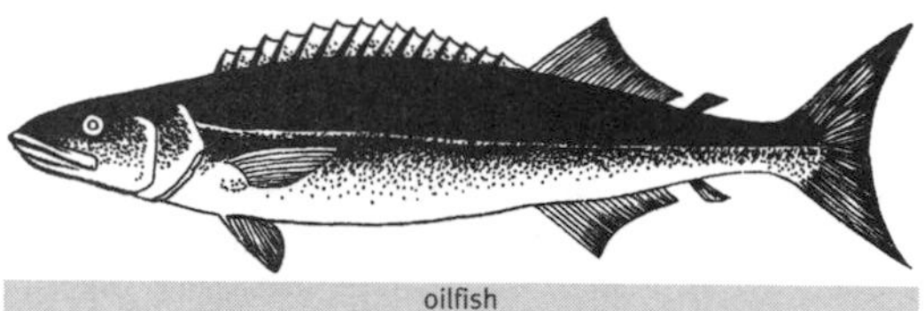

oilfish

oilfish (*Ruvettus pretiosus*) Related to the snake mackerels (Gempylidae), the oilfish reaches a length of 6 feet and is found at depths of about 2,500 feet in all the world's tropical and temperate oceans. They are dark purple-brown, and like many deep-sea species, they have bright green eyes. The oily flesh, which is responsible for its name, can cause diarrhea unless it is grilled to remove the oil. It is a particularly popular food fish in the Mozambique Channel, where Comoro Islanders, East Africans, and Madagascan fishermen catch them regularly. In this region, oilfish inhabit the same waters as the rare, endangered (and inedible) coelacanth (*Latimeria chalumnae*), which when it is caught is usually discarded as a trash fish. The fishery for the oilfish is largely responsible for the drastic decline in the numbers of coelacanths. **See also coelacanth, snake mackerel**

oil spill The accidental discharge of petroleum into the ocean from a ship, a pipeline, or an oil well. The ships required to transport oil have grown larger and larger, and when today's supertankers have accidents of one sort or another—groundings, collisions, breakups—the result can be a catastrophic release of oil into the sea and, occasionally, onto the shore. The largest oil spill in history, however, was not from a ship: it was the intentional discharge of 250 million gallons of oil into the Persian Gulf by the Iraqi army during its retreat from Kuwait after the 1991 Gulf War. The second largest oil spill was not from a ship either. When the Ixtoc offshore oil well in the Gulf of Mexico blew and caught fire in June 1979, 140 million gallons of oil were burned. In terms of quantity of oil discharged (as opposed to the amount of destruction caused by the oil washing ashore), some of the most serious tanker spills have been the *Castillo de Bellver* (78.5 million gallons), the *Amoco Cadiz* (68.7 million gallons), collision of the *Atlantic Empress* and the *Aegean Captain* (49 million gallons), the *Torrey Canyon* (38.2 million gallons), the *Urquiola* (28.1 million gallons), and the *Braer* (25 million gallons). Although the *Exxon Valdez* spilled "only" 10.9 million gallons of oil in Prince William Sound, Alaska, in 1989, the ecological destruction was prodigious. Most oil spills are smaller and less dramatic, and they rarely receive much press coverage.

See also *Amoco Cadiz*, *Atlantic Empress* and *Aegean Captain*, *Braer*, *Castillo de Bellver*, *Exxon Valdez*, Gulf War oil spill, Ixtoc blowout, Santa Barbara oil spill, supertanker, *Torrey Canyon*, *Urquiola*

Ojeda, Alonso de (1466–1515) Spanish conquistador and navigator who sailed with Columbus on his second voyage (1493–1495) and in 1499 led his own expedition to the Indies. With Amerigo Vespucci, he explored the northern coast of South America, but the two were separated, and Ojeda continued his journey alone. He landed on the coast at what is now Suriname, discovered the island of Curaçao, and returned in 1501 to Seville. In 1508, he was named commander of the colony at Cartagena, but while raiding the countryside for gold, he was routed by the Indians and had to be rescued by Diego de Nicuesa. His garrison was attacked again, and he fled for Hispaniola, leaving his men under the command of Francisco Pizarro. Ojeda fled to Cuba, and then to Santo Domingo, where he died in poverty.

Okhotsk, Sea of Northwestern arm of the Pacific Ocean, enclosed by the Siberian coast of Asia on the west and by the Kamchatka Peninsula and the Kuril Islands on the east. It covers some 611,000 square miles, an area roughly the size of Mongolia. The shores are high and rocky, and many rivers flow into this sea. Its average depth is 2,549 feet, and it is 11,069 feet at its deepest point, which is in the southern quadrant, west of the Kuril Islands. The weather in the northern part of the Sea of Okhotsk is severe, with frequent winter gales, but in the south, it is milder because of the proximity to the Pacific Ocean. There are important commercial fisheries for such species as salmon, herring, pollack, flounder, and cod. The California gray whales feed on bottom-dwelling amphipods here before their annual migration to the lagoons of Baja California.

Okinawa Island in the Ryukyu group, south of Kyushu, that is a prefecture of Japan. Its capital is Naha. The 454-square-mile island has a population of well over 1 million. On April 1, 1945, after the bloody battle for Iwo Jima, sixty thousand American troops landed on Okinawa in one of the last great amphibious campaigns of the war in the Pacific. The counterattack involved 355 kamikaze attacks, and participants included Japan's largest battleship, the *Yamato*, which was sunk. The Japanese garrison lost 103,000 of its 120,000 men; the Americans lost 12,000. On June 21, 1945, two months before the atomic bomb was dropped on Hiroshima, the Japanese gave up organized resistance on Okinawa. (They surrendered in August.) From 1945 to 1972, the island was under American military jurisdiction; then it was returned to Japan. The U.S. military bases remaining on the island are a source of deep resentment to the Japanese on Okinawa. The Marine Expo Aquarium on Okinawa, one of the world's largest, exhibits some of the most spectacular marine life, including the only whale sharks (the largest fish in the world) ever maintained in captivity. **See also Ryukyu Islands**

old squaw (*Clangula hyemalis*) A small, graceful, seagoing duck with a long tail, which accounts for its name in Britain: long-tailed duck. In the summer, it aggregates for breeding in loose flocks along the tundras of the circumpolar Arctic, and it winters along the coasts of Europe, North America, and Asia, sometimes straying as far south as California, the Great Lakes, and even Florida. Its summer plumage is brown with white undersides and a dark-capped head, but in the breeding season both the duck and the drake become much whiter around the head and neck. Old squaws are fast, energetic flyers, and deep divers, one having been caught in a net at 200 feet. They are among the most vocal of all sea ducks, whistling and yodeling almost constantly.

old squaw

See also eider, harlequin duck, sea ducks

olivaceous cormorant (*Phalacrocorax olivaceous*) Found from southern North America throughout South America all the way to Cape Horn, this is a small, slim bird that is greenish-black in color with an orange throat patch. It commonly perches in trees.

See also cormorant, shag

ooze Beaches and the littoral zone are often floored with sand, and immediately offshore there may be deposits of mud, silt, rock fragments, and clay ("terriginous sediments") that run off the land, but most of the ocean floor is covered with a substance known as ooze. Composed largely of countless billions of microscopically small one-celled animals known as foraminifera, calcareous ooze covers about half of the seafloor. When composed of the accumulation of millions of years of deposits of the minute calcium carbonate shells of protozoans of the genus *Globigerina*, the resulting sediment is called globigerina ooze. If the ooze is made up of the transparent silica shells of diatoms, it is known as siliceous or diatomaceous ooze. ("Diatomaceous earth" is harvested and used as a fine abrasive, as a base for toothpaste and car polishes, and in insulation.) In some areas of the Pacific, radiolarians predominate in the seafloor. These sediments cover hundreds of millions of square miles of ocean floor, in a blanket that can be several miles thick. Many deepwater basins are floored with reddish or brownish clay, which is formed when the calcium carbonate of the various shells dissolves and combines with atmospheric and volcanic dust. Where currents prevent sediments from accumulating, the floor of the ocean is rock, often covered with manganese nodules.

See also foraminifera, manganese nodules, radiolarian, sand

opah (*Lampris regius*) Widely distributed in warm oceans, the opah, also called the moonfish, is a laterally compressed, deep-bodied fish that can grow to a length of 7 feet and a weight of 600 pounds. It is blue-gray above, rosy below, and completely covered with large white spots. The fins and tail are scarlet. Although it is a tasty food fish, it is so uncommon that it is usually caught by accident. **See also ocean sunfish**

opah

orange roughy (*Hoplostethus atlanticus*) A deepwater fish of temperate waters, fished commercially in South Africa, South Australia, and New Zealand. They reach a maximum length of about 12 inches, and are big-headed, laterally compressed fishes. The name is derived from the bright orange or red color of the body and fins, although there are silver tinges on the flanks. The New Zealand fishery was begun only in 1979; at depths that range from 2,500 to 5,000 feet, it is the deepest commercial fishery in the world. Because of advances in deep-sea trawling, fishes that could not be caught before are now being hauled up in great quantities, and orange roughy was extensively exported to the United States from New Zealand. Orange roughys mature at about 30 years of age, and there are documented records of individuals that have reached 150. They school tightly, so a five-minute tow can fill a trawl net with 10 to 50 tons of fish. This, along with the slow rate of growth and development, makes the orange roughy particularly susceptible to overfishing, and catches have fallen to a fraction of what they were earlier in the decade. Off the Chatham rise of New Zealand, giant squid (*Architeuthis*) are sometimes caught in nets used to catch orange roughys at depths of 2,000 feet or more.

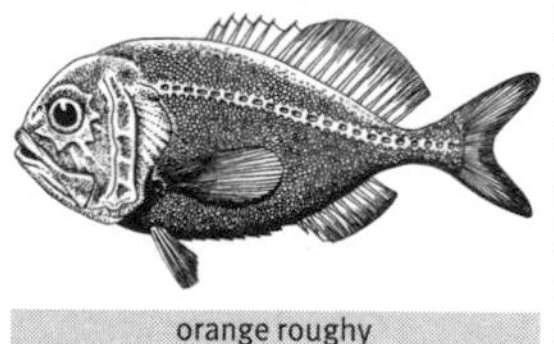
orange roughy

orca: See killer whale

oreo (*Pseudocyttus maculatus*) A flattened, round-bodied, large-eyed fish that is found off the continental shelves of Australia, South Africa, South America, and New Zealand. Like the orange roughy, the oreo was not fished commercially until the 1970s, when large concentrations were discovered in New Zealand waters at depths ranging from 2,500 to 4,000 feet. They are long-lived fishes, maturing at about thirty years of age and living to eighty. Now oreos are the object of a concentrated fishery, with up to 25,000 tons per year being brought up. The unusual name has nothing to do with the cookie; *oreos* in Greek means "hill," and it refers to the warts or bumps on the side of the body of the juveniles. The scientific name of the family is Oreosomatidae ("lumpy-bodied"), and they are related to the fishes known as dories. **See also orange roughy**

Orkney Islands A group separated from the northeast tip of mainland Scotland by the Pentland Firth, the Orkneys consist of some seventy islands, twenty of which are inhabited. The largest is Mainland (also called Pomona), which is divided into East and West Mainland. The main towns are Stromness, reachable by ferry from mainland Scotland, and Kirkwall. Known as the Orcades in ancient times, the islands were occupied by Norse raiders in the eighth century and ruled by Norway and Denmark until 1472, when, along with the Shetlands, they fell under Scottish rule. Although there are only 20,000 Orcadian residents, almost a quarter of a million visitors come to the islands every year for fishing, bird-watching, and exploring archeological sites. The economy of the Orkneys has benefited from the construction of landfall terminals to service the offshore oil fields in Scapa Flow, a sheltered natural harbor that also served as an anchorage for the British North Sea fleet during World War I. When the captured German fleet was interned there in 1918, the ships were scuttled by their German crews.

Ortelius, Abraham (1527–1598) Belgian cartographer; he and Gerardus Mercator were considered the greatest geographers of their age. Ortelius traveled widely, often with his friend Mercator, and in 1570, he produced a seventy-map atlas called *Theatrum Orbis Terrarum* (*Theater of the World*), which was published in Antwerp. Now regarded as more of a publisher than an inspired cartographer, Ortelius incorporated the work of many contemporaneous mapmakers, and was careful to give them credit. The 1587 edition of *Theatrum Orbis Terrarum* contained 103 maps, and was the first modern atlas of the world, even though it showed a huge—and largely nonexistent—southern continent, known as Terra Australis Incognita.

See also Mercator

osprey (*Pandion haliaetus*) Commonly known as the fish hawk, the osprey is one of the most widely distributed birds on earth. It is found around seacoasts, lakes, rivers, and reservoirs throughout all the temperate regions of the world except for South America. It is an easily identifiable bird, with its white underparts, barred tail, and 5-foot wingspan. Unlike the flat, straight wings of the bald eagle—which often steals fish in midair from the osprey—the wings of the osprey are held slightly crooked as it soars or hovers. Ospreys fly over water looking for fish, which they swoop to catch with their talons, holding on with rough protuberances on the underside of their feet. Unless they are bringing the fish back to their nests, which are usually built in tall trees (or on telephone poles in more developed areas), they usually take the fish to a tree to eat it. Like many other birds of prey, ospreys were affected by DDT residues in their food, and during the 1950s, the populations in North America and the British Isles were gravely threatened. They have now made a remarkable comeback, and their huge stick nests are welcome sights in many parts of the world.

osprey

otter trawl The standard bottom trawl, consisting of a large, bag-shaped net that is towed through the water, held open by various ropes, weights, and floats in conjunction with the angled otter boards (also known as doors), which draw apart as the net is pulled over the seabed. Its depth and distance from the bottom can be controlled by floats. The narrow, tapered end of the net, into which the trapped fish are gathered, is known as the cod end. **See also trawling**

outrigger (1) A counterbalancing log of wood rigged from the side of a canoe by native peoples of the Pacific and Indian Oceans; later incorporated into fiberglass sailing boats of the latter twentieth century.

outrigger (2) An extension to each side of the crosstrees of a sailing vessel to spread the backstays.

outrigger (3) A popular private club on Waikiki Beach in Honolulu, specializing in outrigger canoe paddling.

oyster (family Ostreidae) The most famous of bivalves, which have been harvested for food and pearls for millennia. The European oyster (*Ostrea edulis*) has been cultivated for more than two thousand years. The shell is variable in shape, massive, wrinkled, and rough, and reaches a maximum length of 12 inches in some species. The inner surfaces are smooth and white, and lined with nacre, also known as mother-of-pearl. Females produce upwards of 50 million eggs, most of which are eaten, but some of which reach the free-swimming "veliger" stage, after which they attach themselves to rocks, pilings, or other hard surfaces. Oysters are eaten by any number of sea creatures, including the oyster drill (*Urosalpinx cinenea*), starfishes, flatworms, crabs, fishes, ducks and other seabirds, and of course, human beings who farm them in artificial beds on both coasts of the United States, Europe, Australia, New Zealand,

and Japan. Although all oysters can produce pearls, the pearl oyster (genus *Pinctata*) is harvested in the South Pacific. (These are the oysters that are seeded with grains of sand or small mother-of-pearl beads in Japan to produce cultured pearls.) The best natural pearls occur in the Persian Gulf species *Meleagrina vulgaris*. The common edible oyster of eastern North America is *Crassostrea virginica*, and *C. commercialis* is the most important edible Australian species. **See also pearl**

oystercatcher (*Haematopus palliatus*) An unmistakable black-and-white bird with a long, coral-red bill and pale legs, the oystercatcher is named for its habit of chipping various mollusks off the rocks. It can pry open clams, mussels, and oysters with its powerful bill, but it also eats crabs, worms, and other invertebrates that it digs out of the sand. There are six recognized species of oystercatchers found along beaches and rocky coasts throughout the world. Most species exhibit variations of the black-and-white pattern and have white wing patches that are conspicuous in flight, but the black oystercatcher (*H. bachmani*) is all black. Their call is a loud piercing whistle.

oystercatcher

See also plovers, sandpiper

P

Pacific cod (*Gadus macrocephalus*) A somewhat smaller relative of the Atlantic cod, *G. macrocephalus* (*macrocephalus* means "big head") is found along the west coast of North America from northwestern Alaska to Southern California and in the western Pacific from the Chukchi Sea to China. It is a popular food fish, and millions of pounds are caught annually off British Columbia. **See also codfish**

Pacific giant squid (*Moroteuthis robustus*) Although not as large as *Architeuthis,* the Pacific giant squid is a very large squid indeed. At a known maximum length of 19 feet, it is one of the biggest of all squids. It is known mostly from specimens stranded or accidentally caught in fishermen's nets on the Pacific coast of North America, from Southern California to Alaska, and also from corresponding locations in the western North Pacific, from the Aleutians and Siberia to Japan. There are other species of *Moroteuthis* found in Antarctic, South African, and western Australian waters, where they are a preferred food item of sperm whales—as they are in the North Pacific. The tentacular clubs are equipped with two rows of hooks, which might explain the long scars and scratches on sperm whales.
See also giant squid

Pacific Ocean The largest single feature on the planet, covering 64 million square miles (almost a third of the entire surface of the earth) and averaging 14,000 feet in depth. The volume of water in the Pacific has been estimated at 173,625,000 cubic miles—more than twice that of the Atlantic. After weeks of sailing in the stormy Atlantic, Ferdinand Magellan named it *El Pacifico,* "the peaceful one," when he rounded Cape Horn in 1520. The Pacific extends from the Arctic to the Antarctic, and its greatest east-west distance is 11,000 miles, from the coast of Chile to the Malay Peninsula. Surrounded by North and South America on the east, and Asia and Australia on the west, the Pacific contains approximately twenty thousand islands, ranging in size from New Guinea and Borneo (after Greenland, the second and third largest islands in the world) to tiny, uninhabited coral atolls. Most of the islands are concentrated in the south and west, but there are some anomalous groups in the eastern or central Pacific, such as the Galápagos, Easter Island, and the Hawaiian Islands. The Pacific is ringed by deep trenches, one of which, the Mariana Trench, includes the Challenger Deep, at 35,800 feet, the deepest point in all the oceans. Around the perimeter of the Pacific is the "Ring of Fire," where tectonic plate movement creates the instability of such

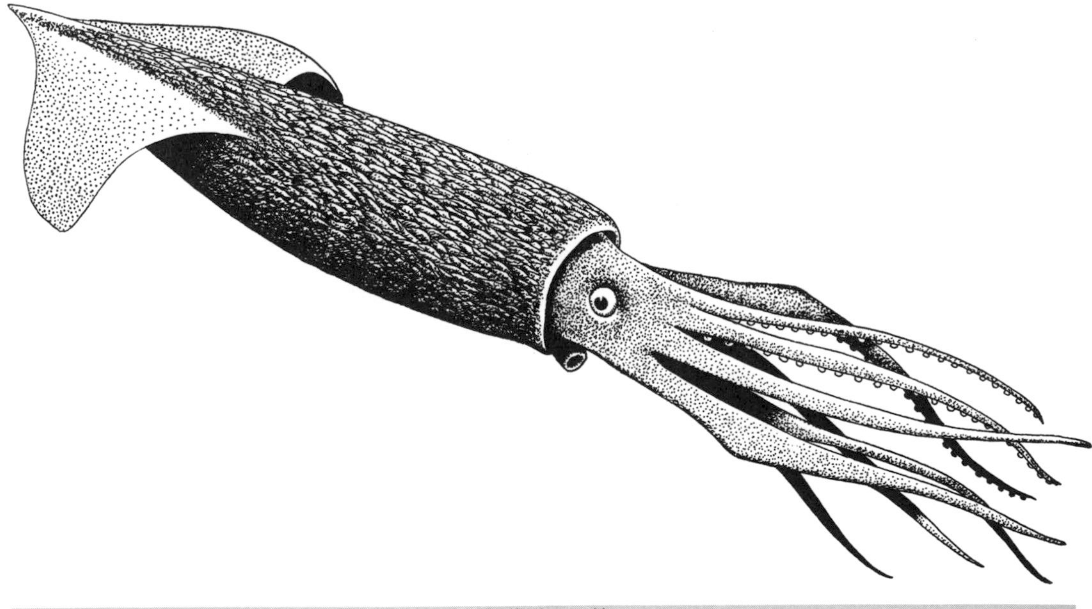

Pacific giant squid

intensely volcanic areas as the Aleutians, the Andes, Kamchatka, Japan, Indonesia, and the Philippines. Many Pacific islands were occupied by Melanesians, Micronesians, and Polynesians before such explorers as Magellan, Drake, Dampier, Tasman, Cook, Bering, and Vancouver "discovered" them and led to their annexation and settlement by various European nations. Sealers and whalers plied the Pacific in the eighteenth and nineteenth centuries, and commercial fishers have been catching salmon, halibut, herring, sardines, tuna, squid, and other edible species since they were first discovered. **See also *various islands;* Atlantic Ocean, Bering, Cook (James), Dampier, Drake, Indian Ocean, Magellan, Mariana Trench, Ring of Fire, Tasman, Vancouver**

Pacific salmon (Genus *Oncorhynchus*) There are six species of salmon from the Pacific: chinook, coho, sockeye, pink, chum, and cherry. All occur on both sides of the ocean, except the cherry salmon, which is found only off Japan. Georg Steller, Vitus Bering's zoologist on the Great Northern Expedition of 1740–1742, was one of the first to identify the different species. All are popular game fishes, and all are commercially fished in great quantities. Like their Atlantic counterparts, Pacific salmon are anadromous, breeding in freshwater, maturing at sea, and then returning to the place they were born to spawn. Pacific salmon species spawn only once and then deteriorate and die. In the process of ascending their natal streams, they undergo drastic physical modifications: their jaws extend into hooks, their backs become humped, and their flesh and internal organs turn to mush. Even though they are becoming weaker and weaker, they fight their way upcurrent, often leaping up waterfalls and transversing rapids to get where they are going. When females arrive at their nesting sites, they dig a hole in the gravel with their tails and lay their eggs, which are then fertilized by the males. After a period of time ranging from a few days to several years, depending upon the species, they swim to the sea. In California, Oregon, Washington, and Idaho, the long history of overfishing, the damming of the rivers up which the salmon must swim, and the destruction of the breeding streams by timber interests has driven many populations of Pacific salmon to the edge of extinction, and those that remain are considered endangered. Only the salmon that breed in Alaska are considered healthy.
See also Atlantic salmon, chinook salmon, chum salmon, coho salmon, pink salmon, sockeye salmon

Pacific white-sided dolphin (*Lagenorhynchus obliquidens*) Lacking a common name, dolphins of the genus *Lagenorhynchus* are popularly known as lags. This 7-foot member of the genus is found in the North Pacific, from Baja California north to Alaska, and east to the waters of Japan, China, and Korea. At sea, they can be easily identified by the sharply hooked, bicolored dorsal fin. They are often seen in huge schools and are famous for their bow-riding and leaping abilities. *L. obliquidens* does well in captivity, and there are few oceanariums on the west coast of the United States or in Japan that do not exhibit these exuberant little dolphins. **See also Atlantic white-sided dolphin, dusky dolphin, hourglass dolphin**

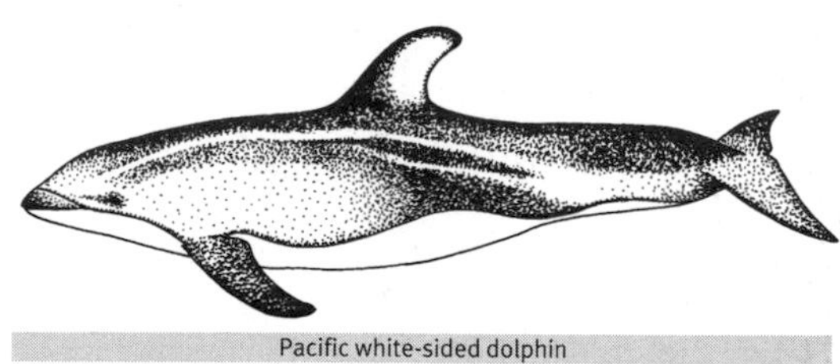
Pacific white-sided dolphin

Padre Island Barrier island that stretches for 113 miles along the Texas coast of the Gulf of Mexico. Often less than 3 miles wide, the island is characterized by large sand dunes, sparse vegetation, and strong winds off the gulf. The shallow Laguna Madre, sometimes 10 miles wide, separates the island from the mainland. There is a causeway to the island at the northern end at Corpus Christi, and another at the southern end. The island was explored and charted by Alfonso Alvarado de Pineda in 1519 and became known as a ship's graveyard after a 1533 hurricane sank a Spanish treasure fleet near the island. It is named for Padre Nicholas Balli, who started the Santa Cruz ranch around 1800. The central portion of the island has been designated Padre Island National Seashore; other islands designated as national parks are Assateague (Maryland) and Fire Island (New York). Hundreds of bird species live on or visit Padre Island, and there are also coyotes, gophers, and jackrabbits. The southern portion, known as South Padre Island, has been developed as one of the most popular tourist destinations in Texas, featuring high-rise hotels, condominiums, restaurants, and every conceivable water activity, including swimming, surfing, sailboarding, parasailing, snorkeling, scuba diving, sport fishing, and dolphin watching. South Padre Island is also the destination for thousands of college students on their spring break.

Palau A group of some 340 islands in the Caroline chain of the western Pacific, the Republic of Palau (sometimes called Pelew or Belau) includes the volcanic islands of Koror (the capital), Babelthuap, and Arakabesan; the coral islands of Angaur, Urukthapel, and Peleliu (scene of some of the fiercest fighting of World War II in the Pacific); and Kayangel Atoll. After its liberation from Japanese occupation, Palau became part of the Trust Territory of the Pacific Islands, administered by the United States under a 1947 mandate from the United Nations. In 1986, the governments of Palau and the United States agreed on the terms of a Compact of Free Association, similar to those reached with other Micronesian Trust members, which allows for virtual independence under a U.S. defense umbrella, but the Palau Compact remained unsigned because a clause forbids the presence of any nuclear weapons on the islands, including visits by ships equipped to carry them. This is unacceptable to the United States—in Palau and anywhere else—which therefore refuses to sign the compact until the clause is rescinded.

See also Caroline Islands

Palawan The second largest province in the Philippines, consisting of 1,768 islands, the largest of which, at 4,500 square miles, is Palawan itself. It is a long, thin island, 270 miles long, and only 24 miles at its widest, oriented on a northeast-southwest axis. If the Philippines is shaped like the Greek letter π then Palawan is the left descender. A chain of mountain ranges runs the entire length of the island, dividing it into two distinct climatic zones. With a total population of 240,000, it is the least densely populated of all the major Philippine islands. There are remnants of a land bridge that once connected Palawan with Borneo; the flora and fauna more closely resemble those of that big island to the southwest than they do the rest of the Philippines.

See also Borneo, Philippines

Palmer, Nathaniel (1799–1877) Sealing captain from Stonington, Connecticut; he went to sea as a youth, first sailed to the South Shetlands in the *Hersilia* in 1819, then returned the following year as part of a large Stonington fleet. Scouting to the south in the 45-ton *Hero*, Palmer sighted land in November 1820 and claimed to have discovered the Antarctic continent, but his claim was preempted by Thaddeus Fabian von Bellingshausen, who had sighted the mainland in January of that year. In 1821, however, he and British sealer George Powell discovered the South Orkneys. Palmer Land, on the Antarctic Peninsula, is named for him, as is Palmer Station, the large American facility on Anvers Island.

See also Antarctic Peninsula, Bellingshausen

Palmyra Atoll A privately owned group of about fifty little islands in the North Pacific, about 1,000 miles southwest of Hawaii, with a combined area of 4 square miles. Administered by the U.S. Department of the Interior, the islands have no permanent inhabitants. Palmyra was first sighted in 1802 by the American ship for which they are named, annexed by the Kingdom of Hawaii in 1862, and then by Britain in 1889. The U.S. government attached the islands to Hawaii and used them for a landing strip during World War II. When Hawaii became a state in 1959, Palmyra was not included.

See also Hawaiian Islands

palolo worm (*Palolo viridis*) Polychaete worm that lives in holes in reefs in the South Pacific and behaves like any other mobile segmented worm, except at the third quarter of the moon in October and November, when the mature worms produce sacs of eggs or sperm (gametes) in the middle and posterior segments, which break off and rise to the surface to release their contents. The tail portion resembles a complete worm and even has eyespots. In Samoa, at the "night of the big gathering," the unruptured tails are gathered in nets; the egg-filled tails of the females are considered the tastiest. The swarming period lasts only a day or two, and the photo-negative anterior portion remains in the coral and regenerates a new posterior, which will swarm at the surface the following year. Palolos are the largest of the polychaetes, with up to a thousand segments; they can reach a maximum length of 10 feet, but most are smaller. There are related species in the Mediterranean and the Caribbean, but their swarming is not as synchronous.

See also featherduster worm, polychaetes, sea mouse

Panama Canal Man-made waterway across the Isthmus of Panama that connects the Caribbean to the Pacific. The idea was broached as early as 1524, when Charles V of Spain ordered a survey to determine if such a waterway could be built. During the U.S. gold rush, anyone who wanted to get by ship from the east coast of the United States to the west had to sail around Cape Horn, a dangerous and time-consuming voyage. Travelers sometimes sailed to Panama and crossed the isthmus on foot, with wagons carrying their goods (to be picked up by another ship on the other side), but this too was dangerous, as the Panamanian jungle was hot and uncomfortable, and yellow fever was rampant. Between 1848 and 1855 the Trans-Panama Railroad was built with U.S. money, and the United States and Britain signed the Clayton-Bulwer treaty, in which they agreed that neither country would acquire exclusive rights to restrict international shipping. In 1878, however, Colombia granted the French a concession to build a sea-level canal under the direction of Ferdinand de Lesseps (builder of the Suez Canal), but poor planning, inadequate financing, and the ravages of yellow fever, malaria, and cholera drove de Lesseps's company into bankruptcy. After ten years and twenty thousand deaths, the French quit. President Theodore Roosevelt orchestrated a plan whereby Panama declared its independence from Colombia and the United States sent warships to lend support. A treaty was signed between the United States and the new country of Panama, and in 1904, work on the canal began. The first three years were spent in the development of construction facilities, surveys, and disease control. American railroad engineer John F. Stevens worked a system of locks into the plans, and introduced trains to carry the vast amount of dirt from the excavation. After spending seven more years and $336 million, the United States opened the Panama Canal on August 15, 1914. After eighty-six years of U.S. operation, in the year 2000 the canal was turned over to Panama. **See also Lesseps**

paper nautilus: See argonaut

Papua New Guinea: See New Guinea

Paracel Islands Strategically important uninhabited islands in the South China Sea, east of Vietnam and southeast of China's Hainan Island. The islands, divided into the Amphitrite and Crescent groups, have no freshwater, and only turtles and seabirds live there. In 1932, they were claimed by French Indochina, and a weather station was built. Japan occupied the Paracels during World War II, but withdrew and renounced its claims. In 1947, Chinese troops occupied Woody Island, but after 1949, both Communist and Nationalist China claimed the islands, as did both North and South Vietnam. The discovery of oil in the South China Sea encouraged South Vietnam to sign contracts with foreign oil companies, at which point China attacked and occupied the islands. The Spratly Islands, also disputed, lie southeast of the Paracels, between Vietnam and the Philippines. To the south of them oil has been discovered in the region of Natuna Island, leading to the signing of a $35 billion contract between the Exxon Corporation and Indonesia. **See also Hainan, Spratly Islands**

paralytic shellfish poisoning During periods of planktonic blooms (times of high concentrations of microscopic organisms in the water) dinoflagellates multiply in large numbers. These often toxic dinoflagellates may be ingested by shellfish; the poisons then accumulate in their digestive glands. Animals and humans may in turn be poisoned by eating poisoned shellfish. Certain species of dinoflagellates are capable of producing some of the most toxic substances known. The two species of dinoflagellates most commonly involved in human intoxications have been *Gymnodinium catenatum,* along the Pacific coast of North America, and *Alexandriam tamarense,* along the eastern coast of North America. Intoxications from these organisms are known as paralytic shellfish poisoning, or PSP. The symptoms, which begin with a tingling or burning sensation, followed by numbness of the lips, gums, tongue, and face, gradually spread. Other symptoms include gastrointestinal upset, weakness, joint aches, and muscular paralysis; death may result. There is no specific treatment or antidote. The poison, variously called paralytic shellfish poison, mussel poison, and saxitoxin, is a complex nonprotein nitrogen-containing compound. Paralytic shellfish poisoning is best avoided by following local public-health quarantine regulations. **See also dinoflagellates, red tide**

parchment worm (*Chaeopterus variopedatus*) A flabby, luminescent, segmented (polychaete) worm that is divided into three regions, with three paddles on the upper side of the middle region. It builds a parchment-like, U-shaped tube in the mud, with a chimney at each end that projects above the ground. The tube is narrower than its occupant at the ends, and the worm, which reaches a length of 10 inches, cannot get out. The paddles keep the water moving through the tube, where the worm produces a mucous net that traps microscopic planktonic organisms. Periodically, the pumping is halted and the net detached and rolled up into a ball, then sent back to the mouth, which is in a groove on the back, and swallowed. Why a worm that spends its entire life in a tube would need to luminesce has puzzled biologists for years, but it appears that if threatened, the worm can release a discharge of luminescent mucous to distract the predator, and then move to the other end of the tube. Parchment worms are found on both coasts of the United States.
See also featherduster worm, palolo worm, polychaetes, tube worms

Parker, Richard (1767–1797) British midshipman and mutineer. After fighting in the American War of Independence (1775–1782), he returned to Britain and tried his hand at making golf balls in Scotland. At the outbreak of the French Revolution (1793–1801) he was back in the navy, but in trouble again; he was court-martialed and discharged. From debtor's prison in Edinburgh he volunteered again for the navy, and in March 1797, he was aboard HMS *Sandwich* in the British Channel port of The Nore. When seamen at Spithead mutinied to protest the pay and conditions of the Royal Navy, Parker led a simultaneous and sympathetic mutiny, but when the Admiralty acceded to all the demands of the Spithead mutineers, Parker and his men obstinately held out. When they were captured, Parker and twenty-four seamen were hanged from the yardarm. **See also mutiny**

parrot fish (family Scaridae) Herbivorous fishes that abound in reef areas of shallow tropical seas around the world. Although many of them are as colorful as their namesakes, their common name comes from the "beak," which is actually their teeth fused together. They eat algae and coral, often scraping into the lime-

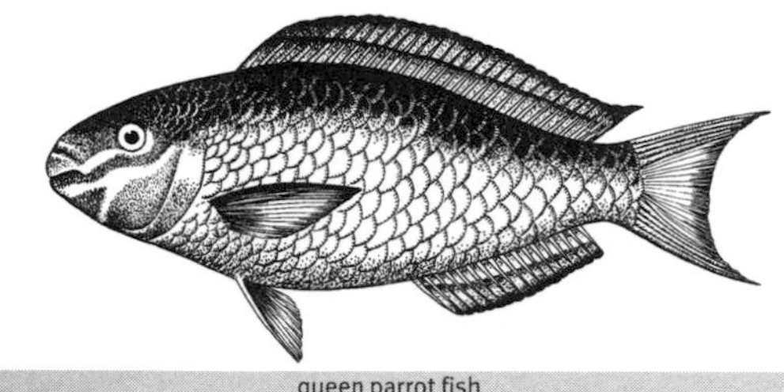
queen parrot fish

stone with their beaks. They grind the limestone into fine sand and are important factors in the production of sediment in reef areas. Like the wrasses, from which they are believed to have evolved, juveniles are colored very differently from adults—they are so different, in fact, that parrot fishes in the "initial" color phase were often regarded as different species from those in the "terminal phase." To further confuse ichthyologists, most parrot fishes undergo a sex change as they mature; drab females become brightly colored males. At night, many species of parrot fishes secrete a transparent veil-like cocoon; it only becomes visible when particulate matter is trapped when it is formed. Shown here is the queen parrot fish (*Scarus vetula*) of the western Atlantic. **See also blue parrot fish, queen parrot fish**

Parry, William Edward (1790–1855) British rear admiral and Arctic explorer, who was hydrographer of the navy from 1823 to 1829. In 1810, he was assigned to the *Alexandria* to protect British interests in the Spitsbergen whale fishery, and in 1818 he served in the Arctic under John Ross on HMS *Alexander.* The following year, in command of the *Hecla* (with Matthew Liddon in the *Griper*), Parry set out to search for the Northwest Passage. After successfully navigating Lancaster Sound and unable to go any farther, Parry organized the first Arctic wintering. He covered the decks with canvas tents, kept a fire going constantly, and banked snow against the ships for insulation. When the ice began to break up in the summer of 1820, he sailed for Melville Island, and although the men could see the Beaufort Sea, the attainment of which would have meant a successful traverse of the Northwest Passage, they could not get through the ice-choked McClure Strait, and they returned to England. Parry led two more expeditions to the Arctic, the last in 1824–1825 in the *Hecla* and the *Fury,* where they became trapped in the ice and the *Fury* was sunk. (Francis Crozier, who was to accompany Sir John Franklin on his disastrous expedition in 1845, accompanied Parry on all three Arctic voyages between 1821 and 1825.) **See also Crozier, Franklin, Northwest Passage**

Patagonian toothfish (*Dissostichus eleginoides*) Found in subantarctic waters north of 55° south, the Patagonian toothfish reaches a length of 6 feet. (A similar species, *D. mawsoni,* known as the Antarctic toothfish, is found only south of the Antarctic Convergence.) Previously known mostly from the stomachs of sperm whales, it has recently become the most sought-after of all Antarctic fish species, and because Antarctic waters are within no nation's exclusive economic zone (EEZ), any nation can fish there. Marketed as Chilean sea bass, black hake, or mero, the toothfish is the target of an intense, uncontrolled international fishery, in which boats from Norway, Argentina, South Africa, and other nations took some 100,000 tons in 1997, more than ten times the legal allotment set by the Convention on the Conservation of Antarctic Marine Living Resources (CCAMLR). The illegal fishing, which only began around 1993, is so intense that the entire population is now threatened with commercial and biological extinction. Toothfishes are caught mostly on longlines (there is also some trawl fishing), which also killed more than 140,000 albatrosses and petrels during the 1996–1997 season. **See also Antarctic Convergence, CCAMLR, longline fishing**

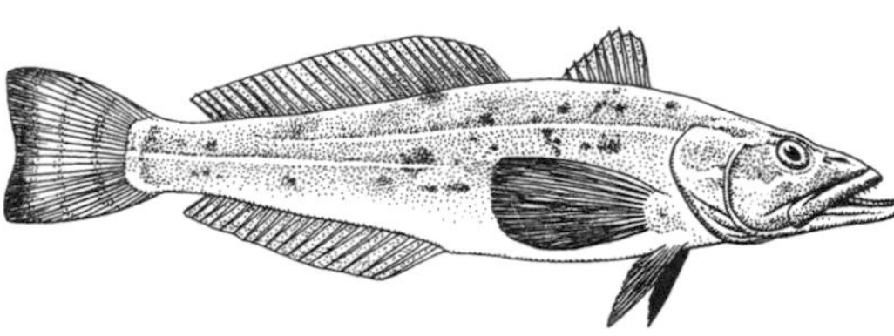
Patagonian toothfish

Payne, Roger (b. 1935) One of the pioneers in twentieth-century whale research. With his wife, Katharine, Payne was one of the first to record and study the songs of the humpback whales, first in Bermuda and then in Hawaii, eventually producing the best-selling record. He also studied the populations of right whales at Peninsula Valdés in southern Argentina, initiating one of the longest-running studies in cetology. He has been affiliated with the World Wildlife Fund, Rockefeller University, and the New York Zoological Society. He is president of the Whale Conservation Institute in Lincoln, Massachusetts, and a scientific adviser to the International Whaling Commission. In addition to many scientific papers and popular articles in such magazines as *National Geographic* and *Animal Kingdom,* he is the editor of *Communication and Behavior of Whales* (1983) and the author of *Among the Whales* (1995).

pea crab (genus *Pinnotheres*) Tiny crabs with pea-shaped bodies that live commensally in bivalve mollusks such as oysters, scallops, clams, and mussels. The oyster crab (*P. ostreum*) is found in the oysters of North America, particularly in Chesapeake Bay. Females, which grow larger than males, are about ½ inch long. **See also crab, oyster**

Peale's dolphin

Peale's dolphin (*Lagenorhynchus australis*) Also known as the blackchin dolphin, this little-known "lag" is found only in the coastal waters of southern South America, from Chile around the Horn to southern Argentina and the Falklands. It is the only lag with an all-black face, and would be fairly easy to identify if the viewer were in the remote locations where this animal lives. Peale's dolphin is named for Titian Ramsey Peale, naturalist aboard the U.S. Exploring Expedition of 1838–1842, who described the first specimen from a sighting off the coast of Patagonia.

See also dusky dolphin, hourglass dolphin

pearl A concretion formed by a mollusk consisting of the same material (nacre, or mother-of-pearl) as the mollusk's shell. Although almost any mollusk can produce a pearl, the best-known ones come from oysters. If and when a foreign substance—such as a grain of sand—gets into its mantle, the animal begins to secrete layers of pearl around it. Pearls can be irregular in shape, or attached to the inner shell of the mollusk, but the most desirable ones are the large and round ones that come from the pearl oyster, *Pinctada martensii.* Pearls have been admired for centuries for their luster and symbolic perfection, and have been incorporated into the jewelry of many cultures. The largest known pearl is an irregularly shaped "baroque" that weighed ⅔ pound. Natural pearls, which can range in color from black to white, as well as cream, blue, yellow, lavender, green, and mauve, are collected in the Persian Gulf, India, and Sri Lanka; various islands in the Pacific and Southeast Asia; and the Caribbean and the Gulf of Mexico. Cultured pearls, which now predominate in the industry, are produced when a foreign object is intentionally introduced into a pearl oyster, in controlled circumstances. Mikimoto Kokichi, the developer of the cultured pearl industry in Japan, learned that a tiny mother-of-pearl bead inserted into the oyster makes for the best pearl. **See also oyster**

pearlfishes (family Carapidae) Slender, knifelike fishes that enter into crevices tail first; those crevices are often in living animals, such as sea cucumbers, clams, oysters, and starfishes. When a pearlfish takes up residence in a sea cucumber—pearlfishes are selective about what species they choose—it sometimes feeds on the host's internal organs. Pearlfishes are also known as cucumber fishes for obvious reasons and as glass eels because they are transparent. They have no scales, they lack pelvic and anal fins, and one species has even lost its pectoral fins. The largest of the family (*Echiodon drummondi*), from British waters, can reach 1 foot in length, but most species are less than half that size. The common pearlfish of the Western Hemisphere is *Carapus bermudensis,* which is found from Bermuda south to Venezuela.

Pearl Harbor U.S. naval base on Oahu, Hawaii, that was attacked by Japanese planes without warning, and without a declaration of war, early Sunday morning, December 7, 1941. (On the same day, the Japanese launched surprise attacks on the Philippines, Malaya, Thailand, Guam, Wake Island, and Hong Kong.) Under Admiral Isoroku Yamamoto, a fleet of aircraft carriers, battleships, destroyers, and support ships had departed in small detachments from Japan, assembled in the Kuril Islands, and sailed for Hawaii. At around 8 A.M., forty torpedo bombers, forty-nine high-level bombers, fifty-one dive-bombers, and fifty-one fighters began a devastating attack on the military installations on Oahu, including Pearl Harbor and Schofield Barracks, the primary army installation. This "first strike" was intended to cripple the American navy, to make a complete Japanese takeover of the Pacific possible. Most of the U.S. Pacific Fleet was in harbor, except for the carriers, which were at sea on a training exercise, and the battleships *Oklahoma, California, West Virginia,* and *Arizona* were sunk at their berths. Five other battleships were badly damaged but remained afloat, and many other ships and harbor facilities were crippled. American casualties exceeded three thousand. President Franklin Roosevelt, upon hearing of the attack, called it "a day that will live in infamy" and declared war on Japan, touching off the Pacific phase of World War II. When the *Arizona* went down, she took 1,102 sailors with her. A memorial structure directly above the sunken hull and her crew commemorates the disaster.

Peary, Robert E. (1856–1920) Born in Pennsylvania and educated at Bowdoin College in Maine, Robert Edward Peary is probably the most controversial figure in all of polar exploration. He joined the U.S. Navy as an engineer in 1881, and in 1886 he crossed Greenland on foot, becoming the first to demonstrate that it was an island. Fully committed to Arctic exploration, he again sailed to Greenland on the research vessel *Roosevelt* in 1905. On foot, he and his party (which included his black servant, Matthew Henson) reached 87°6′ north, setting the record. In August 1908, the *Roosevelt* headed back for Greenland, with Peary determined to reach the North Pole. He set off from Ellsmere Island on March 1 with Henson, seventeen Eskimos, nineteen sledges, and

133 dogs. When he returned to the *Roosevelt* (on April 27), he claimed that he had reached the Pole, with Henson and four Eskimos, on April 6. Peary returned in triumph and was acclaimed as the greatest explorer of all time. However, in a letter to the New York *Herald,* Frederick Cook, a surgeon on Peary's earlier Greenland expedition, claimed that he had reached the Pole almost a year earlier, but had spent the year on Devon Island, not returning until April 15, 1909. Their conflicting claims ignited a battle that still rages, although recent examination of diaries, photographs, meteorological data, and other pertinent material suggest strongly that neither man actually reached the Pole and that both fabricated their accounts. Because Peary's claim was accepted for so long, nobody else tried to reach the Pole on foot until British explorer Wally Herbert did so on April 6, 1969—exactly sixty years after Peary's questionable accomplishment.

See also Cook (Frederick), North Pole

Peleliu One of the many islands of the Republic of Palau, Peleliu is located in the southern portion of the archipelago. A major Japanese stronghold during World War II, it was the scene of some of the most vicious fighting of the entire war. On September 15, 1944, U.S. Marines crossed a barrier reef onto Peleliu, where they expected a two- or three-day battle for the landing strip on the tiny island. But after a week, nearly four thousand marines were dead or wounded. The airfield was in American hands, but the Japanese had mined natural caves into large rooms and connected tunnels with hidden compartments. Fighting continued cave by cave. On November 25, Japanese colonel Kunio Nakagawa sent a radio message to his superiors—"All is over on Peleliu"—and killed himself with a jeweled dagger. American casualties neared 10,000, and virtually every Japanese soldier stationed on Peleliu died there—10,900 in all. **See also Palau**

pelican (*Pelecanus* spp.) Eight species of very large, big-billed birds, with a greatly extensible throat pouch. They feed by plunge-diving into the water from heights of up to 50 feet and scooping up fish. Heavy, slow fliers, they often appear in lines or V formations and sometimes soar to considerable heights. Flying consists of slow flaps, often followed by a long glide, sometimes very close to the surface of the water. Wing beats are sometimes synchronized. Pelicans breed in colonies, sometimes in freshwater lakes or estuaries, but the brown and Chilean pelicans are strictly marine. Most pelicans are white with darker wings or wing tips, but the brown pelican (*P. occidentalis*) is mostly brown. The eastern white pelican (*P. onocratalus*) and the Dalmation (*P. crispus*) are both found in European waters, and are distinguishable because the Dalmatian lacks the black borders on the hind portion of the underwings. A smaller version, seen throughout Asia, is the spot-billed pelican, *P. philippensis.* The pink-backed pelican, *P. rufescens,* is found in Africa, from Namibia around the Cape of Good Hope to the Red Sea; it is also a mostly white bird and is distinguished by a crest at the back of the head. The broad wings of the Australian pelican (*P. conspicillatus*) are boldly marked in black and white. The American white pelican (*P. erythrorhynchos*) is a large white bird with a pouch that can hold up to 4 gallons of water. Unlike the other pelicans, this species does not plunge-dive; instead, groups of them herd fish toward shore to scoop them up. The brown pelican is probably the most spectacular fisherman of all birds. From heights of up to 40 feet, it folds its wings and plummets into the water with a great splash. They are the most marine of all pelicans, found from Cape Hatteras to Brazil, and throughout the Gulf of Mexico and the Caribbean. **See also cormorant, frigate bird, gannet, tropic bird**

white pelican

Pelecaniformes Six families of water birds that differ greatly in looks but have certain anatomical characteristics that justify their inclusion in this order. The cormorants, gannets, tropic birds, frigate birds, pelicans, and anhingas (not seabirds, and therefore not included in this book) all have four webbed toes, with the hind toe pointed forward. Except for the tropic birds, they all have a naked throat (gular) region and nostrils that can be sealed as an adaptation for diving. They are mostly fish eaters, although tropic birds subsist mainly on squid. **See also cormorant, frigate bird, gannet, pelican, tropic bird**

Pelorus Jack A Risso's dolphin (*Grampus griseus*) that accompanied ships crossing Cook Strait, between New Zealand's North and South Islands, from the 1880s to 1912. He would ride the bow wave of steamers, leaping high out of the water as he did so. Such writers as Mark Twain and Rudyard Kipling wrote of Pelorus Jack, and his antics were so popular that when someone took a shot at him, the New Zealand parliament passed a law making it a crime to shoot at him—the first time a law was passed to protect an individual animal. (His name comes from Pelorus Sound, a part of Cook Strait, but a pelorus is also the removable ring around the face of a compass.) **See also Risso's dolphin**

Penang Malaysian island also known as Pulau Pinang, off the northwest coast of the Malayan Peninsula, separated from the mainland by a 2½-mile-wide strait that is spanned by the longest bridge in Asia. The island is 15 miles long and 10 miles wide, with a group of 2,000-foot-high hills in the center. It was discovered by the Chinese admiral Cheng Ho during his 1407–1409 explorations of Malacca. In 1786 Penang, virtually uninhabited, was given to the East India Company by the sultan of Kedah; Captain Francis Light founded a trading company that soon attracted settlers from China, India, Burma, and Sumatra. (Legend has it that in order to clear the forest, Light fired a cannonload of silver dollars into the underbrush and the locals cleared the bush in their frenzy to pick up the coins.) In 1800, Province Wellesley (now Seberang Jaya), the mainland strip across from the island, was added to the British company. In 1826, Penang—then known as Prince of Wales Island—was made a part of the British Straits Settlements along with Singapore and Malacca, and served as an entrepôt for tin and rubber. The island was included when the Malaysian Federation was formed in 1948. Penang Island has a population of approximately .5 million, and for its beaches, scenery, and accommodations is the most popular tourist destination in the country. Its capital, George Town, is also one of the major banking centers of Malaysia; it boasts a sixty-five-story office building. **See also Cheng Ho, Singapore**

penguin (family Spheniscidae) A group of eighteen species of flightless aquatic birds that walk upright. The name may come from the Celtic *pen* ("head") and *gwyn* ("white"), or from the Latin *pinguis,* which means "fat." In English, the name was first applied to the great auk of the North Atlantic, which is now extinct. All but the Galápagos penguin (which lives at the equator) are confined to the Southern Hemisphere, but only a few species actually inhabit the Antarctic area proper. All penguins breed on land and feed at sea, sometimes making long voyages in search of fish, squid, and crustaceans. Their flippers are actually reduced wings, which are adapted for "flying" under water. Overlapping, tightly packed feathers and a thick layer of fat insulate the birds against the cold. They can all walk upright, but they also move by crawling on their bellies or "tobogganing" down slopes of ice or snow. Some are capable of remarkable leaps out of the water. The basic color scheme of penguins is black (or gray) and white, but some species have surprising splashes of yellow or orange. Many penguin species are threatened by overfishing, oil spills, global warming, and tourists, who trample eggs and breeding sites in their attempts to get close to the birds.

See also Adélie penguin, African penguin, chinstrap penguin, emperor penguin, erect-crested penguin, fiordland penguin, Galápagos penguin, gentoo penguin, Humboldt penguin, king penguin, macaroni penguin, Magellanic penguin, rockhopper penguin, royal penguin, Snares penguin, yellow-eyed penguin

periwinkle (*Littorina* spp.) Half-inch-long gastropods with dark brown shells, found on both sides of the Atlantic in great profusion, attached to rocks, pilings, seaweed, marsh grasses, etc. The common periwinkle (*L. littorea*), introduced from Europe in the nineteenth century, has established itself as the most abundant snail in New England. Similar species are found in California, Florida, Alaska, and around the world in temperate waters. They are edible and are used in Western European and Far Eastern cookery. **See also gastropoda**

permit (*Trachinotus falcatus*) A fish that occurs in the western North Atlantic from Massachusetts to Brazil, but the largest concentration is in the waters of south Florida. Although the permit and the common pompano are quite similar in appearance, the permit is less deep in the body. Permits feed like bonefishes, by rooting around in the sand on shallow flats, looking for mollusks, crustaceans, and sea urchins. The permit, which can reach a weight of 53 pounds, makes excellent eating and is a highly desirable game fish. **See also jacks, pompano**

permit

Perón, François (1755–1810) French naturalist who sailed around the world from 1800 to 1804. After taking numerous temperature readings, he became convinced that water became colder with greater depth, and because the floor of the deep ocean must therefore be covered with ice, no life could exist there. Although few traces remain of French attempts to colonize Australia in the early years of the nineteenth century, there are two points in Western Australia (and one in Tasmania) named for François Perón and French naval officer Louis-Claude de Freycinet. Under the command of Nicolas Baudin, a French expedition was dispatched by Napoleon to Australia in 1802. Freycinet and Perón sailed aboard the *Casuarina* and collected 100,000 zoological samples and 2,500 botanical specimens. Perón wrote up the results of this prodigious collecting in *Voyage de découvertes aux Terres Australes,* which was completed by Freycinet in 1816, six years after Perón's death. The scientific name of the southern right whale dolphin is *Lissodelphis peronii,* bestowed by Etienne Lacépède for dolphins observed by Perón off Tasmania.

Perry, Matthew Calbraith (1794–1858) American naval officer and diplomat, whose gruff manner earned him the nickname of "Old Bruin." He is best known for his "opening" of Japan in 1853. He arrived in Sagami Bay with four ships, including the steam frigates *Mississippi* and *Susquehanna,* in June 1853, bearing a letter to the emperor from President Millard Fillmore. He refused to leave until the letter was acknowledged. When the Japanese ordered him to Nagasaki, the only port open to foreigners, he declined and instead waited for five weeks while a special building was erected near the village of Kurihama. Then Perry landed with some 250 marines to face a much larger force of Japanese bowmen, pikemen, and cavalry. He presented the letter to the shogun's envoys, and said he would be back in a year for an answer. However, he only went as far as the China coast and returned in February 1854, anchoring his squadron off Yokohama. In March, the Treaty of Kanagawa was signed between the Americans and the Tokugawa Shogunate, opening certain ports to U.S. ships, providing for the reciprocal return of castaways, and giving permission for the establishment of an American consulate. Although this event marked the breakdown of the wall that had surrounded Japan for three centuries, it was executed particularly in the interests of American commercial ventures, such as whaling, which needed to provision in the western Pacific, and access to the rich sperm whaling region known as the "Japan Grounds." Perry was awarded twenty thousand dollars by Congress, which also paid for the publication of his three-volume *Narrative of the Expedition of an American Squadron to the China Seas and Japan.*

Persian Gulf An almost landlocked arm of the Arabian Sea, covering some 90,000 square miles, and extending from the Shatt el Arab delta to the Strait of Hormuz, which opens into the Gulf of Oman. In the gulf there are many islands, of which Bahrain is the largest. Iran makes up the entire eastern border of the gulf, but on the west are the countries of Kuwait, Saudi Arabia, Bahrain, Qatar, the United Arab Emirates, and Oman. Because it is so critical to transportation, the gulf has traditionally been a source of contention between various powers. By 1835, various sheikdoms agreed to stop harassing British shipping, and in 1907, an international compact placed the gulf in the British sphere of influence. In the 1930s, when the region's vast oil reserves were discovered, international interest was revived. More than half of the world's known oil reserves are found here; only the recently discovered oil deposits in Azerbaijan rival the reserves of the Persian Gulf countries in magnitude. With the closing of the Suez Canal in 1967, the British withdrew from the area, and the United States and the U.S.S.R. attempted to fill the vacuum (the Suez Canal reopened in 1975). The war between Iran and Iraq raged from 1980 to 1988, and foreign tankers in the gulf were attacked by both combatants. When Iraq invaded Kuwait in January 1991, the United States and other countries sent troops, and the Persian Gulf War (January–February 1991) was the result. Iraq's forces were soundly defeated and withdrew, but as they retreated, they set fire to the Kuwaiti oil fields, causing the single greatest oil spill and fire in history. **See also Gulf War oil spill, Suez Canal**

Peruvian beaked whale

Peruvian beaked whale (*Mesoplodon peruvianus*) Between 1976 and 1989, ten specimens of an unknown beaked whale were captured in fishermen's nets or found stranded along the coast of Peru. Described for the first time in 1991, *M. peruvianus* is one of the most recently discovered cetaceans. The largest of the specimens was only 12 feet long, which suggests that it is among the smallest of the beaked whales.

See also Bahamonde's beaked whale, beaked whales, *Mesoplodon*

Pescadores Islands Known as P'eng-hu in Chinese, the Pescadores ("fishermen") were named for the fishermen that Portuguese navigators observed there in the sixteenth century. The archipelago consists of sixty or more small islands west of Taiwan in the P'eng-hu Channel. The islands were settled by the Chinese as early as the seventh century A.D., mostly for the establishment of fishing colonies. When Western powers began to have designs on Taiwan, they settled on the islands. The French arrived in 1884, but after the Sino-Japanese War of 1884–1885, when Taiwan was ceded to Japan, the Pescadores went with it. After World War II, the islands were returned to Taiwan, of which they are now a county. **See also Taiwan**

petrel A general term used for tube-nosed (procellariiform) seabirds, which range in size from the giant petrel (*Macronectes giganteus*) to the gadfly petrel, named for their erratic flight, and the tiny storm petrel, the size of a large swallow. (Albatrosses and shearwaters are also tubenoses.) They come in all possible combinations of black, white, and gray, from the all-white snow petrel (*Pagodroma nivea*) of the Antarctic, to the checkerboard Cape petrel (*Daption capensis*), to the all-dark, great-winged petrel (*Pterodroma macroptera*). There are also diving petrels, which obtain all their food by paddling or "flying" underwater. Petrels spend most of their lives at sea, coming ashore only to nest. Along with the albatrosses and shearwaters, petrels are tubenoses. Their nostrils extend on top of the bill through two horny tubes, associated with salt secretion. The name "petrel" is said to derive from the Italian *petrello,* for the birds' habit of walking on water like Saint Peter.

See also albatross, black-capped petrel, cahow, Cape petrel, diving petrel, gadfly petrel, giant petrel, shearwater, storm petrel

phalaropes (family Phalaropodidae) In almost all species of birds where there is a difference in plumage between males and females, the male is the more brightly colored, and in most cases, he engages in some sort of courtship behavior. The opposite is true for phalaropes: the females are showier than the males, and they do the courting. After mating, the male builds the nest, the female lays the eggs, and then the male incubates them. Phalaropes are the most aquatic of all shorebirds, and can swim as well as they can walk or fly. Found in temperate coastal areas around the world, they have feet that, like those of the much larger coots and grebes ("phalarope" derives from Greek words meaning "coot-footed"), have scalloped membranes that extend from the toes. Sitting on the water, phalaropes spin rapidly, creating a vortex that sucks tiny organisms up toward the surface, where the bird can consume them. There are three species: Wilson's phalarope (*Phalaropus tricolor*) is the largest and least aquatic; the red phalarope (*Phalaropus fulicaria*) is somewhat smaller, and its females are a rich chestnut color; and the red-necked or northern phalarope (*P. lobatus*) is the smallest and most abundant.

Wilson's phalarope

See also shorebirds

Philippines Republic in the southwest Pacific, consisting of more than seven thousand islands, four hundred of which are permanently inhabited. Only about 7 percent are larger than 1 square mile, and only a third have names. The main islands of Luzon (the largest), Mindanao, Samar, Negros, Palawan, Panay, Mindoro, Leyte, Cebu, Bohol, and Masbate occupy 95 percent of the total land area. The islands stretch for 1,100 miles from north to south, and are bounded by the South China Sea to the west, the Philippine Sea to the east, and the Sulu and Celebes Seas to the south. The population of the Philippines is about 75 million, and the population of Manila, the capital, is 11 million. The islands are volcanic in origin, and there are many active volcanoes, the most recent of which was Mount Pinatubo, which erupted catastrophically in 1991, killing hundreds of people, and forcing the abandonment of the U.S. Air Force base at Clark Field. Principal products of this largely tropical country are rice, corn, coconuts, sugarcane, bananas, pineapples, and one of the world's great stands of commercial timber. Magellan discovered the islands in 1521 and claimed them for Spain, which retained possession for the next 350 years. After Admiral George Dewey invaded Manila Bay during the Spanish-American War (1898), the islands were ceded to the United States. This caused the Filipinos to begin a guerrilla war against their new masters, which led to a promise of independence by 1946. But on December 8, 1941, the day after Pearl Harbor, the islands were invaded by Japanese troops, and fierce fighting occurred throughout the islands until the Japanese surrender in 1945. After the fall of Bataan and Corregidor, General Douglas MacArthur withdrew, leaving General Jonathan Wainwright in command. MacArthur returned in October 1944, and after the Japanese fleet was destroyed in the Battle of Leyte Gulf (October 23–26), called the greatest naval engagement in history, the U.S. forces liberated Manila in February 1945. In their defense of the Philippines, the Japanese lost 435,000. The first president of the Philippine Republic was Manuel Roxas y Acuña, elected in 1946. Ferdinand Marcos served from 1965 to 1986, when he was forced to flee because he was suspected of rigging the election and arranging for the 1983 assassination of his opponent, Benigno S. Aquino, Jr. Aquino's widow, Corazon, was declared the winner of the 1986 presidential election, but her administration was hampered by attempts to bring Marcos back, among other things. He died in exile in Hawaii in 1989. Fidel L. Ramos was elected president in 1992, defeating, among other opponents, Imelda Marcos, the ex-president's widow. In 1992, the United States turned over the naval base at Subic Bay, thus ending its military presence in the Philippines.

See also Leyte Gulf, Battle of

Phillip, Arthur (1738–1814) British vice admiral and the first governor of Australia. He served in the Mediterranean and West Indies, then in the Portuguese navy, after which he commanded the *Ariadne* and the *Europe* for Britain during the War of American Independence.

In 1786, he was given command of the First Fleet, eleven ships that sailed in 1787 to convey the first load of 750 convicts to occupy the penal colony in Australia. In the frigate *Sirius*, Phillip arrived in Botany Bay on January 18, 1788, but finding the location unsuitable, he moved on to Port Jackson (which would eventually become Sydney), and the remainder of the fleet followed. After much difficulty, the convicts and their guards settled in, and largely because of Phillip's firm hand, the colony began to prosper. (Subsequent governors, like William Bligh—of *Bounty* mutiny fame—would turn the penal colony into a horror, but Lachlan Macquarie, who served from 1810 to 1821, stressed more humane and sensible treatment for the convicts.) During an altercation between the settlers and the Aborigines, Phillip was speared in the shoulder. Failing health obliged him to return to England in 1792, but in the following years he saw further action and received several promotions. He was made admiral in 1814. Port Phillip Bay in Melbourne is named for him, as is Phillip Island, the site of a breeding colony of blue penguins.

See also Bligh, First Fleet

Phipps, Constantine John (1744–1792) British naval officer, sent in 1773 to search for the Northeast Passage to the Pacific. His two ships were HMS *Racehorse* and HMS *Carcass*, Horatio Nelson serving as midshipman in the latter. When their way was barred by ice north of Spitsbergen, the ships were forced to turn back. Phipps made the first deep-sea sounding. With a weighted line he measured the depth of the North Atlantic between Norway and Iceland, and his measurement of 4,098 feet stood as the depth record until James Clark Ross reached 14,500 feet in Antarctic waters in 1840. Phipps later commanded the 74-gun ship of the line *Courageaux* at the inconclusive battle of the Ushant in 1778.

See also Nelson, Northeast Passage, Ross (James Clark)

Phoenicians The Phoenicians lived on the eastern shore of the Mediterranean, in what is now Syria, Lebanon, and Israel, in a chain of cities rather than a unified nation. Although not now considered the first to sail into the open ocean (the Cycladic Greeks were believed to have reached Spain around 2500 B.C.), around 3000 B.C. they dominated the commercial trade of the Mediterranean, and their long and slender warships, equipped with a ram and sails, were used for exploration and to protect their commercial interests. Using the Pole Star (which the Greeks called the Phoenician Star) for navigation, they traded with Sardinia and Sicily, and by 800 B.C. they had ventured into the Atlantic as far as Tartessus, as the Spanish coast was then known. According to Herodotus, they circumnavigated Africa around 600 B.C., but this accomplishment has not been verified.

phosphorescence A name used for bioluminescence, the glowing condition of the sea—usually a greenish-yellow color—when the surface is broken by a moving vessel, a wave, or an animal like a fish, a squid, or a dolphin. It is believed to derive from minute dinoflagellates, such as *Noctiluca* and *Pyrodinium*, that are stimulated by an oxidation process when they are disturbed. Although the term "phosphorescence" is often used to describe this heatless light, it is more accurately applied to the combustion of the chemical element phosphorus—not found free in nature—which glows in the dark and takes fire spontaneously upon exposure to air. Bioluminescence is altogether different: the generation of light by living organisms, such as the great majority of deep-sea fishes, some sharks, and many squids, shrimps, euphausiids, and even some sea cucumbers and starfishes.

See also bioluminescence, dinoflagellates

photophore Light-producing cells found in bioluminescent fishes, squids, and many other invertebrates. Certain marine creatures are host to bioluminescent bacteria that provide light, but this is considered *extra*—rather than *intra*—cellular. The mechanism by which light is produced in living cells is chemical, involving the oxidation of luciferin and luciferase, which combine to form oxyluciferin, which fluoresces. Jellyfishes are known for their bioluminescence, as are certain worms, crustaceans, and echinoderms, but squids and fishes are the champion light producers of the sea. Many species of squid have lights in or on their bodies and arms, and one species (*Histioteuthis*) has a series of lights around one eye that can serve as a flashlight. *Heteroteuthis*, another small squid, can eject a cloud of luminescent ink, and the very large *Taningia* has the largest light organs of any animal: lemon-sized light organs on two of its arms that it can "strobe" to startle its prey. A large proportion of deep-sea fishes (and many deep-sea sharks) have light organs, and the various species of lantern fishes (*Myctophidae*) are differentiated by the pattern of photophores on their bodies. It is believed that these photophores provide counter-illumination, masking the shadow of the fish and rendering it difficult for predators to see. Many of the anglerfishes are equipped with bioluminescent lures, and some of the viperfishes have lures in addition to rows of photophores on their bodies.

See also anglerfishes, bioluminescence, lantern fishes, *Taningia danae*

Phuket Thailand's largest island, Phuket (pronounced "*pook*-et") covers an area of 210 square miles. It is 500 miles south of Bangkok in the Andaman Sea, on the west coast of the peninsula that Thailand shares with Myanmar and Malaysia. On early maps, the island is sometimes called Junk Ceylon or Cape Salang. It is joined to the mainland by a 730-yard-long bridge. The island is largely flat, but the landscape is occasionally broken by low mountains and valleys. It has a tropical climate with monsoons that blow from December to April. Phuket is known for its production of tin, half of which is mined offshore by dredges and pumps. It also produces rubber, coconuts, and pineapples. With the building of new hotels, restaurants, dive shops, and other tourist facilities in recent years, Phuket has become an important international resort destination. The population of the island is 199,000, mostly Thai Buddhists of Chinese descent, but there are also Thai Muslims, Sikhs, Christians, and a small group of *Chaolay*, or Sea Gypsies.

Piccard, Auguste (1884–1962) Belgian physicist who designed high-altitude and ocean-depth exploratory vehicles. In May 1931, he ascended in a hydrogen-filled balloon to a then-record altitude of 51,793 feet. Later he turned his attention to the depths and developed the bathyscaph, a 10-ton steel sphere that used as its buoyancy device a huge quantity of gasoline, which is lighter than water and, more important, compressible. Piccard tested his bathyscaph (named *FNRS-2*, for the Fonds National de la Recherche Scientifique, the Belgian agency that had funded his endeavors) off West Africa, reaching a depth of 4,554 feet. The first bathyscaph was followed by the *Trieste*, which, with 28,000 gallons of gasoline and Auguste Piccard and his son Jacques aboard, descended to 10,330 feet in the Tyrrhennian Sea in 1953. Unable to afford the expenses of testing and diving, Auguste Piccard sold the bathyscaph to the U.S. Office of Naval Research, which refitted it and took it to the South Pacific. On January 23, 1960, with Jacques Piccard and navy lieutenant Don Walsh aboard, the *Trieste* descended to the bottom of the Mariana Trench, at 35,800 feet the deepest point in the world's oceans. **See also Mariana Trench, *Trieste*, Walsh**

Pigafetta, Antonio (1491–1534) Venetian nobleman who accompanied Ferdinand Magellan on his 1519–1521 epochal voyage around the world. Magellan was killed in the Philippines, but Pigafetta actually completed the circumnavigation in the *Victoria*, which was commanded by Juan Sebastian del Cano. Of the original 270 seamen, only 17 made it back to Spain. One of them was Pigafetta, whose meticulous diary of the voyage is the primary source of our knowledge of Magellan's accomplishments, hardships, and death. It was transcribed for King Charles I as *The First Voyage Round the World by Magellan, Translated from Pigafetta*, and while originals have been lost, several contemporaneous copies have survived. **See also Magellan**

Pillars of Hercules To the north of the Strait of Gibraltar is the imposing headland of Gibraltar rock, and 10 miles to the south is the sentinel rock of Jabal Musa in Morocco. These two guardians of the Atlantic gateway to the Mediterranean have been known as the Pillars of Hercules since ancient times. According to classical mythology, they were erected by Hercules to celebrate the completion of his tenth labor, which was stealing the cattle of the three-bodied monster Geryon and bringing them back to Mycenae. To the Greeks, the twin rocks defined the limits of the inner sea, beyond which one dared not venture. In Plato's description of the lost island of Atlantis (in the dialogues *Timaeus* and *Critias*), he tells us that it was outside the Pillars of Hercules.

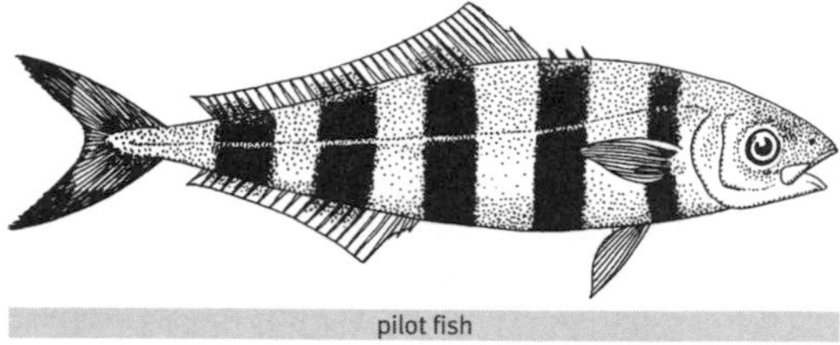

pilot fish

pilot fish (*Naucrates ductor*) The young of this small, boldly striped jack are often found sheltering under jellyfish, but the adults are often seen accompanying some of the larger species of sharks. (It was originally believed that they "piloted" the sharks to their prey, hence their common name.) These 2-foot-long fishes, found in all the world's oceans, also swim with turtles, mantas, tunas, and other large fishes. They feed on scraps from the meals of the larger animals.

See also jacks, remora

pilot whale

pilot whale (*Globicephala* spp.) There are two kinds of pilot whales, differentiated by the length of their pectoral fins. The fin of the longfin (*G. melaena*) is long and graceful, and may be as much as 27 percent of the animal's total length, while in the shortfin (*G. macrorhynchus*), it does not exceed 19 percent. The long-finned species is found predominantly in the North Atlantic, but also in the Southern Ocean. The shortfin is sometimes known as the Pacific pilot whale, although its range overlaps that of the longfin in many areas, including the North Atlantic. They are known as blackfish or potheads, the latter name referring to the bulbous forehead bulge of the males, which can reach a length of 20 feet, compared to a maximum of 17 for females. Pilot whales are gregarious animals and often aggregate in the hundreds. This gregariousness has its downside, however, since they are notorious group stranders—this propensity has been exploited by whalers, who used to drive them ashore en masse and slaughter all of them. (Pilot whales are still killed this way in the Faeroe Islands.) **See also false killer whale, pygmy killer whale**

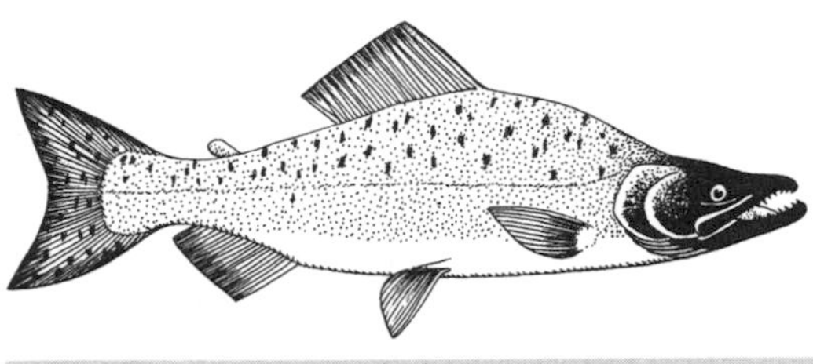

pink salmon

pink salmon (*Oncorhynchus gorbuscha*) Breeding pink salmon males develop an almost grotesque hump of cartilage behind the head, leading to their other name, "humpback salmon." Pinks are the smallest of the Pacific salmons; the word's record is a 13-pounder. They do not manifest the wanderlust of their relatives, often remaining just slightly upstream from the tidewater, instead of battling up hundreds—sometimes thousands—of miles against rapids and waterfalls like the sockeyes or the chinooks. (Chinooks cannot do it anymore; now their rivers are blocked by impassable dams.) They originally bred in streams and rivers of Japan, the Aleutians, and the Mackenzie River delta of northwestern Canada. They have now been introduced into eastern Canadian waters and from the Gulf of St. Lawrence into the Great Lakes. There are frequent reports of hybrid pink and chum salmon. **See also Atlantic salmon, chinook salmon, chum salmon, coho salmon, Pacific salmon, sockeye salmon**

pinnace A small vessel, usually no more than 20 tons and carrying two masts, both square-rigged, that came into use around the sixteenth century. They were often used to take messages between ships, and also to accompany larger ships on early voyages of exploration. The *Squirrel*, a vessel that sank on Sir Humphrey Gilbert's return voyage from Newfoundland in 1583, was a pinnace, and so was the *Charles*, in which Luke Foxe sailed to Frobisher Bay, in 1631. In later usage, the term denoted a ship's boat that was rowed with eight or sixteen oars, and might also be furnished with a mast, and even later, it was applied to ship's boats powered by a small outboard motor.

Pinzón, Martin Alonzo (1440–1493) Spanish navigator who, with his brothers Francisco and Vicente Yañez Pinzón, accompanied Christopher Columbus on his 1492 voyage. Martin Alonzo was master of the *Pinta*, on which Francisco was the pilot, and Vicente was master of the smallest of the three ships, the *Niña*. (Columbus commanded the flagship *Santa Maria*.) It was he who suggested the course change that brought them to the Bahamas on October 12. The *Pinta* left the other ships for six weeks to search for gold and spices, and Martin Alonzo was censured for disloyalty. When the *Santa Maria* was wrecked on Hispaniola, Vicente Pinzón took Columbus aboard the *Niña*. On the return voyage, the *Pinta* and the *Niña* again became separated, but Martin Alonzo reached Palos on the same day (March 15, 1493) as Columbus and Vicente. **See also Columbus**

Pinzón, Vicente Yañez (1460–1523) Younger brother of Martin Alonzo Pinzón; master of the *Niña* on Christopher Columbus's first voyage in 1492. In 1499 he sailed to the east coast of Brazil, and discovered the mouth of the Amazon River. He was named governor of Puerto Rico in 1505. From 1508 to 1509 he explored the Yucatán and Honduras; he may have been the first European to see those places. **See also Columbus**

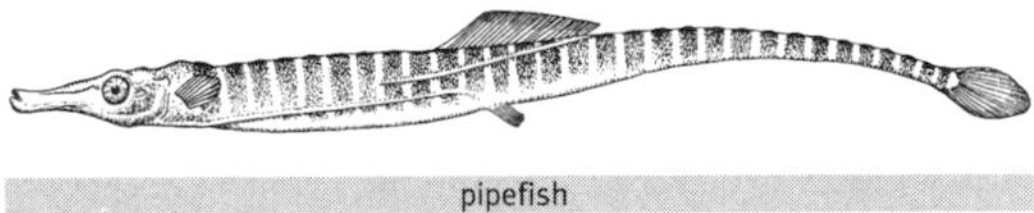

pipefish

pipefish (family Syngnathidae) If you took a sea horse and straightened it out, you would have a pipefish. There are many varieties, but all swim horizontally, while their cousins the sea horses swim head up and tail down. Indeed, with the exception of their swimming orientation and the elongated nature of some of the pipefishes, they are virtually the same. They are armored little fishes that can hang on to coral and plants with their tails; the males incubate the eggs in a brood pouch (although some species of pipefishes have an unprotected area called a brood "patch"); they have small fins that they use by fluttering and tiny, tubelike mouths. The largest of the pipefishes is *Syngnathus californiensis*, which reaches a length of 18 inches; the smallest does not exceed 2 inches in total length. There are more than two hundred species of sea horses and pipefishes, usually confined to shallow water, but some are oceanic, and there are even a few freshwater pipefishes. **See also sea horse**

piracy The act of taking a ship on the high seas from those lawfully entitled to it. Until the birth of regular navies, piracy was endemic among all seafaring nations. Traditionally, pirates overhauled their victims in small, fast ships and captured them by boarding. Piracy flourished in the Mediterranean, particularly along the

North African coast around Morocco, Tunis, and Algiers (the Barbary States), and in British waters until the navy of Henry VIII cleared the seas. The classic age of piracy, however, was the late seventeenth and early eighteenth centuries, when authorized privateers operating in the West Indies and the Indian Ocean often turned to outright piracy. Because piracy was considered an offense against the law of nations, any vessels were permitted to pursue a pirate ship, bring it into port, try the crew (regardless of their nationality), and confiscate the ship. When caught, pirates were often hung or executed in public places as a warning. There are still some areas of the world, such as the South China Sea and parts of Malaysia and Southeast Asia, where pirates board and take ships, often pleasure yachts. **See also Barbary pirates, Blackbeard, Bonnet, buccaneer, Kidd, Morgan, L'Ollonois, privateer, Roberts, women pirates**

Piraeus The port of Athens and the chief port of Greece. The port was fortified by Themistocles after the defeat of the Persians at Salamis in 480 B.C., and the "long walls" were constructed to link the port to Athens. The port itself has three harbors, which were able to dock hundreds of commercial and war galleys. When the Spartans defeated the Athenian navy in the Peloponessian Wars, around 403 B.C., the long walls were destroyed, and Piraeus lost most of its importance until modern times. It is now a shipbuilding and commercial center, the hub of the Aegean ferry network, and one of the main bases of the Greek navy. The current population is about 200,000.

pirates: See buccaneer, piracy, privateer, women pirates

pirate whaling Whaling that violates the restrictions of the International Whaling Commission. In 1978, a South African named Andrew Behr ran the *Tonna,* a converted Japanese trawler, until she was dragged down by one too many whales lashed alongside. Then came the *Sierra,* also owned and operated by Behr, which was rammed and sunk by Paul Watson's *Sea Shepherd* in the harbor at Oporto, Portugal, in 1979. The Japanese hunger for whale meat—legal or illegal—was responsible for many illicit whaling operations, mostly in South America. Aristotle Onassis ran a outlaw whaling company in the 1950s, until public outcry and the Peruvian navy shut him down. The most heinous of all outlaw whalers were the Soviets, who ran wild throughout the whaling grounds of the world during the 1960s, killing every whale they encountered and then filing completely false reports with the IWC, claiming to have conformed to the quotas they had been assigned. For example, when they said they had taken 2,710 humpbacks, they actually killed more than 48,000. They killed nursing mothers with calves, undersized whales of every species, and practically all the right whales of the Southern Ocean. More than any other factor, this egregious travesty of whaling regulations has been responsible for the precipitous decline in some of the world's most endangered whales. In June 1997, a dozen sperm whale carcasses were found floating west of the Azores, with orange radar reflector beacons nearby. So far, the culprits have not been identified. **See also whaling**

pirogue A seagoing, double-ended canoe formed out of the hollowed-out trunk of a tree (sometimes two, lashed together), commonly used in the Gulf of Mexico, the Caribbean, and the west coast of South America during the sixteenth and seventeenth centuries. It was often made of cedar or balsa wood. When travelers in the South Pacific encountered these log canoes, they called them piraguas, based on the South American term. In the waterways of Louisiana, the pirogue is a common flat-bottomed boat, used for fishing and duck hunting.

pistol shrimp (*Alpheus armillatus*) Also known as snapping shrimp, these small (1½-inch) shrimp have a hugely enlarged claw with which they can make a noise loud enough to stun their prey. If one of these animals is in an aquarium, the sound can easily be heard throughout a large room. During World War II, underwater listening devices were disrupted by the constant cacophony of these shrimp snapping their claws. **See also mantis shrimp, shrimp**

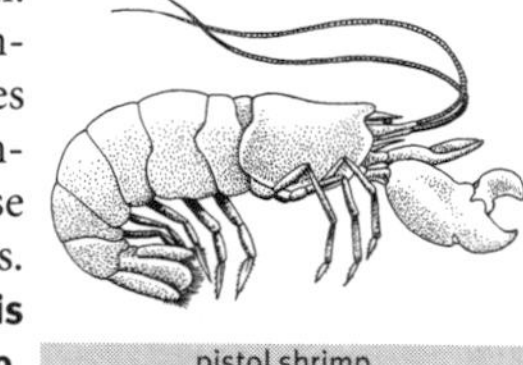
pistol shrimp

Pitcairn Island Discovered by British admiral Philip Carteret in 1767 and named for Robert Pitcairn, the midshipman who first sighted it, this lonely island in the South Pacific has a fascinating history. Its first colonists were the mutineers from the *Bounty,* their Tahitian wives, and several Tahitian men. In 1790, the mutineers, led by Fletcher Christian, put Captain William Bligh over in a small boat, sailed to uninhabited Pitcairn Island, and burned their ship. (The *Bounty*'s remains were discovered in 1957 off the southern end of the island.) The mutineers remained undetected on the island until 1808, when they were discovered by Captain Mayhew Folger in the American sealer *Topaz* and their story was brought back to England. Fletcher Christian, Jack Adams, Matt Quintal, and William McCoy died on Pitcairn, but their English-speaking descendants still live there. They were resettled by missionaries on Tahiti in 1831, but they decided to return to their island. In 1856, because Pitcairn was overcrowded and unable to provide enough food for the 193 inhabitants, they were removed to Norfolk Island, 1,000 miles off the east coast of Australia, but they were unhappy there too, and many of them moved back. The island, with a permanent population of less than 100, is now a British dependency, adminstered by the high commissioner in New Zealand. Of all the inhabited islands in the world, Pitcairn is the farthest from any other human settlements.

See also Bligh, *Bounty*, Christian

Pizarro, Francisco (1476–1541) Spanish conqueror of the Incas of Peru. In 1513, he accompanied Vasco Nuñez de Balboa when he discovered the Pacific, and later formed a partnership with Diego de Almagro and Hernando de Luque to conquer South America. Pizarro and Almagro made two unsuccessful attempts to find the Incas (Luque, who had financed the expeditions, remained in Spain), and Pizarro solicited support from Emperor Charles V of Spain, who appointed him governor and captain-general of a kingdom he had never seen. In 1532, Pizarro landed at Tumbes, on the Bay of Guayaquil in Ecuador, and headed inland, crossing the cordillera of the Andes with 180 men and twenty-seven horses. (Almagro was supposed to meet him later.) At Cajamarca, the Spaniards slaughtered between five thousand and ten thousand Incas, and looted tremendous quantities of gold as well as emeralds and other precious stones. They seized King Atahualpa and held him for a ransom of enough gold to fill a room 22 feet long, 17 feet wide, and 9 feet high, and two smaller rooms were to be filled with silver. While the Incas were scouring the country for the ransom, Pizarro had Atahualpa executed, marking the effective end of the Inca civilization. By this time Almagro had arrived, and he set out for Cuzco, the Inca capital. He had been given the rights to everything south of Pizarro's territory, but there was no gold there, so he returned to Cuzco and claimed it for himself. Once Almagro was defeated and beheaded, Pizarro busied himself constructing new headquarters at Lima (which eventually became the capital of Peru), a settlement he had founded in 1535. On June 26, 1541, Almagro's followers murdered Francisco Pizarro. **See also Balboa**

placoderm The first jawed fishes (gnathostomes), also characterized by head and shoulder girdles that were heavily armored with bony plates. Known first from the early Devonian period (400 million years ago), they were a diverse group that included thirty-five families and more than 270 genera. The largest was the formidable *Dunkleosteus*, which may have reached a length of 30 feet. Although its foreparts were encased in heavy armor, its long, eel-like tail could propel it powerfully through the water. There are no placoderms known after the early Carboniferous period (350 million years ago), and no living fishes show any of the characteristics of these primitive creatures. **See also shark**

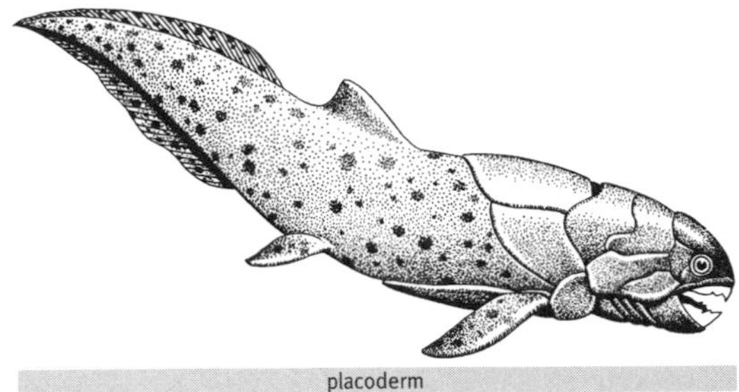

placoderm

plaice (*Pleuronectes platessa*) Long considered one of the more important European food fishes, the plaice is greenish-olive and covered with orange spots. (Since this is a right-eye flounder, the left side—the underside—is white.) It can reach a length of 3 feet and weigh as much as 20 pounds. Lying on the bottom, it can be completely camouflaged, but this does no good when a bottom trawl comes along. It lives and breeds in the North Sea. **See also flounder**

plankton From the Greek word for "wanderers," plankton is composed of the minute organisms that float and drift with tides and currents, unable to move under their own power. Plankton contains plants (phytoplankton) and animals (zooplankton) such as protozoans, larval jellyfishes, crustaceans, cephalopods, and fishes. Directly or indirectly, oceanic plankton provides food for almost all marine animals. The most important forms of phytoplankton are dinoflagellates and diatoms, which, because of their photosynthetic activity, harness the sun's energy and lock it into organic compounds that provide the energy source that forms the basis of the ocean's food pyramids.

See also diatoms, dinoflagellates, zooplankton

plate tectonics The study of modern geology based on the movement of the plates of the earth's crust that unifies theories of continental drift and seafloor spreading. In 1915 Alfred Wegener postulated a single continent ("Pangaea") that existed during the Permian, some 280 million years ago, and that slowly pulled apart into the continents as we know them today. However, he was unable to identify a force that could have made this happen. Fossils in Antarctica and Australia were observed to be similar to those in South America, suggesting that these now-distant landmasses had once been joined, and magnetic anomalies in the earth's crust confirmed its movement. According to the theory, the lighter layer of the crust (the lithosphere) "floats" on the denser layer below (the asthenosphere). The crust is composed of seven major plates and as many as twelve smaller ones. As the plates move (*tectonics* means "shaping" or "assembling"), they slide past each other or converge. The "engine" that moves them is the upwelling of heated material from the core of the planet through the great mountain range that is the Mid-Ocean Ridge—the largest single feature on the planet, which twists around the entire planet like the seams of a baseball. Magma flowing from these volcanic ridges accumulates, pushing away from the ridge and shoving the plates away. The spherical nature of the globe means that some of the continents are moving toward one another while others are moving away. When one plate overrides another, a process known as subduction, the one that sinks is eventually recycled, and the one that is raised up often produces mountain ranges. (The Himalayas were raised up, and are still ascending, because the Indo-Australian Plate collided with the Eurasian Plate.) When two plates encounter each other, mountains may be raised, great trenches may be formed, or earthquakes may occur. Most of the seismic activity of the earth, especially earthquakes and volcanoes, is the direct result of the movement of the earth's plates. The entire Pacific Plate is defined by the "Ring of Fire," the most seismically active region on the planet, with transform faults, fracture zones, and volcanoes ringing the Pacific Ocean.

See also continental drift, Ring of Fire, seafloor spreading, Wegener

Kronosaurus

plesiosaurs Extinct marine lizards (not dinosaurs) identified from fossils from the Late Triassic and Early Jurassic periods, 144 to 230 million years ago. Plesiosaurs had a wide distribution throughout the Old World, and have also been found in South American deposits. Plesiosaurs were inhabitants of the oceans, and had broad, flat flippers to steer with. They were about 15 feet long, with a broad, flat body and a long, flexible neck, like "a snake threaded through the body of a turtle." Like the blowholes of modern whales, the nostrils were set far back near the eyes. There were two types: pliosaurs, with a short neck and an elongated head; and elasmosaurs, where the head was small and the neck extremely long and flexible. Among the largest known plesiosaurs was the 40-foot-long *Kronosaurus*, a short-necked form with a head that was 12 feet long; and *Elasmosaurus*, 43 feet long, almost half of which was neck. (Many references to the Loch Ness monster equate it with a plesiosaur, but the plesiosaurs have been unequivocally extinct for almost 150 million years.)

See also ichthyosaur, Loch Ness monster, marine reptiles

plovers (family Charadriidae) Noisy, boldly patterned little birds with long wings and pigeonlike bills, plovers are found throughout the world except Antarctica. They are mostly denizens of shore or seaside habitats, but there are some that spend their entire lives inland. Those that frequent beaches pursue insects and tiny crustaceans in the wake of receding waves, but one species, the turnstone, overturns rocks and shells to look for food items underneath. Many of them affect a "broken wing" strategy to distract predators from their nests. Like the Eskimo curlew and the passenger pigeon, the golden plover (*Pluvialis dominica*) was nearly eliminated by market hunters in the nineteenth century, but unlike the other species, the golden plover survived. Surfbirds, lapwings, and killdeers are also classified as plovers. **See also Eskimo curlew, golden plover, killdeer**

ringed plover

poacher (family Agonidae) Elongated, big-headed, armored fishes also known as alligator fishes or bullheads. (They are neither alligators nor catfishes, but they are a good recommendation for the use of scientific names.) These bottom dwellers are found in North Atlantic, North Pacific, and southern South American waters. Around the British Isles, the pogge (*Agonus cataphractus*) is a very common fish in waters up to 4,000 feet deep. Although they bear a superficial resemblance to sculpins, they are not related to them.

See also sculpin

Pohnpei Once known as Ponape and also Ascension, Pohnpei is a coral island in the eastern Carolines, now part of the Federated States of Micronesia. It is a hilly island, roughly 129 square miles in area, and surrounded by a barrier reef of many small islets. Because of its lush tropical foliage, it has been called the garden of Micronesia. Many fruits and vegetables are grown there, including oranges, breadfruit, and taro, and the islanders raise pigs and chickens. In the lagoon at Pohnpei is the ancient village of Nan Madol, a group of ninety-two artificial platform islands, surrounded by man-made canals. Although it was a regular stopping point for whalers, Pohnpei was first colonized by

British missionaries in the nineteenth century. After a period of Spanish administration, it was taken over in 1898 by the Germans, who encouraged the production of copra. After World War I, Japan was given a mandate over the island by the League of Nations; during World War II the island was isolated rather than attacked by Allied forces. The island became part of the UN Trust Territories in 1947, and in 1986, it joined the Federated States of Micronesia, whose capital is Palikir, on Pohnpei. To the east of Pohnpei is the island of Pingelap, the home of an anomalous population of color-blind people, who were described by neurologist Oliver Sacks in his 1997 book, *The Island of the Colorblind.*

See also Caroline Islands, Micronesia

polar bear (*Ursus maritimus*) Found only in the circumpolar Arctic, the polar bear is the largest land predator on earth. A full-grown male can stand 11 feet tall on its hind legs and weigh upward of 2,000 pounds. (Females are smaller.) In their harsh habitat, they are kept warm by a thick layer of subcutaneous fat and a thick white coat of hair, which even covers the soles of their feet for insulation and traction on the snow and ice. They feed primarily on seals, ringed seals being the prey of choice, but they will eat fish, birds, and even garbage when they encounter human habitation. There are many records of unprovoked attacks on people, and these animals are considered extremely dangerous. Pregnant females den up in the winter and give birth to one or two cubs approximately a year later. Immature bears stay with their mothers for about two years, avoiding the males, which often kill the cubs. Polar bears are indefatigable roamers, having been found swimming more than 30 miles from land. They are powerful swimmers, but they are also effective runners, and can outrun a reindeer (and certainly a man) on land. Except for *Homo sapiens*, polar bears have no enemies, but humans have wreaked havoc with polar bear populations, from the Inuit hunting them for food and clothing to trophy hunters shooting them for their skins. Although it is illegal to import a polar bear skin into the United States, a smuggled one might be worth ten thousand dollars. In their icy habitat, which covers some 5 million square miles and ignores the political boundaries of Russia, Alaska, Canada, Greenland, Norway, and Finland, there are believed to be between twenty thousand and forty thousand polar bears.

polar bear

Polaris The North Star; the star nearest the celestial North Pole, in the constellation Ursa Minor. It is a very important navigational star, because it always marks due north for an observer. Polaris can be located by following a line upward from the two stars that make up the right end of the Big Dipper.

Polaris The ship that Charles Francis Hall took on an expedition to the North Pole in 1876, which reached 82°11′ north, the farthest north anyone had ever gotten until that time. The U.S. nuclear submarines that can launch nuclear missiles are Polaris submarines, and the intercontinental ballistic missiles (ICBMs) launched by U.S. and British submarines are Polaris missiles.

See also Hall

pole fishing Perhaps the earliest method of commercial fishing. It was from dories with fishing poles that the earliest visitors to Newfoundland caught the plentiful codfish; a variation of this method is employed today in tropical Pacific and Atlantic waters to catch the smaller tuna species, such as albacore, skipjack, and bonito. (The giant bluefin tunas, considered so valuable in Japan, are usually harpooned.) Dories are not used, but the fishermen stand at the rail of a fishing boat, attracting schools by "chumming" (throwing baitfish into the water to excite the object of the fishery to attack anything), and bamboo poles are used with unbarbed, unbaited hooks. The tuna are yanked out of the water by the large crew of fishermen, thrown over their heads onto the deck, killed, and stored belowdecks in freezers. (Some technically advanced countries have replaced the fishermen with "jigging machines" that perform the same function, thus cutting down on the labor costs.)

See also longline fishing, purse seining, trawling

pollack (*Pollachius* spp.) The European pollack (*Pollachius pollachius*) is a member of the cod family that is found in the eastern North Atlantic from Norway to the Bay of Biscay, and also in Icelandic waters and the Mediterranean. It is greenish in color and has a white lateral line and a small chin barbel. It can grow to a length of 40 inches and a weight of 40 pounds, but most are smaller. It is a popular game fish in Europe. Another species (*P. virens*), known as the saithe or coal-

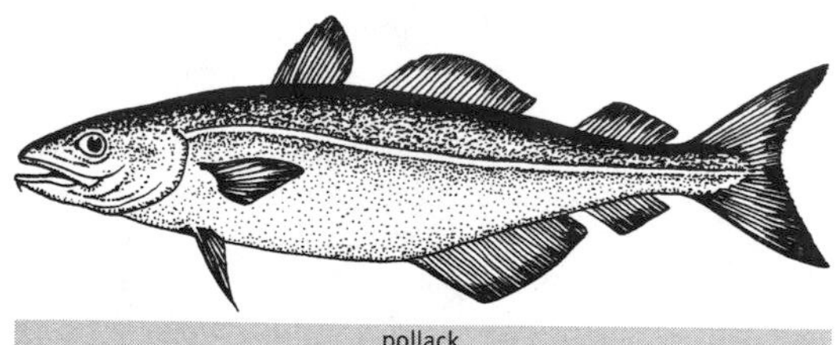

pollack

fish, occurs in the western as well as the eastern North Atlantic, and has a more streamlined shape than its relative. (In Norway, it is known as *seje,* and gave its name to the sei whale, which appears in Norwegian waters at the same time in the spring.) Pollack are popular with sport fishermen because they put up a good fight when hooked, but they are also important commercially. In the area of Georges Bank, some 63,000 metric tons were taken annually until 1993, when the fishery went into a decline. A closely related species is the Alaska pollack (*Theragra chalcogramma*), found in the North Pacific and the Bering Sea, which has the distinction of being the most important commercial fish in the world. In the Bering Sea, Russian, Chinese, Japanese, Korean, and American fleets take tens of millions of tons annually—another disaster in the making.

See also codfish, haddock, hake

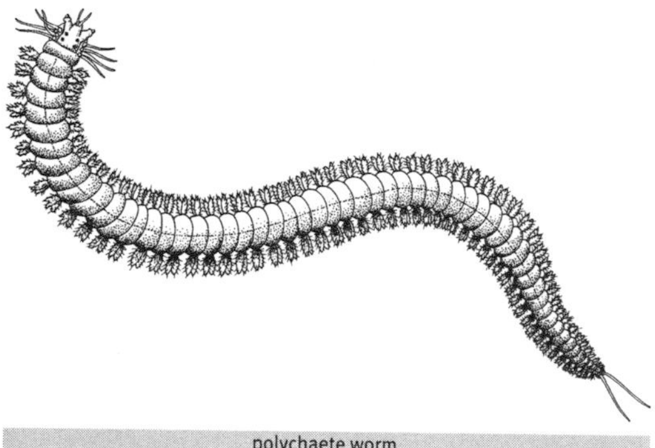

polychaete worm

polychaetes The phylum Annelida incorporates all the segmented worms, including the Oligochaeta (earthworms) and the Hirudinea (leeches), which are found on land and in freshwater, respectively. The marine representatives of the phylum are the polychaetes, known variously as ragworms, bloodworms, fireworms, lugworms, sandworms, scaleworms, fanworms, palolos, sea mice, and, most commonly, bristleworms. Although the polychaetes are usually divided into those that move about (errantia) and those that remain in a tube or burrow (sedentaria), these distinctions are not particularly useful, since many species cross the line. There are more than ten thousand species of polychaete, ranging in size from a tiny fraction of an inch to 10 feet. They are usually elongated, cylindrical, flexible animals that are characterized by a segmented body and the presence of parapodia on the segments. These are paired, fleshy outgrowths that are used for feeding, locomotion, or respiration. These bear bundles of setae (bristles), which can be extended for support. At one end there is a head, often with eyes, antennae, tactile tentacles, and a pair of jaws; at the other end there is an anus and often a pair of projections known as cirri. In between, polychaetes have well-developed digestive, nervous, circulatory, and excretory systems. Their blood is often red, but in some species it is green. Most species breed sexually, by releasing sperm and eggs into the water, but some are hermaphroditic, and others reproduce by breaking into fragments of one of more segments and regenerating the missing portions. Commonly used for bait on both coasts of the United States is the clam worm, *Nereis virens,* which can reach a length of 3 feet and is an iridescent greenish or bluish color. In the reefs and rocks of South Florida and the Caribbean is *Eurythoe complanata,* the orange fireworm, whose toxic bristles can pierce the skin and cause painful wounds. In the South Pacific, the palolo worm (*Palolo viridis*) lives in holes in coral reefs. According to a strict lunar timetable, great breeding swarms appear, an event eagerly awaited by Samoans, for the gamete-laden tail sections of the palolos are a special treat. Featherduster worms (sabellids) live in tubes and extend a feathery plume of gills, which gives them their common name.

See also featherduster worm, palolo worm, parchment worm, sea mouse, tube worms

pomarine skua (*Stercorarius pomarinus*) Somewhat smaller than the great skuas (*Catharacta* spp.), those of the genus *Stercorarius* are similarly aggressive, predatory birds, gull-like in appearance. Like the long-tailed skua (*S. longicaudus*), the pomarine has two tail central feathers that extend beyond the tail, but in this species they are spoon-ended and twisted, which makes for easy recognition. The pomarine skua, commonly called the pomarine jaeger, is circumpolar in the Arctic tundra regions. A brownish-gray bird with white patches on the underside of the wings, a black cap, and yellowish cheeks, the pomarine skua has a wingspan of 48 inches. It falls between the parasitic (Arctic) and the great skua in size, but compared to either of them, its wings are proportionally broader and its bill is more sharply hooked. It comes in light and dark versions (called morphs); the lighter morphs outnumber the darker ones by about 20 to 1. Although the pomarine skua breeds in the high northern latitudes, it roams the world's oceans and has been recorded from the Gulf of Mexico and the West Indies, South Africa, southern Australia, and southern South America. "Pomarine" means "having a covered nostril," and probably refers to the hawklike cere across the base of the upper mandible. (All skuas have the same structure.)

See also Arctic skua, great skua, long-tailed skua, McCormick's skua, skua

pompano (*Trachinotus carolinus*) Placed in the same genus (*Trachinotus*) as the permit, the common pompano is a much smaller fish with a relatively small mouth. (The record pompano weighed 8 pounds, the record permit 53.) Both species are found in the western North Atlantic from Cape Cod to Brazil, and both are excellent food fishes, but some regard the pompano as the most delicately flavored of all fishes. **See also permit**

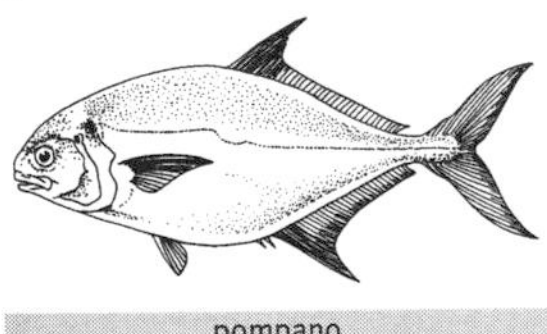

pompano

Ponce de León, Jaun (1460–1521) Spanish explorer, the first European to reach Florida. He served with Columbus on his second voyage in 1493 and returned in 1502 to assist in the conquest of the eastern part of Hispaniola (now the Dominican Republic), where he served under the provincial governor. In 1508–1509, he explored and settled the island of Boriquén (Puerto Rico), founding the colony's oldest settlement, Caparra, near what is now San Juan. From Boriquén he is said to have set out in search of the Fountain of Youth, and in 1513, with three vessels, he sailed through the Bahamas and landed on the North American mainland. Because he landed near modern St. Augustine at the time of the Easter Feast (*Pascua florida*), he claimed the land for Spain and named it La Florida. At the time he believed he had landed on an island. He returned to Spain and was commissioned to subdue the warlike Carib Indians of Guadeloupe and colonize "the island of Florida." His attempt to subdue the Caribs failed, and he returned to Boriquén in 1515; he returned to Spain and sailed again for Florida in 1521. As military governor of Bimini and Florida with permission to colonize those regions, he again set out for Florida, this time with two hundred men, fifty horses, and assorted livestock. He landed on the west coast, near Tampa Bay, and his party was repulsed by the fierce Seminoles. Ponce de León was wounded in the neck by an arrow, and he died after being returned to Cuba. Ponce, Puerto Rico's third largest city, is named for him.

See also Carib Indians, Columbus, Puerto Rico

porbeagle (*Lamna nasus*) Porbeagles are mackerel sharks, smaller and stockier than makos and smaller than great whites; they reach a maximum length of 9 feet and weigh up to 350 pounds. The teeth are unserrated and have small cusps ("lateral denticles") on either side of the main blade. These sharks, usually gray or brownish gray, are characterized by a white patch at the trailing edge of the first dorsal fin and a secondary keel at the base of the tail. Porbeagles are found in the cold temperate waters of the North Atlantic (they are the most popular shark with British anglers), and also in the South Atlantic and South Pacific. The origin of the name is unknown. In the nineteenth and early twentieth centuries, porbeagles were fished extensively for liver oil (an important source of vitamin A) and as food, but their numbers have been greatly reduced and there is no longer a commercial fishery. The salmon shark (*L. ditropis*) is a very similar species, found only in the North Pacific. **See also great white shark, mako shark**

porbeagle

porcupine fish (*Diodon hystrix*) Ordinarily a chubby little spotted fish, when it is threatened, this puffer inflates itself with water (or air, if it is removed from the sea) and becomes a most unappetizing pincushion. Found in warm seas throughout the world, it reaches a length of two feet. **See also sharpnose puffer**

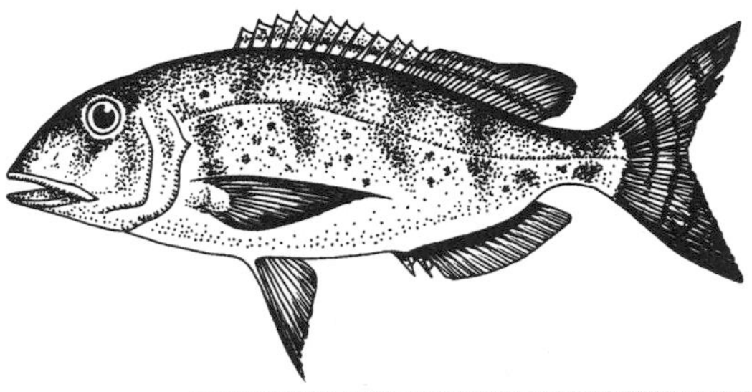

jolthead porgy

porgy (*Calamus* spp.) Laterally flattened fishes with eyes located high on their heads, these fishes of the family Sparidae (commonly called sea breams) are found throughout the temperate waters of the world under many different names. There are about a hundred species, ranging in size from the 1-foot-long porgies to the red steenbras (*Petrus rupestrus*) of South Africa, which can reach a length of 6½ feet. In the western North Atlantic, the 24-inch-long jolthead porgy (*Calamus bajonado*) is characterized by its brassy coloration and its pronounced "Roman" profile. It is found along the coasts from Rhode Island to Brazil, as is the saucereye porgy (*C. calamus*), a smaller fish that lacks the stripes of the jolthead. The sheepshead (*Archosargus probatocephalus*) can grow to 30 inches and is strongly striped with vertical dark bands. Its common name is derived from the prominent incisor teeth that protrude from its mouth. In Florida and Caribbean waters, porgies are popular with recreational fishermen, and they make up a large proportion of commercial catches. In Japan, the red tai (*Pagrus major*), a large porgy, is one of that country's most important food fishes. **See also scup, sparidae, steenbras**

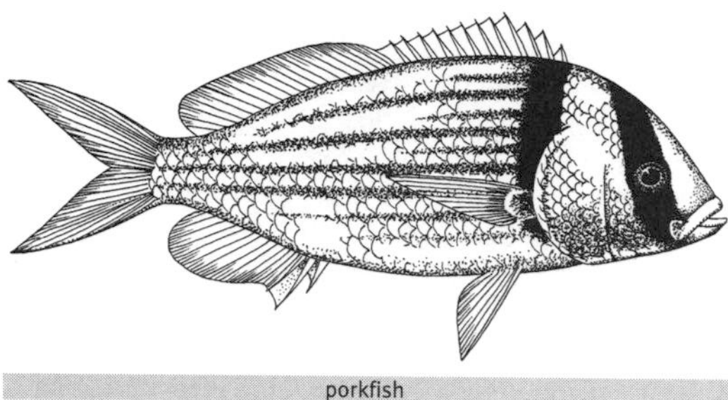
porkfish

porkfish (*Anisotremus virginicus*) A brightly colored, schooling species of grunt, found in the Caribbean, the Bahamas, and Bermuda. Juveniles, which look completely different from the adults, are "cleaner" fishes, picking parasites from the bodies of other fishes.
See also cleaner fishes, grunts

porpoise Although the word is often used interchangeably with "dolphin," it should be applied only to a group of small cetaceans, officially classified as Phocoenidae ("fo-*seen*-a-day") and characterized by their small size (none are larger than 6 feet in length), a beakless profile, and small, spade-shaped teeth. It would be convenient if all dolphins (Delphindae) differed from this description, but some, such as Risso's dolphin (*Grampus griseus*) and the Irriwaddy River dolphin (*Orcaella brevirostris*), have no beaks, and others, such as the Franciscana (*Pontoporia blainvillei*) are as small or smaller than some porpoises. The conical teeth of the dolphins distinguish them from porpoises, but one is not always in a position to examine a small cetacean's teeth. **See also Burmeister's porpoise, Dall porpoise, dolphin, finless porpoise, harbor porpoise, spectacled porpoise, vaquita**

port (1) The left side of a vessel when the viewer is facing the bow (front). The word may be derived from the fact that old-fashioned merchant ships had a loading port on the left side, and warships also had their entry port on that side. Originally, the left side of a vessel was known as the larboard side, but this was officially changed by the British Admiralty in 1844 to avoid confusion with the similar-sounding "starboard."
See also starboard

port (2) A harbor with facilities for berthing ships, the embarkation and disembarkation of passengers, and the loading and unloading of cargo.

Port Royal Now a small fishing village on the southern coast of Jamaica, Port Royal was England's richest New World possession because it was the pirate capital of the West Indies and a center of the rum and slave trades. Pirates such as Henry Morgan were encouraged by the British governors of Jamaica to use it as their base in hopes of dissuading the Spanish from recapturing the island, which they lost in 1655 when Cromwell's ships captured it after failing to take Hispaniola. With a 1680 population estimated at 7,000, Port Royal was the largest English settlement in the New World. On June 7, 1692, a massive earthquake hit Port Royal, dumping two-thirds of the city into the Caribbean, and a tsunami followed, killing two thousand people. Afterward, another three thousand died of injuries, disease, and famine. In 1981, marine archaeologists began the excavation of the submerged portions of Port Royal, producing a fascinating and detailed picture of life in a seventeenth-century pirate stronghold.
See also Jamaica, Morgan

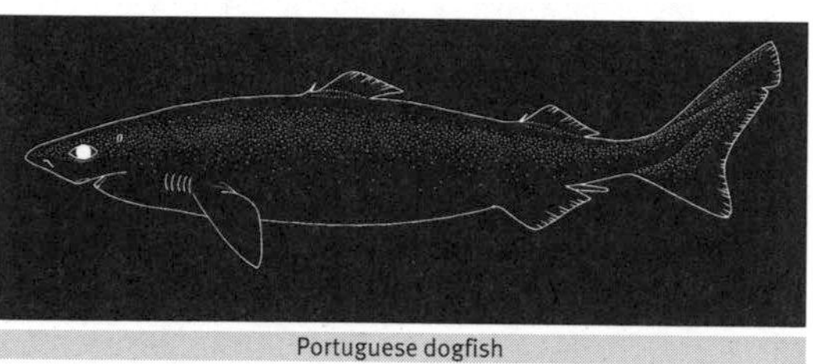
Portuguese dogfish

Portuguese dogfish (*Centroscymnus coelolepis*) Probably the deepest-living of all sharks, this species has been recorded at 12,000 feet in the North Atlantic. It has also been reported off Europe and North Africa, in the South China Sea, and off New Zealand. It reaches a length of about 4 feet, and is fished commercially in Portuguese waters. **See also dogfish**

Portuguese man-of-war (*Physalia physalis*) A colony of differentiated cells associated to form a functioning "animal." The lovely blue and pink iridescence of the gas-filled float belies the deadly nature of the stinging tentacles that dangle beneath it, sometimes 100 feet long. The stinging cells (cnidocytes) contain long, threadlike nematocysts, which consist of a harpoon-tipped organelle that can inject a neurotoxin into the victim's flesh. The toxin is powerful enough to kill a swimmer, and the animals are almost as dangerous when they are dead and

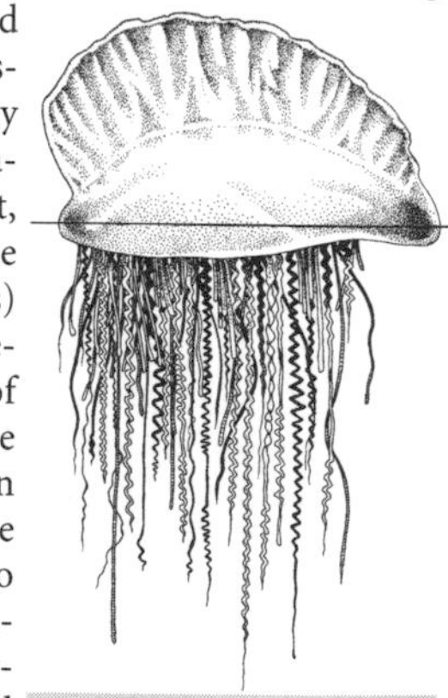
Portuguese man-of-war

washed up on the beach as when they are floating in the sea. The Portuguese man-of-war eats fish up to the size of a mackerel, first stinging the prey to death, then hauling it up under the float and eating it with the specially adapted feeding polyps. A small fish called *Nomeus* lives with apparent impunity in the tentacles, and has even been observed nibbling on them. Known as the Portuguese man-of-war fish, *Nomeus* may function as a lure to attract larger fish into the deadly web of its host. **See also cnidarian, jellyfish, siphonophore**

portulan chart Also "portolan" and, in Italian, *portolani;* sailing chart first used in the Mediterranean in the thirteenth century. Usually drawn on sheepskin or goatskin, the charts depended on the compass rose (showing the points of the compass) and rhumb lines (lines followed by a vessel sailing on one course or a wind blowing continuously in one direction) and covered the entire skin. The coastline was shown, but no details of what lay beyond, and distances between points were often indicated. They were drawn in black ink, but red was sometimes used to indicate major harbors, with colored banners to show each one's allegiance. The portulan chart is descended from—and easily confused with—the *portolano,* a book of sailing directions describing coasts and ports, anchorages and hazards. **See also compass**

Poseidon Originally the Greek god of earthquakes, whose dominion was later extended to the sea. The son of Chronos and Rhea and the brother of Zeus and Hades, Poseidon's palace was in the depths of the sea, although he occasionally visited Olympus. He traveled in a chariot accompanied by dolphins and sea monsters, and the waves became smooth as he approached. He was married to Amphitrite and among his children by her were Triton, Rhode, and Benthesicyme. The symbol of Poseidon's power was the trident, a three-pointed spear, and also the dolphin. A memorable fifth-century B.C. bronze shows him about to hurl his trident. In the *Odyssey,* he is hostile to Odysseus and prevents him from returning home. In the Roman pantheon, he was known as Neptune.

posh A term referring to sumptuously comfortable or luxurious surroundings, supposedly derived from the letters POSH (Port Out, Starboard Home), which were printed on the tickets of first-class passengers on the P&O Line, so they would be assigned cabins on the port side of the vessel going out and on the starboard side coming home, thus avoiding the glare of the sun in the Indian Ocean. Unfortunately, there is no evidence whatsoever to substantiate this story.

Potemkin Russian battleship of the Black Sea fleet; the scene in 1905 of a famous mutiny. Encouraged by agitators, the men refused to eat meat that they claimed was full of maggots. The ship's second-in-command, Commander Ippolit Giliarovsky, declared that this amounted to mutiny and ordered that a number of men be selected at random, covered with a tarpaulin, and shot. At that, the crew mutinied, took the ship, and killed Giliarovsky, Captain Golikov, the ship's surgeon (who had insisted that the meat was edible), and most of the eighteen other officers. Flying the red flag, the mutineers took the ship to Odessa during one of the largest strikes in Russian history, and during the riots, Cossacks killed some six thousand—including those on the Richelieu Steps, an incident shown dramatically in the "Odessa Steps" segment of Sergei Eisenstein's 1925 film, *The Battleship Potemkin.* The film tells the story of the ship, the riots, and the Cossack attack, and is generally considered to be one of the best movies ever made. Unable to convince other Russian crews to mutiny, the *Potemkin* left Odessa and sailed around the Black Sea until she was scuttled by her crew near the Romanian port of Constanza. The great battleship was raised and towed unceremoniously back to Odessa, where Tsar Nicholas II had her renamed *Pantelymon,* or "Low Peasant." In 1919, at the completion of her uneventful service in World War I, she was sunk by her officers in Sebastopol harbor.

See also mutiny

Pribilof Islands The Pribilofs include the major islands of St. Paul and St. George and two islets in the Bering Sea; they are about 180 miles north of the Aleutian chain. They were first visited in 1786 by Gavril Pribylov, a Russian sea captain, who reported that there were fur seal rookeries there. The islands were transferred to the United States with the purchase of Alaska in 1867. The fur seals have been the main focus of the islands since their discovery. During the 1880s, sealing rights were leased to the Alaska Commercial Company, which allowed other nations to harvest the seals. Signed by the United States, Canada, Japan, and Russia in 1911, the North Pacific Sealing Convention abolished pelagic sealing (killing seals at sea, where no discrimination between males and females was possible), and mandated that all skins were to be shared. This treaty was ended when Japan withdrew in 1941, but in 1957, the original signatories created the North Pacific Fur Seal Commission, which permitted limited hunting. In 1979, the Pribilof Aleuts received $8.5 million in compensation for the treatment they received by the U.S. federal government from 1870 to 1946, and the discriminatory United States ended the seal harvest and withdrew from the islands. Another $20 million was provided to help the Aleuts develop their own tourist and commercial fishing industries. The Aleuts still take some four thousand young male fur seals annually.

See also Aleutian Islands

pricklebacks (family Stichaeidae) A group of eel-like, elongated fishes, related to the blennies, with long anal and dorsal fins. There are thirty-one genera and some sixty species, found throughout the colder waters of the northern Atlantic and Pacific. The monkeyface prickleback (*Cebidichthys violaceus*) reaches a length of 2 feet, and is considered a delicacy by California anglers. The decorated warbonnet (*Chirolophis polyactocephalus*) has numerous plumelike filaments on its head; it is found from Washington to Alaska. In the North Atlantic, the 3-foot-long wrymouth (*Cryptocanthodes maculatus*) burrows into muddy bottoms, and has a mouth that faces upward. **See also blenny**

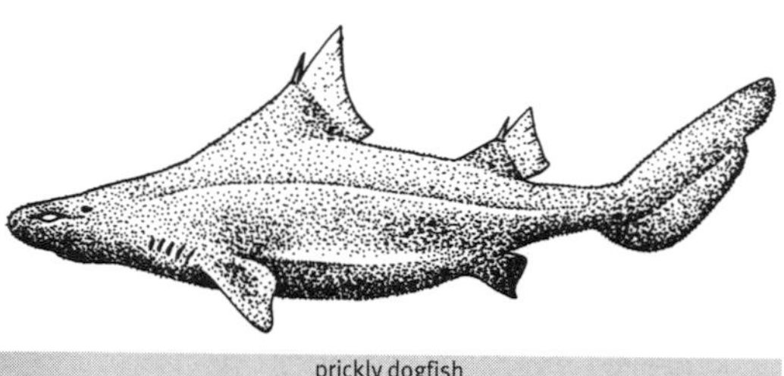

prickly dogfish

prickly dogfish (*Oxynotus centrina*) Though differing somewhat in morphology, all these bizarre sharks have small heads, very rough skin, and two high, wide-based dorsal fins with a prominent spine in the middle. While most sharks are more or less cylindrical in cross-section, the prickly dogfish, or oxynotids, are triangular with a flattened base. Their teeth are different in the upper and lower jaws: the uppers are small and narrow, while the lowers are compressed and bladelike. They probably live on the bottom, eating crabs and other small invertebrates. There are four species in the genus, each found in a distinct area. The Mediterranean and eastern Atlantic version (*O. centrina*) is the largest, reaching a length of 58 inches; the sailfin roughshark (*O. paradoxus*) is a smaller, narrower-finned species found in the eastern North Atlantic from Scotland to North Africa. In Australian and New Zealand waters, *O. bruniensis*, known as the roughshark, reaches a length of 28 inches. The Caribbean roughshark (*O. caribbaeus*) is the smallest (19 inches) but has the most exaggerated dorsal fins and a color pattern of broad dark bands on a lighter background.

See also bramble shark

Prince Edward Island Before the arrival of the Europeans, the Micmac Indians called this crescent-shaped island Abegweit, or "cradled on the waves." Jacques Cartier "discovered" the island in 1534, and in 1603, Samuel Champlain arrived and called it Île Saint Jean. It was known by that name until the British took over in 1799 and named it after Edward, Duke of Kent, commander of the British forces in America and the father of Queen Victoria. In 1867, PEI became the seventh province of Canada. The island is about 140 miles long, 40 miles wide at its widest point, and covers some 2,185 square miles. It is separated from New Brunswick and Nova Scotia by the Northumberland Strait, which was spanned by a pedestrian and automobile bridge in 1997. The island is a popular vacation destination for Canadians; its year-round population of 140,000 swells to 700,000 in the summer. On the northwestern coast is Malpeque Bay, whose shallow inshore waters are ideal for the growth of oysters. **See also Cartier**

Prince of Wales Island (1) The largest of the Torres Straits islands, in the Arafura Sea between the tip of the Cape York Peninsula of Queensland and the southern shore of Papua New Guinea. (The other islands are Wednesday, Thursday, Friday, Goode, Hammond, and Horn.) After repairing the *Endeavour* in August 1770, James Cook sailed farther up the east coast of Australia, rounding Cape York and hoping that Luis de Torres was right when he said there was a way through. There was, of course, and on his westward way into the Arafura Sea, he rounded and named Prince of Wales Island. (The other Torres Strait islands were not surveyed and named until Matthew Flinders circumnavigated Australia in 1802.) Prince of Wales Island covers about 70 square miles of rocky, wooded terrain, with no arable land and no source of freshwater. With the decline of the market for pearl shell (trochus), most of the islanders migrated to adjacent Thursday Island, the administrative capital of the islands, or to Queensland, and the population on Prince of Wales now numbers only a few dozen. **See also Torres Strait**

Prince of Wales Island (2) Island of southeast Alaska, at 2,770 square miles the largest of the Alexander Archipelago, which includes Admiralty, Kupreanof, Baranof, Chichagof, Kuiu, Mitkof, Wrangell, Revillagigedo, and many smaller islands. (The city of Sitka is on Baranof, and Ketchikan is on Revillagigedo.) The entire archipelago is the Tongass National Forest, but the U.S. government has allowed lumber interests to clear-cut vast tracts of the 17 million acres of hemlock, spruce, and cedar to the point where Prince of Wales Island, once the most heavily forested of all the islands, is virtually stripped bare of its original growth.

See also Alexander Archipelago

Prince Patrick Island Island in the Parry Islands group of the Northwest Territories, Canada, with its western shore facing the Beaufort Sea. It is about 150 miles long and 20 to 50 miles wide, and consists of 6,119 square miles of snow- and ice-covered sandstone. It was discovered by Francis Leopold McClintock (1819–1907) during an expedition in search of Sir John Franklin, whose two ships, *Erebus* and *Terror*, had disappeared while searching for the Northwest Passage. McClintock, who developed the art of traveling by sledge, discovered Prince Patrick Island in 1853, naming it for the third son of Queen Victoria. In 1857, on his fourth expedition in search of answers to the mystery of John Franklin's disappearance, he found the remains of the crew members on King William Island.

See also Beaufort Sea, Franklin, McClintock

prion (*Pachyptila* spp.) Also known as whalebirds because they feed on krill, the same food that sustains the baleen whales, prions are small, fast-flying seabirds of the Southern Ocean. These little birds feed in the same manner as the whales, but on a considerably smaller scale: taking in mouthfuls of crustacean-packed water, trapping the food on the tiny serrations on the sides of its bill, and expelling the water. Of the four or five species, the broad-billed and thin-billed prions are the largest, and the fairy prion is the smallest. Prions used to be incredibly abundant, but the introduction of egg-eating rats and cats to their nesting islands has greatly reduced the populations.

fairy prion

See also fulmar, krill, petrel, shearwater

Pritchard, Walter (1886–1935) Although little known today, Walter "Zarh" Pritchard was celebrated during his lifetime as the first (and only) painter to actually work underwater. Born in India and schooled in Scotland, Pritchard was fascinated by the sea, and as a child, he was enraptured by the subaqueous sights of the cold waters of the Firth of Forth. He moved to London, where he sold a painting to Sarah Bernhardt; then to California, where he set up shop in Santa Barbara, painting in the waters around Catalina Island. Around 1904, he headed to Tahiti because of the warm, clear waters and brightly colored corals there. At first, he made quick sketches that he would finish in his studio, but later he took advantage of diving suits and underwater breathing apparatus, and actually sat on a rock on the bottom, painting on oil-soaked canvas with oil paints. His paintings were chosen by art collectors (the director of the Louvre); royalty (Prince Albert of Monaco and the Crown Prince of Japan); and such scientists as William Beebe and David Starr Jordan, one of California's leading ichthyologists. Pritchard had exhibitions at galleries around the world, as well as at the New York Aquarium and the American Museum of Natural History. By the time he died in 1936, he was rich and famous, but since then his unique painting style has faded from popularity. Nevertheless, he deserves a place alongside Jules Verne, William Beebe, Jacques Cousteau, and others who have sought to communicate the actual appearance of the underwater world.

privateer An armed, privately owned vessel that operated in time of war against the trade of an enemy. (The name is used for both the ships and the men who sailed them.) In Britain, they were commissioned by "letters of marque" that were issued by the Admiralty, authorizing ships to cruise in search of enemy merchant vessels and seize them. Francis Drake was a privateer on his voyage around the world (although he did not have a letter of marque), and so was John Paul Jones before he became a regular naval officer. The great age of privateering was from 1589 to 1815, during which time the ships were considered lawful auxiliaries to regular navies. The purpose of the privateer was to disrupt the enemy's commerce, and all the resulting booty belonged to the the ship that took it. This inevitably led to a blurring of the line between privateering and outright piracy, in which private, armed vessels, often with no authorization whatsoever, would seize ships friendly or unfriendly for their cargoes. Privateering was an important element in naval warfare until it was abolished by the Declaration of Paris in 1856, but because the United States did not sign the declaration, shipping during the Civil War was heavily affected by Union and Confederate privateers. The actions of privateers against the commercial shipping of the enemy is considered the strategic forerunner of the German submarine assaults on merchant ships during World Wars I and II.

See also letter of marque, piracy

Project FAMOUS Joint project between France and the United States in 1973–1974 to explore the underwater Mid-Atlantic Ridge. It was described as "the most ambitious exploration effort ever to be undertaken with deep-sea submersibles; three of the world's deepest diving craft are to make some 60 penetrations this summer into the so-called navel of the world—the volcanic rift valley beneath the Mid-Atlantic." Scientists participating in Project FAMOUS (French-American Mid-Ocean Undersea Study) became the first to view the heretofore hidden wonders of the mid-Atlantic seafloor. For this cooperative venture, bottom charts were updated by American and French hydrographers; the British contributed a 7-ton, side-scan sonar system named GLORIA (Geological LOng-Range Inclined Asdic), and the U.S. Naval Research Laboratory added a new underwater photography system called LIBEC (LIght BEhind Camera). Woods Hole provided a camera sled they named ANGUS (Acoustically Navigated Geological Undersea Surveyor), which could be controlled from the surface, and which would prove to be indispensable in the collection of photographic data. The most important elements in the project, however, were the three submersibles: *Alvin* of Woods Hole, fitted with a new titanium hull that withstood test pressures of 22,500 feet, far greater than the depth of the rift; the French *Archimede,* a 69-foot-long bathyscaph that had been to 32,000 feet in the Pacific; and the newly built *Cyana,* a descendant of Jacques Cousteau's diving saucer (*le soucoupe plongeante*), and at a length of 20 feet, a little smaller than *Alvin.* Like its predecessor *Trieste, Archimede* employed a gasoline-filled hull, which accounted for her immense size, but *Alvin* and *Cyana* relied on the weight of their ballast to bring them down; compared to the ponderous *Archimede,* they were as handy as bicycles. Altogether, the three submersibles made a total of forty-seven dives to depths of approximately 3,000 meters (9,842½ feet). On August 2, 1973, some 400 miles southwest of the Azores, *Archimede* carried the first human beings to the floor of the Atlantic rift valley. (Project FAMOUS turned out to be the last voyage of the *Archimede;* she was too large, too unmaneuverable, and too expensive to run.) **See also *Alvin*, Ballard, Mid-Atlantic Ridge, *Trieste***

protective coloration The color or pattern of an animal that affords it protection from its predators and the opportunity to conceal itself from potential prey. The most common kind of protective coloration is cryptic resemblance, where the animal resembles its environment, in color, texture, or both. In the sea, many fishes are colored (or even shaped) like rocks or corals, including the stonefishes, frogfishes, and scorpionfishes. One of the best examples is the sargassum fish (*Histrio histrio*), a small frogfish that has evolved to resemble the sargassum weed in which it lives. In addition to those animals that inherently resemble their environment, there are those that can change color—in the case of certain octopuses, texture as well—to imitate a new or different environment. Coral reef fishes are often brightly colored so that they do not stand out in the fragmented, changing light and shadow of the reefs, and there are some fishes and squid that can mimic plants and corals, in deployment as well as coloration. **See also bioluminescence, countershading, mimicry**

Ptolemy (A.D. 90–168) Claudius Ptolomaeus was probably born in Egypt of Greek parentage. He worked at the library in Alexandria from A.D. 127 to 150. A mathematician and astronomer, he was also responsible for the foundations of cartography as we know it today. He knew that the world was round and invented a number of projections whereby an area on the curved surface of the earth could be represented on a flat plane. His eight-book *Guide to Geography,* usually known as *Ptolemy's Geography,* is his most important work; it includes an atlas of the world as it was then known. It includes a list of eight thousand places with a grid that shows their latitude and longitude. (His calculations were off because he accepted Posidonius's estimate that the earth's circumference was 18,000 miles instead of the 25,000 miles more accurately estimated by Eratosthenes, and he therefore estimated a degree of latitude as 50 miles, instead of its true length of 60 miles.) With the fall of the Roman Empire, Ptolemy's great work was lost, and it was not discovered again until 1400, when a copy was found in Constantinople. Ptolemy's work formed the basis for the mapping of all the new discoveries of the fifteenth and sixteenth centuries. He also postulated a vast continent to the south, Terra Australis Incognita, which showed up on subsequent maps and led navigators on a merry search for a great southern landmass that was supposed to balance the continental landmasses of the Northern Hemisphere.

Puerto Rico Commonwealth of the United States, located in the Caribbean, about 1,000 miles southeast of Miami. Puerto Rico ("rich port") measures about 100 miles in length and 35 miles across. When Columbus landed there in 1493, he named it San Juan Bautista (St. John the Baptist). Before the arrival of the Spanish explorer Juan Ponce de León in 1508, the island was inhabited by the peaceful Arawak Indians, but disease and the hardships of slavery eliminated them altogether, so the Spanish imported African slaves to work the sugar and coffee plantations. In the nineteenth century there were a few ineffectual uprisings against the Spanish, but this island has never waged a war for independence. During the Spanish-American War, U.S. troops occupied the island, and by the Treaty of Paris, which ended the war, Puerto Rico was ceded to the United States. The island was made a commonwealth in 1952. The ongoing struggle for Puerto Rican independence was marked by the assassination attempt on President Harry Truman in 1950 and an armed attack on the U.S. House of Representatives in 1954. Puerto Ricans, who do not pay federal taxes and cannot vote in national elections, are divided in their opinions about statehood, independence, and the status quo. The island, with a population of 3.5 million, is now a major center of Caribbean commerce, finance, and tourism, and San Juan is one of the world's busiest cruise ship ports. **See also Arawak Indians**

puffer (family Tetraodontidae) There are about ninety species of puffers found throughout the world's temperate and warm seas. Most grow only to about 1 foot in length, but some species can be as long as 3 feet. When pulled from the water, puffers swallow enough air to make them swell up like a balloon. Thrown back in the water, they float upside down until they can expel the air and return to their normal shape and swimming position. To avoid predators, they can also swallow water while beneath the surface. Puffers can be differentiated from porcupine fishes (which are also able to inflate themselves with air or water) by their teeth: puffers have a sharp beak with a division in the center, while porcupine fishes (Diodontidae) have a solid beak with fused teeth. In some puffer species, the liver, ovaries, and gut contain a neurotoxin that is so powerful that it can cause serious illness or death if eaten. In Japan, puffers are called *fugu;* it is considered heroic to eat their flesh, which has to be prepared by a specially trained chef. **See also fugu, mimicry, porcupine fish**

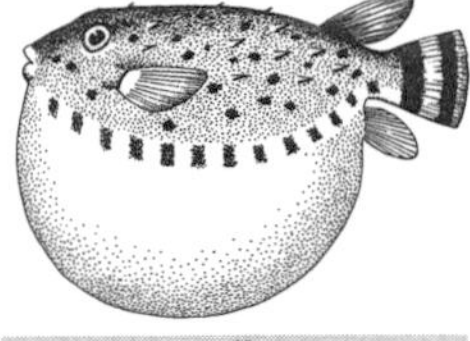
puffer

puffin (*Fratercula* spp.) There is one species of puffin in the North Atlantic and two in the North Pacific. All three are black-and-white birds that stand a little over a foot high on bright red feet, and they all develop huge, brightly colored beaks during the breeding season. The North Atlantic version (*F. arctica*) nests in burrows in Iceland, Greenland, Spitsbergen, Jan Mayen, the Shetlands, the Faeroes, Norway, and Sweden. After the chicks hatch, the parents fly out to sea to catch fish to feed their offspring, which they can carry crosswise in their beaks. Unlike most other alcids, puffins are quite agile on land, and they stand upright with comic dignity. Their flight, a straight line with wingbeats so rapid that the wings cannot be seen, distinguishes them from other birds. The horned puffin (*F. corniculata*) of the North Pacific breeds along the shores of the Chukchi and Bering Seas, southeast to Glacier Bay and British Columbia, and is characterized during the breeding season by a yellow bill with a red tip. The tufted puffin (*F. cirrhata*), abundant in Alaska, has a white, clownlike visage, with shaggy, straw-colored plumes that spring from behind the eye and droop to the nape.

See also Alcidae family, auklet, murre, razorbill

puffin

Puget Sound Named by Captain George Vancouver for his lieutenant, Peter Puget, in 1792, Puget Sound is an arm of the North Pacific, connected to the ocean by the 100-mile-long Strait of Juan de Fuca. The sound is deep and navigable by large ships, and many important ports take advantage of this, including Victoria on Vancouver Island and Bellingham, Seattle, and Tacoma on the Washington mainland. Ferries are the mode of transportation between the mainland and the islands in Puget Sound, such as the San Juans (Lopez, San Juan, and Orcas), and Whidbey Island.

See also Juan de Fuca Strait, San Juan Islands, Vancouver, Vancouver Island

Purchas, Samuel (1575–1626) English cleric and friend of Richard Hakluyt, whose collection of manuscripts fell into his hands at Hakluyt's death in 1616. In 1582, Hakluyt had published *Divers Voyages touching on the Discovery of America*, and in 1589, the *Principall Navigations, Voyages and Discoveries of the English Nation within these 1500 Years.* Purchas published *Hakluytus Posthumous, or Purchas his Pilgrimes* in 1626. Purchas was evidently a poor editor, for he mutilated many of Hakluyt's accounts, but he did add several new voyages, which partially compensates for his garbling of Hakluyt's texts. **See also Hakluyt**

purse seining A commercial surface fishery technique that depends on the schooling inclinations of certain fishes. Fishes usually caught by this method include mackerel, herring, sardines, and various species of tuna, including yellowfin and albacore. When a school is sighted, it is surrounded by a curtain or wall of netting (often deployed by a small motorboat) that is buoyed at the surface and weighted at the bottom. When the school is enclosed, the bottom of the net is "pursed" by a line drawn through the purse rings, and the net is closed at the bottom, like an upside-down drawstring purse. The fish are then "brailed" (removed by dip net) or dumped on deck. In the late 1970s, purse seiners in the eastern tropical Pacific were "setting on dolphins," which meant finding schools of tuna that, for unknown reasons, swam beneath large aggregations of spinner and spotter dolphins. Both tuna and dolphins were gathered into the purse seines, and while the tuna fishermen benefited enormously, millions of dolphins died unnecessarily.

See also gill net, longline fishing, otter trawl, tuna-porpoise problem

pygmy killer whale (*Feresa attenuata*) Although the common name of this 8½-foot dolphin is derived from the shape of the skull and teeth, the pygmy killer whale may be even more fierce and aggressive than its larger namesake. In captivity, where the pygmy killer has been known to attack and kill other dolphins in its tank, it is probably the most aggressive dolphin toward humans. On the rare occasions pygmy killers have been encountered by divers in the wild, they have demonstrated no such inclinations. These animals are not common anywhere, but they have been spotted in the West Indies, around Florida, and in the Mediterranean, the Indian

pygmy killer whale

Ocean, the South Atlantic, and the eastern tropical Pacific. At sea, it is easy to confuse the pygmy killer with a similar species, the melon-headed whale (*Peponocephala electra*). **See also false killer whale, killer whale, melon-headed whale**

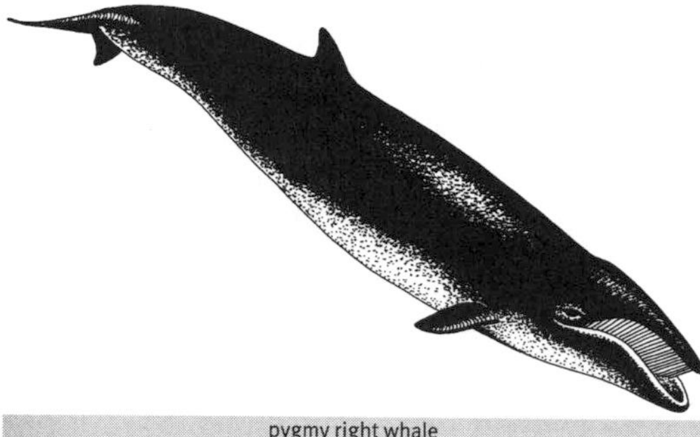
pygmy right whale

pygmy right whale (*Caperea marginata*) The smallest of the baleen whales, and one of the least-known cetaceans in the world. There are fewer than one hundred records of this animal, and almost nothing is known of its natural history. It is found only in the Southern Hemisphere, having been reported from South Africa, Australia, Tasmania, New Zealand, and various islands in the Southern Ocean, including the Falklands. The pygmy right whale can reach a length of 21 feet, and it has relatively long baleen plates, similar to those of the right whale, but it has a prominent dorsal fin, which its larger cousin does not.
See also right whale

pygmy shark (*Euprotomicrus bispinatus*) One of the smallest of all sharks, the pygmy reaches a maximum length of 10½ inches. It is blackish in color, with strongly marked white-edged fins. Not much is known about its habits, but it has been collected near the surface, always at night. It is believed that it descends to great depths during the day, probably to more than 5,000 feet, well beyond the range of midwater trawls. Catches have occurred randomly throughout the temperate and tropical waters of the world, always far offshore. **See also cigar shark, cookie-cutter shark, green dogfish**

pygmy sperm whale

pygmy sperm whale (*Kogia breviceps*) Because of the shape of their skulls, two small-toothed whales are believed to be related to the giant sperm whale: the pygmy sperm whale and the dwarf sperm whale (*K. simus*). The two little whales look very much alike, and both have sharp teeth in both jaws—unlike the sperm whale, which has erupted teeth only in its narrow lower jaw. They also have falcate dorsal fins, which the sperm whale lacks. Their biology is poorly known, but they have been reported from almost all the oceans of the world, except for the high polar latitudes. Both species have mysterious "bracket marks" behind the head that look very much like gills, but it is difficult to imagine why a whale would develop superficial similarities to a shark. **See also dwarf sperm whale, sperm whale**

Pytheas (fourth century B.C.) Greek navigator and astronomer who voyaged from the Mediterranean around the western and northern coasts of Europe, visiting Britain (and maybe even Iceland) and giving the ancient Greeks their first knowledge of these distant lands. Sailing out through the Pillars of Hercules (the Strait of Gibraltar), he began his voyage from Gades (Cádiz) in Spain, sailed past France and England, and certainly reached Scotland. He also sailed into the Baltic, perhaps as far as the Elbe estuary. Pytheas was also the first Greek navigator to recognize that the periodical fluctuations of the tides were connected with the phases of the moon. Although his writings have not survived, contemporaneous accounts by the geographer-historians Polybius and Strabo confirm his accomplishments.

Q

quadrant From the Latin *quadrans,* the fourth part, a nautical instrument for measuring the altitude of the sun. It was also called a backstaff. At noon, the sun would be directly overhead at the equator, so however far the sun deviates from that position gives a pilot a measurement of how far north or south he has sailed. Used during the sixteenth century, it gained its name because the user had the sun behind him when taking an observation; the cross-staff, which it replaced, required its user to look directly into the sun. (The term "quadrant" is also employed to indicate the quarters of the magnetic compass; each quadrant consists of 90 degrees.) **See also astrolabe, compass, sextant**

quahog (*Mercenaria mercenaria*) Also known as the hard-shelled clam, the quahog (pronounced "*co*-hog") is found along the Atlantic coast of North America, from Cape Cod to the Gulf of Mexico. The broad, thick shells of full-grown adult quahogs can be 6 inches long. From the shells' annual growth rings it has been determined that quahogs can live for more than two hundred years, placing them among the longest-lived animals known. The quahog is the clam of New England clam chowder, and in its smaller sizes, it is served as the littleneck or cherrystone clam. The name *Mercenaria* comes from the shells' use in making Indian money, or wampum. **See also clam**

quay A projection, usually made of stone, along the boundaries of a harbor to provide accommodation for ships to lie alongside, for the loading and unloading of cargo or for the embarkation and disembarkation of passengers. When a quay extends beyond the boundary of the harbor into the water, it is called a mole.

queen angelfish (*Holacanthus ciliaris*) So named because of the blue-ringed black crown on its forehead, the queen angel is one of the most spectacular of all Caribbean reef fishes. (The blue angel, *H. isabelita,* lacks the crown.) The young have five light blue vertical bars on the body, two of which border a dark bar through the eye. It can reach a length of 18 inches. **See also angelfishes**

queen angelfish

Queen Charlotte Islands North-south archipelago off the coast of western British Columbia, consisting of the large islands of Graham and Moresby and many smaller ones, covering a total of 3,700 square miles. The main settlement is Masset, on Graham Island. Approximately half of the 5,000 inhabitants of the Queen Charlottes are Haida Indians, descendants of the occupants of the islands when the Spanish explorer Juan Peréz first visited in 1774. James Cook arrived in 1778, but it was George Dixon who surveyed the islands in 1787 and named them for his ship and for the queen consort of King George III. (On his first voyage in 1769 Cook called the harbor on South Island, New Zealand, "Queen Charlotte's Sound," the same, minus the apostrophe, as the body of water between these Canadian islands and the northern tip of Vancouver Island.)

Queen Elizabeth Launched in 1938, the *Queen Elizabeth* was 1,031 feet overall, 12 feet longer than her sister ship, the *Queen Mary.* She had not been completed when World War II broke out, and by 1940, she had been refitted as a troopship with a capacity of 8,200 passengers. Between April 1941 and April 1943, she transported Allied troops wherever they were needed, often carrying as many as sixteen thousand men and ignoring safety precautions in the interest of the war effort. Hitler promised a huge cash reward and the Iron Cross to any U-boat commander or Luftwaffe pilot who sank the ship. After the war, she was fitted out as a Cunard passenger liner; she made her first voyage in October 1946, with a passenger capacity of about two thousand. After a distinguished career, she was withdrawn from service in 1969 and sold to American speculators, who stationed her in Port Everglades, Florida, as a tourist attraction and convention center. When fire safety requirements could not be met, she was sold to a Hong Kong investor, who planned to turn her into "Seawise University." Under suspicious circumstances, she caught fire in Hong Kong harbor in 1972 and was scrapped. The much smaller *Queen Elizabeth 2,* popularly known as *QE2,* was built in 1967, christened by her namesake in 1969, and used to carry troops to the Falkland Islands in 1982. The last of the great luxury liners afloat, *QE2* continues in service as a cruise ship. In 1998, the Norwegian shipbuilding company Kvaerner ASA sold the Cunard line to Carnival Cruise Lines of Miami. **See also *Queen Mary***

Queen Mary Planned as a replacement for the *Mauretania*, the *Queen Mary* was intended to be the epitome of luxurious cruise-ship design. In 1932, after Cunard merged with White Star Line, *Queen Mary* was launched, and for several years she transported passengers across the Atlantic in style while setting speed records for east- and westbound crossings. She was 1,000 feet in length, displaced 80,774 tons, and was capable of speeds in excess of 30 knots. In 1940, she and her newly built sister ship, the *Queen Elizabeth*, were converted to troop carriers, and they spent the war shuttling between Europe, Australia, and the United States with large contingents of troops aboard. On October 2, 1942, while approaching Scotland on a zigzag course to avoid U-boats, the *Queen Mary* collided with her destroyer escort, HMS *Curacao*, and sliced it in half, killing almost three hundred of her crew. After the war, during which the *Queens* had traveled more than 600,000 miles and transported nearly a million soldiers, the *Queen Mary* was handed back to Cunard White Star and refitted for her original purpose as a luxury liner. By the 1950s, however, transatlantic airplane travel was becoming more popular, and cruise ships, no matter how fast, could not compete with eight-hour crossings. (The normal duration for an Atlantic crossing was four days.) *Queen Mary* made her last trip in 1967, then was sold to an American company for £1,230,000 and taken to Long Beach, California, as a museum, hotel, and convention center, where she resides today.

See also *Queen Elizabeth*

queen parrot fish (*Scarus vetula*) What is it about the colors blue, yellow, and green that causes people to name fishes "queen" when they display these colors? The queen parrot fish, found in the West Indies and Bermuda, begins life as a drab, grayish creature with a faint white stripe on the flanks, but as it metamorphoses from female to male, it assumes the bright color scheme of the other queens, angelfish and triggerfish. It reaches a maximum known length of 20 inches.

See also parrot fish

Queensland grouper (*Protomicrops lanceolatus*) In East African and Indo-Pacific waters, the giant grouper is one of the largest of all reef fishes. It has been known to achieve a weight of 880 pounds, and may grow even larger. The common and scientific nomenclature of this fish is completely muddled; it is sometimes referred to as the genus *Epinephelus*, and throughout its range, it is called brindle bass, garrupa, jewfish, and giant grouper. **See also giant sea bass, jewfish**

queen triggerfish (*Balistes vetula*) Like all triggerfishes, the queen has a spine before the dorsal fin that can be erected and locked into the upright position by a triggerlike mechanism. While some of the Pacific triggerfishes are among the most colorful fish in the ocean, most Caribbean varieties are drab in color. The exception is the queen trigger, with its blue, green, and yellow coloration, accented by a vivid pattern of stripes.

queen triggerfish

See also humuhumunukunukuapua'a, triggerfish

Quemoy Now known as Jinmen or Chin-men, Quemoy is an island in the Taiwan Strait, 150 miles west of Taiwan but within a stone's throw of mainland China. It is the largest of a group of twelve islands collectively known as the Quemoy Islands. The hilly island is 51 square miles in area, with a largely agricultural population of 50,000. It was occupied in 1949 by the Nationalist Chinese when they fled the mainland. In 1958 Quemoy and Matsu (150 miles to the north) were subjected to heavy artillery bombardment by the Communist Chinese. In support of the Nationalists, the United States dispatched the Seventh Fleet to patrol the strait, and further hostilities were avoided. It remains under the jurisdiction of Taiwan.

Quirós, Pedro Fernández de (1565–1615) Born in Portugal, Quirós became Spanish in 1580 when the two crowns were united under Philip II. He emigrated to Peru, where in 1595 he was appointed chief pilot to Álvaro de Mendaña, who was setting out on an expedition to colonize the Solomon Islands, which he had discovered thirty years earlier. Mendaña failed to find the Solomons and died on Santa Cruz Island. Quirós returned to Manila, discovering Ponape in the Caroline Islands on the way. In 1605, he set out again to find Terra Australis Incognita. When he landed on a previously undiscovered island, he believed he had found the mysterious southern continent, and in honor of his king, who was also Archduke of Austria, he named it Austrialia del Espiritu Santo. It wasn't a continent; it was only a 1,500-square-mile island in the New Hebrides (now Vanuatu), which he couldn't find again after he was blown off the leeward shore. When Quirós returned to Mexico, he left the expedition under the command of Luis Váez de Torres, who discovered the Torres Strait, between the southern shore of New Guinea and the Cape York Peninsula of Australia. **See also Mendaña, Torres**

R

Rabaul Town on the island of New Britain in the Bismarck Archipelago, now part of Papua New Guinea. With one of the finest harbors in the world, Rabaul was chosen by the Japanese to be their major naval and air base for the projected invasion of Australia during World War II. In January 1942, they captured the island with twenty thousand troops, and massacred or imprisoned the Australian garrison stationed there. Japanese planes and ships deployed for the battles of Guadalcanal, New Georgia, and Bougainville were based at Rabaul. For the Americans, the destruction of the base was one of the most important objectives of the Solomon Islands campaign—indeed, of the entire war in the Pacific—and they devoted enormous amounts of men and matériel to this mission. By late 1943, the base was completely destroyed. In 1994, the volcanoes of Rabaul erupted with no warning, rediucing large portions of the town to ashes—again.
See also Bougainville Island, Guadalcanal, New Britain

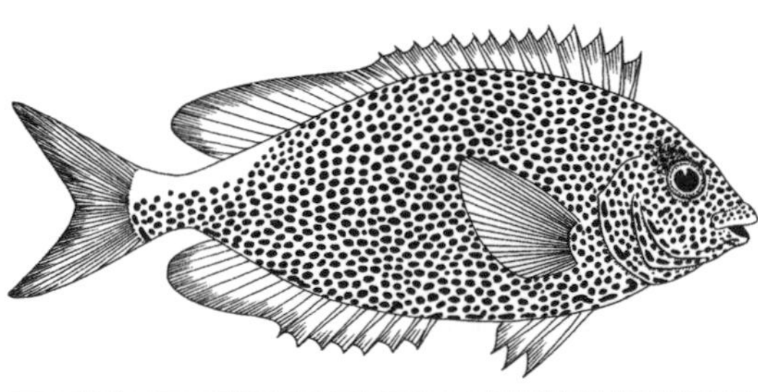
coral rabbitfish

rabbitfishes (family Siganidae) Obviously named for their distinctive mouths, rabbitfishes nibble and scrape algae off rocks with their buck teeth. In Australia, these 1-foot-long fishes are known as spinefeet, because of the venomous spines in their pelvic, dorsal, and anal fins, which contain a highly toxic venom. There are about twenty-eight species of rabbitfishes in the waters of the Indo-Pacific region. Most are colored in an arrangement of blue and yellow, but one species, commonly known as the foxface (*Siganus vulpinus*), is bright yellow with a white head and strong black markings.

radar System for detecting the position, motion, and nature of a remote object by means of radio waves reflected from its surface. Radar is an "active" sensing device, meaning that it generates electromagnetic energy, as opposed to a "passive" device, which can only recognize objects by their reflections. (Vision and hearing are passive senses.) Because the speed of radio waves is known, the time it takes for the echo to bounce back can be translated into distance from the object. The direction can be read by the direction in which the transmitting aerial points at the time of transmission. As the aerial revolves to scan around a ship, information is presented two-dimensionally on a cathode ray tube, giving a complete picture of all surrounding radio-reflective objects. Although the principles were understood as early as the 1880s, when James Clerk Maxwell formulated the general equations about the electromagnetic field, serious work on radar was not performed until the 1930s. Originally known as RDF (for "radio direction finding"), its current name comes from the acronym for RAdio Detecting And Ranging. During World War II, it was found to have many military applications, such as the detection of U-boat periscopes (radar cannot detect objects under water, because radio waves cannot be transmitted through water) and the location of enemy ships. Many ships, buoys, and other floating objects are equipped with metal "radar reflectors" to make them more visible to radar receivers. Advances in technology have led to more uses of radar, such as air traffic control, weather forecasting, remote environmental sensing, and speed measurements of everything from automobiles to pitched baseballs. Radar beams were reflected back from the moon in 1946, and from Venus in 1958, thereby opening a completely new field: radio astronomy. With more and more powerful transmitters, the Ballistic Missile Early Warning System uses radar, as does the system that tracks orbiting satellites. **See also sonar**

radiolarian Single-celled animals composed of radiating threads of protoplasm supported by a perforated central skeleton. The skeletons, which come in astounding shapes resembling snowflakes, urns, hats, and spheres-within-spheres, are made of silica, and when they fall to the bottom—which they do in countless billions—they form the "radiolarian ooze" that blankets some 3 million square miles of the floor of the equatorial Pacific and Indian Oceans. Like foraminiferans, radiolarians trap and digest microscopic plants and animals, such as diatoms, flagellates, and even copepods in their young stages. **See also diatoms, dinoflagellates, foraminifera, sediments**

Rae, John (1813–1893) Scottish physician and Arctic explorer. He was employed by the Hudson's Bay Company in 1846 to survey the Boothia Peninsula, and after charting 700 miles of its coastline, he was able to show that it was not an island. He spent a year in the Arctic with twelve men, wintering in a stone house and shooting and fishing for food. In 1848, he joined Sir John Richardson in the search for the Franklin expedition, which had been missing since 1845, and in 1851, he returned to the Arctic to explore the coastline south of Victoria Island, for which he received the Gold Medal of the Royal Geographic Society. In 1854, while mapping the west coast of the Boothia Peninsula, he met with some Eskimos who were able to tell him that Franklin had starved to death, and he also obtained some pieces of silver plate that were stamped with the name or initials of Franklin and other officers. When he returned to England, he collected the ten-thousand-pound reward that the Admiralty was offering for information that would confirm Franklin's fate.

See also Franklin, Kane, McClintock, McClure, Ross (John)

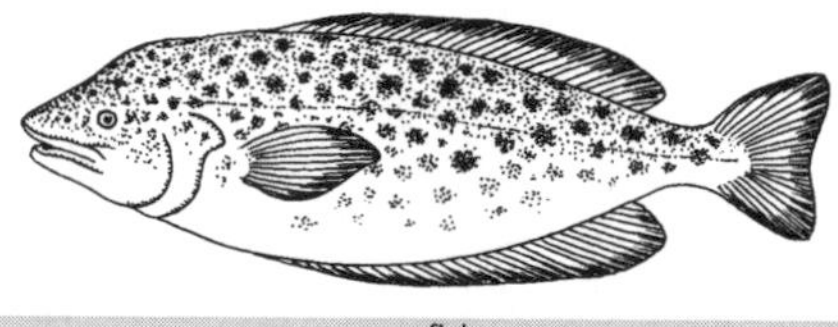

ragfish

ragfish (*Icosteus aenigmaticus*) A little-known species from the North Pacific, the ragfish lacks scales and fin spines, and its skeleton is largely cartilage. Their insubstantial structure may be responsible for their common name because they droop noticeably when carried. There may be two species of ragfish—*I. aenigmaticus,* which does not exceed 2 feet in length, and the much larger *Acrotus willoughbyi,* which may get to be 7 feet long—but they also may represent different stages of the same species. Sperm whales captured off the Queen Charlotte Islands of British Columbia during the 1930s frequently had ragfishes in their stomachs.

ragged-tooth shark: See sand tiger shark

rainbow runner (*Elagatis bipinnulata*) A sleek, streamlined relative of the jacks and pompanos, the rainbow runner gets its common name from its bright greenish-blue back and a succession of lateral yellow and blue stripes. It is an inhabitant of tropical waters around the world and is a popular game fish. The world's record, caught off Baja California in 1991, weighed 37 pounds.

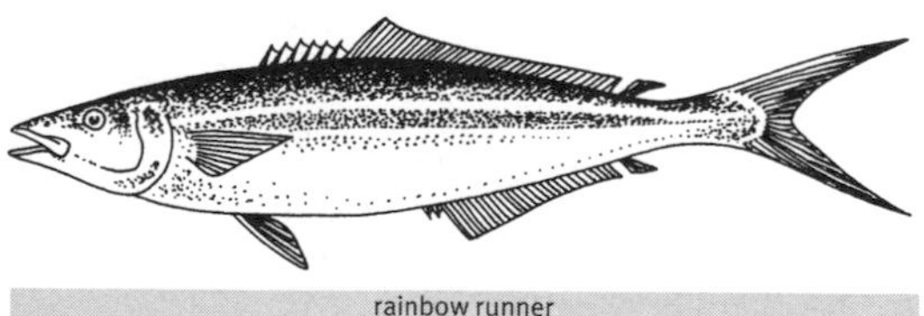

rainbow runner

See also jacks, pompano

Raleigh, Walter (1553–1618) British adventurer, explorer, and courtier. He served in the Huguenot army in France in 1569, and was probably in Paris at the time of the St. Bartholomew's Day massacre in 1572. When his half brother Humphrey Gilbert obtained a patent from Queen Elizabeth to "discover and take possession of remote lands," Raleigh enthusiastically joined the expedition, but it became merely a piratical venture against the Spanish that met with little success. When he returned to England, he immediately became a favorite of the queen and grew rich from her generosity. After suppressing an Irish revolution in Munster, he was given 40,000 acres there, and introduced the potato as a crop. Raleigh then helped to finance Gilbert's 1583 voyage, in which Gilbert discovered and took possession of Newfoundland for England. Although he conceived and organized the unsuccessful 1584 attempts to colonize America, the closest he ever got to North America himself was when one of the settlements on Roanoke Island (in present North Carolina) was named the "Citie of Raleigh." Appointed vice admiral of Devon in 1588, he took no part in the battle with the Spanish Armada. At court he was falling under the shadow of Robert Devereux, the Earl of Essex, so he led a campaign to Portugal to foster a revolt against Phillip II of Spain, which failed miserably. In 1592, Elizabeth had him thrown into the Tower of London for secretly marrying Elizabeth Throckmorton, one of her maids of honor, but he bought his release with profits from an earlier privateering voyage in which he had invested. He led a group to South America to search for gold in 1595, and although he did not find the fabled El Dorado, he wrote *The Discoverie of Guiana,* one of the finest travel narratives of the Elizabethan period. In 1596, he went with Devereux in an unsuccessful attempt against Cádiz, and he was Devereux's rear admiral in a 1597 expedition to the Azores. When the queen died in 1603, Raleigh's star fell precipitously. He was accused of plotting against her successor, James I, and was again thrown into the Tower, where he remained until 1616. He was released when he promised the king that he would find for him a gold mine in Guiana, and he sailed to the mouth of the Orinoco, where he took ill. The king had promised the Spanish ambassador that he would execute Raleigh if his forces clashed with the Spanish there, and since they did (Raleigh's son Walter was killed in the battle), he was arrested and executed upon his return. The capital of the state of North Carolina is named for him, but it is not the site of the original settlement. **See also Gilbert**

Rangiroa Coral island in the Tuamotu Archipelago; part of French Polynesia. Rangiroa, the most populated of the group, has the only airport, with regularly scheduled flights from Tahiti and Bora-Bora. Only Kwajalein in the Marshall Islands has a larger atoll; Rangiroa's is 48 miles long and 15 miles wide, big enough to enclose the entire island of Tahiti. Encircling the lagoon are more than two hundred coral outcroppings, known as *motus*. The island is home to some 2,000 people, almost all of whom live in the two settlements of Avatoru and Tiputa. Most visitors come for the marine life in the lagoon, where there are myriads of fishes, sharks, mantas, and morays, and to ride the onrushing current through Tiputa Pass.

See also French Polynesia, Tuamotu Archipelago

Rarotonga The largest of the Cook Islands, Rarotonga is a 26-square-mile volcanic island in the South Pacific, some 2,100 miles northeast of New Zealand. It was the first island visited by the mutineers of the *Bounty* in October 1789, but they found no place for a settlement and sailed on to Tonga and eventually to Pitcairn. The arrival of Fletcher Christian and his crew marked the European discovery of the island, but it was obviously not reported. The "official" discovery of Rarotonga occurred in 1823, when the Reverend John Williams of the London Missionary Society landed there and used it as a base to Christianize the adjacent islands. The fringing coral reef makes the island appealing to divers, and the construction of an airport in 1973 led to a great increase in tourism. **See also *Bounty,* Christian, Cook Islands, Pitcairn Island**

rattail: See grenadier

rat-trap fish

rat-trap fish (*Malacosteus niger*) This 6-inch-long fish is also known as the loose-jaw, because its mouth has no floor or walls, eliminating water resistance as it snaps up fishes that may be as large or larger than it is. Rat-trap fishes have a comma-shaped light organ under the eye that glows red, an unusual color, since the photophores of most other deep-sea fishes glow with a blue-green light. Recent studies have shown that this species can emit red light from the photophore to illuminate its prey, and it can see the red image because its retinas contain a chlorophyll derivative. No animal can produce chlorophyll—the ability to do so defines plants—so *Malacosteus* appears to obtain the substance from eating fishes that have eaten copepods, which in turn have eaten chlorophyll-containing phytoplankton. How the chlorophyll is incorporated into the fish's retina is unknown. **See also bioluminescence**

rays (order Batoidei) Flattened, cartilaginous fishes that are mostly bottom dwellers, but some species, such as the mantas, are creatures of midwater and the surface. The pectoral fins have evolved into broad, winglike appendages, which the animal flaps for locomotion. In most forms, the eyes are on the dorsal surface, the mouth and multiple gill slits are on the underside, and there is a long, whiplike tail. Most rays feed on shellfish on the bottom, but the mantas are filter-feeders, straining planktonic organisms from the water. Because they lie on the bottom, rays have developed a variety of defense mechanisms, including electricity (the electric rays or torpedoes) and assorted poisonous stingers and barbed spines (the stingrays). Skates (Rajoidei) are closely related to the rays, and are usually smaller and more rounded. The guitarfishes (Rhinobatoidei) are more sharklike in form, but they are classified with the rays because their mouths and gill slits are on the underside of the head. Sawfishes too, although decidedly sharklike in shape, have their mouths and gill openings on the bottom.

See also guitarfish, manta, sawfish, skates, torpedo

razorbill (*Alca torda*) Sometimes known as the razor-billed auk, this member of the family Alcidae is a smaller version of its extinct larger relative, the great auk. The razorbill is a competent flyer; the great auk, with its tiny flippers, could not fly at all. The razorbill gets its name from its laterally flattened bill, which has a white band toward the tip and a white line that extends from the base of the bill to the eye. The bird is black above and white below; it has a wingspan of about 27 inches. (The guillemot, which is about the same size and coloration, has a much narrower bill.) When the bird swims, its tail is cocked rather high. Found only in the North Atlantic, razorbills nest in colonies on cliffs in North America, Greenland, France, Norway, Sweden, Finland, Russia, Iceland, the Faeroes, and the British Isles.

razorbill

See also great auk, murre

red algae (*Rhodophyta*) Red algae live in deeper water than the green and brown seaweeds. They come in forms ranging from flat sheets to branches, and often feel slippery to the touch. While other seaweeds live in shallow coastal waters, some red algae inhabit waters down to 650 feet, largely because of their ability to use available light more efficiently for photosynthesis. Red algae have been harvested for years in Asia for human consumption, with the annual harvest worth more than $500 million. In addition, red algae are important ingredients in fertilizer and animal feeds, and because of their unusual jelling properties, they are a source for agar and carrageenan, which are used as thickeners in many products. Agar, which is extracted from *Gelidium* and *Gracilaria,* is the preferred solid medium for growing bacteria in the laboratory, and carrageenan, which comes from the red alga commonly called Irish moss (*Chrondus crispus*), is used in the manufacture of ice cream, toothpaste, infant formulas, and puddings, such as blancmange.

red crab (*Gecarcoidea natalis*) Every year on tiny Christmas Island, some 220 miles south of Java, hundreds of millions of large, bright-red crabs—some measuring 1 foot between outstretched claws—come ashore. They mate in burrows excavated by the males, and twelve days later, each female gives birth to thousands of tiny red crabs. Hundreds of thousands die during their attempt to return to the sea because they dry out or because they are run over by trains or cars. After about three weeks at sea, the baby crabs return to the island by the millions, overwhelming the small settlement, and climbing through windows, under doors, and into every nook and cranny. During the dry season, they are hardly visible, living in burrows in the forests, and dining on the leaf litter on the ground. **See also crab**

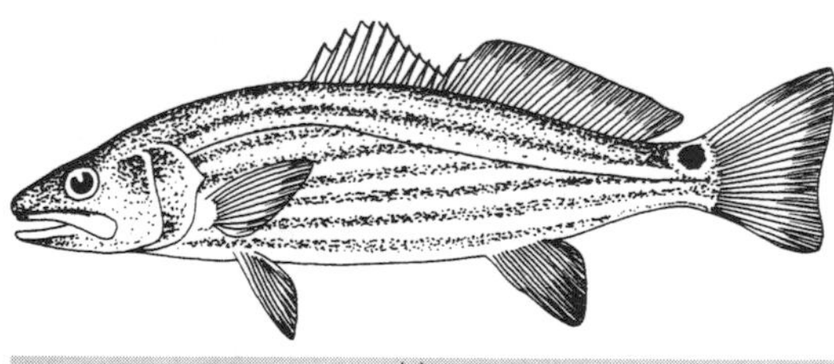

red drum

red drum (*Sciaenops ocellatus*) Also known as channel bass, redfish, and redhorse, the red drum is found in the western Atlantic from Maine to the Gulf of Mexico, in schools over sandy or muddy bottoms. They are popular game fish and can exceed 4 feet in length. The record, taken off North Carolina in 1984, weighed 94 pounds. They are considered tasty food fishes at under 10 pounds, when they are known as puppy drum, but the flesh of the larger specimens is not as desirable. Around 1986, the popularity of the Cajun recipe for blackened redfish caused an enormous increase in the Gulf of Mexico fishery for this species, and in 1987, 5 million pounds were landed in Louisiana. The fishery crashed, and the popularity of the species declined proportionally; nonetheless, the red drum is now considered an endangered species. **See also black drum**

red-footed booby (*Sula sula*) The smallest of the boobies, with a wingspan that does not exceed 40 inches, the red-footed booby comes in a variety of color morphs: all-white, brown, and a assortment of mixed plumages, such as the white-tailed brown morph and the white-headed brown morph. In all cases, the bill is pale bluish-gray, the feet and legs bright red. Unlike other boobies that nest on rocks or cliffs, the red-footed booby nests in trees. It is found throughout the Caribbean as far south as Trinidad, throughout the tropical Pacific and the Indian Ocean, and also in western Mexico and the Galápagos. The name "booby" comes from the Spanish *bobo,* which means "stupid," and refers to the birds' reluctance to move when approached by people.

See also blue-footed booby, brown booby, gannet, masked booby

red tide A discoloration of seawater, caused by a concentrated bloom of dinoflagellates. It is not always red, but may show up as brown, pink, violet, orange, yellow, green, or blue. On the Gulf Coast of Florida and Walvis Bay, South Africa, the red tide is caused by *Gymnodinium brevis,* which often results in massive fish kills. When clams and oysters ingest the dinoflagellate *Gonyaulax catenella,* the toxins do not harm the shellfish, but accumulate in their tissues. When the shellfish are eaten by people, the concentrated neurotoxins are so powerful that death can result, a phenomenon known as paralytic shellfish poisoning (PSP). Most outbreaks of PSP have occurred along the northwest coast of North America and the coasts of northwestern Europe, and in Japan.

See also dinoflagellates, paralytic shellfish poisoning

remora (*Remora remora*) With a dorsal fin modified into a sucker disk, the 24-inch-long remora attaches itself by suction to various larger fishes and cetaceans

remora

and hitches a ride. As the host feeds, these suckerfishes dash out to eat the scraps. Of the seven genera, some are host-specific, such as the shark remora (*Echeneis naucrates*), the whale remora (*R. australis*), and the spearfish remora (*R. brachyptera*). They also hitchhike on turtles and, occasionally, boats. The Latin word *remora* means "delay" or "hindrance," because the fishes were thought to slow ships down when they attached themselves to the hull. **See also pilot fish**

requiem sharks Another name for the Carcharhinidae, a large family of sharks. Although there are many records of these sharks attacking people, their name does not derive from "requiem," meaning a mass for the dead. The French *requin* means "shark," and the name of this group of sharks is believed to be a corruption of that. **See also carcharhinid sharks**

Resolution A common ship's name in British naval history; the best known was a Whitby collier of 461 tons and 110 feet overall that was originally christened *Marquis of Granby* before being renamed HMS *Resolution* and assigned to Captain James Cook for his second exploration voyage around the world. In 1772, with Captain Tobias Furneaux commanding *Adventure,* Cook headed south and, on January 13, 1773, became the first man to cross the Antarctic Circle. They headed for Tahiti and Tonga, and then New Zealand, Easter Island, Cape Horn, and South Georgia, but failing to find the southern continent, Cook returned to England, arriving on July 29, 1775. In 1776, Cook took *Resolution* in search of the Northwest Passage, this time accompanied by Captain Charles Clerke in HMS *Discovery.* They visited Van Diemen's Land (Tasmania), Tonga, Tahiti, and the Sandwich Islands (Hawaii), and then headed north to Alaska. They spent the winter of 1777 in Hawaii, then went north again, returning to Hawaii in February 1778. They sailed on February 4, but a sprung foremast in *Resolution* caused them to turn back, and in a confrontation with the Hawaiians, Cook was killed. Clerke assumed command of *Resolution* and attempted again to find the Northwest Passage, but he died at sea, and the ship arrived back in England on August 22, 1780, after a voyage that had lasted more than four years. **See also Cook (James)**

Réunion One of two islands in the Indian Ocean, east of Madagascar and 110 miles apart, known as the Mascarene Islands. The other is Mauritius. Réunion is about 40 miles long and 30 miles wide, covering an area of 970 square miles. It is mountainous, with the Piton des Neiges reaching a height of 10,000 feet. It was first settled in the seventeenth century by French colonists, who named the island Bourbon and imported East African slaves to work the sugar and coffee plantations. Later, Malays, Chinese, and Indians from Malabar were brought in as indentured laborers, which has given the island its mixed (creole) population. Réunion was a French colony until 1946, when its status was changed to overseas department. (Mauritius, once a part of the British Commonwealth, is now an independent republic.) The total population is around 630,000. The dodo of Mauritius was exterminated in the seventeenth century, and the same fate befell its close relative, the Réunion solitaire, sometimes called the white dodo. The Mauritius dodo was grayish brown with perky white tailfeathers; the Réunion solitaire, a flightless bird of the same size and shape, was pure white.
See also dodo, Madagascar, Mauritius

Revillagigedo Islands Uninhabited archipelago in the Pacific, belonging to Mexico, approximately 300 miles southwest of Cabo San Lucas, the southern tip of the Baja Peninsula. These volcanic islands (and the island in southeast Alaska on which the city of Ketchikan is located) were named for Juan Francisco de Güemes y Horcasitas, the Count of Revilla Gigedo, who was the viceroy of New Spain from 1746 to 1755. The largest island is Socorro, which is 24 miles long and 9 miles wide. In 1952 a volcano appeared on the island of San Benedicto, 40 miles north of Socorro. Attempts are being made to reintroduce the Socorro dove (*Zenaida graysoni*), which has been extinct in the wild since 1978. The humpback whales that make an annual migration from Hawaii to southeast Alaska sometimes detour to the Revillagigedos. (They are identifiable by the individual patterns on the underside of their flukes.)
See also humpback whale

Reynolds, Jeremiah N. (1799–1859) American lecturer and author. While a newspaperman in Ohio, he became enthusiastic about John Cleves Symmes's theories that the earth was hollow and could be entered by ship from the "holes in the poles." Reynolds and Symmes went on a nationwide lecture tour in 1828 to drum up interest in an exploratory voyage to the north, but when Symmes died the next year, Reynolds abandoned his crackpot theories and promoted a general scientific expedition instead. Congress turned down his request for funding, so he organized his own excursion to the South Pacific. In 1829, the ships *Seraph, Annawam,* and *Penguin* sailed to the coast of Chile, where they accomplished little, and because the crews began to desert, they were forced to return. Reynolds remained in Chile, where he heard the stories of a white whale that sank ships. In 1839, in *The Knickerbocker,* America's leading literary magazine, he published "Mocha Dick, the White Whale of the Pacific," a story that is known to have influenced Herman Melville. (Mocha is an island off the coast of Chile, and has nothing to do with the whale's color.) Reynolds became a banker in Texas, where he died. (His plans for scientific investigation were adopted by U.S. Navy in 1836, and turned into the U.S. Exploring Expedition of 1838–1842, under Lieutenant Charles Wilkes.)

See also Melville, Mocha Dick, Symmes, U.S. Exploring Expedition

Rhodes The largest of the Greek islands known as the Dodecanese, and the easternmost part of all modern Greece. (*Dodecanese* means "twelve islands," but there are quite a few more.) Rhodes is separated from Turkey by the Strait of Marmara. The Minoans had outposts on Rhodes during the fourteenth and thirteenth centuries B.C., but when the Minoan civilization collapsed, Rhodes became a powerful, independent kingdom in its own right. In 480 B.C. Rhodes supported the Persians at Salamis, but two years later, it became a member of the Dorian League led by Athens. Following the death of Alexander the Great in 323 B.C., Rhodes was ruled by Ptolemy I of Egypt. In 294 B.C., Rhodes withstood a massive siege by the Macedonians and erected the famous Colossus in 282 B.C. to commemorate their success. The Saracens invaded the island in the fifth century; the Arabs two centuries later. In subsequent years, the Crusaders, the Turks, and the Genoese laid claim to the island, and in 1309, the Genoese sold the island to the Knights of St. John of Jerusalem (the Knights Hospitallers), who fortified the city of Rhodes and remained there until 1522, when they capitulated to Suleiman the Great. Rhodes remained under Turkish rule until 1912, when the Italians wrested it away during the battle for Libya. When the Italians surrendered to the Allies in 1943, Rhodes became a bloody battleground for the fighting between the Germans and the British, and after the war, the Dodecanese were united with Greece. The capital city is also called Rhodes—or *Ródhos,* as the Greeks say. Because of the richness of its history, Rhodes is one of the most popular tourist destinatons in Greece.

Rhodes, Colossus of One of the Seven Wonders of the Ancient World, the Colossus of Rhodes was a gigantic bronze statue erected shortly after 300 B.C. by the citizens of Rhodes to celebrate their successful defense of the city against the powerful Macedonians. Designed and sculpted by Chares of Lindos, a pupil of Lysippus, it was a 110-foot-tall hollow bronze statue of Helios, the god of the sun. At that height, it would have been far too small to straddle the harbor as some interpretations show. (The harbor is 1,300 feet across.) We do not know what it looked like, but it was probably a naked man with one arm upraised. It fell after standing for only fifty-six years when an earthquake hit Rhodes about 225 B.C. The statue broke at the knees, tumbling onto the ground. It lay there, a mass of twisted metal and piles of stones (which had been used to weight it during its construction), until A.D. 653 when the Arabs invaded Rhodes, disassembled the Colossus, and sold the bronze for scrap to a Syrian, who is said to have required nine hundred camels to transport the pieces.

See also Rhodes

ribbonfish (*Desmodema polysticum*) Like the oarfish to which it is related, the ribbonfish is a long, flattened fish, silvery in color, with bright red fins. The juveniles, which look nothing at all like the adults, are heavily spotted. Several other species of long, flattened fishes are also commonly referred to as ribbonfishes, including the scalloped ribbonfish (*Zu cristatus*), whose unusual generic name comes from Zu, the Babylonian god of storms, because the only known specimens were washed ashore after a storm. Juveniles of *Zu cristatus* have streamers on their dorsal fins that may be five times longer than the fish itself. In South Africa, the cutlass fish (*Trichiurus lepturus*) is sometimes called ribbonfish.

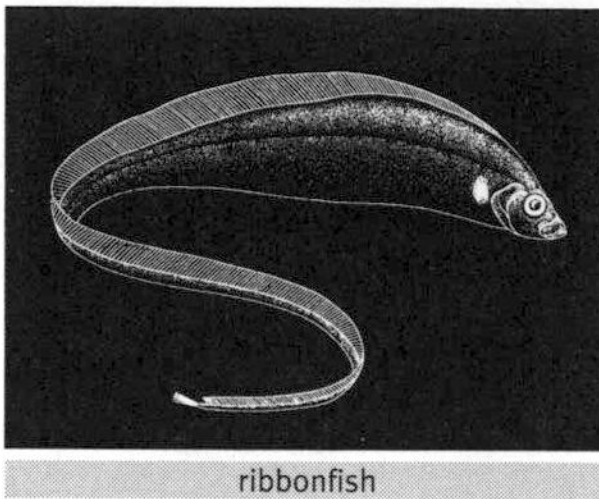

ribbonfish

See also cutlass fish, oarfish

ribbon seal (*Histriophoca fasciata*) Probably the most strikingly colored of all pinnipeds, the ribbon seal is found only in the Arctic North Pacific, from Alaska to northern Japan. An estimated 250,000 live around the Sea of Okhotsk. They are creatures of the ice, and show a remarkable indifference to the approach of predators, including polar bears, Arctic foxes, or humans. Ribbon seals are born white, but after a couple of weeks, the white coat, called the lanugo, is shed. The "ribbons" that characterize the adults do not appear until they are between two and four years old.

ribbon seal

See also harbor seal, harp seal, ringed seal

ribbon worms (phylum Nemertea) Marine worms that range in size from threadlike species, a fraction of an inch long, to the bootlace worm (*Lineus longissimus*) of the North Sea, which has been reported at a stretched-out length of 100 feet. Most, however, are considerably smaller: long, narrow, carnivorous worms that live in crevices in rocks or burrow into the sand. Most have a tubular proboscis that is everted for feeding and can be as long or longer than the worm itself. The proboscis is covered with a sticky mucus that entangles the prey; in some species, it is armed with a sharp spike (stylet) that pierces the prey and injects a toxic secretion. If a ribbon worm is damaged, a new worm can regenerate from any portion that contains even a portion of one of the lateral nerve cords. Reproduction usually takes place externally, but some species reproduce by fragmentation, where a large specimen may break up into as many as twenty segments, each of which grows into a new worm. Found under rocks in the colder waters of the world, from Siberia to South Africa, is *L. ruber,* a dark reddish or greenish ribbon worm that may reach a length of 10 inches. **See also polychaetes**

Rickover, Hyman (1900–1986) American admiral who served in World War II as head of the Bureau of Ships, and in 1946 was assigned to the nuclear research facility at Oak Ridge, Tennessee. He was responsible for the construction of the USS *Nautilus,* the world's first atomic submarine, commissioned in 1954. Rickover's career was marked by controversy, because he was Jewish and also because of his autocratic and outspoken nature. He later became chief of the Naval Reactors Branch of the Atomic Energy Commission, and was in charge of the development of the missile-firing submarines of the *Polaris* project.

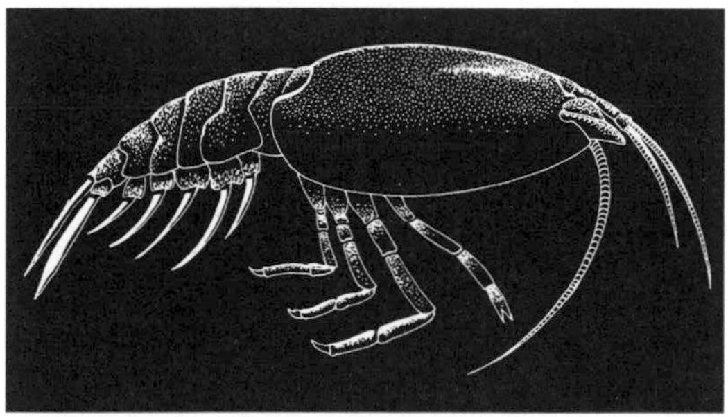

rift shrimp

rift shrimp (*Rimicaris exoculata*) Discovered in 1985 at the 11,000-foot-deep Trans-Atlantic Geotraverse (TAG), these 2-inch-long shrimp occur in dense schools in the immediate vicinity of hydrothermal vent sites. They have no eyes (their scientific name can be translated as "rift shrimp without eyes"), but on their dorsal surface there is a pair of light-sensitive organs just below the skin. Since they live in total darkness, the ability to "see" is probably unnecessary, but these organs may be useful in detecting the faint light emitted by the vents. Like many other hydrothermal vent animals, rift shrimps do not breathe oxygen, but subsist on sulfides dissolved in the water or scraped off the sides of the mineral chimneys.

See also hydrothermal vents, tube worms

right whale (*Eubalaena glacialis*) Because they produced plentiful oil and whalebone (baleen), were slow swimmers and easy to kill, and floated when dead, these animals became known as the "right" whales to hunt. Because of their inshore breeding habits, right whales were often the first hunted when settlers arrived in a new area. Almost as soon as they were discovered, they were eliminated from the waters of Cape Cod, Alaska, South Africa, Japan, Australia, and New Zealand; only the lack of arable land and settlers in Patagonia saved the right whale population there. Some cetologists believe there are two distinct species, the northern (*E. glacialis*) and the southern (*E. australis*), but they are similar enough to be included here as a single species. Right whales are thickset, heavy animals with no dorsal fin; they measure up to 60 feet in length and weigh as many tons. They have extremely long baleen plates (up to 10 feet long), and feed by "skimming" through shoals of small crustaceans, allowing the water to enter through the opening between the two "sides" of the baleen plates and pass out through the plates while the food items are trapped in the inner fringes. Right whales are born with "callosities" on their heads, the patterns of which remain constant throughout their lives, making identification of individual whales possible. There are some 750 right whales that breed in the protected bays of Peninsula Valdéz in southern Argentina, and perhaps 5,000 in the world's oceans. (The preexploitation population has been estimated at 200,000 animals.) The North Atlantic right whale, which breeds off the coasts of Georgia and Florida and feeds in the Gulf of Maine and the Bay of Fundy, numbers only a couple of hundred individuals, making it one of the rarest large animals in the world.

right whale

See also baleen, bowhead whale, whaling

ringed seal (*Phoca hispida*) Creatures of the Northern Hemisphere ice pack, ringed seals have a completely circumpolar distribution. They are the smallest of the earless seals (Phocidae), rarely reaching 4½ feet in length. Females give birth to 2-foot-long pups in caves or tunnels in the snow, a most unusual habitat, even for seals. Throughout their range, they are hunted by Eskimos who use the meat, blubber, and skins, and they are the favorite prey of polar bears, who wait for them to poke their heads out of breathing holes in the ice, then swipe them out and kill them. There are several subspecies of ringed seals, distinguished primarily by distribution. **See also harbor seal, harp seal, ribbon seal**

ringed seal

Ring of Fire Popular name for a belt of intense earthquake and volcanic activity that follows the outer limits of the Pacific Ocean basin. It is believed to be the definition of the Pacific Plate, whose motion creates the seismic activity. The "ring," which includes most of the active volcanoes on earth, incorporates the west coasts of North and South America as well as New Zealand, Indonesia, the Philippines, Japan, Kamchatka, and the Aleutian Islands. Major volcanic areas that are not in this 30,000-mile-long ring are the Hawaiian Islands, the volcanoes of the Mediterranean and the Caribbean, and those along the Mid-Atlantic Ridge, such as Iceland and the Azores. Some of the most destructive seismic events in history have taken place along the Ring of Fire, including major earthquakes in Japan, Chile, Indonesia, Alaska, and Mexico that have killed hundreds of thousands of people and caused incalculable property damage. The Ring of Fire contains some of the largest cities in the world, such as Tokyo and Los Angeles, and a major catastrophe still looms as a strong possibility. **See also Aleutian Islands, Hawaiian Islands, Iceland, Kamchatka, Krakatau, plate tectonics**

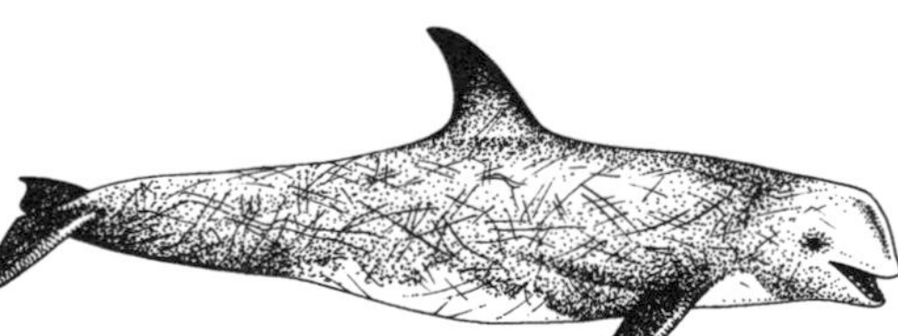
Risso's dolphin

Risso's dolphin (*Grampus griseus*) Easily identifiable at sea because of its light coloration, high dorsal fin, and beakless profile, Risso's dolphin can be found throughout the world's tropical and temperate waters. They vary in color from light to dark gray, but most tend toward the lighter color, and they are always found heavily scratched and scarred, which is thought to be a result of fighting among members of the species, or perhaps of trying to eat clawed squids. Risso's dolphins have teeth only in the lower jaw, and the forehead (more accurately known as the melon) has a unique vertical crease. The maximum length is about 13 feet. Amenable to training, they are often used in oceanarium shows, particularly in Japan. One of the most famous of all dolphins was Pelorus Jack, a Risso's dolphin that accompanied steamers in New Zealand waters at the end of the nineteenth and the early twentieth centuries. **See also bow riding, Pelorus Jack**

roaring forties The region in the Southern Ocean between the latitudes of 40° south and 50° south, where the prevailing wind blows strongly from the west. Sailing ships in the Australian trade would make for this area after rounding the Cape of Good Hope, so the westerly gales could speed them onward. A similar belt of powerful winds can be found between 40° north and 50° north in the North Atlantic.
See also Cape of Good Hope

robber crab (*Birgus latro*) Also known as coconut crabs, robber crabs are big, powerful, and dangerous inhabitants of South Pacific islands. They may measure 40 inches from head to tail and weigh as much as 35 pounds. Their claws are strong enough to open a coconut—or sever a finger. They are adept at climbing palm trees, and are said to be able to cut the coconuts loose so that they fall to the ground. Robber crabs are nocturnal and spend the daylight hours in holes in the ground. Although their gill chambers have been adapted to breathe atmospheric oxygen, they return to the water to lay their eggs. **See also crab**

Roberts, Bartholomew (1682–1722) "Black Bart" was a British pirate of rather unusual habits: he never drank anything stronger than tea, went to bed early, and did not work on the Sabbath. Despite these habits—or because of them—Roberts was probably the most successful pirate who ever lived. He began his career at the age of thirty-seven, and plied his trade off the coast of Africa and in the West Indies. He is said to have captured as many as four hundred vessels. His flag showed a figure of himself with a sword in his right hand and each foot standing on a skull. He was killed in a battle off the coast of West Africa with Chaloner Ogle, a British naval officer who had been sent after him, and who was knighted for having brought down Batholomew Roberts. Black Bart was not yet forty when he died; he lived what he called "a merry life, but a short one." **See also piracy, privateer**

Robinson Crusoe Novel written by Daniel Defoe in 1719, about a sailor who is shipwrecked and marooned on an uninhabited island for twenty-seven years and rescues his only companion, Friday, from a cannibal's feast. Defoe based his novel on the true story of Alexander Selkirk (1676–1721), a British seaman who in 1703 joined the crew of a privateer commanded by Thomas Stradling for a voyage in the South Seas. After a violent disagreement with Stradling, Selkirk asked to be put ashore on the island of Juan Fernández, 400 miles west of Chile. He lived alone on the island for four years, subsisting on fruits, vegetables, and goat meat, when he was found by the privateers *Duke* and *Duchess*, commanded by Woodes Rogers. When he returned to England in 1711, Selkirk told his story to Richard Steele, who published it in *The Englishman* in 1713. This account was seen by Daniel Defoe, who wrote *Robinson Crusoe*, published in 1719. Several movies have been made, including a 1964 version entitled *Robinson Crusoe on Mars*. **See also Juan Fernández Islands, Rogers, Selkirk**

Rockall Uninhabited island in the North Atlantic, some 220 miles west of the Outer Hebrides and 180 miles from St. Kilda. It was annexed by Great Britain in 1955, and has now assumed a new importance because of the European Economic Zone 200-mile limits for fishing and oil exploration. Rockall is the only visible part of the Rockall Bank, a vast submerged plateau between the British Isles and the Mid-Atlantic Ridge. About 70 feet above sea level, it is a "stack" with no possible landing sites. During the early crossings of the Atlantic, it may have been sighted by other explorers like Martin Frobisher, who identified it as the "Sunken Land of Buss" but never managed to find it again. In time, when its location was verified, it began to appear on maps as Rokol, and later Rockall. British ornithologist James Fisher visited the island by boat in 1949 and by helicopter in 1955. The following year, he published a book about the history and natural history of this lonely rock, in which he listed the twenty-one species of birds that have been spotted on and around the island.

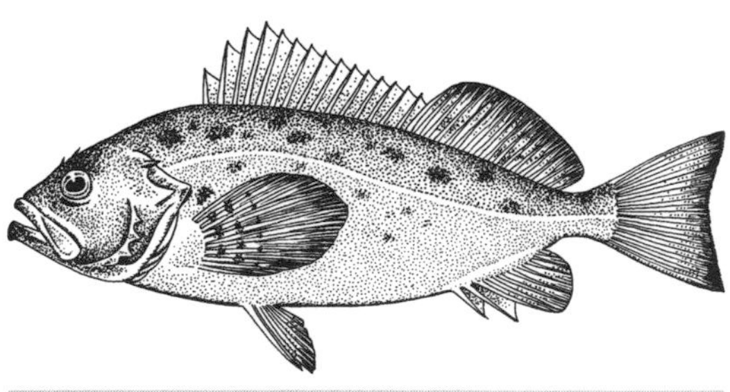

rockfish

rockfish (*Sebastes* spp.) Large family of fishes related to the scorpion fishes, but without the poisonous spines. They range in length from 1 to 3 feet, and are usually characterized by mailed, spiny cheeks and fanlike pectorals. Off the coast of California there are more than sixty species that are extremely popular with anglers; they are also the most important commercial catches in the state. Included are the bocaccio (*S. paucispinis*), the yellowtail rockfish (*S. flavidus*), the widow rockfish (*S. entomelas*), the vermilion rockfish (*S. miniatus*), the blue rockfish (*S. mystinus*), and the most important rockfish of all, the Pacific ocean perch (*S. alutus*). In the North Atlantic there is only one species, *S. marinus*, known variously as redfish, ocean perch, or rosefish, which is found from Spitsbergen, Iceland, and Greenland to the Gulf of St. Lawrence. A trawl fishery in the Gulf of Maine reached its maximum in 1942, and although the population has been steadily declining, the fishery continues. **See also scorpion fish**

rockhopper penguin (*Eudyptes chrysocome*) One of the species known as crested penguins (genus *Eudyptes*), the rockhopper is one of the smallest of the penguins; only the little blue penguin is smaller. It is a chubby little bird, with an orange-brown bill, pink feet, and a bright yellow stripe that does not meet above the bill but ends in long, bright yellow plumes that project laterally. The name comes from its habit of traveling on land by bouncing along with both feet held together. It often slides or bounces down banks and cliffs to get to the sea, and then jumps feet first into the water, unlike other species, which walk or dive into the sea. It is estimated that the total population of rockhoppers is about 3.7 million pairs.

rockhopper penguin

See also erect-crested penguin, fiordland penguin, macaroni penguin, royal penguin, Snares penguin

Rogers, Woodes (1679–1732) In 1708, a syndicate of Bristol merchants named Woodes Rogers commander of a world-ranging expedition in pursuit of the privateers who were posing a threat to their commerce. With a letter of marque as his authorization, he sailed the *Duke* and *Duchess* to the South Seas (with William Dampier aboard as navigator). Although they took few prizes, when they arrived at Juan Fernández Island in 1709, they found Alexander Selkirk, who had been living there for four years, ever since he had asked to be put ashore because of a disagreement with Captain Thomas Stradling of the ship *Cinque Ports*. (It was Selkirk's story that Daniel Defoe turned into *Robinson Crusoe*.) Off the coast of Mexico, Rogers intercepted the Manila galleon *Nuestra Señora de la Encarnación Desengano*, and brought her cargo of silks, gold, and gems back to England, where it was sold for £148,000.

Rogers wrote *A Cruising Voyage Round the World* in 1712, and in 1717, this ex-privateer was named royal governor of the Bahamas, with a specific mandate to stamp out piracy. (The British had abandoned their settlement at Nassau on New Providence because the French and Spanish raided it so often, and the pirates had taken over.) Only Charles Vane resisted; the rest of the pirates surrendered because the Crown was prepared to pardon them. Many of the pirates gave up their wicked ways, but Blackbeard continued his depredations along the American coast until he was captured and hanged in 1718. Rogers served two terms as governor of Nassau, where he died in 1732.

See also Blackbeard, Dampier, privateer, *Robinson Crusoe*, Selkirk, Vane

Roggeveen, Jacob (1659–1729) Dutch navigator who led an expedition to the South Pacific in 1721, sponsored by the Dutch West India Company for the purpose of establishing a commercial presence there. During the 1687 voyage of buccaneer Edward Davis, William Dampier had reported a large island along latitude 27° south that might be the coast of the great southern continent Terra Australis Incognita, and Roggeveen proposed to search for "Davis Land" by sailing west from Chile. He entered the Pacific through the Le Maire Strait, visited the Juan Fernández Islands, and on Easter Sunday 1722 he discovered an island that he called Paasch Eyland, Easter Island. He and his men marveled at the gigantic stone idols they saw, but the locals—who were completely naked, but covered from head to toe with elaborate tattoos—provided no explanation for their meaning or manufacture. When the natives crowded them too closely, Roggeveen's sailors panicked and shot a dozen of them, but peace was restored before the three Dutch ships sailed. After a week, Roggeveen headed west and spent a month in the Tuamotus. Discouraged by his failure to find the southern continent, he returned to Holland by way of Batavia (now Djakarta).

rogue wave A huge surface wave, not to be confused with the seismically generated tsunamis, which can reach great heights and do enormous damage if and when they reach land. Rogue waves—also known as freak waves—are caused by a poorly understood synchronicity of sinusoidal wave movements that result in a single wave that can be more than 100 feet high from the trough to the peak. Occasionally, the wave is preceded by a trough that resembles a deep moat. Reports of anomalous waves 60 to 80 feet in height are not uncommon; in 1933, while sailing through a South Pacific storm, the U.S. Navy tanker *Ramapo* encountered (and survived) the largest wave ever measured, 112 feet high as calculated by the ship's officers. In 1942, while ferrying Allied troops across the North Atlantic, the *Queen Mary* was slammed broadside by a mountainous wave that burst her pilothouse windows 90 feet up and lay her so far over that she was within a few degrees of sinking. The Italian passenger liner *Michelangelo* was struck by a huge wave in 1966 that tore a hole in the bow, submerged the bridge, and killed three people. Off the southeastern coast of Africa, between Durban and Port Elizabeth, there have been a disproportionate number of rogue waves; in 1973, the cargo-carrier *Bencrauchan* was cracked and disabled, and several weeks later, the *Neptune Sapphire* was broken in half in the same area. In 1974, the Norwegian tanker *Wilstar* fell into a deep trough and hit so hard that steel beams and plates were folded and the bow bulb of the ship ripped off. Sebastian Junger's *The Perfect Storm* (1997) details the sinking of the Gloucester fishing vessel *Andrea Gail* and the loss of her crew of six by a wave estimated at more than 100 feet high. **See also tsunami, wave**

Rongelap Atoll in the Marshall Islands, approximately 125 miles east of Bikini. British explorer Samuel Wallis happened upon the atolls of Rongelap and Rongerik in 1767, on his way from Tahiti (which he had discovered) to Tinian. Along with the rest of the Marshalls, Rongelap has belonged to Germany, Britain, Japan, and then the United States. When the United States detonated the "Bravo" thermonuclear (hydrogen) device at nearby Eniwetok on March 1, 1954, the explosion was far greater than expected—a thousand times more powerful than the bombs that had been dropped on Hiroshima and Nagasaki—and the fallout covered a much larger area than had been predicted. The unfortunate crew of the Japanese fishing boat *Lucky Dragon,* just east of the test site, saw a gigantic fireball and then a mushroom cloud that rose 21 miles into the sky. Everything on Rongelap, downwind of the explosion, was covered by 2 inches of radioactive ash, and two days later, the sixty-four residents were evacuated to Lae, south of Kwajalein. Many of the Rongelap residents contracted radiation-induced diseases, and many women who were pregnant gave birth to deformed or stillborn babies. An attempt was made in 1985 to move people back to Rongelap, but tests showed that the atoll was still contaminated and uninhabitable.

See also Bikini, Eniwetok, Kwajalein, Marshall Islands

Rooke, George (1650–1709) British admiral of the fleet, considered the greatest seaman of his age. He commanded the *Deptford* at Bantry Bay during France's 1698 attempt to invade Ireland; at Barfleur in 1692 he burned thirteen of the French ships that arrived to try to reinstate James II on the throne of England. During the War of the Spanish Succession (1702–1713) he was commander-in-chief of the combined Anglo-Dutch fleet that failed to capture Cádiz and Barcelona, but did take Vigo and Gibraltar. The battle of Gibraltar led to Rooke's downfall: when his friends in Parliament claimed that it was a greater victory than the Duke of Marlborough's at Blenheim, the assertion so infuriated the Whigs in power that they had Rooke put on the shelf, and he was never given another ship.

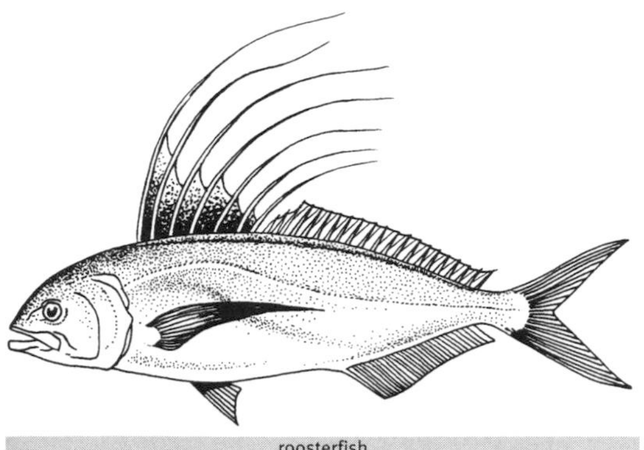

roosterfish

roosterfish (*Nematistius pectoralis*) A relative of the jacks, the roosterfish is a rather ordinary-looking, silvery-blue fish, except for the seven towering dorsal spines that it can raise or tuck into a groove on its back. It is found only on the Pacific coast of the Americas, from Southern California to Peru, usually close to shore. It averages 2 feet in length and 15 pounds in weight, but occasional giants are taken by sport fishermen. The world-record rod-and-reel roosterfish, taken off Baja California in 1960, weighed 114 pounds.

See also jacks

rorqual From the Norwegian *rörhval,* meaning "grooved whale"; any of five whale species that are characterized by long pleats on the underside of the throat. The rorquals are, in descending order of size, the blue whale, fin whale, sei whale, Bryde's whale, and minke whale. These long, graceful animals eat small organisms, such as krill and herring, by catching them in the hairy fringes of their relatively short baleen plates. All except the sei whale feed by swimming through a school of prey with their capacious mouths opened wide—often greatly distending the throat pouch as well—and taking in a huge mouthful of food and seawater. The sei whale is a "skimmer," swimming through shoals of food with its mouth open and trapping the food items in its baleen fringes, in the manner of right whales and bowheads. All the rorquals, again in descending order of size, were hunted commercially, and the larger ones now exist in drastically reduced populations. Only the minke, which was once considered too small to hunt, has increased in numbers, because the larger rorquals, which competed with it for food, were removed from the food chain.

See also blue whale, Bryde's whale, fin whale, minke whale, sei whale, whaling

Ross, James Clark (1800–1862) British naval officer who carried out magnetic surveys in the Arctic and the Antarctic. When he was seventeen, he traveled with his uncle, Sir John Ross, on an Arctic voyage, and in 1831, he became the first man to set foot on the magnetic North Pole. In the ships *Erebus* and *Terror* he led the British Antarctic Expedition of 1839–1842, seeking the magnetic South Pole, which he failed to reach, but he discovered the Ross Sea, Victoria Land, and, one of the most remarkable discoveries of the nineteenth century, the Ross Ice Shelf. He also took the deepest soundings to date, reaching 14,500 feet in Antarctic waters. In recognition of his exploits, he was knighted by Queen Victoria in 1847.

See also Ross (John), Ross Sea, Ross seal, Ross's gull

Ross, John (1777–1856) British Arctic explorer, who first sailed to the Arctic in 1818 in the ships *Isabella* and *Alexander.* When he reached Lancaster Sound, he refused to sail any farther, claiming to have sighted an impenetrable chain of mountains. His colleagues did not agree, and when he returned to England, he was refused any further support. It was not until 1829 that he was able to return to the Arctic in search of the Northwest Passage, this time on the paddle steamer *Victory.* He discovered and surveyed the Boothia Peninsula and King William Island, but his ship became bound in the ice, and it was four years before a rescue was effected. During this enforced stay, his nephew (and second-in-command), James Clark Ross, located the magnetic North Pole. He was knighted upon his return to England in 1833, and in 1850 he commanded a privately financed expedition to search for Sir John Franklin.

See also Franklin, Northwest Passage, Ross (James Clark)

Ross Ice Shelf Originally named the "Victoria Barrier" by James Clark Ross when he discovered it in 1841, the Ross Ice Shelf is the world's largest body of floating ice. It covers approximately 200,000 square miles, an area about the size of France. It lies at the head of the Ross Sea, between spectacular clifflike escarpments on the west and the grounded Antarctic ice sheet on the east. The ice shelf is fed by the Byrd, Beardmore, Nimrod, Scott, and Amundsen glaciers as they move inexorably to the sea. The shelf has had great historical importance, having served as the gateway for such expeditions as Roald Amundsen's 1911 attainment of the South Pole, Robert Falcon Scott's epic 1911–1912 journey (only to learn that Amundsen had beaten him to the Pole), and Robert Byrd's three "Little America" expeditions of 1928–1941.

See also Ross (James Clark), Ross Sea

Ross Island Approximately 45 miles from east to west and the same from north to south, Ross Island is irregularly cross-shaped but difficult to measure because it is usually incorporated into the Ross Ice Shelf. It is separated from the Antarctic continent by McMurdo Sound. The island is dominated by two volcanoes, Mount Erebus (12,250 feet high, and the world's southernmost active volcano), and Mount Terror, named for the ships in which James Clark Ross sailed to the Antarctic in 1839–1843 (and the same ships that Sir John Franklin sailed on his ill-fated search for the Northwest Passage in 1845.) Scott's 1902 expedition established winter quarters at Hut Point on this island but the men actually lived aboard *Discovery,* which was frozen into the ice. Shackleton, who arrived six years later, could not reach Scott's hut, so he built his own at Cape Royds. Ross Island is the home of New Zealand's Scott Base and America's McMurdo, the largest facility in the Antarctic, which contains, among other things, a landing strip on the ice.

Ross Sea The southernmost extension of the Pacific Ocean, which makes a deep indentation into the Antarctic continent. The sea covers some 370,000 square miles and lies between Cape Adare on the west and Cape Colbeck on the east. The sea was first penetrated in 1841 by James Clark Ross in the *Erebus* and *Terror,* as he sailed past the Ross Ice Shelf. It remained unexplored until 1895, when the Norwegian vessel *Antarctic,* searching for new whaling grounds, put a party ashore at Cape Adare. The nutrient-rich waters support abundant planktonic life, which in turn supports large populations of fish, seals, whales, and birds.

See also Ross (James Clark), whaling

Ross seal (*Ommatophoca rossi*) Named for James Clark Ross, who explored the Antarctic from 1839 to 1843, the Ross seal has a proportionally small head and well-developed vibrissae (whiskers). Its color pattern is streaked rather than spotted, and there are longitudinally oriented stripes on the neck and throat. Essentially solitary, it inhabits the pack ice all around the Antarctic. Its vocalizations consist of clicks, gurgles, and a sound that has been described as "chugging."

Ross seal

See also Ross (James Clark), Ross Sea

Ross's gull (*Rhodostethia rosea*) Named for Sir John Ross, who explored the Arctic in the early decades of the nineteenth century, this small, delicate gull is generally found only in the high Arctic. It breeds in Siberia, and has been spotted as far north as 89°20′ north, almost at the North Pole. For those in a position to see it, it can be identified by its white head, pale gray back, black ring around the neck, wedge-shaped tail, and, in summer, its all-over pink coloration. To the excitement of birdwatchers, this species sometimes shows up far from its normal range, in such places as Ohio and Massachusetts.

Ross's gull

See also ivory gull

rough-toothed dolphin

rough-toothed dolphin (*Steno bredanensis*) This dolphin gets its name from the striations on its teeth, ridges that are too small to be seen with the naked eye. *Steno* is a dark purplish-brown on the back and lighter on the undersides. It is often marked with pinkish circular scars, making it possible to refer to this animal as a "purple dolphin with pink polka dots." (The scars are believed to be caused by the cookie-cutter shark, which takes circular bites from the living dolphin.) The rough-toothed dolphin reaches a length of 8 feet, and although it is found throughout the world's tropical and temperate waters, it is common nowhere. It is believed to be the "smartest" of all the dolphin species, and in captivity it performs even better than the more common bottlenose. It is—perhaps because of its "intelligence"—often temperamental and difficult, so despite its trainability, it is rarely seen in oceanarium shows.

See also bottlenose dolphin, cookie-cutter shark, ichthyosaur

royal penguin (*Eudyptes schlegeli*) One of the species known as "crested penguins" (genus *Eudyptes*), the royal penguin breeds only on subantarctic Macquarie

Island and adjacent islets. Royals differ from other crested penguins in that their throat is white, whereas the other species have a black head and a white chest and belly. The total population is around 850,000 pairs.

See also erect-crested penguin, macaroni penguin, rockhopper penguin, Snares penguin

Russo-Japanese War (1904–1905) Conflict that grew out of the designs of Russia and Japan on Manchuria and Korea. Although the Japanese tried to negotiate a treaty based on spheres of influence, the Russians rejected the offer, confident that they could easily defeat the Japanese and that a victory would divert the threat of internal revolution in Russia. Japan broke off diplomatic negotiations and on February 8, 1904, attacked the Russian naval squadron at Port Arthur (now Lü-shun in China) and bottled up the fleet. Following a succession of decisive Japanese land victories, the Russian commander surrendered Port Arthur, and the outnumbered Japanese troops took Mukden. On May 27–29, 1905, at Tsushima, Admiral Heihachiro Togo's fleet destroyed the Baltic Fleet commanded by Admiral Zinovy Rozhestvensky, and the Russians, whose country was in the throes of the 1905 revolution, came to the peace table at Portsmouth, New Hampshire. The Treaty of Portsmouth (negotiated between August 9 and September 5, 1905), with President Theodore Roosevelt as mediator, granted Japan control of the Liaotung Peninsula (with Port Arthur) and the South Manchurian railroad, and its control of Korea was recognized. It was the first time in the modern era that a European power had been defeated by an Asian country.

Rut, John (fl. 1527) British navigator (birth and death dates unknown), one of the few to venture across the Atlantic in the half century after John Cabot's 1497 voyage. In 1527, Rut was commissioned by Henry VIII to find the Northwest Passage. In the *Mary of Guilford* (his other ship, the *Sampson*, was lost in a storm), he cleared the Scilly Islands and made it to Newfoundland (or maybe Labrador), where thirty years before Cabot had reported that he could catch cod by lowering a basket. He saw no fewer than fifty Spanish, French, and Portuguese fishing boats. (He reported this to the king in the first letter from North America to Europe, sending it with an English ship that was returning with a load of cod.) Rut evidently had no relish for northern exploration, for after snooping around ineffectually for a couple of weeks among the icebergs and finding no suitable anchorages, he coasted southward to the Caribbean, where he encountered the Spanish captain Gines Navarro, who drove him off. He headed for Puerto Rico, where he reprovisioned and headed back to England, arriving in the spring of 1528. Rut found nothing new, and although he passed right by it, he missed the opportunity to enter the Gulf of St. Lawrence eight years ahead of Cartier.

See also Cabot (John), Cartier, Northwest Passage

Ryder, Albert Pinkham (1847–1917) American marine and landscape painter, born in New Bedford, Massachusetts, who spent most of his life in New York City's Greenwich Village. His paintings, which are often heavily overpainted, are usually small and dark; most are devoted to scenes of moonlight and the sea. His vision was bad and he could not focus on details or stand bright lights. He also cared little for technique, and most of his works have deteriorated badly, but despite his limitations, he is considered one of nineteenth-century America's most important painters. Among his best-known pictures, all in the collections of major museums, are *Toilers of the Sea, Death on a Pale Horse,* and *The Flying Dutchman.* **See also Flying Dutchman**

Ryukyu Islands An archipelago 400 miles southwest of Kyushu (the southernmost of Japan's main islands), the Ryukyus consist of fifty-five volcanic islands, including Okinawa, and coralline islets. The original inhabitants of the islands resemble the Japanese, but their language is completely different from Japanese. Yankee whalers on the "Japan Grounds" in the mid-nineteenth century could not pronounce the Japanese name for the islands and called them "Loo-choo." This name has lasted into this century; another accepted name for the islands is Luchu. After Japan's surrender in 1945, the islands came under the control of the Americans, whose administrative government was based at Naha, the capital of Okinawa. Successive treaties returned sovereignty to the Japanese, but the United States still maintains a large military presence on Okinawa. **See also Okinawa**

S

sailfish

sailfish (*Istiophorus platypterus*) The sailfish gets its common name from the towering dorsal fin, which is peppered with dark spots. Previously, the Atlantic and Pacific sailfishes were believed to be separate species, but they are now considered the same. They grow larger in the Pacific, however, and the record for a Pacific sailfish is 221 pounds, while the Atlantic record is only 128. This is one of the world's most popular game fishes, and its acrobatics endear it to saltwater anglers. It is usually considered the fastest fish in the ocean; its speed is enhanced by its ability to lower its dorsal fin into a groove on its back to reduce water resistance. Estimates of its speed range from 50 to 65 mph. The sailfish is believed to use its erected sail underwater to herd prey species into a ball so that it can slash at them more effectively with its bill. **See also spearfish**

St. Barthélemy Island in the Windward chain of the Lesser Antilles, commonly known as St. Barts. With the northern half of St. Martin, it is an *arrondissement* of Guadeloupe, itself an overseas *département* of France. St. Barts and St. Martin, 150 miles northwest of Guadeloupe, were sighted by Columbus in 1493, but not colonized. Around 1648, St. Barts was occupied by the French, who sold it to Sweden in 1748. (St. Barts is the only island in the Caribbean ever to fly the Swedish flag.) Returned to France in 1877, it is still considered a colony. The rocky soil of the 8-square-mile island never supported plantations worked by slaves, so the population is mostly of Swedish and French extraction. The capital—and only town—is Gustavia, named for a king of Sweden. The only industry is tourism. **See also Guadeloupe**

St. Croix Columbus visited this island in 1493 and christened it Santa Cruz ("sacred cross"), but French settlers later modified the name. (It is now pronounced "Saint Croy," a French-American amalgamation.) The largest of the American Virgin Islands, it encompasses a total of 84 square miles and is home to some 55,000 Cruzans. In the early seventeenth century, English and Dutch settlers vied for control of the island's profitable sugar plantations, but the British prevailed, at least until they were ousted by the Spanish. The Knights of Malta acquired the island in 1651 and sold it in 1665 to the French West India Company. In 1733 the islands were bought by Denmark, which sold them to the United States in 1917. The two major settlements, Christiansted (the capital) and Frederiksted, with their Dutch-style houses, are holdovers from Danish rule. Tourism is now St. Croix's predominant industry, with many cruise ships docking in its harbors, but there is also a Hess oil refinery on the south coast. In 1989, St. Croix was devastated by Hurricane Hugo, which destroyed 90 percent of its buildings, left 22,000 homeless, and brought about prolonged looting and violence.

See also American Virgin Islands

Saint Elmo's fire An electrical discharge that creates a brushlike glow at the tip of the extremities of pointed objects such as the mastheads and yardarms of a ship, or the steeple of a church. It also occurs on the wing tips and propellers of airplanes flying in dry snow or in the vicinity of thunderstorms. Sailors regarded its appearance as a favorable omen, signaling the end of bad weather.

St. Eustatius First settled by the Dutch West India Company in 1636, Sint Eustatius is a tiny island of only 8 square miles. Southeast of Saba and Sint Maarten, the other Dutch possessions in the Windward Islands, it is, with them, considered part of the Netherlands Antilles, the rest of which (Aruba, Bonaire, and Curaçao) are hundreds of miles to the south. Its capital and only town is Oranjestad on the west coast. The island is dominated by the Quill, a cone-shaped extinct volcano in the south. During the seventeenth century, the island served as an entrepôt for Dutch traders in the Caribbean; its warehouses so filled up with casks of sugar and rum that the merchants were forced to build dikes to reclaim land from the sea, much as they had done in their home country, so they could build more storage. As a free port, Oranjestad allowed ships of any nation to trade there, and on November 16, 1776, Sint Eustatius became the first foreign power to acknowledge the sovereignty of the newly independent United States. In 1779, 25 million pounds of sugar were shipped from Oranjestad. The island changed hands numerous times and became a hotbed of illegal trading and smuggling, a base for pirates and privateers. The British admiral George Rodney sacked the island in 1781 to subdue the pirates, and afterward the British abandoned it. Its commercial importance long faded into history, Sint Eustatius—called Statia by its 2,000 inhabitants—is struggling to establish a tourist industry, but it has scrubby vegetation and poor beaches. **See also Aruba, Bonaire, Curaçao**

St. Helena Located some 1,200 miles from the west coast of Africa in the South Atlantic, St. Helena is the tip of an extinct volcano that is a visible part of the Mid-Atlantic Ridge, the undersea mountain range that bisects the Atlantic Ocean. (Ascension Island and Tristan da Cunha, two other South Atlantic surface manifestations of the Mid-Atlantic Ridge, are both dependencies of St. Helena.) The island was discovered in 1502 by the Portuguese navigator João da Nova Castella and named for the mother of the Roman emperor Constantine. In 1669 the British East India Company claimed the island, and in 1815, Napoleon was exiled there. He died at Jamestown, the capital, in 1821. Long an important way station for sailing ships, St. Helena's fortunes faded with the introduction of steam, and it is now largely supported by British subsidies.

See also Ascension Island, Mid-Atlantic Ridge, Tristan da Cunha

St. John The smallest of the three U.S. Virgin Islands, St. John is only 9 miles long and 5 miles wide. It is 5 miles east of St. Thomas and 35 miles north of St. Croix. When Columbus arrived in 1493, the island was inhabited by the warlike Carib Indians, but they were quickly exterminated. In the eighteenth century, when the Danes owned it, the island was devoted to sugar and cotton plantations. When the slaves revolted in 1733, British and Danish troops were brought in. They failed to put down the rebellion, and after an eight-month standoff, French troops arrived from Martinique. Many of the rebels committed suicide rather than be returned to slavery; slavery was not actually abolished until 1843. The population now is only 3,500, mostly concentrated in the towns of Cruz Bay and Coral Bay. After 1950, when most of St. John had been abandoned, Lawrence Rockefeller bought half of it and turned it into the Virgin Islands National Park. Since then, the island has remained largely undeveloped, and much of it has reverted to second growth forest. The national park covers some 13,000 acres, and more than a third of it is offshore, so it includes the reefs and marine life. Hotels on St. John attract thousands of tourists each year.

See also American Virgin Islands

St. Kilda Cluster of small, remote, rocky islands 110 miles west of the Scottish mainland and 41 miles from the Outer Hebridean island of North Uist. St. Kilda consists of the islands of Hirta, Dun, Soay, and Boreray, all of which are unoccupied by people but which support sheep, wrens, mice, and gray seals. The largest population of northern gannets (*Sula bassana*) in the world, some fifty thousand pairs, breeds on the steep cliffs of St. Kilda, and there are also large numbers of skuas, fulmars, kittiwakes, puffins, and guillemots. The islands were owned by the MacLeods of Skye, and rent was paid once a year in the form of feathers, wool, grain, and dairy products. Hirta had sustained human inhabitants for perhaps two thousand years, but in 1930, the inhabitants were evacuated at their own request, and now all the islands are held as a nature reserve under the National Trust for Scotland. (The British army maintains a radar station there now.) There is no saint named Kilda, and the name is thought to be derived from Khilta, the Gaelic name for Hirta. A district of Melbourne, Australia, is called St. Kilda; there is also a small town north of Adelaide called St. Kilda. Still another St. Kilda exists near the town of Dunedin, on South Island, New Zealand. **See also Hebrides**

St. Kitts and Nevis In 1493, when Christopher Columbus arrived at these islands in the Lesser Antilles, he named one after St. Christopher, his patron saint, but it was later shortened to St. Kitts by the British settlers who landed there in 1623. "Nevis" is derived from Columbus's use of *las nieves* ("the snows") to describe that island's cloud-capped peak. A British-French rivalry lasted until the 1783 Treaty of Versailles, which ended the American Revolutionary War and the hostilities between England and France, ceded the islands to Britain. As part of an alliance with Nevis and Anguilla, St. Kitts joined the British-sponsored West Indian Federation in 1958, but after it was dissolved in 1962, the islands sought independence, and Anguilla broke from the others. The Federation of St. Kitts and Nevis became an independent nation in 1983. The American statesman Alexander Hamilton (1755–1804) was born on Nevis, the illegitimate son of a Scottish planter and the estranged wife of a merchant. Agriculture and manufacturing account fôr most of the island's economy, but tourism is increasing. **See also Anguilla**

St. Lawrence, Gulf of A marginal sea lying at the mouth of the St. Lawrence River in eastern Canada. It covers about 93,000 square miles and includes Anticosti Island, the Magdalen Islands, Prince Edward Island, and Cape Breton Island. The two outlets to the North Atlantic are the Strait of Belle Isle between Newfoundland and Labrador, and the Cabot Strait, between Newfoundland and Nova Scotia.

See also Anticosti Island, Cape Breton Island, Magdalen Islands, Newfoundland, Prince Edward Island

St. Lawrence Island A rocky, barren island in the Bering Sea, south of the Bering Strait. The island, 90 miles long and 22 miles wide, belongs to the United States and is inhabited by Yupik Inuit, who engage in fishing, walrus hunting, and whaling. It was visited by Vitus Bering in 1728 (along with the Diomedes); during the middle of the nineteenth century, it became known to American whalers as one of the best areas to find bowhead whales. In 1879, during a particularly harsh winter (in a region where harsh winters are the norm), two-thirds of the native population either starved or froze to death. Some attribute this catastrophe to the whalers' introduction of alcohol to the islanders. Eskimo whalers from the villages of Gambell and Savoonga still take an occasional bowhead.

St. Lucia The original settlers on St. Lucia were Arawak Indians, who were displaced by the Caribs; when the first Englishmen arrived on the island in 1605, they were intimidated by the Caribs and fled. Only by making a treaty with the Caribs were the French able to remain on the island after 1650. European politics and wars meant that the island shuttled between French and British ownership until 1871, when it was officially designated one of the Windward Islands of Britain. For much of the nineteenth century, the major crop was sugarcane, but it was replaced by bananas and cacao. In 1958, St. Lucia joined the short-lived West Indian Federation, but when the federation was dissolved in 1962, the island voted for independence as a member of the British Commonwealth, achieved in 1979. Like many Caribbean nations, St. Lucia depends heavily on tourism. The population is approximately 152,000, and the capital is Castries. Soufrière ("sulfur producer") is a town nestled in the crater of the dormant volcano Qualibou, which last erupted in 1766 and still has many bubbling mud pits and steaming sulfur springs. (The Soufrières of Montserrat, St. Vincent, and Guadeloupe are active volcanoes.) The island grows many tropical fruits such as mangoes, guavas, coconuts, papayas, and breadfruit, and is the home of the St. Lucia parrot (*Amazona versicolor*), once endangered but now fully protected.

See also Guadeloupe, Montserrat, St. Vincent and the Grenadines

St. Martin Like Hispaniola and New Guinea, the island of St. Martin in the Leeward West Indies is shared by two governments. The south is the Dutch side, Sint Maarten, with its capital at Philipsburg; on the northern side is the French portion, with its capital at Marigot. Columbus sighted the island in 1493, but did not recommend colonization, and it was not until 1648 that the Dutch West India Company banded together with the French to keep the Spanish and the British from claiming it. Along with Saba and Sint Eustatius, Sint Maarten is part of the Netherlands Antilles, as are Aruba, Bonaire, and Curaçao, 500 miles away off the coast of Venezuela. Like its neighbors St. Barts and Anguilla, St. Martin is a popular Caribbean tourist destination. **See also Aruba, Bonaire, Curaçao, St. Eustatius**

St. Peter and St. Paul Rocks Located nearly on the equator between South America and West Africa, this Brazilian-owned group consists of five tiny islets, the largest of which measures only 400 by 200 feet. Like the Azores, Ascension Island, and Tristan da Cunha, the rocks are visible manifestations of the great Mid-Atlantic Ridge that snakes down the middle of the ocean floor. But unlike the other islands, which are composed of extrusive basaltic volcanic material, St. Peter and St. Paul are made of peridotite, an olivine-rich igneous rock formed at considerable depths below the surface and believed to have been lifted up by massive faulting. When Charles Darwin visited this remote outpost on the *Beagle* in 1832, he observed that from a distance the islands appeared a brilliantly white color, because of the "dung of a vast multitude of seafowl." The *Challenger* tied up but did not attempt a landing in 1873; her crew did row ashore and go fishing. Because of their unique nature, the rocks are often visited by oceanographic expeditions, including the *Albatross* (1885), the *Meteor* (1925), and the Woods Hole vessel *Chain* in 1963. The Brazilians erected an untended lighthouse there in 1930, but it failed, and now only a skeletal ruin remains.

See also Ascension Island, Azores, *Beagle*, *Challenger*, Tristan da Cunha

St. Pierre and Miquelon The only French colony in North America, St. Pierre and Miquelon are two small islands south of Newfoundland, in the Gulf of St. Lawrence. The population consists mostly of fishermen who live near the capital of St. Pierre on that island. Because of their proximity to the Grand Banks, long one of the most productive codfishing grounds in the world, the islands were colonized by Basques, Normans, and Bretons from France at the beginning of the seventeenth century. The islands were taken by the British in 1713, retaken by France in 1763, captured twice more by Britain, and finally ceded to France in 1814, with the provision that they remain unfortified. The 6,000-odd residents speak French and are French citizens. The introduction of offshore fishing zones have made the islands a source of friction between Canada and France since 1976, but an agreement was reached in 1992, granting St. Pierre and Miquelon an exclusive economic zone of 3,607 square nautical miles.

St. Thomas One of the three U.S. Virgin Islands, 75 miles east of Puerto Rico and 40 miles north of St. Croix in the Caribbean. St. Thomas was sighted in 1493 by Columbus, colonized by the Dutch in 1657, and then by the Danes, as part of the Danish West India Company. Sugar was grown until slavery was abolished, and in 1917, the United States bought the island from Denmark, for $25 million. St. Thomas covers 32 square miles and is one of the most developed islands in the Caribbean; its population numbers about 51,000, most of whom live in or around the capital, Charlotte Amalie. There are no wells, so drinking water has to be brought in from Puerto Rico. Many cruise ships dock in the spacious harbor, releasing hundreds of thousands of passengers per year to shop for jewelry, clothes, and souvenirs. In 1995, Hurricane Marilyn struck St. Thomas with such force that it destroyed a quarter of all the houses on the island and caused an estimated $3.5 billion worth of damage.

See also American Virgin Islands

St. Vincent and the Grenadines A nation in the Lesser Antilles, consisting of the large island of St. Vincent (134 square miles) and the Grenadines: Bequia, Balliceau, Mayreau, Mustique, Isle à Quatre, Petit St. Vincent, and Union Island. Like St. Lucia, Montserrat, and Guadeloupe, St. Vincent has a volcano, which is known as Soufrière; it blew violently in 1812, 1902, and 1979. (The 1902 eruption occurred concurrently with the devastating eruption of Mount Pelée on Martinique, which killed thirty thousand.) The island was inhabited by Carib Indians when Columbus arrived in 1498, and no Europeans lived there until the middle of the eighteenth century, when the French made a treaty with the Caribs that allowed them to settle there. In 1763 the island was ceded to Britain, but the Caribs joined the French in ousting the British ten years later. Only the arrival of British troops in 1796 subdued the Caribs, and the remaining Indians were deported to Honduras. Cotton was the main crop until the British switched to sugarcane and brought in African slaves to work the plantations. St. Vincent was made a part of the British Windward Island colony in 1871; that arrangement was terminated in 1958 when the island joined the West Indian Federation. In 1969, the islands became a self-governing state in the British Commonwealth, and they achieved independence in 1979. Agriculture still dominates the economy, but tourism is on the rise, especially catering to visiting sailboats. The St. Vincent parrot (*Amazona guildingii*), one of the world's most endangered birds, was almost extinct before dedicated conservationists rescued it.

Saipan Capital of the Northern Mariana Islands in the western Pacific. Discovered by Magellan in 1521, the Marianas were sold to the Germans in 1898 and seized by the Japanese in 1914. During World War II, the Japanese had an air base on Saipan, but when the island was taken by American troops in 1944, it became a major air facility for the bombing raids on Japan. On September 30, 1638, *Nuestra Señora de la Concepción*, the largest of all the Manila galleons, ran into a reef and sank off the southern tip of Saipan. The cargo of the *Concepción* was located in 1988, and storage jars, glass beads, pottery fragments, and more than thirteen hundred pieces of gold jewelry have been recovered so far. The Manila galleons, which carried treasure from China and Japan to Mexico (and thence across the land to Veracruz and ultimately, if they were not wrecked, to Spain), often stopped in the Marianas for water.

See also Manila galleon, Mariana Islands

Sakhalin Island Island in the Sea of Okhotsk, 580 miles long and about 100 miles across at its widest point. Its weather is harsh, and its inhabitants are engaged mostly in fishing, lumbering, or coal mining. Since the mid-nineteenth century, ownership of Sakhalin has been the subject of dispute between Japan and Russia, but after World War II, Sakhalin, together with the Kuril Islands, came under Soviet control.

See also Kuril Islands; Okhotsk, Sea of

Salamis Greek island in the Saronic Gulf of the Aegean, just southwest of Piraeus, the port of Athens. During the Persian Wars (490–479 B.C.), after the death of Darius, his father, Xerxes continued Persia's attempts to conquer Greece. After the Greeks lost the battle of Thermopylae on land, the Athenian leader Themistocles drew up his fleet of triremes in the Salamis Channel (between the island of Salamis and the Greek mainland), but they appeared to be trapped when the Persians massed their ships in the bay of Eleusis to the north. The very narrowness of the channel led to the defeat of the Persians, for they were attacked and destroyed by Themistocles's war galleys as they tried to enter. On September 20, 480 B.C., after several hours of fighting, the Persians had lost some 200 ships out of a total of 1,200, while the Greeks lost only 40 of 450. With this defeat, the Persians retreated and never attempted to invade Greece again. The period immediately following the Battle of Salamis was the height of Athenian culture, which included the age of Pericles, the building of the Parthenon, and the inspired teaching of Socrates, Plato, and Artistotle.

salinity The quantity of dissolved salts in seawater. Most of the sea's salinity results from sodium chloride, the same substance as table salt. The average salinity of the ocean is about 35 parts per thousand (ppt), but some bodies of water are saltier. The Red Sea, affected by high evaporation rates, has a salinity of 43 ppt, while the Baltic, fed by numerous rivers and melting snows, has a salinity of 7 ppt. Saltier water is heavier, so it sinks below less salty water. Because the salt in seawater speeds the rate of water loss from cells, humans cannot drink it, especially if they are dehydrated to begin with.

sally lightfoot (*Grapsus grapsus*) A multicolored, square-bodied crab, often dark red and sky blue or solid red with white speckles; it has black pincers with red patches and cream-colored fingertips. It is found on rocky shores and jetties at the water's edge, from Florida and the West Indies to Brazil, Baja California to Chile and the Galápagos. The name comes from their habit of skipping lightly over rocks and lava.
See also crab

salmon, Atlantic: See Atlantic salmon

salmon, Pacific: See Pacific salmon

salmon shark: See porbeagle

salp (*Thaliacea*) Transparent, barrel-shaped animals, supported by conspicuous hooplike bands, sometimes joined in chains or clusters. Salps are classified as tunicates, with the sea quirts and lancelets. Although there are thought to be several different species (e.g., *Salpa* and *Thalia*), identification is difficult because of individual variations. Salps are usually found in offshore waters, but sometimes they come closer to shore. They reproduce by budding: an outgrowth of the parent breaks off and becomes a new animal. One species (*Pyrosoma*) is bioluminescent and emits light when touched. **See also sea squirts, tunicate**

saltwater crocodile (*Crocodylus porosus*) The largest of the crocodiles, and the largest of all living reptiles, the saltwater crocodile grows to a known length of 23 feet, and may get even larger. It is also the most dangerous of all crocodiles, with a deserved reputation as a man-eater. "Salties" occur in Southeast Asia, Indonesia, Malaysia, and northern Australia, where they move freely from island to island, but they can also swim great distances in the sea, and have been found in the Solomon Islands, New Guinea, Fiji, Japan, and the Cocos (Keeling) Islands, more than 600 miles southwest of Java, the nearest large island. They are not confined to saltwater; they also appear in freshwater rivers, lakes, and swamps, where, depending upon their size, they feed on crabs, turtles, snakes, shore- and wading birds, buffaloes, domestic livestock, wild boars, monkeys—and occasionally people. **See also marine reptiles**

salvage In maritime law, the proportion of the value of a ship or its cargo paid by the owner or the insurance company to those who have saved the ship or cargo when it was in danger of being lost. (No salvage can be claimed by the crew who saves its own ship.) Offers of salvage may be refused as long as an officer or the agent of the owner remains aboard, but a derelict vessel that has been deserted or abandoned is fair game. "Salvage" also applies to the underwater recovery of a ship or its contents. The recovery of treasure from sunken ships (or the remains of the ships themselves) is also considered salvage, but there are often contradictory claims of ownership in these cases. If the owner of a sunken or derelict vessel is known, the salvor does not become the owner of the property upon salving it, but the owner may reclaim its property upon payment of the salvage fee. **See also Law of the Sea**

Samoa, American An unincorporated territory of the United States in the central Pacific Ocean, comprising

saltwater crocodile

the islands of Tutuila, Aunu'u, Ta'u, Olosega, and Ofu, Rose (an uninhabited coral atoll), and Swain Island, 280 miles to the north. Pago Pago, on Tutuila, is the capital. The Samoans are a Polynesian people, believed to have inhabited these islands three thousand years ago and to have, around the year 1200, traveled throughout the Pacific, colonizing Hawaii, Tahiti, and New Zealand. The islands were first sighted in 1722 by the Dutch navigator Jacob Roggeveen, and Bougainville landed there in 1768. After a series of tribal wars, the 1899 Treaty of Berlin gave Western Samoa to Germany and the islands east of longitude 171° west to the United States. Western Samoa became an independent country in 1962, but American Samoa is administered by the U.S. Department of the Interior. The population of 55,000 American Samoans are U.S. nationals, not citizens, and even adults cannot vote in presidential elections. The primary industry of the territory is tuna fishing, and canneries are the largest employers. Fagatle Bay, 163 acres off the southwest coast of Tutuila, is the only tropical coral reef to be designated a national marine sanctuary.

See also Samoa, Western

Samoa, Western The first independent nation in the South Pacific, Western Samoa consists of two main islands, Upolu and Savai'i, and seven smaller islands. Apia, the capital, is on Upolu. The Dutch navigator Jacob Roggeveen sighted the islands in 1722. By 1830 the London Missionary Society had arrived, and until 1873, Samoan chiefs fought bitter wars for supremacy. The British, Germans, and Americans also wanted to dominate Samoa, but Britain withdrew to pursue her interests in Tonga and the Solomons. More tribal warfare brought the three powers back to intervene, and in 1899, a treaty gave Western Samoa to Germany, and the islands east of longitude 171° west to the United States. Robert Louis Stevenson (1850–1894), the author of *Treasure Island, Kidnapped, The Strange Case of Dr. Jekyll and Mr. Hyde,* and many other works of fiction and poetry, spent six weeks in Western Samoa in 1889. In 1890, he returned to his estate near Apia, where he died four years later. He is buried on the side of Mount Vaea. After World War I, when New Zealand occupied Samoa, an influenza epidemic killed 8,500 Samoans. After World War II, Samoa became a UN Trust Territory; it voted for independence in 1962. The current population of Western Samoa is about 164,000.

See also Samoa, American

Samos One of the Sporades Islands of Greece, Samos is in the Aegean only 2 miles from Turkey, from which it is separated by the Mykale Strait. It is a fertile, mountainous island that covers some 181 square miles and supports a population of approximately 32,000. It was inhabited in the eleventh century B.C. and later colonized by Ionian Greeks. By the sixth century B.C. it was a commercial and maritime power, and it is said that the poet Anacreon and the fabulist Aesop lived there. Pythagoras, the philosopher-mathematician, was born on Samos, but he spent most of his life in exile in Italy. According to Herodotus, the tyrant Polycrates had a 3,000-foot-long tunnel dug by slaves to carry water, but the two teams did not meet, and they had to solve the problem with a U-turn. Samos remained a loyal ally of Athens during the Peloponnesian War, but the island declined after it fell out of Athenian hands after 323 B.C. It was owned by a Genoese trading company in the fourteenth century A.D., and was later captured by the Ottoman Turks. It was a semi-independent principality until it joined Greece in 1913. The capital of Samos is Vathí, and its second city is Pythagorion. It is a popular destination for tourists interested in its ancient ruins, statuary, and beaches. **See also Aegean Sea**

sampan A typical small and light boat of inshore Asian waters and rivers. There are two types: the harbor sampan, which usually has an awning over the center and is powered and steered by a single oar, and the coastal sampan, with a single mast and a junk-type sail.

See also junk (1)

San Blas Islands A group of some three hundred islands off the Caribbean (northern) coast of Panama, originally known as the Mulatas Islands. Most of the tiny islands are nameless sand spits, but some are occupied by Cuna Indians, who are descended from the Caribs. A small percentage of them are "white Indians," albinos who are not permitted to intermarry. On the windier islands, the Indians live in thatched-roof huts, to keep off the *chitras,* or biting sand fleas. The island of Porvenir, the administrative seat of the region, has an airstrip for planes bringing tourists from Panama City. The Cuna women are famous for their *molas,* appliquéd panels that are extremely popular with tourists. Each *mola* is actually the front or back half of the traditional women's blouse, but they are often framed and used as wall decorations. The colors and patterns are particularly bright and can represent anything from brain corals (a popular motif) to birds, mythological figures, and geometrical designs. **See also Caribbean Sea**

sand Small mineral particles in the form of loose rounded or angular grains, formed by the extensive activity of ocean or river currents, wind, waves, or glacial movement. Most sand comes from the land and is carried to the ocean by rivers. The most common material in beach sand is quartz, which comes from quartzite rocks. Quartz is abundant in rocks, it is not readily worn down, and it is insoluble in water. Sand may also be composed of grains of volcanic rock or glass (as in the black sand beaches of Hawaii), coral (as in the pink beaches of Bermuda), or seashells ground small. Sandstone is formed by the cementing together of grains of sand, usually with calcium carbonate as the bonding agent; worn-down sandstone again becomes sand. Most of the world's sand is found along beaches and underwater in areas close to shore (and also in deserts, of course), but much of the ocean floor is covered by mud, ooze, and sediments that have accumulated over the eons. **See also ooze, sediments**

sandalwood (*Santalum album*) Any of several fragrant tropical woods, especially *S. album,* an evergreen native to India, where it is used for joss sticks and in various funeral rites and religious ceremonies. The yellow aromatic oil distilled from the wood is added to perfumes, soaps, candles, and medicines. Some twenty-five species of *Santalum* are distributed throughout Hawaii and other South Pacific islands, where in many cases they have been harvested almost out of existence.

sandbar shark (*Carcharhinus plumbeus*) Although it might be confused with the dusky shark or the bull shark, the sandbar shark can be identified by its very high, triangular dorsal fin. Sandbars are common throughout the coastal temperate waters of the world, in northern and southern Africa, the east and west coasts of North America, Japan, China, and parts of Australasia. It is viviparous, giving birth to up to fourteen pups per litter. There is no record of a sandbar attack on a swimmer or diver. Because of its abundance, this shark is fished commercially for its meat, leather, liver, and fins. The maximum size is 8 feet. **See also bull shark, dusky shark**

sandbar sharks

sand crab (*Ocypode albicans*) Familiar to beachgoers along the Atlantic coast of North America from Long Island southward, these are the little crabs seen scuttling swiftly along the sand, sideways and on tiptoe. (*Ocypode* means "swift-footed.") The maximum size is 2 inches across the sand-colored shell. Also known as ghost crabs because they are so hard to see, they are adapted to life on the beach and will drown if they are held under water. These seminocturnal little crabs spend the day in burrows, avoiding bright sunny days because their shadow gives them away to predatory birds. **See also crab**

sand diver (*Synodus intermedius*) One of the lizardfishes, the sand diver is a voracious 18-inch-long carnivore that is characterized by a lizardlike head, a mouthful of sharp teeth, and a long, cylindrical body. It can be seen resting on the bottom, or sometimes buried in the sand, with only its eyes and the top of its scaly head showing. Sand divers are found in shallow waters on both sides of the Atlantic. Two related species are the galliwasp (*S. foetens*) and the rockspear (*S. synodus*). **See also lizardfish**

sand dollar (*Echinarachnius parma*) A thin, disk-shaped sea urchin covered with many reddish-brown, close-set, feltlike spines, which can reach a diameter of 3 inches. On the upper surface are five petal-shaped loops of tube feet that serve as gills, and the mouth is in the center of the lower surface. They are found on sand bottoms in water up to a mile deep, off the Atlantic and Pacific coasts of North America. They feed on fine particles of organic matter, and are eaten by bottom-feeding fishes. The skeleton of the sand dollar, called the test, is usually a bleached white flattened disk, with a five-pointed flowerlike design that marks the location of the loops of tube feet. Pentamerous radial symmetry—five points arranged around the center—is characteristic of the body plan of all echinoderms, including starfishes (in which it is most visible), and also sea lilies and feather stars, brittle stars and basket stars, and sea cucumbers.

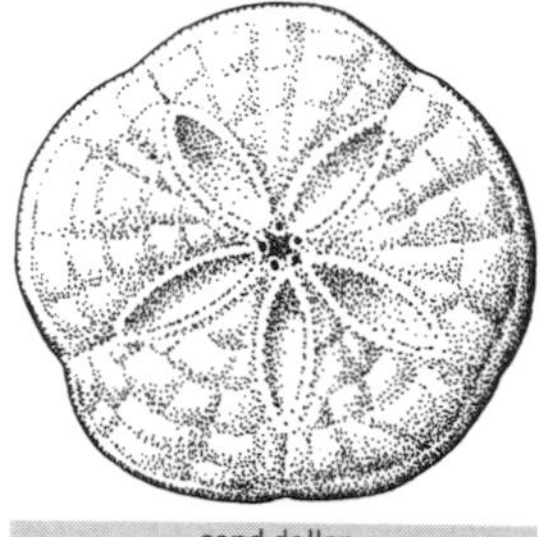
sand dollar

See also basket star, brittle star, echinoderms, sea cucumber, starfish

sanderling (*Calidris alba*) Among the smallest and most visible of all the sandpipers, sanderlings are only about 8 inches long. With flitting, toylike steps, they dart after receding waves, catching mollusks and shrimps from the wet sand. They nest on Arctic islands, migrate into the Southern Hemisphere, and can be found on almost every beach in the world. These plump little birds can easily be recognized by the flashing white stripe on the wings. **See also sandpiper, shorebirds**

sanderling

sandpiper There are many species of sandpipers—perhaps as many as eighty—most of which are characterized by a long bill and generally uniform plumage. The turnstones (named for their habit of turning over stones to look for food), long-legged sandpipers (Tringinae), short-legged sandpipers (Eroliinae), and snipes (including the woodcock) make up this large group. Curlews, which are long-legged sandpipers, are identifiable by their long curved bills, and the short-legged varieties—known in America as peeps for the sounds they make—are commonly seen running along beaches picking up tiny food items from the sand exposed by receding waves. Most of these birds are migratory, and some travel enormous distances every year.

See also Eskimo curlew, sanderling, turnstone

sand tiger shark

sand tiger shark (*Carcharias taurus*) A shark with many scientific names; it has also been known as *Odontaspis* and *Eugomphodus*. Because of its "sharky" look and its adaptability to captivity, the sand tiger shark is common in commercial aquarium displays. It swims slowly with its mouth open, displaying an impressive set of jagged, curved teeth. Despite its appearance, however, the sand tiger, which can reach a length of 10 feet, is a sluggish, inoffensive creature that feeds on small fishes and swims close to the bottom, usually in shallow water close to shore. It is found on both sides of the North Atlantic and in the Mediterranean. The sand tiger shark is oviparous, which means that the young are born live from an egg that hatches within the female, but they are also oviphagous ("egg-eating"). One embryo in each oviduct survives, after it eats its unborn siblings. At birth, the surviving cannibals are about 3 feet in length. Throughout the world's temperate shallow waters, there are species closely related to the sand tiger. (None of them are related to the tiger shark, *Galeocerdo cuvier.*) The South African version is commonly known as the ragged-tooth and the Australian is the gray nurse. In the early history of spearfishing in Australia, gray nurses were killed frequently by divers who wanted to conquer—and brag about killing—such a terrifying-looking beast, despite its harmless nature.

See also tiger shark

Sanibel Island Forming a gentle east-west arc in the Gulf of Mexico off Fort Myers, Sanibel is one of Florida's premier tourist attractions. Where most barrier islands run parallel to the coasts they protect, Sanibel juts out into the gulf. Its longest shore faces south, athwart the currents that move northward, creating a convenient barricade for the shells that wash ashore. The island, 14 miles long but only 3 miles wide, is a shell collector's paradise, the most popular place in the United States for would-be malacologists. When Juan Ponce de León arrived here in 1513, he was driven off by the Caloosa Indians, and when he returned with a much larger landing party in 1521, he was shot through the neck with an arrow and died shortly thereafter in Cuba. The island remained a popular haunt for pirates like José Gaspar and "Calico Jack" Rackham during the eighteenth century. By the 1880s, it had become known as a fishermen's haven, particularly for tarpon. In the 1930s, development arrived on both Sanibel and Captiva, its immediate neighbor to the north. Until 1963, the only access to the island was by ferry or private boat, but a causeway from Fort Myers has increased traffic and tourism exponentially.

San Juan Islands Archipelago of 172 islands in Puget Sound, at the Canadian border east of Juan de Fuca Strait. The largest islands are Orcas, San Juan, and Lopez. Some of the other islands are occupied, often by single families, but many are uninhabited rocks. The San Juans, originally occupied by the Lummi Indians, were "discovered" by Spanish and British explorers in the late eighteenth century but not settled until the 1850s. In the "Pig War" in 1859 (an American shot and killed a marauding British-owned pig), British and American settlers almost came to blows, but a compromise was reached where joint occupation was accepted pending nonpolitical arbitration, and the German kaiser decided in favor of the Americans in 1872. San Juan, with the village of Friday Harbor, is the busiest of the islands, with a fishing fleet, a whaling museum, and the University of Washington Oceanographic Laboratory. Orcas (named for the orcas, or killer whales, that are frequently seen around the San Juans) is smaller and less populous; Lopez is even smaller and popular with bicyclists. Residents and visitors get to and from the islands by ferry.

See also Juan de Fuca Strait, Puget Sound

Santa Barbara Islands Sometimes called the Channel Islands (but not to be confused with the *English* Channel Islands), the Santa Barbara Islands extend for 150 miles off the coast of Southern California. There are two groups, separated by the San Pedro Channel: the Channel Islands of San Miguel, Santa Rosa, Santa Cruz, and Anacapa; and the Outer Channel Islands of Santa Barbara, San Nicolas, Santa Catalina, and San Clemente. Many were visited by the Portuguese navigator Juan Rodríguez Cabrillo in 1542, and he is reputed to be buried on one of them. San Miguel, Santa Rosa, Santa Cruz, Santa Barbara, Anacapa, and their surrounding waters are part of the Channel Islands National Marine Sanctuary, which was dedicated in 1980. The islands are frequented by seals, sea lions, and a variety of seabirds. The larger islands, such as Santa Rosa, Santa Cruz, Santa Catalina, and San Clemente, are inhabited, mostly by sheep and cattle ranchers, but they are also popular destinatons for visitors. The Tuna Club on Santa Catalina was California's most celebrated big-game fishing club, occasionally hosting such angling luminaries as Zane Grey. **See also national marine sanctuary, Santa Barbara oil spill**

Santa Barbara oil spill On January 28, 1969, at Unocal's "Platform A," 3 miles off the coast of Santa Barbara, a drilling operation resulted in an uncontrolled flow of oil from a deep reservoir through oil-bearing sands. Some 3.2 million gallons (79,000 barrels) were released into the Pacific Ocean. Three days after the spill began, winds and currents drove the oil ashore, oiling more than 100 miles of shoreline; by the fourth day, it had spread to the Channel Islands of Santa Rosa, Santa Catalina, and Anacapa. Eventually, the spill covered more than 800 square miles and reached all the way to the Mexican border. Marine and terrestrial plants were destroyed; marine mammals, seabirds, fishes, and invertebrates were oil-soaked and killed. Bills were introduced in the California and federal legislatures to create oil-well-free zones, and in 1972, the National Marine Sanctuaries Act was passed, leading to the establishment of the Channel Islands National Marine Sanctuary in 1980. **See also oil spill**

Santoríni Island in the Aegean, known in ancient times as Thera, that was subjected to a monumental volcanic explosion around 1500 B.C., destroying most of the island and leaving only a ring of three fragments (Thira, Thirasía, and Aspronísi) surrounding a huge caldera. In the center of the caldera are two recent cones, Néa Kaméni and Palaía Kaméni, which are the manifestations of more recent volcanic eruptions. The discovery of Minoan buildings and artifacts at Akrotíri on Santoríni (closely related to the Minoan settlements of Knossos, Phaistos, and Zakros on the island of Crete, some 75 miles away) has led many to suggest that Santoríni is somehow connected to the legend of Atlantis, the fabled city that Plato described as having disappeared into the sea around 9000 B.C. In two of his dialogues, Plato said that Atlantis vanished into the Atlantic, and while most of Santoríni was vaporized, Akrotiri and all of the island of Crete, including Knossos, are still above the surface of the Aegean. Because of its spectacular views and history, Santoríni is one of the most popular destinations for Aegean cruise ships, and the capital city of Firá (also Thirá) buzzes with excitement night and day.

See also Aegean Sea, Atlantis, Crete

São Tomé and Príncipe Island republic in the Gulf of Guinea, straddling the equator, about 150 miles off Gabon in West Africa. São Tomé (pronounced "san tomay") covers about 330 square miles, and 90 miles to the north is Príncipe, with an area of about 55 square miles. The population is around 144,000. The islands were discovered by Portuguese navigators around 1471, during early exploratory voyages down the west coast of Africa. The Portuguese settled the islands with slaves, and by about 1483, yams, rice, fruit, and sugar were being grown to supply passing ships. During the sixteenth century, São Tomé was briefly the world's largest producer of sugar, but it was soon eclipsed by Brazil. The Dutch took the islands in 1641, but were expelled by the Portuguese in 1740. Under Portuguese control, São Tomé served as a staging area for slavers heading for Brazil, but when slavery was abolished in the nineteenth century, the islanders began to grow cocoa and soon became the world's largest producer of chocolate. When cocoa prices fell after World War I, the economy declined, and a nationalist movement took root. Guerrilla wars were fought throughout the 1960s, and when the Portuguese dictatorship of António Salazar fell in 1974, power was handed over to the Liberation Movement of São Tomé. Many landowners fled to Portugal, and in 1975, independence was granted to the Republic of São Tomé and Príncipe. The country, one of the smallest in the world, depends largely on foreign aid.

Sarawak Malaysian state on the island of Borneo, bordered on the northeast by Sabah and Brunei and on the south by the Indonesian province of Borneo (Kalimantan). In 1841, Sarawak was ceded by the sultan of Brunei to James Brooke, an Englishman who became the rajah of the independent state. It was made a British protectorate in 1888, but remained under the control of the Brooke family. Along with the rest of the island, Sarawak was occupied by the Japanese in 1942, and after the war, the Brookes turned their fiefdom over to the British. In 1963, after a series of revolts, Sarawak joined the new Malaysian Federation as its largest state, joining with Sabah to become East Malaysia.

See also Borneo, Brooke, Brunei

sardine (*Sardinops sagax*) The name "sardine" is commonly used for several different fishes. The Pacific sardine was so heavily fished in California in the 1930s and '40s that the fishery completely collapsed. (John Stein-

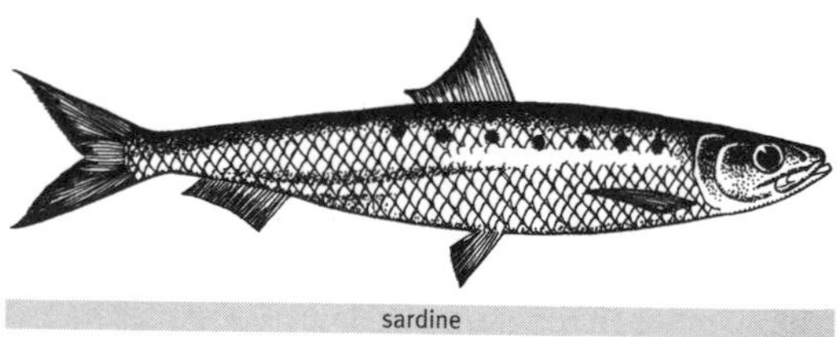

sardine

beck's 1945 novel *Cannery Row* was about this fishery.) In Australia, *S. neopilchardus,* known as pilchard, is harvested for food; and in South Africa, the "sardine" is *S. ocellata,* also called pilchard. The Spanish sardine (*Sardina pilchardus*) is eaten in great quantities in Spain and Portugal. All sardines are silvery fishes that are less than 1 foot long, aggregate in huge schools, and are usually caught in purse seines.

See also anchovy, Atlantic herring

Sardinia Italian island in the Mediterranean, separated from the French island of Corsica by the Strait of Bonifacio. The island covers some 9,302 square miles; it is slightly smaller than Sicily. Throughout the island are the *nuraghi,* truncated cones of huge basaltic rocks that were built in prehistoric times. The Phoenicians and the Carthaginians settled Sardinia before the Romans arrived in 238 B.C., and used it for the production of salt and grain. The kingdoms of Pisa and Genoa fought for this island (and Corsica as well), until Pope Boniface VIII gave Sardinia to the house of Aragón, which subsequently passed it along to Spain. Spain ceded it to Austria, but in 1717 the Spanish regained it by force. In 1720, the island was awarded to Victor Amadeus II of Savoy (who named himself king of Sardinia). It has been part of Italy since 1835. Although the island has some agricultural resources, it is mountainous, stony, and unproductive. Much of it is given over to pasturage for sheep and goats; wheat, barley, grapes, olives, cork, and tobacco are also grown. Troubled economically, Sardinia has a low per capita income and a high unemployment rate. Tourism has become increasingly popular, and visitors can view the colorful village festivals and buy traditional crafts. The population of Sardinia is approximately 1,660,000; some 220,000 live in Cagliari, the capital.

See also Corsica, Mediterranean Sea

Sargasso Sea First described by Christopher Columbus in 1492, the Sargasso Sea is a roughly elliptical area of the North Atlantic east of the Bahamas, where a powerful eddy causes sargassum weed to collect in vast quantities and float on the surface. The sea ranges in depth from 5,000 to 23,000 feet, and is characterized by weak currents, low precipitation, and high evaporation, which combine to produce a body of water that is relatively plankton-free. The Sargasso Sea is believed to be the primary breeding place of the common eel (*Anguilla anguilla*), which then disperses to America and Europe. The Bermuda islands are in the Sargasso Sea.

sargassum fish (*Histrio histrio*) Almost completely invisible in the seaweed it inhabits, the sargassum fish is a species of frogfish that is the same mottled color as the weed, with fins that are split, forked, and filamented to increase its camouflage. Like other frogfishes, the sargassum fish is equipped with a "fishing pole," a lure atop its head that it wiggles to attract potential prey, which consists of shrimps and other fishes. The prehensile pectoral fins of the 7-inch-long fish are especially adapted for "crawling" through the weed; it rarely swims. It is best known among the sargassum patches in the western Atlantic, but it also found in floating weeds of the tropical western Pacific. **See also anglerfishes, frogfish**

sargassum fish

sargassum weed (*Fucus natans*) A genus of free-floating brown algae, also known as gulfweed or rockweed, characterized by hollow, berrylike floats, called pneumatocysts, that keep it floating at the surface. An entire community of specially adapted fishes and crustaceans makes its home in the patches of weed. Early travelers believed that sargassum weed would entangle their ships, but this is not so. **See also algae, sargassum fish**

Sark, Isle of One of the Channel Islands, Sark (Sercq in French) is actually two islands, Great Sark and Little Sark, connected by a 100-yard-long isthmus (the Coupée) that is only 6 feet wide in some places. Much closer to France than to England, it has been considered part of England since the Norman invasion in 1066. Along with Alderney, Sark is included in the bailiwick of Guernsey. The total area is 2.1 square miles, and the population is around 500. Sark is Europe's last remaining fiefdom, ruled by seigneurs or dames under a charter granted by Queen Elizabeth I in 1565. The Carteret family were seigneurs during the eighteenth century; the seigneury subsequently changed hands several times. It was held by Sybil Hathaway, the Dame of Sark, from 1927 until her death in 1974, when it passed to her son, Michael Beaumont, the twenty-second seigneur. Cars are not allowed on the island, and only the seigneur may keep a female dog. **See also Alderney, Channel Islands, Guernsey, Jersey**

saury (*Scomberesox saurus*) A small offshore Atlantic fish with moderately elongated jaws. Sauries congregate in huge schools, and ten of thousands of the 1-foot-long fishes can sometimes be seen skittering along the surface. They are found in the warm and temperate waters of the world, where they are preyed on by

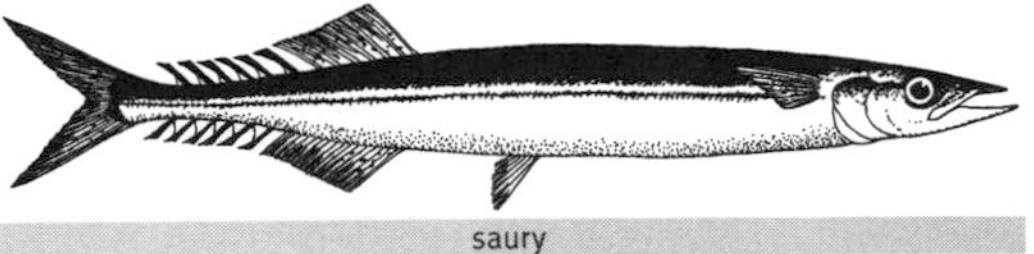

saury

any and every fish that can catch them. The Pacific saury (*Cololabis saira*) demonstrates the same habits and characteristics from California to Japan.

See also ballyhoo, halfbeak, needlefish

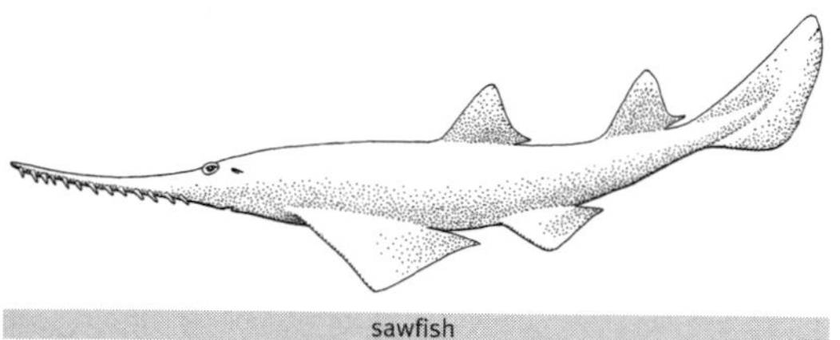

sawfish

sawfish (*Pristis pectinata*) With its nostrils and gill slits on the underside of its head, the sawfish is classified as a ray, but it looks more like a shark with a saw on the end of its nose. There are several species of sawfishes throughout the world, and one (*P. microdon*) that is found primarily in freshwater rivers in Australia and Indonesia. The saw, which has about twenty pairs of "teeth" on it—actually modified dermal denticles—is used to slash at schooling fishes to stun or impale them. (Sawfishes actually have numerous small, blunt teeth in their jaws.) Stories of sawfish attacking whales are apocryphal, but they can inflict nasty wounds in swimmers if provoked. Found throughout the world's shallow subtropical waters, the largest of the sawfishes can reach a length of 22 feet. **See also rays, sawshark**

sawshark (*Pristiophorus spp.*) Sawfishes are rays that look like sharks; sawsharks actually are sharks, defined by the location of their gill slits on the side of the head. The flattened rostrum is armed with toothlike denticles, and their teeth are more pointed and sharklike than those of the sawfishes. Sawsharks reach a length of 5 feet, and with the exception of the Caribbean species *P. schroederi,* they are all found in Indo-Pacific waters. They are deepwater sharks, found at depths of 200 feet or more. All sawsharks have a pair of barbels located about midway on the underside of the "saw." There is one species, *Pliotrema warreni,* that has six gill slits and is commonly known as the sixgill sawshark. (Most shark species have five on each side, three have six, and one has seven.) **See also frilled shark, sawfish, sixgill shark**

scales The thin epidermal layer of fishes contains mucus cells, which secrete the slippery, slimy substance so familiar to fishermen. Beneath this layer, cartilaginous fishes (sharks and rays) have tiny toothlike projections called dermal denticles (or placoid scales), while the bony fishes have scales, thin overlapping plates embedded in the inner layer of the skin. Some of the more primitive fishes, like the gars, have diamond-shaped scales attached to one another by joints, but most other fishes have scales that are either cycloid (more or less circular in outline) or ctenoid (from the Greek *ctenos,* "comb"), which means that the posterior margin is comb- or toothlike. Most of the spiny-rayed fishes (Acanthopterygii) have ctenoid scales. When the scales are so loosely fastened that they can be easily rubbed off, as with many of the herrings, they are termed "deciduous." Many structures in fishes are derived from scales, such as the lancets or blades of surgeonfishes, the sharp belly scutes of herrings, the armorlike plates of sea horses and pipefishes, the spikes of porcupine fishes, and the encasements of trunkfishes. Some fishes, like tarpon, have enormous scales, while those of tunas and other scombroids are minute, and in eels the scales are so tiny that they are invisible to the naked eye. **See also dermal denticles**

scallop (family Pectinidae) Bivalve mollusks with ribbed, fan-shaped shells, scallops are found throughout the world's oceans, from the shallows to depths of 10,000 feet. There are more than four hundred species, whose shells can be white, yellow, red, purple, or variegated combinations. They do not crawl or burrow, and they do not have a large "foot" like clams, but with their highly developed adductor muscle, they can open and close their shells rapidly and move themselves through the water by jet-propelled jumps. On the outer edge of the mantle there are numerous blue eyes. Scallops are hermaphroditic, each individual containing both the orange roe and a whitish testis. The bay scallop (*Aequipecten irradians*), with a shell 3 inches across, is harvested off New England and eastern Canada, but the deep-sea scallop (*Placopecten magellanicus*) can be 8 inches across. In the wild, starfish are the scallop's normal predators, but men have harvested them for thousands of years, eating the adductor muscle and, in some countries, the roe. During the Middle Ages, the shell became a religious emblem, and nowadays, it is most familiar as the symbol of the Shell Oil Company.

scallop

See also clam, mollusk, oyster

Scammon, Charles Melville (1825–1911) American whaling captain and author, born in Pittston, Maine. He sailed for California in 1850 and in 1852 commanded the San Francisco brig *Mary Helen* on a sealing and whaling voyage. In 1856, in command of the ship *Lenore,* he visited Magdalena Bay, one of the areas where the gray whales mate and give birth. Two years later, he brought the *Boston* over the barrier sandbar of the previously unexplored lagoon called Ojo de Liebre ("Jackrabbit Springs"), now known as Scammon's Lagoon. In 1859,

Scammon breached the third of the major breeding grounds of the gray whales, called Ballenas or San Ignacio Lagoon. Although he was directly responsible for the initiation of lagoon whaling, he also made a major contribution to whale conservation when he returned to San Francisco in 1872 and published *The Marine Mammals of the Northwestern Coast of North America, Together with an Account of the American Whale-Fishery,* which is considered one of the most important books ever written on the biology of the gray whale and the history of the fishery.

scamperdown whale

scamperdown whale (*Mesoplodon grayi*) This beaked whale, also known as Gray's beaked whale, was long considered an inhabitant of Southern Hemisphere waters only, particularly around New Zealand—until one showed up in the Netherlands in 1927. This suggests that we know little about where beaked whales live, only where they die. In addition to the characteristic pair of teeth in the lower jaw of the male, males and females have a row of tiny teeth in the upper jaw.

See also beaked whales, *Mesoplodon*

Scapa Flow A huge expanse of water in the Orkney Islands, sheltered by the islands of Hoy, Flotta, South Ronaldsay, and Burray, which was used by the Royal Navy as an anchorage for the fleet during both world wars. It was from Scapa Flow that the Grand Fleet sailed into the Battle of Jutland (May 31, 1916), and where the German High Seas Fleet was interned and scuttled itself on June 21, 1919. In October 1939, while the British fleet was anchored at Scapa Flow, the German submarine *U-14* slipped into Scapa Flow undetected and sank the battleship *Royal Oak* with a loss of 833 lives. **See also Jutland, Battle of; Orkney Islands**

Schmidt, Johannes (1877–1933) Danish oceanographer and ichthyologist who began his career as a botanist, writing his Ph.D. dissertation on mangroves. In 1904, at the behest of the International Council for the Study of the Sea, he made a voyage aboard the *Thor* to study the food fishes of the North Atlantic. When he found an eel larva (leptocephalus) in deep water west of the Faeroes, he began to study the then completely unresolved question of eel migration. (The larva was the first ever found outside the Mediterranean.) From 1905 to 1920, he scoured the Atlantic studying the eels that he captured, and finally, on an expedition aboard the *Dana* in 1922, he established that eels migrate to the Sargasso Sea, where they breed and die. After passing through the leptocephalus phase, their offspring metamorphose into "elvers" and are carried by currents to disperse in either European or North American streams and rivers. From 1924 to 1930, Schmidt led a series of expeditions aboard the *Dana,* making detailed observations of marine life, many of which were published in the *Dana Reports,* issued by the Zoological Museum of the University of Copenhagen. **See also eel, leptocephalus**

schooner A sailing vessel rigged with fore-and-aft sails on two or more masts, and originally with square topsails on the foremast. Properly speaking, a schooner has only two masts, but four-, five-, and even seven-masted schooners have been built. They were largely used for the coasting trade, and notably for fishing on the Grand Banks of Newfoundland.

See also codfish, Grand Banks

Schouten, Willem Cornelisz (1527–1625) Dutch East India Company captain who sailed as pilot on Jakob Le Maire's *Eendracht* on a 1616 expedition to discover a new way from the Atlantic to the Pacific Ocean. With Schouten's brother Jan as captain of the *Hoorn,* they sailed to Patagonia, but there the *Hoorn* burned to the waterline. Then they proceeded south to the Strait of Magellan, where they discovered Staten Island (which they believed was part of the great southern continent), and the strait that separates that land from the tip of Tierra del Fuego, which they named the Le Maire Strait. (Farther south, separating the Antarctic Peninsula from Petermann Island and the Argentine islands, is the Le Maire Channel.) Schouten and Le Maire were the first Europeans to sight Cape Horn, which they named Hoorn, after Schouten's birthplace. In the Pacific, they reached the Juan Fernández Islands, and then found several small islands between Samoa and Fiji, which they named the Hoorn Islands. The group, now known as the Horn Islands, includes Futuna and Alofi. When they arrived in Batavia (now Djakarta), Jan Pieterszoon Coen, the governor-general of the Dutch East India Company, refused to believe they had found a route to the Pacific, confiscated their ship, and sent them in disgrace back to Holland.

See also Cape Horn, Le Maire, Staten Island, Wallis and Futuna

Scilly Isles More than 150 islands, islets, and rocks, located 30 miles off Land's End, the western tip of southern England. Five of them are inhabited: St. Mary's, Tresco, St. Martin's, St. Agnes, and Bryher. Warm, rainy weather is characteristic of the Scillies, and the residents are famous for their flower gardens. To mark the rocky coasts, there are numerous lighthouses and lightships, but the isles have been the scene of numerous shipwrecks, including the flagship and three other ships of Sir Cloudesley Shovell, which were returning home in 1707 from a successful engagement with the French when they ran onto the rocks in a storm and were wrecked. Any number of other ships have wrecked in the dangerous rocks of the Scilly Isles, but the most recent—and infamous—disaster was the 1967 wreck of the *Torrey Canyon,* a 63,000-ton supertanker that ran into Pollard Rock in clear weather, spilling 38 million gallons of crude oil into the sea off Cornwall.

See also Shovel, *Torrey Canyon*

Scombridae A large group of fishes that are characterized by their streamlined, spindle-shaped bodies, pointed head, and tapered tailstock. The caudal (tail) fin is lunate or crescent-shaped. The two dorsal fins are separate, and tailward of the second dorsal and the anal fin are a series of little finlets. In some species, such as the tunas, there is a slot into which the spiny dorsal fin fits when it is lowered, adding to the streamlining. Scombrids are among the fastest swimmers in the ocean, and some have been clocked at speeds up to 50 mph. They are all important food fishes.

See also albacore, Atlantic mackerel, bonito, wahoo

Scoresby, William, Jr. (1789–1857) The son of a whaling captain of the same name, William Scoresby, Jr., went to sea at the age of ten aboard his father's ship *Resolution.* Before he returned to sea, he entered the University of Edinburgh, where he studied chemistry, natural philosophy, and anatomy. In 1810, when he was twenty-one, he was made captain of the *Resolution* and began his illustrious whaling career. He sailed the "West Ice" (Greenland and Spitsbergen) for sixteen years, becoming the most successful whaler in history, but he will be remembered not only for his deeds, but for his words. In 1820, he wrote one of the most influential books ever written on the history and practice of whaling, *Account of the Arctic Regions with a History and Description of the Northern Whale-Fishery.* The whales they hunted were polar whales (now known as bowheads), but when Herman Melville wrote *Moby-Dick* in 1851, he gratefully acknowledged the contributions of Scoresby, whom he caricatured as "Charley Coffin."

See also *Moby-Dick,* whaling

scorpionfish (family Scorpaenidae) There are hundreds of species of scorpionfishes, widely distributed throughout the tropical and temperate waters of the world. Most of them spend their time on the bottom, hence the name "rockfish." Many are armed with venomous fin spines, which they use defensively. Although most species are hard to see because of their coloration and a variety of flaps or projections on the head and body, the lionfish (*Pterois volitans*) is a dramatic-looking (and deadly) fish with striking brown and white coloration and flaring pectoral fins that swims slowly around Indo-Pacific reefs, as if daring anything to attack it.

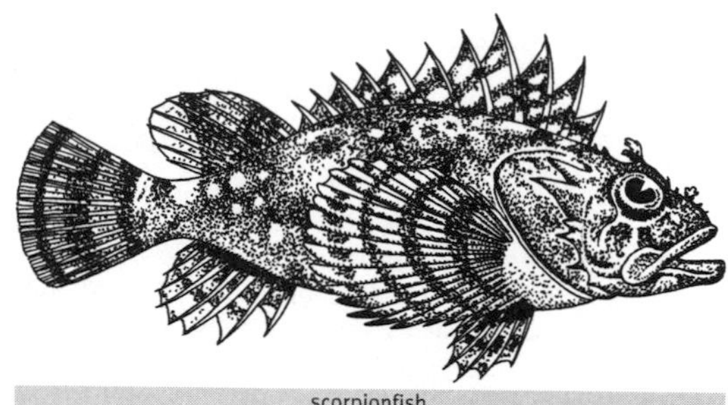

scorpionfish

See also lionfish, stonefish

scoter (*Melanitta* spp.) There are three species of scoters, large, seagoing ducks that are found throughout the northern reaches of the Northern Hemisphere. The surf scoter (*M. perspicillata*) is the smallest of the three, identifiable by the drake's black coloring with white markings on the head. (As with most ducks, female scoters are brown.) Its range (Labrador south to the Gulf of St. Lawrence; Alaska to Baja California) overlaps that of the white-winged scoter (*M. fusca*), which is easily identified by the white speculum (patch of color on the wing) in both sexes. The common scoter (*M. nigra*) breeds in the Aleutians and northern Canada and winters as far south as California and South Carolina. It is all black except for the yellowish bump at the base of the bill.

surf scoter

See also eider, harlequin duck, old squaw, sea ducks

Scott, Robert Falcon (1868–1912) British naval officer and Antarctic explorer whose first expedition, in 1901–1904, was aboard the *Discovery.* With Ernest Shackleton and Edward Wilson, Scott sledged to the record southern latitude of 82°17′. Upon his safe and triumphant return to England, Scott was promoted to captain. He then decided to lead an expedition to the Antarctic in the *Terra Nova,* this time to study the Ross Sea and to reach the South Pole. In October 1911, they started overland for the Pole, with motorized sledges, ponies, and dogs. The motors broke down, the ponies were unsuited for polar travel and had to be shot, and the dogs were sent back. Scott's party of twelve began to haul their heavily burdened sledges toward the Pole. By December 31, seven were sent back, leaving Scott, Edward Wilson, H. R. Bowers, L. E. G. Oates, and Edgar Evans to continue. They reached the Pole on January

18, 1912, only to learn that Roald Amundsen had beaten them by about a month. The weather on their 800-mile return journey was terrible, and they all froze to death before they could reach the food and fuel that had been left for them at a supply depot. Scott is considered the quintessential British hero, even though his expedition failed because of poor planning and organization.

See also Amundsen, *Discovery*, Shackleton, South Pole

scrimshaw The carving done by American whalemen on whale bones and teeth or, less frequently, on the tusks of walruses. Between bouts of frantic whaling and trying-out activities, whalemen had plenty of spare time, and they produced what has been called "the only folk art in America." The favorite surface was the large ivory teeth of the sperm whale, which were carved with images of ships, women, whaling, patriotic scenes, and many other subjects. First the teeth were smoothed and polished, then the design picked out with a sharp tool, often no more sophisticated than a jackknife, and then ink or lampblack was rubbed into the etched surface.

See also trypot, tryworks, whaling

Scripps Institution of Oceanography (SIO) Originally known as the Marine Biological Association of San Diego and based in Coronado, the SIO moved to its current location in La Jolla in 1904. Its first benefactors were Ellen and E. W. Scripps, and University of California professor William E. Ritter was named the first director. In 1912, the facility was renamed the Scripps Institution for Biological Research. Thomas Wayland Vaughan, appointed director in 1924, oversaw the acquisition of the first research vessel, the building of new laboratories, and an aquarium, which was named for him. Later, under the direction of oceanographer Harald Sverdrup, SIO became part of the University of California and acquired the research vessel *E. W. Scripps*. During World War II, Scripps scientists provided a great deal of useful information for the U.S. Navy on sonar and sound in the sea, and in 1952, physicist Hugh Bradner invented the wet suit there. In 1956, under director Roger Revelle, graduate education was initiated at Scripps, and its administrative functions were transferred to the University of California at San Diego. Scripps was a part of the Joint Oceanographic Institutions Deep Earth Sampling program (JOIDES) that did such important work sampling the ocean floor from 1968 to 1983 with the research drill ship *Glomar Challenger*. SIO is now one of the country's leading institutions for the study of climatic change, ocean engineering, and marine biology. The campus occupies sixty-seven buildings on 230 acres, and employs more than twelve hundred people.

See also Woods Hole Oceanographic Institution

scuba An acronym for "Self-Contained Underwater Breathing Apparatus." In 1943, Jacques-Yves Cousteau of the French navy and Émile Gagnan, a civilian engineer, perfected a "demand regulator," which, when attached to a tank of compressed air, allowed a diver to descend beneath the surface and breathe without being tethered to the surface by an air hose. (They originally called their device an Aqua-Lung; the patents from the sales of the equipment made Cousteau a very rich man, enabling him to finance his numerous television films.) The basic idea had been proposed in 1865 by Benôit Rouquayrol and Auguste Denayrouze, whose device consisted of a canister filled with compressed air, released by a regulator valve, with another valve that removed the exhaled air. Jules Verne employed the Rouquayrol-Denayrouze apparatus in *Twenty Thousand Leagues under the Sea* (1870), when Captain Nemo and Professor Arronax leave the submarine *Nautilus* and take a walk on the ocean floor. **See also Cousteau**

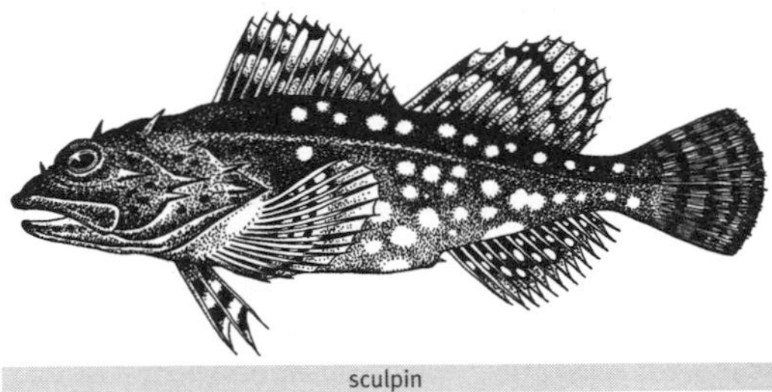

sculpin

sculpin (family Cottidae) Big-headed, bottom-dwelling fishes, with eyes mounted atop their heads and sharp spines on the gill covers. There are some two hundred species, found throughout the colder waters of the world but predominantly in the North Pacific. Many of the smaller species (up to 4 inches in length) live in tide pools, but there are also many species that inhabit freshwater streams and lakes. The largest of the sculpins is the cabezone (*Scorpaenichthys marmoratus*) of the North Pacific, which reaches a length of 2½ feet and a weight of 25 pounds. The most common American marine sculpin is the staghorn sculpin (*Leptocottus armatus*), found in shallow waters from Alaska to Baja California. **See also cabezone**

scup (*Stenotomus chrysops*) The scup, a relative of the porgies, is found in the western North Atlantic from Nova Scotia to Florida. Although they average about 10 inches in length, an occasional 2-footer is caught. In summer, they school in inshore waters but they move into deeper water in winter. They are caught by recreational fishermen and until recently were the target of commercial fisheries off southern New England and New Jersey. The catch has dropped markedly in recent years, from a high of 9,800 tons in 1981 to a low of 3,700 tons in 1989.

See also porgy, Sparidae

scurvy A nutritional disorder caused by a lack of vitamin C. The symptoms are swollen and bleeding gums with loosened teeth, soreness and stiffness of the joints, bleeding under the skin, and anemia. Toward the end of the sixteenth century, scurvy was the major cause of death of seamen on long voyages. It was not until 1753 that James Lind, a Scottish physician, recognized that scurvy could be prevented by the ingestion of the juices of oranges, lemons, and limes, but vitamin C was not isolated as the specific antiscorbutic until 1932. Although limes were the preferred preventative—thus the name "limeys" for British sailors—oranges and lemons are actually more effective.

See also Lind

Scylla and Charybdis Two navigational hazards in the Strait of Messina, which, according to Greek legend, Odysseus had to pass through on his voyage from Troy. Scylla, a nymph, was seen bathing by the sea god Glaucus, who fell in love with her but whose advances she repelled. Glaucus appealed to Circe for a love potion, but instead she gave him a poisonous mixture that turned Scylla into a frightful monster, rooted to a rock. She retained her lovely voice, however, and sang enticing songs to passing mariners. Charybdis was a dangerous whirlpool on the opposite side of the strait. She had been an avaricious woman who stole the oxen of Hercules and was punished by Zeus by being turned into a whirlpool. The phrase "between Scylla and Charybdis" is now used to denote a passage of great danger.

See also Maelström

sea anemone (order Actiniaria) A sea polyp that generally resembles a flower, with a crown of tentacles surrounding an oral disk atop a columnar body. The tentacles are equipped with stinging cells (nematocysts) that are used to immobilize or kill their prey, which ranges in size from small planktonic organisms to good-sized fishes. The tentacles draw the victim into the central body cavity, which is almost completely filled with a giant digestive gland. When severely disturbed, some species can protrude acontia (singular: acontium), special stinging structures, through the mouth or pores in the column, which are heavily armed with nematocysts. Although they appear rooted in place, sea anemones are capable of movement, gliding slowly on the muscular pedal disk that makes up the base. Some communal species do not get much larger than a couple of inches wide, but the tropical species *Stoichactis* can have a tentacle disk that is 3 feet across. They have long life spans, with some known to live almost a hundred years. Although some sea

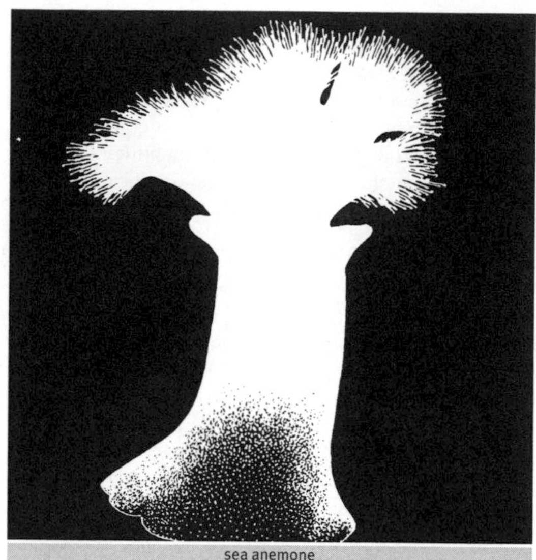

sea anemone

anemones are found at great depths (the deepest was hauled up from 30,000 feet), most species inhabit tide pools or tropical coral reefs. The tropical species are especially colorful, with tentacles that range from ivory to pink, yellow, and red. They have the ability to retract the tentacles completely, which gives them the appearance of a soft blob. They can reproduce sexually or, in some cases, by budding a new individual from the base of the column. There are many instances where a symbiotic or commensal relationship is established between a sea anemone and another creature; perhaps the best known is the clownfish (*Amphiprion*) that lives amid the waving poisonous tentacles, immune to the venom, and feeds on the crumbs from the anemone's meal. Members of a genus of hermit crabs (*Eupagurus*) "adopt" their own sea anemone and affix it to the shell they have moved into. When they change shells, they take the sea anemone along.

See also cnidarian, hermit crab, nematocyst

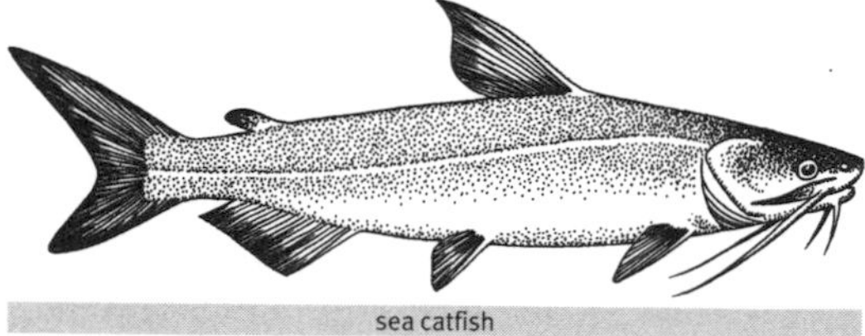

sea catfish

sea catfish (*Arius felis*) Most catfishes are freshwater denizens, but the sea catfish and the gafftopsail catfish (*Bagre marinus*) live in salt or brackish water. The sea catfish, at a maximum length of about 1 foot, is common in shallow waters from Cape Cod to the

Gulf of Mexico. As the female lays her eggs, the male picks them up in his mouth and incubates them there until they hatch. After hatching, the young continue to use the male's mouth for refuge until a month passes and they are able to go off on their own.

See also gafftopsail catfish

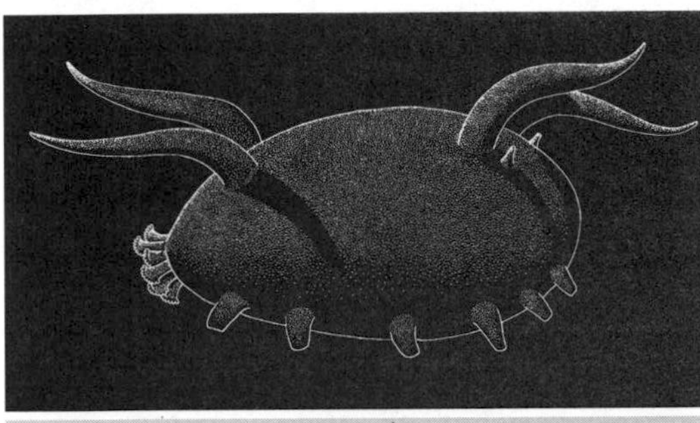

sea cucumber

sea cucumber The class Holothuriodea includes some twelve hundred species, the commonest form of which has the characteristic cucumberlike shape. At one end, modified tube feet (podia) form a circle of tentacles around the mouth; the anal opening at the other end is used for discharging wastes and respiration. Water is taken into the body cavity and oxygen is removed by branching organs known as respiratory trees. Most sea cucumbers move slowly along the substrate by using five rows of podia on the underside, but there are some species that can swim, and a few (such as *Pelagothuria natatrix*) that spend their entire lives floating and swimming in the water column. It is not clear how swimming holothurians feed, but the bottom dwellers take in quantities of sand and other matter to extract nourishment, expelling the sand when they have completed their feeding. Because they move slowly, sea cucumbers cannot easily escape from predators; for defense, they can either discharge a mass of sticky tubules or expel their internal organs completely. Some species of pearlfishes (*Carapus* spp.) live commensally inside of sea cucumbers, coming out only to feed. Sea cucumbers are also eaten by people, particularly in Asia, where they are known as *bêche-de-mer* ("shovel of the sea"), or *trepang*. **See also echinoderms, pearlfishes**

sea ducks Several species of ducks spend most of their lives near the sea, but because there are so many borderline cases, such as ducks that breed inland and live along the shore, they are usually not included in discussions of seabirds. The most marine of all ducks are the eiders, scoters, old squaws, and harlequin ducks. All sea ducks are excellent divers, living on mollusks and other animal matter that they pick from the bottom. Their bills are strong and well adapted to crushing the shells. As with most ducks, the drake is strikingly colored while the female is quite drab. The sea ducks included here are all birds of the Arctic or subarctic.

See also eider, harlequin duck, Labrador duck, old squaw, scoter

sea eagle (*Haliaeetus* spp.) Also known as fish eagles, these large, broad-winged birds of prey are found throughout the world's coastal areas, except for South America. Although they are powerful fliers, they also soar a great deal, with the wings held horizontally. The bald eagle, Steller's sea eagle, and the white-tailed sea eagle inhabit subarctic latitudes, while the white-bellied sea eagle (*H. leucogaster*) is distributed throughout the Australasian region. There is an African species (*H. vocifer*), a Madagascar variety (*H. vociferoides*), one (*H. sanfordi*) that is restricted to the Solomon Islands, and one, Pallas's sea eagle (*H. leucoryphus*), found mostly inland in central Asia. When fish are not readily available, sea eagles will take diving birds, small mammals, lizards, snakes, crabs, and turtles, but they are also carrion eaters. They build large stick nests high in trees, or on cliffs if trees are not available.

See also bald eagle, Steller's sea eagle, white-tailed sea eagle

Sea Empress Norwegian-owned, Liberian-registered oil tanker that ran aground on February 15, 1996, at the entrance to the harbor at Milford Haven, Wales, one of Britain's premier deepwater harbors. In gale-force winds, the *Sea Empress* discharged nearly 30,000 tons of crude oil, most of which was forced out to sea. Harbor tugs were unable to hold the ship against the strong winds and tides, and she grounded several more times before she was finally towed across the Irish Sea to Belfast. By February 21, the *Sea Empress* had released some 65,000 tons (24 million gallons) of crude oil off the coast of Wales. Cleanup efforts succeeded in recovering much of the oil, and large areas were sprayed with chemical dispersants to break it up. Nevertheless, thousands of birds died and 120 miles of shoreline were seriously polluted. Even though the *Sea Empress* spilled almost twice as much oil as the *Exxon Valdez,* the lightness of the oil and the rapid responses by local and government agencies saved the area from a major ecological disaster. **See also oil spill, supertanker**

sea fan (*Gorgonia*) Invertebrate animals, also known as horny corals, characterized by a flattened, fanlike structure that consists of a colony of tiny polyps that grow upon one another in a radiating, treelike fashion. Spicules of lime give the colony its support structure. Sea fans are most often yellow, pink, brown, or purple. They can grow to about 2 feet in height, and are found mostly in the warm, shallow waters of the Atlantic and Pacific, particularly near or on coral reefs.

See also coral, coral reef

seafloor spreading The movement of the ocean floors outward from an underwater mountain system that encircles the globe like the seams of a baseball. Some parts of this 40,000-mile-long oceanic ridge contain clefts or rifts that may be a mile deep and 30 miles wide, from which liquid rock (magma), rising from the earth's mantle, wells up to create the seafloor. As this basaltic magma hardens, it forms a new crust that is moved outward in both directions from the ridge. This is believed to be the agent that causes the movement of the tectonic plates and therefore the continents. That the seafloor can be dated as youngest at the mid-ocean seams and oldest at the continental shelves is strong supporting evidence for the theory of seafloor spreading, first proposed by oceanographer Harry Hess in the 1960s, and later verified by R. J. Dietz, J. Tuzo Wilson, Bruce Heezen, Walter Pitman, Frederick Vine, and others. It is known that the earth's poles switch positions every million years or so, so when analysis of the magnetic properties of the seafloor showed alternating, symmetrical bands of reversed polarity on either side of the ridges, it was clear that the seafloor was spreading outward, as if on matched conveyor belts. Some of the moving seafloor material subducts under the plates, but other material is rammed into the continental plates, causing mountains to be thrown up.

See also continental drift, Mid-Atlantic Ridge

sea hare (*Aplysia* spp.) Marine gastropods with a greatly reduced internal shell and prominent tentacles that resemble a rabbit's ears. Although like their namesakes they are gentle creatures, they can eject a slimy purple ink if roughly handled. Most sea hares are about a foot in length, but a species on the west coast of North America (*A. vaccaria*) can get to be 30 inches long and weigh 35 pounds, making it the world's largest gastropod. They are found in warm shallow waters, where they feed on seaweeds. **See also gastropoda, nudibranch**

sea horse (*Hippocampus* spp.) Looking like an underwater chess knight, the sea horse has a tiny, tubular mouth at the end of a long snout and a head that is angled like that of a horse. (The heads of the closely related pipefishes are a continuation of the fish's body.) Like chameleons, sea horses can move their eyes independently. They swim upright, using their prehensile tails to fasten themselves to plants or coral branches. Female sea horses lay their eggs in a pouch in the male's belly, where they are fertilized and incubated. The male then "gives birth" to the tiny babies, which resemble miniature adults. They are found throughout the world's warm, shallow waters. In many areas they are overfished and endangered.

See also leafy sea dragon, pipefish

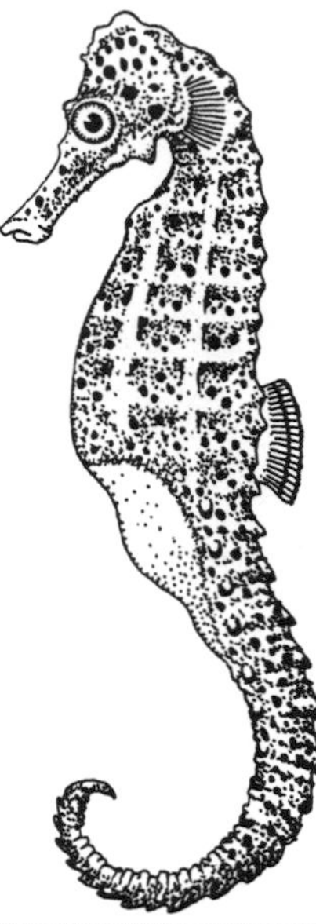
sea horse

sea ice Ice that forms when saltwater freezes. While freshwater freezes at 32°F, saltwater freezes between 28°F and 29°F. Because ice is less dense than water, it floats; ice begins to form in the ocean from the surface downward. The youngest ice (known as "frazil" ice) begins as a sort of icy slush, and then circular floes with slightly raised edges ("pancake" ice) are formed; "young" ice is unhummocked and about a foot thick. The younger ice floes are about 6 feet thick, but older ice can be twice that thickness. Even though it is frozen saltwater, sea ice is usually devoid of salt; as the ice forms, the salt is expelled from it as brine. After a year, sea ice is salt-free enough to be melted for drinking. Overland travelers to the North Pole had to traverse the sea ice, which would often crack open into wide "leads" or pile up in high-pressure ridges. It is conceivable that at any given time, the North Pole would be under open water.

See also ice, iceberg, ice formation, North Pole

sea krait (*Laticauda* spp.) Sea kraits are sometimes known as amphibious sea snakes because they return to land to bask, mate, and lay their eggs. Named for kraits, the terrestrial, venomous Asiatic snakes, they are found in scattered locations throughout the Indo-Pacific region. They are banded, beautifully colored, cylindrical snakes with a flattened tail (*Laticauda* means "broad tail") that reach a maximum known length of 7 feet. They are very venomous, but even when handled, they rarely bite humans.

See also sea snakes

sealing The catching of seals, sea lions, and elephant seals for their fur, blubber, and meat. Some of the earliest exploration of the Southern Hemisphere was conducted by sealers searching for more fur seals and elephant seals to kill for their fur and oil, respectively. During the eighteenth century, fur seals were killed along the east coast of South America, and also the Juan Fernández Islands and the Galápagos. Shortly after James Cook discovered South Georgia in 1775, the sealers arrived in pursuit of the elephant seals. (The first whaling station on South Georgia was named Grytviken, meaning "cauldron bay," because of the oil pots the whalers found there.) Sealing captain William Smith discovered the South Shetland Islands in 1819, attracting sealers from many nations. The hunting was so intense that by 1829, the fur seals had been eliminated. The same sorry story was repeated throughout the Southern Ocean; wherever fur seals were found, they were slaughtered nearly to extinction. In the North Pacific, where the northern fur seals were discovered during Vitus Bering's 1741 voyage, Russian and American sealers were soon competing for the seals of the Pribilofs, the Aleutians, and the Kuril Islands. (It was Russian sealers visiting the western Aleutians who completely eliminated Steller's sea cow, only twenty-seven years after it had been discovered.) Sealing has also played an important part in the history of the North Atlantic, especially Newfoundland, where Norwegian and Canadian sealers engaged in the bloody slaughter of baby harp seals as early as the sixteenth century. Except in Siberia and the Pribilofs, where hunting by native peoples is still allowed, most sealing is finished, largely because of its anachronistic nature, but also because of the shortage of seals. Harp seals, however, are still being killed in large numbers in the Gulf of St. Lawrence (Canada) and Norway.

See also elephant seals, sealskin, Steller's sea cow

seals and sea lions The pinnipeds ("fin feet") of the world can be divided into two types: the Phocidae, known as the earless seals (sometimes known as the true seals); and the Otariidae, the eared seals, which are characterized by small but visible pointed ear flaps. The Phocids cannot rotate their hind flippers forward; they hunch themselves over land or ice with the strong nails on their forelimbs. The Otariids can use all four limbs for locomotion. Fur seals, like sea lions, are Otariids. The walrus is neither seal nor sea lion, but intermediate between the two groups, since it has no visible ear flaps but can rotate its rear flippers forward.

See also *individual species;* walrus

sealskin Among the world's most luxurious furs because of its densely packed hairs, sealskin has been made into garments for millennia. Eskimos have always relied on sealskin for its tough and waterproof leather, and because the skins with the fur on were extremely warm, they were used extensively for outer parkas, pants, and boots. By the middle of the nineteenth century, seal fur coats were the height of fashion in Europe and China, and to serve this market, the sealers harvested the seals with remorseless efficiency. Harp seal pups, known as whitecoats, were clubbed to death before they were two weeks old, so that their molt did not begin and spoil the fluffy white fur. **See also harp seal**

sea moth (family Pegasidae) Armored little fishes (maximum length: 5 inches) that resemble flying gurnards or poachers, but are more closely related to sea horses and pipefishes. Even though the sea moth has an elongated snout like the sea horses and pipefishes, the mouth is not at the end of it, but rather on the underside of the head. The common name is obviously derived from the very broad pectoral fins, and the generic name (*Pegasus*) from the winged horse of mythology. They live on the bottom in the waters of the Pacific, from Hawaii throughout Micronesia and Melanesia and across the Indian Ocean to the east coast of Africa.

See also flying gurnard, pipefish, poacher, sea horse

seamount An isolated or comparatively isolated elevation of the deep-sea floor of approximately 1,000 meters (3,280 feet) or more. They are found in all major ocean basins; more than fifteen hundred have been counted in the Pacific alone. They are believed to have been formed volcanically, but the exact process is not known. Seamounts vary in size from relatively small, conical peaks to massive structures such as the Great Meteor Seamount in the northeast Atlantic, with a base diameter of 68 miles and an elevation of 13,000 feet above the seafloor. The area of its flat summit platform is nearly 1,250 square miles, larger than the state of Rhode Island. A sea peak is a seamount with a pointed summit, and guyot is a seamount with a flattened, relatively smooth top.

sea mouse (*Aphrodita hastata*) A polychaete worm that is covered in what appears to be a coat of silky felt, but is actually extended bristles. For the most part, sea mice live just below the surface of the sand or mud, and appear quite lethargic, although they can scuttle quickly when disturbed. If the "fur" is cut and lifted, it will be seen that the worm has a series of overlapping disk-shaped plates (scales) on the back; it is actually a large scaleworm in which the scales are concealed by the long bristles. Sea mice, which reach a length of 9 inches, are found from the Gulf of St. Lawrence to Chesapeake Bay. **See also polychaetes**

Sea of Cortez: See Gulf of California

sea otter (*Enhydra lutris*) The smallest of all marine mammals, sea otters do not exceed 5 feet in length, including tail. They spend a great deal of time floating on their backs, feeding and grooming themselves. They are among the few tool-using mammals, breaking open shellfish with rocks. Once believed to be almost

sea otter

extinct, the sea otter has made a remarkable comeback. In the past it was hunted for its luxurious fur, from northern Japan all around the islands of the North Pacific, as far south as Baja California. By 1938 it was considered commercially extinct—that is, hunting was no longer worthwhile. Once it was fully protected, "seed populations" were discovered in Alaska and Siberia, and the animal was reintroduced into its former range, where it is now fully protected. There may be as many as 200,000 sea otters alive today.

Sea Peoples Collective name for the mysterious pirates who invaded Anatolia, Syria, Palestine, Cyprus, and Egypt around the thirteenth century B.C. Although their actual identity is unknown, they have been tentatively identified as the Bronze Age Greeks known as the Ekwesh (the Achaeans of Homer?), Tyrrhenians, Sherdens (from Sardinia), the Siculi (from Sicily), and the Philistines from Palestine. It is considered possible that the Sea Peoples were responsible for the downfall of Mycenaean Greece. The Egyptians defeated them in 1187 B.C. off the western delta of the Nile, and again in 1180 B.C. off Cyprus, but their forces and governments were so depleted by the massive effort involved that the Egyptians soon fell to the Philistines and the Assyrians.

sea robin (*Prionotus carolinus*) Because they make loud noises by vibrating the muscles attached to their air bladder, these homely fishes have been named after a familiar songbird. The lower portion of the large pectoral fins consists of two or three enlarged rays that are uses for detecting food. Like the gurnards, these 1-foot-long fishes "walk" along the seafloor. There are nineteen species in the Atlantic and two in the Pacific.

See also flying gurnard

sea serpent Ever since men began to sail the open ocean, there have been reports of frightening monsters eager to sink ships and gobble up hapless sailors. Olaus Magnus (1490–1557), the influential Catholic archbishop of Sweden, wrote an account of northern peoples in which all sorts of sea serpents were described and illustrated, and authors who followed him, including Gesner, Topsell, and Aldrovandi, reproduced his drawings and descriptions. By the nineteenth century, many of these creatures had been relegated to fantasy, but still the stories persisted. Strange carcasses that washed up on the beach were often described as sea serpents, but these frequently turned out to be the decaying remains of basking sharks, which decompose in a way that suggests a long-necked vertebrate. (The "Stronsa Beast" of 1808 turned out to be a basking shark, as did the "dinosaur" that was hauled aboard a Japanese fishing vessel off New Zealand in 1977.) Others were fishes like the oarfish or decomposed whales. There have been some sightings that are still unexplained, like the 1817 Gloucester, Massachusetts, "serpent" that hundreds of people saw, and for which there is still no explanation, but many stories can now be taken as references to the giant squid (*Architeuthis*). This 60-foot-long apparition with gigantic eyes and snakelike arms was unknown both to science and to sailors until the middle of the nineteenth century. Although many species of sea snake are venomous, they rarely grow to more than 5 feet in length and are therefore not the basis for stories about sea monsters.

See also basking shark, cadborosaurus, giant squid, Loch Ness monster, oarfish

sea sickness A feeling of unease, nausea, and general discomfort, attributed to the motion of a vessel at sea, usually during rough weather. At its worst, sea sickness produces waves of nausea, vomiting, and the feeling that the victim would be better off dead. ("Motion sickness" can also occur on airplanes, and in cars, buses, and trains.) The principal cause of the disturbance is the effect of motion on the semicircular canals of the inner ear, although other factors can contribute, such as inadequate ventilation, fumes, or noxious odors. There is no universal cure, but fresh air, medications, or combinations of medications, antihistamine patches, and acupressure wristbands sometimes work for some people.

yellow-bellied sea snake

sea snakes (family Hydrophiidae) Sea snakes are poorly known, but most species live in Indo-Pacific waters, largely between the peninsulas of Southeast Asia and the waters of northern Australia. (There are none in the Atlantic or the Mediterranean.) They tend to live in the shallow waters around continental shelves or islands, and there are very few sightings in water that is more than 50 fathoms deep. They are adapted for an aquatic existence and can remain underwater on a single breath for two hours or more. Most are 2 to 5 feet long, and have laterally flattened tails, which they use like oars to propel themselves through the water. Many sea snakes are very venomous. They use their poison to subdue their prey, which consists mainly of small fishes; although the venom certainly is powerful enough to injure or even kill a person, most bites occur when the snakes are being removed from fishermen's nets. There are, however, occasional reports of unprovoked attacks in Asian coastal waters by the beaked sea snake (*Enhydrina schistosa*), whose venom is twice as toxic as that of the Indian cobra. The yellow-bellied sea snake (*Pelamis platurus*) has the widest distribution of any species, having been reported from the eastern and western Pacific, almost always swimming at the surface. This species sometimes aggregates in huge groups—one such group was 10 feet wide and 60 miles long—but the reason for this behavior is unknown.

See also sea krait, sea serpent, yellow-bellied sea snake

sea squirts (tunicates) Bottom-dwelling marine animals with a primitive, hollow dorsal nerve chord. Their bodies, which resemble potatoes more than animals, are encased in a thick tunic composed of a material that resembles cellulose, and there are two siphons, one for inhaling and the other for exhaling. The inhalant siphon takes in water, and the gills extract suspended particles of food. They have no head, no eyes, no mouth. The heart lies in a loop of the gut, and there is a very simple circulatory system. Some species live individually, others in colonies. (Salps, which are related to sea squirts, usually form long chains.) All sea squirts are hermaphroditic (having both male and female organs), and in some species the embryos develop inside the animal, while in other species both sperm and eggs are ejected for outside fertilization. The embryos are tiny, free-swimming, tadpolelike creatures that attach themselves permanently to the substrate. Some species reproduce by budding: a fingerlike projection growing near the base breaks off and settles to become a new individual. They are in all oceans, from the intertidal zone to the great depths, and they attach themselves to everything from rocks and pillings to ships' hulls and the backs of crabs. **See also lancelet, salp, tunicate**

sea turtle Several species of large marine turtles are found in the tropical and subtropical oceans of the world. They all have flippers for forelimbs, lightweight shells, and heads that cannot be drawn into their shells. They spend most of their lives in the water, but the females come ashore to dig a nest in the sand and lay their eggs. All species are declining in numbers, because the eggs are frequently "harvested" for human consumption, and because the meat is also eaten. The shell of the hawksbill turtle is the source of tortoiseshell, still popularly used for decorative objects, even though plastic has replaced most uses. The green turtle, highly valued for soup, has been known to weigh 1,000 pounds, but most today are less than half that size. The loggerhead is a large-headed, chiefly carnivorous species, that occasionally comes ashore to bask in the sun. The Pacific leatherback is the largest and the fastest of the sea turtles; an 8½-foot-long specimen weighed 1,908 pounds. The Atlantic leatherback, which ranges from the Gulf of Mexico and the Caribbean to the British Isles in the north, and in the south to Argentina and South Africa, is smaller. Leatherbacks have no visible shell, but instead a series of longitudinal ridges on the back and underside. The smallest of the sea turtles are the ridleys. The 2-foot-long Kemp's ridley breeds only in the Gulf of Mexico, and the slightly larger Pacific ridley is found in the Indian and Pacific Oceans. Kemp's ridley is seriously endangered, and there may be no more than fifteen hundred left in the world. The flatback turtle is found only in northern Australia, and it is the only species of sea turtle that is not considered threatened or endangered. The United States and 115 other countries have banned the import or export of sea turtle products. **See also flatback turtle, green turtle, hawksbill turtle, Kemp's ridley turtle, leatherback turtle, loggerhead turtle**

sea urchin (class Echinoidea) An echinoderm enclosed in a thin, brittle shell covered with movable spines. The word "urchin" is derived from the Latin *erecius*, meaning "hedgehog." Although they look very unlike sea stars, sea urchins have the same fundamental body plan: the five-pointed (pentamerous) radial symmetry expressed in the five rows of minute holes through which the tube feet project. The tube feet (podia) are longer than those of the sea stars, because the sea urchins have to be able to extend them beyond the spines in order to move around. They also move by using the spines. The mouth, which is on the underside, is composed of an elaborate set of five teeth (another example of five-pointed symmetry), which are arranged radially and worked by a complex set of muscles and plates known as Aristotle's lantern. Some species do not exceed a couple of inches in diameter (including spines), but the slate-pencil urchin (*Heterocentrotus mammilatus*), named for its thick spines, can reach a span of a foot. The spines of some species are poisonous, but all are dangerous because even those that are not equipped with venom glands can pierce the skin of an unwary bather and cause a nasty infection.

needle-spined sea urchin

See also echinoderms, sand dollar, starfish

sea wasp: See box jelly

seaweed Generic term for plants that grow in the ocean; usually algae of one form or another. The more highly developed types have a basal disk known as a holdfast, as well as fronds of varying length or shape. The simplest seaweeds are the blue-green algae (*Cyanophyta*), which are found in shallow water near the shore and grow in sheets, filaments, or branching fronds. The largest of these is the bright green sea lettuce (*Ulva lactuca*), which grows in 3-foot-long sheets, often with ruffled edges. Green fleece (*Codium fragile*) is a bushy weed with tubular branches. Brown algae (*Phaeophyta*) are the most common; they grow at depths of 50 to 75 feet and take the shapes of brushes, whips, or ferns. Rockweed or bladderwrack (*Fucus* spp.) is a tough, leathery brown algae with flattened fronds and air bladders at the tips. The red seaweeds (*Rhodophyta*) are fernlike and grow at the greatest depths, perhaps 200 to 600 feet down, where they can photosynthesize blue and violet light. Agar, which is a vegetable gelatin obtained from the red seaweed *Gelidium*, is probably the most valuable seaweed product, since it is an important element in the cultivation of bacteria for study. Irish moss, also known as carrageen, is a source of gelatin that is used in cosmetics, jellies, candy, salad dressing, cooking, and food preparation. Obtained from large kelps, algin is a gelatinous material that can be treated to produce alginates, which are used as stabilizers in dairy products, puddings, ice creams, and various bakery and confectionery goods. Many seaweeds are rich in iodine and potassium and are used in the manufacture of fertilizers, medicines, and minerals. In some North Atlantic coastal areas, seaweeds are used directly as fertilizers. In Japan, the red seaweed *Poryphyra* is grown commercially in large quantities, dried, and used to make *nori*, which is used to wrap and flavor various dishes. (In Europe *Poryphyra* is known as laver or dulse.) Sargassum weed, also known as gulfweed or rockweed, is a free-floating brown algae, characterized by hollow, berrylike floats that keep it floating at the surface. It occurs primarily in the Sargasso Sea.

See also algae, blue-green algae, brown algae, kelp, red algae, sargassum weed

sediments Except for those areas where the elemental igneous rock pokes through, the floor of the ocean is covered with layers of sediment of varying thickness and composition. In some places, the sediments can be 40,000 feet thick, but the average is less. Near-shore sediments may consist of terrestrial material that is washed into the sea by rivers or floods, or material that flows into the sea from volcanoes, but the greatest proportion of the seafloor is covered with ooze—clay that contains minute particles of rock that drifted far from shore before they fell to the bottom, combined with the remains of organic substances like calcium carbonate (which forms the calcareous ooze) or silica (which forms the siliceous ooze). (Calcareous ooze is composed largely of clay and the skeletons of foraminiferans, siliceous ooze mostly of clay and the skeletons of radiolarians.) For the most part, the sediments remain where they fall, but infrequently, underwater landslides and massive turbidity currents redistribute the bottom materials. Not to be discounted are the seafloor inhabitants, such as the holothurians, that annually redistribute hundreds of tons of sediment by passing it through their digestive systems.

See also foraminifera, ooze, radiolarian, sand, turbidity currents

sei whale (*Balaenoptera borealis*) At a maximum length of 60 or more feet, this is among the largest of the rorquals, exceeded only by the blue and fin whales. It can be differentiated from the fin whale because it lacks the white on the right lower jaw. While other rorquals, such as the blue and fin whales, take in mouthfuls of small organisms and then expel the water through the fringes of the baleen plates, the sei whale is a "skimmer" and swims through schools of krill with its mouth open, allowing the food items to become trapped in its silky baleen fringes. Sei whales are found in both hemispheres. After the Antarctic whalers had practically wiped out their larger relatives, they began to kill the sei whales. What may have been a pre-exploitation population of 300,000 around the world has been reduced to perhaps 75,000. The name comes from the Norwegian *seje*, the coalfish that appeared off the coast of Norway every spring, along with these whales. It is pronounced "say."

See also blue whale, Bryde's whale, fin whale, International Whaling Commission, rorqual,whaling

Selkirk, Alexander (1676–1721) British seaman who in 1703 joined the crew of the privateer *Cinque Ports*, commanded by Thomas Stradling, with William Dampier as navigator, for a voyage in the South Seas. After a violent disagreement with Stradling, Selkirk asked to be put ashore on Juan Fernández Island. He lived alone on

sei whale

the otherwise uninhabited island for four years, subsisting on fruits, vegetables, and goat meat, until he was found by the privateers *Duke* and *Duchess,* commanded by Woodes Rogers, and with Dampier again aboard as navigator. Selkirk sailed aboard the *Duke* for two years, privateering along the coasts of Peru and Chile, and returned to England in 1711. There he told his story to Richard Steele, who published it in *The Englishman* in 1713. This account was seen by Daniel Defoe, who wrote the novel *Robinson Crusoe* (published in 1719), based on Selkirk's adventures.

See also *Robinson Crusoe*

sergeant major (*Abudefduf saxatilis*) Although there are several species of sergeant majors found throughout the shallow waters and coral reefs of the tropics, they all look alike, with their grayish-yellow body color and bold vertical stripes. These little fishes (maximum length: 6 inches) often venture into the open ocean, but at the first sign of danger, they retire to the shelter of the reef.

sergeant major

See also damselfishes

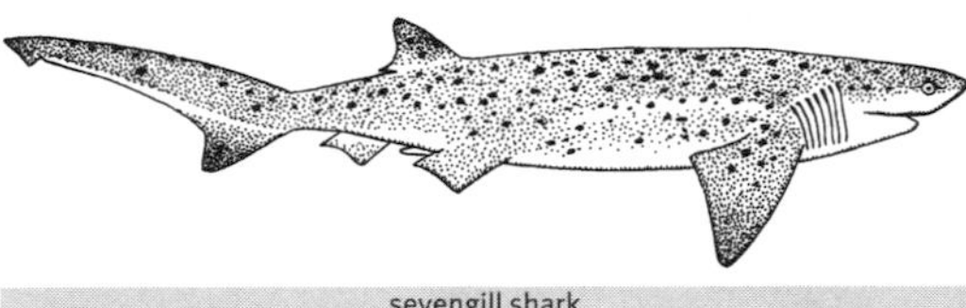
sevengill shark

sevengill shark (*Notorhynchus cepedianus*) Most sharks have five gill slits on each side of the head, but there are two with six and one with seven. The sevengill shark is found throughout the world's temperate waters, where it feeds on fishes and small sharks. It is an aggressive predator, and in captivity it has been known to attack divers. It has been recorded at 10 feet in length, but may grow larger. Like its relative the sixgill shark, it has only a single dorsal fin, placed back near the tail.

See also frilled shark, sixgill shark

sextant A navigational instrument for the measurement of vertical and horizontal angles at sea, developed independently in England and America in 1731 from the quadrant, which could measure angles only up to 90°. The need for an additional 30° arose from the need to make lunar observations in order to determine a ship's longitude, even after the chronometer had been perfected. The modern sextant consists of a triangular frame, the bottom of which is a graduated arc of 60°. A telescope is attached horizontally to the plane of the frame. A small mirror is mounted perpendicular to the frame at the top of a movable index arm, which swings along the arc. In front of the telescope is the horizon glass, half transparent and half mirror. The image of the sun or other body is reflected from the index mirror to the mirror half of the horizon glass and then into the telescope. If the index (or image) arm is then adjusted so that the horizon is seen through the transparent half of the horizon glass, with the reflected image of the sun lined up with it, the sun's altitude can be read from the position of the index arm and the arc. By reference to navigational tables, the geographical position can then be determined.

See also astrolabe, Global Positioning System (GPS), quadrant

Seychelles Island republic in the Indian Ocean, approximately 600 miles north of Madagascar and 1,000 miles east of Kenya. The group consists of about one hundred islands and coralline islets. The largest of these is Mahé, where 90 percent of the population of 75,000 lives. The islands were first explored by Vasco da Gama in 1502, but were not settled until 1756, when French planters from Mauritius brought their slaves over and the French annexed the islands, which they called the Séchelles. in 1810, France surrendered the islands to the British, who administered them as a part of Mauritius until 1903, when they became a Crown Colony. Labor-intensive crops like cotton and grains were grown until the British abolition of slavery in 1830, after which coconuts, vanilla, and cinnamon became common. In 1976, the Seychelles became an independent republic within the British Commonwealth. In 1977 James Mancham, the country's first president, was ousted by Prime Minister France-Albert René, who claimed that Mancham was under Soviet influence. A coup against René in 1981 failed, but attracted international attention because a group of South African mercenaries entered the country posing as rugby players. Since the opening of Mahé International Airport in 1971, tourism has become the islands' most important industry. A very controversial law was passed in 1995 promising immunity from criminal prosecution and extradition to anyone investing $10 million in the Seychelles.

See also da Gama, Indian Ocean, Madagascar

Shackleton, Ernest (1874–1922) Ernest Shackleton joined Robert Falcon Scott's first expedition to the Antarctic (1901–1904) and was one of those who reached the farthest south of anyone until that time. Invalided home after an attack of scurvy, Shackleton decided to lead an expedition of his own to the South Pole. Arriving on the *Nimrod* in 1908, he scaled the Beardmore Glacier, then reached a point 97 miles from the Pole before he had to turn back. In 1914 he sailed again in the *Endurance,* with the intention of crossing the continent on foot from the Weddell Sea to the Ross Sea via the South Pole, a distance of 2,050 miles. Before they could even set out, the *Endurance* was trapped in the pack ice, and after drifting for nine months, the ship was crushed by the closing ice. The expedition floated on ice floes for another five months until Shackleton led his men in small boats to Elephant Island in the South Shetlands. From there, in a ship's boat, Shackleton and five men sailed 800 miles to South Georgia, eventually crossing the mountainous island on foot. In 1921 he sailed again to the Antarctic aboard the *Quest,* but he died of a heart attack on the island of South Georgia, where he is buried at Grytviken.

See also Scott, South Georgia, South Pole

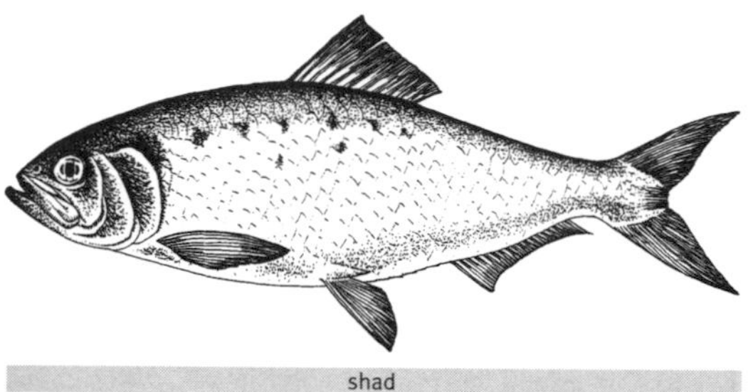

shad

shad (*Alosa sapidissima*) The largest member of the herring family (Clupeidae), the shad grows to a length of 30 inches. It is anadromous, meaning that it ascends coastal freshwater streams and rivers to spawn. Most fishing takes place during the winter in the southern United States, and in the spring in rivers like the Potomac, the Susquehanna, the Delaware, and the Hudson. At sea, huge schools of shad feed on plankton. Its eggs are served as the popular shad roe. A recent study has shown that shad can detect sounds at a much higher frequency than was ever believed possible, which may enable them to hear the ultrasonic clicking of hunting dolphins and thereby avoid capture.

See also alewife, Atlantic herring, menhaden

shag (*Phalacrocorax aristotelis*) The great cormorant of European coastal waters, the shag is a blackish-green bird that develops a wispy crest during the breeding season. It is found on rocky coasts throughout northwest Europe and also in Morocco, Tunisia, Cyprus, and Turkey. It feeds almost exclusively on eels. The common cormorant, a somewhat larger bird that shares much of its range, eats mostly flatfish.

See also common cormorant, cormorant

shagreen Because the skin of sharks is composed of toothlike processes known as dermal denticles, the skin has a rough, abrasive texture. Shagreen, or untanned sharkskin, was used in place of sandpaper in the finishing of wood. The nature of the leather also made it useful for those items where a good grip was required, such as oars, which water might make slippery, or sword handles, where blood might do the same. The word is believed to have come from the Persian *saghari,* which meant horsehide in which seeds were imbedded in the tanning process; when the skin dried and the seeds fell out, the hide had a granular texture and was used for sword hilts. Samurai swordsmen used the skin of sharks and rays for sword handles, not only for its nonslip character, but also because the granular nature of the leather was particularly beautiful to them. The French word *chagrin,* which also refers to "rough skin," has become a metaphor for vexation, disappointment, or humiliation.

See also sharkskin

shark A marine vertebrate with a cartilaginous skeleton and skin composed of dermal denticles, that breathes through multiple gill slits. There are some 350 species of sharks, ranging in size from the 8-inch-long cigar shark to the whale shark, which is at 50 feet the largest fish in the world.

See also *individual species*

shark attack Although the number of people attacked by sharks worldwide every year is considerably fewer than those attacked by pigs or stung to death by bees, "shark attack" remains one of the most horrific word combinations in the English language. Of the 350-odd shark species, only a small percentage have been implicated in unprovoked attacks on swimmers or divers, but these attacks are usually so hysterically inflated by the media that it often appears that the shark menace is the worst thing about the ocean. We do not know why some sharks attack some people, but it is rarely out of hunger. It may be because the shark is protecting its territory, because the shark mistakes the human for a more familiar prey item, or because the shark is stimulated by splashing or movement in the water that suggests the "prey" is injured. An average of between fifty and sixty shark attacks are reported per year, with the results ranging from death (in between five and ten cases), to abrasions from the shark's rough skin. Those species known to attack humans are the bull shark, great white, mako, tiger shark, great hammerhead, oceanic whitetip, blue shark, and various carcharhinids such as the dusky shark, the Galápagos shark, the bronze whaler, and the gray reef shark. Regardless of its size or appearance, almost any shark will retaliate if it is threatened or provoked. The culprit in the movie *Jaws,* a great white shark, was probably responsible for more fear of the ocean than anything since the iceberg that sank the *Titanic.*

See also *individual species*; *Jaws*, Shark Attack File

Shark Attack File Started by the U.S. Office of Naval Research in 1958, the Shark Attack File was an attempt to collect data on all historical attacks and document new ones as they occurred. The file was begun when the American Institute of Biological Sciences established a Shark Research Panel at the Smithsonian Institution and Cornell University. Originally housed at the Smithsonian, the files were moved to the Mote Marine Laboratory in Sarasota, Florida, where in 1974 David Baldridge produced a computerized statistical analysis entitled *Shark Attack: A Program of Data Reduction and Analysis.* The file was then transferred to the University of Rhode Island, and finally, to the Florida Museum of Natural History, where it is now under the auspices of the American Elasmobranch Society and curated by George Burgess. **See also Mote Marine Laboratory, Springer**

shark cartilage products Because it is commonly believed that sharks do not get cancer, those who would find out why hit on the idea that since sharks have cartilage where vertebrates like us have bones—and we *do* get cancer—it must be the cartilage that makes the difference. Unscrupulous entrepreneurs began marketing "shark cartilage pills," which they claimed would ward off cancer. Even if this were true (which it isn't), taking the cartilage internally makes about as much sense as eating the sawdust from a redwood tree so you will become taller. **See also cartilaginous fishes**

shark repellent During World War II, a "shark chaser" was developed, ostensibly to protect downed airmen and the survivors of torpedoed ships. It consisted of copper acetate (a component of decomposing shark meat), suspended in a black soluble substance that was supposed to keep the sharks away and hide the seaman at the same time. It didn't work. Various prods and spearguns have been sporadically successful, and a secretion from the Moses sole, a Red Sea flatfish, was shown to be highly repellent to sharks. It was too difficult and expensive to synthesize, however, and most divers were unwilling to carry a live sole with them. Various other chemicals have been tried over time, but a "surrounding cloud" disperses too quickly and a "squirt" presupposes a proximity to the shark that might be problematic even if the device worked. Because sharks are known to be extremely sensitive to electrical fields, the shark "POD" (protective oceanic device) was recently developed in South Africa. It is a battery-powered device that, strapped to the diver's tank, emits an electrical field strong enough to keep sharks at bay. It has been tested among feeding white sharks (by Australian divers Ron and Valerie Taylor) and appears to work. Still, the best way to avoid shark attacks is to stay out of the ocean. **See also shark attack, Taylor**

shark's fin soup An exotic delicacy of Chinese cuisine; only the first dorsal, pectorals, and lower lobes of the tail fin can be used in its preparation. The skin is removed, and all muscle tissue as well, leaving only the inner cartilaginous fin rays (ceratorichia) to soak. The fibers are boiled in water for hours, then the water is changed, and the fibers boiled again and again, a process that may take as long as five days. The glutinous mass that results is served in a thick broth of chicken stock seasoned with soy sauce, ginger root, onions, vinegar, mushrooms, and other ingredients, and after the soup is drunk, the gelatinous fibers of the shark's fin are eaten.

sharkskin The skin of sharks is covered with dermal denticles, tiny toothlike projections. Like teeth, the denticles differ from species to species and so can be used to differentiate one species from another. (In most species, the skin is rough only if rubbed against the grain of the denticles, but in the basking shark, the denticles are rough in all directions.) Because of these denticles, the skin of sharks may abrade the skin of a human swimmer if the two come into contact. This quality, however, also makes sharkskin useful as an abrasive; in some places it was called shagreen and used in place of sandpaper. The denticles can be removed by grinding them down or by chemical means, where the entire denticle is extracted from the skin. Thus treated, sharkskin makes excellent leather and has become popular for shoes, small accessories (wallets and belts), and bookbinding. **See also dermal denticles, shagreen, shark**

Sharpe, Bartholomew (1650–1688) English buccaneer who served under Henry Morgan in the piratical raid on Panama in 1671. He then seized a Spanish ship and cruised the west coast of South America looking for ships to capture. William Dampier left Sharpe's band of pirates to return to England, but the rest of his crew continued to pillage until they captured the Spanish ship *Rosario,* on which they found an atlas of the South Seas. Sharpe became the first Englishman to round Cape Horn; he presented the atlas to Charles II, in time to be acquitted of the charges of piracy that awaited him in England. He was appointed captain in the Royal Navy, but the lure of piracy was too strong, and he captured a Dutch vessel and deserted. He was last heard of commanding a pirate colony on the Caribbean island of Anguilla. **See also buccaneer, piracy**

sharpnose puffer (*Canthigaster rostrata*) A 5-inch-long fish, found on both sides of the Atlantic, that feeds on vegetable matter, such as sea grasses, and also sponges, crabs, mollusks, worms, starfishes, and algae. Puffers have tough, prickly skin, sharp teeth, and the ability to inflate themselves with water or air when threatened. **See also porcupine fish, puffer**

shearwater (*Puffinus* spp.) Graceful, gull-like birds of the order Procellariiformes that are found throughout the ice-free waters of the world. The common name is derived from their habit of flying stiff-winged along

shearwater

the troughs of waves. Shearwaters are slender-billed, drab-colored birds whose average wingspan is around 45 inches. Not known as ship followers, shearwaters feed on krill and small fishes that they pluck from the surface, but they occasionally dive in pursuit of their prey. They nest in vast burrows on offshore islands and coastal hills throughout the world; after spending the day at sea, they return to land under cover of darkness. Many of the seventeen species undertake long migrations; they are among the champion long-distance travelers of all birds. For example, the short-tailed shearwater (*P. tenuirostris*) breeds on islands in southern Australia and spends the rest of the year circumnavigating the Pacific, as far north as the Bering Sea off Alaska and Siberia. Along with the albatrosses, storm petrels, and diving petrels, shearwaters are tubenoses. Their nostrils extend on top of the bill through two horny tubes, employed in the secretion of salt.

See also albatross, giant petrel, greater shearwater, Manx shearwater, petrel, short-tailed shearwater

sheathbill (*Chionis* spp.) All-white, chicken-sized birds of Antarctic and subantarctic islands, sheathbills are terrestrial scavengers, feeding on corpses, placentas, and anything else they can find. They are often seen around penguin rookeries, where they are adept at stealing eggs. Because of their feeding habits, their white plumage is usually stained with blood or dirt. Their feet are not webbed, and they are agile runners on rocky terrain. They are the only shorebirds of the Antarctic. There are two species, the snowy sheathbill (*C. alba*) and the lesser sheathbill (*C. minor*), which is somewhat smaller and has a black bill. **See also penguin, skua**

sheathbill

shell collecting Because of their beauty and variety, the shells of marine gastropods have long been prized by collectors. Many, however, do not realize that the shell is made by the living snail, and that its owner must be killed before it can be added to a collection. (There are, of course, many collectors who restrict their acquisitions to shells washed up on the beach.) Most books about marine gastropods devote little space to the animals, emphasizing only the shells. The emphasis on shell collecting has been so strong that the animals themselves are erroneously referred to as "shells." Thus the cone snails (Conidae) are popularly known as cone *shells;* the Muricidae as murex *shells;* the cowries (Cypraeidae) as cowrie *shells;* and so on.

Shepherd's beaked whale

Shepherd's beaked whale (*Tasmacetus shepherdi*) Because both males and females have functional teeth in the upper and lower jaws—in addition to the pair of much larger teeth at the end of the male's lower jaw—this beaked whale has been placed in its own genus, *Tasmacetus.* It has never been seen alive, and it is known only from specimens that have stranded in New Zealand, South Australia, Argentina, and Chile.

See also beaked whales, *Mesoplodon*

Shetland Islands A group of about a hundred islands north of Scotland, including Mainland (the largest, where both the administrative capital of Lerwick and the fishing port of Scalloway are located) and, to the north, the islands of Yell and Unst. (Unst is the northernmost point of the British Isles.) In the eighth and ninth centuries, the islands were invaded by Norsemen, who ruled them until 1472, when the islands were claimed by the Scottish crown. In the late nineteenth century, Shetlanders were in great demand to ship out on British whaleships bound for the Greenland fishery. The population of the Shetlands is approximately 22,000 people and 300,000 sheep. The main industries are fishing and sheep raising; the wool is knitted into the world-famous Shetland and Fair Isle sweaters. The islands are also the home of Shetland ponies, which once were used in coal mines but are now raised as pets and for children. The discovery of oil in the North Sea completely changed the economy of the Shetlands, and in the 1970s a major oil terminal was built at Sullom Voe on Mainland. Pipelines reach from the oil rigs to the terminal, and the oil is loaded into tankers in the deep water of Yell Sound. In January 1993, the supertanker *Braer* ran aground in Lerwick, dumping 25 million gallons of oil into the Atlantic. A week later, the tanker broke up, discharging the remainder of her 85,000-ton cargo. **See also *Braer,* oil spill**

ship From the Old English *scip,* the generic name for seagoing vessels. Originally personified as masculine, by the sixteenth century ships were almost universally considered feminine. In strict maritime usage, a ship is a three-masted, square-rigged vessel with a bowsprit, but despite this narrow definition the term is widely used for other types of seagoing vessels. The first ships were propelled with oars, but by the eleventh century, sail had been introduced. The introduction of gunpowder around the fourteenth century meant that sailing ships could engage in battle, instead of merely transporting fighting men who fought aboard vessels with swords and spears. From the seventeenth to the nineteenth centuries, there was little change in the design of warships, except that they became larger and larger. Merchant ships paralleled the warships, and both changed dramatically with the introduction of steam power around 1800. From that point on, warships, merchant ships, and passenger vessels improved significantly. The introduction of the internal combustion engine and the diesel engine changed the very nature of shipbuilding yet again. A few types of ships—particularly submarines and icebreakers—now employ nuclear power.

shipworm: See teredo

shorebirds As used in the United States, "shorebirds" is the equivalent of the British "waders." It refers to sandpipers, plovers, avocets, stilts, and oystercatchers—birds that usually frequent beaches and mudflats of the seashore, but are sometimes found inland. Most shorebirds are long-legged, long-billed, and brownish, but some, like the oystercatcher and the avocet, are strongly marked. The smallest of the shorebirds is the least sandpiper; the largest is the long-billed curlew. Species such as the golden plover and the Eskimo curlew once migrated in huge flocks, but they were so extensively hunted for food that the golden plover came close to extinction, and it appears that the Eskimo curlew was eliminated from the face of the earth.

See also avocet, curlew, killdeer, oystercatcher, plovers, sanderling, sandpiper

short-finned squid (*Illex illecebrosus*) Reaching about 18 inches in length, the short-finned squid is named for its small tail fins. It is found in enormous numbers in the western North Atlantic, ranging from Cape Hatteras to Greenland and Iceland. Like many other squid species, *Illex* is at the base of the food pyramid, serving as a major food source for various fishes, seabirds, whales, and dolphins. It is also the object of large commercial fisheries by Canadians and Americans. This is one of the species popularly marketed as calamari.

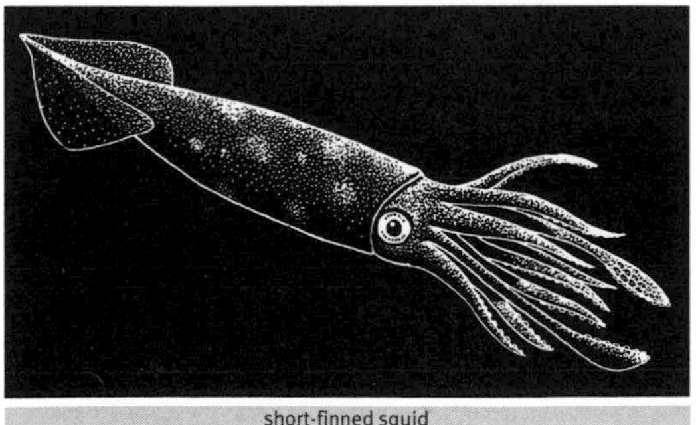

short-finned squid

See also calamari, long-finned squid, squid fishing

short-tailed albatross (*Diomedea albatrus*) Once abundant in the North Pacific, this white-bodied albatross was brought to the verge of extinction by Japanese feather hunters in the late nineteenth century. It used to breed on the Bonin, Izu, Ryukyu, and Pescadores Islands, but it has now been found only on Tori-shima, some 350 miles south of Tokyo. In 1903, the volcano on Tori-shima erupted, killing all the Japanese hunters living there. By that time, an estimated 5 million birds had been killed. At one time, the short-tailed albatross was believed to be extinct, but there are now about 250 of these magnificent birds left. With its 84-inch wingspan, it is one of the larger albatrosses. Adults are mostly white, with black wing tips, a tawny crown, and a pinkish bill with a blue tip. Because of its rarity, and because it does not normally follow ships, the short-tailed albatross, the largest flying bird of the North Pacific, is hardly ever seen. (The more common North Pacific albatross nowadays is the Laysan, which breeds on Midway and Laysan Islands.) On his 1740–1741 voyage with Commander Vitus Bering, zoologist Georg Wilhelm Steller became the first European to describe this species, and it used to be known as Steller's albatross.

See also albatross, black-footed albatross, Laysan albatross, Steller, wandering albatross

short-tailed shearwater (*Puffinus tenuirostris*) Sooty-brown above, with paler underparts, and a wingspan that may reach 39 inches, the short-tailed shearwater is one of the great distance travelers among birds. Known as the muttonbird in Australia because it was collected in huge numbers to feed people, the short-tailed shearwater breeds in burrows on islands in the Bass Strait (between Tasmania and South Australia), and when the young are about ten weeks old, they take off in a clockwise direction around the Pacific Ocean, passing Japan, Siberia, Alaska, and California before heading back across the trackless ocean to Australia. The annual flight covers approximately 20,000 miles. Although the "harvest" has been curtailed by the Australian government, muttonbirds are still being collected from the burrows and packaged as Tasmanian squab.

See also albatross, petrel, shearwater

Shovell, Cloudesley (1650–1707) English admiral who distinguished himself in the War of the Grand Alliance

(1689–1697), and the War of the Spanish Succession (1701–1714), when he captured the important Spanish ports of Vigo, Gibraltar, and Barcelona. Returning home from an attack on the French at Toulon in 1707, his flagship *Association* and three other ships were blown onto the rocks of the Scilly Isles and wrecked. All nine hundred of his men died, but Shovell managed to get ashore, where he was murdered by a local woman for an emerald ring he was wearing.

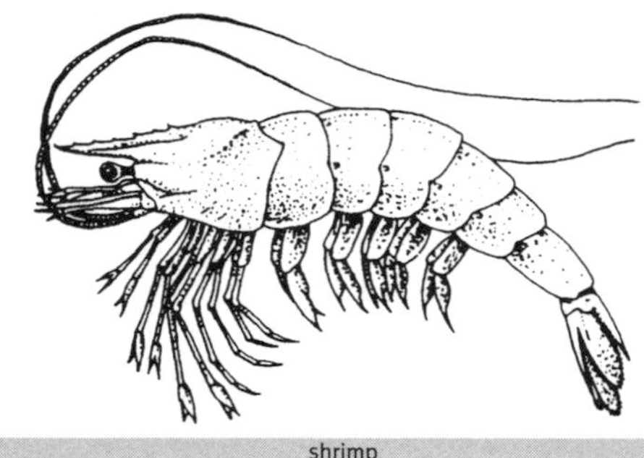

shrimp

shrimp Any of a large number (about two thousand species) of decapod crustaceans characterized by a laterally flattened body, a flexible abdomen, and a fanlike tail. The appendages are modified for swimming, and there is a pair of long, whiplike antennae. Shrimps differ from lobsters and crabs in that they are primarily swimmers, as opposed to crawlers. They can swim forward by using their abdominal appendages (known as swimmerets) and backward by flexing their tails. They are gregarious and swim near the ocean floor in large schools, making them particularly susceptible to trawl fishing. Females lay large numbers of eggs—from fifteen hundred to half a million, depending on the species—and carry them on their abdominal appendages. They range from the size of a mosquito to *Macrobrachium carcinus* of the Philippines, which can reach a length of 14 inches. Many species of deepwater shrimps are scarlet in color, and many have bioluminescent organs. With the exception of these and some tropical species, shrimps are grayish in color, but they turn an appetizing pink when cooked. The predominant fishery in the Gulf of Mexico is for the 6-inch brown shrimp (*Penaeus aztecus*), whose catch exceeds the total of all finfish combined. (In 1993, Texas shrimpers landed 74 million pounds.) The gulf also supports fisheries for the white shrimp (*P. setiferus*) and the pink shrimp (*P. duorarum*). In Europe, the predominant fishery is for *Crago septemspinosus*, which is known there as a prawn, but there is no scientific difference between a prawn and a shrimp. There are equally important fisheries in South America, Africa, and Australasia. (The prawn fishery in Queensland is Australia's most valuable.) Shrimps can be differentiated from the euphausiids, which they closely resemble, because the latter's gills are exposed. **See also euphausiids, mantis shrimp, pistol shrimp, rift shrimp**

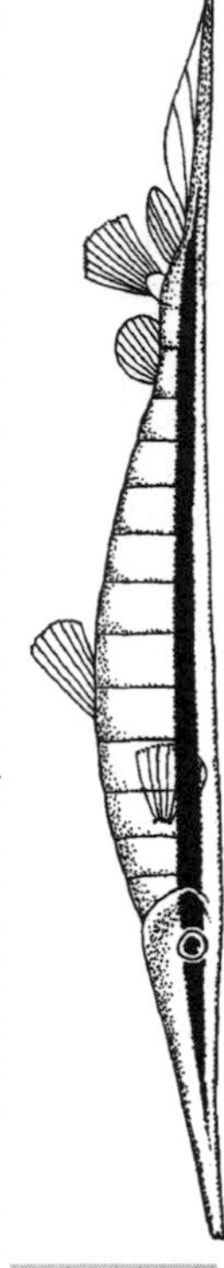

shrimpfish (*Aeoliscus punctulatus*) Because these 6-inch-long fishes are so laterally flattened and the bony plates on the ventral surface fused into a sharp, razorlike edge, they are sometimes known as razorfishes. With an extended tubular snout, a dorsal spine that extends to the rear, and a body that looks as if it is jointed (but isn't), *Aeoliscus* looks like a shrimp in fish's clothing. To add to its oddity, it swims vertically, head down. Its fins have all migrated to its ventral surface, and it moves by fine adjustments of these fins to position itself directly over items of food, which are usually amphipods, copepods, and other midwater crustaceans. Shrimpfishes are restricted to the tropical Indo-Pacific region, but their close relatives, the snipefishes, are worldwide in distribution. The snipefishes (Macrorhamphosidae) also have a pointed snout and swim head down, but they lack the body plates and their dorsal spine extends from the back, not from the tail area. Because of their tubular mouths, both families are classified with the sea horses and pipefishes. **See also pipefish, sea horse**

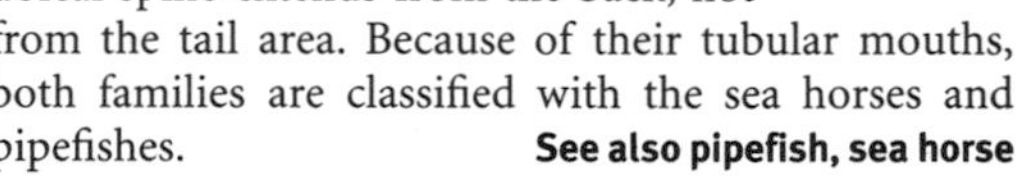

shrimpfish

shy albatross (*Diomedea cauta*) Sometimes known as the white-capped albatross, this species has a dark dorsal surface from wing tip to wing tip, and a white belly and white underwings that are bordered in black. It can be distinguished from other albatrosses by the diagnostic "thumbprint" on the leading edge of the underwing. Shy albatrosses breed in the Australasian region and disperse throughout the southern oceans. The species is considered "vulnerable" in Australia because large numbers are caught by longliners who trail their baits behind the boats; the albatrosses swallow the baits, become hooked, and drown. Tasmanian researchers outfitted several shy albatrosses with time-depth recorders and learned to their surprise that these birds were competent divers, able to reach depths of 25 feet in pursuit of squid and fishes.

See also albatross, black-browed albatross, wandering albatross

Sicily Italian island in the Mediterranean, separated from Italy by the Strait of Messina. The largest island in the Mediterranean, it is roughly triangular and covers some 9,925 square miles. It is almost entirely mountainous, and is dominated in the east by 10,900-foot-high Mount Etna, Europe's largest and most active volcano. Its rich soil has made agriculture its primary activity, and Sicily produces an abundance of wheat, barley, maize, olives, citrus fruits, and wine grapes. The first known inhabitants were the Siculi (for whom the island is named), followed by the Phoenicians, Carthaginians, Greeks, and Romans. After the fall of Rome, Sicily passed into the hands of the Vandals, then the Goths and the Byzantines. The Arabs finally took the island in the ninth century, after raiding it for two centuries, but the Normans replaced them and Roger I became the first king of Sicily. In the twelfth and thirteenth centuries, the island was a part of the Kingdom of the Two Sicilies, and by the eighteenth century it was ruled by the Bourbons. During the nineteenth century, it was the major center of the Italian revolutionary movement; in 1860 it was liberated from the Bourbons, and in 1861, it was made a part of the Kingdom of Italy. During World War II, after the defeat of the Germans in North Africa, the Allies made large-scale amphibious landings on Sicily from North African bases in July 1943, and after heavy fighting, the German and Italian troops withdrew to the Italian mainland, leaving the island in the hands of the Allies. In recent years, Sicily has been interested in increasing tourism, but the real or imaginary presence of the Mafia (which originated in nineteenth-century Sicily) has been a substantial deterrent. The capital is Palermo, and other important centers are Catania and Messina. The island's population is around 5 million.

Siebe, Augustus (1788–1872) German inventor of the first diving suit in 1819, and widely recognized as the father of diving. His "suit" consisted of a metal helmet and breastplate, attached to a loose watertight jacket. Under the jacket, the diver wore a suit that reached to the armpits. Air was pumped down from the surface through a flexible metal tube attached to the helmet. The constant stream of air escaped between the jacket and the combination suit, and the pressure kept the water from entering the jacket from below. As long as the diver remained on his feet, the system worked, but if he tripped, his open suit soon filled with water, and unless he was pulled up quickly, he would drown. In 1830, Siebe introduced a "closed" version, which had an air-regulating outlet built into the helmet, and with some modifications, this is the "dress" used today for "hard-hat" diving. **See also Cousteau, scuba**

silky shark (*Carcharhinus falciformis*) Circumtropical in distribution, the silky shark is best known from the western North Atlantic (Cape Cod to Brazil) and the eastern Pacific (Baja California to Peru). It is a large, slender, dark-colored carcharhinid shark that reaches a maximum length of 10 feet. Although it has not been implicated in attacks on man, it is considered potentially dangerous. **See also carcharhinid sharks**

Singapore Republic in Southeast Asia, consisting of the island of Singapore and fifty-eight small adjacent islets, at the southern tip of the Malay Peninsula. Singapore Island, which covers 210 square miles, is almost completely urbanized, but it is the only urbanized island that is an independent nation. (Manhattan and Hong Kong Islands, more or equally urbanized, are parts of larger entities; Venice is a city.) Its population, mostly of Chinese descent, but including Malays, Tamils, and Sikhs, numbers around 2.9 million. With its natural deepwater harbor and its strategic location on the Strait of Malacca between the Indian Ocean and the South China Sea, Singapore has become the busiest port in Southeast Asia. Before the arrival of Portuguese and Dutch colonists in the sixteenth and seventeenth centuries, Singapore (which had been known as Tumasik) was an outpost for pirates and fishermen. The British captured Malacca from the Dutch in 1795 and established their presence in Southeast Asia. In 1819, Sir Stamford Raffles (1781–1826), of the British East India Company, signed the treaty with the Sultan of Johore that made Singapore the center of the British colonial empire in Southeast Asia, leading to the establishment in 1826 of the Straits Settlements, the Crown Colony that included Singapore, Penang, and Malacca. During the nineteenth century Singapore was one of the world's leading ports for the export of tin and rubber. On January 8, 1941, Japanese planes bombed Singapore, and the city surrendered in February 1942. After the war, the Straits Settlements were dissolved, Penang and Malacca joined the Malayan Union, and Singapore became a separate Crown Colony. It was annexed to Malaysia in 1963, but separated by mutual agreement in 1965. The People's Action Party (PAP), under the leadership of Lee Kuan Yew, has controlled Singapore since then and has brought the country to an unprecedented level of prosperity, often at the expense of civil rights. In 1995, Singapore-based Barings Bank collapsed after trader Nicholas Leeson ran up more than $1 billion in losses on the Singapore Monetary Exchange. Along with Hong Kong, Taiwan, and South Korea, Singapore is known as one of the "Four Little Dragons of Eastern Asia," the newly industrialized economies that have emerged to rival Japan and China.

See also British East India Company, Penang

siphonophore Floating hydrozoan colonies in which several kinds of polyps and a variety of medusalike individuals are combined into a single, functioning "animal" that swims or drifts, and dangles an array of stinging tentacles. Siphonophores (the word means "siphon bearer") have component polyps that are devoted to reproduction, but the means by which this colony reproduces is not fully understood. There are many kinds of siphonophores (*Hippopodius, Physophora, Porpita, Muggiaea,* etc.), but the best known are the purple sailor (*Velella*) and the Portuguese man-of-war (*Physalia*). With the aid of their balloonlike sail, siphonophores float on the surface, but jellyfishes, with no such air bladder, live fully submerged.

See also jellyfish, Portuguese man-of-war

sirenians A group of herbivorous aquatic mammals sometimes known as sea cows. They breathe through nostrils at the end of their muzzle, unlike whales and dolphins, who breathe through blowholes on the top of their heads. There are three species of manatee, one dugong, and Steller's sea cow, which became extinct in the eighteenth century. The sirenians spend their entire lives in the water and cannot move on land. Their placid disposition and near-shore distribution has made them particularly susceptible to human predation. **See also dugong, manatee, Steller's sea cow**

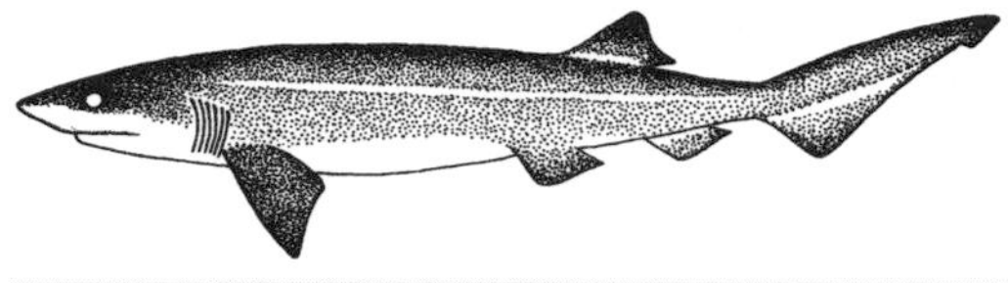

sixgill shark

sixgill shark (*Hexanchus* spp.) There are two species of sixgills: *H. griseus,* the bluntnose sixgill; and *H. vitulus,* the bigeye sixgill, which is believed to inhabit deeper waters. The upper jaw teeth are pointed sharply toward the side of the mouth, but the lower teeth are much larger and comb- or sawlike. Females grow larger than males, and the largest on record was 15 feet long. (There are unsubstantiated reports of individuals more than 20 feet in length.) In Bermuda waters, sixgills have been photographed from submersibles at depths up to 2,000 feet. They are often seen in aquariums, but they do not eat in captivity and have to be force-fed. There is one species of sawshark (*Pliotrema warreni,* not related to the sixgills) that has six gill slits on either side of its head.

See also frilled shark, sawshark, sevengill shark

skates Flat-bodied cartilaginous fishes that are generally rounded in form. They have large pectoral fins that extend from the snout to the base of the tail, and some species have a projecting snout that is helpful in identification. Many have spiny or thornlike structures on the dorsal surface, and most have two small dorsal fins on the tail but virtually no caudal fin. The mouth and gill openings are on the underside. Skates vary in size from the hedgehog skate (*Raja erinacea*), which does not grow larger than 20 inches, to the giant skate (*R. binoculata*), which can be 8 feet long. All skates lay eggs, many of which are the leathery cases known as mermaid's purses. Rays have large winglike pectoral fins, often pointed at the tips, and a long, whiplike tail, with one or more venomous spines at the base. Unlike skates, most rays give birth to live young.

See also manta, stingray

skeleton shrimp (*Caprella* spp.) Not a shrimp at all, *Caprella* is an elongated amphipod, with the thorax fused to the head. Less than ½ inch long, the body of this little creature is composed of elongated cylindrical segments that bear two pairs of hook-ended legs near the head and three more pairs at the rear. Thus equipped, it is one of the dominant small predators of the lower intertidal zone. Its movements are slow and methodical, rather like those of a praying mantis, and it perches on seaweeds, hydroids, and corals while stalking its tiny prey. **See also amphipod**

skimmer (family Rhynchopidae) There are three species of skimmers: the American black (*Rhynchops niger*), the African (*R. flavirostris*), and the Indian (*R. albicollis*). All are strongly marked black-and-white birds with a long red bill; they are the only birds whose lower bill is longer than the upper. Skimmers get their common name from their feeding technique, which consists of flying low over the water with their knifelike lower mandible skimming the surface to flip up small fish and other organisms, which are quickly swallowed. Because they fly so close to the water, their wing strokes never break the bird's horizontal plane. Skimmers usually feed in the evening, when their prey comes closer to the surface, and their eyes have a vertical pupil that can be narrowed to a slit as protection against bright sunlight during the day. They roost in flocks on open beaches, and from a distance, their cries sound like the baying of hounds. **See also gull, tern**

black skimmer

skipjack (*Katsuwonus pelamis*) Also known as the skipjack tuna, ocean bonito, and striped tuna, this small tuna is one of the world's most important food fishes. Skipjacks may reach a length of 3 feet, but most are smaller. The world's record weighed 41 pounds and was caught off Mauritius. They are found worldwide in tropical and subtropical waters, often aggregating in schools that may number as many as 50,000. In the Atlantic, they frequently associate with blackfin tuna

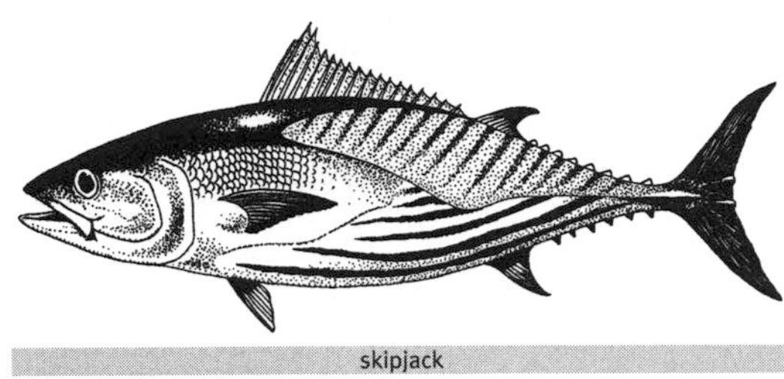

skipjack

(*Thunnus atlanticus*), and in the Pacific and Indian Oceans, they school with the yellowfin tuna (*Thunnus albacares*). The skipjack is a popular game fish, and it is the mainstay of the California tuna fishery. It is of tremendous commercial importance in Japan, Hawaii, and the Caribbean. It is marketed frozen, salted, and fresh, and when canned, it is known as light-meat tuna. **See also tuna, yellowfin tuna**

skua (family Stercorariidae) There are two genera in this family: *Catharacta*, the large skuas, and *Stercorarius*, the smaller ones. Some species, particularly those in the North Atlantic, are called jaegers, a German word that means "hunters." They are gull-like in overall structure, but more robust and powerful, as befits birds of prey. Some species are strongly "klepto-parasitic," which means they steal food from other birds, often on the wing, but skuas also feed on the chicks and eggs of other birds, and will kill juvenile birds or injured adults. Skuas, which are largely brownish in color, are extremely widespread, and occur in the higher latitudes of both hemispheres. They breed on the ground, normally laying two eggs. The adults are fiercely protective of their chicks, and will attack anything—including people—that comes too close. The name "skua" is probably onomatopoetic, derived from their cry.
See also Arctic skua, great skua, long-tailed skua, McCormick's skua, pomarine skua

Skye, Isle of The largest and most northerly of the Inner Hebrides, Skye is separated from the mainland by a ferry ride of only a couple of hundred yards. The island is 50 miles long, but so deeply indented that no part is more than 5 miles from the sea. Much of the island is moorland, and crofters farm small parcels of land. Fishing was once a mainstay of the economy, but it has declined, and whisky distilling has increased. Tourists flock to this spectacular island, which is dominated by the Cuillin Hills. In 1995 a bridge joined the island to the mainland, rendering the ferries obsolete. Dunvegan Castle, ancestral home of clan MacLeod, was built in the ninth century and has been occupied longer than any other house in Scotland. The current resident is John MacLeod of MacLeod, leader of the clan. After World War II, Gavin Maxwell, the author of *Ring of Bright Water*, bought the tiny island of Soay, just off Skye, and initiated a basking shark fishery. It was not a financial success, but he wrote *Harpoon Venture* (1952) about the experience. **See also basking shark, Hebrides**

slave trade During the fifteenth and sixteenth centuries, the exploration of the West African coast by Portuguese navigators resulted in the capture and exploitation of Africans as slaves. The Spanish conquerors of Mexico and Central and South America attempted to enslave the native peoples that they encountered and robbed, but the Indians escaped, revolted, or died, increasing the demand for Africans to replace them. Bartolomé de Las Casas, the Indians' most vociferous defender, suggested that Spanish settlers bring black African slaves with them when they came to the Indies. The trade in Africans was so lucrative that British, Dutch, French, Spanish, and Portuguese companies stationed agents all along the West African coast to barter for slaves to send to the West Indies and America. No one knows how many African slaves were transported across the Atlantic, but estimates run as high as 1 million. Conditions aboard the slave ships (known as blackbirders, as were the captains of the ships) were worse than intolerable; stench and filth prevailed, and the slaves were chained and packed so tightly belowdecks (where there were no sanitary facilities whatsoever) that they often suffocated or starved to death, whereupon they were summarily tossed overboard. With the development of the sugar plantations, so many African slaves were brought to the West Indies that the population of the islands soon became predominantly black. African slaves were introduced into the American colonies as early as 1616, when the first shipload arrived in Virginia. The raising of tobacco, sugar, rice, and, much later, cotton meant that there was an ever increasing need for workers, especially in the southern colonies. The slave trade was often triangular: ships from Britain would transport trade goods to the west coast of Africa, where they would be exchanged for slaves. The slaves were then brought to the West Indies or to the colonies of North and South America, where they would be traded for rum and sugar to be brought back to England. **See also middle passage**

Slocum, Joshua (1844–1909) From 1895 to 1898, after an adventurous career as a merchant captain, Joshua Slocum sailed alone around the world in the 35-foot, 12-ton sloop *Spray*. He departed from Boston, passed through the Strait of Magellan, to Australia and South Africa, spending time wherever it pleased him, since his was not an attempt at speed. In 1900, *Sailing Alone around the World* was published, a classic of sailing literature. In 1909, at the age of sixty-five, he planned a voyage down the Amazon in the *Spray*, but he left Martha's Vineyard in a storm and was never seen again. **See also Chichester**

slump In geology, the downward movement of rock debris, usually initiated by the removal of some supporting or buttressing material. Underwater, it refers to the sporadic downslope displacement of rock or sand that has built up in a canyon or a continental slope. If and when an underwater earthquake causes the slumping displacement of great quantities of rock or sediment, the water rushing into the resulting chasm is likely to produce a tsunami.

See also sediments, tsunami

small-spotted catshark (*Scyliorhinus canicula*) Also known as the sandy dogfish and the rough houndshark, this is a catshark with small spots. It is the most plentiful catshark in the eastern Atlantic and the Mediterranean, and is fished heavily for food in Britain, where it is sold as flake, rock eel, or rock salmon, and makes up a large proportion of the "fish" in fish and chips. It is also caught in large numbers for laboratory study, and because it adapts well to captivity, much experimental work with living sharks is performed with this species. (The electroreceptor capabilities of specialized cells in sharks was first observed in *S. canicula.*) The largest measured was 39 inches long, but most are smaller.

See also catsharks

Smeerenburg Dutch for "Blubbertown," Smeerenburg was intended to be a permanent settlement on the tiny northwestern Spitsbergen island of Amsterdamøya. Because bowhead whales were so plentiful in Spitsbergen waters, the Dutch believed they could save valuable time by wintering over instead of making the long round-trip journey every year. Although the whalers had been working on Amsterdam Island since 1619, they did not intentionally spend the winter there. (In the mid-1620s, nine men were stranded there and died.) Six Englishmen survived the winter in 1631, so the Dutch set up shop in Smeerenburg in 1632. It consisted of a small village with tryworks and other facilities for processing whale oil and baleen (whalebone). Despite what they considered adequate provisioning and precautions, the first overwintering party died in 1634, and the same thing happened the following year. The experiment was then abandoned.

See also bowhead whale, Jan Mayen, Spitsbergen, whaling

smelt (*Osmerus mordax*) Resembling miniature salmon with a small, fleshy adipose fin behind the dorsal, smelts reach a maximum length of 14 inches. The American smelt, sometimes known as the rainbow smelt, is found in great profusion from the Gulf of St. Lawrence to Virginia. This silvery fish with an olive-green back is anadromous, living in the sea and entering freshwater to breed, but a permanent population introduced into the Great Lakes has become so numerous that it supports a commercial fishery. The European smelt (*O. eperlanus*) is a somewhat smaller species, reaching only 8 inches in length. Freshly caught smelts smell like cucumbers. A species (*Thaleichthys pacificus*) from the Pacific Northwest is called candlefish because its flesh was so oily that Indians attached a dried fish to a stick and used it as a torch. The capelin (*Mallotus villosus*) is a species of smelt. **See also capelin**

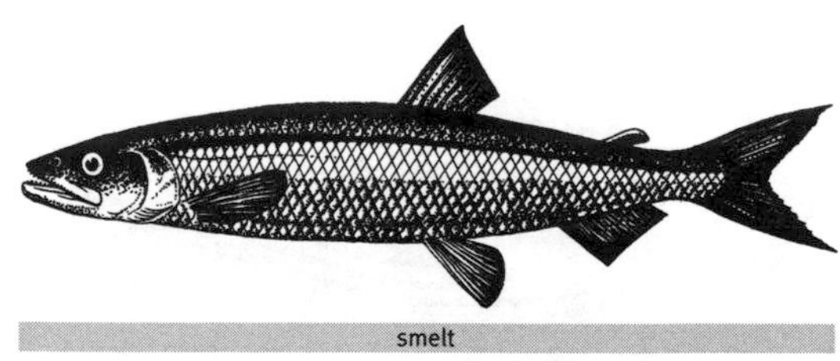

smelt

Smith, J. L. B. (1897–1968) James Leonard Brierley Smith was born in Graaf-Reinet and educated in South Africa, but he took his Ph.D. in chemistry at Cambridge. He returned to teach chemistry at Rhodes University College at Grahamstown, but maintained a passionate interest in ichthyology. He published papers on new fish species in South Africa, and had soon established a reputation as an expert on fishes; in fact, he was one of the very few ichthyologists, amateur or professional, in all of South Africa. He built a small laboratory at Knysna, and he was there in 1938 when he received a letter from Marjorie Courtney Latimer describing and illustrating an unusual fish that she had seen in a market in East London. Smith recognized it as a coelacanth, a fish that was supposed to have gone extinct some 65 million years earlier, and he was very anxious to examine the specimen. Unfortunately, it had been so poorly preserved that only the skin and the skull were saved. When he described this "living fossil," *Latimeria chalumnae* became one of the most significant discoveries in all of zoology, and Smith almost immediately became one of the most important figures in world ichthyology. Seeking more specimens, he posted broadsides all along the South African coast, but World War II interfered, and it was not until 1952 that another coelacanth was found, this time in the Comoros. More began to appear—all in the Comoros, which belonged to France—and Smith felt that he was no longer in charge of the discovery that had catapulted him to worldwide scientific and popular recognition. He had terminal cancer and took his own life. Often with his wife, Margaret, as coauthor, he published more than two hundred scientific papers, as well as the important *Sea Fishes of Southern Africa* (1949), *Old Fourlegs, The Story of the Coelacanth* (1956), *The Fishes* *of the Seychelles,* and a book of articles entitled *High Tide* (1968). Margaret MacDonald Smith was instrumental in starting the J. L. B. Smith Laboratory at Rhodes University in Grahamstown, now one of South Africa's most important ichthyological institutions.

See also coelacanth, Comoro Islands

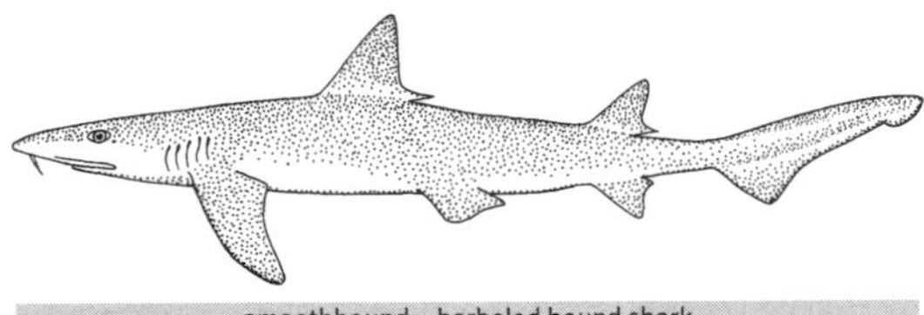

smoothhound—barbeled hound shark

smoothhound sharks (family Triakidae) Known variously as smoothhounds, smooth dogs, or smooth dogfishes, these are small- to medium-sized sharks (none longer than 5 feet), with slender bodies, large oval eyes, and dorsal fins without spines. In general form they resemble the carcharhinids, but their dentition is closer to that of the nurse sharks. There are about thirty species of smoothhounds inhabiting shallow to moderately deep water in all oceans. After the spiny dogfish (*Squalus acanthias*), the most numerous shark on the American Atlantic coast is the 4-foot-long dusky smoothhound, *Mustelus canis.* Similar species are found in California (*M. californicus*), Florida (*M. norrisi*), western Europe and the Mediterranean (*M. asterias* and *M. mustelus*), eastern Asia (*M. griseus*), New Zealand (*M. lenticulatus*), and Australia (*M. antarcticus*). Most of the smoothhounds are gray or brownish, but the leopard shark (*Triakis semifasciata*), which is spotted like its namesake, is also a smoothhound.

See also leopard shark, spiny dogfish

Snail: See gastropoda, mollusk

snake eel (*Ophichthus* spp.) Fishes with a snakelike body that are probably responsible for reports of sea snakes in the West Indies. They have cylindrical, muscular bodies, sharp noses, and a habit of burrowing in the sand, either headfirst or tailfirst. Garden eels inhabit permanent burrows. Snake eels lack the impressive teeth of the morays and reach a maximum length of 4 feet.

snake eel

See also garden eel, moray, sea snakes

snake mackerel (*Gempylus serpens*) A 5-foot-long deep-sea predator, the snake mackerel gets its common name from its elongated body and the little finlets at the base of the caudal fin, which are reminiscent of the mackerels, though it is not related to them. The eyes are large, suggesting a deep-water habitat, and the fanglike teeth are numerous and sharp. The snake mackerel hunts in large packs, and during their feeding frenzies, they are caught in large numbers by commercial fishermen using handlines. A closely related species, *Thyristes atun,* is known as *snoek* in South Africa and *barracouta* in New Zealand.

snake mackerel

See also cutlass fish, lancet fish

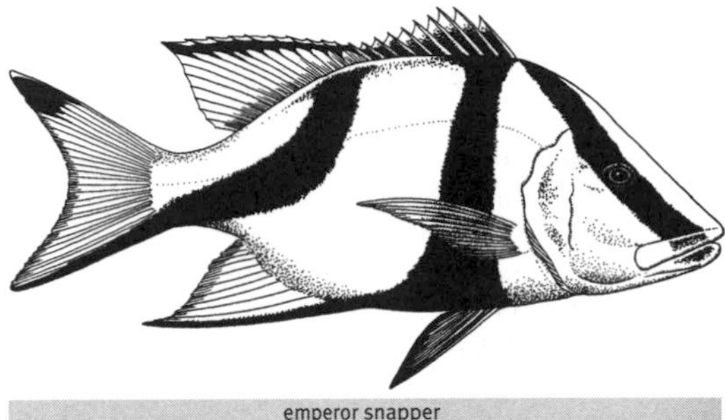

emperor snapper

snappers (family Lutjanidae) There are more than a hundred species of snappers found in shallow tropical seas around the world. (The family name is from *ikan lutjang,* the common Malayan name for these fishes; they are called snappers because they snap and bite when caught.) Many of these long-faced fishes are commercially important, but some are also responsible for ciguatera, human food poisoning caused by toxic buildup in fish. Snappers are nocturnal hunters, feeding on smaller fishes that they catch with their large canine teeth. Some are drab-colored, but the emperor snapper (*Lutjanus sebae*) of the Indo-Pacific is a white fish with bold mahogany markings, and the blue-lined threadfin snapper (*Symphorhicthys spilurus*) has, in addition to the horizontal stripes that give it its common name, long golden streamers on its fins. In North America, the most valuable is the red snapper (*L. campechanus*), a 30-inch-long, bright red fish that inhabits the deeper waters of the Gulf of Mexico. Fished almost exclusively with handlines, the red snapper is an extremely popular food fish. The mutton snapper (*L. analis*) is another popular American food fish, but it is also fished for sport. In Florida waters the mangrove snapper (*L. griseus*), also known as the gray snapper, is a very popular sport fish, usually caught in or around mangrove stands. The schoolmaster (*L. apodus*) is the most common of the Caribbean snappers, found from coral reefs to tide pools. The largest of the family is the cubera snapper (*L. cyanopterus*), which ranges from Florida and Cuba south to Brazil in the western Atlantic, measuring 4 feet in length and weighing as much as 120 pounds. The yellowtail snapper (*Ocyurus chrysurus*) is bright yellow with blue stripes and a red eye, a solitary fish that spends most of its time about 20 feet above the bottom, unlike most other snappers, which hug the bottom. **See also ciguatera**

Snares penguin (*Eudyptes robustus*) Found only on Snares Island south of New Zealand, this yellow-crested penguin reaches a height of 24 inches. They forage by making shallow dives near the shore and keep in contact with one another by short, barking calls. The population on the island is estimated at 66,000 birds. **See also erect-crested penguin, fiordland penguin, macaroni penguin, rockhopper penguin, royal penguin**

snipe eel (*Nemichthys scolopaeus*) Where the common eel (*Anguilla*) has a blunt, rounded head, *Nemichthys* has beaklike upper and lower jaws that do not close. Exactly how the snipe eel feeds with this arrangement is not clear, but it may entangle the antennae of shrimps in the tiny, backward-pointing teeth in the jaws. Snipe eels, which can reach a length of 5 feet, have 670 vertebrae—more than any other vertebrate. Blackish-brown in color, they are found at depths up to 6,000 feet in all oceans. Related species are the avocet eel (*Avocettina infans*), with an even more exaggerated gape, and the bobtail snipe eel (*Cyema atrum*), a dartlike creature that does not get to be more than 6 inches long. **See also eel, leptocephalus, moray**

snipe eel

snipefish (family Macrorhamphosidae) Related to sea horses, pipefishes, and shrimpfishes, the snipefishes are small tropical and semitropical fishes that are characterized by a compressed body with bony plates on each side of the head, a very long second dorsal spine, and a long, tubelike snout that is probably used to suck up tiny organisms. There are three genera, *Centriscops, Macrorhamphosus,* and *Notopogon,* which are found from moderate depths to deep water in the Indo-Pacific from California to the east coast of Africa. In New Zealand they are known as bellowsfishes. **See also pipefish, sea horse, shrimpfish**

snipefish

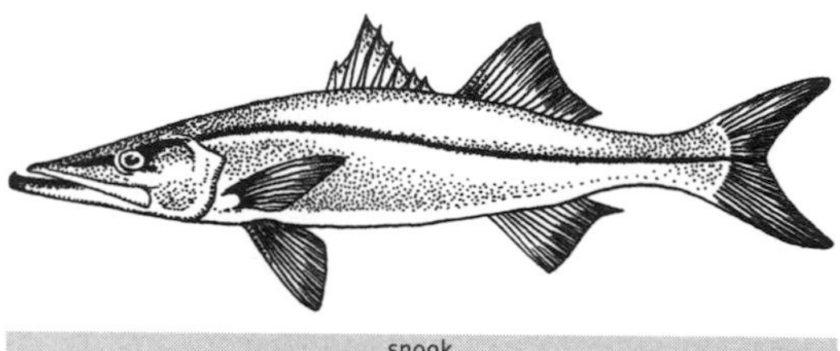
snook

snook (*Centropomus undecimalis*) A popular Florida game fish, the snook is a silvery fish with a prognathous lower jaw and a prominent black lateral line. Once fished commercially, they now can only be taken by sport fishermen. Thirty-pounders are not uncommon, and the record is 53 pounds. Because of their dependence on inshore shallow waters, estuaries, and mangrove swamps as nurseries, snooks are threatened by coastal development that is destroying their habitat.

snorkel Originally *schnorkel,* the tube attachment that provided the air supply for submerged German submarines during World War II. The tube was fitted with a flap at the top that prevented the entry of seawater in rough weather or when submerged. It not only supplied air for the crew to breathe, but also the oxygen required to run diesel engines. The English version is also applied to the breathing tube designed along the same principles, which enables swimmers to breathe with their faces in the water while swimming or diving.

snow petrel (*Pagodroma nivea*) The world's only all-white petrel, snow petrels are found in the vicinity of the Antarctic pack ice. They breed on South Georgia, the South Shetlands, the South Orkneys, and other Antarctic islands. The call of this lovely little bird is a raucous caw. **See also petrel, South Georgia, South Shetland Islands**

snow petrel

soapfish (*Rypticus saponaceus*) There are several species of soapfish (family Grammistidae), whose name is derived from their response when they are attacked or handled: a thick mucus covers their skin like a soapy froth. The mucus and the froth are toxic to predators. Soapfishes, which are related to the sea basses, grow

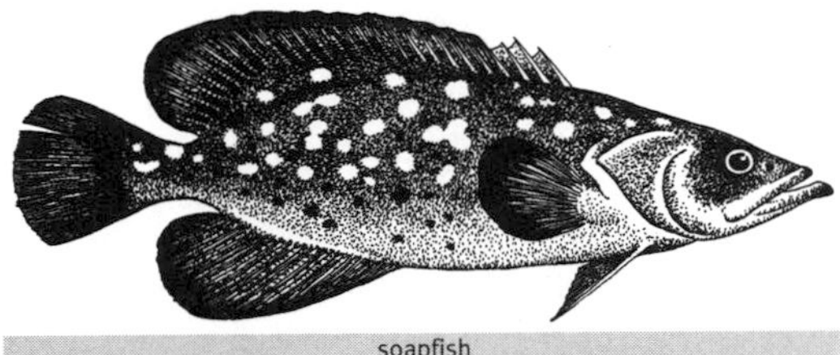

soapfish

to a maximum length of 12 inches. The several species are found on both sides of the Atlantic.

See also giant sea bass

sockeye salmon (*Oncorhynchus nerka*) The unusual name of this Pacific salmon has nothing to do with its eyes; it is a corruption of the Amerind *sukkai.* It is also known as the red salmon because its flesh becomes a deep red when canned, and in its breeding colors, it has a lobster-red body and a green head. Sockeyes are among the longest travelers of all salmon species: they may wander 2,000 miles from their parent river into the ocean, and upon their return they may travel as much as 1,000 miles upstream in rivers like the Columbia and the Salmon in Idaho. Before they begin their epic journey, sockeyes remain in lakes and rivers for as much as five years. *O. nerka* is the number one salmon in the cannery industry of Alaska.

See also chinook salmon, coho salmon, Pacific salmon, pink salmon

soft corals (Alcyonacea) Also known as octocorals because of the eight-tentacled polyps, soft corals do not excrete a limestone skeleton but are strengthened by fused or unfused spicules. They are common in shallow, warm seas, and are often colorful and sometimes luminescent. Dead man's fingers (*Alcyonium digitatium*) is a common species on both sides of the North Atlantic, and other common species are the organ-pipe coral, which has emerald-green polyps emerging from brick-red tubes, and the blue corals, which get their coloration from iron salts. Soft corals cannot be made into jewelry, but they are very photogenic and often become the highlight of an underwater photographer's diving expedition. The gorgonians (sea fans, sea pens, and sea plumes) are classified with the soft corals to differentiate them from the "true," or "stony," corals, but they take a different form.

See also coral, dead man's fingers, gorgonians

soldierfish (family Myripristidae) Differentiated from the squirrelfishes by the absence of an opercular spine, soldierfishes are also small, averaging under 1 foot in length; they are nocturnal and usually reddish in color. They are schooling fishes and are sometimes caught commercially. **See also squirrelfish**

sole (family Soleidae) Mostly right-eyed flatfishes with an elongated teardrop shape and eyes that are small and close together. The 18-inch European sole (*Solea solea*), also known as Dover sole, is an extremely popular food fish. They are caught at night when they leave their hiding places in the sand to search for worms, brittle stars, and other bottom-dwelling animals. The hogchoker (*Trinectes maculatus*), found in American waters, is a very small sole, rarely exceeding 8 inches in length. Its name is supposed to have come from its rough scales that choked hogs that attempted to eat some on the beach. **See also flounder, plaice, turbot**

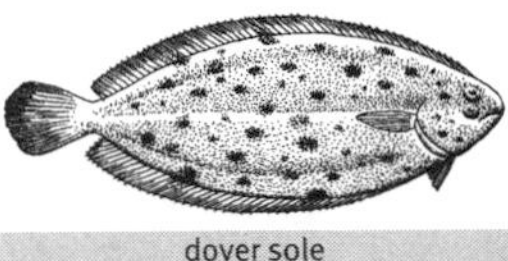

dover sole

Solomon Islands A self-governing nation in the southwest Pacific, which became independent in 1978; it consists of Guadalcanal, Malaita, New Georgia, Choiseul, Santa Isabel, San Cristóbal, the Santa Cruz Islands, and nearly a thousand smaller islands. (Bougainville and Buka are part of Papua New Guinea.) The total population is around 340,000, and the administrative capital is Honiara, on Guadalcanal. In 1568 the Spanish explorer Álvaro de Mendaña was the first European to visit, but his attempts at colonization failed, largely because of the hostility and anthropophagous inclinations of the natives. (Mendaña found alluvial gold on Guadalcanal, and thinking he had discovered Terra Australis Incognita, which contained the biblical King Solomon's gold mines, he named the islands accordingly.) Assorted settlers and missionaries came to the Solomons throughout the eighteenth, nineteenth, and twentieth centuries, and at one time or another, they were under the control of the Germans, the British, or the Australians. During World War II, Japanese forces occupied Choiseul, New Georgia, Ysabel, and Guadalcanal, leading to some of the fiercest fighting of the war.

See also Guadalcanal, Mendaña, New Georgia, Terra Australis Incognita

sonar Originally developed for the detection of icebergs after the *Titanic* disaster of 1912, the name "sonar" is an acronym derived from SOund NAvigation Ranging. (It was known as "asdic" in the United Kingdom, but the origin of this word is unclear.) Sonar uses the transmission of an energy pulse from an array of transducers that emit high-frequency sounds, and then detects the echoes reflected back by an object. The accurate measurement of the time between the emission of a pulse and the return of the echo gives the range of the object. By using a narrow sweeping pulse beam, a bearing can also be obtained, and in the case of a moving object, such as a ship, the direction of movement can also be determined. (The same principles are employed by toothed whales and dolphins in their echolocation). "Doppler sonar" relies upon the relative speed of the target and the receiving station because there is a change in frequency when the observed and the observer are in motion relative to each other, a phenomenon known as the Doppler effect. "Side-scan sonar" consists of a horizontally towed array of transducers that broadcast sound beams along a wide vertical plane, giving a more complete picture of the seabed or objects. Sonar is also used by the fishing industry to detect schools of fish, and it is the basis of all echo-sounding equipment used to determine the depth of water under a ship. **See also echolocation, radar**

sooty albatross (*Phoebetria* spp.) The ranges of the two species of sooty albatross, the dark-mantled (*P. fusca*) and the light-mantled (*P. palbebrata*), overlap in the Southern Ocean. Both are sooty brown, darker on the head. The light-mantled version has a frosty gray patch that extends from the nape to the lower back. The wingspan averages around 7 feet. They breed on South Atlantic islands like Tristan da Cunha and the Gough Islands, and also in the southern Indian Ocean on Amsterdam, St. Paul, Marion, Kerguelen, and Crozet Islands. The light-mantled species breeds in South Georgia, the sooty does not. Like all albatrosses, sooties are superb fliers, following ships for hours without any visible effort. At sea, they may be mistaken for the southern giant petrel, a shorter-winged, less graceful bird with a much heavier bill. **See also albatross, giant petrel, wandering albatross**

sooty shearwater (*Puffinus griseus*) A wide-ranging, dark grayish-brown bird of the Pacific and Atlantic Oceans, the sooty shearwater aggregates loosely in huge flocks, swooping low over the waves and dipping down to pluck food items from the water. Somewhat larger than the short-tailed shearwater, with a wingspan that can reach 50 inches, the sooty has darker underwings. It is also much more widely distributed, and can be seen at sea throughout the world's oceans except in the high Arctic. **See also Manx shearwater, shearwater, short-tailed shearwater**

SOS The internationally recognized signal from a ship in distress, adopted in 1908. The three letters were chosen because they were easy to read and to tap out in Morse code: three dots, three dashes, three dots. It does not stand for "Save Our Ship" or "Save Our Souls."

SOSUS SOSUS stands for SOund SUrveillance System In the 1950s, the U.S. Navy employed a network of bottom-mounted hydrophone arrays to monitor the movement of Soviet warships, particularly nuclear submarines. The end of the cold war meant that the monitoring of warships was no longer necessary, and in 1990, the navy turned over the equipment and data to the National Oceanic and Atmospheric Administration (NOAA) for the collection and processing of nonmilitary data. Since 1991, when it began monitoring volcanic and hydrothermal activity in the North Pacific, NOAA has reported deep-sea earthquakes and actual evidence of seafloor spreading. SOSUS is also capable of listening to and recording the sounds of large whales, enabling researchers to learn more about their daily activities and migration patterns than ever before. **See also seafloor spreading**

Southampton Island The largest island at the entrance to Hudson Bay, separated from the mainland of the Northwest Territories by Roes Welcome Sound. The island is about 210 miles long and 220 miles wide, and covers some 15,913 square miles. It was discovered in 1613 by Thomas Button as he made a futile search for the elusive Northwest Passage in the *Discovery*, the very same ship from which Henry Hudson had been abandoned in an open boat two years earlier. Button named the island after his patron, Henry Wriothesley, the third earl of Southampton—to whom William Shakespeare may have written the sonnets. The coastal waters of the island are known for Arctic char fishing.

South China Sea Surrounded by the Chinese mainland, the island of Taiwan, and the Taiwan Strait on the north; Vietnam on the west; Malaysia, Sumatra, and the island of Borneo on the south; the Philippines (Palawan and Luzon) on the east, the South China Sea covers about 1.5 million square miles. Surrounding the Indochinese peninsula (Thailand, Cambodia, Laos, and Vietnam) are two large embayments, the Gulf of Thailand and the Gulf of Tonkin. In the southwestern quadrant, from the Gulf of Siam to the Java Sea, is the Sunda Platform, one of the largest sea shelves in the world, where the water is about 200 feet deep. To the north is a deep basin, with depths of more than 18,000 feet. Monsoons affect most of the area in the summer, and the region is subject to violent typhoons. **See also East China Sea, Yellow Sea**

southern elephant seal: See elephant seals

southern fur seal (*Arctocephalus* spp.) There are eight species in the genus *Arctocephalus* (which means "bear-headed"), mostly found south of the equator, but also on Guadalupe Island off the coast of Baja California and in the Galápagos Islands. They are eared seals (Otariids), as are the sea lions, but all southern fur seals have a sharply pointed snout, while many of the Southern Hemisphere sea

southern fur seal

lions have shorter, stubbier muzzles. The species of southern fur seals, differentiated by minor skeletal differences and geography, are the Galápagos (*A. galapagoensis*), Guadalupe (*A. townsendi*), Juan Fernández (*A. philippi*), South American (*A. australis*), South African and Australian (*A. pusillus*), New Zealand (*A. forsteri*), Antarctic (*A. gazella*), and subantarctic (*A. tropicalis*). In recent years, the depleted populations of fur seals have rebounded, probably because of the cessation of hunting, and also because the baleen whales have been so reduced in numbers that more food is available for the fur seals. **See also sealing, seals and sea lions**

Southern Ocean Sometimes called the Antarctic Ocean, the Southern Ocean surrounds the Antarctic Continent and comprises the southern portion of the Atlantic, Pacific, and Indian Oceans. It has no rigid northern boundaries, but it may be defined by the Antarctic Convergence, the northern limit of cold Antarctic surface water. Most geographers now recognize the Southern Ocean, because it serves conveniently to identify a distinct body of water with its own physical characteristics and a common fauna. Components of this ocean are the Ross, Bellingshausen, and Weddell Seas, which border on the Antarctic continent and are iced over in winter, producing thousands of icebergs; and the infamous Drake Passage, which separates the tip of South America from the Antarctic Peninsula. The Southern Ocean is especially rich in nutrients, especially in the area of the Convergence, and it supports an abundance of planktonic organisms, which in turn support a large and diverse fauna ranging from squids and seabirds to seals and whales. Seal hunters performed much of the early exploration of the islands while they virtually extirpated their quarry; they were followed by the whalers who began the onslaught on the great whales of the Southern Ocean that led to their near extinction as well.

See also Antarctic Convergence, Drake Passage

southern right whale dolphin (*Lissodelphis peronii*) Found in a circumpolar distribution throughout the Southern Ocean, *L. peronii* is a 7-foot-long, elongated teardrop of a dolphin, its sleek silhouette uninterrupted by a dorsal fin. Because of its sharply demarcated white face, Herman Melville called it the "mealy-mouthed porpoise." It differs from its northern relative in its coloration and also in its morphology: it is more flattened from top to bottom, an adaptation that is thought to give it the stability it needs because it lacks a dorsal fin. It was named for François Peron, a French naturalist who described it from a specimen sighted south of Tasmania.

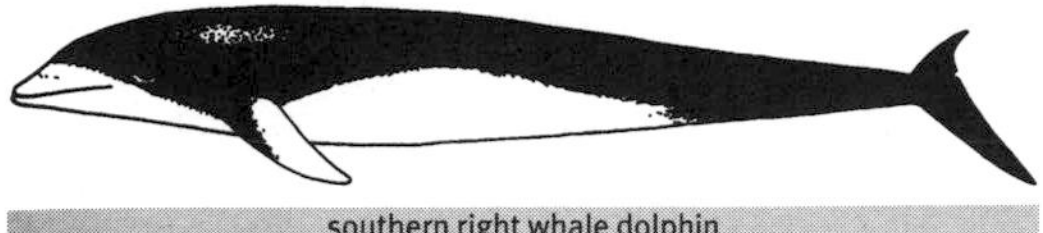
southern right whale dolphin

See also northern right whale dolphin

southern sea lion (*Otaria byronia*) Males of this species are distinguished by their short, upturned muzzle, massive head, and neck with a mane that gives them their English common name. (They are known as *lobos marinos*—"sea wolves"—in Spanish.) The bulls can weigh as much as 750 pounds, but the females are much smaller, lighter in color, and more graceful. The southern sea lion, also known as the South American sea lion, inhabits both coasts of southern South America, from Peru to Cape Horn on the west, and as far north as Uruguay on the east. Like most of the pinnipeds, this species is primarily a fish eater, and it can often be seen in shallow inshore waters. Southern sea lions were hunted extensively by sealers in the nineteenth century, and although they are now protected, their populations are declining.

southern sea lion

See also southern fur seal

South Georgia A roughly crescent-shaped island, 105 miles long and 18 miles across at its widest point, South Georgia is located in the Southern Ocean, almost 1,300 miles east of Cape Horn and 800 miles east of the Falkland Islands. An island of snow-capped mountains, it has been described as the "Alps in mid-ocean." The highest point is Mount Paget, 9,625 feet high. The island was first sighted in 1675 by Antoine de la Roché, but was first explored in 1775 by James Cook, who reported the abundance of elephant and fur seals. British sealers arrived shortly thereafter, and continued to kill the pinnipeds for their valuable pelts throughout the nineteenth century. With the arrival of the Norwegian captain Carl Anton Larsen in 1904, whaling began in and around the island, and continued without interruption until 1965. The Norwegians built a whaling station at Grytviken and the British built one at Leith Harbour, for the processing of blue and fin whales that were caught off the ice edge. When Ernest Shackleton's *Endurance* was crushed by the ice in 1915, he and his men sailed in a longboat across the open ocean for 800 miles to South Georgia, which they then crossed on foot to reach the British whaling station at Stromness. (Shackleton died on South Georgia in 1922 and is buried there.) During the whaling period, the whaling stations were peopled by settlements of whalers, but when whaling closed down, the communities were abandoned. South Georgia is uninhabited except for a detachment of Royal Marines and a caretaker for the museum at Grytviken, and the pinnipeds and penguins. There are also reindeer, the descendants of those introduced by the Norwegians around 1910 to provide fresh meat. **See also Cook (James), Shackleton**

South Orkney Islands Part of the British Antarctic Territory, the South Orkneys are composed of two large islands, Coronation and Laurie, and many smaller islets. The islands were first sighted by sealing captains George Powell and Nathaniel Palmer in December 1821, and generations of fur seals were killed by those who followed them. The islands are now are uninhabited by humans (except for a British scientific station on Signy Island), but they are occupied by breeding penguins, seabirds, and seals. Like the South Shetlands, these islands were named for their more hospitable counterparts north of Scotland.

south polar skua: See McCormick's skua

South Pole The southern end of the earth's axis, at latitude 90° south. Unlike the North Pole, which is located in the middle of the Arctic Ocean, the South Pole is on the Antarctic continent and therefore reachable overland. From the late eighteenth through the nineteenth centuries, many explorers sailed to the Antarctic, among them James Cook, Thaddeus von Bellingshausen, Jules Dumont d'Urville, Charles Wilkes, and James Clark Ross, but it was not until the early years of the twentieth century that the conquest of the Pole became a possibility. In 1898, Adrien de Gerlache, a Belgian naval officer, explored the Antarctic mainland, followed by Otto Nordenskjöld in 1902 and Jean-Baptiste Charcot in 1903. Robert Falcon Scott led the British Antarctic Expedition in 1901–1903, and although he, Ernest Shackleton, and Edward Wilson got farther south (82° south) than anyone had ever done, they did not reach the Pole. Shackleton tried again in 1909, but he also failed. On December 14, 1911, Roald Amundsen and four fellow Norwegians became the first human beings to stand at the South Pole. Captain Scott and his four companions arrived there a month later, and on their disastrous return journey, they all died.

See also Amundsen, Bellingshausen, Cook (James), Dumont d'Urville, North Pole, Ross (James Clark), Scott, Shackleton, Wilkes

South Sandwich Islands A dependency of the Falkland Islands, the South Sandwich Islands are located north of the Weddell Sea, 500 miles southeast of South Georgia. These glacier-covered, uninhabited islands were sighted and named by James Cook in 1775. (Cook also named the present Hawaiian Islands the Sandwich Islands after Lord Sandwich, but the name stuck only to these inhospitable islets.) Cook reported that fur seals inhabited the islands, which brought the sealers, who almost eliminated the seals. Argentine forces occupied the South Sandwiches in 1976, but were removed after the Falkland Islands War in 1982.

South Sea Bubble A series of stock speculations that failed disastrously in 1720. When Robert Harley founded the South Seas Company in 1711, investors believed that the treaty following the War of the Spanish Succession (1702–1713) would open enormous trade possibilities with Spanish America, especially in slaves. King George I became a governor of the company, which greatly enhanced its prestige, and a wild boom followed, where stock was guaranteed a 6 percent interest rate, and shares that were issued at one hundred pounds in January 1720 were worth a thousand pounds in August. The company was so successful that its offer to take over a large part of the national debt was accepted by Parliament. By the end of the year, however, shares had dropped to £124, and six months later they were practically worthless. A few investors made vast fortunes trading the stock, but the great majority lost heavily, and many were ruined. A committee of inquiry found that the chancellor of the exchequer and two other government ministers had invested heavily in the stock and accepted bribes, and they were sent to prison. When the bubble burst, Robert Walpole was made chancellor of the exchequer and reorganized the company so that its trading rights with Spain were sold, and until it was dissolved in 1853, the South Sea Company was primarily involved with the Greenland whale fishery.

South Shetland Islands Part of the British Antarctic Territory just to the west of the tip of the Antarctic Peninsula in the South Atlantic Ocean, these islands are now uninhabited but once served as bases for Antarctic sealing and whaling activities in the eighteenth and nineteenth centuries. The largest of the South Shetlands are Livingston, King George, and Elephant, but the best known is Deception, an active volcano that last erupted in 1970. Its harbor at Port Foster was one of the centers for British Antarctic whaling during the early decades of the twentieth century. It was from Elephant Island that Shackleton made his epic crossing of the Drake Passage to South Georgia that resulted in the successful rescue of all his men after their ship *Endurance* was crushed and lost in 1916. The islands are now the home of millions of penguins, petrels, seals, and sea lions.

See also Antarctic Peninsula, Drake Passage

Sowerby's beaked whale (*Mesoplodon bidens*) Known only from the North Atlantic, Sowerby's beaked whale is named for James Sowerby (1752–1822), the British naturalist who described the first specimen, which was stranded in Scotland in 1800. Its specific name, *bidens*,

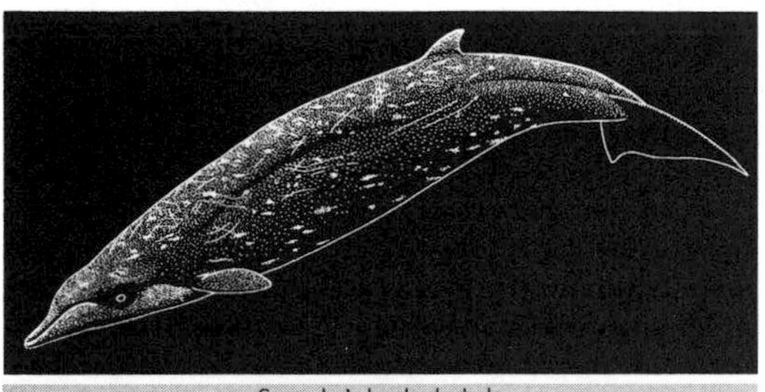
Sowerby's beaked whale

means "two teeth," a characteristic of all adult male *Mesoplodons*. These whales have been reported from strandings in Norway, the British Isles, Newfoundland, and Massachusetts. It was the first beaked whale to be described. **See also beaked whales, *Mesoplodon***

spadefish (family Ephippidae) From tropical and subtropical seas, spadefishes take their common name from their flattened, compressed body, which may be as deep as it is long. The long-finned spadefish (*Platax teira*) has extended dorsal and anal fins that are most pronounced in the juvenile stages. Adults reach a maximum length of 3 feet. The Atlantic spadefish (*Chaetodipterus faber*) is essentially an inhabitant of warm Florida and Caribbean waters, but they sometimes range as far north as Massachusetts. A similar species (*C. zonatus*) is found from Southern California to Mexico. **See also batfish**

spadefish

Spanish-American War Conflict lasting less than a year between Spain and the United States, fought ostensibly for Cuban independence, but in reality to satisfy U.S. expansionists. Cuban insurgents were rioting in 1897 when President William McKinley dispatched the battleship *Maine* to Havana, to protect U.S. citizens. On February 15, 1898, the battleship exploded and sank in Havana harbor, taking 268 crew members to the bottom. Americans accused the Spanish of sinking the ship, the jingoistic "yellow press" trumpeted "Remember the *Maine*," and by April, the two countries were at war. In May, a squadron under the command of Commodore George Dewey steamed into Manila harbor in the Philippines and soundly defeated the Spanish fleet there. In Cuba, the fleet under the command of Admiral Pascual Servera was destroyed by American ships, and seventeen thousand American troops came ashore to capture the city of Santiago. Colonel Theodore Roosevelt, who as assistant secretary of the navy had been one of the most fervent advocates of a free Cuba, led his Rough Riders to victory at San Juan Hill. Manila and Puerto Rico were occupied without a shot being fired, and the war was over. Signed on December 10, 1898, the Treaty of Paris freed Cuba (but left it under U.S. control), and Guam, Puerto Rico, and the Philippines were ceded to the United States for reparations of $20 million. **See also Cuba, Guam, *Maine*, Philippines, Puerto Rico**

Spanish Armada: See Armada, Spanish

Spanish mackerel

Spanish mackerel (*Scomberomorus maculatus*) The Spanish mackerel can be distinguished from other mackerels by the presence of bronze or yellow spots on the flanks, and the lack of stripes. It is a popular game fish in the western North Atlantic, from Chesapeake Bay to the Gulf of Mexico, and reaches a maximum weight of 20 pounds. **See also Atlantic mackerel, Scombridae**

Spanish Main A term used mostly by romantic writers to describe the mainland of South America from the sixteenth through the eighteenth centuries, following the voyages of Columbus and the Spanish conquistadors. Buccaneers were attracted to these areas by reports of gold and other valuable treasures being shipped back to Spain. Originally the term referred to the northeast coast of the continent, from the Orinoco River to the Isthmus of Panama, but it later came to include the Caribbean. Originally it referred to the land, but it eventually came to mean the sea. **See also buccaneer, piracy**

Sparidae About one hundred species of perchlike fishes with deep bodies, large heads, steep foreheads, and small mouths. There are prominent spines on the fins, and the scales and lateral line are well developed. They are found around the world in most tropical and temperate waters, but there are a large proportion in the region of southern Africa. Among the better-known species are sea bream, musselcrackers, porgies, pinfish, steenbras, and scup. **See also porgy, scup, steenbras**

spearfish (*Tetrapterus* spp.) Spearfishes look so much like juvenile marlins that some ichthyologists consider them just that. For the most part, they are recognized as separate species, although the distinction between them and the smaller marlins is not clear, and they have been placed in the same genus, *Tetrapterus*. (The black and blue marlins belong to the genus *Makaira*.) Like the striped marlin, spearfishes are slender, light-bodied

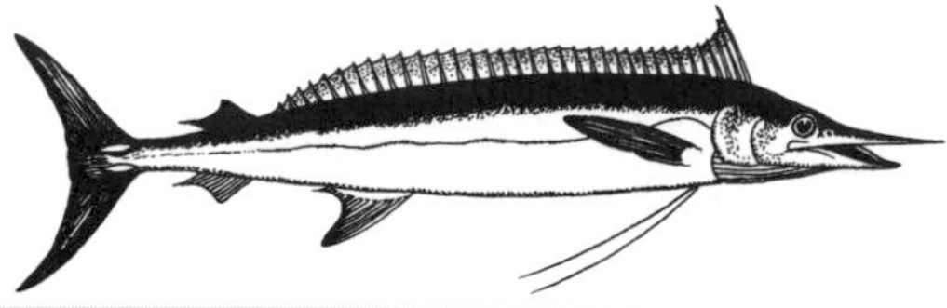
shortbill spearfish

fishes, but they have a much shorter bill; in the shortbill spearfish (*T. angustirostris*), it is barely longer than the lower jaw. The dorsal fin is pronounced for most of its length, unlike that of the marlins, which consists of a high anterior element, diminishing to a low spiny ridge for most of its length. Spearfishes are sought-after game fishes, but they are not common anywhere. The largest ones weigh about 90 pounds. The longbill spearfish (*T. pfluegeri*) is found in the North and South Atlantic, the shortbill in the Pacific and Indian Oceans. There may be a separate species (*T. belone*) that is restricted to the Mediterranean.

See also sailfish, striped marlin, white marlin

spearfishing The practice of fishing while swimming underwater, armed with a spear. The earliest devices were simply sharpened poles, or poles with a point on the end, but now rubber slings are employed to shoot the spear farther from the diver. There are also spear "guns," powered with CO_2 or explosive cartridges. Originally, divers simply held their breath when they hunted, but scuba gear has enabled divers to reach greater depths and to remain underwater much longer.

See also scuba

spectacled cormorant (*Phalacrocorax perspicillatus*) When Vitus Bering's ship was wrecked on the Commander Islands in 1741, he and his crew found many unusual animals there. Bering's zoologist, Georg Steller, eventually described a species of sea eagle, the sea otter, sea lions, and a gigantic sirenian, subsequently known as Steller's sea cow, that became extinct at the hands of Russian sealers only twenty-eight years after it was discovered. The crew also encountered a very large cormorant (12 to 15 pounds) with wings so short that it could hardly fly. Bering's crew killed these hapless birds for food, and when the Russian sealers came, they did the same. By the mid-nineteenth century, Steller's spectacled cormorant was extinct. **See also Bering, cormorant, shag, Steller, Steller's sea cow**

spectacled eider (*Somateria fischeri*) Unlike the common eider, which is very numerous, the spectacled eider is uncommon. It is among the smallest of the eiders (only Steller's is smaller), but with its black-rimmed white spectacles, it is the most easily recognized. Like all other eiders, the ducks are brown where the drakes are strongly marked, but even in the brown females, the "spectacles" can be seen as light-colored patches around the eyes. Although it was known that spectacled eiders bred along the Russian coast of the Arctic Ocean, the wintering grounds were not discovered until 1995, when biologists found nearly 75,000 pairs—believed to be the entire breeding population—packed into dense rafts in the icy Arctic Ocean between St. Lawrence and St. Matthew Islands in the Bering Sea.

See also eider, king eider, sea ducks, Steller's eider

spectacled porpoise

spectacled porpoise (*Australophocœna dioptrica*) Black on the dorsal surface with a pronounced white eye-ring and black lips, this little porpoise is easy to identify in the vicinity of Cape Horn or Tierra del Fuego. The spectacled porpoise was previously known only from southern South American waters, but there are now several records from the New Zealand region, suggesting a circumpolar distribution. There is pronounced sexual dimorphism in this species: males have a greatly exaggerated dorsal fin, whereas the females have a stubby, rounded little fin.

See also harbor porpoise, porpoise

Spee, Maximilian von (1861–1914) German vice admiral who served with distinction in West Africa and with the German contingent during the Boxer Rebellion in China in 1900. He was appointed to command the German East Asian squadron in 1912. He kept the heavy cruisers *Scharnhorst* and *Gneisnau,* plus the three light cruisers *Emden, Nürnberg,* and *Leipzig,* in the China Sea, until it became evident that he was outnumbered by British and Japanese ships. After Japan declared war on Germany (August 22, 1914), he headed across the Pacific to South America, shelling the French base at Papeete, Tahiti, en route (September 22). On November 1, 1914, at Coronel, off Chile, he defeated British admiral Sir Christopher Cradock, sinking the cruisers *Good Hope* and *Monmouth* with all hands. After rounding Cape Horn, he was met by Admiral Doveton Sturdee with the battle cruisers *Invincible* and *Inflexible* and the light cruisers *Kent, Cornwall,* and *Glasgow.* Von Spee tried to make a run for it, but he was chased and caught by the British on December 8, 1914, who succeeded in sinking the *Scharnhorst,* with 765 men (including von Spee), the *Gneisnau,* the *Nürnberg,* and *Leipzig.* The pocket battleship *Graf Spee*, scuttled by her crew in Montevideo harbor in 1939, was named for him.

See also Coronel, Battle of; Falkland Islands, Battle of the

sperm whale (*Physeter macrocephalus*) Although this is probably the most familiar of all the great whales (Moby Dick was a sperm whale), it is still one of the least known of all large animals. Bull sperm whales can be 60 feet long and weigh as many tons, but females do not exceed 45 feet. *P. macrocephalus* is the only great whale with a single nostril at the front of its head; all the others have paired blowholes located much farther back. In the sperm whale's head there is a huge reservoir of clear oil that may be used to focus and resonate sounds. We do not know what this animal does with its 20-pound brain—probably the largest of any animal that ever lived—but it probably has to do with sound production for hunting and communication. Sperm whales are among the deep-diving champions of the mammalian world, able to dive to 10,000 feet and hold their breath for an hour and a half. They feed primarily on squid, but it is not known how they are able to catch enough of these elusive cephalopods to maintain their high metabolism. Ambergris, a peatlike, waxy substance, forms in the intestinal tract of some sperm whales, but we do not know why. In groups that can number two to fifty animals, sperm whales swim in all the world's oceans, but only adult males are found in the high polar latitudes. Although they were heavily hunted during the eighteenth, nineteenth, and twentieth centuries, sperm whales are now protected throughout the world, and the population is believed to be more than 1 million. **See also ambergris, pygmy sperm whale, sperm whaling**

sperm whale

sperm whaling It is said that the first Nantucket sperm whalers came across a pod of sperm whales while hunting right whales, but whatever the impetus, sperm whaling began in New England around 1715. Soon Yankee whalers were scouring the oceans for the "spermacetty whale," the source of the finest oil for lubrication and lighting. From small boats lowered from square-rigged whaling ships, the whales were harpooned, killed, and then flensed out alongside the ship. The blubber was boiled down in brick tryworks on deck, and then casked for transport to the home ports of New Bedford, Nantucket, Sag Harbor, Mystic, and numerous other New England and Long Island towns. The best source of information on the sperm whale fishery is still Herman Melville's 1851 novel, *Moby-Dick.* During the height of the New England sperm whale fishery—say, from 1830 to 1870—the whalers probably did not kill more than 15,000 whales. In contrast, during the 1960s, when Soviet and Japanese factory ships were operating in the North Pacific, they were taking 25,000 sperm whales *every year.* **See also whaling**

spider crab (*Macrocheira kaempferi*) One of several species of spider crabs (family Majidae), the Japanese giant spider crab is the largest of the family and the largest of the arthropods (joint-legged animals) as well. Slow-moving like most spider crabs, *Macrocheira* lives off the southeastern coast of Japan, in waters down to 175 feet. The biggest have been measured at 16 feet from the tip of one outstretched claw to another. But with a body that is only 18 inches across, it is mostly spindly arms, and a crab this size would weigh only about 40 pounds. Spider crabs often decorate their shells with algae, sponges, and other organisms, which they affix with a sticky secretion from the mouth. The kelp crab (*Pugettia producta*), a spider crab of the kelp beds of California, is only ½ inch wide. **See also crab**

spinnaker A three-cornered lightweight sail that is normally set forward of a yacht's mast, with or without a boom to increase sail area while running before the wind. The name is said to be derived from "spinxer," a word coined in 1866 because the sail was first introduced by the British yacht *Sphinx.* In recent years, with the introduction of newer, stronger fabrics, spinnakers have become much larger and much more colorful.

spinner dolphins

spinner dolphin (*Stenella longirostris*) This slender, graceful little dolphin is named for its habit of leaping from the water and spinning on its long axis, sometimes as many as six times before splashing down. The reason for this behavior is unknown (males, females, and juveniles do it), but it is thought to serve some communication function. Spinners are characterized by a long narrow snout and a perky, vertical dorsal fin that hooks so far forward in some forms that it looks as if it has been put on backward. They are about 7½ feet long and do not weigh more than 150 pounds. A tropical species, spinners are also caught in tuna nets, but not in such large numbers as the spotter dolphins. Mixed schools of spinners and spotters are sometimes encountered, and they are often trapped together in the nets.

See also spotter dolphin, tuna-porpoise problem

spinner shark (*Carcharhinus brevipinna*) Like the spinner dolphin, this shark has a habit of leaping from the water and rotating on its long axis. Where the dolphins are believed to spin so they can reenter the water with a greater sound and thereby communicate their position to other dolphins, the shark spins as it is feeding below the surface and continues right out of the water. Spinners are large (to 9 feet) gray sharks with black fin tips, small eyes, and a long pointed snout. They are found in the inshore waters of every temperate ocean except the eastern Pacific. Because of its coloration, it is easy to confuse this species with the blacktip shark (*C. limbatus*), but the spinner has much smaller eyes and smaller teeth. It feeds on small fishes and is not considered dangerous to man.

See also blacktip shark, carcharhinid sharks

spiny dogfish (*Squalus acanthias*) The common spiny dogfish has a spine in front of each of the dorsal fins that is attached to a venom gland and can cause a nasty wound. Spiny dogfishes, also known as piked dogfishes, are abundant in the temperate waters of the North Atlantic and Pacific, southern South America, South Africa, Australia, and New Zealand. They form very large schools, sometimes numbering in the thousands; they might be the most abundant of all living shark species. The spiny dogfish can reach a maximum length of 5 feet, but most are smaller. They are opportunistic feeders, sometimes feeding on fish trapped in nets, which makes them extremely unpopular with fishermen. They are also unpopular with biology students who have to dissect them in comparative anatomy classes. They are not a popular food item in America, but in England the flesh is marketed as "greyfish" or "flake" and is widely used as the "fish" in fish and chips.

See also dogfish, smoothhound sharks

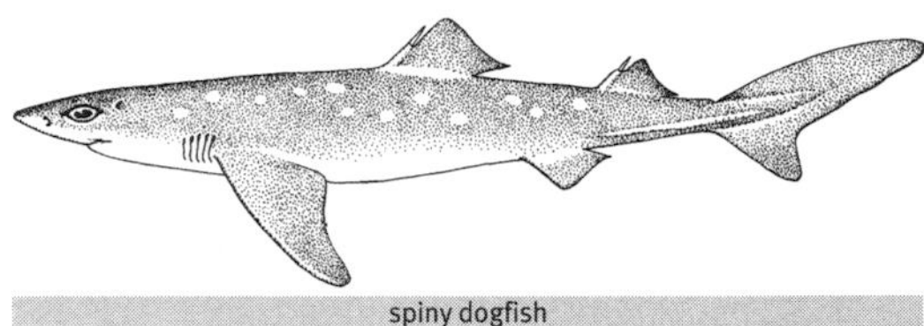
spiny dogfish

spiny lobster (*Palinurus argus*) Decapod crustacean with a spiny carapace, a broad tail fin, large antennae, and no large claws. Spiny lobsters are inhabitants of subtropical and warm temperate waters: *P. argus* from Bermuda to the West Indies, and the California version (*P. interruptus*) from the Pacific coast of North America. The Australian rock lobster (*P. ornatus*), and the South African rock lobster (*Jasus lalandei*) are popular fare in their respective countries, but the abdomen meat of the South African species is marketed around the world as "lobster tail." Because they lack the powerful claws of the American lobster, spiny lobsters are often collected by divers and spearfishermen, who refer to them as "bugs." **See also lobster**

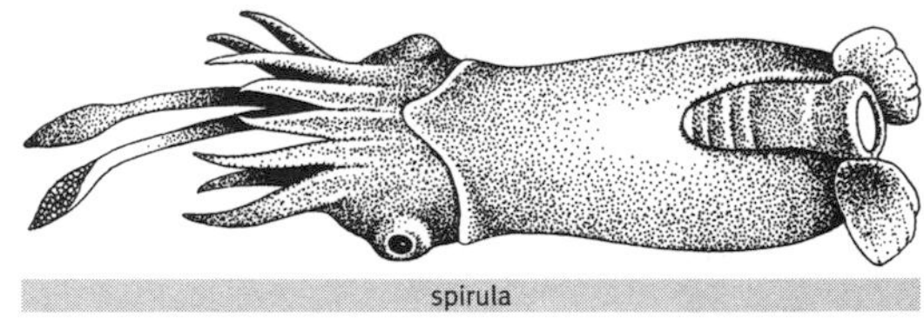
spirula

spirula (*Spirula spirula*) A small cephalopod that is neither squid nor cuttlefish, spirula is a true pelagic species, occurring worldwide in tropical waters, usually at depths of 300 to 1,600 feet, but it has been collected at 3,000 feet. It is believed to be the only descendant of the ammonites, extinct by the Cretaceous period, whose coiled shells reached 10 feet in diameter. About 3 inches long, adults are whitish in color with rust-red markings. They move about in large schools, swimming in head-down position, with their tiny fins and light organ uppermost. Spirula is unique in having an interior, spiral shell shaped like a ram's horn, with a series of gas-filled chambers. These shells are very buoyant, and they are often found washed ashore in great numbers. **See also ammonite, cephalopod**

Spitsbergen A Norwegian island group (also known as Svalbard, which means "cold coast") consisting of Spitsbergen, North East Land, Edge Island, Barents Island, Prince Charles Foreland, and several smaller islands. The Dutch explorers Willem Barents and Jacob van Heemskerck discovered the islands in June 1596 and reported great numbers of whales in the icy waters. First the Dutch sent whalers to these remote islands, and by 1611, the British did the same; they were followed by French, Norwegian, Swedish, and Danish ships. The rivalry for bowhead whales—known then as polar whales, or the Mysticetus—was fierce, and after several failed attempts to have crews overwinter there to be ready to hunt as soon as the ice began to break up, the Dutch established Smeerenburg ("blubbertown") on Amsterdam Island, a community devoted entirely to the processing of whales. Since the decline of whaling around 1800—the whales were completely eliminated—the islands have been the site of coal mines and weather stations.

sponge (phylum Porifera) There are some five thousand species of sponges; all are marine except for about twenty freshwater species. It is still not clear if an individual sponge is one animal or a colony. However they are defined, they are the most primitive of multicellular animals, and they lack organs or organ systems. Inside the central water-filled chamber (the spongocoel), there are whiplike structures that create currents to draw in water. These currents bring planktonic and detritic food in through numerous perforations, where a jellylike substance (the mesoglea) contains free-moving cells, called amoebocytes, that absorb food particles and oxygen from the water, and eject water and waste products through the oscula, or large openings. The amoebocytes also produce spherical eggs that swim by means of cilia until they find a suitable site for attachment. Almost all sponges have a skeleton of microscopic glassy spicules, whose form and arrangement are diagnostic. Sponges live in all seas and at all depths, from the intertidal zones to the abyss. The glass sponges (*Hexactinella*) live only in very deep water; the Calcarea have limy (calcaraeous) spicules, and the Demospongiae are the more familiar, shallow-water species. In these, the skeleton is composed of spongin, a flexible, plasticlike material. Sponges come in all colors and all shapes and sizes. There are organ-pipe sponges, tubular sponges, finger sponges, elephant's ear sponges, egg sponges, beard sponges, boring sponges, fig sponges, and a glass sponge (*Euplectella*) from the deep Pacific that is commonly known as Venus's flower basket. A living sponge is not at all like the object we use in the bath, which is, in fact, the skeleton of the sponge. Although living sponges are still commercially collected in the Mediterranean and Florida, most kitchen and bath sponges nowadays are synthetic.

sponge crab (*Dromia erythropus*) A family of primitive crabs that camouflage themselves with pieces that they cut from living sponges. Some carry the pieces around, while others settle into the hole they have made, pulling the piece back over them. One member of the family does the same thing with a bivalve shell, holding it over its body with its fifth pair of thoracic legs, which are bent over the back. **See also crab**

spotter dolphin

spotter dolphin (*Stenella attenuata*) The spotter is found throughout the tropics, often in huge schools. For reasons that are still poorly understood, the dolphins and the tuna swim together and are therefore trapped together in the nets used to catch tuna. Prior to 1972, millions of dolphins were drowned in the nets, but since 1982, the kill quota for U.S. fishermen has been 20,500 annually. Foreign fishing fleets often ignore restrictions, and continue to trap and kill dolphins in their nets. The spotter is one of the fastest of all dolphins, having been clocked in excess of 20 mph over a measured course. There are now thought to be several subspecific populations of spotters, discovered only when scientists were able to examine and compare specimens caught in the tuna nets.
See also purse seining, spinner dolphin, tuna-porpoise problem

Spratly Islands Uninhabited reefs in the South China Sea, located midway between the Philippines and Vietnam. Itu Aba is the largest at 90 acres; Spratly Island itself is 1,500 feet long. France held the islands from 1933 to 1939, but Japan occupied them during the war and used them as a submarine base. After the war, China established a garrison on Itu Aba, which was claimed by the Chinese Nationalists after they had fled the mainland. South Vietnam occupied the islands in the 1970s; then Taiwanese troops moved onto Itu Aba. When the

Exxon Corporation signed a $35 billion contract with Indonesia to explore and develop the oil reserves in the Natuna area, south of the Spratlys in the South China Sea, various countries reconfigured their exclusive economic zones in an attempt to include these islands. Many of the Spratly Islands are now claimed by Vietnam, China, Taiwan, Malaysia, and the Philippines and manned by small military detachments.

Springer, Stewart (1906–1991) With no formal training, Stewart Springer taught himself to be a shark biologist, and went on to become one of the most influential and respected scientists of his time. He began as a collector of sharks for laboratories and industry, and gradually moved up through the ranks until he was recognized as an expert. In 1942 he worked on the development of "Shark Chaser," a compound that was supposed to protect U.S. Navy fliers that were downed at sea during World War II, and also wrote many of the survival manuals that were used. After a stint with Shark Industries, a company that caught sharks for the vitamin A in their livers, Springer went to work for the U.S. Fish and Wildlife Service, and in 1958, he was named to the Shark Research Panel, sponsored by the Office of Naval Research and the American Institute of Biological Sciences, which created the original Shark Attack File. He retired in 1970, but remained a research associate at the Mote Marine Laboratory in Sarasota, Florida. As an indication of the respect with which he was held, his fellow scientists gave the name *springeri* to a snipe eel, a pipefish, a crab, a mollusk, and a stingray.

See also Shark Attack File, shark repellent

Spruance, Raymond (1886–1969) American admiral, a 1908 graduate of the U.S. Naval Academy, and a member of the Naval War College staff from 1931 to 1933 and again from 1935 to 1938. He served as second-in-command in task force operations in the Marshalls and Wake Island in 1942, but he distinguished himself by his tactics and strategy during the Battle of Midway (June 3–6, 1942) and the Battle of the Philippine Sea (June 19–20, 1944). At Midway, he commanded the aircraft carriers *Enterprise* and *Hornet*, six cruisers, and nine destroyers in the decisive battle against Admiral Yamamoto. Spruance was named chief of staff to Admiral Chester Nimitz in June 1942, and was named deputy commander-in-chief of the Pacific Fleet. He was in charge of the occupation of the Gilbert Islands and the invasion of the Marshall Islands in 1943. He commanded the forces employed in the capture of Saipan, Guam, and Tinian in the Marianas, battles that led to the Japanese realization that they could not win the war in the Pacific. Further successes were the operations involving Iwo Jima and Okinawa. In 1945, Spruance succeeded Nimitz as commander-in-chief of the Pacific Fleet, and at the war's end he was named president of the Naval War College. He was ambassador to the Philippines from 1952 to 1955.

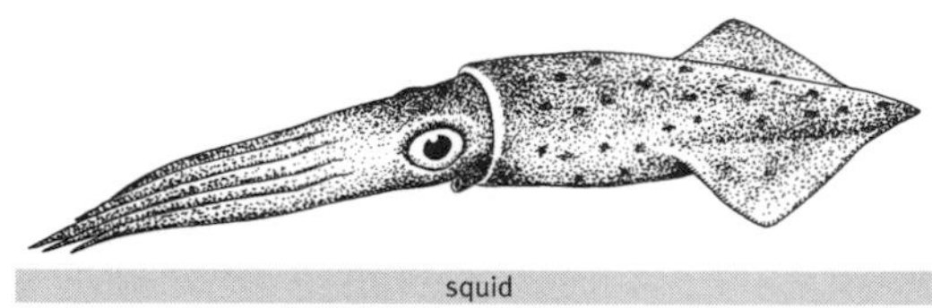

squid

squid Cephalopods ("head-feet") that are characterized by eight grasping arms and two longer tentacles. The arms and tentacles are equipped with suckers, and in some species the suckers are equipped with claws. They have elongated, tubular bodies that are strengthened by an internal, flattened, plume-shaped shell, known as the pen or gladius. They range in size from less than ¾ of an inch to 57 feet. All squid are equipped with chromatophores, which enable them to change color with astonishing speed, and many are also equipped with photophores, which means they can light up. One common defense mechanism is the release of a cloud of ink in the shape of the squid (a "pseudomorph") that attracts the predator and enables the squid to dart away. Squid move by a sort of "jet propulsion," where they take water into the body cavity and then eject it forcibly from a funnel or siphon that they can move in any direction. There are more than seven hundred known species of squid, and they are so numerous that they serve as the primary food source for many of the ocean's predators, such as beaked whales, dolphins, fishes, sharks, and seabirds. In addition, squid are being caught in ever-increasing quantities to feed humans.

See also *individual species;* chambered nautilus, cuttlefish, octopus, squid fishing

squid fishing With many species being caught in many locations, squid in general constitute one of the world's most important fisheries. The Japanese lead the way, catching mostly the common squid (also known as the flying or red squid), *Todarodes pacificus,* in vast quantities. (In 1996 the Japanese catch of this species was more than half a million tons.) Throughout the world's oceans, the other species heavily fished are the long-finned squid (*Loligo pealei*), the California market squid (*L. opalescens*), the short-finned squid (*Illex illecebrosus*), the Argentine squid (*I. argentinus*), and the Humboldt squid (*Dosidicus gigas*). Commercial boats employ lights—underwater and topside—to attract the squids, and "jigs" to catch them. Until about 1950, handlines were used, but then the Japanese, the leaders in squid-fishing technology, developed automated jigging machines, which deploy numerous jigs concurrently. The jig itself is a spindle-shaped lure with two or three rows of barbless hooks in a ring around the end away from the eye with which it is fastened to the line. Either by hand or by machine, the lures are "jiggled" in the water to attract the squid.

See also Humboldt squid, Japanese flying squid, long-finned squid, short-finned squid

squirrelfish (family Holocentridae) During the daylight hours, squirrelfishes stay hidden in cracks and crevices in the coral, coming out only at night to feed. As befits nocturnal fishes, they have particularly large eyes. Many, but not all, of the hundred-odd species are red in color, and all squirrelfishes have a long spine on the gill cover, while the closely related soldierfishes (Myripristidae) do not. In the squirrelfishes, this spine is venomous. The various species are found throughout the world, in warm shallow waters. They are highly regarded as food fishes, but because they are small, spiny, and difficult to handle, they have little commercial value. **See also bigeye, cardinal fishes, soldierfish**

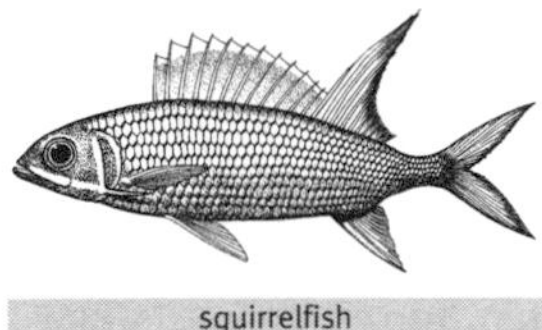
squirrelfish

Sri Lanka Island republic southeast of the southern tip of India, from which it is separated by the 20-mile-wide Palk Strait. The ancient Greeks knew the island as Taprobane, and the Arabs called it Serendib. The pear-shaped island covers 25,332 square miles and has a population of 18.5 million. The capital is Colombo, and other major cities are Batticaloa, Kurunegala, Kandy, Trincomalee, and Jaffna. Lourenço de Almeida (son of Francisco de Almeida, the first viceroy of Portuguese India) landed there in 1505, and by 1619 the Portuguese controlled most of the island. The Kandyans enlisted the help of the Dutch to oust the Portuguese, and Ceylon, as it was then known, remained under the control of the Dutch East India Company until 1802, when the island became a British Crown Colony. Under British rule, Ceylonese plantations grew coffee, tea, and rubber on the island's rich and fertile soil. The island is also famous for its gemstones, particularly rubies and sapphires. It is the world leader in the production of black tea, cinnamon, and graphite. After several constitutions, Ceylon was granted independence in 1948; its name was changed to Sri Lanka ("resplendent isle") in 1972. Sinhalese make up about three-quarters of the population, but the Tamils, mostly in the north, have been fighting for a separate homeland since 1981, and violent confrontations, guerrilla warfare, and civilian bombings continue, despite efforts by the United Nations and India to bring the warring factions to the peace table. **See also Dutch East India Company**

starboard The right side of a vessel as the viewer faces the bow. It is probably a corruption of "steer-board," the board or oar that projected into the sea from this side, and that was used to steer the vessel before the invention of the hanging rudder. Its opposite used to be "larboard," but this was changed to "port" because of the possible confusion when the two words were said or shouted. **See also port (1)**

starfish Also known as sea stars, these echinoderms most commonly have five arms, but there are species with ten to twenty arms, and the sun star (*Heliaster*) might have as many as fifty. Starfishes are found throughout the world's oceans, from the shallows and tidepools to depths of 20,000 feet. The largest starfish (*Pisaster brevispinus* of the U.S. Pacific Northwest) can be more than 2 feet across. The mouth is on the underside, and when the animals feed—they are efficient predators of scallops, clams, mussels, oysters, and snails—they force open the shells with constant pressure, and then evert their stomachs to envelop the prey, bringing their stomachs to the food instead of vice versa. (More primitive species feed by sweeping organic particles into their mouths, which are on the underside of their bodies.) Having no gills, starfishes breathe through special structures on the skin. Most can regenerate a new arm if one is lost, and there are some species that can grow a whole new animal from a small piece of arm.

See also brittle star, crown-of-thorns starfish, echinoderms

ten-armed sea star

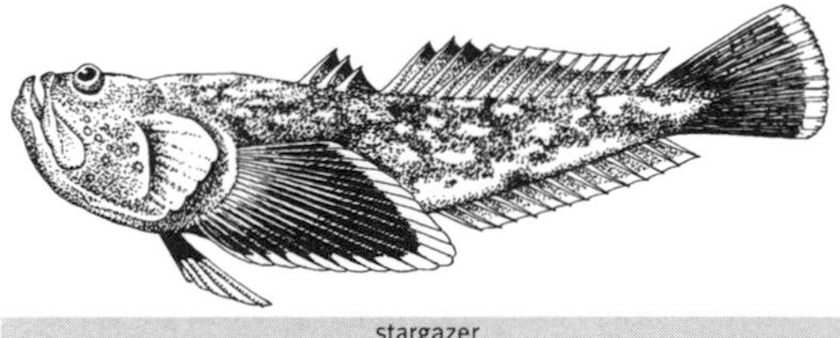
stargazer

stargazer (*Uranoscopus scaber*) Found in the Mediterranean and the eastern North Atlantic from Spain to Senegal, the stargazer is a 10-inch-long fish with a big head, a nearly vertical mouth, and eyes on the top of its head that seem to be staring upward. A leafy appendage on the tip of the lower jaw can be wiggled to attract prey fishes. It is equipped with powerfully venomous spines that protrude from each gill cover and, behind each eye, a electrical organ that can impart an unpleasant shock to a person handling the fish out of the water. There are similar species in Asian waters and also in the tropical Indo-Pacific.

See also scorpionfish, stonefish, weever

Staten Island (1) The least densely populated of the five boroughs of New York City, Staten Island (formerly known as the borough of Richmond) is home to more than 400,000 people, only 5 percent of the people of New York. The diamond-shaped island is 15½ miles long and 7 miles wide; it covers some 64 square miles. It is separated from New Jersey by the Arthur Kill, which is spanned by two bridges: the Outerbridge Crossing and the Goethals Bridge; and by the Kill Van Kull, which is spanned by the Bayonne Bridge. Staten Island can be reached by ferry from lower Manhattan, and from Brooklyn by the Verrazano-Narrows Bridge, the longest suspension bridge in the world when completed in 1964. In 1524, Giovanni da Verrazano was the first European to visit the island. Henry Hudson dropped anchor there in 1609 and named it Staaten Eylandt, for the States General of the Netherlands. The first Dutch settlement was in 1661, and for many years the island remained culturally separated from the rest of the city. During the War of Independence, Staten Island was occupied by the British and served as an important staging area for troops preparing for the Battle of Long Island in 1776. Compared to the rest of New York City, Staten Island is less industrialized and more rustic. Dissatisfied with their generally uncomfortable relationship with New York City, in 1993 a majority of the residents of Staten Island proposed secession and the establishment of an independent city. The idea was rejected by city officials, but it has been repeatedly reintroduced.

See also Long Island, Manhattan

Staten Island (2) Discovered by Willem Schouten and Jakob Le Maire on their 1616 voyage in the *Hoorn* (an attempt to discover a way around the tip of South America), this Staten Island was also named for the governing body of the Netherlands. Staten Island lies off the toe of the island of Tierra del Fuego at the southern tip of South America. Separated from Tierra del Fuego by the Le Maire Strait, it is a mountainous island, about 50 miles long, shaped by many bays and peninsulas. Numerous sailing ships attempting to round the Horn were wrecked off this island because of the strong currents, powerful westerly winds, and perpetual cloud cover. All attempts to populate the barren island have failed, and even a prison was transferred from Staten Island to Ushuaia. Staten Island's only edifice is an unmanned lighthouse.

See also Le Maire, Tierra del Fuego

steelhead (*Oncorhynchus mykiss*) Like the salmons, the steelhead is an anadromous, migratory fish that is born in freshwater, spends its adult life in the sea, and returns to the place of its birth to spawn and die. It is actually a rainbow trout that goes to sea and in the process grows considerably larger than its landlocked relative. (Rainbow trout belong to the same species, *O. mykiss.*) The record steelhead (which is also the record rainbow) was a 42-pounder caught in Alaska in 1970. Steelheads are a favorite object of fly fishermen in northern California, Oregon, Washington, British Columbia, and Alaska. They enter the North Pacific from Alaska and Japan, and are heavily fished by Japanese gillnetters. The common name in English is simply "steelhead"—not steelhead trout or steelhead salmon. The confusion between trout and salmon is nowhere more evident than here: rainbow trout belong to the same genus (*Oncorhynchus*) as the Pacific salmon, while the Atlantic salmon belongs to the same genus (*Salmo*) as the brown trout. (Other trouts—all freshwater fishes—are classified in the genus *Salvelinus.*)

See also Atlantic salmon, Pacific salmon

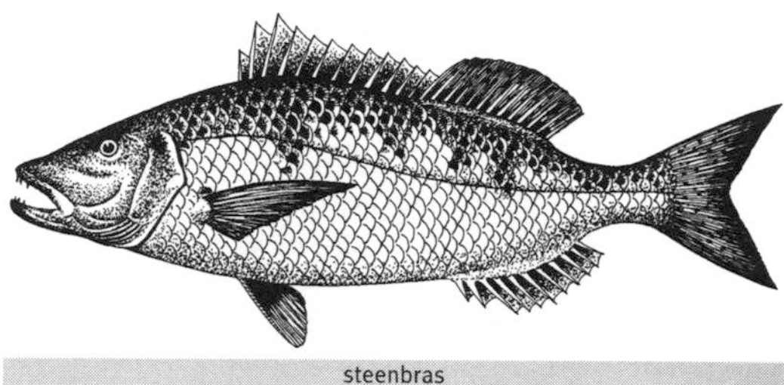

steenbras

steenbras (*Petrus rupestris*) One of the largest of the sea breams (Sparidae), the red steenbras is a powerful, predatory fish that is found in the deeper inshore waters of eastern South Africa. It is a bronzy red in color, often with yellow undersides. At a length of 6 feet and a weight of 150 pounds, the steenbras is one of South Africa's premier game fishes. Spearfishermen also seek out this species, and there are undocumented reports of attacks on divers. (These seem unlikely, because the normal diet of the steenbras consists of small fishes, octopuses, and squid.) The flesh makes excellent eating, but the liver is poisonously high in vitamin A and causes the hair of those who eat it to fall out.

See also Sparidae

Stejneger's beaked whale (*Mesoplodon stejnegeri*) Pronounced "*sty*-ne-ger's," this species was named for Leonhard Stejneger, who found the first skull on Bering Island in 1883. (Stejneger was also Georg Steller's biographer.) The "tusks" of this saber-toothed beaked whale may be more than 5 inches in height, and as with all *Mesoplodon* species, they appear only in the males. To date, it has been recorded only from the North Pa-

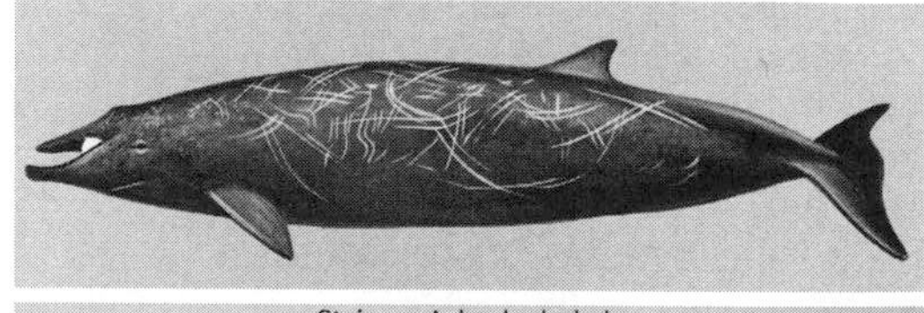

Stejneger's beaked whale

cific, with the majority of the strandings from the Aleutian Islands. **See also beaked whales, *Mesoplodon***

Steller, Georg Wilhelm (1709–1746) German-born naturalist and mineralogist who accompanied Vitus Bering on his epochal 1741–1742 voyage from Kamchatka to the Aleutian Islands and Alaska. Bering died when their ships were wrecked in a storm in the western Aleutians in 1742, but the survivors built a new vessel from the remains of the wrecked ones and sailed back to Kamchatka. Steller lived to write up his observations of the voyage (which were published posthumously), including the first scientific description of the sea otter, the fur seal, the northern jay, a sea lion, and a sea cow—the last three of which were named for him. Steller's eider, a seagoing duck, was also named for him. He was probably the first zoologist to identify the different species of Pacific salmon. His description of a "sea monkey" remains a cryptozoological mystery to this day.
See also Bering, Steller's sea cow, Steller's sea lion, Steller's sea monkey

Steller's eider (*Polysticta stelleri*) The smallest of the eiders, Steller's is a diving duck that feeds on mollusks and crustaceans by diving to the bottom in fairly shallow water. It lives and breeds in the Arctic, from Siberia to Alaska. The world population, historically around half a million, has been reduced to perhaps 200,000, largely because of loss of habitat and expansion of human habitation, particularly around Point Barrow, Alaska. **See also eider, Steller**

Steller's sea cow (*Hydrodamalis gigas*) A relative of the dugong and the manatee, this huge and ungainly sirenian is now extinct. It was discovered in the Aleutians by the shipwrecked sailors on Bering's 1741–1742 voyage. Bering died there, but those sailors who survived ate the meat of the sea cows. By 1769, only twenty-seven years after its discovery, the sea cow had been completely eliminated, killed off for food and leather by the Russian sealers who stopped off at these islands. It probably reached a length of 30 feet, and may have weighed as much as 4 tons. Its skin was thick, dark, and wrinkled, and its forelimbs were attenuated, lacking fingers, nails, or hooves. It fed on seaweed that it crushed between the bony plates in its otherwise toothless jaws. There is some speculation that it lived and fed in the shallows because it was too buoyant to dive.
See also dugong, manatee, sirenians, Steller

Steller's sea eagle (*Haliaetus pelagicus*) Found above and around the inshore waters of northeastern Asia, Steller's sea eagle is a large, black bird, with snow-white shoulder bands, white thighs, and white tail. The beak, much the heaviest of any sea eagle's, is slate-gray in juveniles but turns bright yellow in adults. A dark relative, which completely lacks the white shoulder patches, occurs only in Korea.
See also bald eagle, Steller, white-tailed sea eagle

Steller's sea lion (*Eumetopias jubatus*) The largest of all the eared seals (Otariidae). As with others of this family, the males attain a much greater size than the females: bulls can weigh as much as a ton, while females do not exceed 700 pounds. Found only in the North Pacific, Steller's sea lion breeds in the Pribilofs, the Aleutians, and the coast of North America as far south as northern California. Extremely wary of people, these animals are difficult to approach on land. Until the 1980s the estimated population was some 300,000 animals, but in recent years, for reasons that

Steller's sea lion

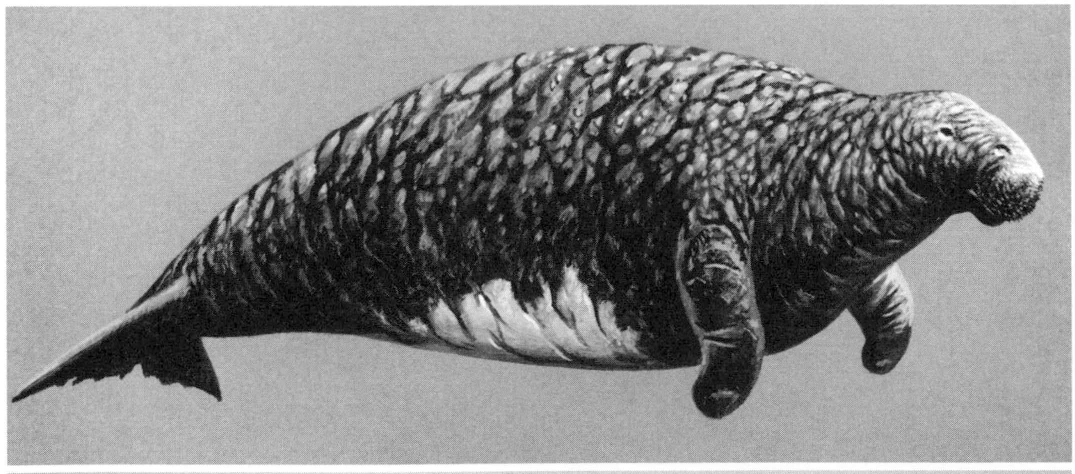
Steller's sea cow

are not understood, the population has declined at an alarming rate. As of 1997, the National Marine Fisheries Service has listed Steller's sea lion as "endangered."

See also California sea lion, northern fur seal, Steller

Steller's sea monkey In August 1741, while participating in Vitus Bering's Great Northern Expedition, Georg Wilhelm Steller observed an animal that was about 6 feet long, with erect ears on a doglike head. On the upper and lower lips, it had whiskers that hung down, "which made it look almost like a Chinaman." It was swimming so close to his boat that Steller said he could have "touched it with a pole." He did not see any front flippers, but he noticed a vertical tail with a large upper lobe, like that of a shark. He later wrote that its behavior reminded him of that of the "Danish sea monkey" (*Simia marina danica*) that had been described in Konrad Gesner's 1587 *Historia animalium.* Steller tried to collect it by shooting it, but he missed. An experienced observer, he is not likely to have confused this creature with a sea otter or a fur seal, both of which he saw shortly thereafter, and to this day, the identity of the sea monkey remains a mystery. (Leonhard Stejneger, Steller's biographer, believed that it was a fur seal, which Steller had not seen up to that time.)

See also northern fur seal, sea otter, Steller

stern slipway A 1925 invention that completely changed the nature of industrial whaling. Previously, whale carcasses were processed onshore or alongside the whaleship, but it was obvious that if the whale could be brought onto the deck, it could be reduced to its component parts much more efficiently. The first (completely impractical) solution was a slipway in the bow, which meant that the ship could not be under way when the whale was being hauled aboard. While the stern seemed an obvious solution, a way had to be devised to circumvent the steering gear and rudder, which were usually located in that position. Norwegian whaleman Petter Sørlle designed a system where the steering mechanism would be split in two, leaving the middle available for the slipway. From 1926 onward, all whaling factory ships were equipped with a gaping hole in the stern, capacious enough for a 100-ton blue whale. **See also catcher boat, factory ship**

stinging corals (*Millepora* spp.) Also known as fire corals, these are true corals that have nematocysts (stinging cells) like those of anemones, jellyfishes, and the Portuguese man-of-war, whose sting can be extremely painful. Stinging corals secrete limestone skeletons and the colonies are similar in appearance to other, harmless, branched corals, so the safest thing to do is avoid touching all corals. Both divers and corals will benefit from this policy because the polyps are fragile and easily damaged by handling. **See also nematocyst**

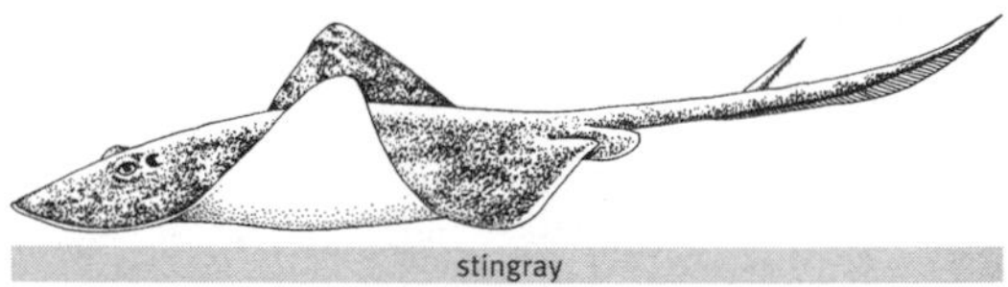

stingray

stingray (family Dasyatidae) Stingrays are flattened, cartilaginous fishes that lie on the bottom when they aren't swimming gracefully. When caught or stepped on, a stingray lashes its tail, trying to impale the offender with one of its spines. The spines are equipped with poison glands. The Atlantic stingray (*Dasyatis sabina*) measures about a foot across its wings, while the giant stingaree of Australia and Hawaii (*Plesiobatis daviesi*) can be 7 feet across. All rays give birth to live young. There are some stingrays that live exclusively in freshwater, in the rivers of southeastern Asia, Africa, and South America. **See also eagle ray, manta**

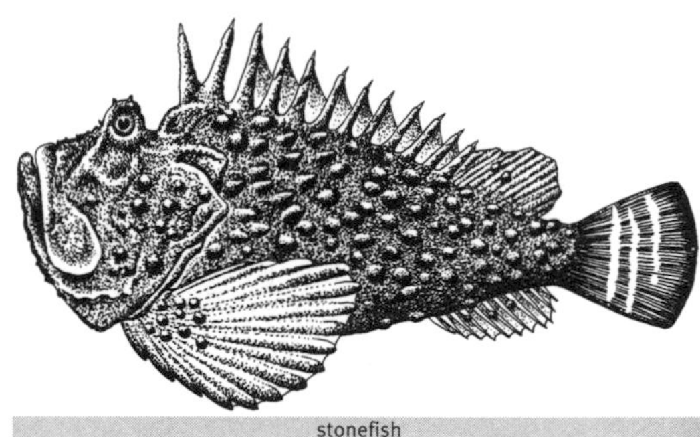

stonefish

stonefish (*Synanceja horrida*) The deadliest fish in the world, the stonefish has venom glands located near the base of its hypodermiclike dorsal spines. The species closely resembles a small rock as it lies half buried in the sand. A person accidentally stepping on this fish is injected with the venom, which can be fatal in less than two hours. There are several species of stonefishes found throughout temperate and tropical waters of the world. **See also scorpionfish**

storm petrel (family Hydrobatidae) The smallest of the tube-nosed birds, the swallow-sized storm petrels are a diverse group of perhaps twenty species, which are extremely difficult to differentiate at sea. They all have slender bills and legs that seem too long for their bodies. Found throughout the world, most are dark brown or black with a conspicuous white rump patch. Some species, like Wilson's, are migratory, while others, such as the Galápagos, remain in a particular region. Along with the albatrosses, shearwaters, and diving petrels,

storm petrel

storm petrels are tubenoses. Their nostrils extend on top of the bill through two horny tubes. The common name comes from the old sailor's belief that they appeared before a storm; they are called petrels for their habit of "walking" on the water, as Saint Peter is said to have done. (*Petrello* is "little Peter" in Italian.) Sailors once called them "Mother Carey's chickens," from the Latin *mater cara,* which means "dear mother" and refers to the Virgin Mary, the guardian of all seamen.

See also albatross, British storm petrel, diving petrel, Leach's storm petrel, petrel, shearwater, Wilson's storm petrel

storm surge A rise in the normal water level in a particular part of the ocean, caused by a combination of high winds and low pressure. Storm surges take the form of onrushing walls of water and can occur over hundreds of miles of coastline and last for several hours. Typically, storm surges can cause the water to be 10 feet deeper than normal, and if one peaks at high tide, extensive flooding can result. An intense surge in 1953 breached many of the dikes of the Netherlands, flooded about 15,000 square miles of land, forced 600,000 from their homes, and drowned 2,000 people. In conjunction with a typhoon that roared up the Bay of Bengal in 1737, a 40-foot storm surge flooded the city of Calcutta and killed 300,000. In 1876, in the part of India that is now Bangladesh, the ebbing Meghna River was blocked by a storm surge, and 28,000 square miles of the delta were flooded, drowning more than 100,000.

See also Bengal, Bay of; hurricane; typhoon

stranding At one time or another, every species of cetacean, from the largest blue whale to the smallest harbor porpoise, has stranded on a beach and died. Some strand singly, others in groups, usually depending upon the relative gregariousness of the species: cetaceans that form large groups are likely to strand en masse, those that swim singly or in pairs normally strand likewise. The most notorious group-stranders are toothed whales (odontocetes) such as sperm whales, pilot whales, and false killer whales. Most of the eighteen species of beaked whales (which are odontocetes) are known only from stranded specimens. Various explanations have been offered for this phenomenon, including parasites in the brain; a malfunction of the whales' navigational systems; being chased by predators; a protest against whaling; a memory of where water used to be; a migraine headache; and even suicide. When the "leader" of a group of whales or dolphins strands, the rest of the group sometimes follows him or her onto the beach. Attempts to put them back in the water often result in the whales beaching themselves again in another location. In his *Historia animalium,* Aristotle wrote, "It is not known for what reason they run themselves aground on dry land; it is said that they do so at times and for no obvious reason." He said that some twenty-three centuries ago, and we are no closer to solving the mystery than he was.

See also beaked whales, false killer whale, pilot whale, sperm whale

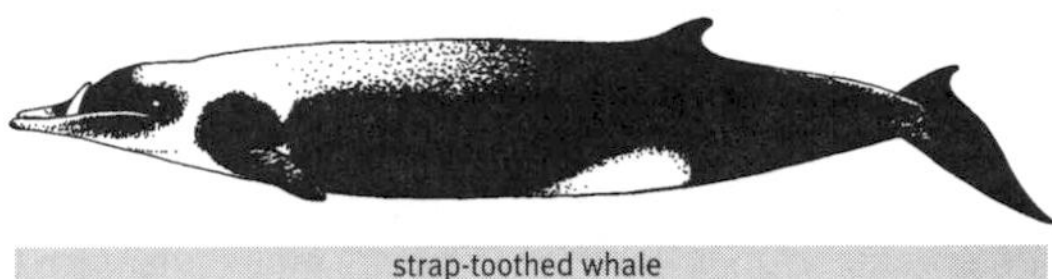
strap-toothed whale

strap-toothed whale (*Mesoplodon layardii*) In the juveniles of this species, the teeth look like those of any other beaked whale, but as the males mature, the teeth grow up and arch over the upper jaw until they form an arch that limits the opening of the mouth. How an animal can eat if it cannot open its mouth very wide—if at all—is another of the mysteries surrounding the beaked whales. This is a Southern Hemisphere species, having stranded in New Zealand, Tasmania, South Africa, and Uruguay. **See also beaked whales, *Mesoplodon***

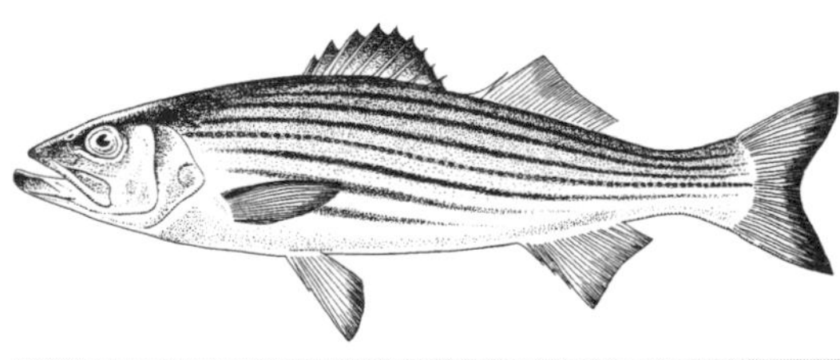
striped bass

striped bass (*Morone saxatilis*) A favorite fish of surf casters, whose catch often outweighs that of commercial fishermen, stripers are found along the Atlantic coast from the Gulf of St. Lawrence to Florida, and in California, where the species was introduced as a sport fish in 1879. Also known as rockfish, they grow to a length of 6 feet and a weight of 78 pounds. Stripers are anadromous, which means they spawn in freshwater, then return to the sea. A 20-pound female may lay 1 million eggs, a 50-pounder 5 million. The most popular spawning areas for stripers on the east coast of the United States are the Hudson River and the smaller rivers adjacent to Chesapeake Bay. Young fishes remain in the streams until they are two years old and then swim to the ocean. When stocks declined in the 1980s, stringent management measures were adopted from Maine to Virginia to rebuild the stocks, and they are now back to the previous high levels.

See also wreckfish

striped dolphin

striped dolphin (*Stenella coeruleoalba*) From its scientific name (*coeruleoalba* can be translated as "blue-white"), one can get an idea of the coloration of this species. It is strongly marked, with a prominent pattern of dark blue stripes on a grayish or white ground. It is the largest member of the genus *Stenella* and can reach a length of 10 feet. There is some controversy about the number of valid species, but however many there are, striped dolphins are found throughout the temperate and tropical offshore waters of the world. These are the dolphins shown in the famous "dolphin frieze" of the palace of Knossos on the Aegean island of Crete.

See also spinner dolphin, spotter dolphin

striped marlin (*Tetrapterus audax*) Not nearly as massive as its larger cousins the blue and black marlins, the striper is still a very popular game fish. Found in the tropical and temperate waters of the Indian and Pacific Oceans, it can be recognized by its high first dorsal fin and of course the stripes, which are pale blue or lavender but fade with the death of the fish. When hooked, this fish takes to the air or "tailwalks" across the water. The rod-and-reel record is 494 pounds for a fish caught off New Zealand in 1986. Commercial longliners are causing severe depredations of the populations, as with most other billfishes.

See also black marlin, blue marlin, longline fishing

Stromboli Known as the "lighthouse of the Mediterranean" because of its constant eruptions, Stromboli is one of the actively volcanic islands in the Lipari group off the northeastern coast of Sicily. The group, which used to be known as the Aeolian Islands, includes Lipari and Vulcano, named for Vulcan, the Roman god of fire. The 5 square miles of Stromboli are dominated by the eponymous mountain, which has been spewing lava into the air since ancient times. It has a number of vents, but the main crater is located 600 feet below the 3,000-foot-high summit, which allows for observation in relative safety. In 1907 the mountain blew and shattered windows throughout the town; five years later, another eruption catapulted large blocks of lava into the air, some landing as far as ½ mile from the eruption. In September 1930, the mountain produced a mushroom cloud that rose 2 miles high and again sent huge lava blocks, one of which weighed 30 tons, into the air. An incandescent landslide rolled down the mountain, killing six people. Incredibly, some 350 people now live on the island, mostly catering to the tourists who come to see the volcano. At the conclusion of Jules Verne's 1864 novel *Journey to the Center of the Earth*, the travelers escape from the fiery core of the earth by riding a stream of molten lava up the shaft of Stromboli.

See also Aeolian Islands, Vulcano

sturgeon (*Acipenser* spp.) There are several species of sturgeon, primitive fishes characterized by five rows of bony plates that run from head to tail, a skeleton that is mostly cartilage, an upturned snout, and barbels before the protrusible, tubelike mouth on the underside of the head. In all species, the upper lobe of the tail is longer than the lower. They are usually anadromous (living in coastal waters and ascending streams and rivers to breed), but there are some species that spend their entire lives in lakes or rivers. Sturgeon are among the largest of the bony fishes, and the record is said to be a 24-foot-long Russian sturgeon, or beluga (*Huso huso*), that was caught in the Volga River in 1827 and weighed 3,249 pounds. This record is undocumented, and fish of 10 to 12 feet are considered large. Various species are found in North American rivers, including the Mississippi and the Hudson. They are popular food fishes in Europe, and the flesh can be eaten fresh or smoked. Sturgeon eggs are served as caviar throughout the world, but most of the world's supply comes from the Black and Caspian Seas. Although sturgeon roe can be

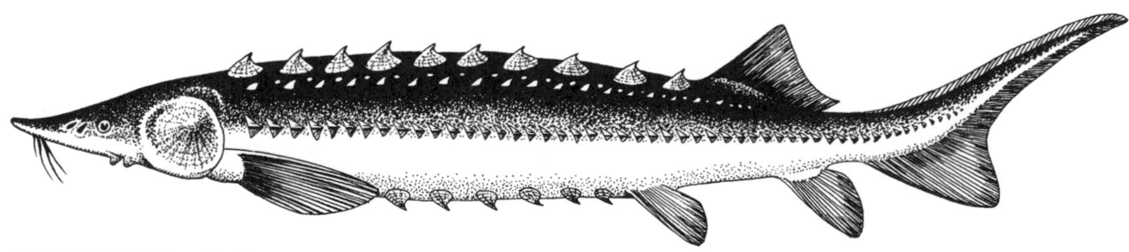
Atlantic sturgeon

stripped from a living fish, it is mostly collected from females that have been killed. A single female can produce 3 million eggs. After the eggs are removed from the egg membrane, they are lightly salted to remove the water from them, and then the brine is separated before they are canned and shipped for consumption. In addition to the slaughter of sturgeon for caviar, dams, pollution, and overfishing have greatly reduced their numbers.

submarine A naval vessel capable of operating completely submerged. In 1620, a Dutchman named Cornelis van Drebbel (1572–1633) exhibited in the Thames River a greased-leather boat that was powered downward by rowers until it was awash. The first submarine intended for military purposes was the egg-shaped, one-man *Turtle,* invented by David Bushnell (1742–1824) in 1775. In 1800, Robert Fulton developed the *Nautilus,* a submarine that could remain underwater for six hours. During the U.S. Civil War (1861–1865), the Confederate navy used several submarines named *Hunley* (after their inventor), but the only one that succeeded in blowing up a Union vessel blew itself up as well. John Holland and Simon Lake worked independently on their own designs for submarine boats, and in 1900, Holland succeeded in selling his ideas to the U.S. Navy. But it was Lake's *Argonaut* that was the first submarine to navigate in the open sea, when it traveled from Norfolk, Virginia, to New York in 1897. The "E-boats," developed in 1912, were the first diesel-powered submarines, and the first to cross the Atlantic. World War I saw the use of submarines by the Americans, the British, and the Germans, the beginning of submarine warfare as we know it. The underwater warships (the German "U-boat" stands for *unterseeboot,* or undersea boat) were armed with torpedoes, and depth charges were employed against them. By World War II, submarines and other warships were equipped with sonar, which made it possible to locate an underwater vessel. Submarines have no windows, and have to rely on sound to locate their adversaries. Only the periscope, raised above the surface, enables the crew of the submarine to see other ships. The advent of nuclear power allowed submarines to go faster and remain submerged longer. The development of missile-firing submarines meant that the vessels could now target land sites as well as ships, perhaps the most profound change in the long history of undersea warfare.

See also Fulton, Holland, Lake, *Nautilus,* submersible, U-boat

submarine canyons Fan-shaped valleys with steep walls and a V-shaped cross-section, carved into the continental slopes and elsewhere into the ocean floor. There are prominent canyons off the northeastern United States (Hudson Canyon), California (La Jolla, Monterey, Coronado), the Bahamas, Corsica, the French Riviera, the Congo, Sri Lanka, and New Zealand (Kaikoura). The longest of all are in the Bering Sea. Although many canyons are associated with rivers, no logical explanation can be offered as to how, say, the Hudson Canyon was formed by the Hudson River. Rivers may bring heavy, silt-laden waters into the river mouth and then into the ocean, but the agent that makes this carve out a submerged canyon is not apparent—unless they are simply drowned riverbeds, left over from a time when the level of the sea was much lower. (This explanation has now been rejected by geologists.) It is now believed that canyons are created by turbidity currents, but how the process works is not obvious.

See also seafloor spreading, turbidity currents

submersible Underwater vessels intended for nonmilitary applications. Although based on earlier submarine technology, submersibles were intended for research and exploration, not warfare, and were therefore outfitted with windows or portholes so the occupants could see out. The first operational submersible was developed by Otis Barton in 1926, when he presented his plans to William Beebe, who wanted to explore the depths. The bathysphere that Barton designed was a cast-steel ball, less than 5 feet in diameter and with three quartz portholes, that was lowered on a steel cable into the depths. Barton and Beebe got to 3,000 feet in 1935, a record that would not be broken until Barton took his improved "benthoscope" to 4,500 feet off the California coast in 1949. In 1960, Jacques Piccard and Don Walsh descended in the bathyscaph *Trieste* to the bottom of the Mariana Trench in the southwest Pacific off Guam—at 35,800 feet, the deepest possible point in the world's oceans. Submersible development then focused more on research, and the *Alvin* was developed at Woods Hole. Since she was launched in 1964, this stubby little sub has been the workhorse of submersible technology: finding a lost H-bomb off Spain, exploring the Mid-Atlantic Ridge, locating the first hydrothermal vents, and allowing the first people in seventy-five years to see the *Titanic.* **See also *Alvin,* Barton, bathyscaph; bathysphere, Beebe, submarine, *Trieste***

Suez Canal The 100-mile-long canal that connects the Mediterranean and the Red Sea across the Isthmus of Suez in Egypt. Designed by the French diplomat Ferdinand de Lesseps (1805–1894), it was begun in 1859. When it officially opened ten years later, it provided the shortest maritime route between the Mediterranean and the Indian Ocean and the western Pacific. It was built under the aegis of a joint French-Egyptian stock company, but when the khedive of Egypt experienced financial difficulties in 1875, he sold his 44 percent share to the British government at the instigation of British prime minister Benjamin Disraeli. Under pressure from Egypt, Britain withdrew from Suez in 1954, and Egyptian president Gamal Abdal Nasser nationalized the canal in 1956. In what became known as the Suez Crisis, British and French troops invaded Egypt and retook the canal on October 29, 1956. A UN emergency force replaced the invaders, and the canal was opened to all shipping except that of Israel. The canal was blocked by the Egyptians during the Arab-Israeli War of 1967 and remained closed until 1975. In the meantime, the development of immense bulk carriers and supertankers has cost the canal a significant proportion of the tonnage that formerly used it. In order for these very large ships to use the canal, it would have to be widened and deepened. **See also Lesseps**

Sumatra The second largest of the Indonesian islands, 1,110 miles long and separated from the Malay Peninsula by the Malacca Strait, and from the island of Java by the narrow Sunda Strait. Much of the eastern half of the island is swampland, and much of the interior is impenetrable jungle. It is Indonesia's richest island: 70 percent of the nation's oil, coal, gold, silver, rubber, pepper, coffee, tea, and sugarcane come from Sumatra. As with most of the "East Indies," Hindus settled here first, then Arabs, and Marco Polo arrived around 1292. By the sixteenth century, Europeans had begun to plunder the island's riches, particularly spices. After the Japanese occupation during World War II, the island was incorporated into the new republic of Indonesia. The Atjehnese people have waged constant guerrilla warfare against the government since the 1950s, seeking independence for Sumatra. **See also Indonesia, Java**

summer flounder (*Paralichthys dentatus*) Commonly known as fluke, this is one of the left-eye flounders. The basic color of the left side is brownish or greenish, but this species is especially adept at changing color to match its background. Even though it can reach a weight of 20 pounds, 2- to 5-pounders are more likely, and they are caught in great numbers by sport fishermen from bridges, jetties, and small boats. It is also the object of a commercial fishery along the Middle Atlantic coast of the United States.

See also flounder, plaice, winter flounder

supertanker A ship designed to transport large quantities of liquid cargo, especially oil. Tankers were originally small ships, built to carry loads of 5,000 gallons of oil, but increased consumption of oil and gas by industry and automobiles led to the development of larger and larger ships. Tankers were employed to transport oil through the Suez Canal, but when it closed in 1967, a way had to be found to carry large cargoes around the Cape of Good Hope. Very large crude carriers (VLCCs) of between 250,000 and 275,000 tons were considered the only economic way to transport bulk oil by sea. Ultra-large crude carriers (ULCCs) followed, approaching 800,000 tons. Supertankers consist of a number of separate oil containers built into the hull, running the full width of the ship but separated by narrow compartments. Some of these leviathans can be 1,500 feet long, and crew members have to ride bicycles to get from one point on deck to another. (The *Exxon Valdez* was 987 feet long—longer than three football fields.) A fully loaded supertanker traveling at its top speed of 15 mph takes three miles to come to a complete stop. Safety and steering problems have plagued these ships from their inception, and groundings, collisions, and other accidents have led to some of the worst environmental disasters in history, as millions of gallons of oil are disgorged into the ocean and onto the beaches.

See also *Amoco Cadiz, Atlantic Empress* and *Aegean Captain, Braer, Exxon Valdez, Torrey Canyon, Urquiola*

surfing The sport of riding cresting waves on a "board." Although the origins of the sport are lost, when James Cook arrived in Hawaii in 1778, he noticed that some Hawaiians paddled out to meet him on koa wood boards. The islands were definitely the birthplace of surfing, and for a long time, the sport was restricted to those areas—such as the beach at Waikiki—where the waves were conducive to riding shoreward. Duke Kahanamoku (1890–1968), a Hawaiian gold-medal winner in swimming at the 1912 and 1920 Olympics, was the man who single-handedly popularized surfing. Boards were made of wood until the 1950s, when fiberglass was introduced. The stronger, smaller, and lighter boards that resulted led to a revolution in technique: instead of simply standing up on the board and coasting toward shore, surfers could now engage in intricate maneuvers on the face of the wave. Surfing is now popular around the world, wherever there is the essential shore break, and although California, Florida, New Jersey, Mexico, Australia, New Zealand, and South Africa boast excellent surfers and surfing beaches, Hawaii remains the ne plus ultra of surfing destinations—especially the north shore of the island of Oahu.

surfperches (family Embiotocidae) Found only in the Pacific, surfperches are small (the largest is 18 inches long), compressed fishes, generally oval in shape. They inhabit the surf zone along sandy and rocky coasts, and they give birth to live young, which is unusual for a saltwater fish. There are about two dozen species, many of which are popular with anglers, particularly the black perch (*Embiotoca jacksoni*) and the walleye surfperch (*Hyperprosopon argenteum*).

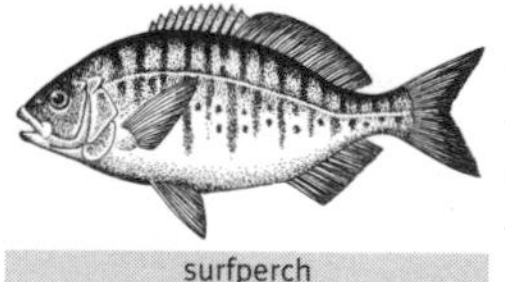
surfperch

surgeonfishes (family Acanthuridae) Sometimes called tangs, surgeonfishes are named for the sharp, erectile spines on either side of the caudal peduncle, which resemble scalpels. When raised, these spines point forward, and the fish uses them to slash at attackers. Surgeonfishes are compressed fishes with small scales and eyes placed high on the head. They feed by nibbling on algae and other marine botanicals. Most surgeonfishes appear oval because of the identical rounded shape of the dorsal and anal fins, but there are some with unusual projections, like the forehead spike of the unicorn fish, or the long dorsal filament of the moorish idol. **See also blue tang, moorish idol, unicorn fish**

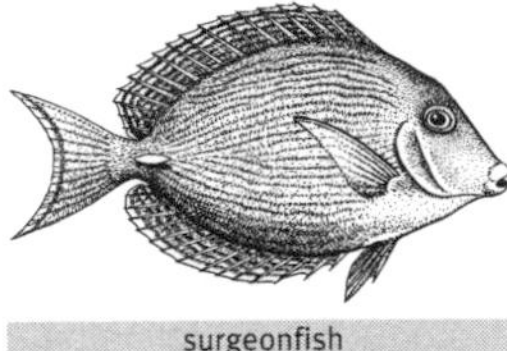
surgeonfish

Surtsey Lava island off the southern coast of Iceland that arose from the sea in 1963. The crew of a passing fishing boat first noticed steam coming from the sea, and the rotten-egg smell of hydrogen sulfide was discernible on the Icelandic mainland, 20 miles away. In water 500 feet deep, an underwater volcano, part of the Mid-Atlantic Ridge that gives all of Iceland its volcanic nature, had erupted, pushing a mountain of lava toward the surface until it burst through in a cataclysm of steam and fire. The island, now about 1 square mile in area, was in a state of sporadic eruption until 1967, when it began to cool. Surtsey is just southwest of the island of Heimaey, where a volcano erupted ten years later, partially obliterating the village of Vestmannaeyjar. The volcano and the island were named for Sutur, the Norse fire giant who is to come from the south at the end of the world and burn everything up.
See also Heimaey, Iceland, Mid-Atlantic Ridge

swallow-tailed gull (*Larus furcatus*) Along with the smaller, darker lava gull (*L. fuliginosus*), the swallow-tailed gull breeds only in the Galápagos Islands. The body and wings are pale gray, the beak is black with a white tip, and the feet and eye-rings are red. The prominently forked tail is white in adults, black-edged in juvenile birds. Swallow-tailed gulls are largely nocturnal (with large eyes to show for it), catching fish and squid at night.
See also Galápagos Islands, gull, lava gull

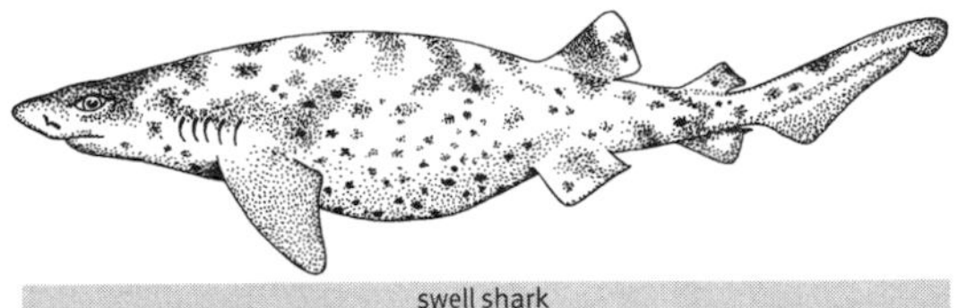
swell shark

swell shark (*Cephalocyllium ventriosum*) Several species of catshark are known as swell sharks. Found around the world, they are characterized by a robust body and a much smaller second dorsal fin. These 4-foot-long, strongly marked sharks spend the daylight hours resting in caves, then emerge at night to feed on small fishes and invertebrates. They all have the habit of distending their stomachs by taking in water or air. This remarkable ability, unique among the sharks, is believed to be a defense mechanism, but it is difficult to understand how a shark that is bobbing at the surface with a bellyful of air is protecting itself. They may blow themselves up like balloons to wedge themselves in caves or crevices, thus keeping larger predators from pulling them out. There are swell sharks found in the kelp beds of California, from Monterey Bay southward, and others that occur off southeast Asia (*C. fasciatum*), New Zealand (*C. isabellum*), Australia (*C. laticeps*), and southwest India (*C. silasi*). **See also catsharks, puffer**

swim bladder: See gas bladder

Symmes, John Cleves (1780–1829) John Cleves Symmes, the son of an American Revolutionary officer and member of the Continental Congress of the same name, developed the theory that the earth was hollow and "habitable within," that it could be entered from a hole located at each pole by ships that would sail in upside-down. By 1825, he had attracted the attention of newspaperman Jeremiah N. Reynolds (1799–1859), who went on a nationwide lecture tour, but distanced himself from Symmes's crackpot "Holes in the Poles" theory. The project was shelved until 1836, when the American whaling and sealing interests managed to win congressional support for a scientific expedition, which eventually became the U.S. Exploring Expedition of 1838–1842, led by Charles Wilkes. It is not unlikely that Jules Verne's *Journey to the Center of the Earth*, published in 1864, was based on Symmes's ideas.
See also Reynolds, U.S. Exploring Expedition, Wilkes

T

Tahiti One of the Society Islands of French Polynesia. It is composed of two volcanic cones (Tahiti Nui and Tahiti Iti), connected by an isthmus. The terrain is jagged and mountainous, rising to Mount Orohena at 7,333 feet. The island is 33 miles long and fringed by coral reefs and shallow lagoons. Samuel Wallis, commander of the *Dolphin,* discovered Tahiti in 1767, and named it King George III Island, but its native name, pronounced "Otaheite," was quickly adopted and slightly modified. Bougainville later visited the island and claimed it for France, and in quick succession, two famous British navigators dropped their anchors there: Captain James Cook in 1769 and Captain William Bligh in 1788. The Royal Society, for which Cook named the islands, had sent him to Tahiti to observe the transit of Venus, and Bligh took the ill-fated *Bounty* there to collect breadfruit seedlings to bring back to the West Indies. The first permanent European settlers in Tahiti were members of the Protestant London Missionary Society, who helped the royal Pomare family gain control of the entire island. In 1843, Queen Pomare IV was forced to agree to a French protectorate, and in 1880, the island became a French colony. The capital of Tahiti, and of all French Polynesia, is Papeete. The French painter Paul Gauguin (1848–1903) arrived in Papeete in 1891, but found it too built up and crowded, and so removed himself to Mataiea on the south coast.

Taiwan Island in the Pacific Ocean, about 100 miles off the coast of China across the Taiwan Strait. (Both the strait and the island used to be named Formosa, the Portuguese word for "beautiful.") The island is about 245 miles long and 90 miles at its widest, covering some 14,000 square miles. It is a mountainous island, with a large proportion of the population of 21.5 million relegated to the cities and the coastal plains. Taipei is the capital and largest city, with a population of 2.64 million. The Portuguese were the original Europeans to visit the island, but the Dutch and Spanish were the first settlers, around 1625. The Dutch were ousted in 1661 by Chinese refugees from the deposed Ming Dynasty, and by 1683 the Manchus had taken control of the island. In 1886 Taiwan became a province of China, but in 1894, after the Sino-Japanese War, it was ceded to Japan, along with the Pescadores. During World War II, Taiwan served as a major staging area for Japan's invasion of Southeast Asia. After the defeat of the Japanese, the island was returned to China, which was then governed by the Nationalists. When the Communists conquered China in 1949, the Nationalist government under Chiang Kai-shek fled to Taiwan, claiming that they were the legitimate government of all of China. The Chinese Communists planned an invasion of Taiwan in 1950, but were thwarted when U.S. president Harry Truman ordered the Seventh Fleet to the Formosa Strait. Communist shelling of the islands of Quemoy and Matsu brought about a security treaty between Taiwan and the United States, again avoiding an invasion. In 1971, the United Nations voted to turn over the "China seat" to the People's Republic of China, ousting Taiwan and leaving it unrepresented in the UN, the International Monetary Fund, and the World Bank. Taiwan persists in its claims that it is the rightful government of China, but it seems unlikely that the Beijing government is going to hand over its country—the most populous in the world—to the government-in-exile in Taiwan. **See also Quemoy**

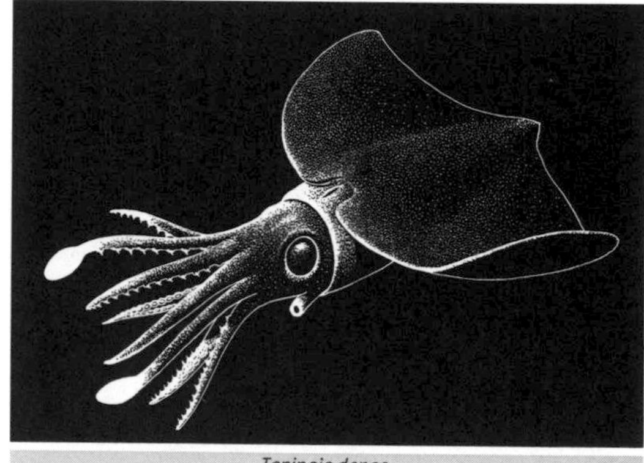

Taningia danae

Taningia danae Very large squid that has the largest light organs of any known animal on two of its arms. Named for the Danish ichthyologist Aage Vedel Tåning, *Taningia* has no common name. It has eight appendages instead of the ten that most squid species possess, and on the ends of two of these are lemon-sized (and lemon-colored) photophores that can be flashed at will. Each of the photophores is equipped with a black, eyelidlike membrane that can be opened and closed. The function of these stroboscopic flashers is unknown, but they may be used to startle their prey or to confuse a predator. (The strobes probably have the opposite effect on a 60-foot sperm whale.) On the suckers of its arms, *Taningia* has retractable claws, like those of a cat. Little is known about this species, which can reach a length of 7 feet and a weight of 135 pounds, but it has been collected—mostly from the stomachs of sperm whales—from all the subpolar, temperate, and tropical oceans of the world, making it one of the most widely distributed of all squids.

See also bioluminescence, giant squid

Tanna Island in the Republic of Vanuatu in the southwest Pacific, 25 miles long and 12 miles wide. Although not the largest of the islands (Espiritu Santo, Malakula, and Éfaté are larger), Tanna is the most fertile and the most populated, with almost 20,000 inhabitants. The southern part of the island is dominated by Mount Yasur, a 1,000-foot-high volcano that actively sizzles and spits fire. Tanna has many beaches, often alternating black and white sand, and is a popular tourist destination for Australians and New Zealanders. During World War II, many men from Tanna were brought to Éfaté to work on American military bases, and they were mightily impressed with the abundance of goods and matériel. They came to believe that a savior whom they called Jon Frum would return to their island bringing all sorts of riches, and they built wooden hangars, tin-can radio installations, and airstrips to facilitate his arrival. (The name Jon Frum is derived from the idea that "Jon *from* America" would come to be their salvation.) The "cargo cults" were semireligious groups that believed in the return of Jon Frum. They adopted the red cross seen on ambulances as the symbol, and today, the island is dotted with red crosses surrounded by picket fences, and there is a Jon Frum church topped by a red cross. **See also Vanuatu**

Tarawa Capital of the island republic of Kiribati in the western Pacific. Kiribati became independent in 1979. Originally the Gilbert Islands, the group includes the widely separated Phoenix Islands, the Line Islands, and Banaba (formerly Ocean) Island. Tarawa Atoll was selected as the main objective of the 1943 assault by U.S. Marines in their central Pacific campaign because it contained an important Japanese airfield. In one of the bloodiest battles of the entire war, the marines took the island, establishing the amphibious assault as a method that would be used repeatedly in the defeat of the Japanese in the Pacific. **See also Gilbert Islands, Kiribati**

Tarpon (*Megalops atlanticus*) Considered primitive fishes, tarpon are placed in the family Elopidae, which also includes the ladyfish. Like eels, these fishes pass through a transparent, ribbonlike larval phase, known as the leptocephalus. One of America's premier sport fishes, the Atlantic tarpon can be found in the inshore and offshore waters of the Atlantic Ocean. It lives in shallow water, feeding on smaller fishes, crawfish, and shrimp. The "silver king"—named for its silver-dollar scales—is sought by fishermen from bridges, jetties, or skiffs, and as soon as the hook is set, the fish takes to the air in acrobatic, twisting leaps, sometimes rising 10 feet out of the water. They are not considered edible, and in recent years, more have been released than taken to be mounted by taxidermists. The world's record tarpon, caught in the eastern South Atlantic off Sierra Leone, weighed 283 pounds. In the Indo-Pacific and the coasts of eastern Africa, there is a smaller version, known as the oxeye tarpon (*M. cyprinoides*).

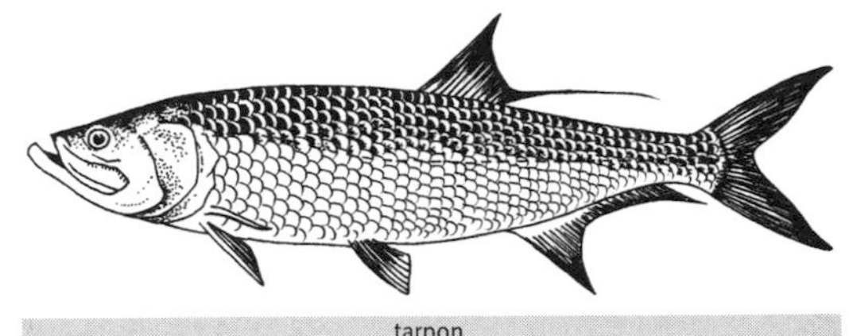

tarpon

See also ladyfish, leptocephalus

Tasman, Abel Janszoon (1603–1659) Generally regarded as the greatest of the Dutch navigators and explorers, Tasman made his first voyage for the Dutch East India Company in 1634, when he sailed to what is now the island of Ceram in Indonesia. He later visited Japan, Formosa, Cambodia, and Sumatra, and was selected in 1642 by Anthony van Diemen (governor general of the company) to explore the Southern Hemisphere to determine if there really was a Terra Australis Incognita, an undiscovered southern continent. In the *Heemskerk* and *Zeehaen*, Tasman departed from Batavia (now Djakarta), sailed to Mauritius in the Indian Ocean, and then headed west until he encountered the island that now bears his name, which he called Van Diemen's Land. He then sailed eastward, discovered the South Island of New Zealand (which he christened Staten Landt), and then headed eastward for Chile, but found the islands of Tonga and New Guinea instead. The Tasman Sea, between eastern Australia and New Zealand, is named for him, as is Australia's only island state.

See also Dutch East India Company, Tasmania

Tasmania Discovered in 1642 by the Dutch navigator Abel Tasman, and named Van Diemen's Land after Anthony van Diemen, governor-general of the Dutch East India Company in Batavia (now Djakarta) from 1636 to 1645. James Cook visited the island in 1777, and in 1803, Britain took possession and established Australia's third penal colony there, after Sydney and Norfolk Island. The island was inhabited by Aborigines, but they were methodically killed off by the settlers; the last full-blooded Tasmanian, a woman called Truganini, died in 1876. Around 1806, with four Aborigine rowers, James Kelley circumnavigated the island in forty-nine days, and discovered the two harbors that would become the notorious penal colonies of Port Davey and Macquarie

Harbor. At this time, right whales were observed in the estuary of the Derwent River, and bay whaling became Tasmania's first industry; the oil was shipped to England. Because of public opposition, the penal colony was shut down in 1853, and Van Diemen's Land was made a separate colony. (Tasmania was renamed in 1856.) The island, some 26,000 square miles in area, is separated from the mainland by the Bass Strait. It was incorporated into the Commonwealth of Australia in 1901, and is now one of the seven states, with Hobart as its capital. The Tasmanian wolf, also known as the Tasmanian tiger or thylacine (*Thylacinus cynocephalus*), is (or was) a large marsupial predator, about the size of a German shepherd. It was found in the dense forests, where it fed on wallabies and birds. Because it was also thought to be a sheep killer, it was hunted under a bounty system, and although there are occasional (usually unverified) reports of animals deep in the bush, it is believed to be extinct. **See also Tasman, transportation**

Tasman Sea The arm of the southwestern Pacific Ocean between Australia and New Zealand. It covers about 900,000 square miles and merges with the Coral Sea to the north and the Southern Ocean to the south. Abel Tasman, for whom the sea is named, navigated it in 1642, and James Cook was among the first to explore the Australian and New Zealand coastlines. The Great Barrier Reef is in the Tasman Sea, as are Lord Howe and Norfolk Islands. Sydney, Australia, is the largest city on the Tasman Sea.

tautog

tautog (*Tautoga onitis*) Most wrasses are warm-water fishes, but the tautog is an exception; it is found in the cool waters from Massachusetts to New Jersey. It is a heavy-bodied fish that can weigh as much as 20 pounds, but most are smaller. Tautogs cruise over shallow areas searching for mussels and crustaceans, which they crush with their flat, rounded teeth. Because of their size and inshore habitat, tautogs are eagerly sought by New England spearfishermen. **See also cunner, wrasse**

Taylor, Ron (b. 1934) and Valerie (b. 1936) Australia's most famous diving and filmmaking couple, Ron and Valerie Taylor, both born in Sydney, have been making underwater films since they were married in 1963. Accomplished divers, they were Australian spearfishing champions before they exchanged their spears for cameras. Ron made the first film ever shot of the great white shark, off South Australia's Dangerous Reef, and then both he and Valerie served as camera crew for Peter Gimbel's much-acclaimed 1971 feature, *Blue Water, White Death: The Search for the Great White Shark.* They shot the live shark footage for *Jaws*, and they have produced and photographed innumerable television films on the underwater world, and written hundreds of magazine articles and several books, all dedicated to the appreciation and preservation of the marine environment. In 1986, Valerie was named Ridder (Knight) of the Golden Ark, a Dutch knighthood, for her conservation efforts. **See also Gimbel, *Jaws***

Tegetthof, Wilhelm von (1827–1871) Admiral who commanded the Austrian fleet at the Battle of Lissa (1866) in the Adriatic, where his flagship *Ferdinand Maximillian* rammed and sank the Italian flagship, *Re d'Italia*, in one of the first battles between armored ships. Because of the Austrian success at Lissa, the ram became a prominent feature on warships for the next forty years. After von Tegetthof's successes, the Austrian navy decreed that there would always be an ironclad that bore his name. The ship sailed to Novaya Zemlya by the Austrian explorers Julius von Payer and Karl Weyprecht in 1872 (when they discovered Franz Josef Land) was named *Admiral Tegetthof.*

Tenerife The largest of the Canary Islands, off North Africa, Tenerife has an area of 795 square miles and a total population of 658,000. In the northeastern sector is Santa Cruz de Tenerife, the major port and capital, with a population of 220,000. It was here in 1797 that Admiral Horatio Nelson lost his arm and, at the Los Rodeos airport in 1977, that two jets collided, killing 582. The larger, southwestern portion of the island is composed almost entirely of the dormant volcano known as Pico de Teide (Teide Peak). At a height of 12,185 feet, it is Spain's highest mountain, and after Mauna Loa and Mauna Kea on the island of Hawaii, it is the third largest shield volcano on earth. (A shield volcano is a dome-shaped volcano, formed by centuries of eruptions of basaltic lava.) Tiede last erupted in 1909. The northern climate is conducive to agriculture, but the south is much drier, and camels are common beasts of burden. **See also Canary Islands**

teredo (*Teredo navalis*) Commonly known as shipworms, teredos are actually bivalve mollusks (clams of the family Teredinae) that are notorious for the damage they cause in the timbers of wooden ships, docks, wooden pilings, etc. They occur in all seas, but are most common in warmer waters. In their life span of between two and ten months, some species can reach a

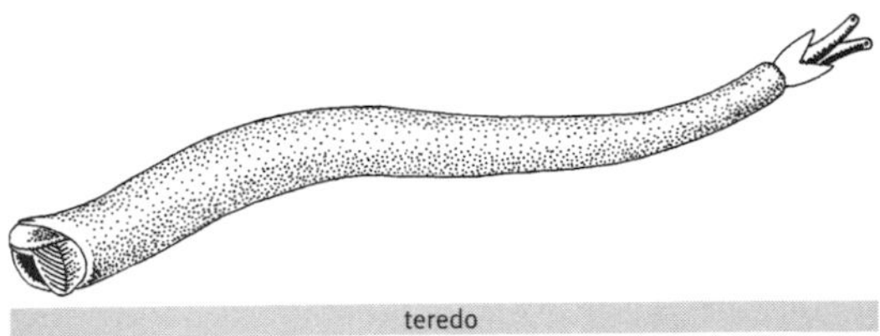
teredo

length of 36 inches. The shells of teredos enclose only a small portion of the anterior end of the body; the rest is elongated and wormlike, hence the nickname. The ridged and roughened edges of the shells are rotated back and forth to rasp away at the wood, providing food for the clam and creating tunnels in which the shipworm lives. After creating the tunnel, the shipworm remains there, eating food brought by currents created by the cilia of the gills. The twisted tunnels weaken the wood to such a degree that it can collapse and disintegrate. Copper sheathing was used to protect the hulls of ships until the development of antifouling paints in the twentieth century. **See also clam, ship**

tern (*Sterna* spp.) Closely related to gulls, terns are usually smaller (although there are some that are larger than some gulls) and daintier and fly in a more graceful, bouncy manner. In general, they have sharp, fine bills and are more slender than gulls, with narrower, more pointed wings. They spend a great part of the day on the wing, often with the bill pointed down. Their legs are short and set far back, and although their feet are webbed, they seldom swim. All terns have forked tails, and many have greatly elongated outer tail feathers. Terns are found throughout the world's marine environments, but many of them reside in the tropics or the more temperate latitudes. Those that do visit the polar regions do so in the summer, and often migrate vast distances to warmer winter climates. They fish by plunge-diving, often from considerable heights, catching fish near the surface. There are some thirty-two species, most of which are white and gray with a black cap. (The Inca tern, with its all-dark coloration, and the all-white *Gygis alber* are prominent exceptions.) The largest is the Caspian tern (*S. caspia*), with a 55-inch wingspan; the smallest is the fairy tern of Australasia (*S. nereis*), with a span of 19 inches. Noddies belong to a different genus (*Anous*) but are often included with the terns. **See also Antarctic tern, Arctic tern, Caspian tern, common tern, Inca tern, noddy, white tern**

Terra Australis Incognita The Unknown Land to the South, postulated by early geographers who believed that because the earth was spherical, there had to be a southern continent to balance the known northern landmasses and keep the planet from wobbling. It was contrived by Ptolemy in the second century A.D. and later incorporated into maps such as Ortelius's 1570 *Theatrum orbis terrarum* ("Theater of the World"). After the Portuguese rounded Africa and showed that men could cross the equator without burning up, it became possible to reach the missing continent, and even when men had sailed around Cape Horn, they believed there was a great landmass to the south. The prevailing winds and currents, however, tended to force ships northward along the west coast of South America after they rounded the Horn, so they were usually as far north as Peru before they could head west again. When Álvaro de Mendaña discovered the Solomon Islands, some 200 miles east of New Guinea, he named them because he expected to find King Solomon's mines on this outpost of Terra Australis. In his 1519–1522 circumnavigation of the world, Francis Drake expected to pass through the Strait of Magellan and sail along the coast of the southern continent, presenting gifts to any rulers he encountered. Pedro Fernández de Quirós landed in the New Hebrides in 1595, believed that the largest island was Terra Australis, and accordingly named it Australia del Espíritu Santo. Abel Tasman sailed eastward from Batavia (Djakarta) in 1642, passed south of Australia, and fetched up first on the island that would eventually be named for him, then in New Zealand, and he too believed he had found the great southern continent. By the time James Cook sailed in 1768, he was under express orders to find the "Great South Land," even though it seemed to be moving farther and farther south. He did, of course, discover Australia and circumnavigate New Zealand, but Tasman had already sailed south of Australia, demonstrating that it was not the long-sought landmass. Cook never saw the Antarctic continent, which was not sighted until 1821 and not delineated until the twentieth century, but he—and all those who preceded him in the search for Terra Australis Incognita—would not have been surprised to find that it really is there. They *would* have been shocked to learn that it is a gigantic, uninhabited, ice-covered desert. **See also Cook (James), Drake, Mendaña, Ortelius, Ptolemy, Quirós, Tasman**

Tew, Thomas (d. 1695) In 1692, Thomas Tew was offered a share in the 70-ton sloop *Amity* by a group of Bermuda merchants, for a privateering voyage along the coast of Africa. This didn't much appeal to him, so he changed course and headed for the East, where in 1693, he captured a treasure ship in the Mocha Fleet of the Grand Mogul Aurangzab, then took his loot to Madagascar to divide it up. His plunderings had netted him more than £100,000, and another fortune in ivory, spices, and silks. With his associate John Avery, he returned in triumph in the *Amity* to Newport, Rhode Island, in 1694, and journeyed to New York, where he was received by Governor Benjamin Fletcher. In 1695, when he engaged an East Indiaman in the Red Sea, he was shot in the gut and killed. **See also piracy, privateer**

Thomson, Charles Wyville (1830–1882) Scottish medical doctor and oceanographer, successor to Edward Forbes (who died at the age of thirty-nine in 1854) as professor of natural history at the University of Edinburgh. C. Wyville Thomson sailed on a dredging expedition aboard *Lightning* in 1868 to study undersea life in the Atlantic from the Faeroes to Gibraltar, and this expedition proved so successful that the following year, with the *Porcupine* and the *Shearwater* at his disposal, Thomson surprised everyone—including himself—by dredging living creatures from the North Sea at depths of 15,000 feet. In 1872, he was chosen to lead the scientific team aboard HMS *Challenger* on her epochal three-year oceanographic voyage around the world. Under Thomson's direction, a team of naturalists, chemists, and artists, equipped with the latest equipment, hauled up their nets and dredges and established conclusively that there was life at the ocean's greatest depths. He was knighted in 1876, after the *Challenger* returned to England. The results of the expedition, published in a fifty-volume report, appeared in 1895, thirteen years after Thomson's death. **See also *Challenger***

Thorfinn Karlsefni (active c. 1002–1007) Icelandic trader who came to Greenland; married Gudrid, the widow of Thorstein Ericsson (Leif's brother); and set out for Vinland, the land recently discovered by Leif Ericsson. They landed at Helluland ("Land of Flat Stones," possibly Baffin Island), continued south to Markland ("Land of Trees," probably Labrador), and eventually reached Vinland ("Land of Vines," almost certainly Newfoundland). They built a settlement in 1004, and Gudrid gave birth to Snorri, the first white child born in North America, but after much quarreling and fighting with the Skraelings (local Eskimos), they abandoned Vinland and headed back toward Greenland. One of their ships was lost in the Irish Sea, but the other, commanded by Thorfinn, reached Greenland safely in 1006. **See also Eric the Red, Leif Ericsson, Vinland**

threadfins (family Polynemidae) Threadfins are named for the two-part pectoral fin, the lower portion of which consists of several threadlike rays that are used as feelers for finding food on the bottom. Also known as threadfishes, they occur worldwide in warm, shallow waters, and are characterized by a protruding snout, giving them a piglike profile, more exaggerated than that of a mullet. The foot-long barbu (*Polydactylus virginicus*) inhabits the sandflats from New Jersey to Uruguay. Most of the thirty-odd species are less than 18 inches in length, but the giant threadfin (*Eleutheronema tetradactylum*), found off India, grows to be 6 feet long. The juvenile forms of certain jacks—unrelated to the polynemids—such as *Alectis ciliaris* and *A. indicus*, have extremely long dorsal and anal fin rays that they lose as adults. Because of these streamers, these fishes are sometimes called threadfins too. **See also jacks, mullet**

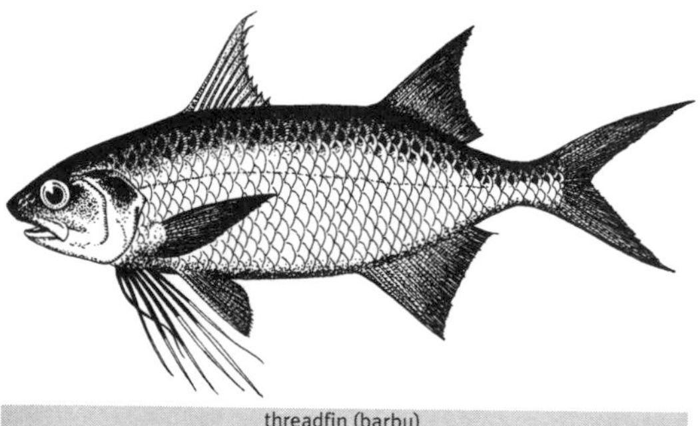

threadfin (barbu)

thresher shark

thresher shark (*Alopias* spp.) A large shark with a tail that can be as long again as the shark itself. Including the tail, threshers have been measured at 20 feet. They evidently use the whiplike tail to herd and stun the smaller fishes that are their prey, but this feeding method has never been directly observed. There are three species of thresher sharks, generally similar, but differentiated either by range or by some physical characteristic. The common thresher (*A. vulpinus*) is the most widely distributed, being found around the world in temperate and tropical seas. The pelagic thresher (*A. pelagicus*) is an uncommon offshore species; the bigeye thresher (*A. superciliosus*) inhabits greater depths and has larger eyes to compensate for the reduced light. All are susceptible to longline fisheries, and they are often the "shark" served in restaurants. **See also shark**

Thursday Island The administrative and commercial center of the Torres Strait Islands, located just off the Cape York Peninsula of northern Queensland, Australia. Some 2,600 live on the 1.25-square-mile island, engaging in pearl and *bêche-de-mer* (sea cucumber) fishing. The naming of Thursday Island ("T.I." to Australians) remains a mystery. Between James Cook's passage through the Torres Strait in 1770 and Captain Owen Stanley's in 1848, several navigators had the opportunity to bestow a name upon the little island. In June 1798, as he sailed in an open boat through the Torres Strait on the way to Timor, William Bligh named Wednesday and Sunday Islands (just to the south of Torres Strait), but curiously, he made no mention of Thursday Island. Evidently, Captain Stanley, surveying in the *Rattlesnake*, named Thursday and Friday Islands, but later the Admiralty switched the names.

See also Torres Strait

tide The periodic rise and fall of the sea as a result of the gravitational pull of the sun and the moon. The largest tides, "spring tides," occur when the sun and the moon are in line and act together; the smallest, "neap tides," take place when the sun and moon are at right angles to each other and do not exert a combined attraction. In each case, because the moon is so much nearer, it exerts two and a half times the pull of the sun. There is usually a period of six hours and twenty minutes between high and low water, and therefore, the total time between successive high (or low) tides is twelve hours and forty minutes. But in some places, the interval between two successive tides is less than twelve hours, while in others, such as the China Sea, the interval can be more than twenty-four hours. Inland seas, such as the Mediterranean, are virtually tideless, because the entrance, the Strait of Gibraltar, is too narrow to allow the influx or outflow of sufficient tidal water. When certain conditions apply, such as narrowing channels, the inrushing tides can reach great heights, such as the 20-foot tides in the Gulf of Tonkin and the 70-footers in the Bay of Fundy, the highest in the world. The inrushing tide is known as a tidal bore, but so-called tidal waves, which are the result of seismic disturbances and are properly known as tsunamis, have nothing to do with tides. **See also current, tsunami**

Tierra del Fuego Discovered by Ferdinand Magellan in 1520, Tierra del Fuego ("Land of Fire," named for the fires made by the natives) is an archipelago at the southern tip of South America. It is composed of the island of Tierra del Fuego, with an area of 18,530 square miles, the largest in South America; five medium-sized islands; and hundreds of small islands, islets, and rocks, separated from the mainland by the strait. When it was first discovered, there were small bands of Ona, Yaghan, and Alakaluf peoples living there, but by the nineteenth century, they had been killed off by disease introduced by sailors and settlers. After its discovery, it remained unmapped until 1831–1836, when Robert FitzRoy conducted the first thorough survey in the *Beagle* (with Charles Darwin aboard as naturalist). Tierra del Fuego is now divided between Chile on the west and Argentina on the east. The Argentine town of Ushuaia is the southernmost permanent settlement in the world, with a population of approximately 12,000. (Puerto Williams in Chile is farther south but is mostly a naval base.)

tiger shark (*Galeocerdo cuvier*) With its short, squared-off snout, long upper tail lobe, and vertical stripes, the tiger shark is probably the easiest to identify of all the carcharhinids. Newborn tiger sharks, which are born alive, are heavily striped and spotted, but the markings fade as the shark matures. Tigers are also among the largest sharks, the longest ones having been measured at 18 feet. The coxcomb-shaped teeth are also diagnostic; they have a deep notch on the outer margin. Tigers are cosmopolitan sharks, found throughout the world's tropical and subtropical waters, and their range of food items is enormous and opportunistic. They will eat whatever they come across, including mollusks, sea turtles, fishes, skates, rays, other sharks, porpoises, birds, garbage, human remains, and, occasionally, living people. Their stomachs have also been found to contain inedible items, such as cans, bottles, license plates, rolls of tar paper, and armor. Tigers are considered among the most dangerous of sharks, and there are many documented records of their unprovoked attacks on swimmers and divers.

tiger shark

See also carcharhinid sharks, shark attack

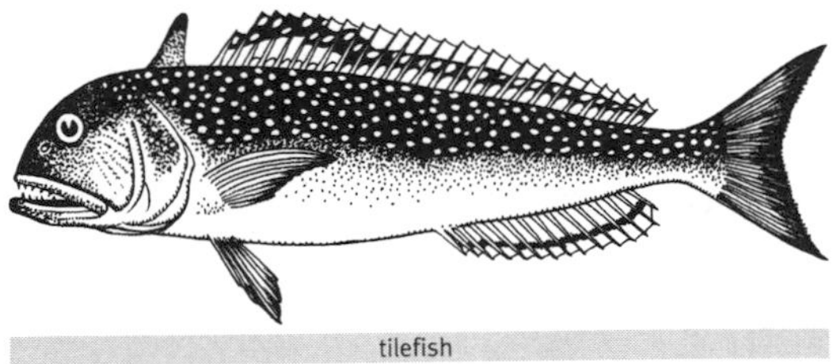

tilefish

tilefish (*Lopholatilus chamaeleonticeps*) Shortly after this 3-foot-long fish was discovered by Nantucket cod fishermen in 1879, it appeared on the way to becoming a popular food fish. But before the tilefish could be fished commercially, reports began to come in of immense rafts of these fishes floating dead on the surface. For hundreds of miles, the seas were covered with dead tilefish, their numbers estimated in the billions. Examination of the dead fishes showed no signs of disease or parasites; they appeared to have simply died. (Subsequent analysis suggested that an anomalous influx of cold Arctic water had lowered the temperature beyond that which the fish could tolerate.) By 1882, no tilefish at all were seen, and it was believed that this fish, which had been known for only three years, had become extinct. It was not until 1892 that small numbers were caught again. Although the fish has made a biological comeback, its popularity as a food has declined, and a directed commercial longline fishery, beginning in the 1970s, has succeeded in lowering the population again, but not in reviving great interest in the tilefish.

Timor A long, narrow island in the Indonesian Lesser Sunda group, originally exploited in 1520 by Portuguese traders, who came for sandalwood. By 1620, the Dutch had established a beachhead at Kupang, and the Portuguese had moved up into the mountains. Treaties of 1860 and 1914 effectively divided the island between Portugal and the Netherlands. Dutch Timor became part of Indonesia in 1950, but the Timorese people fought to establish their independence and rid themselves of Indonesian rule. After a war in which tens of thousands of combatants were killed and the population was decimated by famine and disease, Indonesian troops were driven out in late 1999, and East Timor declared itself an independent republic.

See also Indonesia

Tinian One of the Mariana Islands in the southwest Pacific, immediately south of Saipan. The Marianas were discovered by Magellan in 1521 and remained Spanish possessions until 1898, when they were sold to Germany. The Japanese seized the islands in 1914, but in 1944, Tinian was taken by American troops and turned into an important base for bombing attacks on Japan. The B-29s that dropped the atomic bombs on Hiroshima and Nagasaki took off from Tinian.

See also Magellan, Mariana Islands

Tirpitz, Alfred von (1849–1930) German grand admiral, who joined the Prussian navy at age sixteen, in the era of wooden ships. In 1879, he was given command of a torpedo boat and saw the great possibilities of this type of weapon. By 1892, he was chief of staff of the navy, in charge of working out the tactics of Germany's battle fleet. His plans for expansion were accepted by the kaiser, and Germany set out to build the world's largest navy, which would consist of forty-one battleships, twenty battle cruisers, and forty light cruisers. The expansion of the fleet contributed to the enmity between Britain and Germany and is considered a major cause of World War I. Tirpitz began the construction of submarines, but his advice was systematically ignored when Germany entered the war, and by 1916, impatient with the navy's refusal to follow his recommendations for unrestricted submarine warfare, he resigned.

Titanic The most infamous shipwreck in history. At 882 feet in length and displacing 46,328 tons, the "unsinkable" RMS *Titanic* was the largest ship ever built when she was launched on May 31, 1911. She left Southampton for New York on April 10, 1912, but four days later she hit an iceberg off the Grand Banks of Newfoundland and sank. The iceberg ripped a hole in the hull, rupturing several of the watertight compartments, but to this day it is not clear why the *Titanic* sank so quickly—or why she sank at all. With far too few lifeboats for the 2,200 passengers and crew, chaos ensued as the liner settled into the icy sea and broke in half; 1,500 people died. Rockets were fired from the sinking ship, and SOS signals sent out on the new Marconi wireless system, but the only ship close enough was the *Carpathia,* which arrived four hours after the accident and picked up some seven hundred survivors who were in *Titanic*'s lifeboats. Some of America's richest people died in the disaster, and Captain E. J. Smith chose to go down with his ship. The hull, broken in two, remained lost (the exact coordinates of the sinking were not known) until 1985, when an expedition led by Woods Hole oceanographer Robert Ballard, using side-scanning sonar and remote-controlled underwater television cameras, located the wreck some 13,000 feet below the surface. The following year, Ballard descended in the submersible *Alvin* and became the first person to look upon the *Titanic* in seventy-four years. Although there was a strong feeling that she should remain unmolested, salvage laws permitted anyone who could get down to the wreck to remove whatever they could, and the *Titanic* has been looted and desecrated. In 1998, the movie *Titanic,* directed by James Cameron, won eleven Academy Awards, and became the largest grossing movie in history, taking in more than $1 billion.

toadfish

toadfish There are several different fishes that are commonly called toadfishes; all are slow-moving bottom dwellers with large mouths and many sharp teeth. The head is broad and heavy, and tapers to a long, slender tail. They are particularly well named, since they not only have a broad head like that of their namesakes, but they also make toadlike croaking noises. They usually lie in crevices, burrows, or half buried in the sand. Toadfishes are ill-tempered and quick to bite, and they are also equipped with two venomous dorsal-fin spines and one on each gill cover, which are used for defense. The 10-inch-long oyster toadfish (*Opsanus tau*) is found in the shallow coastal waters of eastern North America, and the venomous toadfish (*Thalassophryne* spp.) is found in the American tropics. **See also midshipman**

Tonga The only independent kingdom in the Pacific, Tonga consists of 150 islands divided into three groups, Tongatapu, Vava'u, and Ha'apai. The population in 1996 was 103,466; the natives are Polynesian in character, and most closely related to the Samoans. A few of the islands are volcanic, but most are coral atolls. The Dutch navigator Jakob Le Maire discovered the islands in 1616, and Abel Tasman, another Dutchman, visited in 1643. When James Cook arrived in 1773, he named the archipelago the Friendly Islands. The London Missionary Society succeeded in stamping out the heathen religion of the Tongans and converted King George Topou I to Christianity in 1862. A British protectorate from 1900, Tonga was ruled by the beloved Queen Salote Tupou III from 1918 to 1965. She was succeeded by her son, who now rules as King Taufa'ahau Tupou IV.
See also Cook (James), Le Maire, Tasman

tonguefish (family Cynoglossidae) Teardrop- or tongue-shaped flatfishes that have their eyes on the left side of the head, a small mouth, and no fin spines and no ribs. In American waters, these foot-long fishes have no commercial value, but several Asiatic species are highly prized, especially in Japan. **See also flounder, plaice, sole**

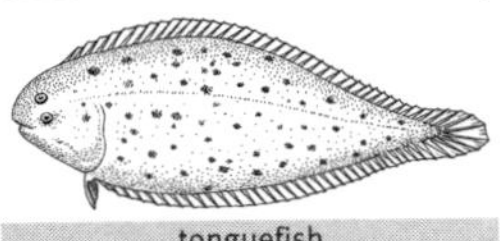

tonguefish

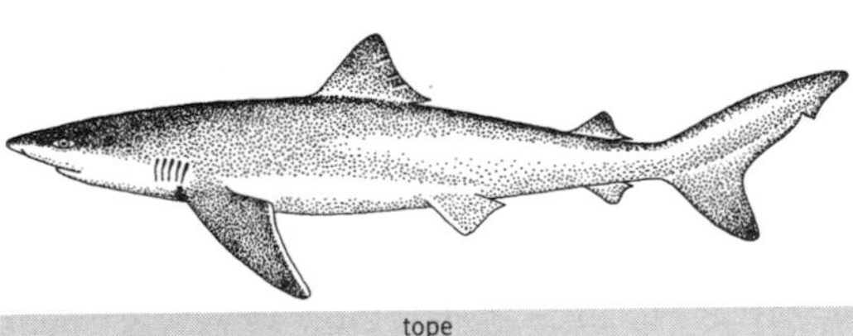

tope

tope (*Galeorhinus galeus*) Also known as the school shark, the tope is an abundant shark that is found throughout the temperate coastal and insular waters of the world. In California waters, it is known by the unfortunate name of soupfin shark, although shark's fin soup is consumed mostly in Asian countries. Topes are heavily fished around the world, with the most intensive activity occurring off Uruguay, Argentina, South Africa, southern Australia, and California, but there are signs that overfishing is affecting the catches. Before World War II, the liver of this species was one of the primary sources of vitamin A, but it has since been synthesized. Nowadays the meat is prepared fresh, frozen, or salted. Although not considered a game fish by the International Game Fish Association, the tope, which grows to a length of 6 feet, is a popular sport fish in Britain. **See also shark's fin soup**

Torcello This island in the lagoon of Venice was first settled in A.D. 452 by refugees from the mainland. In the seventh century the bishopric was moved there. During the tenth century, Torcello had about 10,000 inhabitants and was the richest and most powerful island in the lagoon. It began to decline when the bishopric was moved to the Rialto in 811. In the sixteenth century, malaria struck the island, and those who survived moved to Murano or Venice. The island was nearly deserted then, and has not changed much. The cathedral of Torcello was founded in 639 and reconstructed in 1008. It contains mosaics of the Madonna and the Last Judgment. The island's population is now around 60 people, mostly fishermen and market gardeners.
See also Burano, Murano, Venice

torpedo (1) Small rays of the families Torpedinidae, Narkidae, or Temeridae, known for their ability to produce electrical shocks. They are found worldwide in tropical and temperate, mostly shallow, waters, but occasionally as deep as 3,000 feet. The electrical capability, used by the ray for defense and for prey capture, is generated by two organs of modified muscle tissue, one on each side of the head. **See also electric ray**

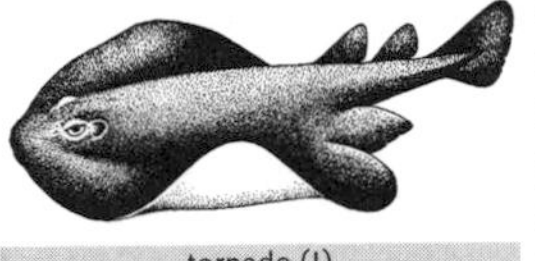

torpedo (1)

torpedo (2) A naval weapon, used principally by submarines and aircraft. The first torpedoes were explosive charges moored in the sea, which later became known as mines. Then explosive charges were attached to long spars projecting from the bows of small craft ("spar torpedoes"), such as those used by the Confederate submarine *Hunley* during the U.S. Civil War. The Whitehead torpedo, developed around 1865, was the first self-propelled, buoyant missile, and was powered by a small reciprocating engine run by compressed air. By 1878, most of the world's navies had Whitehead torpedoes, and the first torpedo casualty was a Turkish revenue cutter that was sunk by the Russians in that year. Directional accuracy was achieved in 1885 when John Howell developed the gyroscope to control the vertical rudder. By 1914, torpedoes were 17 to 22 feet long and carried as much as 600 pounds of explosives. During the world wars, the Japanese developed torpedoes powered by liquid oxygen, and the German U-boats had electrically driven torpedoes. Whereas older torpedoes had to travel in a straight line, modern torpedoes utilize homing devices that allow them to change course to seek out their target, responding to sounds made by the target and following until they make contact. The newer torpedoes are powered by electric motors, but some, designed for effectiveness against deep-diving nuclear submarines, have turbines driven by solid propellant. Torpedoes can now be fitted with nuclear warheads. **See also submarine**

Torres, Luis Váez de (d. 1613) Portuguese navigator, second in command to Pedro Fernández de Quirós on the expedition that sailed from Callao (Peru) in 1605 and discovered the New Hebrides Islands (now Vanuatu) the following year. Quirós believed he had found the southern continent, but it was actually the island of Espiritu Santo in the New Hebrides. Unable to locate the island after his ship, the *Capitana,* had been blown to leeward, Quirós headed back to Mexico, but Torres continued to explore the southwest Pacific. He reached the Louisiade Archipelago at the eastern tip of New Guinea, then sailed through a maze of reefs and shoals between the southern shore of New Guinea and the promontory that is the northern tip of Australia—through the strait that now bears his name. From there he went to the Moluccas and then to the Philippines, where he died. The Spanish authorities did not disclose the existence of the Torres Strait, which might explain why Abel Tasman, who approached it some forty years later, believed it was only a bay, and never tried to penetrate it.

Torres Strait Passage between the northern tip of the Cape York Peninsula of Queensland, Australia, and the southern shore of Papua New Guinea. It was first traversed by Luis Váez de Torres in 1606, but it was not named until 1802, when Matthew Flinders circumnavigated Australia. When James Cook sailed through in 1770, he did so very cautiously, not only because of the dangerous reefs and shoals, but also because he didn't trust Torres's description. The strait contains several islands, known collectively as the Torres Strait Islands, and consisting of Prince of Wales Island (named by Cook in 1770), Wednesday, Thursday, Friday, Horn, Hammond, and Goode. The Torres Strait Islands are part of Queensland and are administered by a central council on Thursday Island.

Torrey Canyon Carrying a cargo of crude oil from Kuwait, the 63,000-ton supertanker *Torrey Canyon* ran aground on the Scilly Isles, off Land's End, Cornwall, England, on March 18, 1967, in the world's first massive pollution event. In a hurry to reach Milford Haven, Wales, Britain's premier deepwater port, the captain tried to steer between the "Seven Stones" reef and the main islands, and ran onto the rocks. The ship's container compartments were holed, and 38 million gallons of oil polluted the beaches of Cornwall and Brittany. At that time, there was no plan to deal with such a catastrophe, and the first move was to try to salvage the ship. After ten days of futile pushing and shoving, during which another 3 million gallons poured out, it was decided to bomb the ship to set the remaining oil afire. Flames 300 feet high shot into the air, and a mile-high plume of oily smoke blanketed southern England. Tens of thousands of seabirds were killed by the oil that permeated their lungs or their feathers. (The oil killed 80 percent of the twelve thousand puffins that lived in the Scilly Isles, and the population has not recovered.) Across the Channel, the Bretons had a little more time to prepare for the oil that was approaching their beaches, and they used powdered chalk to bind the oil into particles, which sank to the bottom. (Unless treated, oil is lighter than water, so it floats.)

See also *Amoco Cadiz, Braer, Castillo de Bellver, Exxon Valdez,* oil spill, supertanker

Tortola Largest of the British Virgin Islands, Tortola is 12 miles long but only 3 miles wide. Running lengthwise on the 21-square-mile island is a high, unbroken ridge, with its highest point at Mount Sage, 1,781 feet above sea level. Road Town, with a population of about 8,000, is the largest settlement and the administrative center of the British Virgin Islands. On his second voyage in 1493, Columbus came upon the Virgin Islands—which he named for the eleven thousand martyred virgin followers of Saint Ursula—but finding no gold there, he sailed on. Perhaps it was he who named Tortola for the *tórtola,* or turtle dove. Dutch settlers colonized the island first, but the British drove them off and claimed the island in 1672. Sugar plantations dominated the economy until 1838, when the British abolished slavery and many settlers returned to England. The island languished until the advent of tourism in the middle of the twentieth century, but it is now one of the Caribbean's most popular destinations, especially for blue-water sailors; Road Town specializes in yacht charters.

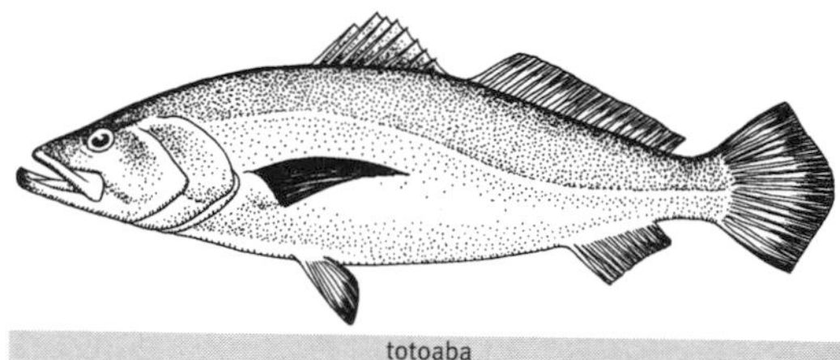

totoaba

totoaba (*Cynoscion macdonaldi*) Also spelled "totuava," this member of the drum family (Sciaenidae) is found only in the Sea of Cortez (Gulf of California), between Baja California and Mexico. A popular food fish, it reaches a maximum length of 6 feet and weight of 225 pounds. Earlier in the century, it was caught for its large air bladder, which was used in the preparation of soups in Asian countries. Before the construction of dams on the Colorado River, fresh, muddy water used to flow into the sea, providing the necessary spawning conditions for the totoaba, but now these conditions have greatly deteriorated. In addition, commercial fishing—first with hook and line, then with dynamite, and finally, with gill nets—has reduced its numbers to a dangerously low level, and it is believed to be the only marine fish on the brink of extinction.

See also gill net, white sea bass

Toussaint-Louverture, François-Dominique (1744–1803) A self-educated slave who led the 1791–1803 uprising in Haiti and is regarded as one of the outstanding heroes in black history. In 1793, François-Dominique Toussaint adopted the surname L'Ouverture, meaning "opening," because of his fast-moving attacks. Although he professed allegiance to France and then to Napoleon, he was devoted only to freeing his people from the yoke of slavery. Late in 1793, when the British occupied all of Haiti's coastal cities and allied themselves with the Spanish in the east, Toussaint drove the British out and then put down an attempted coup by the mulatto generals Pétion, Rigaud, and Boyer. Intent upon freeing the slaves, Toussaint invaded Spanish-held Santo Domingo in 1801 and named himself governor-general for life of the entire island of Hispaniola. Concerned about French colonial ambitions in the West Indies, Napoleon sent General Charles Leclerc in 1802 to deal with Toussaint, but the rebel army resisted so strongly that the best Leclerc could get was a truce. Toussaint was seized, however, and brought to France in chains, where he died in a dungeon. In 1804, Haiti became the second independent nation in the Western Hemisphere, and two of Toussaint's generals, Jean-Jacques Dessalines and Henri Christophe, were the first and second emperors.

See also Dominican Republic, Haiti

trade winds Winds that blow from the northeast in the Northern Hemisphere and from the southeast in the Southern, and originate on the equatorial sides of the horse latitudes, which are two belts of high air pressure, one lying between 25° north and 30° north of the equator and the other at the same latitudes south of the equator. The high air pressure in these belts forces air along the equator in an area known as the doldrums. As the air converges in the doldrums, it rises high over the earth, recirculates poleward, and sinks back toward the earth's surface at the horse latitudes, completing a cycle. The air does not move directly north or south because it is deflected by the rotation of the earth. Trade winds get their name from the regularity with which they blow, assisting ships that used to carry trade around the world in the days that preceded steam propulsion. **See also doldrums, horse latitudes**

Trafalgar, Battle of Fought on October 21, 1805, off the eponymous cape off the southwest coast of Spain, the Battle of Trafalgar was one of the last major encounters of sailing fleets. After failing to contain the French fleet under Admiral Pierre de Villeneuve in the Mediterranean, Admiral Horatio Nelson returned to England, but he came back in his flagship *Victory,* joined the British fleet, and encountered the combined French and Spanish fleets as they were attempting to reenter the Mediterranean after a rendezvous in the Caribbean. Nelson's plan was to attack the enemy in two columns, one led by himself in the *Victory,* the other by Admiral Cuthbert Collingwood in the *Royal Sovereign.* Before the battle, Nelson gave his famous signal: "England expects that every man will do his duty." Although Britain's twenty-seven ships were outnumbered by the enemy's thirty-three, Nelson's audacious plan of attack allowed his ships to triumph; no fewer than twenty enemy ships surrendered or were taken. Nelson did not live to celebrate his victory, because he was hit by a sniper's bullet from the French ship *Redoubtable* and lived only a few hours more. Except for the Battle of Navarino Bay, Trafalgar was the last major combat fought entirely under sail.

See also Navarino Bay, Battle of; Nelson

transatlantic cable The first attempt at spanning the Atlantic with a telegraph cable took place in 1857, when the American frigate *Niagara* met the British man-of-war *Agamemnon* in mid-ocean and attempted to splice their cables together. The cable snapped. Cyrus Field's New York, Newfoundland, and London Telegraph Company tried it again the next year, with the same results. When it was realized that the two-ship method was impracticable, they looked for a single ship large enough to carry 2,700 miles of cable that weighed between 5,000 and 7,000 tons. The U.S. Civil War interrupted the program, but in 1865, the *Great Eastern,* up to that time the largest vessel ever built, was converted into a cable ship, and on July 27, 1866, the cable was successfully brought from Ireland to Heart's Content, Newfoundland.

See also *Great Eastern*

transportation Prior to the American Revolution, British authorities had used certain colonies, particularly Maryland and Georgia, for the internment of convicted criminals who could not be maintained in the jails of England, but afterward, they had to seek elsewhere. Thousands had been housed on "hulks," decommissioned and rotting warships anchored in the Thames, on which the conditions and the smell were abominable. Punishment in Georgian England was intended to be as uncomfortable as possible, and the prisoners were often permanently shackled, flogged regularly, and fed rotten and inedible food. The government sought other locations in which to dispose of these unwanted souls, and although various places in Africa were suggested, the new, and far distant, colony of Australia was chosen. For crimes of robbery, forgery, murder (but also for offenses no more serious than stealing a piece of cloth), Englishmen and -women were sentenced to "transportation" to New Holland (Australia) for periods of seven years, fourteen years, or life. Sir Joseph Banks, who had been James Cook's botanist, suggested Botany Bay as a site, and on May 13, 1787, the First Fleet under Captain Arthur Phillip set out for Australia. Upon arrival in Sydney Harbor (Botany Bay was deemed unsuitable for a settlement) on January 26, 1788, they established the first permanent settlement on the Australian continent. Shortly thereafter, other convict settlements were founded on Norfolk Island and Van Diemen's Land (now Tasmania).

See also Botany Bay, First Fleet, Norfolk Island, Tasmania

trawling Bottom-fishing by towing a cone-shaped net, kept open at the front and closed at the rear. In the earliest version, the wide mouth was spread by being towed by two boats. Then came the beam trawl, where the mouth of the net was opened by a horizontal wooden beam, and finally, the otter trawl, where the mouth is kept open by "otter boards," flat metal plates that shear outward when towed through the water. Metal or plastic floats attached to the upper lip of the net's mouth keep it open, and steel balls on the bottom allow it to follow the contours of the seafloor. Today's bottom trawl, operated from a diesel-powered trawler, may be 400 feet long, with an opening that may be 200 feet wide between the wings. The smaller, closed part of the trawl is the "cod end." This is the primary method used to catch demersal (bottom-dwelling) species such as flounder, plaice, sole, and various crustaceans, such as shrimp, but of course, such a fishery also entraps many "trash fish" that are discarded. Stern trawlers replaced the side-working trawlers around 1950, and factory trawlers are now in operation that can freeze or process the catch on board, or even transport it "wet" (on ice, but not frozen) back to port. Midwater trawls are also used to fish between the bottom and the surface; the depth at which the trawl is operated can be regulated by the length of the towing warps and the speed of the towing vessel. In recent years, bottom-trawling has been identified by scientists and conservationists as a primary factor in the destruction of the seafloor, including not only the fishes but every other living thing on the bottom.

See also drift-net fishing, gill net, longline fishing, otter trawl, purse seining

tremoctopus (*Tremoctopus violaceus*) Once thought to be one of the smaller octopuses, tremoctopus has been measured at 6 feet in total length. For its habit of autonomously releasing part of its luminous spotted web as a decoy when threatened, it is sometimes known as the blanket or handkerchief octopus. As with the unrelated argonaut (*Argonauta argo*), the sexes are strongly dimorphic: the tiny males are ½ inch in length to the female's 5 feet, but the hectocotylized (breeding) arm is larger than the entire animal. Tremoctopus is a tool-using octopus. Males and juvenile females usually carry small pieces of the stinging tentacles of the Portuguese man-of-war (*Physalia*) as a defense against predators, or perhaps to immobilize their own prey.

See also argonaut, octopus, Portuguese man-of-war

Trieste Submersible commissioned by Auguste Piccard in 1953 and named for the Adriatic seaport where she was built. The 50-foot-long flotation hull was designed to hold 28,000 gallons of gasoline, and suspended below it was the passenger compartment, a 10-ton, forged-steel chamber whose inside diameter was 7 feet. The portholes were 6-inch-thick truncated cones made of a newly developed shatterproof plastic called Plexiglas. The bathyscaph descended by taking on water ballast into the forward and aft tanks and by valving off small quantities of gasoline. She could be moved forward or backward by propellers located on top of the hull, but only in a very limited way. In August 1953, off her home base of Castellammare in southern Italy, with Piccard and his son Jacques aboard, *Trieste* made her first successful manned descent: 26 feet down to the bottom of the harbor. There followed a succession of deeper dives in the Mediterranean from 1953 to 1956: 3,540 feet in the Bay of Naples; 10,300 feet in the Tyrrhenian Sea. In 1958, Auguste Piccard, no longer able to afford the upkeep on his own, sold the *Trieste* to the U.S. Office of Naval Research (ONR) but stayed on as a consultant. (Jacques Piccard was the primary pilot.) They were going to send the submersible down 35,800 feet, to the bottom of the Challenger Deep in the Mariana Trench, the deepest possible dive in the world. The sphere that had been forged in Italy was not considered strong enough for such a dive, so a new one was manufactured by the Krupp steel works in Germany. New instrumentation was designed and built in Switzerland, and the revised electronics were supplied by the U.S. Navy. Record after record fell as the bathyscaph made practice dives: 18,150 feet; 24,000 feet; and finally, on January 23, 1960, with Lieutenant Don Walsh as copilot, Jacques Piccard and the *Trieste* landed on the bottom at 5,966 fathoms (35,800 feet). Immediately a foot-long flatfish moved out of the way of the descending steel monster, refuting once and for all Edward Forbes's smug declaration that no life could exist in the depths. At the very bottom of creation, in water 7 miles deep, there was a fish, and, even more startling, a fish with *eyes*. It is unfortunate that *Trieste* was not equipped with an external camera. Shortly after her historic dive, Torben Wolff of the Zoological Museum of Copenhagen published a note in *Nature* in which he disputed the identification of the fish. Despite the "eyes," he suggested that it was not a fish at all, but probably a sea cucumber, "perhaps related to the bathypelagic, cushion-shaped *Galatheathauria aspera*, which is almost a foot long and oval in outline."

See also Piccard, sea cucumber, submersible, Walsh

triggerfish (family Balistidae) Including the filefishes, there are some 120 species of Balistidae. The most obvious identifying characteristic is the first dorsal spine, which can be locked upright by a small second spine and released by a trigger. In the filefishes, the "trigger" spine is located just above the eye. Triggerfishes have strong, chisel-like teeth that are well adapted for crunching through the hard parts of invertebrates such as crabs, mollusks, and sea urchins. The eyes, located high on opposite sides of the flattened head, can function independently. Many triggerfishes are drab in color, but some, like the Indo-Pacific clown triggerfish, are patterned with elaborate and colorful designs.

See also filefish, humuhumunukunukuapua'a, queen triggerfish

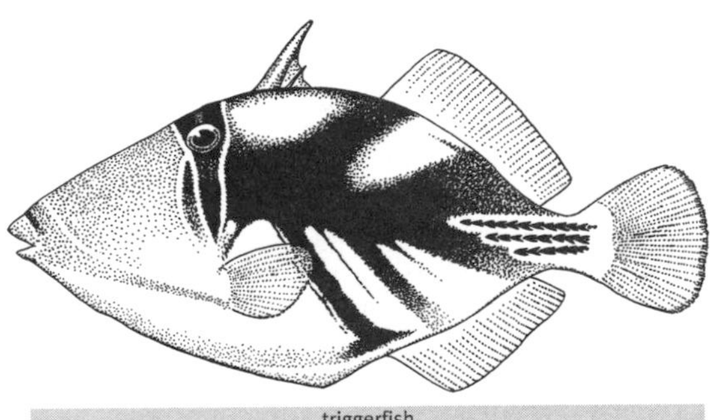

triggerfish

trilobite Extinct marine invertebrates, relatives of the arthropods (joint-legged animals), that were characterized by a flattened, oval body with a pair of longitudinal furrows that divided the shell into three sections—hence the name "trilobite," which means "three-lobed." The animals were also divided into three sections: the head, the thorax, and the tail. They ranged in length from less than 1 inch to 2 feet, and most known species had a pair of antennae and many pairs of legs, which corresponded to the segments of the body. The upper shell, which was usually segmented, was thicker than the lower and is therefore better represented in the fossil record. From these fossils, we have been able to determine that trilobites had eyes, the first such organs of any creature on earth. Trilobites, which became extinct

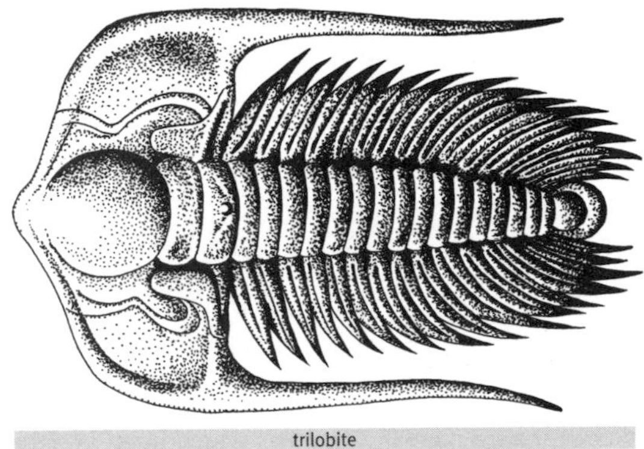

trilobite

in the Permian period some 245 million years ago, were extremely plentiful, and represent more than half of the known fossils from the Cambrian period, 500 million years ago. The reason for their extinction is not known, but it may have to do with the rise of cephalopods and fishes that fed on them. Today's horseshoe crabs (Chelicerates) are similar to trilobites.

See also horseshoe crab

trimaran A vessel with a single central hull and floats on either side, propelled by sails or mechanical power. It is a further development of the catamaran, a twin-hulled vessel, which was probably based on outrigger canoes. Trimarans are twentieth-century innovations, largely used as ocean racing yachts or cruising vessels. Because of their lightness and stability, they are considerably faster than a comparably sized single hull. In 1999, the Defense Evaluation and Research Agency (DERA) of the British Ministry of Defense announced the commission of a 300-foot-long trimaran warship demonstrator (RV *Triton*) that they claim "represents probably the most significant design shape in warship hulls since the advent of the Ironclads."

See also catamaran, outrigger

trireme A warship of the Mediterranean, powered by three banks of oars. Originally developed by the Greeks to add more speed and power to the two-tiered bireme of the Phoenicians, triremes were built to a maximum length of 135 feet, and a beam of 20, but they lay quite low in the water, usually with no more than 8 feet of freeboard above the waterline. The two hundred–man crew consisted mostly of rowers in three tiers: the upper rank (*thranites*), the middle (*zygites*), and the lower tier (*thalamites*). Although triremes are mentioned in various accounts, we do not know exactly how the three tiers of oarsmen were positioned. Triremes were outfitted with a long ram and sometimes carried archers and soldiers for combat. During the Persian Wars (490–479 B.C.) the Athenian leader Themistocles defeated the Persians with his fleet of triremes. The Romans later elaborated on the three-tier arrangement by building quinqueremes (five tiers) and even septiremes (seven tiers). Triremes remained in operation until around A.D. 1200, when they were rigged as sailing vessels with two masts and lateen sails, the oars being used only for extra mobility.

See also bireme, Salamis

Trinidad and Tobago Republic in the Caribbean consisting of two islands immediately north of the Orinoco River delta of Venezuela. Trinidad is the larger, at 1,864 square miles; Tobago, 21 miles to the northeast, is 116 square miles. Trinidad was visited by Columbus in 1498 on his third voyage, but finding no gold there, he claimed it for Spain and moved on. When the Spanish arrived, the islands were inhabited by some 35,000 peaceful Arawaks and their nemeses, the warlike Caribs, but disease and enslavement soon reduced the Indian populations almost to zero. Dutch raiders came in 1640, followed by the French in 1677, and the British formally annexed the islands in 1797. When Britain abolished slavery in 1834, Trinidadian landowners imported laborers from India. (Today, Indians make up 45 percent of Trinidad's population.) The islands were joined politically in 1888. In 1958, they became part of the West Indies Federation, which was dissolved in 1962. The Republic of Trinidad and Tobago achieved its independence in 1976. Trinidad has the largest pitch lake in the world, providing asphalt for export, and there are also large petroleum and natural gas deposits. The high standard of living is due to its thriving export economy and a healthy tourist industry. Calypso, the music that has spread to so many West Indian islands, originated in Trinidad, as did the steel drum that is representative of Caribbean music. Trinidad's pre-Lenten carnival, Canbouley, is spectacular.

tripletail (*Lobotes surinamensis*) Because its dorsal and anal fins are large and well developed, they give—to some people—the appearance of additional tails. Found throughout the inshore warm temperate and tropical waters of the world, these relatives of the sea basses can reach a length of 3 feet and a weight of 50 pounds, but most are much smaller. Juvenile tripletails resemble leaves and drift with the current before they mature. These fishes have the curious habit of swimming on their sides or with their heads down.

See also giant sea bass

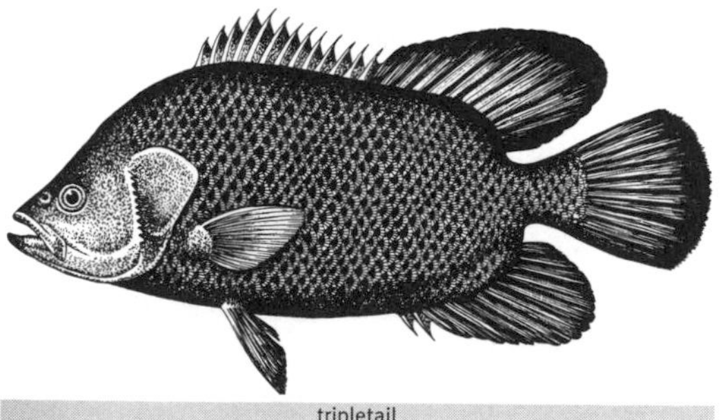

tripletail

tripod fish

tripod fish (*Bathypterois* spp.) Deepwater species that make a tripod of the right and left ventral fin and the lower lobe of the tail fin to perch high off the bottom. They are found at great depths, up to 18,000 feet, and we know of their existence only from occasional specimens brought up in trawls or from rare photographs. Most have tiny eyes (some have no eyes at all), so they must depend on their other senses to find food. In some of the eighteen known species, the "landing gear" is longer than the fish itself; a 10-inch fish can raise itself 1 foot off the bottom.

See also *Ipnops murrayi*

Tripolitan War (1801–1805) The first war fought by the United States after its formation. In 1785, when pirates of the Barbary Coast (Mediterranean North Africa, including Tripoli, Algiers, and Tunis) captured two American ships and made slaves of their crews, the Americans tried to bargain for their freedom. When this failed, Thomas Jefferson proposed the creation of an armed fleet to rescue them; it was the beginning of the U.S. Navy. On May 14, 1801, the American flagpole at the U.S. embassy in Tripoli was chopped down, and the Pasha of Tripoli declared war on the United States. The pasha demanded twenty thousand dollars a year in tribute, which the United States refused to pay. Instead, in 1803 it sent a small force to blockade Tripoli. One of the ships was the first commissioned for the new navy, the frigate *Philadelphia*, William Bainbridge commanding. When she ran aground and was captured and her crew imprisoned, Commander Edward Preble sent Stephen Decatur to burn the *Philadelphia* to prevent the enemy from using her guns against the American squadron. Decatur's success made him America's first naval hero. A settlement with the pasha was negotiated in 1805, but it did not end the threat of Barbary piracy, since the pirates continued their raids on American shipping during the Napoleonic Wars and the War of 1812. In 1815, a squadron under Decatur forced the Dey of Algiers to sign a treaty renouncing U.S. tribute, and the war was officially over. **See also Barbary pirates**

Tristan da Cunha A group of small volcanic islands in the South Atlantic; along with Ascension and Saint Helena, they are the visible southern manifestations of the Mid-Atlantic Ridge. Of the islands, which include Gough, Nightingale, and Inaccessible, only Tristan is inhabited. Discovered by the Portuguese admiral Tristão da Cunha in 1506, the island was visited by explorers, whalers, and sealers, but the first inhabitants came from St. Helena in the nineteenth century. (The name is pronounced "Tristan d' *Coon*-ya.") It was annexed by Great Britain in 1816, and became a dependency of St. Helena in 1936. The long-dormant volcano on Tristan da Cunha erupted in 1961, and the entire population of the island was evacuated to England. In 1963, the 198 inhabitants voluntarily returned to Tristan.

See also Ascension Island, Mid-Atlantic Ridge, St. Helena

Trobriand Islands Low-lying coral islands in the southwest Pacific, some 90 miles north of the easternmost tip of New Guinea; part of the country of Papua New Guinea. The 20,000 Trobrianders, who are of Melanesian stock, subsist on agriculture, particularly the growing of yams. The intricate trading systems of the islanders were discussed in Polish anthropologist Bronislaw Malinowski's classic study, *Argonauts of the Western Pacific* (1922).

tropic bird (*Phaethon* spp.) There are three species of tropic bird—the red-billed (*P. aethereus*), the red-tailed (*P. rubricauda*), and the white-tailed (*P. lepturus*)—

white-tailed tropic bird

which are similar to one another except for the differences identified by their common names. All are medium-sized, highly aerial seabirds, mostly white in color, with two tail feathers that may be as long or longer than the body. They feed by hovering with fluttering wings, and then diving with their wings half closed, like a gannet. As their name implies, tropic birds are creatures of the tropics, found throughout the South Pacific, South Atlantic, and Indian Oceans, often hundreds of miles from land.

See also frigate bird, tern

True's beaked whale

True's beaked whale (*Mesoplodon mirus*) This small beaked whale (maximum length: 17 feet) is found primarily in the North Atlantic from Nova Scotia to Florida, but there also seems to be a population in South African waters. Only the males have teeth, which are located only at the extremity of the lower jaw. The scientific name *mirus* means "amazing" or "wonderful."

See also beaked whales, *Mesoplodon*

Truk One of four island states that make up the United Federation of Micronesia (the others are Pohnpei, Kosrae, and Yap), Truk is a 39-square-mile island in the Caroline group of the western Pacific. First reported in 1583 by the Spanish explorer Álvaro de Saavedra, Truk was sold to Germany in 1899 and transferred after World War I to Japan, which heavily fortified it and used it as a naval base. In February 1944, Truk Lagoon was attacked by U.S. Task Force 58, a group of nine aircraft carriers and related cruisers, destroyers, and submarines. More than four hundred Japanese planes were wiped out and between fifty and sixty ships were believed sunk. Those two days of devastating air assault created what is known today as the Ghost Fleet of Truk Lagoon. Now named Chuuk, it is a popular destination for divers and underwater photographers.

See also Micronesia, Pohnpei, Yap

trumpetfish (*Aulostomus maculatus*) With its attenuated body and terminal mouth, the trumpetfish aligns itself with sea grasses and coral as a camouflage device. It has even been known to swim along with larger fishes in an attempt to approach unsuspecting prey. Related to the pipefishes, sea horses, and cornetfishes, trumpet-

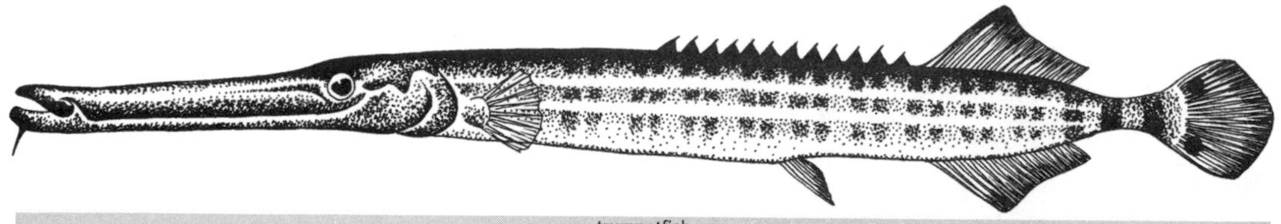
trumpetfish

fishes are found in the warm, shallow waters of the Caribbean and the Indo-Pacific regions.

See also pipefish, sea horse

trypot A large iron cauldron used aboard whaling ships in the nineteenth century for the boiling down of blubber, the process known as trying out. Trypots were also used for rendering the fat from seals and penguins. The Norwegian whalers who arrived in South Georgia in 1904 named the first whaling station there Grytviken, "Cauldron Bay," for the trypots that had been left behind by the sealers. **See also tryworks, whaling**

tryworks First used on shore, tryworks were brick furnaces that incorporated iron trypots and were used for the boiling of whale blubber to extract the oil, which was then ladled into casks. The fires were started with wood, but then whaleskin was used, which meant that the whales were being boiled down in fires fueled by themselves. In the seventeenth- and eighteenth-century British and Dutch whale fisheries, the blubber was packed into casks and "tried-out" when the whalers returned to their home ports. Then around the middle of the eighteenth century, American sperm whalers began to build brick tryworks on the decks of their ships, completely changing the way whaling was conducted, because the ships did not have to return before the blubber spoiled and so could remain at sea for much longer periods of time. **See also trypot, whaling**

tsunami In Japanese *tsunami* means "harbor wave." In the past, they were called tidal waves, although they have nothing to do with tides. They are sometimes known as seismic sea waves, but this too is a misnomer, because it is possible to have a tsunami with no earthquake (seismic) activity. Submarine and terrestrial landslides can cause tsunamis, and so can cosmic collisions, such as comets or meteors that hit the ocean. Most tsunamis, however, *are* caused by seismic activity, as when an earthquake beneath the sea causes the seafloor to subside, or an undersea volcano erupts and creates an impulsive force that uplifts the water column. Rising landmasses, such as the birth of the Icelandic island of Surtsey in 1963, also displace a great amount of water, which may cause tsunamis. Seismic waves are formed concentrically, radiating in all directions from the source of the disturbance, like the ripples caused by a pebble thrown into a shallow pool. Most ordinary waves are wind-generated and have a short period (the length of time required for two successive crests to pass a given point) and a short wavelength (the distance from crest to crest). Depending upon the magnitude of the event that triggers the tsunami and the depth of water in which it is formed, the period of a tsunami may be ten hours, and its wavelength anywhere from 60 to 600 miles. Moreover, the water may move at speeds of up to 600 miles an hour. (The huge breakers that surfers ride in Hawaii are wind-driven, and although they may crest at 30 feet, they are separate, unsupported entities, with no mass of water behind them, and may slow down to a wavelength from 60 mph to 40 mph.) Waters rushing in to fill the original "hole," which may be thousands of miles from any land, may cause the tide to recede at first, but the ocean's struggle to regain its equilibrium creates an even greater disturbance, and it is this backlash that is the tsunami. Gravity acts to flatten out a wave at the surface (even a large tsunami may be only a couple of feet high and in deep water will not be felt by a ship), but it is not slowed down until it reaches progressively shallower water, when all that water, which has been traveling at high speed for thousands of miles, piles up as the wave approaches land, giving it its height and causing it to run up on any land that is in its path. Billions of tons of fast-moving water, moving at speeds of up to 600 mph, provide the incredible destructive force of a tsunami. An earthquake in Chile in May 1960 destroyed much of the town of Chiloe, but the resulting tsunami, which reached the Hawaiian Islands fourteen hours later, after crossing the Pacific at 450 mph, was 30 feet high and caused an estimated $24 million worth of damage. The 1883 eruption of the Indonesian volcano of Krakatau produced a wave 130 feet high that drowned 36,000. In July 1998, an earthquake off the north shore of Papua New Guinea generated a wave 50 feet high that roared ashore, obliterating villages and killing more than 2,200 people. **See also wave**

Tuamotu Archipelago About eighty atolls scattered over 600 square miles in the eastern Pacific. Rangiroa, the second largest coral atoll in the world, is 46 miles across; Fakarava, which commercially is the Tuamotus' most important island, spreads out for 38 miles. The group also includes Raroia, the island on which the *Kon-Tiki* made her Polynesian landfall after sailing across the Pacific from Peru in 1947, and also Mururoa, the atoll where the French conducted their nuclear testing for thirty years. The Tuamotus differ dramatically from the mountainous Society Islands 200 miles to the southwest, since they consist only of low coral islands and atolls. At one time, these islands were called the Dangerous Islands because hundreds of ships had been wrecked on the reefs and atolls. The islands were discovered by the Spanish navigator Pedro Fernández de Quirós in 1606; they became dependencies of Tahiti in 1847, and were annexed by France in 1881. Now they are part of French Polynesia, along with the Society Islands, the Austral Archipelago, and the Marquesas.

See also Heyerdahl, Marquesas Islands, Mururoa, Quirós, Rangiroa

tubeshoulder

tubeshoulder (*Searsia koefoedi*) A deep-sea fish that has a special subdermal sac that connects to an external tube above and anterior to the gill covers, through which the 1-foot-long fish can eject a luminous fluid. The tiny squid *Heteroteuthis dispar* has a comparable mechanism, by which it ejects a bioluminescent ink cloud. Tubeshoulders also have ventral light organs.

See also bioluminescence, *Heteroteuthis dispar*

tube worms (phylum Pogonophora) The tube worms known as Pogonophores ("beard-bearers") are characterized by a long and slender body encased in a self-secreted tube. Most of the known species are less than 2 feet long and less than ⅛ inch in diameter. They are usually found in soft marine sediments, buried except for the gills (the "beard" of the name), which protrude above the ocean floor. They have no mouth or digestive tract, and they take all nourishment through the surface of the body. In 1977, at the Galápagos Rift hydrothermal vent site, at a depth of 8,000 feet, tube worms were discovered that were so unusual they completely upset our conception of these animals—in fact, they completely upset our conception of life on earth. Where all other living creatures depend on the photosynthetic cycle—where the light from the sun nourishes the plants, which nourish the smaller animals, and so on up the food chain—the rift tube worms (and the clams, crabs, shrimp, and mussels that were found alongside them) subsisted on hydrogen sulfide, processed by symbiotic bacteria in their bodies. These vent tube worms—now considered a separate phylum, the Vestimentifera—can reach a length of 10 feet and have a crown of blood-red, feathery plumes that extend from a white tube 2 inches in diameter. The rift tube worm (*Riftia pachyptila*) is found in dense clusters in the vicinity of hydrothermal vents around the world.

See also hydrothermal vents, rift shrimp

tucuxi

tucuxi (*Sotalia fluviatilis*) Because it is a nondescript little dolphin from a part of the world rarely visited by travelers, the tucuxi (pronounced "too-*cux*-ee," after an Indian word) is one of the least-known dolphins. There is some confusion about how many species there are, since its range includes most of the lakes and rivers of northern South America, as well as the coastal waters of Venezuela, Guyana, and Suriname. The tucuxi shares its riverine habitat with the Amazon River dolphin (boutu), and its coastal waters with the bottlenose dolphin. It is not difficult to differentiate the Amazon river dolphin from the boutu, because the boutu has a long snout and a low dorsal ridge—and may be pink in color—but tucixis look very much like smallish bottlenoses. The major difference is in the shape of the dorsal fin: where the tucuxi has a low, triangular fin, the bottlenose has one that is high and gracefully curved.

See also Amazon River dolphin, bottlenose dolphin, freshwater dolphins

tuna Any of a number of streamlined fishes included in the family Scombridae with the similarly shaped mackerels. Tunas are characterized by beaklike, nonprotrusible jaws, a sickle-shaped tail on a narrow tailstock, and a series of small finlets on the dorsal and ventral surfaces before the tail. They have a high proportion of red muscle, which is conducive to continuous activity, and they are among the fastest and most powerful of all fishes. Through a system of heat-exchangers in the blood and muscles, tunas can increase performance by maintaining a higher body temperature than the surrounding water. To reduce drag, the fins can be folded into slots, the eyes are flush with the head, and the scales are tiny or lacking altogether. Most tunas are dark blue or black above and silvery below, often iridescent. The largest of the tunas is the bluefin, which can be 12 feet long and weigh 1,500 pounds; most species are smaller. Collectively they are among the world's most important food fishes, and all of them are fished commercially.

See also *individual species*

tuna-porpoise problem For reasons that are still unknown, yellowfin tuna aggregate under schools of spotter and spinner dolphins. In 1966, government biologists learned that California tuna fishermen were trapping both dolphins and tuna in their nets, removing the dolphins, and dumping some of them overboard dead. (Even though the cetaceans being caught were commonly known as dolphins, the problem was referred to as "tuna-porpoise," because the fishermen call the animals porpoises.) It was estimated that from 1960 onward, when the fishermen converted from pole-and-line fishing to purse seines, between 3 million and 5 million dolphins were killed. The (U.S.) Marine Mammal Protection Act of 1972 made it illegal to harm any cetacean, but the tuna fishermen lobbied for an exemption, and continued to kill dolphins in staggering numbers—more than 300,000 in 1972. The fishermen continued to set their nets "on dolphins," until they were sued in federal court by a consortium of conservation groups and forced to suspend their entire fishing operations. They were allowed to commence again only if they could abide by strict quotas imposed by the government, and they agreed to kill no more than twenty thousand dolphins per year. Later, American canners refused to accept tuna caught with dolphins, which led to the concept of "dolphin-safe tuna," but foreign fishing fleets did not abide by American restrictions, necessitating an embargo of foreign tuna caught with dolphins. **See also purse seining, spinner dolphin, spotter dolphin, yellowfin tuna**

tunicate (phylum Urochordata) Tunicates are invertebrate animals whose larvae possess a primitive notochord that they lose as adults, suggesting a common ancestor with the backboned animals. They have a heart and a network of blood vessels and nerves, but the distributive functions are performed by cells known as macrophages. There are more than two thousand classified species, found worldwide and at every depth. Most tunicates (the name is derived from the Latin *tunic,* referring to their flexible protective covering) attach themselves to the bottom or some other object, but the sea grapes (*Molgula*) are free-floating. The grape- or vase-shaped sea squirts are common in shallow waters, where they affix themselves to rocks, seaweeds, pilings, or jetties, and squirt water from two tubelike openings on the dorsal surface if stepped on or otherwise disturbed. The phylum Urochordata ("tail-chord") also includes the tiny pelagic animals known as appendicularians, which never metamorphose from the tadpolelike larval stage. The thaliacea are transparent, pelagic tunicates that group themselves into colonies, the most spectacular of which is *Pyrosoma* ("fire-body"), which moves slowly through the water as a glowing, bioluminescent column that can be 4 feet long. Other colonial thaliacians are known as salps, which aggregate in even larger colonies. Some tunicates reproduce by budding; a fingerlike projection growing near the base breaks off and settles to become a new individual. **See also salp**

turbidity currents Underwater currents that flow because of horizontal differences in density resulting from suspended sediment, affecting the deep circulation of the oceans. The mechanics of these currents can be tested experimentally, but the actual events have never been observed, and much of the study of turbidity currents is hypothetical or speculative. These powerful currents may be responsible for the creation of submarine canyons, but the way in which this might happen is unclear. Broad fans at the lower end of these canyons suggest a gradually decreasing sediment flow, but whether the current was triggered by an earthquake, a landslide, a slump, a tsunami, or some other tectonic event that caused a great wedge to be carved out of the continental slope is uncertain. In 1929, an earthquake around the Grand Banks initiated a vast landslide, which rolled over the seafloor at 50 mph, severing telegraph cables as it went. More than 60,000 square miles (an area approximately the size of Michigan) of sediment was distributed onto the old ocean floor. **See also sediments, slump, submarine canyons**

turbot (*Scophthalmus maximus*) One of the left-eye flounders, which means that the eyes are on the left side of the head and the fish rests on the bottom on its right side. The turbot is found on sandy, muddy, shell and gravel bottoms, usually in shallow water. The upper (eyed) side is speckled with gray, brown, or black, and varies with the color of the seabed. It is found throughout the Black Sea, the Mediterranean, and the Baltic. The turbot is fished heavily; its firm white flesh makes it popular. Another European species is *Psetta maxima.* The Turbot War, where Canada seized a Spanish fishing boat in 1995 over quotas, was fought over the Greenland halibut, *Reinhardtius hippoglossoides,* which is also known as turbot. **See also flounder**

Turks and Caicos Islands British dependency in the West Indies, south of the Bahamas, consisting of Grand Turk and Salt Cay (the Turks); and Providenciales and East, West, North, South, and Middle (or Grand) Caicos (the Caicos). The origins of the names are unknown, but "Turks" may be derived from a species of cactus whose flower resembles a Turkish fez; and *cai icoco* may be the name of a type of plum tree. In 1513, searching for the Fountain of Youth, Ponce de Léon found no rejuvenating waters in these islands, only Indians, and he departed. British settlers from Bermuda arrived in 1678, and at first the islands were placed under the government of the Bahamas, but in 1874, they were annexed to Jamaica. They remained so until

1962, when they were designated a separate Crown Colony. Tourism, especially scuba diving, is the mainstay of the islands' economy.

Turner, J. M. W. (1775–1851) Joseph Mallord William Turner was born in London and studied at the Royal Academy School. After years of teaching, he took a trip to the Continent and began to paint the landscapes that would make him one of England's foremost artists. After he became financially secure, he lived a reclusive life with his father and painted in a highly original style that turned more and more abstract as he sought to capture light, space, and the elemental forces of nature on canvas. Although he painted all sorts of landscapes, he was particularly interested in nautical subjects, such as shipwrecks, fires, slave ships, sunsets at sea, and whaling ships, all of which he infused with a luminosity that was most unusual for the period and foreshadowed the Impressionists. His painting of the HMS *Téméraire,* a ship made famous at the Battle of Trafalgar in 1805, being towed to her last berth is considered one of his greatest works.

turnstone (*Arenaria interpres*) Sometimes known as the ruddy turnstone because of its chestnut-colored back, this is a strongly marked, pugnacious little shorebird that feeds by overturning shells, pebbles, and seaweed, and even digging holes to get at small mollusks, crustaceans, and insects. Turnstones breed in the high Arctic tundra surrounding the North Pole, and migrate southward to North America, Africa, Asia, New Zealand, and Australia. To taxonomists they are a problem, since they have been variously classified as plovers, sandpipers, or in a family by themselves. There is also a darker variety, known as the black turnstone (*A. melanocephala*).

turnstone

See also plovers, sandpiper, shorebirds

Tuvalu Independent Commonwealth nation in the western Pacific, made up of nine low coral atolls scattered over .5 million square miles. The islands were visited by Captain John ("Foul Weather Jack") Byron in 1764 in his search for Terra Australis Incognita. Formerly the Ellice Islands, a component of the Gilbert and Ellice protectorate, Tuvalu gained its independence in 1978. The islands are not self-supporting, and a large proportion of their subsistence comes from trust funds set up by Australia, New Zealand, and the United Kingdom. **See also Gilbert Islands, Kiribati**

Twenty Thousand Leagues under the Sea One of the great adventure novels written by Jules Verne (1828–1905), the father of science fiction. (He also wrote *Journey to the Center of the Earth* and *Around the World in Eighty Days.*) In this book we meet Professor Arronax, his servant Conseil, and harpooner Ned Land, who, while investigating a series of mysterious ship sinkings, are captured and taken aboard the *Nautilus,* a submarine manufactured and commanded by Captain Nemo, a self-styled recluse from topside civilization. While on their journey—the title refers to the distance traveled while under the sea, not the depth—the undersea travelers sail around the world underwater, experiencing such wonders as an underwater volcano, a stroll through Atlantis, a battle with a giant squid, the underside of an iceberg, and a trip through a tunnel from the Red Sea into the Mediterranean. The novel was made into the first underwater feature film by J. E. Williamson in 1917, and then again in 1954 by Walt Disney Studios, whose production was directed by Richard Fleischer. **See also Verne**

typhoon Known as hurricanes in the western Atlantic, typhoons (sometimes called cyclones) are severe tropical storms of the western Pacific. (The name comes from the Chinese *tai-fun,* meaning "big wind.") Where the winds of a hurricane spiral inward counterclockwise, those of a typhoon spin clockwise. Characterized by howling winds and torrential rains, typhoons can cause incalculable damage. The 1881 typhoon in Indochina was responsible for 320,000 deaths. The funnel shape of the Bay of Bengal makes it particularly susceptible to killer typhoons, and history is filled with these disasters. In October 1737, a typhoon hit Calcutta, causing a 40-foot storm surge and killing 300,000. During an 1876 cyclone in India, the ebbing Megna River was blocked by a storm surge, and 28,000 square miles of the delta were flooded, drowning more than 100,000. Then in 1970, a typhoon swept over the same area (now Bangladesh) and may have killed as many as .5 million. Another typhoon and storm surge drowned 130,000 more Bangladeshis in 1991. In December 1997, supertyphoon Paka roared across the southwest Pacific, and winds of 236 mph were measured at Andersen Air Force Base on Guam, the highest surface-velocity winds ever recorded. **See also cyclone, hurricane, storm surge**

Tyrian purple The deep purple coloring agent used to dye the woolen cloth of the ancient Romans. By 68 B.C., the city of Tyre, in what is now Lebanon in the eastern Mediterranean, had come under Roman rule. To make the precious dye, murex snails were collected and the animals removed from the shells and boiled to extract the dye. The mixture was exposed to the sun, and the cloth plunged into it when the desired hue was reached. The color was used exclusively for the clothing of the royalty and aristocracy, hence the term "royal purple." **See also murex snail**

U

U.S. Exploring Expedition One of the first "official" U.S. government scientific expeditions was the Great United States Exploring Expedition of 1838–1842. Under the command of forty-year-old Lieutenant Charles Wilkes, the six-ship squadron was originally commissioned to search for the hollow center of the earth, which John Cleves Symmes had determined could be entered from an entrance under the South Pole. However crackbrained this "Holes in the Poles" idea sounds now, it was enough to get President John Quincy Adams to encourage Congress to support the venture. A more realistic, commercial objective was the search for new sealing and whaling grounds in the South Pacific, but Symmes's theory was the nominal impetus for the expedition. Consisting of the flagship *Vincennes* and the *Porpoise, Peacock, Relief, Flying Fish,* and *Sea Gull,* the expedition departed from Hampton Roads, Virginia, on August 18, 1838. Most of their "exploring" would take them to the South Pacific and the Antarctic, and the complement of scientists—known as "the scientifics"—gave the expedition an oceanographic importance far beyond its stated or subliminal warrants. On board were mineralogist James Dwight Dana, philologist Horatio Hale, naturalists Titian R. Peale and Charles Pickering, botanists William Brackenridge and William Rich, conchologist Joseph Couthouy, and artists Alfred Agate and Joseph Drayton. On the outward leg of the voyage the squadron put into Madeira for provisions, passed the bulge of Africa, and stopped briefly at the Cape Verdes en route to Rio de Janeiro, where the botanists collected some fifty thousand specimens. Wilkes maintained the scientific nature of the voyage by seeing to it that hydrographic and meteorological data were collected regularly, and they traced the course of the Gulf Stream by taking the temperature of the water. From Rio, they sailed south to round Cape Horn, passed Tierra del Fuego, and entered the South Pacific. Their legacy remains; the "scientifics" set the stage for all scientific expeditions to follow. **See also Wilkes**

U-boat Abbreviation for the German *unterseeboot* ("undersea boat"). U-boats were originally intended by the Germans to attack British civilian supply ships during the early days of World War I. Their efforts were enormously successful and came within a whisker of isolating Britain from her suppliers, stopped only by the intervention of the United States. The Cunard liner *Lusitania* was sunk off Ireland by *U-20* on May 7, 1915, and among her 1,198 casualties were 128 Americans. Once the United States entered the war, Germany began unrestricted attacks on merchant shipping and, by April 1917, had sunk 430 Allied vessels. By the end of the war, there were 140 U-boats in action, but the introduction of the convoy system and American destroyers greatly reduced their effectiveness. Hitler's Germany, though expressly forbidden to build submarines, began to do so anyway, and by 1939, they had fifty-seven in operation. On October 8, 1939, *U-47* slipped into the British anchorage at Scapa Flow in the Orkneys and torpedoed and sank the battleship *Royal Oak,* which had 833 men aboard. In what came to be known as the Battle of the Atlantic, "wolf packs" of U-boats attacked Allied vessels; during 1942, they sank more than 6 million tons of shipping. By 1943, with the introduction of the *schnorkel* that enabled the subs to remain submerged longer, they almost succeeded in severing Britain's Atlantic lifeline. The Allies countered with radar and attacked the U-boats with Liberator bombers, and by the end of the war, 785 German U-boats (out of a total of 1,162) had been sunk, scuttled, or captured. **See also submarine**

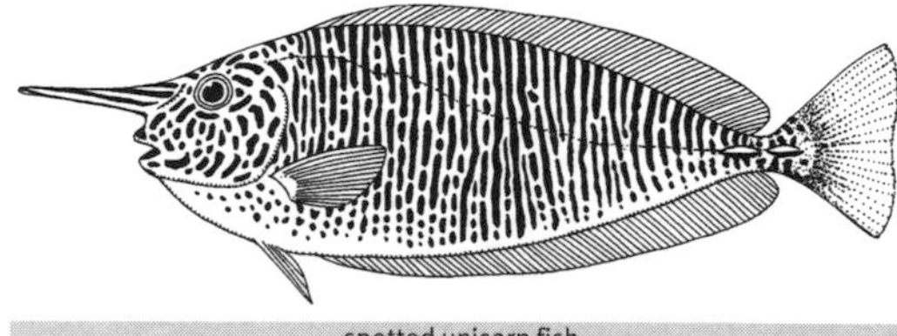

spotted unicorn fish

unicorn fish (*Naso brevirostris*) Other surgeonfishes have one spine on each side of the base of the tail, but the unicorn fishes of the genus *Naso* have two. In addition, they have a "horn" that projects straight out from the head. In some species, the horn is just a little bump, but in others it is quite pronounced. Unicorn fishes are found throughout the Indo-Pacific region. One species (*N. tuberosus*) is called the humpnose unicorn fish, because instead of a horn it grows a large protuberance on its snout, making it look not unlike a miniature beluga. Another fish, completely unrelated to these surgeonfishes, is also known as the unicorn fish. It is *Lophotus fiski,* a relative of the oarfishes and ribbonfishes. It is very little known, but it has a long, flexible dorsal spine that originates from its snout, which protrudes beyond the mouth. **See also oarfish, surgeonfishes**

upside-down jellyfish (*Cassiopea andromeda*) *C. andromeda* is one of the "many-mouthed jellyfishes" (or Rhizostomeae, which means "root-mouthed"). Members of this species spend their time closely packed together, lying on their backs on the bottom, with their voluminous mouth lobes exposed. This species occurs in shallow lagoons, intertidal sandbanks, or mudflats around the Caribbean and the Indo-Pacific. Like corals, these jellyfishes have zooxanthellae in their tissues, so they are capable of photosynthesizing sugars from the carbon dioxide produced by their hosts. Instead of trapping prey in its tentacles, *Cassiopea* lives on the products of photosynthesis in its tissues, exposed to sunlight because of its upside-down position. In Greek mythology, Cassiopeia was the mother of Andromeda, both of whom were incorporated into this jellyfish's scientific name. **See also jellyfish, zooxanthellae**

upwelling The upward movement of colder water, usually along the western coasts of continents, when warmer surface water is displaced from the shoreline by the prevailing winds. Upwellings, which occur on the northern coasts of western North and South America as well as northwestern and southwestern Africa, bring up nutrient-rich waters, which produce a plankton bloom that feeds large numbers of fish and birds. Because of these upwellings, the Humboldt Current off the Peruvian coast supports huge populations of anchovies, which in turn support the great colonies of cormorants, gannets, and pelicans that populated and contributed to the guano islands. When the winds shift (often as a result of El Niño), the upwellings diminish, the plankton decreases, and the number of anchovies decreases dramatically. If the rising water is colder than the ambient air, fog results, as in northern California.
See also El Niño, guano islands, Humboldt Current

Urquiola Supertanker that ran aground on the northwest coast of Spain on May 12, 1976, discharging 28 million gallons of crude oil into the harbor and bay of La Coruña. The ship ran aground while entering the harbor, but as tugs approached to pull her off, there was an explosion belowdecks and the ship caught fire. Oil on the water burst into flames, and crewmen who had jumped overboard were killed. The fire burned for eighteen hours, and some 30,000 gallons of oil washed ashore. It covered 130 miles of beaches with thick sludge, killing much of the invertebrate life of the shore and littering the beaches with dead fish.
See also oil spill, supertanker

Ushant French island (Île d'Ouessant) off the coast of Brittany; the most westerly point of France. Ten miles off the mainland, it divides the entrance to the English Channel from the Bay of Biscay, and in the days of sail, it was an important landmark for ships heading out of the Channel. Now a very powerful lighthouse, the Phare de Créach, identifies the entrance to the Channel. The island is about 5 miles long and 2 miles wide, and the population of around 800 are mostly fishermen and farmers. On October 14, 1747, an important battle of the War of the Austrian Succession was fought about 200 miles to the south, but the battle was given the name of the island. In 1794, the Battle of the Glorious First of June, also known as the Battle of Ushant, was the first great naval engagement of the French Revolution. On that date, the British fleet under Lord Howe attempted to intercept a grain convoy from the United States that was being escorted by a French squadron. Although the grain got through, the British claimed victory (and named the battle) because they captured six of the French ships.
See also English Channel

V

vampyroteuthis (*Vampyroteuthis infernalis*) Not exactly a squid and not exactly an octopus, the "vampire squid from hell" has eight tentacles and two wispy filaments instead of the grasping tentacles of the squids. It is a deepwater species that gets to be about a foot in length and has the ability to wrap its cirrate arms over its head so that it effectively turns itself inside out. The reason for this behavior is a mystery. It is a warm sepia brown in color, with startling, sky-blue eyes that are proportionally the largest eyes of any animal for its size. It was previously known only from dead animals that had been captured in nets, but in recent years, the living animal has been photographed from a submersible.

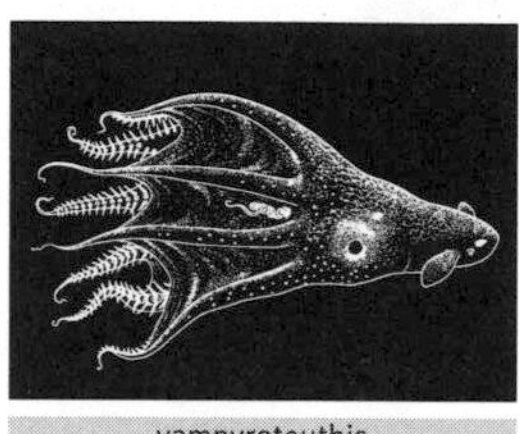
vampyroteuthis

See also cirrate octopus, octopus, squid

Vancouver, George (1757–1798) British navigator and explorer who sailed with James Cook on his second (1772–1775) and third (1776–1780) voyages. In 1791 he set out for the Pacific Northwest to survey the coastline and to take over the territory that had been assigned to England by the Nootka Convention that resolved the difficulties of sovereignty with Spain. En route, he rounded the Cape of Good Hope, investigated parts of Australia and New Zealand, visited Tahiti and Hawaii, and arrived in North America in 1792. For three years, Vancouver meticulously explored and surveyed the American coast from Alaska to San Francisco, and he circumnavigated the island that now bears his name. He named Puget Sound after his lieutenant, Peter Puget, and the Strait of Georgia after George III. George Vancouver died shortly after his return to England in 1795, but his brother and Peter Puget published his three-volume *Voyage of Discovery to the North Pacific Ocean and round the World* in 1798.

Vancouver Island The largest island in western North America, Vancouver is 285 miles long and 30 to 80 miles wide. Heavily forested, it is separated from the mainland of Canada by the Queen Charlotte, Georgia, and Johnstone Straits, and from the United States by the Strait of Juan de Fuca. The capital is Victoria, and the other major cities are Nanaimo, Port Alberni, and Esquimalt. (There are also two cities called Vancouver: one is the provincial capital of British Columbia on the Canadian mainland; the other is in the state of Washington, opposite Portland, Oregon.) Before the arrival of the Europeans, the island was inhabited by southern Kwakiutl and Nootka Indians, but by the eighteenth century, both Spain and Britain claimed it. It was sighted in 1774 by Juan Peréz, but in 1778, James Cook was the first to land there. In 1789, the English trader John Meares built a fort on Nootka Sound, which was taken by the Spanish, but after Captain George Vancouver circumnavigated and mapped the island in 1792, the Spanish gave up all claims on the northwest coast. The island was held by the Hudson's Bay Company until it was made a Crown Colony in 1849; it was annexed to British Columbia in 1866. Lumbering and wood processing are the major industries, but there are also coal, gold, copper, and iron ore mines. The island has many large parks, and its beautiful scenery and abundant wildlife (killer whales, eagles, salmon) make it popular with visitors.

See also Cook (James), Vancouver

Vane, Charles (d. 1719) One of the pirate leaders (another was Blackbeard) who were based in the pirate colony of New Providence in the Bahamas, and who ranged the eastern seaboard of the American colonies from the Caribbean and Florida to Maine, preying on merchant shipping. When Woodes Rogers, the ex-privateer who in 1717 had been appointed governor of the Bahamas in an effort to eradicate piracy, arrived in Nassau, Vane welcomed him with a fireship. Later his crew replaced him with "Calico Jack" Rackham because he proved to be a coward in an encounter with a French warship. After he was rescued from a shipwreck in Honduras, he was hanged in Jamaica. At the gallows, according to Daniel Defoe, "he showed not the least Remorse for the Crimes of his past Life."

See also Blackbeard, Rogers

Vanuatu Discovered in 1606 by the Portuguese navigator Pedro Fernández de Quirós, this independent republic in the southwestern Pacific consists of thirteen large islands (the largest is Espirítu Santo) and many smaller ones. The capital is Vila, on the island of Éfaté. Rediscovered by Louis de Bougainville in 1768, the islands were charted and named the New Hebrides by James Cook in 1774. British and French settlers began arriving in the 1860s. Conflicts between British and French interests were resolved by the establishment in 1881 of a joint naval commission to administer the islands. After the "sandalwooders" of the nineteenth century stripped the islands of these aromatic trees, they began kidnapping the islanders to work the sugar and cotton plantations of Queensland, Australia. There was no Japanese invasion of the New Hebrides during World War II, and the island became a major Allied base. The island of Tanna was the site of the Jon Frum cargo cult, where believers dressed in homemade uniforms and built airstrips and dummy equipment in anticipation of the arrival of air cargo. Because so much of the land was owned by foreigners, the islanders held their own elections, and in 1980 declared themselves the Republic of Vanuatu. Vanuatu covers 5,700 square miles and has a population of 177,000. The major industries are copra production, tuna fishing, manganese mining, and cattle raising. Vanuatu refused to condemn France's 1995 resumption of nuclear testing at the Tuamotu atoll of Mururoa, maintaining that it was a French internal matter.

See also Greenpeace, Mururoa, Tanna

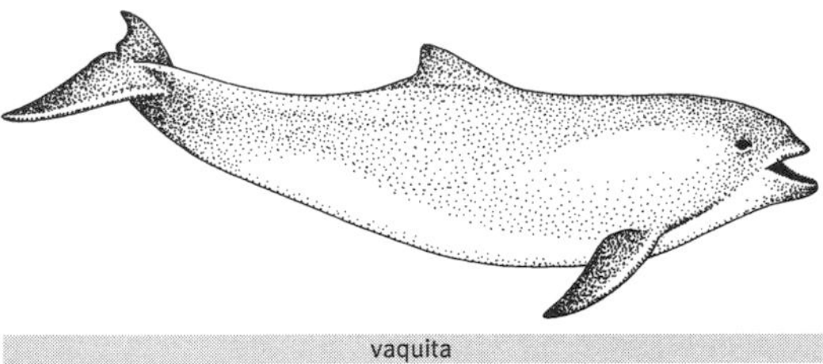

vaquita

vaquita (*Phocoena sinus*) Also known as the cochito, this close relative of the harbor porpoise is found only in the Sea of Cortez (Gulf of California), between Baja and mainland Mexico. At a maximum length of 4½ feet, it is even smaller than the harbor porpoise, and may be the smallest of all the cetaceans. It is similar in coloration to the harbor porpoise, although the mouth-to-flipper stripe is not as distinct and the dorsal fin is more sharply curved. Directed hunting and entanglement in gill nets (in the totoaba fishery) have seriously reduced the population of vaquitas. With only a few hundred remaining, they are among the most endangered of all marine mammals.

See also harbor porpoise, totoaba

Vasa A 200-foot-long, 64-gun wooden warship built for King Gustavus Adolphus of Sweden. On August 10, 1628, just after she was launched in Stockholm Harbor, a sudden gust of wind caused her to keel over, and water began to rush into the open cannon ports. She sank with her sails raised and her flags flying. Attempts to raise the ship in the seventeenth century were unsuccessful (only some of the cannons were salvaged), but in 1961, five years after she was rediscovered by Anders Franzen, the almost-complete hull was brought to the surface from a depth of 110 feet. Archaeologists found a treasure trove of historical artifacts, including carvings, textiles, coins, equipment, and skeletons of some of the fifty crew members who were drowned in the accident. To keep the waterlogged wood from drying too fast, the hull was continuously sprayed with water and a preservative solution from 1962 to 1972. The remains of the ship, the sculptures, and other artifacts are now housed in their own museum on an island in Stockholm Harbor.

See also marine archaeology

Vaterland *Vaterland*—"Fatherland" in German—was designed in 1911 to be the world's largest ship; she was launched in 1913. She measured 908 feet overall, compared to *Titanic*'s 882 feet. In her sea trials off Norway, she achieved a speed of 26.3 knots, faster than any other ship of the period. She crossed the Atlantic on her maiden voyage, but she was set to return to Germany when in July 1914, the war in Europe broke out, and when the *Lusitania* was sunk by German torpedoes on May 7, 1915, *Vaterland* was interned in New York. When the United States entered the war in 1917, American forces seized the ship; she was renamed *Leviathan* and transformed from luxury liner to troopship. On one crossing, she carried 14,416 soldiers—more human beings than had ever before sailed on a single ship. After the war, *Leviathan* was converted back to a liner and became the flagship of the United States Lines. In the ensuing years, she traversed the Atlantic many times as a popular cruise ship, and in 1938, having begun to age noticeably and proving too costly to maintain, she was retired and broken up in Scotland.

See also *Lusitania, Titanic*

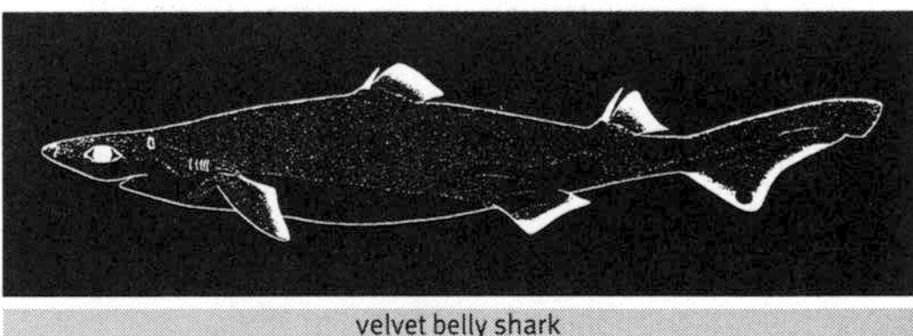

velvet belly shark

velvet belly shark (*Etmopterus spinax*) One of the so-called lanternsharks, named for their bioluminescent capabilities. This 2-foot-long species is common in the eastern Atlantic, from Norway to the Gulf of Guinea, along the continental slopes. Accounting for its name, the little shark's belly is black, but it can glow with a strong greenish color.

See also green dogfish, lanternsharks

Venice An island city at the head of the Adriatic Sea, Venice is not only surrounded by water, but water runs through its very landscape, as it does nowhere else on earth. Now moored in a crescent-shaped lagoon that stretches for 32 miles, the island was originally composed of islets, mudflats, and sandbanks that were united by its early settlers into a single entity, with canals replacing the water that once separated the little patches of land. The canals are crossed by more than four hundred footbridges. A series of permanent sandbanks (*lidi*) protect the shallow waters of the lagoon from the south, but there are gaps that allow the passage of the tides and the city's maritime traffic. The city includes all the islands inside the *lidi*, such as Murano, Burano, Torcello, and the *lidi* themselves, one of which is named Lido, the singular of *lidi*. Every year between October and April, the city is inundated by high tides (*l'acqua alta*) that pour through the three entrances to the lagoon, threatening the city's fragile foundations. Originally settled by mainland Venetii fleeing from the invading Lombards, the islanders elected their first doge (from the Latin *dux*, "leader," also the root of "duke") in 727. Byzantines controlled the islands until 800, when Venice fell under the control of the Italian king Pepin. By the ninth century, the group of islands had proudly proclaimed itself the city of Venice. Under Doge Domenico Contarini (1043–1070) a new church was built for Saint Mark, the patron saint of the city. (In 828, two merchants had stolen the saint's body from Alexandria and brought it to Venice.) From the eleventh to the fourteenth centuries, Venetians dominated the Adriatic, and supplied ships, food, and money to the Crusaders. In 1348, along with most of Europe, Venice was decimated by the Black Death, and in its weakened state, it lost Dalmatia to the Hungarian king Ludwig I. Throughout the fourteenth century, the increasingly powerful Serenissima ("Serene City") expanded its sphere to the mainland, controlling Padua, Vicenza, Verona, Belluna, and Friuli. Constantinople's conquest by the Turks in 1479 marked the end of trade with the Levant, and the discovery of America transferred commercial power to Spain and Portugal. Although Venice had extended her influence through the eastern Mediterranean to Corfu and Cyprus, the Turks soon took these islands, but were prevented from further expansion when the Venetians joined with Spain and Rome to defeat the Turkish fleet at the Battle of Lepanto in 1572. Another plague in 1630 further contributed to Venice's decline. The city was taken by Napoleon in 1797 and handed over to Austria the following year. After Prussia defeated Austria at Sadowa in 1866, Venice was absorbed into the Kingdom of Italy. A railroad bridge was built in 1847, and in 1882, an aqueduct was built to bring water from the mainland (which the Venetians call *terraferma*). The city of Venice is world-famous for its architecture, painting, music, festivals, bridges, and canals. Once known as the "queen of the sea," the city has begun to decline for many of the reasons that made it famous in the first place. Tourists flood Venice, much as the threatening high waters do, causing a strain on the island's limited resources. Venice has practically no sewer system; the canals are supposed to carry the wastes away. There are no cars—only motorboats and gondolas are allowed—but pollution is widespread because of the acid-spewing factories on the mainland. Venice's watery passages are contributing to the sinking of the city; her artworks and buildings are decaying, and the population of the historic center (100,000 and dropping), perhaps unwilling to remain in a city that has become a museum of itself, is deserting the city.

See also Burano, Murano, Torcello

Verne, Jules (1828–1905) French author, generally considered by many to be the originator of science fiction. Born in Nantes and trained as a lawyer, Verne became interested in the theater and wrote librettos for operettas, but he soon turned to writing novels, usually about extraordinary journeys. His first was *Five Weeks in a Balloon* (1863), followed by *A Journey to the Center of the Earth* (1864), *From the Earth to the Moon* (1865), and, in 1870, what is considered his masterpiece, *Twenty Thousand Leagues under the Sea. Around the World in Eighty Days* was published in 1873, *The Mysterious Island* in 1875, and *Michael Strogoff* in 1876. During his lifetime, Verne's novels were extremely popular, and they have continued so to this day, largely through the medium of movies. *Twenty Thousand Leagues under the Sea* was filmed twice: first in 1917 by J. E. Williamson, and then in 1954 by Walt Disney Studios.

See also *Twenty Thousand Leagues under the Sea*

Verrazano, Giovanni da (1480?–1528?) Italian navigator and explorer in the service of François I of France. He sailed west from Madeira in 1524 seeking a passage to the Pacific, but encountered the impenetrable coast of North America instead and explored from North Carolina to Maine, and perhaps even Newfoundland. He anchored *La Dauphine* in "a very pleasant place" that was later to become New York Harbor, at the mouth of a large river that would later be named for Henry Hudson. Verrazano then sailed east from New York, rounded Long Island, and sailed into Narragansett Bay, where he is said to have named "Rode Island" because it reminded him of the Greek island of Rhodes. On a later voyage (1528?) he explored the West Indies and was killed by natives somewhere in the Caribbean. Some four centuries later (1964), the Narrows would be spanned by the Verrazano-Narrows bridge from Brooklyn to Staten Island—at 4,260 feet, the longest suspension bridge in the United States.

Vespucci, Amerigo (1454–1512) Italian navigator in the service of Spain in whose honor America was named. Born in Florence, Amerigo Vespucci entered the service of the Medici family and moved to Seville in 1492. He was probably there when Columbus returned from his first voyage in 1493. (There is an account of a 1497 crossing that would have put him on the American continent a year before Columbus's 1498 landfall in Venezuela, but this is evidently a fiction.) Vespucci was appointed navigator on the 1499 passage of Alonso de Ojeda to the West Indies, but the two became separated, and Vespucci sailed alone along the coast of South America, becoming the first European to see the mouth of the Amazon River. Believing he had found Asia, Vespucci returned to Spain in 1500, eager to return to the New World. The Spanish would not send him (or perhaps the ships they offered were unsatisfactory), so he sailed under the Portuguese flag instead. In 1501, he reached Baía de Guanabara (Rio de Janeiro), then went south and discovered Rio de la Plata. Upon his return to Lisbon in 1502, Vespucci reported that he had found a "new world" that was not Asia at all. A 1507 map by the German cartographer Martin Waldseemüller, based on Vespucci's accounts, included the word AMERICA prominently on South America. This is believed to be the first use of the word "America," and although it is disputed by some historians, it may be the reason the Americas—North and South—are so named. Vespucci's accomplishments are so poorly documented that his very explorations are the subject of ongoing debates: Did he make three or four voyages? Are his letters genuine or forgeries? Was America named for him or for a Nicaraguan tribe known as the Amerrique? Even if America was not named for him, the Italian three-masted sailing ship was, and the airport at Florence is called Amerigo Vespucci. **See also Ojeda**

Victoria Island After Baffin and Ellesmere, Victoria is Canada's third largest island, covering an area of 81,000 square miles, about the same size as Kansas. To the south is the fabled Northwest Passage. The island was discovered by Thomas Simpson in 1838 and named for Queen Victoria. It is sparsely inhabited by nomadic Eskimos, and there is a joint United States–Canadian weather station at Cambridge Bay. Since April 1999, when the vast territory of Nunavut was separated from the Northwest Territories and placed under control of the native peoples, Victoria Island has been administratively divided in half, with the western portion remaining in the Northwest Territories.

Vikings Scandinavian raiders of the eighth to tenth centuries, who sailed southward in search of land, wealth, and fame. The early Swedish Vikings were primarily traders who crossed the Baltic and made their way along the Vistula, the Dneiper, and the Volga and eventually reached Constantinople. They founded the city-states of Novgorod and Kiev, and one Swedish tribe, the Rus, gave Russia its name. From Norway and Denmark, however, came the Vikings that were dedicated to plunder. The best shipbuilders and sailors in the world at that time, they took their *knorrs* (longboats) to the coasts of western Europe and the Atlantic, particularly England and Ireland. By 868 they had conquered nearly half of England, until the rampaging tide was stemmed by King Ethelred of Wessex and his brother Alfred, who became known as Alfred the Great. They also landed in northern France, plundered Paris in 845, and passed through the Straits of Gibraltar into the Mediterranean, pillaging some parts of North Africa, the Balearic Islands, and the coast of Italy. At first, they burned and raided and returned home, but in time, they began to settle those lands, which were so much richer than those they had left. The Vikings discovered Iceland, and sailing from there, around 930, a Norwegian sailor named Gunnbjorn was the first European to set foot on the island of Greenland. And Eric the Red, who had been found guilty of manslaughter in Norway, fled to Iceland in 984, and the following year landed on a frozen land that he named Greenland. In 986, an Icelander named Bjarni Herjolfsson sighted the North American continent, and around the year 1000 Leif Ericsson, the son of Eric the Red, became the first European to set foot on it, first at Baffin Island (which he named Helluland), then at Labrador (Markland), and finally, at Newfoundland, which he named Vinland. Stories of the Vikings reaching Rhode Island or Minnesota are generally considered folk tales or hoaxes.
See also Greenland, Iceland, Leif Ericsson, Newfoundland, Vinland

Villiers, Alan (1903–1982) British mariner and marine author, born in Melbourne, Australia. He went to sea at the age of fifteen and spent the next five years in square-rigged ships. In 1923, while he was working as a proofreader for the local newspaper in Hobart, Tasmania, the whaling factory ship *Sir James Clark Ross* arrived, and Villiers signed on as a laborer, promising his editor he would take notes. These notes became *Whaling in the Frozen South,* considered the definitive description of Antarctic whaling up to that time. In 1931, he joined the crew of the four-masted bark *Parma,* which twice won the grain race from Australia to England. In 1934 he purchased the Danish sail training ship *Georg Stage,* renamed her *Joseph Conrad,* and sailed around the world with a crew of cadets. During World War II he served as an officer in the volunteer reserve, and commanded a squadron of landing craft during the invasions of Sicily and Normandy. In 1956, he sailed a replica of the *Mayflower* across the Atlantic. Villiers served as consultant for the ships used in such movies as John Huston's *Moby Dick* (1956) and Peter Ustinov's *Billy Budd* (1962). A prolific author—and a stickler for correctness in nautical terminology—he also wrote *The Making of a Sailor, Whalers of the Midnight Sun, Wild Ocean, The Coral Sea, Windjammer, Falmouth for Orders, The Quest of the Schooner Argus, The Way of a Ship, Captain James Cook, The New Mayflower, The*

Battle of Trafalgar, Give Me a Ship to Sail, and *Men, Ships, and the Sea.*

Vinland The name given by Leif Ericsson to the landfall he made around A.D. 1000. Ericsson had sailed from Iceland heading for Greenland, but missed the large island because he sailed too far south. Bjarni Herjolffsson, another Icelander, evidently made the same off-course journey around 986, but he only saw the land and never set foot on it. Ericsson reported three different locations: Helluland, Markland, and Vinland. The first ("Flat Rock Land") is believed to have been Baffin Island; Markland ("Forest Land") was probably Labrador; and Vinland was almost certainly Newfoundland, specifically the location known as L'Anse aux Meadows. Recent archaeological excavations at this site have revealed the remains of eight turf-walled houses, one of which was a longhouse, 72 feet long. Ericsson wintered over at Vinland, loaded a cargo of grapes and timber, and returned to Greenland. His reports led to other voyages to Vinland; Thorfinn Karlsefni led one of these, accompanied by his wife, Gudrid; their son Snorri is believed to have been the first European child born in the New World. Karlsefni's settlers were met by hostile "Skraelings" (either Eskimos or Algonquin Indians) and were eventually forced to abandon their settlement and return to Greenland.

See also Eric the Red, Greenland, Leif Ericsson, Thorfinn Karlsefni, Vinland Map

Vinland Map In 1965, Yale University revealed the heretofore unsuspected existence of a map, dated 1440, that showed, among other things, Bjarni Herjolffsson's Vinland. (It also depicted Greenland as an island, a fact that was not known until 1650, but then, "Greenland" appears with two deep fjords on the east coast, which exist on Baffin Island—which, of course, *is* an island.) Published fifty years before Columbus's first voyage, the map was further evidence that the Norsemen had reached North America long before any other Europeans. When it was published—as *The Vinland Map and the Tartar Succession*—it created a great furor, especially among Italian Americans, who wanted Columbus to retain the title of discoverer of America. The map seemed to "prove" that the Icelanders had discovered Vinland, but some experts questioned its authenticity, citing certain cartographic anomalies and claiming that some of the inks on the map could be dated to the twentieth century. Others maintain that the map is genuine, and the "proof" that it is a forgery—or at least, an altered version of an authentic map—is inconclusive. Whether the Vinland Map proves it or not, there is little question that the first Europeans to land in North America were the Norsemen.

See also Columbus, Leif Ericsson, Vinland

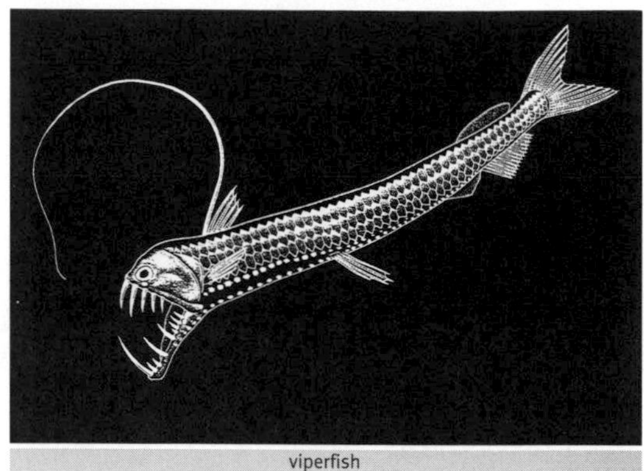

viperfish

viperfish (*Chauliodus sloanei*) Lurking at depths of 8,000 feet or more during the day, the foot-long viperfish moves closer to the surface at night in search of prey. Like many of the similar dragonfishes, *Chauliodus* has rows of photophores along its body, but unlike most of them, it also has scales. Its fangs, too long to fit in its mouth, extend beyond the eyes. It is able to open its jaws to an angle greater than 90° and can engulf fishes larger than itself. **See also dragonfish**

Virgin Gorda Island in the British Virgin Islands named "Fat Virgin" because early Spanish sailors likened its shape, narrow at both ends and swollen in the middle, to a reclining woman. The island is 7 miles long and supports a population of approximately 5,000 people. It is visible from Tortola, just across the Sir Francis Drake Channel. The vegetation on the 8-square-mile island is mostly scrub and cactus, except where hotel landscapers have introduced tropical foliage. On the southern end of the island there is a curious assemblage of house-sized granite boulders that make up The Baths, a popular spot for swimming and snorkeling. Virgin Gorda is now largely devoted to tourism, with several hotels, restaurants, and Little Dix Bay, a luxurious resort complex.

See also British Virgin Islands, Tortola

Vizcaíno, Sebastián de (1550–1615) Spanish explorer aboard the Manila galleon *Santa Ana* in November 1587 when she was looted by the British privateer Thomas Cavendish off Baja California; he led the survivors to safety. He returned in 1596 to lead an expedition to Baja California in an unsuccessful attempt to plant a colony there. In 1602 he was named chief pilot for New Spain, and sailed with three ships, *San Diego, Santo Tomás,* and *Tres Reyes,* northward along the coast from Acapulco, stopping at and naming—among other things—San Diego and Monterey Bays, San Clemente Island, Carmel, and Santa Barbara. One of his ships, under Juan Martin de Aguilar, became separated from the others and sailed as far north as Cabo Blanco in Oregon (and may have entered the Columbia River), but Aguilar had to turn back when his crews became ill with cold and scurvy. Admiral Vizcaíno voyaged across the Pacific in 1611 to search for the fabled (and nonexistent) islands of Rica de Ojo and Rica de Plata, and to establish trade with Japan, but did not succeed. Vizcaíno Peninsula, Desert, and Bay on the Pacific side of the Baja Peninsula are named for him.

See also Cavendish, Gulf of California, Manila galleon

Voss, Gilbert (1918–1989) Renowned cephalopod biologist. Even before he went to college, Gilbert Voss had served in the Merchant Marine and in the U.S. Coast Guard. He entered the University of Miami in 1947, graduated in 1951, and went on to George Washington University for his Ph.D. He returned to the University of Miami and taught there until his death in 1989. Under his tutelage, an entire generation of cephalopod biologists was trained, including Clyde Roper, Richard Young, and Roger Hanlon, all of whom are now leaders in the field. Voss was equally at home at sea and in the laboratory, and he published some one hundred scientific papers, mostly about squid and octopuses. He served as editor of the prestigious *Bulletin of Marine Science*, and also wrote popular articles on cephalopods for *National Geographic* and *Sea Frontiers*.

Vulcano Named for Vulcan, the Roman god of fire, this island in the Lipari group (once known as the Aeolian Islands) off the northeast coast of Sicily has given its name to all of the exploding mountains of the world. Since its last major eruption in 1888–1890, numerous towns have arisen on the slopes of what is considered, after Vesuvius, to be Italy's most dangerous volcano. The highest point on the island is 1,640 feet above sea level. There are no fewer than four potentially active locations: the *gran cratere* of Vulcano itself, the crater Vulcanello, the Lentia complex, and the Fossa cone. Where once there were only small houses, there are now hotels and villas, and while there are only about 500 permanent residents of the island, another 10,000 tourists arrive during the summer.

See also Aeolian Islands, Stromboli

W

Waddell, James (1824–1886) Confederate naval officer who took the raider *Shenandoah* to the whaling grounds, intent upon capturing or sinking Union whaleships. In October 1864, he captured the Maine whaler *Aliana* off Dakar, took her crew prisoner, and scuttled the ship. In the Okhotsk Sea he captured three more ships, but when he was told that the war was over, he refused to believe it and increased his efforts. In Alaskan waters he caught and burned thirteen more vessels, and by the time he quit, he had destroyed a total of twenty-four Yankee whaleships—most of them after the surrender at Appomatox on April 9, 1865. In November he arrived in Liverpool, but the British would not arrest him, since they believed he was doing his duty as a Confederate officer. He returned to Maryland, where he was put in charge of the flotilla that protected the Chesapeake Bay oyster beds.

wahoo (*Acanthocybium solanderi*) Known as *ono* in Hawaii, the wahoo is distributed throughout the warm waters of the world, where it tends to be a loner or to travel in small groups. It is classified with the mackerels and tunas (family Scombridae), and is reputed to be one of the fastest fish in the sea, attaining speeds of up to 50 mph. A favorite game fish that fights courageously when hooked, the wahoo averages 15 to 20 pounds, but the record rod-and-reel catch weighed 155 pounds.

See also frigate mackerel, Scombridae

Wake Island Atoll in the central Pacific, 2,300 miles west of Honolulu, Wake is composed of three low coral islets, Wake, Wilkes, and Peale. Wake was named for Captain William Wake, an Englishman who visited in 1796, and Wilkes and Peale for two prominent members of the U.S. Exploring Expedition of 1841, Charles Wilkes and Titian Ramsey Peale. The islands were annexed by the United States in 1899 for use as a cable station. In 1938, Pan American Airlines built a seaplane base there, and on December 8, 1941, the day after their attack on Pearl Harbor, the Japanese attacked Wake, but were repulsed by coastal defense guns and aircraft. They struck again on December 23 and overran the island. U.S. planes repeatedly bombarded Wake from 1942 until the Japanese surrender in 1945. One of the casualties of this occupation was the flightless Wake Island rail (*Gallirallus wakensis*), the only land bird on the island. Starving Japanese troops, cut off from any supplies, hunted down and ate every single one.

See also U.S. Exploring Expedition

Wallis, Samuel (1728–1795) British naval officer who made a 'round-the-world voyage from 1766 to 1768 in HMS *Dolphin.* He departed England in company with the *Swallow,* Philip Carteret commanding, but they were separated after passing through the Strait of Magellan and went on their own exploring cruises. Wallis headed more to the northwest and discovered Tahiti (which he named King George III Island), while Carteret sailed farther south, discovering Pitcairn, New Ireland, and the Admiralty Islands. Wallis also came upon two groups of islands now known as the Wallis and Futuna Islands of French Polynesia. Wallis set the

wahoo

stage for the epochal first voyage of Captain James Cook, which would commence in August 1768, only three months after Wallis returned to England in the *Dolphin.* (Carteret returned in February 1769.) It was Tahiti, discovered by Wallis, that the Royal Society decided would be the perfect spot to observe the transit of Venus that was to occur in June 1769 and that would be of great value in determining the earth's distance from the sun. The observation of the transit of Venus was the reason for Cook's first voyage.

See also Carteret, Cook (James), New Ireland, Tahiti

Wallis and Futuna Two island groups, some 125 miles apart, in the southwestern Pacific, between Fiji in the west and Western Samoa in the east. The Wallis Islands (Îles Wallis in French) were named for their discoverer, British navigator Samuel Wallis, who spotted the island of Uvea and its surrounding coral islets in 1767. The islands were occupied by the French in 1842. During World War II, they were garrisoned by six thousand American troops. By popular referendum, Wallis and Futuna became part of France's overseas territory in 1959. Uvea, also called Wallis Island, occupies the middle of an atoll 30 miles across; it has a population of 8,000 people—and 25,000 pigs. The islands are very French in character, and being so far off the beaten track, they entertain few tourists. They grow enough food to be self-sufficient. Futuna (not to be confused with an island of the same name in Vanuatu) and its accompanying atoll, Alofi, are recently emerged volcanic islands, as evidenced by their hot springs and steaming vents. Futuna and Alofi were first sighted in 1616 by the Dutch navigators Willem Schouten and Jakob Le Maire, who named them the Hoorn Islands, after their home port in the Netherlands. (Earlier in the same voyage, the same explorers named Cape Horn for the same town in Holland.) The total population of Futuna is around 5,000, mostly concentrated in the settlement of Sigavé. Marist missionaries converted virtually the entire population of Wallis and Futuna in the nineteenth century, and they are now all practicing Roman Catholics.

See also Fiji; Le Maire; Samoa, Western; Schouten; Wallis

walrus (*Odobenus rosmarus*) There are two subspecies of walrus, the Atlantic and the Pacific, both inhabitants of the moving pack ice and rocky islands of the Arctic. They have no visible earflaps, so they are not Otariids, and they can rotate their hind limbs forward, so they are not Phocids. Extremely gregarious, they often cluster together in large groups. Males are larger than females, and both sexes have tusks. The largest bulls can weigh 3,500 pounds. Feeding underwater, they use a powerful suction to eat the clams and other mollusks that make up their diet. They may employ the tusks as "guide rails" as they suck up clams from the muddy bottom, but they do not dig with them. (They do occasionally pull themselves along on the ice with their tusks, however.) Walruses have been exploited for their ivory tusks, tough hides, and blubber oil for thousands of years, and although most commercial hunting has ceased, native peoples still kill them in large numbers for their ivory, which they carve into trinkets for the tourist trade.

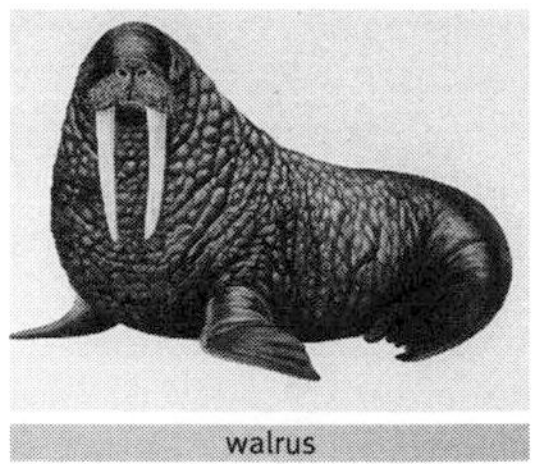

walrus

Walsh, Don (b. 1931) Retired U.S. naval officer, now president of International Maritime Inc. (IMI), an Oregon-based marine consulting practice. In January 1960, Lieutenant Walsh and Jacques Piccard piloted the bathyscaph *Trieste* to the deepest place in the ocean, 35,800 feet down in the Challenger Deep of the Mariana Trench. After retiring from the navy with the rank of captain, Walsh joined the University of Southern California as founder and the first director of the Institute of Marine and Coastal Studies, and was also a professor of ocean engineering. In 1989, IMI formed a joint venture in the Soviet Union, Soyuz Marine Service, a diving services company that still exists in Russia. In addition to working underwater operations in the mid-1960s, Walsh was one of the first oceanographers in the United States to work on remote sensing of the oceans from satellites and aircraft. He also has participated in polar operations in the Arctic and Antarctic, beginning in the late 1950s. The Walsh Spur in the Antarctic is named for his contributions to the U.S. Antarctic Program. He was appointed by Presidents Carter and Reagan to the National Advisory Committee on Oceans and Atmosphere and was a member of the Law of the Sea Advisory Committee for the State Department.

See also Mariana Trench, *Trieste*

Walvis Bay Discovered by Bartolomeu Dias in 1486, Walvis Bay (*Walvisbaai* in Afrikaans, which means "whalefish bay") is one of the few protected harbors on the South Atlantic coast of southern Africa. With its 5-mile-long natural breakwater, the bay was a breeding ground for southern right whales, so it attracted many Yankee whalers. Britain annexed the bay in 1878 to control the rich guano deposits, and in 1884 incorporated it into their Cape Colony. The Germans claimed Walvis Bay as part of their colony of Southwest Africa, but in 1916 they were ousted by British troops, ending Germany's presence in southern Africa. Walvis Bay had been part of the Union of South Africa since 1910, but became part of Namibia when that country achieved its independence in 1994. **See also Dias, right whale**

wandering albatross (*Diomedea exulans*) The largest of the albatrosses—in fact, the largest of all flying birds, with a wingspan that may approach 12 feet. (Besides other albatrosses, the Andean condor is its only rival, with a wingspan that may be 10 feet across.) Wanderers breed on numerous subantarctic islands, including Crozet, Heard, Kerguelen, Gough, South Georgia, Auckland, Campbell, and Macquarie, laying a single, large red-speckled egg. The downy young are fed by one parent or the other until the adult birds leave them sitting on the nest for the entire winter. Juveniles pass through various brownish color stages, becoming lighter and lighter until, by the age of about 10, they become pure white with black wing tips and trailing edges. (Adult females are all white, but with a dark cap.) Like all other albatrosses, wanderers spend almost their entire lives at sea, and most of that in the air. They are magnificent and masterly flyers, soaring tirelessly on stiff, outstretched wings for days or even weeks. Supremely graceful in the air, wanderers are awkward on land, but they alight on the water to feed on squid and small fishes. They are habitual ship followers in the Southern Ocean. Almost as large is the royal albatross (*D. epomophora*), which breeds only on the outlying islands of New Zealand.

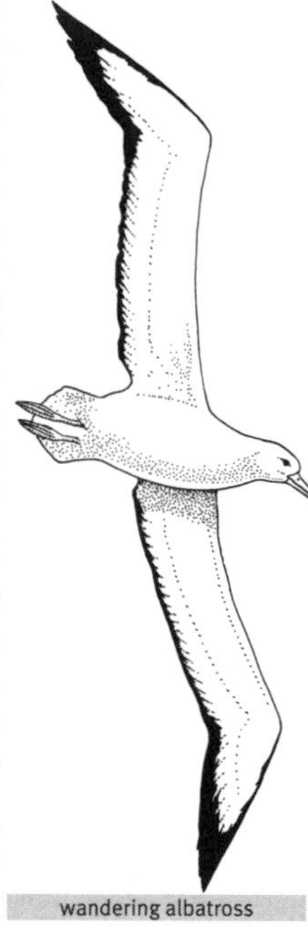
wandering albatross

See also albatross, black-browed albatross, black-footed albatross, Laysan albatross, sooty albatross

warm-blooded fishes Almost all fishes are cold-blooded (poikilothermic), which means that their body temperature is the same as that of the surrounding water. There is, however, a group of large predatory fishes that are equipped with circulatory system modifications that warm the blood, and therefore the muscles, making them faster swimmers and more efficient predators. Warm-blooded fishes retain the heat because of a web of tiny veins and arteries called a *rete mirabile,* or "miracle net." The *rete* acts as a heat-exchanger, circulating the cooler arterial blood from the heart past the warm, oxygen-rich blood from the gills. Heat passes from the veins to the adjacent arteries and then to the warm muscles. It seems to be advantageous for the muscles to be warmer—think of an athlete "warming up." Fishes that employ this system are certain tunas, especially the bluefin; the broadbill swordfish; and the mackerel sharks: the great white, mako, and porbeagle.

See also bluefin tuna, broadbill swordfish, mackerel shark, mako shark, porbeagle

water The most abundant liquid on earth, a tasteless, odorless substance that is colorless in small amounts, but takes on a bluish tinge in large quantities. It is a compound of two parts hydrogen and one part oxygen: H_2O. In its solid form (ice) and its liquid form, it covers more than 70 percent of the planet. (The Pacific Ocean, our world's largest feature, covers one-third of the earth.) It is the stuff of life on earth; protoplasm is a suspension of a number of substances in water, and every living thing is made of protoplasm. Water is also the home of most of the earth's living things, perhaps nine-tenths of whom live in water. Most creatures in water breathe dissolved oxygen, not the oxygen of H_2O. Water is one of the few substances that occurs in all three states—solid, liquid, and gas—at ordinary temperatures. To see all three states simultaneously, put an ice cube into a pot of boiling water. Most substances expand when heated and contract when cooled, but water contracts and becomes more dense when it is heated; when cooled, it expands. As it freezes, it enlarges greatly. Ice is less dense than water, and therefore it floats. Pure water at sea level freezes at 32°F; the effect of salts in water is to lower the freezing point, so seawater freezes at 28°F or 29°F. When salt water freezes, however, the dissolved salts are rejected, so even sea ice is nearly fresh. In its solid state, water behaves strangely: a glacier, which is made of ice, can flow like a river. Hail, frost, and snow are all forms of ice. Water boils and assumes its gaseous form (steam) at 212°F, but evaporation causes water molecules to rise above the land and sea and form clouds. The sea contains about 326 million cubic miles of water. Each cubic mile holds, among other things, 160 million tons of salt, 20 million tons of magnesium chloride and magnesium sulfate, 9.5 million tons of calcium chloride, and 4 million tons of potassium sulfate. Water and its movements—rain, snow, evaporation, waves, tides, currents, gyres, upwellings—affect all life on earth.

See also glacier, ice, iceberg, tide, wave

water pressure Pressure is the force exerted on a surface, and water pressure is the force exerted by water on objects in it, the equivalent of one atmosphere (atm) per square inch (psi), sometimes given as 14.7 lb/in^2. Ocean pressure is directly proportional to depth and acts in all directions (Pascal's Law), not only downward. As an organism descends in water, it has to deal with the air pressure on the water, as well as the pressure exerted downward by the water, which weighs approximately eight hundred times as much as air. (A cubic foot of water weighs 62.4 pounds.) The pressure increases by 1 atm per 10 meters (32.8 feet), so at a depth of 33 feet, the pressure is 1 atm; at 66 feet it is 2 atm; at 100 feet, it is 3 atm, and so on. In the abyssal zone (about 5,000 feet down) the pressure is the equivalent of 21 atm, or 7,364 pounds per square inch, and in the deep trenches (the hadal zone), it can reach 1,000 atm, or 14,714 psi—more than 7 tons. Because their tissues consist mostly of water, which is uncompressible, creatures of the depths are unaffected by pressure. The gas bladders of some, however, expand as pressure decreases. When these animals are hauled up in a net, their organs expand violently, usually killing them. But there are many creatures—sperm whales, for example—that are equipped with lungs, yet are able to dive to great depths and surface rapidly. In this case, the lungs contract as the animal dives, and the oxygen is forced into the myoglobin of the muscle tissue. Without mechanical assistance (Aqua-Lungs pumping compressed air, for instance), human beings can dive to about 300 feet. Beyond that, their lungs collapse from the pressure. High pressure forces the nitrogen in air into the bloodstream, and a rapid ascent causes this gas to bubble out again, and if it ends up in the joints or the brain, it can cause the often fatal disease known as the bends. To avoid this, divers may ascend slowly, according to certain tables, or find themselves in a decompression chamber, which gradually lowers the pressure until it is equal to that at the surface.

See also bends, the; gas bladder; sperm whale

waterspout A small tornado that occurs at sea. A funnel-shaped cloud is formed at the base of a cumulus-type cloud, and the updraft extends downward to the water's surface, where it picks up spray. Waterspouts are most frequent in tropical regions, but they can occur in the higher latitudes as well. The highest waterspout ever measured occurred off New South Wales, Australia, in 1898, and was 5,014 feet high—nearly a mile.

Watson, Paul (b. 1950) One of the most visible, outspoken, and controversial of all marine conservationists. A founding member of Greenpeace, he was active in the first whaling-interference voyages of the *Phyllis Cormack* and later, the *Rainbow Warrior.* He led the Greenpeace mission to Newfoundland in 1975 to protest the merciless bashing of harp seal pups. When the methods of Greenpeace seemed too tame for him, he formed the Sea Shepherd Conservation Society (with the support of Cleveland Amory). In 1978, he took the ship *Sea Shepherd* to Portugal, where he rammed and incapacitated the pirate whaler *Sierra* in 1979. He blew up two Spanish catcher boats in Vigo harbor, and scuttled two Icelandic whaling ships in Reykjavik harbor in 1980. He returned to Newfoundland and spray-painted baby harp seals red, so their white coats would be useless to furriers. Despite arrests by the Canadian and Norwegian authorities, he continues his activities on behalf of whales, seals, and other species that he believes are being exploited. **See also Greenpeace, pirate whaling**

wave The oscillations of the sea caused by wind blowing along the surface and moving in the direction in which the wind blows. The particles of water in a wave do not move forward in a horizontal direction, but rise and fall below the surface. If the force of the wind causes the crest of the wave to overbalance and break, however, then the water particles do move forward. The height of a wave depends on the wind's strength and the length of the fetch, the distance that the wind has been blowing over the open sea; the speed of waves depends on wind strength, up to a maximum speed of about 25 knots. The maximum height of a wave from trough to crest is about 45 feet, except in the center of a hurricane, where much greater heights have been recorded. **See also current, tide, tsunami**

waved albatross (*Diomedea irrorata*) Because it breeds only on Hood (Española) Island in the Galápagos, this albatross, with a yellow-tinged white head and a dark back marked with wavy lines, is unlikely to be confused with any other species. As with all albatrosses, the parents of the single chick hunt food for their young at sea and bring it back in the form of a rich, oily liquid that they transfer to the chick by regurgitation. The maximum wingspan of the waved albatross is 93 inches.

See also albatross, Laysan albatross, wandering albatross

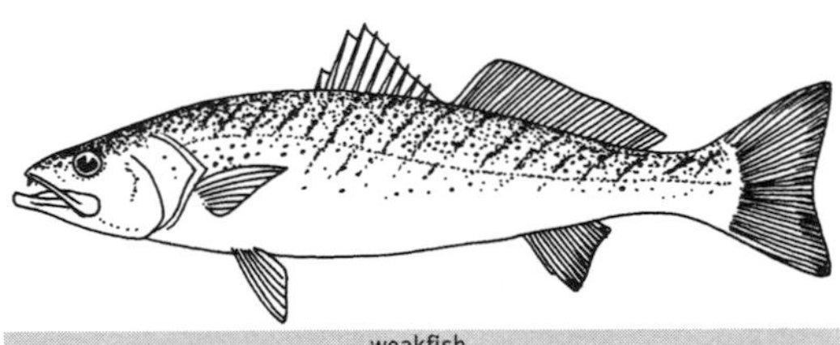

weakfish

weakfish (*Cynoscion regalis*) Also known as squeteague, the weakfish is a species of sea trout that inhabits the western North Atlantic from Massachusetts to Florida, where it can be found in the surf zone and in bays or estuaries. The common name is derived from the "weakness" of the membrane of the mouth, which is easily torn when the fish is hooked. It ranges in weight from 3 to 20 pounds, and because of the delicate flavor and texture of the flesh, it is fished commercially.

weasel shark (family Hemigaleidae) Medium-sized sharks, ranging up to 4 feet in length. They resemble the carcharhinids and houndsharks, but differ from

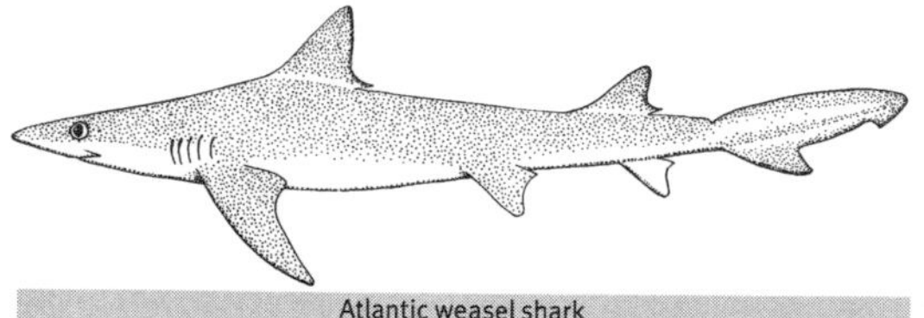
Atlantic weasel shark

the houndsharks in having a precaudal pit before the base of the upper lobe of the tail, and from the carcharhinids in having a spiral intestinal valve rather than a scroll-type. There are seven nominal species, found in the continental and offshore waters of the Atlantic and the Pacific Oceans.

See also carcharhinid sharks, smoothhound sharks

Weddell, James (1787–1834) Born in Ostend (then in the Netherlands, now in Belgium), Weddell was a British explorer and sealer for whom the Weddell Sea and the Weddell seal were named. He commanded the sealing brig *Jane* on three Antarctic voyages between 1819 and 1824, visiting South Georgia and the South Shetlands, and he discovered and named the South Orkneys. On his third (1822–1824) trip in the *Jane* and the cutter *Beaufoy,* he sailed farther south than anyone had done before (74°15′ south), entering the sea that was later named for him.

Weddell Sea A deep embayment of the Antarctic coastline, southeast of the Antarctic Peninsula. Its 3 million square miles are usually covered with thick pack ice. James Weddell originally named it the George IV Sea, but in 1900, the British Admiralty decided to honor its discoverer. **See also Antarctic Peninsula**

Weddell seal

Weddell seal (*Leptonychotes weddelli*) With its fat body, small head, and benign expression, the Weddell seal is easily recognized—if you happen to be in the Antarctic. These seals probably live farther south than any other mammals, and they spend a great deal of time under the ice, breaking through it with specially developed canine teeth. Weddell seals, which were named for the Antarctic explorer James Weddell, are among the most accomplished divers of all pinnipeds. They can remain submerged for well over an hour and descend to almost 2,000 feet in pursuit of fish or squid.

See also crabeater seal, Weddell

weever (*Trachinus* spp.) Weevers are elongated, bottom-dwelling fishes that have venom glands at the base of the first dorsal fin and also at the base of a strong spine on the gill cover. Probably as a warning to potential predators, the first dorsal in many species is jet black. In shallow European waters they lie buried in the sand with only the mouth, eyes, and dorsal spines protruding, and woe betide the swimmer who steps on one. The powerful poison, similar in composition to some snake venoms, can cause great pain, and even death. A person picking up one of these fishes is likely to get stabbed with one of the cheek spines as the fish thrashes. The lesser weever (*T. vipera*) only reaches a length of 6 inches, but is considered the most dangerous of the European weevers because of its inshore habitat and powerful venom.

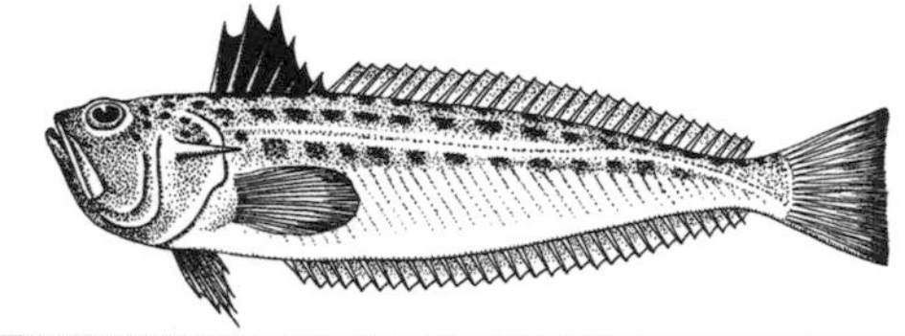
lesser weever

Wegener, Alfred (1880–1930) German meteorologist, geologist, and explorer who first proposed the theory of continental drift. In 1912, observing the jigsaw lineup of the western African and eastern South American continents, as well as that of North America and western Europe, he suggested that some 300 million years ago, from a single continent that he called Pangaea, the present continents began the movement that has resulted in their present positions. Wegener was unable to identify the force that could move continents, however, and his ideas were ridiculed by most scientists during his lifetime and up to the mid-1950s, when the spreading of the seafloor was identified as the geological engine that could—and still does—move the tectonic plates of the earth's crust. Wegener died while on an expedition to Greenland to measure ice thickness and did not see his theory vindicated. The meteorological institute in Bremerhaven is named for him.

See also continental drift, plate tectonics

West Indies When the first Europeans headed west in search of the Indies, those (like Columbus) who went southward encountered an archipelago composed of large and small islands, now known collectively as the West Indies or, sometimes, the Antilles. Each group has its own geographical or political designation. The Bahamas are the most northerly, with some of them located due east of Florida. The Greater Antilles are a group of larger islands, including Cuba, Jamaica, Puerto Rico, and Hispaniola (Haiti and the Dominican Republic), and the Lesser Antilles comprise the Leeward Islands (the U.S. and the British Virgin Islands, Anguilla,

St. Barts, St. Kitts and Nevis, Antigua and Barbuda, Guadeloupe, Dominica); the Windward Islands (Martinique, St. Lucia, St. Vincent and the Grenadines, Grenada); Barbados; and Trinidad and Tobago. The Netherlands Antilles, consisting of Aruba, Bonaire, and Curaçao, are located close to the northern coast of Venezuela. Before the arrival of the Europeans, many of the islands were inhabited by Arawak and Carib Indians, but within a short time, they were eliminated by disease, exploitation, or murder. In 1958, Britain influenced the various islands to form the West Indian Federation, but internal dissension caused its breakup in 1962. Some of these islands are now independent nations, while others are dependencies or territories of France, Great Britain, the Netherlands, or the United States. **See also Anguilla, Antigua and Barbuda, Aruba, Barbados, Bonaire, Cuba, Curaçao, Dominica, Grenada, Guadeloupe, Hispaniola, Jamaica, Martinique, Puerto Rico, St. Barthélemy, St. Kitts and Nevis, St. Lucia, St. Vincent and the Grenadines, Trinidad and Tobago**

Weyprecht, Karl (1838–1881) Austrian Arctic explorer. With Julius von Payer (who had been part of an earlier expedition that camped on a Greenland ice floe for six months), he took an expedition to Novaya Zemlya in 1871, sponsored by Count Wilczek. Lieutenants Weyprecht and Payer, of the Austrian navy and army, respectively, led another expedition in 1872, intending to drift to the North Pole from Siberia, but their ship, the *Admiral Tegetthof,* became trapped in the ice north of Novaya Zemlya, and they drifted for almost a year before they sighted land on August 30, 1873. It was a previously unknown group of islands that they named Franz Josef Land, after their emperor. (They called the first island Wilczek, after their patron.) Always hopeful that the *Admiral Tegetthof* would break free of the ice, they explored and mapped the islands for another year, until it became obvious that they were running out of food and that the ship was permanently trapped. In open boats they headed south through the ice until they were picked up by a Russian fishing boat off Novaya Zemlya on August 24, 1874.

whale Large marine mammals that spend their entire lives in the water, breathe by means of nostrils on top of the head, and are equipped with pectoral fins (flippers), and horizontal tail fins (flukes). Living whales (order Cetacea) are divided into two groups; the Odontoceti, or toothed whales, which include the sperm whale, pygmy and dwarf sperm whales, the beluga and narwhal, the beaked whales, and all dolphins and porpoises; and the Mysticeti, or baleen whales, which include the rorquals (blue, fin, sei, Bryde's, and minke whales), right whale and bowhead, gray whales, and humpbacks. Whales and dolphins are found in all the world's oceans, from pole to pole, and many species were the object of directed fisheries that nearly exterminated them. **See also *individual whales and dolphins***

whaleboat A graceful open boat, pointed at both ends so that it could be rowed (or towed) in either direction, as necessary. It had no rudder and was steered by an oar over the stern. Originally launched from land, whaleboats were later hung from the davits of a whaleship and lowered in pursuit of whales. According to their size, whaleships carried six to eight boats, each with its own designated crew and harpooner. **See also whaleship**

whalebone: See baleen

whale catcher: See catcher boat

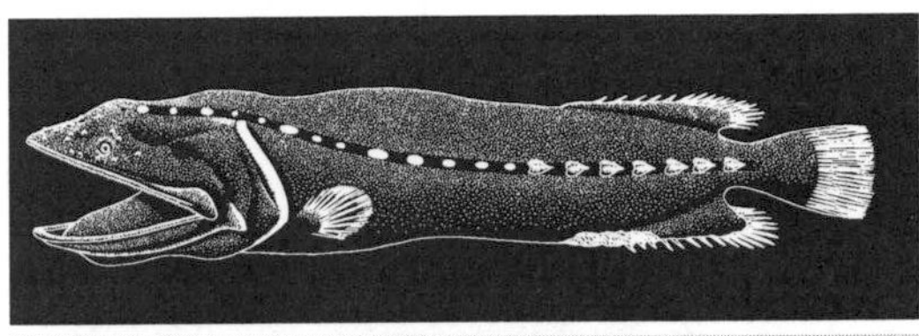

whalefish

whalefish (family Cetomimidae) Perhaps the huge mouth, reminiscent of a baleen whale's, is responsible for the name of this family, but otherwise these 1-foot-long, deepwater fishes bear no resemblance to cetaceans. (The scientific name *Cetomimus* means "whale mimic.") There are about thirty-five species, some known as flabby whalefishes. Many have a red mouth and jaws, and instead of a lateral line, they have either a series of vertical lines equipped with sensory organs or a row of raised papillae. Little is known about them. They have no scales, no pelvic fins, and no ribs, but some have bioluminescent organs.

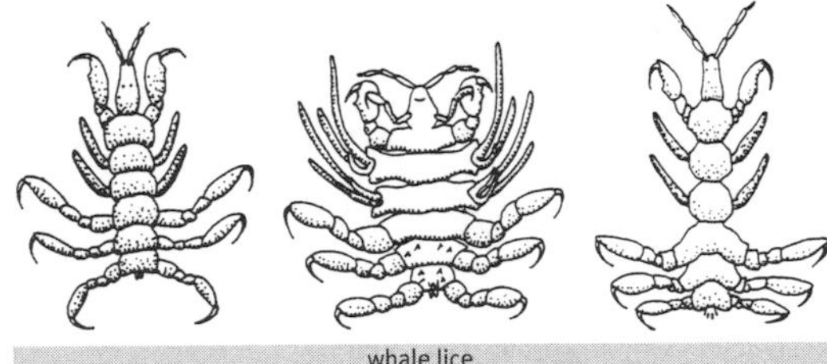

whale lice

whale lice (family Cyamidae) Whale lice are dorsoventrally flattened amphipods that are adapted to live on whales, where they feed on sloughed-off scraps of skin. These whitish little creatures, about the size of a human fingernail, grip the skin of the whale with tiny little hooks on the end of their legs. Host-specific whale lice have evolved for different whale species, such as *Cyamus boopis,* which lives on the humpback whale; *C. gracilis* (right whale); *Neocyamus physteris* (sperm whale); and *C. scammoni,* the most familiar whale louse because it can be observed on gray whales that are often seen by whale watchers in Baja California.

See also amphipod, gray whale

whale oil From the sixteenth to the nineteenth centuries, this clear or brownish oil was used to make soap and as a fuel for lighting. In the twentieth century, it was a component in the manufacture of margarine and nitroglycerine. It can also be utilized for quenching of steel, as a leather dressing, and in the manufacture of paints and varnishes. It was obtained by stripping the blubber layer from the whales, cutting it into pieces, and boiling the pieces in large vats or pressure cookers. With the elimination of commercial whaling, petroleum products have taken the place of whale oil.

See also whaling

whaler sharks A term loosely applied to various carcharhinid sharks in Australia, mostly because of their habit of following whaling ships. Only the bronze whaler (*Carcharhinus brachyurus*) retains this name; the black whaler (*C. obscurus*) is now known as the dusky shark, the silvertip whaler (*C. albimarginatus*) as the silvertip shark, the river whaler (*C. leucas*) as the bull shark, and the blue whaler (*Prionace glauca*) as simply the blue shark. All were previously considered dangerous to man, but only the bull shark deserves that reputation, having been implicated in numerous attacks.

See also bull shark

whale shark (*Rhincodon typus*) A gigantic shark, reaching 50 feet in length. It is the largest shark, and also the largest fish in the world. (Only a few species of whales grow larger.) It is grayish or brownish in color, and covered with a geometric pattern of spots and vertical stripes. It is harmless to humans, and although it has hundreds of tiny teeth, it feeds on planktonic organisms, which it filters through its gill-rakers. It is found throughout the tropical waters of the world, but in the 1990s a large population was located off Ningaloo Reef, Western Australia, and seen to be feeding on the minute spawn of breeding corals. Scuba divers often hitch rides on these slow-moving and inoffensive giants.

See also basking shark

whaleship A square-rigged ship, especially equipped for the hunting and processing of whales; also called a whaler. It was first developed for the eighteenth-century Dutch and British Arctic bowhead fishery, and later modified for the Yankee whalers out of such ports as New Bedford and Nantucket. Whaleships were typically 100 to 150 feet long, and averaged about 300 tons. Each had a complement of whaleboats, hung on davits over the side, which were lowered when whales were sighted. In earlier days, the whale carcasses were processed on shore, but Yankee ingenuity soon provided onboard tryworks, which consisted of brick fireplaces enclosing huge iron cauldrons, so that the blubber could be boiled down at sea. The ships were bluff-bowed and ungraceful but highly efficient, and had plentiful space below for the stowage of casks of whale oil. When the hold was filled, the ships would offload at ports like Lahaina, Maui, or San Francisco, and continue on their whaling voyages, sometimes remaining at sea for as long as five years.

See also Nantucket, New Bedford, tryworks, whaleboat, whaling

whale watching Commercial enterprise devoted to the observation of living whales at sea, usually fairly close to shore. For years, people had watched the migrating gray whales from the cliffs of Southern California

whale shark

around San Diego, but watching from boats is said to have begun around 1975, when a fisherman in Massachusetts took people out to see the humpbacks of Stellwagen Bank. Whales in New England waters proved to be exceptionally abundant—and watchable—and soon an industry was established, with boats from Boston, Provincetown, and Gloucester taking passengers to see humpbacks, fin whales, right whales, sperm whales, and various dolphins. The humpbacks that migrated annually from southeast Alaska to Hawaii and back became another popular object of whale watchers in both locations, as did the killer whales that lived all year round in Puget Sound. The annual gray whale migration is watched from boats in Southern California, and also in the destination lagoons of Baja California, where the activity is closely monitored by the Mexican government. Whale watching has now become a worldwide activity, and areas where whales were once killed are now available for more benign interactions. Whale watchers seek sperm whales in the Azores; right whales in South Australia and South Africa; blue whales, fin whales, and belugas in Canada; fin whales and minke whales in Norwegian waters; and, probably most surprising, humpbacks in the Bonin Islands of Japan.

whaling The organized killing of whales for commercial purposes. Beginning around A.D. 1000, the Basques pioneered whaling in the Bay of Biscay, eventually moving west across the Atlantic to Newfoundland. In the search for the Northwest and Northeast Passages, explorers from England and the Netherlands came across huge stocks of whales, which led to their slaughter in the waters of Greenland, Spitsbergen, and eastern Arctic Canada. By 1650, American settlers had begun killing right whales in the inshore waters of Massachusetts, and by 1715, these intrepid Yankee whalers had discovered the sperm whale. British vessels transporting convicts to Australia in 1788 opened the Pacific Ocean to whalers, and soon whaleships from Old and New England were scouring the seven seas. In 1867, the Norwegian whaler Svend Foyn invented the grenade harpoon, which made it possible to kill blue and fin whales that could not be taken with a hand-thrown harpoon. The factory ship soon freed the whalers from land stations by enabling them to process whales on the high seas. In 1855, Charles Scammon discovered the breeding lagoons of the California gray whale in Baja California, and these whales were hunted almost to extinction. During the first third of the twentieth century, millions of whales were killed in the Arctic and the Antarctic for their bone, oil, and meat. The International Whaling Commission (IWC) was formed in 1949, ostensibly for the purpose of controlling whaling, but it was in reality a "whalers' club" that oversaw the systematic destruction of the world's whales. It was not until the 1970s that conservationists around the world began to protest this needless slaughter, and in 1982, the IWC passed a moratorium on commercial whaling. Eskimos and other aboriginal peoples may still kill whales for subsistence reasons, and a few countries, such as Norway and Japan, continue to take a small number per annum, but by and large, killing whales for profit seems to have ceased.

See also bowhead whale, gray whale, International Whaling Commission, sperm whale

whimbrel (*Numenius phaeopus*) Formerly known as the Hudsonian curlew, the whimbrel breeds in northern Canada, especially along the shores of Hudson Bay. It winters along the coasts of western North America from California to Chile, and is also found in Eurasia, Africa, and Australia. While market gunning has decreased the numbers of most other curlews throughout the world (the Eskimo curlew was driven to extinction), the whimbrel has taken advantage of bird refuges and a decrease in hunting and actually become more numerous. They often travel in enormous flocks, flying high in V formations or skimming low across the water. The long, downward-curving bill is used to probe the mud or sand for shellfish. **See also curlew, Eskimo curlew**

whimbrel

white-beaked dolphin (*Lagenorhynchus albirostris*) The white-beaked dolphin is the largest of the lags (the common name for members of the genus *Lagenorhynchus*), reaching a length of 10 feet. As its name implies, it has a black head and a white beak. It has grayish patches on the flanks and the dorsal surface of the tail stock, but it is not as crisply marked as the other North Atlantic lag, the white-sided dolphin. It is sometimes seen in groups of a thousand or more, but more commonly in schools of between thirty and fifty. Like the other lags, it is acrobatic and fond of jumping. It is shy of ships, however, and does not bow-ride, which means there has been little observation of this species in the

white-beaked dolphin

wild. It is found in the waters of Greenland, Norway, Newfoundland, and Labrador, and occasionally as far south as Cape Cod.

See also Atlantic white-sided dolphin

white-bellied storm petrel (*Fregetta grallaria*) Large as storm petrels go, the white-bellied breeds in Southern Hemisphere islands such as Gough, Lord Howe, Kermadec, Juan Fernández, and Tristan da Cunha. Like other storm petrels, it patters over the water with its feet dangling and its wings raised, but this species also splashes into the water and then springs clear. It accompanies ships, flying ahead of them more often than behind. Some ornithologists regard the black-bellied storm petrel, which gets its name from the line that runs from breast to tail, as a subspecies or even a color morph of the white-bellied.

See also Leach's storm petrel, storm petrel, Wilson's storm petrel

White Island Called Kvitøya in Norwegian, this lonely speck of land east-northeast of Nordaustlandet, Spitsbergen's eastern island, would be recognized only as a habitat for walruses were it not for its unfortunate place in the history of North Pole exploration. In July 1897, Major Salomon Andrée, a Swedish engineer, took off in a gigantic hot-air balloon from Danes Island (Danskøya) in northwestern Spitsbergen with two other men and was never heard from again. Thirty-three years later, the Norwegian sealing vessel *Isbjörn* landed on White Island and found the remains of Andrée's camp, his diary, and the bodies of the three explorers. Andrée's balloon had floated almost due east over all of Spitsbergen, then over the icy sea until it crashed. The three men hiked over the ice until they got to White Island, where they died. In the camp, a camera was found with undeveloped film that showed the men and the half-deflated balloon on the sea ice. Although their bodies and the diary were found, it was not clear how the men died. (There is a White Island in Canada, just north of Hudson Bay; another off the Antarctic continent; and two off New Zealand, one off North Island and another off South Island.) **See also Spitsbergen**

white marlin (*Tetrapturus albidus*) Smaller than the blue or black marlin, the white marlin rarely reaches a length of 10 feet. It has rounded pectoral and dorsal fins, while the dorsals of the larger marlins are falcate. Its coloration is dark blue or green, with pale lavender vertical bars that can be seen in the living fish, but fade quickly with death. It is found only in the Atlantic from the Gulf of Maine to Brazil, where it is a top-rated sport fish. White marlins average about 50 pounds, but the record is a 181-pounder, caught off Brazil in 1979.

See also black marlin, blue marlin, striped marlin

White Sea An inlet of the Barents Sea between the Kola and Kanin Peninsulas in northeastern Siberia. The White Sea (Beloye More) covers approximately 37,000 square miles and it is over 1,000 feet deep at Kandalashka Bay in the south. Near the mouth of the Dvina River is Archangel, its chief port. A canal system 140 miles long connects the White Sea to the Baltic at St. Petersburg. In the 1990s, it was revealed that the Soviet navy had been using the White Sea as a dumping ground for spent nuclear reactors. **See also Barents Sea**

white sea bass

white sea bass (*Cynoscion nobilis*) Not a true sea bass, *C. nobilis* is more closely related to weakfishes and the totoaba. Sometimes known as corvina, it can be found in the Pacific from Alaska and California as far south as Chile, and it is a popular game fish. Adults average around 20 pounds, but the world's record, taken off Mexico in 1953, weighed 83 pounds.

See also totoaba, weakfish

White Star Line The popular name of the Ocean Steam Navigation Co. Ltd., one of the great transatlantic steamship lines. It was consolidated after the failure of the Aberdeen White Star Line of clippers to Australia, and bought by Thomas Ismay in 1867. Over the years Ismay ordered four ships to be built by Harland and Wolff of Belfast. The first of these was the *Oceanic*, which made her maiden voyage in 1871. When the *Olympic* was launched in 1911, she was the largest ship in the world. The next year, *Olympic*'s sister ship, the *Titanic*, was launched. Making what was surely the most unhappy debut of any passenger ship, *Titanic* collided with an iceberg in the North Atlantic on April 14, 1912, and sank, taking 1,489 passengers and crew members down with her. A third sister ship, the *Britannic*, was sunk by a mine in the Aegean during World War I, but in spite of these disasters, White Star continued until 1934, when it was merged with Cunard.

See also *Britannic*, Ismay, *Titanic*

white-tailed sea eagle (*Halieaetus albicilla*) Found mostly in and around Greenland, this large fishing eagle (wingspan: 8 feet) is sometimes known as the gray sea eagle because its plumage is gray all over, except for the tail, which is white. Sea eagles can sit for hours on a dead branch waiting for a fish to rise, at which point they swoop down and capture it with their powerful talons. This species has been recorded as a rare visitor to Maine and Massachusetts. **See also bald eagle, Steller's sea eagle**

white-tailed sea eagle

white tern (*Gygis alba*) The only all-white tern, it has large black eyes, black feet, a black bill, and a wingspan of about 34 inches. It is also called the fairy tern, a name that probably comes from the translucent appearance of the wings when the bird is overhead. It has a circumequatorial distribution throughout tropical seas and is found in the Seychelles, Madagascar, the Mascarene Islands, and throughout the South Pacific. In Australia it is known as the white noddy. **See also noddy, tern**

white-tipped reef shark

white-tipped reef shark (*Trianodon obesus*) Inhabiting only the Indo-Pacific region, this slender, 5-foot-long shark is easily recognizable by its short, broad snout and the white-tipped dorsal and caudal fins that give it its common name. It spends most of its time in caves, coming out to hunt fishes, eels, lobsters, and crabs. Because of its unaggressive nature, it is a favorite with skin divers, especially in Hawaii, where it is known as *lalakea*. It should not be confused with the much larger (and much more dangerous) oceanic whitetip, *Carcharhinus longimanus*. **See also oceanic whitetip shark**

Whydah A 100-foot-long three-masted galley, loaded with ivory, indigo, and gold coins, the *Whydah* was captured off Cuba by the pirate captain Black Sam Bellamy, who made it his flagship. Heading for Massachusetts in April 1717, she ran aground off Wellfleet. The ship was wrecked, and the crew and cargo remained lost for 267 years. It was discovered in 1984 by treasure hunter Barry Clifford, who had been searching for it for almost ten years. Clifford retrieved from it cannons, pewter tableware, navigational instruments, gold bars, and silver and gold coins. In addition, the wreck of the *Whydah* yielded nearly four hundred pieces of gold that had been worked into jewelry by Africans living in what is now Ghana and Ivory Coast. The ship is also believed to have carried 4 or 5 tons of unworked gold and silver, but this "mother lode" has not been found.

Wight, Isle of In the English Channel separated from the English mainland by the Solent and the Spithead Channel, the Isle of Wight is 23 miles long and 13 miles wide. It was probably first settled by the Romans in A.D. 47; it was later occupied by the Jutes, then annexed to the Kingdom of Wessex in 661. When William conquered England in 1066, he bestowed the island on his lieutenant, William Fitz-Osbern. Always worried about invasions from France, the British built coastal forts on the island at Cowes, Sandown, Freshwater, and Yarmouth. During the English Civil War, King Charles I escaped to the island and was imprisoned in Carisbrooke Castle from 1647 to 1648, before being taken up to London to be executed. Queen Victoria maintained a seaside residence at Osborne House, near the famous yachting center at Cowes. The island is now a popular resort, with a permanent population of around 126,600. **See also English Channel**

Wild, Frank (1874–1930) A Yorkshireman who joined the Royal Navy at twenty-six. He later volunteered for Robert Scott's 1901–1904 Antarctic expedition, where he became friends with Ernest Shackleton. He sailed aboard the *Nimrod* in 1907 and, with Shackleton, Jameson Adams, and Eric Marshall, made the greatest of all sledge journeys, discovering the Beardmore Glacier and reaching a point only 97 miles from the South Pole. Under Douglas Mawson, Wild was the leader of the Western Party (1911–1914), which mapped 310 miles of new Antarctic coastline. As deputy to Shackleton on the *Endurance* in 1916, Wild was left in charge of the twenty-two men on Elephant Island, as the 22-foot longboat *James Caird* sailed for South Georgia with Shackleton and five men aboard. Wild saw the men through the terrible winter, where they lived in overturned boats and ate penguins and seals until they were rescued. He again served with Shackleton aboard the *Quest* in 1922; when Shackleton died, Wild assumed command and completed the voyage. His seafaring days over, he moved to Africa, where he died of pneumonia. **See also Mawson, Scott, Shackleton**

Wilkes, Charles (1798–1877) American naval officer and explorer, who joined the U.S. Navy as a midshipman in 1818, after three years in the merchant service. In 1836, when Congress approved a national expedition to the southern Atlantic and Pacific Oceans with a view to promote the whale fishery, Wilkes was appointed commander. Six ships were placed at his disposal, staffed with scientists of every kind. They visited the South Pacific, then headed for the Antarctic. In January 1840, they sighted land, and were the first to name the Antarctic continent. After sailing along the coast for 1,500 miles along the region that would later be named Wilkes Land, they headed north, where they surveyed the Strait of Juan de Fuca and the coast of western

North America. Instead of being honored for his leadership and surveying skills, Wilkes was court-martialed for having exceeded his authority. He was acquitted, and in 1861, he was given command of the *San Jacinto* in the Union navy. When he stopped the British mail ship *Trent* and removed two Confederate commissioners, he so far exceeded his authority that he was court-martialed again and was this time convicted of disobedience, insubordination, and conduct unbecoming an officer. He was suspended from duty for a year, and then given command of a squadron operating against Confederate raiders in the West Indies and the Bahamas. He retired in 1866 with the rank of rear admiral. **See also U.S. Exploring Expedition**

Willoughby, Hugh (d. 1554) British soldier and adventurer of unknown origin, who first appears in 1553, making a voyage in search of the Northeast Passage. Willoughby was aboard the *Bona Esperanza*, Cornelius Durforth was captain of the *Bona Confidentia*, and Richard Chancellor captained the *Edward Bonaventure*. They sailed north to the Lofoten Islands, and then to the coast of northern Norway, where a great storm blew up and separated the *Bona Esperanza* and the *Bona Confidentia* from Chancellor's *Edward Bonaventure*. They landed on the snowbound coast of Lapland, and although they all died, Willoughby's diary was found a year later by Russian fishermen. Richard Chancellor waited to rendezvous with Willoughby, but when he didn't appear, Chancellor went on to Archangel and was taken overland to see Tsar Ivan IV (Ivan the Terrible) in Moscow. This led to the founding of the Muscovy Company, designed to stimulate trade between Russia and England. Willoughby's journey is detailed in Richard Hakluyt's 1583 *Principall Navigations, Voyages and Discoveries of the English Nation within these 1500 Years*. **See also Chancellor, Hakluyt, Muscovy Company**

Wilson's storm petrel (*Oceanites oceanicus*) A small, sooty-brown petrel with a 15-inch wingspan and a conspicuous white rump, found in all oceans except the North Pacific. It breeds on Southern Hemisphere islands such as South Georgia, the Crozets, Kerguelen, the Falklands, and Tierra del Fuego, then disperses in large numbers to the North Atlantic. It is identifiable by its long legs, which protrude well beyond the tail when raised, and the feet, which have bright yellow webs. A ship follower, it has a habit of "walking" on the water with legs outstretched and wings held high. It is one of the few birds that breeds in the Southern Hemisphere and spends the winters in the north; the sooty and great shearwaters are others. Some ornithologists suggest that Wilson's storm petrel may be the most numerous bird in the world.

See also British storm petrel, petrel

windowpane flounder (*Scophthalmus aquosus*) A thin-bodied, left-eyed flounder, distributed along the northwest Atlantic continental shelf from the Gulf of St. Lawrence to Florida. The eyed side is brownish, reddish, or grayish, mottled with small irregular blotches on the head, side, and fins. The blind side is white. Although some are caught as a by-catch of other fisheries, the windowpane has little commercial value.

See also flounder, plaice, sole

Windward Islands The southern group of the Lesser Antilles in the Caribbean, curving south from the Leeward Islands toward Venezuela, the Windward Islands consist of the independent countries of Dominica, Grenada, St. Lucia, and St. Vincent and the Grenadines, and the French island of Martinique. The Windwards are all volcanic islands, and in fact, almost all of them have a volcano named La Soufrière, "the sulfur producer." (The volcano on Martinique, Mount Pelée, erupted in 1902, completely eradicating the town of St. Pierre and killing all of its thirty thousand residents.) The Windward Islands are all exceptionally fertile and scenic, and, with the exception of economically deprived Dominica, all are major Caribbean tourist attractions. **See also Dominica, Grenada, Martinique, St. Lucia, St. Vincent and the Grenadines**

winter flounder (*Pseudopleuronedes*) Also known as the blackback or lemon sole, this right-eye flounder can be 23 inches long and is found in the western North Atlantic from Labrador to Georgia. This important food fish is more easily caught in the winter when it comes into shallower water, hence its name. As with most commercially harvested fishes on the east coast of the United States, the numbers of winter flounder are in decline.

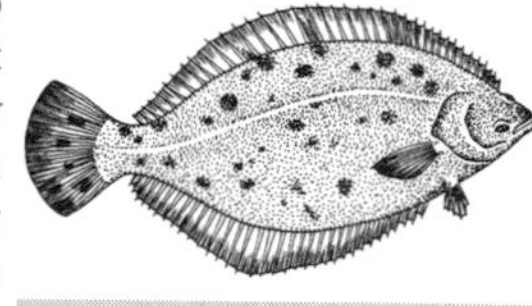
winter flounder

See also English sole, flounder, plaice

wobbegong (*Orectolobus ornatus*) Several species of flattened, tasseled sharks that are found mostly in Australian waters, hence their aboriginal common name. The banded wobbegong of northern Australia and New Guinea reaches a maximum length of 10 feet, but most are smaller. Like the nurse sharks to which they are related, wobbegongs spend their time motionless on the bottom, their tassels and mottled coloration

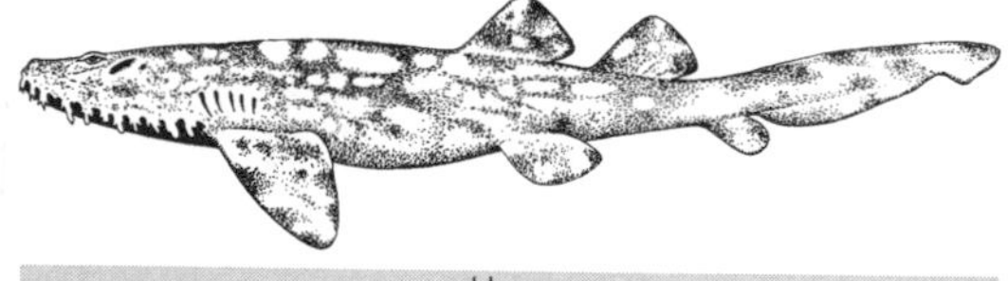
wobbegong

camouflaging them until they burst forth to engulf an unwary fish, octopus, lobster, or crayfish. They are considered harmless to man unless they are provoked, when they will bite aggressively.

See also nurse shark, zebra shark

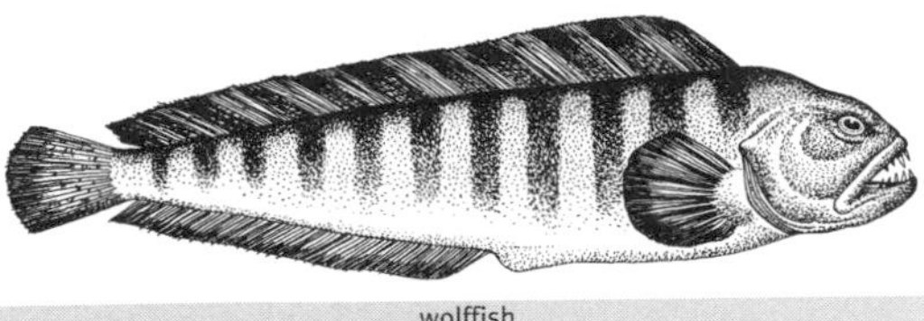

wolffish

wolffish (*Anarhichas lupus*) With its large head and bared canine teeth, the wolffish is an ominous and dangerous predator. In addition to their canines, they have massive grinding teeth to crush the shellfish that they eat. There are several species, some known as catfish, distributed throughout the North Atlantic and the North Pacific. They include the spotted catfish (*A. minor*) of European waters, which can reach a length of 7 feet, and the jelly cat (*A. latifrons*), which is smaller and generally used for bait. On the west coast of North America, *A. ocellatus* is known as the wolf-eel and can grow to a length of 6 feet. Because of their ferocious visage, wolffish are a favorite of underwater photographers; indeed, they are as nasty as they look and can inflict severe bites. In New England waters, there is a minor fishery for wolffishes, which, despite their appearance, make good eating.

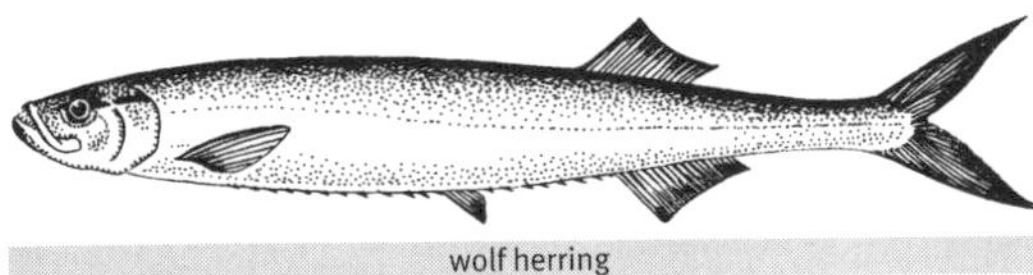

wolf herring

wolf herring (*Chirocentrus dorab*) A fish that is closely related to the other herrings and looks very much like them, except that it can get to be 10 feet long. As with the other herrings, the scales on the underside of the belly meet in a knifelike ridge. Unlike most other bony fishes, it has a spiral valve in the intestine, a feature usually restricted to sharks and rays. Widely distributed in Indo-Pacific and East African waters, it is a voracious predator with large fanglike teeth.

See also Atlantic herring

women pirates Although much less common than their male counterparts, there were several female buccaneers. The best known were Mary Read and Anne Bonny, both of whom dressed in men's clothes and sailed with "Calico Jack" Rackham. Bonny was the illegitimate child of an Irish lawyer and a housemaid, who was taken to Charleston, South Carolina, as a child, then married to a man who took her to the pirates' lair in New Providence in the Bahamas. Read had been wearing men's clothes for some time when she fell in with Rackham. After they were captured off Jamaica in 1720, Rackham was hung, but the women avoided the death sentence because both of them were pregnant at their trials.

See also buccaneer, privateer

Woods Hole Oceanographic Institution (WHOI) Located on the tip of a peninsula at the southwestern corner of Cape Cod on Buzzard's Bay, Massachusetts, WHOI is a private, nonprofit corporation dedicated to research and education in the ocean sciences. Spencer Baird, assistant secretary of the Smithsonian Institution, opened the first laboratory at Woods Hole in 1875, and Harvard naturalist Louis Aggasiz was responsible for starting the Marine Biological Laboratory there in 1888. In 1930, WHOI was founded with Harvard ichthyologist Henry Bryant Bigelow as its first director. Woods Hole scientists were active in mapping the ocean floor; in 1977 they were the first to identify the animals of the hydrothermal vents, and they located the wreck of the *Titanic* in 1985. Today WHOI employs 850 people, and in addition to its buildings and laboratories, it maintains three oceangoing research vessels—the *Knorr*, the *Oceanus*, and the *Atlantis* (the second Woods Hole ship of that name)—as well as the research submersible *Alvin*. The National Marine Fisheries Service (NMFS) maintains offices and a public aquarium at Woods Hole.

See also Scripps Institution of Oceanography

Wrangel Island Russian island in the Arctic Ocean, between the East Siberian and the Chukchi Seas. The barren, frozen island covers some 2,800 square miles and is occupied by polar bears, seals, lemmings, numerous seabirds, and a Russian weather station. The island was the object of an unsuccessful quest by Baron Ferdinand Petrovich von Wrangel (1796–1870). Captain Henry Kellett of the British navy probably sighted the island in 1846, but it was discovered in 1867 by Thomas W. Long, captain of the New London, Connecticut, whaler *Nile*, who proposed that it be named Wrangel Land. (The passage between the island and the mainland is now known as Long Strait.) In 1879, American explorer George Washington De Long believed he could reach the North Pole by sailing north along what he believed to be the large landmass of Wrangel Land, and then sledging to the Pole. In the *Jeannette* he passed north of Wrangel Island, proving that it was an island and not part of the northern landmass. (The *Jeannette* drifted for twenty-two months until she was crushed in the ice, and in their escape in small boats, many of the crew died, including De Long. It was the wreckage of the *Jeannette*, washed ashore in Greenland, that prompted Fridtjof Nansen to try to drift to the Pole in the *Fram* in 1893–1896.) Vilhjalmur Steffanson, a Canadian explorer, tried to secure the island for England in 1921 by establishing a colony of sympathetic Eskimos there, but the Russians ousted them and established the first permanent colony there in 1926. (One of the smaller islands and the town on it in the Alexander Archipelago of southeast Alaska were also named for Wrangel, but an extra "l" was added to the spelling.)

See also *Fram*, Nansen

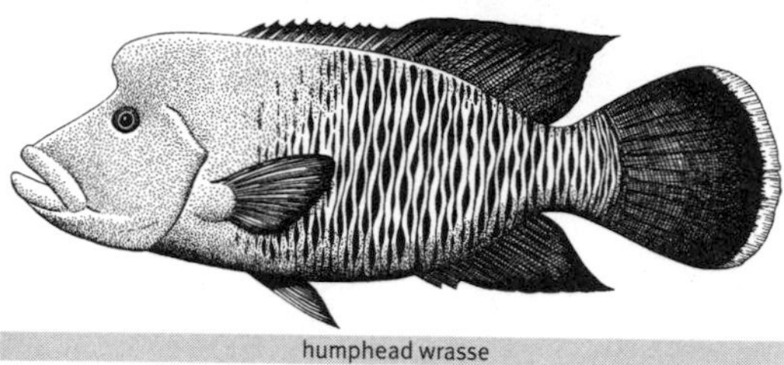

humphead wrasse

wrasse (family Labridae) Any of three hundred–odd species of thick-lipped, large-scaled fishes, often with protruding canine teeth in the front of the jaw. They range in size from the cleaner wrasses, which seldom exceed 2 inches in length, to the gigantic humphead (Napoleon) wrasse (*Cheilinus undulatus*), which can be 8 feet long and weigh more than 400 pounds. Juvenile wrasses are frequently differently colored than adults, and in some species, the males and females are strikingly different. Sex reversal has been demonstrated for many members of the Labridae; they begin their lives as females, but are able to alter their sex, often acquiring the gaudier coloration of the male. Most wrasses are diurnal; they retire for the evening as darkness approaches and resume activity the following morning. They are found throughout the world's tropical and temperate waters, but few species are of economic importance.

See also bluehead, cleaner fishes, hogfish, tautog

wreckfish (*Polyprion americanus*) Named for its affinity for sunken shipwrecks, the wreckfish is one of the sea basses, related to the groupers, perches, and striped bass. (It is sometimes called the stone bass.) Inhabiting both sides of the Atlantic and the Mediterranean, it reaches a length of 6 feet and a weight of about 75 pounds. It hangs around floating logs to prey on other fishes, but it has also been found at depths of up to 2,500 feet. **See also grouper, striped bass**

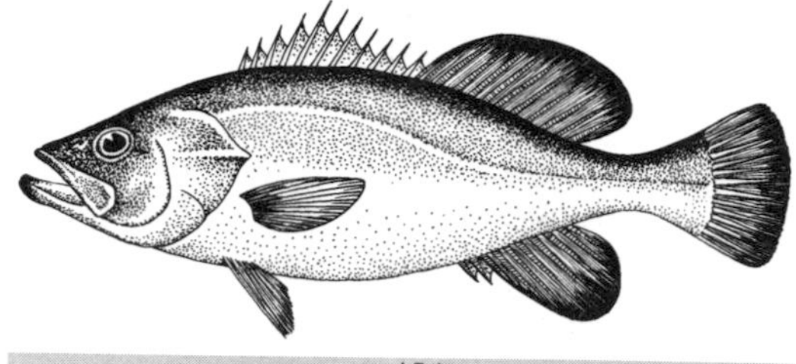

wreckfish

Yamamoto, Isoroku (1884–1943) Japanese admiral, born Isoroku Takano; his name was changed when he was adopted by a prominent family in his home district. He graduated from the Japanese Naval Academy in 1900, took part in the Battle of Tsushima in the Russo-Japanese War (1904–1905), and spent the years 1917–1919 at Harvard. A trained pilot, he became Japan's chief proponent of the aircraft carrier. Although originally opposed to joining the Axis powers in 1938, he believed that a powerful first strike would cripple the American Pacific Fleet, and he planned the surprise attack on Pearl Harbor in December 1941. But the American carriers were at sea at the time, and in subsequent battles, such as Midway (June 3–7, 1942), his forces were defeated by the very air power that he advocated. Yamamoto deployed his fleet in support of Guadalcanal and the Solomons, but in early 1943, these entrenched strongholds fell to American forces. Because they had intercepted the Japanese codes, American fighters located the aircraft in which Admiral Yamamoto was traveling and shot it down over Bougainville.
See also Bougainville Island; Midway, Battle of; Pearl Harbor; Russo-Japanese War

Yamato Japanese battleship; with its sister ship *Musashi,* the largest battleship ever built. Under great secrecy, work on these ships was begun in 1937, and they were launched in 1939. They were equipped with nine 18-inch guns that could fire a broadside able to hit a target 27 miles away, and one hundred antiaircraft guns. Neither ship ever had the opportunity to use its great firepower, however. The *Yamato* was present at Midway and Leyte Gulf but spent all her time chasing American ships. In a final grand gesture, she was sent on a suicide mission to draw the American carriers away from Okinawa and then beach herself. She never reached Okinawa; 270 miles away she was attacked from the air by fighters, and by torpedo bombers from the carrier *Yorktown,* and she sank on April 7, 1945. (The *Musashi* was sunk at Leyte Gulf in 1944.)

Yap One of the Caroline Islands in the western Pacific, now one of four states making up the Federated States of Micronesia. (The others are Kosrae, Pohnpei, and Truk.) The first Europeans to visit Yap were probably the early sixteenth-century Portuguese navigators. The island was controlled by Spain until 1899, when it was transferred to Germany. The Germans made Yap a center for underwater cable communications and divided the islands into ten administrative units that are still maintained. Because Japan sided with the Allies in World War I, she was awarded these and other Pacific islands, but the island was taken by U.S. forces in 1944. To denote wealth and status, the Yappese display gigantic stone disks outside of their houses. Some of these "fei stones," holed in the center to make them transportable, are greater in diameter than a man's height.
See also Caroline Islands, Micronesia, Pohnpei, Truk

yellow-bellied sea snake (*Pelamis platurus*) A brilliantly colored marine snake, widely distributed throughout the Pacific from east to west. It is shiny black above, with a chrome yellow belly, a pattern that turns into black spots on the yellow flattened tail. These snakes form gigantic aggregations that may be miles in length, but the reason for this behavior is not known. The yellow-bellied sea snake is one of the few vertebrates in the world with no natural enemies; even such predators as seabirds, fishes, and sharks that are willing to eat other species of sea snakes will not eat *P. platurus.*
See also sea snakes

yellow-eyed penguin (*Megadyptes antipodes*) The feathers of the face are pale yellow tipped with black, and there is a lighter yellow mask around the straw-colored eyes. The long, slender bill is cream-colored with a reddish-brown tip. The yellow-eyed penguin has the smallest population of any penguin species, and the approximately 5,000 individuals live only on the southeastern coasts of South Island, New Zealand, and on the Stewart, Campbell, and Auckland Islands. In recent years, loggers and farmers have introduced predators, such as cats, dogs, pigs, stoats, and ferrets, and the species is considered endangered. **See also penguin**

yellowfin tuna (*Thunnus albacares*) With its extremely long, bright yellow second dorsal and anal fins, the yellowfin is easily differentiated from other tunas. It is one of the larger of the species, reaching a maximum weight of 388 pounds. Hundreds of thousands of tons are caught annually, making it one of the most important commercial tunas. Because of its unexplained inclination to associate with schools of spotter and spinner dolphins, yellowfins were caught by "setting on dolphins," which means drawing a purse seine around a school of dolphins at the surface, then closing the net

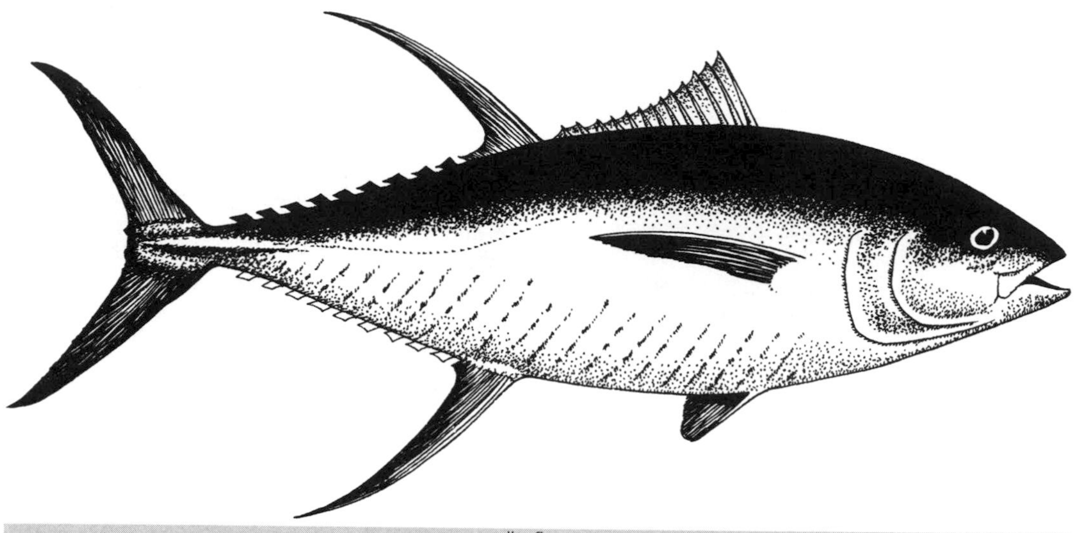

yellowfin tuna

on both the dolphins and the tuna. This resulted in a large catch of tunas, but also in the death of millions of dolphins. This method is now carefully controlled for American tuna fishermen, but other nationals, such as Mexicans and Ecuadorians, are under no such restrictions, and thousands of dolphins are still being killed.

See also purse seining, tuna-porpoise problem

Yellow Sea Arm of the North Pacific Ocean between China and Korea, marked at its northern end by two bays, the Gulf of China (*Po Hai* in Chinese) and Korea Bay. The boundary between the Yellow Sea and the East China Sea is not clear. It covers approximately 160,000 square miles. The name refers to the yellowish, silty water that results from the discharge into the sea of such major rivers as, from China, the Yellow (*Huang Ho*) and the Yangtze and, from Korea, the Yalu. The Yellow Sea is shallow, never more than 260 feet deep, and it is a heavily utilized fishing area for boats from China and North and South Korea. Tsingtao, Shanghai, and Tientsin are on the Yellow Sea in China; in Korea, the main ports are Inchon and Nampo.

See also East China Sea

yellowtail (*Seriola lalandi*) The confusion of common names is nowhere more evident than with this fish. In California it is known as yellowtail; in South Africa it is called amberjack, in Australia yellowtail kingfish, and in New Zealand yellowtail kingfish or kahu. In all those places, however, it is still *S. lalandi,* a legendary game fish. The fins and tail are yellowish, and there is sometimes a bronzy stripe along the mid-side of the body. Because it sometimes aggregates in huge schools, it is caught commercially by purse seiners, and it is popular with anglers and spearfishermen as well. Throughout its range, it is considered an excellent food fish. The world-record yellowtail (or kahu), caught in New Zealand in 1984, weighed 114 pounds.

See also amberjack

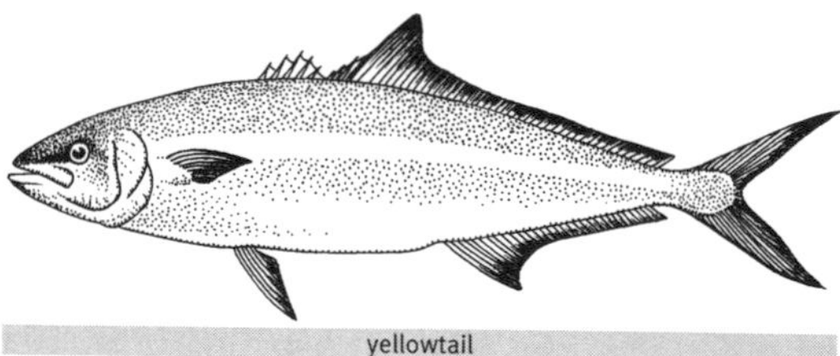

yellowtail

Z

Zacynthus Also spelled Zákinthos, Zakyntos, or Zante, the southernmost of the Greek islands in the Ionian Sea, just east of the Peloponnese. Like the other Ionian Islands, Zacynthus was occupied by Romans, Normans, Ottoman Turks, and Venetians before 1801, when the island was taken by the French under the Emperor Napoleon, and then by the British after Napoleon's defeat. In 1863, after the war for Greek independence, Zacynthus was united with Greece. Considered one of the most beautiful of the Greek islands, its beaches have been cluttered with commercial food stands and cheap lodgings. The loggerhead turtle (*Caretta caretta*), which breeds on the beaches of Zacynthus (and on Cephalonia), is also being threatened by the intrusive presence of packaged tourism. Zacynthus has been subjected to several earthquakes: in 1514, 1893, and most recently, August 12, 1953, when one hundred people were killed, and many of the standing buildings on the island were destroyed.

Zanzibar Two islands, Zanzibar and Pemba, 22 miles off the coast of East Africa; now part of Tanzania. Originally settled around A.D. 1000 by Hadimu villagers from the mainland, the island remained undisturbed until Vasco da Gama arrived in 1499. En route to India, the Portuguese gained control of Zanzibar, as well as most of the East African coast. In 1698, Arabs from the state of Muscat and Oman gained control of the island; they held it into the nineteenth century. In 1832, Sultan ibn Sayyid moved his court to Zanzibar and made it the center of a lucrative slave and ivory trade, as the Omanis mounted slave raids and ivory-collecting caravans to the mainland and sold the goods to visiting traders. Britain gained control of the island in 1870 and abolished the slave trade. In 1963, Zanzibar became independent, but a year later, the Arab government was overthrown by African rebels, and Tanganyika, Zanzibar, and Pemba agreed to merge as Tanzania. Now the major exports of Zanzibar are cinnamon, cloves, and coconuts.

Zealand Known to the Danes as Sjælland, Zealand is the largest of the islands that make up most of Denmark, and the seat of the capital of the country, Copenhagen. (The largest part of Denmark is on the Jutland Peninsula, the northern end of which, Vendsyssel-Thys, is actually an island connected to the mainland by various bridges; the other large islands include Funen to the west of Zealand, Lolland and Falster to the immediate south, and Langeland and Bornholm.) Zealand is between the Kattegat and the Baltic Sea, separated from Sweden by the Øresund. (Across the Øresund from Copenhagen is the Swedish city of Malmö.) The island covers some 2,715 square miles, and including Copenhagen's 1.6 million residents, the population is around 1.972 million. Other important centers on the island are Roskilde, Helsingør (Elsinore), Slagelse, Næstved, and Hillerød. The northern part of the island, indented by the many-branched Isefjord, has many beaches, while the south is largely devoted to dairy farming and wheat growing.

zebra mussel (*Dreissena polymorpha*) Zebra mussels are inhabitants of freshwater and not a sea species, but their proliferation in recent years is causing major problems for ships, power plants, water treatment facilities, and harbors, which are all sea-related. Originally native to the Black, Caspian, and Azov Seas, the mussels first appeared in Lake St. Clair near Detroit in 1988, probably transported as ballast. Since then, they have spread to the St. Lawrence River system and all the Great Lakes, and are threatening every state west of the Mississippi. They can extend their range by commercial and recreational boats, amphibious airplanes, scuba equipment, and fishing gear. Zebra mussels are D-shaped, striped bivalves, about the size of a human thumbnail. Clustering together in colonies of hundreds of thousands per square yard, they can clog and eventually close the openings of underwater pipes. To combat this invasion, everything from chlorination, chemicals, ozone, heat, ultraviolet radiation, antifouling paints, sonic vibrations, electric shock, and parasites have been tried, but nothing works very well, and the mussels keep on proliferating. **See also mussel**

zebra shark (*Stegastoma fasciatum*) Also called the leopard shark in Australian waters, this monotypical (only one species in the genus) shark has a broadly rounded head, plentiful spots, and longitudinal ridges on the body like the whale shark, to which it is not related. (Juveniles are striped black and white, like a

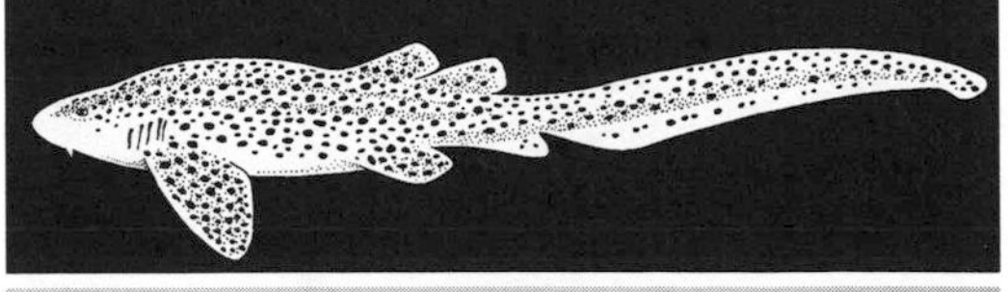

zebra shark

zebra, accounting for the common name.) It grows to a length of 11 feet, almost half of which is the extremely long upper lobe of the caudal fin. Zebra sharks sometimes lie on the bottom. They are unaggressive and considered harmless. **See also nurse shark, wobbegong**

zooplankton That part of the plankton that is composed of living animals, as opposed to the phytoplankton, which is composed of plants. Virtually every known animal phylum is represented in the sea, and many of them have planktonic larvae. Examples are the developing larvae of mussels, clams, polychaete worms, crabs, lobsters, and starfish. Some of these can swim, but not well enough to prevent them from being swept along by currents in the water. **See also plankton**

zooxanthellae Symbiotic protozoans or dinoflagellates that are pigmented (usually yellowish-brown), and are therefore able to conduct photosynthesis. In coral polyps, they utilize the carbon dioxide, nitrates, and phosphates produced as waste products by their hosts to manufacture sugars, amino acids, and lipids to supplement the growth of the corals. Because reef coral polyps depend on the photosynthesis of the zooxanthellae, corals must grow in clear water with good light. Photosynthesis takes place during the daylight hours, and the coral polyps come out to feed at night. Zooxanthellae can live outside of corals, but corals must have these protozoans aboard or they die. The giant clam (*Tridacna gigas*) also has zooxanthellae inside the tissue of its mantle lips and siphons. **See also coral**

A Note about the Author

Richard Ellis is the author of ten previous books, including *The Book of Sharks; The Book of Whales, Dolphins and Porpoises; Men and Whales; Monsters of the Sea; Great White Shark* (with John McCosker); *Deep Atlantic; Imagining Atlantis;* and *The Search for the Giant Squid.* He is also a celebrated marine artist whose paintings have been exhibited in museums and galleries around the world. He has written and illustrated articles for numerous magazines, including *Audubon, National Geographic, Discover,* and *Scientific American.* He lives in New York City.

A Note on the Type

This book was set in Minion, a typeface produced by the Adobe Corporation specifically for the Macintosh personal computer and released in 1990. Designed by Robert Slimbach, Minion combines the classic characteristics of old-style faces with the full complement of weights required for modern typesetting.

Composed by North Market Street Graphics

Printed and bound by Quebecor Printing, Fairfield, Pennsylvania